Math Study:

Your Roadmap to Success in Math

The Carson Math Study System is designed to help you succeed in your math course. You will discover your own learning style, and you will use study strategies that match the way you learn best. You will also learn how to organize your course materials, manage your time efficiently, and study and review effectively.

Your Notebook

Notes (see pages xviii–xix):

Section 1.7   9/20

We can simplify an expression by combining like terms.

p. 74 def. Like terms: constant terms or variable terms that have the same variable(s) raised to the same powers.

Ex 1) 2x and 3x are like terms.
Ex 2) 5x and 7y are not like terms.

Homework (see pages xix–xxii):

Section 1.5 Homework   9/21

#1 – 15 odd

1. $5^2 + 3 \cdot 4 - 7$
$= 25 + 3 \cdot 4 - 7$
$= 25 + 12 - 7$
$= 37 - 7$
$= 30$

Quizzes/Tests (see page xxv):

Chapter 2 Quiz   10/10

For 1-4, simplify.

4/4 = 100%
Nice work!

1. $|-8|$ $= 8$ ✓

2. $|9|$ $= 9$ ✓

3. $-15 + 5$ $= -10$ ✓

4. $-8 + -6$ $= -14$ ✓

Study Materials (see pages xxiii–xxv):

Chapter 2 Study Sheets   9/10

To graph a number on a number line, draw a dot on the mark for the number.   ex) Graph -2,

$-4 \;-3 \;-2 \;-1 \;\; 0 \;\; 1$

The absolute value of a positive number is positive.   ex) $|7| = 7$

The absolute value of a negative number is positive.   ex) $|-12| = 12$

Your Learning Style

To discover what kind of learner you are, complete the Learning Styles Inventory on page xvi (or in MyMathLab). Then, in the textbook, watch for the Learning Strategy boxes and the accompanying icons that provide ideas for maximizing your own learning style.

3 = Often applies   **2** = Sometimes applies   **1** = Never or almost never applies

——	**1.** I remember information better if I write it down or draw a picture of it.
——	**2.** I remember things better when I hear them instead of just reading or seeing them.
——	**3.** When I receive something that has to be assembled, I just start doing it. I don't read the directions.
——	**4.** If I am taking a test, I can "visualize" the page of text or lecture notes where the answer is located.
——	**5.** I would rather the professor explain a graph, chart, or diagram to me instead of just showing it to me.
——	**6.** When learning new things, I want to do it rather than hear about it.
——	**7.** I would rather the instructor write the information on the board or overhead instead of just lecturing.
——	**8.** I would rather listen to a book on tape than read it.
——	**9.** I enjoy making things, putting things together, and working with my hands.
——	**10.** I am able to conceptualize quickly and visualize information.
——	**11.** I learn best by hearing words.
——	**12.** I have been called hyperactive by my parents, spouse, partner, or professor.
——	**13.** I have no trouble reading maps, charts, or diagrams.
——	**14.** I can usually pick up on small sounds like bells, crickets, frogs, or distant sounds like train whistles.
——	**15.** I use my hands and gesture a lot when I speak to others.

Your Learning Strategies

LEARNING Strategy

AUDITORY · TACTILE · VISUAL Developing a good study system and understanding how you best learn is essential to academic success. Make sure you familiarize yourself with the study system outlined in the To the Student section at the beginning of the text. Also take a moment to complete the Learning Styles Inventory found at the end of that section to discover your personal learning style. In these Learning Strategy boxes, we offer tips and suggestions on how to connect the study system and your learning style to help you be successful in the course.

LEARNING Strategy

AUDITORY · TACTILE · VISUAL In the To the Student section, we suggest that when taking notes, you use a red pen for definitions and a blue pen for rules and procedures. Notice that we have used those colors in the design of the text to connect with your notes.

Definition   ***Absolute value:*** A given number's distance from 0 on a number line.

For example, the absolute value of 5 is 5 because the number 5 is 5 units from 0 on a number line. Likewise, the absolute value of -5 is 5 because -5 is also 5 units from 0 on a number line.

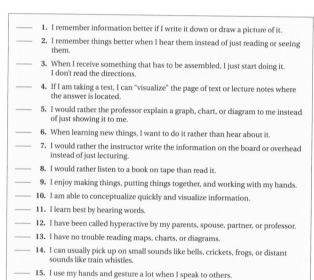

The absolute value of a number n is written $|n|$. The examples just mentioned translate this way:

The absolute value of 5 is 5.   The absolute value of -5 is 5.

Symbolic form: $|5| = 5$ $|-5| = 5$

What about the absolute value of 0? Since there are no units between 0 and itself, the absolute value of 0 is 0.

Rule   **Absolute Value**

The absolute value of every real number is either positive or 0.

Elementary Algebra

Third Edition

Tom Carson

Bill Jordan
Seminole State College of Florida

PEARSON

Addison
Wesley

Boston   Columbus   Indianapolis   New York   San Francisco   Upper Saddle River
Amsterdam   Cape Town   Dubai   London   Madrid   Milan   Munich   Paris   Montréal   Toronto
Delhi   Mexico City   São Paulo   Sydney   Hong Kong   Seoul   Singapore   Taipei   Tokyo

Editorial Director: Christine Hoag
Editor in Chief: Maureen O'Connor
Executive Editor: Cathy Cantin
Executive Project Manager: Kari Heen
Senior Project Editor: Lauren Morse
Project Editor: Katie DePasquale
Editorial Assistant: Jonathan Wooding
Media Producers: Ceci Fleming and Vicki Dreyfus
Software Development: Mary Durnwald, Tanya Farber, and Eileen Moore
Senior Development Editor: Dawn Nuttall
Production Manager: Ron Hampton
Senior Design Supervisor: Andrea Nix
Art Director: Heather Scott
Production Services: Pre-Press PMG
Executive Marketing Manager: Michelle Renda
Marketing Manager: Marlana Voerster
Marketing Assistant: Margaret Wheeler
Prepress Services Buyer: Caroline Fell
Manufacturing Manager: Evelyn Beaton
Senior Manufacturing Buyer: Carol Melville
Senior Media Buyer: Ginny Michaud
Manager, Cover Visual Research and Permissions: Karen Sanatar

Cover photo: © Shutterstock/Hemisferic, City of Arts and Sciences, Valencia, Spain

Photo credits can be found on page vi.

Library of Congress Cataloging-in-Publication Data

Carson, Tom, 1967–
   Elementary algebra / Tom Carson, Bill Jordan.—3rd ed.
     p. cm.
   ISBN 978-0-321-57729-0
   1. Algebra—Textbooks. I. Jordan, Bill. II. Title.
   QA152.3.C37 2011
   512.9—dc22
                                  2009020867

1 2 3 4 5 6 7 8 9—WC—14 13 12 11 10 09

Addison-Wesley
is an imprint of

www.pearsonhighered.com   ISBN-10: 0-321-57729-9
                                                ISBN-13: 978-0-321-57729-0

Contents

Photograph Credits

Preface

Welcome to the third edition of *Elementary Algebra*! Revising this series has been both exciting and rewarding, and it has given us the opportunity to respond to valuable instructor and student feedback. It is with great pride that we share with you both the improvements and additions to this edition as well as the hallmark features and style of the Carson/Jordan series.

Elementary Algebra, Third Edition, is the second book in a series that includes *Prealgebra*, Third Edition; *Intermediate Algebra*, Third Edition; and *Elementary and Intermediate Algebra*, Third Edition. This text is designed to be versatile enough for use in a standard lecture format, a self-paced lab, or even in an independent study format. Written in a relaxed, nonthreatening style, we have taken great care to ensure that students who have struggled with math in the past will be comfortable with the subject matter. Throughout the text, we explain *why* an algebraic process works the way it does, instead of just showing how to follow the process. In addition, problems from science, engineering, accounting, health, the arts, and everyday life link algebra to the real world. A complete Study System— including a Learning Styles Inventory, integrated Learning Strategy boxes, and a plan for developing a notebook as a personalized organization study tool—is also provided to guide students to success in the course. (See page xiii.)

Changes to the Third Edition

This revision includes refinements to the presentation of the material as well as the addition of examples, exercises, Your Turns, Instructor's Notes, and updated applications throughout the text.

The *To the Student* material within this Preface (pages xiii–xxv) has been extensively revised. In addition to revising the discussion about why mathematics is important and what it takes to succeed, a table comparing behaviors of strong and weak students has been added. Also, the explanation of the notebook system now includes sample notes, homework, and study materials.

Each Your Turn is now numbered to match its example. Additionally, examples are now correlated to corresponding groups of exercises in each exercise set. The authors have added section references to both the section Review Exercises and Cumulative Reviews. In the Annotated Instructor's Edition, the answers for the Practice Test exercises also have a reference to the corresponding section.

The text has been reorganized so that the topic of systems is presented earlier in the course. Former Chapter 8, *Systems of Equations*, now follows Chapter 4, *Graphing Linear Equations and Inequalities*, and is now Chapter 5, *Systems of Linear Equations and Inequalities*. Additionally, a new appendix, *Mean, Median, and Mode*, has been added to the text.

To provide students with frequent opportunities to review what they have learned, Cumulative Review exercises now appear after every chapter, starting with Chapter 2.

The exercise sets have been carefully revised in response to user and reviewer feedback. Changes have been made to reflect the reorganization of topic coverage, to improve grading and pairing, to offer extra practice in problem solving, and to ensure a strong correlation between exercises and examples to help students master the math. Algebra Pyramid icons, used throughout the text to help students differentiate between constants and variables, expressions and equations and inequalities, have been added to each set of end-of-section Review Exercises in addition to the Chapter Review Exercises and the Cumulative Review exercises. They have also been added to the start of relevant sections.

Finally, new supplemental resources have been added to the Carson/Jordan series to provide additional support for students and instructors: *Worksheets for Classroom or Lab Practice, Chapter Test Prep Video* (in *Video Resources on DVD*), and enhanced topic coverage in *MyMathLab.*

Key Features

Study System   A study system is introduced in the To the Student section on pages xviii–xxv and is then reinforced throughout the text. The system recommends color codes for taking notes. The color codes are consistent in the text itself: red for definitions, blue for procedures and rules, and black for notes and examples. In addition, the study system presents strategies for succeeding in the course.

Learning Styles Inventory   A Learning Styles Inventory is presented on page xvi to help students assess their particular learning styles. Throughout the book, Learning Strategy boxes offer study tips keyed to specific learning styles.

Learning Strategy Boxes   Learning Strategy boxes with icons depicting the learning style appear where appropriate in the text to offer advice on how to use the study system effectively and how to study specific topics based on a student's individual learning style (see pages 3, 107, and 414).

The Algebra Pyramid   An Algebra Pyramid is used throughout the text to help students see how the topic they are learning relates to the big picture of algebra—particularly focusing on the relationship between constants, variables, expressions, and equations and inequalities (see pages 3, 92, and 408). In the end-of-section Review Exercises, Chapter Review Exercises, and Cumulative Review Exercises, an Algebra Pyramid icon indicates the level of the pyramid that correlates to a particular group of exercises to help students determine what actions are appropriate with these exercises, for example, whether to "simplify" or "solve" (see pages 238, 326, and 683).

Chapter Openers   Like the Algebra Pyramid, chapter openers are designed to help students see how the topics in the upcoming chapter relate to the big picture of the entire course. The chapter openers give information about the importance of the topics in each chapter and how they fit into the overall structure of the course (see pages 1, 91, and 169).

Connection Boxes   Connection boxes help students understand how math concepts are related and build on each other (see pages 92, 212, and 422).

Your Turn Practice Exercises   The Your Turn practice exercises found after most examples give students an opportunity to work problems similar to the examples they have just seen. This practice step makes the text more interactive and provides immediate feedback to help students develop confidence in their problem-solving skills (see pages 5, 97, and 294).

Real, Relevant, and Interesting Applications   The application problems in the examples and exercises reflect real situations in science, engineering, health, finance, the arts, and other areas. These real-world applications illustrate the everyday use of basic algebraic concepts and encourage students to apply mathematical concepts to solve problems (see pages 98, 211, and 302).

Thorough Explanations   Great care is taken to explain not only *how* to do the math but also *why* the math works the way it does, how concepts are related, and how the math is relevant to students' everyday lives. Knowing all of this gives students a context in which to learn and remember math concepts.

Problem-Solving Method   A five-step problem-solving method is introduced on page 94 of Section 2.1 with the following steps:

1. Understand
2. Plan
3. Execute
4. Answer
5. Check

Application examples throughout the text follow these steps, modeling the problem-solving process clearly and concisely (see pages 113, 175, and 188).

Warning Boxes   Warning boxes alert students to common mistakes and false assumptions that students often make and explain why these are incorrect (see pages 201, 309, and 413).

Of Interest Boxes   Throughout the text, Of Interest boxes offer a unique perspective on the mathematics. With historical notes and fun facts, these are designed to enhance the learning process by making math concepts interesting and memorable (see pages 20, 129, and 412).

Puzzle Problems   Mathematical brainteasers, often solved without a formulaic approach, appear at the end of selected exercise sets to encourage critical thinking (see pages 104, 438, and 463).

Collaborative Exercises   These exercises, which appear once per chapter, encourage students to work in groups to discuss mathematics and use the topics from a particular section or group of sections to solve a problem (see pages 135, 195, and 282).

Calculator Tips   The relevant functions of calculators (scientific or graphing, depending on the topic) are explained and illustrated throughout the text in the optional Calculator Tips. In addition, an occasional calculator icon in the exercise sets indicates that the problem is designed to be solved using a calculator, though one is not required (see pages 16, 44, and 277).

Review Exercises   Since continuous review is important in every mathematics course, this text includes Review Exercises at the end of each section's exercise set. These exercises review previously learned concepts not only to keep the material fresh for students, but also to serve as a foundational review for the discussion in the upcoming section (see pages 36, 302, and 418).

Chapter Summaries and Review Exercises   An extensive Summary at the end of each chapter provides a list of defined terms referenced by section and page number, a list of important formulas from the chapter, and a summary of key concepts with examples. A set of Chapter Review Exercises is also provided. Answers to all Chapter Review Exercises are given in the back of the book (see pages 156–163, 228–236, and 319–330).

Chapter Practice Tests   A Practice Test follows each set of chapter review exercises. The problem types in the practice tests correlate to the short-answer tests in the Printable Test Bank. This is especially comforting for students who have math anxiety or who experience test anxiety (see pages 89, 164, and 236). Answers to all Practice Test Exercises are provided in the back of the book. Students can also watch a video of an instructor working through each chapter test exercise solution in the Chapter Test Prep Videos, available on the *Video Resources on DVD,* in *MyMathLab,* and on **YouTube** (search "CarsonElemAlg" and click on "Channels").

Cumulative Reviews   Cumulative Review Exercises now appear after every chapter starting with Chapter 2. These exercises help students stay current with all the material they have learned and help prepare them for midterm and final exams (see pages 166 and 472). Answers to all Cumulative Review Exercises are provided in the back of the book.

Supplements for *Elementary Algebra*, Third Edition

Student Supplements

Student's Solutions Manual
- By Emily Keaton
- Contains complete solutions to the odd-numbered section exercises and solutions to all of the section-level Review Exercises, Chapter Review Exercises, Practice Tests, and Cumulative Review Exercises.

ISBN-10: 0-321-62281-2
ISBN-13: 978-0-321-62281-5

Worksheets for Classroom or Lab Practice
- Extra practice exercises for every section of the text with ample space for students to show their work.
- These lab- and classroom-friendly workbooks also list the learning objectives and key vocabulary terms for every text section, along with vocabulary practice problems.

ISBN-10: 0-321-62817-9
ISBN-13: 978-0-321-62817-6

Video Resources on DVD Featuring
Chapter Test Prep Videos
- Short lectures for each section of the text.
- Complete set of digitized videos on DVD for student use at home or on campus.
- Chapter Test Prep Videos allow students to watch an instructor work through step-by-step solutions for all of the Chapter Test exercises from the textbook.
- Ideal for distance learning or supplemental instruction.
- All videos include optional English and Spanish subtitles.

ISBN-10: 0-321-62818-7
ISBN-13: 978-0-321-62818-3

MathXL® Tutorials on CD
This interactive tutorial CD-ROM provides algorithmically generated practice exercises that are correlated at the objective level to the exercises in the textbook. Every practice exercise is accompanied by an example and a guided solution designed to involve students in the solution process. Selected exercises may also include a video clip to help students visualize concepts. The software provides helpful feedback for incorrect answers and can generate printed summaries of students' progress.

ISBN-10: 0-321-62280-4
ISBN-13: 978-0-321-62280-8

Instructor Supplements

Annotated Instructor's Edition
- Includes answers to all exercises, including Puzzle Problems and Collaborative Exercises, printed in bright blue near the corresponding problems.
- Useful teaching tips are printed in the margin.
- A ★ icon, found in the AIE only, indicates especially challenging exercises in the exercise sets.

ISBN-10: 0-321-62134-4
ISBN-13: 978-0-321-62134-4

Instructor's Solutions Manual
- By Emily Keaton
- Contains complete solutions to all even-numbered section exercises, Puzzle Problems, and Collaborative Exercises.
- Available on the Instructor's Resource Center.

Instructor's Resource and Support Manual
- Includes resources designed to help both new and adjunct faculty with course preparation and classroom management.
- Offers helpful teaching tips specific to the sections of the text.
- Available on the Instructor's Resource Center.

Printable Test Bank
- By Laura Hoye, *Trident Technical College*
- Contains one diagnostic test per chapter, four free-response tests per chapter, one multiple-choice test per chapter, a mid-chapter check-up for each chapter, one midterm exam, and two final exams.
- Available on the Instructor's Resource Center.

TestGen®
TestGen (www.pearsonhighered.com/testgen) enables instructors to build, edit, print, and administer tests using a computerized bank of questions developed to cover all the objectives of the text. TestGen is algorithmically based, allowing instructors to create multiple, but equivalent, versions of the same question or test with the click of a button. Instructors can also modify test bank questions or add new questions. Tests can be printed or administered online. The software and testbank are available for download from Pearson Education's online catalog.

PowerPoint® Lecture Slides
- Present key concepts and definitions from the text.
- Available in MyMathLab or can be downloaded from the Instructor's Resource Center.

MyMathLab MyMathLab® Online Course (access code required)

MyMathLab is a text-specific, easily customizable online course that integrates interactive multimedia instruction with textbook content. MyMathLab gives you the tools you need to deliver all or a portion of your course online, whether your students are in a lab setting or working from home.

- **Interactive homework exercises,** correlated to your textbook at the objective level, are algorithmically generated for unlimited practice and mastery. Most exercises are free-response and provide guided solutions, sample problems, and learning aids for extra help.
- **Personalized Study Plan,** generated when students complete a test or quiz, indicates which topics have been mastered and links to tutorial exercises for topics students have not mastered. Instructors can customize the available topics in the study plan to match their course concepts.
- **Multimedia learning aids,** such as video lectures, animations, and a complete multi-media textbook, help students independently improve their understanding and performance.
- **Assessment Manager** lets you assign media resources (such as a video segment or a textbook passage), homework, quizzes, and tests. If you prefer, you can create your own online homework, quizzes, and tests that are automatically graded. Select just the right mix of questions from the MyMathLab exercise bank, instructor-created custom exercises, and/or TestGen test items.
- **Gradebook,** designed specifically for mathematics and statistics, automatically tracks students' results, lets you stay on top of student performance, and gives you control over how to calculate final grades. You can also add offline (paper-and-pencil) grades to the gradebook.
- **MathXL Exercise Builder** allows you to create static and algorithmic exercises for your online assignments. You can use the library of sample exercises as an easy starting point, or you can edit any course-related exercise.
- **Pearson Tutor Center** (www.pearsontutorservices.com) access is automatically included with MyMathLab. The Tutor Center is staffed by qualified math instructors who provide textbook-specific tutoring for students via toll-free phone, fax, email, and interactive Web sessions.

MyMathLab is powered by CourseCompass™, Pearson Education's online teaching and learning environment, and by MathXL®, our online homework, tutorial, and assessment system. MyMathLab is available to qualified adopters. For more information, visit www.mymathlab.com or contact your Pearson sales representative.

MathXL MathXL® Online Course (access code required)

MathXL® is an online homework, tutorial, and assessment system that accompanies Pearson's textbooks in mathematics or statistics.

- **Interactive homework exercises,** correlated to your textbook at the objective level, are algorithmically generated for unlimited practice and mastery. Most exercises are free-response and provide guided solutions, sample problems, and learning aids for extra help.
- **Personalized Study Plan,** generated when students complete a test or quiz, indicates which topics have been mastered and links to tutorial exercises for topics students have not mastered. Instructors can customize the available topics in the study plan to match their course concepts.
- **Multimedia learning aids,** such as video lectures and animations, help students independently improve their understanding and performance.
- **Gradebook,** designed specifically for mathematics and statistics, automatically tracks students' results, lets you stay on top of student performance, and gives you control over how to calculate final grades.
- **MathXL Exercise Builder** allows you to create static and algorithmic exercises for your online assignments. You can use the library of sample exercises as an easy starting point or use the Exercise Builder to edit any of the course-related exercises.
- **Assessment Manager** lets you create online homework, quizzes, and tests that are automatically graded. Select just the right mix of questions from the MathXL exercise bank, instructor-created custom exercises, and/or TestGen test items.

MathXL is available to qualified adopters. For more information, visit our website at www.mathxl.com or contact your Pearson sales representative.

Acknowledgments

Many people gave of themselves in so many ways during the development of this text. Mere words cannot contain the fullness of our gratitude. Though the words of thanks that follow may be few, please know that our gratitude is great.

We would like to thank the following people who have given of their time in reviewing this and previous editions of this textbook. Their thoughtful input was vital to the development of the text.

Marwan Abu-Sawwa, *Florida Community College at Jacksonville*
Jannette Avery, *Monroe Community College*
Elizabeth Barrow, *Charleston Southern University*
Valerie Beaman-Hackle, *Macon State College*
Sandra Belcher, *Midwestern State University*
Annette M. Burden, *Youngstown State University*
Nancy Carpenter, *Johnson County Community College*
Carol Cheshire, *Macon State College*
Susi Curl, *McCook Community College*
Sharon Edgmon, *Bakersfield College*
Beth Fraser, *Middlesex Community College*
Kathy Garrison, *Clayton College and State University*
Haile Haile, *Minneapolis Community and Technical College*
Nancy R. Johnson, *Manatee Community College*
Tracey L. Johnson, *University of Georgia*
Ryan H. Kasha, *Valencia Community College–West Campus*
Jeffrey Kroll, *Brazosport College*
Peter Lampe, *University of Wisconsin–Whitewater*
Lonnie Larson, *Sacramento City College*
Sandra Lofstock, *California Lutheran University*
Stephanie Logan, *Lower Columbia College*

John Long, *Jefferson Community College*
Lynn Maracek, *Santa Ana College*
Rogers Martin, *Louisiana State University*
Elizabeth Morrison, *Valencia Community College*
Linda Murphy, *Northern Essex Community College*
Charles Odion, *Houston Community College*
Karen Pagel, *New Mexico State University*
Janis Orinson, *Central Piedmont Community College*
Lourdes Pajo, *Pikes Peak Community College*
Merrell Pepper, *Southeast Technical Institute*
Patrick Riley, *Hopkinsville Community College*
Reynaldo Rivera, *Estrella Mountain Community College*
John Rochowicz Jr., *Alvernia College*
Rebecca Schantz, *Prairie State College*
Terri Seiver, *San Jacinto College–Central*
James Smith, *Columbia State Community College*
Kay Stroope, *Phillips County Community College*
John Thoo, *Yuba College*
Bettie Truitt, *Black Hawk College*
Judith A. Wells, *University of Southern Indiana*
Joe Westfall, *Carl Albert State College*
Peter Willett, *Diablo Valley College*
Tom Williams, *Rowan-Cabarrus Community College*

We would like to extend a heartfelt thank-you to everyone at Pearson Education for giving so much to this project. We would like to offer special thanks to Cathy Cantin and Greg Tobin, who believed in us and gave us the opportunity; to Dawn Nuttall for her insightful comments during the development of this edition; to Jonathan Wooding, Katie DePasquale, Lauren Morse, and Kari Heen, for keeping us on track; and also to Marlana Voerster and Margaret Wheeler for their encouragement and for working so hard to get us "out there."

A very special thank-you to Ron Hampton, whose keen eyes and editorial sense were invaluable during production; and to Timothy Rodes, Laura Hakala, and all of the fabulous people at Pre-Press PMG for working so hard to put together the finished pages.

To Ceci Fleming, Vicki Dreyfus, Peter Silvia, and all the people involved in developing the media supplements package, we are so grateful for all that you do. A special thank-you to Laura Hoye, who created the excellent Printable Test Bank, and to Emily Keaton for her work on the solutions manuals. Thank you to Gary Williams, Elizabeth Morrison, Perian Herring, and Paul Lorczak for their wonderful job of accuracy checking the manuscript and page proofs. A big thank-you goes to Elizabeth Morrison for her additional help reviewing art during production.

Finally, we'd like to thank our families for their support and encouragement during the process of developing and revising this text.

Tom Carson

Bill Jordan

To the Student

Why Do I Have to Take This Course?

Often, this is one of the first questions students ask when they find out they must take an algebra course. What a great question! But why focus on math alone? What about English, history, psychology, or science? Does anyone really use *every* topic of *every* course in the curriculum? Most jobs do not require that we write essays on Shakespeare, discuss the difference between various psychological theories, or analyze the cell structure of a frog's liver. So what's the point? The issue comes down to recognizing that general education courses are not job training. The purpose of those courses is to stretch and exercise the mind so that the educated person can better communicate, analyze situations, and solve problems, which are all valuable skills in life and any job.

Professional athletes offer a good analogy. A professional athlete usually has an exercise routine apart from their sport designed to build and improve their body. They may seek a trainer to design exercises intended to improve strength, stamina, or balance and then push them in ways they would not normally push themselves. That trainer may have absolutely no experience with their client's sport, but can still be quite effective in designing an exercise program because the trainer is focused on building basic skills useful for any athlete in any sport. Education is similar: it is exercise for the mind. A teacher's job is like that of a physical trainer. A teacher develops exercises intended to improve communication skills, critical-thinking skills, and problem-solving skills. Different courses are like different types of fitness equipment. Some courses may focus more on communication through writing papers, discussion, and debate. Other courses, such as mathematics, focus more on critical thinking and problem solving.

Another similarity is that physical exercise must be challenging for your body to improve. Similarly, mental exercise must be challenging for the mind to improve. Expect course assignments to challenge you and push you mentally in ways you wouldn't push yourself. That's the best way to grow. So as you think about the courses you are taking and the assignments in those courses, remember the bigger picture of what you are developing: your mind. When you are writing papers, responding to questions, analyzing data, and solving problems, you are developing skills important to life and any career out there.

What Do I Need to Succeed?

☑ **Adequate Time**   To succeed, you must have adequate time and be willing to use that time to perform whatever is necessary. To determine if you have adequate time, use the following guide.

> **Step 1.** Calculate your work hours per week.
> **Step 2.** Calculate the number of hours in class each week.
> **Step 3.** Calculate the number of hours required for study by doubling the number of hours you spend in class.
> **Step 4.** Add the number of hours from steps 1–3 together.

Adequate time: If the total number of hours is below 60, then you have adequate time.

Inadequate time: If the total number of hours is 60 or more, then you do not have adequate time. You may be able to hang in there for a while, but eventually, you will find yourself overwhelmed and unable to fulfill all of your obligations. Remember, the above calculations do not consider other likely elements of life such as commuting, family, recreation, and so on. The wise thing to do is cut back on work hours or drop some courses.

Assuming you have adequate time available, choosing to use that time to perform whatever is necessary depends on your attitude, commitment, and self-discipline.

☑ **Positive Attitude**   We do not always get to choose our circumstances, but we do get to choose our reaction and behavior. A positive attitude is choosing to be cheerful, hopeful, and encouraging no matter the situation. A benefit of a positive attitude is that it tends to encourage people around you. As a result, they are more likely to want to help you achieve your goal. A negative attitude, on the other hand, tends to discourage people around you. As a result, they are less likely to want to help you. The best way to maintain a positive attitude is to keep life in perspective, recognizing that difficulties and setbacks are temporary.

☑ **Commitment**   Commitment means binding yourself to a course of action. Remember, expect difficulties and setbacks, but don't give up. That's why a positive attitude is important, it helps you stay committed in the face of difficulty.

☑ **Self-Discipline**   Self-discipline is choosing to do what needs to be done—even when you don't feel like it. In pursuing a goal, it is normal to get distracted or tired. It is at those times that your positive attitude and commitment to the goal help you discipline yourself to stay on task.

Thomas Edison, inventor of the lightbulb, provides an excellent example of all of these principles. Edison tried over 2000 different combinations of materials for the filament before he found a successful combination. When asked about all his failed attempts, Edison replied, "I didn't fail once, I invented the lightbulb. It was just a 2000-step process." He also said, "Our greatest weakness lies in giving up. The most certain way to succeed is always to try just one more time." In those two quotes, we see a person who obviously had time to try 2000 experiments, had a positive attitude about the setbacks, never gave up, and had the self-discipline to keep working.

Behaviors of Strong Students and Weak Students

The four requirements for success can be translated into behaviors. The following table compares the typical behaviors of strong students with the typical behaviors of weak students.

Strong students ...	Weak students ...
are relaxed, patient, and work carefully.	are rushed, impatient, and hurry through work.
almost always arrive on time and leave the classroom only in an emergency.	often arrive late and often leave class to "take a break."
sit as close to the front as possible.	sit as far away from the front as possible.
pay attention to instruction.	ignore instruction, chit-chat, draw, fidget, etc.
use courteous and respectful language, encourage others, make positive comments, are cheerful and friendly.	use disrespectful language, discourage others, make negative comments, are grumpy and unfriendly. Examples of unacceptable language include: • "I hate this stuff!" (or even worse!) • "Are we doing anything important today?" • "Can we leave early?"
ask appropriate questions and answer instructor's questions during class.	avoid asking questions and rarely answer instructor's questions in class.

(continued)

Strong students ...	Weak students ...
take lots of notes, have organized notebooks, seek out and use study strategies	take few notes, have disorganized notebooks, do not use study strategies
begin assignments promptly and manage time wisely, and almost always complete assignments on time.	Procrastinate, manage time poorly, and often complete assignments late.
label assignments properly and show all work neatly.	show little or no work and write sloppily.
read and work ahead.	rarely read or work ahead.
contact instructors outside of class for help, and use additional resources such as study guides, solutions manuals, computer aids, videos, and tutorial services.	avoid contacting instructors outside of class, and rarely use additional resources available.

Assuming you have the prerequisites for success and understand the behaviors of a good student, our next step is to develop two major tools for success:

1. **Learning Style:** complete the Learning Styles Inventory to determine how you tend to learn.
2. **The Study System:** This system describes a way to organize your notebook, take notes, and create study tools to complement your learning style.

Learning Styles Inventory

What is *your* personal learning style?

A learning style is the way in which a person processes new information. Knowing your learning style can help you make choices in the way you study and focus on new material. Below are fifteen statements that will help you assess your learning style. After reading each statement, rate your response to the statement using the scale below. There are no right or wrong answers.

3 = Often applies   **2** = Sometimes applies   **1** = Never or almost never applies

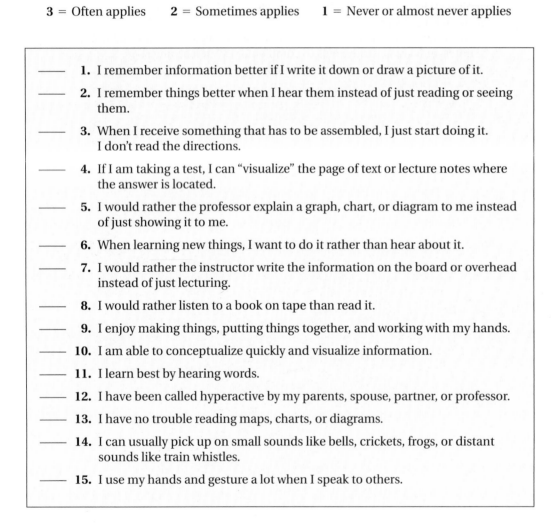

—— **1.** I remember information better if I write it down or draw a picture of it.

—— **2.** I remember things better when I hear them instead of just reading or seeing them.

—— **3.** When I receive something that has to be assembled, I just start doing it. I don't read the directions.

—— **4.** If I am taking a test, I can "visualize" the page of text or lecture notes where the answer is located.

—— **5.** I would rather the professor explain a graph, chart, or diagram to me instead of just showing it to me.

—— **6.** When learning new things, I want to do it rather than hear about it.

—— **7.** I would rather the instructor write the information on the board or overhead instead of just lecturing.

—— **8.** I would rather listen to a book on tape than read it.

—— **9.** I enjoy making things, putting things together, and working with my hands.

—— **10.** I am able to conceptualize quickly and visualize information.

—— **11.** I learn best by hearing words.

—— **12.** I have been called hyperactive by my parents, spouse, partner, or professor.

—— **13.** I have no trouble reading maps, charts, or diagrams.

—— **14.** I can usually pick up on small sounds like bells, crickets, frogs, or distant sounds like train whistles.

—— **15.** I use my hands and gesture a lot when I speak to others.

Determining Your Learning Style Write your score for each statement beside the appropriate statement number below. Then add the scores in each column to get a total score for that column.

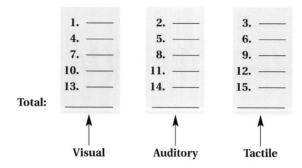

The largest total of the three columns indicates *your* dominant learning style.

Visual learners learn best by *seeing*. If this is your dominant learning style, then you should focus on learning strategies that involve seeing. The color coding in the study system (see page xviii) will be especially important. The same color coding is used in the text. Draw lots of diagrams, arrows, and pictures in your notes to help you see what is happening. Reading your notes, study sheets, and text repeatedly will be an important strategy.

Auditory learners learn best by *hearing*. If this is your dominant learning style, you should use learning strategies that involve hearing. After getting permission from your instructor, bring a tape recorder to class to record the discussion. When you study your notes, play back the tape. Also, when you learn rules, say the rule over and over. As you work problems, say the rule before you do the problem. You may also find the videotapes beneficial in that you can hear explanations of problems taken from the text.

Tactile (also known as kinesthetic) learners learn best by *touching or doing*. If this is your dominant learning style, you should use learning strategies that involve doing. Doing lots of practice problems will be important. Make use of the Your Turn exercises in the text. These are designed to give you an opportunity to do problems that are similar to the examples as soon as the topic is discussed. Writing out your study sheets and doing your practice tests repeatedly will be important strategies for you.

Note that the study system developed in this text is for all learners. Your learning style will help you decide what aspects and strategies in the study system to focus on, but being predominantly an auditory learner does not mean that you shouldn't read the textbook, do lots of practice problems, or use the color-coding system in your notes. Auditory learners can benefit from seeing and doing, and tactile learners can benefit from seeing and hearing. In other words, do not use your dominant learning style as a reason for not doing things that are beneficial to the learning process. Also, remember that the Learning Strategy boxes presented throughout the text provide tips to help you use your personal learning style to your advantage.

Index of Learning Strategies Learning strategies keyed to visual, auditory, and tactile learners can be found on the following pages:

Visual	Auditory	Tactile
3, 6, 93, 107, 137, 156, 164, 172, 265, 276, 414, 423, 439, 479, 507, 660, 716	3, 6, 93, 110, 156, 164, 186, 480, 507, 716, 735	3, 6, 93, 110, 156, 164, 212, 244, 307, 414, 423, 507, 691, 701

This Learning Styles Inventory is adapted from *Cornerstone: Creating Success through Positive Change*, Sixth Edition, by Moody/Sherfield © 2011. Reprinted by permission of Prentice Hall, Inc., Upper Saddle River, NJ.

The Study System

Organize the notebook into four parts using dividers as shown:

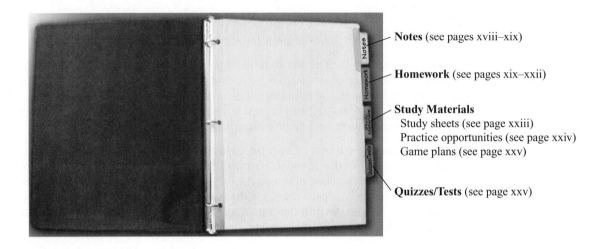

Notes (see pages xviii–xix)

Homework (see pages xix–xxii)

Study Materials
  Study sheets (see page xxiii)
  Practice opportunities (see page xxiv)
  Game plans (see page xxv)

Quizzes/Tests (see page xxv)

Notes

- Use a color code: **red** for definitions, **blue** for rules or procedures, and **pencil** for all examples and other notes.

- Begin notes for each class on a new page (front and back for that day is okay). Include a topic title or section number and the date on each page.

- Try to write your instructor's spoken explanations along with the things he or she writes on the board.

- Mark examples your instructor emphasizes in some way to give them a higher priority. These problems often appear on quizzes and tests.

- Write warnings your instructor discusses about a particular situation.

- Include common errors that your instructor illustrates, but mark them clearly as errors so that you do not mistake them for correct.

- To speed note taking, eliminate unnecessary words like "the" and use codes for common words like $+$ for "and" and $\therefore$ for "therefore." Also, instead of writing complete definitions, rules, or procedures, write the first few words and place the page reference from the text so that you can copy from the text later.

Sample Notes with Color Code

Include title.

Include date.

Definition in red with textbook page reference.

Procedure in blue with textbook page reference.

Section 1.7   9/20

We can simplify an expression by combining like terms.

p. 74   def. Like terms: constant terms or variable terms that have the same variable(s) raised to the same powers.

Ex 1) $2x$ and $3x$ are like terms.

Ex 2) $5x$ and $7y$ are not like terms.

Consider $2x + 3x$

$2x$ means two x's are added together.

$3x$ means three x's are added together.

$2x \;+\; 3x$

$= x + x + x + x + x$

We have a total of five x's added together.

$= 5x$

We can just add the coefficients.

p. 74 Procedure: To combine like terms, add or subtract the coefficients and keep the variables and their exponents the same.

Ex 1) $7x + 5x = 12x$

Ex 2) $4y^2 - 10y^2 = -6y^2$

Homework

This section of the notebook contains all homework. Use the following guidelines whether your assignments are from the textbook, a handout, or a computer program like MyMathLab or MathXL.

- Use pencil so that mistakes can be erased (scratching through mistakes is messy and should be avoided).

- Label according to your instructor's requirements. Usually, at least include your name, the date, and assignment title. It is also wise to write the assigned problems at the top as they were given. For example, if your instructor writes "Section 1.5 #1–15 odd," write it that way at the top. Labeling each assignment with this much detail shows that you take the assignment seriously and leaves no doubt about what you interpreted the assignment to be.

- For each problem you solve, write the problem number and show all solution steps neatly.

Why do I need to show work and write all the steps? Isn't the right answer all that's needed?

- Mathematics is not just about getting correct answers. You really learn mathematics when you organize your thoughts and present those thoughts clearly using mathematical language.
- You can arrive at correct answers with incorrect thinking. Showing your work allows your instructor to verify you are using correct procedures to arrive at your answers.
- Having a labeled, well-organized, and neat hard copy is a good study tool for exams.

What if I submit my answers in MyMathLab or MathXL? Do I still need to show work?

Think of MyMathlab or MathXL as a personal tutor who provides the exercises, offers guided assistance, and checks your answers before you submit the assignment. For the same reasons as those listed above, you should still create a neatly written hardcopy of your solutions, even if your instructor does not check the work. Following are some additional reasons to show your work when submitting answers in MyMathlab or MathXL.

- If you have difficulties that are unresolved by the program, you can show your instructor. Without the written work, your instructor cannot see your thinking.
- If you have a correct answer but have difficulty entering that correct answer, you have record of it and can show your instructor. If correct, your instructor can override the score.

Sample Homework:
Simplifying Expressions or Solving Equations

Suppose you are given the following exercise:

For Exercises 1–30 simplify.

   1. $5^2 + 3 \cdot 4 - 7$

Your homework should look something like the following:

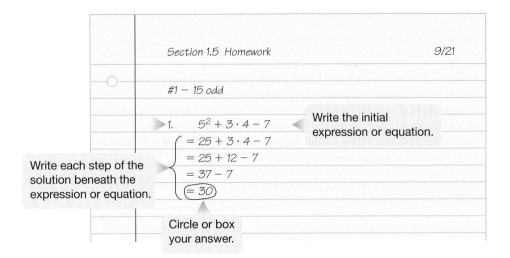

Sample Homework: Solving Application Problems

Example: Suppose you are given the following two problems.

For Exercises 1 and 2, solve.

1. Find the area of a circle with a diameter of 10 feet.
2. Two cars are traveling toward each other on the same highway. One car is traveling 65 miles per hour and the other is traveling at 60 miles per hour. If the two cars are 20 miles apart, how long will it be until they meet?

Your homework should look something like the following:

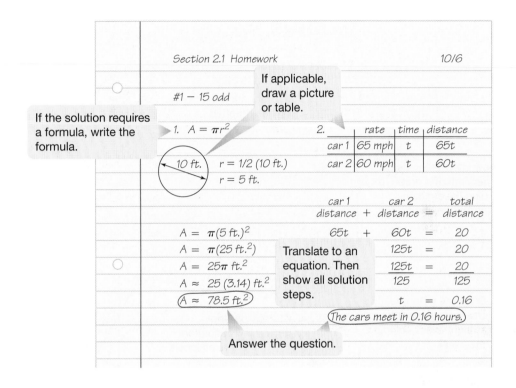

Sample Homework: Graphing

Example: Suppose you are given the following two problems.

For Exercises 1 and 2, graph the equation.

  1. $y = 2x - 3$ **2.** $y = -2x + 2$

Your homework should look something like the following:

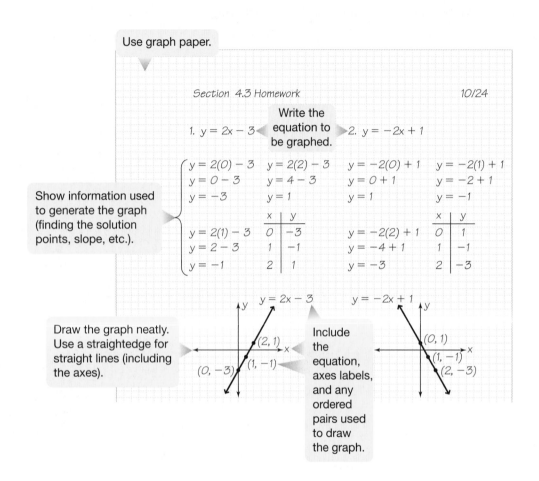

Use graph paper.

Section 4.3 Homework   10/24

Write the equation to be graphed.

1. $y = 2x - 3$ 2. $y = -2x + 1$

Show information used to generate the graph (finding the solution points, slope, etc.).

$y = 2(0) - 3$ $y = 2(2) - 3$ $y = -2(0) + 1$ $y = -2(1) + 1$
$y = 0 - 3$ $y = 4 - 3$ $y = 0 + 1$ $y = -2 + 1$
$y = -3$ $y = 1$ $y = 1$ $y = -1$

x	y
0	-3
1	-1
2	1

$y = 2(1) - 3$
$y = 2 - 3$
$y = -1$

$y = -2(2) + 1$
$y = -4 + 1$
$y = -3$

x	y
0	1
1	-1
2	-3

$y = 2x - 3$ $y = -2x + 1$

(2, 1)
(1, -1)
(0, -3)

(0, 1)
(1, -1)
(2, -3)

Draw the graph neatly. Use a straightedge for straight lines (including the axes).

Include the equation, axes labels, and any ordered pairs used to draw the graph.

Study Materials

This section of the notebook contains three types of study materials for each chapter.

Study Material 1: The Study Sheet   A study sheet contains *every* rule or procedure in the current chapter.

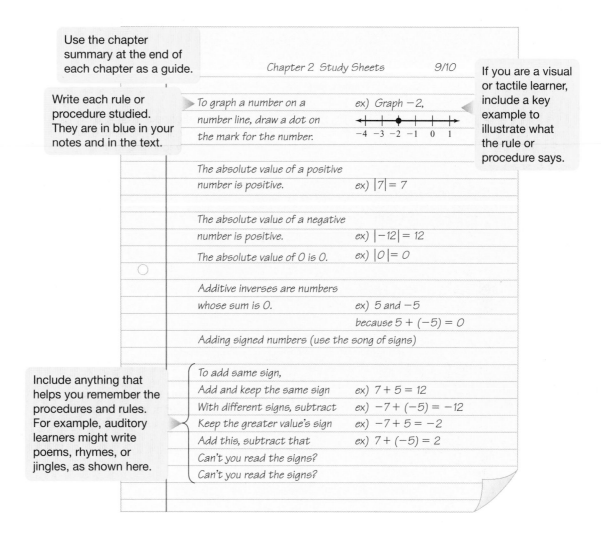

Use the chapter summary at the end of each chapter as a guide.

Write each rule or procedure studied. They are in blue in your notes and in the text.

If you are a visual or tactile learner, include a key example to illustrate what the rule or procedure says.

Include anything that helps you remember the procedures and rules. For example, auditory learners might write poems, rhymes, or jingles, as shown here.

Chapter 2 Study Sheets   9/10

To graph a number on a number line, draw a dot on the mark for the number.   ex) Graph -2,

The absolute value of a positive number is positive.   ex) $|7| = 7$

The absolute value of a negative number is positive.   ex) $|-12| = 12$

The absolute value of 0 is 0.   ex) $|0| = 0$

Additive inverses are numbers whose sum is 0.   ex) 5 and -5
because $5 + (-5) = 0$

Adding signed numbers (use the song of signs)

To add same sign,
Add and keep the same sign   ex) $7 + 5 = 12$
With different signs, subtract   ex) $-7 + (-5) = -12$
Keep the greater value's sign   ex) $-7 + 5 = -2$
Add this, subtract that   ex) $7 + (-5) = 2$
Can't you read the signs?
Can't you read the signs?

Study Material 2: The Practice Test If your instructor gives you a practice test, proceed to the discussion of creating a game plan.

If your instructor does not give you a practice test, use your notes to create your own practice test out of the examples given in class. Include only the instructions and the problem, not the solutions. The following sample practice test was created from examples in the notes for Chapter 2 in a prealgebra course.

Chapter 2 Practice Test   10/10

For # 1 and 2, graph on a number line.

1. 4

2. −3

For #3 and 4, simplify.

3. $|-8|$

4. $|9|$

For #5 and 6, find the additive inverse of the number.

5. 13

6. −15

For #7 − 10, add.

7. $13 + (-9)$ 8. $-20 + (-6)$

9. $-15 + 8$ 10. $3 + (-24)$

For each example in your notes, write the directions and the problem but not the solution.

After working through the practice test, use your notes to check your solutions.

Study Material 3: The Game Plan   The game plan refines the study process further. It is your plan for the test based on the practice test. For each problem on your practice test, write the definition, rule, or procedure used to solve the problem.

The sample shown gives the rule or procedure used to solve each problem on the sample practice test on the previous page. The rules and procedures came from the sample study sheet.

Multiple problems that use the same rule or procedure can be grouped together.

Write the rule or procedure used to solve the problems on the practice test.

Chapter 2 Game Plan   *9/10*

#1 and 2: Draw a dot on the mark for the number.

#3 and 4: The absolute value of a positive number
is a positive number.
The absolute value of a negative number
is a negative number.
The absolute value of 0 is 0.

#5 and 6: Additive inverses are numbers whose
sum is 0.

#7 – 10: To add same sign,
Add and keep the same sign
With different signs
Subtract and keep the greater value's sign
Add this, subtract that
Can't you read the signs?
Can't you read the signs?

Quizzes/Tests

Archive all returned quizzes and tests in this section of the notebook.

- Midterm and final exams questions are often taken from the quizzes and tests, so they make excellent study tools for those cumulative exams.
- Keeping all graded quizzes and tests offers a backup system in the unlikely event your instructor should loose any of your scores.
- If a dispute arises about a particular score, you have the graded test to show your instructor.

Chapter 2 Quiz   *10/10*

For 1-4, simplify.

4/4 = 100%
Nice work!

1. $|-8|$ $= 8$ ✓

2. $|9|$ $= 9$ ✓

3. $-15 + 5$ $= -10$ ✓

4. $-8 + -6$ $= -14$ ✓

Foundations of Algebra

1

" Three people were at work on a construction site. All were doing the same job, but when each was asked what the job was, the answers varied. 'Breaking rocks,' the first replied. 'Earning my living,' the second said. 'Helping to build a cathedral,' said the third. "

—Peter Schultz,
German businessman

" Build up your weaknesses until they become your strengths. "

—Knute Rockne (1888–1931)
Notre Dame football coach

In Peter Schultz's quote, each of the three people answered the question correctly. Notice that the first two people were rather narrowly focused, whereas the third person had a greater vision and appreciation of the finished structure and his or her place in the construction process. Your curriculum and the courses in it are like a building, with the units or chapters like construction materials. Try to keep the big picture in mind as your courses and curriculum build. In the chapter openers of this book, we discuss how the mathematics in the chapter fits into the bigger picture of the entire course structure to help you see the "cathedral" we are building.

In this chapter, we review the foundation of algebra, which is arithmetic. More specifically, we review number sets, operations of arithmetic, properties of real numbers, and the evaluation and simplification of expressions. This review is by no means a complete instruction of arithmetic. Since a solid foundation is important to support the rest of our building, if you encounter a topic in this chapter that is a particular weakness for you, it is important that you consult with your instructor or other more complete sources for extra practice.

1.1 Number Sets and the Structure of Algebra

Objectives

1. Understand the structure of algebra.
2. Classify number sets.
3. Graph rational numbers on a number line.
4. Determine the absolute value of a number.
5. Compare numbers.

Objective 1 Understand the structure of algebra. Learning mathematics is like learning a language. When we learn a language, we must learn the alphabet, vocabulary, and sentence structure. Similarly, mathematics has its own alphabet, vocabulary, and sentence structure. In this section, we begin the development of the foundation of algebra with an overview of its components and structure. The basic components are **variables** and **constants**.

Definitions *Variable:* A symbol that can vary in value.

Constant: A symbol that does not vary in value.

Variables are usually letters of the alphabet, such as x or y. Usually, constants are symbols for numbers such as $1, 2, \frac{3}{4}$, and 6.74. However, constants can sometimes be symbols such as e or the Greek letter π, each having special numeric values. Variables and constants are used to make **expressions**, **equations**, and **inequalities**.

Definition *Expression:* A constant, a variable, or any combination of constants, variables, and arithmetic operations that describes a calculation.

Examples of expressions:
$$2 + 6 \quad 4x - 5 \quad \frac{1}{3}\pi r^2 h$$

Definition *Equation:* A mathematical relationship that contains an equal sign.

Examples of equations:
$$2 + 6 = 8 \quad 4x - 5 = 12 \quad V = \frac{1}{3}\pi r^2 h$$

Connection Think of expressions as phrases and equations as complete sentences. The expression $2 + 6$ is read "two plus six," which is not a complete sentence. The equation $2 + 6 = 8$ is read "two plus six is eight." Notice that the equal sign translates to the verb *is*, which makes the sentence a complete sentence.

Definition *Inequality:* A mathematical relationship that contains an inequality symbol
$$(\neq, \; <, \; >, \; \leq, \text{ or } \geq).$$

Inequality Symbols and Their Translations

Symbolic Form	Translation
$8 \neq 3$	Eight is not equal to three.
$5 < 7$	Five is less than seven.
$7 > 5$	Seven is greater than five.
$x \leq 3$	x is less than or equal to three.
$y \geq 2$	y is greater than or equal to two.

The algebra pyramid shown illustrates how variables, constants, expressions, equations, and inequalities relate. At the foundation of algebra and our pyramid are constants and variables, which are used to build expressions, which in turn are used to build equations and inequalities.

The Algebra Pyramid

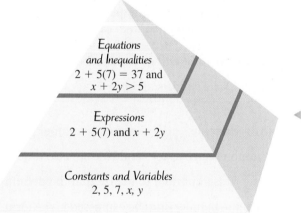

Equations and Inequalities
$2 + 5(7) = 37$ and $x + 2y > 5$

Expressions
$2 + 5(7)$ and $x + 2y$

Constants and Variables
$2, 5, 7, x, y$

Note: During this course, we move back and forth between expressions, equations, and inequalities. When we transition from one to the other, we use the algebra pyramid as a quick visual so that it's clear what we are working on.

Objective 2 **Classify number sets.** Numbers can be placed into different categories using **sets**.

Definition **Set:** A collection of objects.

Braces are used to indicate a set. For example, the set containing the numbers 1, 2, 3, and 4 would be written $\{1, 2, 3, 4\}$. The numbers 1, 2, 3, and 4 are called the *members* or *elements* of this set.

Procedure **Writing Sets**

To write a set, write the members or elements of the set separated by commas within braces, $\{\ \}$.

Example 1 Write the set containing the first five letters of the alphabet.

Answer $\{A, B, C, D, E\}$

Your Turn 1 Write the set containing the first four months of the year.

Answer **Your Turn 1**

$\{January, February, March, April\}$

Numbers are classified using number sets. The set of *natural numbers* contains the counting numbers 1, 2, 3, 4, ... and is written {1, 2, 3, ... }. The three dots are an *ellipsis* and indicate that the numbers continue forever in the same pattern. The set of *whole numbers* contains all of the natural numbers and 0 and is written {0, 1, 2, 3, ... }. The set of *integers* contains all of the whole numbers and the opposite (or negative) of every natural number and is written { ..., −3, −2, −1, 0, 1, 2, 3, ... }. Notice that the integers continue forever in both directions.

A number line is often useful in mathematics. The following number line is marked with the integers. Note that the positive numbers are to the right of 0 and the negative numbers are to the left.

Although we can mark and view only a portion of the number line, the arrows at the ends indicate that the line and the numbers on it continue forever in both directions. There are other types of numbers that include the integers and all numbers between them. If we traveled along the number line forever in both directions, we would encounter every number in the set of *real numbers*.

Some of the real numbers that are not integers are the **rational numbers** that contain every real number that can be expressed as a ratio of integers.

Definition

Rational number: Any real number that can be expressed in the form $\frac{a}{b}$, where a and b are integers and $b \neq 0$.

Note: In the definition for a rational number, the notation $b \neq 0$ is important because if the denominator b were to equal 0, the fraction would be undefined. The reason it is undefined will be explained later.

For example, $\frac{3}{4}$ is a rational number because 3 and 4 are integers. Note that numbers such as 0.75 and 75% are also rational numbers because they can be written as $\frac{3}{4}$. It is important to realize that the definition of a rational number does not state that the number *must* be expressed in the form $\frac{a}{b}$. Rather, if a number *can* be expressed in the form $\frac{a}{b}$, it is a rational number. For example, the number 5 is a rational number because it can be expressed as $\frac{5}{1}$.

Example 2   Determine whether the given number is a rational number.

a. $\frac{2}{3}$

Answer   Yes, because 2 and 3 are integers.

b. 0.4

Answer   Yes, 0.4 is a rational number because it can be expressed as $\frac{4}{10}$ and 4 and 10 are integers.

c. $0.\overline{6}$

Answer   A bar written over a decimal digit indicates that the digit repeats without end. So $0.\overline{6} = 0.66666\ldots$, and we say that these decimal numbers are nonterminating decimal numbers. Usually, we encounter these numbers as quotients in certain division problems, such as when we write certain fractions as decimals. In this case, $0.\overline{6}$ is the decimal equivalent of $\frac{2}{3}$. Because $0.\overline{6}$ can be expressed as the fraction $\frac{2}{3}$, it is a rational number.

Note: All nonterminating decimal numbers with repeating digits are rational numbers because they can be expressed as fractions with integers in both the numerator and denominator.

d. 3

Answer   Yes, because 3 can be thought of as $\frac{3}{1}, \frac{6}{2}, \frac{9}{3}$, etc. All integers are rational numbers.

Your Turn 2 Determine whether the given number is a rational number.

a. $\frac{4}{7}$ **b.** 0.56 **c.** $0.\overline{2}$ **d.** -7

Not all numbers can be expressed as a ratio of integers. One such number is π (pronounced "pie"). Because the exact value of π cannot be expressed as a ratio of integers, it is categorized as an **irrational number**.

Definition   *Irrational number:* Any real number that is not rational.

Some other irrational numbers are $\sqrt{2}$ and $\sqrt{3}$. (Square roots are explained in more detail in Section 1.5.) Because an irrational number cannot be written as a ratio of integers, if a calculation involves an irrational number, we must leave it in symbolic form or use a rational number approximation. We can approximate π with rational numbers such as 3.14 and $\frac{22}{7}$. Any decimal representation of an irrational number is a nonrepeating, nonterminating decimal number. For example, $0.1010010001\ldots$ is irrational.

The rational and irrational numbers make up the **real numbers**.

Definition   *Real numbers:* The union of the rational and irrational numbers.

The following figure illustrates how the number sets relate in the real number system.

Real Numbers

| **Rational Numbers:** Real numbers that can be expressed in the form $\frac{a}{b}$, where a and b are integers and $b \neq 0$, such as $-4\frac{3}{4}$, $-\frac{2}{3}$, 0.018, $0.\overline{3}$, and $\frac{5}{8}$. | **Irrational Numbers:** Any real number that is not rational, such as $-\sqrt{2}$, $-\sqrt{3}$, $\sqrt{0.8}$, and π. |

Integers: $\ldots, -3, -2, -1, 0, 1, 2, 3, \ldots$

Whole Numbers: $0, 1, 2, 3, \ldots$

Natural Numbers: $1, 2, 3, \ldots$

Objective ❸ Graph rational numbers on a number line. Number lines can be useful tools when comparing numbers or solving certain arithmetic problems. Let's review how to graph a number on a number line.

Example 3 Graph on a number line.

a. $1\frac{3}{5}$

SOLUTION The number $1\frac{3}{5}$ is located $\frac{3}{5}$ of the way between 1 and 2, so we divide the space between 1 and 2 into 5 equal divisions and place a dot on the 3rd mark.

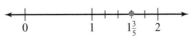

b. −2.56

SOLUTION Because −2.56 means $-2\frac{56}{100}$, we could divide the space between −2 and −3 into 100 divisions and count to the 56th mark. Because this is tedious to do, we gradually use smaller and smaller sections of the number line to graph the number.

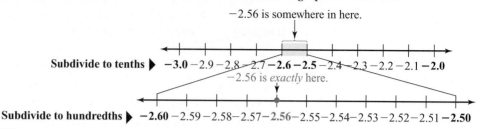

−2.56 is somewhere in here.

Subdivide to tenths ▶ −3.0 −2.9 −2.8 −2.7 −2.6 −2.5 −2.4 −2.3 −2.2 −2.1 −2.0

−2.56 is *exactly* here.

Subdivide to hundredths ▶ −2.60 −2.59 −2.58 −2.57 −2.56 −2.55 −2.54 −2.53 −2.52 −2.51 −2.50

Your Turn 3 Graph on a number line.

a. $2\frac{3}{4}$

b. −1.78

Objective ④ Determine the absolute value of a number. The word *value* indicates how much something is worth or its magnitude. In mathematics, **absolute value** indicates the magnitude of a number.

Definition ***Absolute value:*** A given number's distance from 0 on a number line.

For example, the absolute value of 5 is 5 because the number 5 is 5 units from 0 on a number line. Likewise, the absolute value of −5 is 5 because −5 is also 5 units from 0 on a number line.

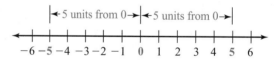

The absolute value of a number n is written $|n|$. The examples just mentioned translate this way:

The absolute value of 5 is 5.   The absolute value of −5 is 5.

Symbolic form: $|5| = 5$ $|-5| = 5$

What about the absolute value of 0? Since there are no units between 0 and itself, the absolute value of 0 is 0.

Rule **Absolute Value**

The absolute value of every real number is either positive or 0.

Answers   Your Turn 3

a.

1   2 $2\frac{3}{4}$ 3

b.

−1.8 −1.78 −1.76 −1.74 −1.72 −1.7

Example 4 Simplify.

a. $|-4.5|$

Answer $|-4.5| = 4.5$ because −4.5 is 4.5 units from 0 on a number line.

b. $\left|\dfrac{3}{8}\right|$

Answer $\left|\dfrac{3}{8}\right| = \dfrac{3}{8}$ because $\dfrac{3}{8}$ is $\dfrac{3}{8}$ of a unit from 0 on a number line.

Your Turn 4   Simplify.

a. $\left|-2\dfrac{4}{5}\right|$

b. $|12.8|$

Objective ⑤ Compare numbers.   Number lines can also be used to determine which of two numbers is greater. Because numbers increase from left to right on a number line, the number farthest to the right will be the greater of two numbers.

Rule Comparing Numbers

For any two real numbers a and b, a is greater than b if a is to the right of b on a number line. Equivalently, b is less than a if b is to the left of a on a number line.

Consider the following number line where we compare 2 and 8.

8 is farther right than 2.

0 1 2 3 4 5 6 7 8 9

Because the number 8 is farther to the right on a number line than the number 2, we say that 8 is greater than 2, or in symbols, $8 > 2$. Or we could say that since 2 is to the left of 8, $2 < 8$.

Example 5   Use $=$, $<$, or $>$ to write a true statement.

a. 4 ⬚ -4

Answer $4 > -4$ because 4 is farther to the right on a number line than -4.

b. -2.7 ⬚ -2.5

Answer $-2.7 < -2.5$ because -2.7 is farther to the left on a number line than -2.5.

c. $\left|3\dfrac{5}{6}\right|$ ⬚ $3\dfrac{5}{6}$

Answer $\left|3\dfrac{5}{6}\right| = 3\dfrac{5}{6}$ because the absolute value of $3\dfrac{5}{6}$ is equal to $3\dfrac{5}{6}$.

d. $|-0.7|$ ⬚ -1.5

Answer $|-0.7| > -1.5$ because the absolute value of -0.7 is equal to 0.7, which is farther to the right on a number line than -1.5.

Answers Your Turn 4

a. $2\dfrac{4}{5}$ **b.** 12.8

Answers Your Turn 5

a. $-15 > -21$ **b.** $-4.1 < 0$

c. $2\dfrac{5}{6} > 2\dfrac{1}{4}$ **d.** $|-9.5| = 9.5$

Your Turn 5   Use $=$, $<$, or $>$ to write a true statement.

a. -15 ⬚ -21 **b.** -4.1 ⬚ 0 **c.** $2\dfrac{5}{6}$ ⬚ $2\dfrac{1}{4}$ **d.** $|-9.5|$ ⬚ 9.5

1.1 Exercises

Note: Exercises marked with a ★ represent challenging exercises.

1. Define a set in your own words.

2. Define rational numbers in your own words.

3. Explain the difference between a rational number and an irrational number.

4. Explain why it is incorrect to say that absolute value is always positive.

For Exercises 5–10, answer true or false.

5. Every rational number is a real number.

6. Every real number is a rational number.

7. Every whole number is an integer.

8. Every real number is a natural number.

9. A number exists that is both rational and irrational.

10. Zero is a rational number.

For Exercises 11–20, write a set representing each description. See Example 1.

11. The days of the week

12. The last ten letters of the English alphabet

13. The vowels of the English alphabet

14. The states in the United States that do not share a border with any other state

15. The natural numbers that are multiples of 5

16. The even natural numbers

17. The odd natural numbers greater than 7

18. The even integers greater than or equal to 16

19. The integers greater than or equal to −6 but less than −2

20. The integers greater than −2.1 and less than $\frac{3}{4}$

For Exercises 21–30, determine whether each number is a rational number or an irrational number. See Example 2.

21. $-\frac{4}{5}$

22. $\frac{1}{4}$

23. 9

24. −12

25. π

26. $\dfrac{\pi}{4}$ **27.** -0.21 **28.** -0.8 **29.** $0.\overline{6}$ **30.** $0.\overline{13}$

For Exercises 31–38, graph each number on a number line. See Example 3.

31. $-\dfrac{5}{6}$ **32.** $5\dfrac{1}{2}$ **33.** $2\dfrac{3}{8}$ **34.** $-\dfrac{2}{5}$

35. -3.5 **36.** 7.4 **37.** 2.45 **38.** -7.62

For Exercises 39–48, simplify. See Example 4.

39. $|23|$ **40.** $|6|$ **41.** $|-2|$ **42.** $|-8|$ **43.** $|-5.7|$

44. $|-4.5|$ **45.** $\left|-3\dfrac{1}{8}\right|$ **46.** $\left|2\dfrac{3}{5}\right|$ **47.** $|0|$ **48.** $|-67.8|$

For Exercises 49–76, use $=$, $<$, or $>$ to write a true statement. See Example 5.

49. $9 \quad 3$ **50.** $2 \quad 7$ **51.** $4 \quad -3$ **52.** $-6 \quad 5$ **53.** $-7 \quad -8$

54. $-19 \quad -7$ **55.** $-8 \quad 0$ **56.** $0 \quad -5$ **57.** $6.2 \quad 4.5$ **58.** $2.63 \quad 3.75$

59. $-4.3 \quad -7.6$ **60.** $-3.5 \quad -3.1$ **61.** $2\dfrac{2}{3} \quad 2\dfrac{3}{4}$ **62.** $3\dfrac{5}{6} \quad 3\dfrac{1}{4}$ **63.** $5.8 \quad |-5.8|$

64. $|-4.1| \quad 4.1$ **65.** $7.3 \quad |-8.7|$ **66.** $|-10.4| \quad 3.2$ **67.** $|4.31| \quad |-4.31|$ **68.** $|-0.59| \quad |0.59|$

69. $\left|-6\dfrac{5}{8}\right| \quad 5\dfrac{3}{8}$ **70.** $4\dfrac{2}{9} \quad \left|4\dfrac{5}{9}\right|$ **71.** $|-4| \quad |-2|$ **72.** $|-10| \quad |-8|$ **73.** $|29.5| \quad |-29.7|$

74. $|-5.36| \quad |5.76|$ **75.** $\left|-\dfrac{2}{3}\right| \quad \left|-\dfrac{4}{3}\right|$ **76.** $\left|-\dfrac{9}{11}\right| \quad \left|-\dfrac{7}{11}\right|$

★ *For Exercises 77–80, list the given numbers in order from least to greatest.*

77. $-2.56, 5.4, |8.3|, \left|-7\dfrac{1}{2}\right|, -4.7$ **78.** $2.9, 1, -12.6, |-1.3|, -9.6, \left|-2\dfrac{3}{4}\right|$

79. $0.4, -0.6, 0, 3\dfrac{1}{4}, |-0.02|, -0.44, \left|1\dfrac{2}{3}\right|$ **80.** $1.02, -0.13, -4\dfrac{1}{8}, -2\dfrac{1}{4}, |-1.06|, -2, |0.1|$

1.2 Fractions

Objectives

1. Write equivalent fractions.
2. Write equivalent fractions with the LCD.
3. Write the prime factorization of a number.
4. Simplify a fraction to lowest terms.

In this section, we review some basic principles of **fractions**.

Definition **Fraction:** A quotient of two numbers or expressions a and b having the form $\dfrac{a}{b}$, where $b \neq 0$.

Connection We learned in Section 1.1 that if a number can be expressed in the form $\dfrac{a}{b}$, where a and b are integers, it is a rational number. So fractions that have integer numerators and denominators are rational numbers.

For example, $\dfrac{3}{4}$ is a fraction. The top number in a fraction is called the *numerator*, and the bottom number is called the *denominator*.

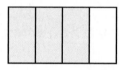

$$\dfrac{3}{4} \quad \longleftarrow \text{Numerator} \\ \quad \longleftarrow \text{Denominator}$$

We can use fractions to indicate a part of a whole. For example, the rectangle to the right has been divided into four equal pieces, three of which are shaded. So the shaded region represents $\dfrac{3}{4}$ of the rectangle.

Objective 1 **Write equivalent fractions.** Now let's review how to write *equivalent fractions*, which are fractions that represent the same amount. For example, $\dfrac{1}{2}$ and $\dfrac{2}{4}$ are equivalent. We can see that they represent the same amount on a ruler.

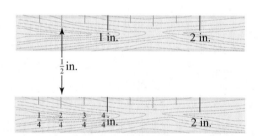

We can rewrite $\dfrac{1}{2}$ as $\dfrac{2}{4}$ by multiplying its numerator and denominator by 2.

$$\frac{1}{2} = \frac{1 \cdot 2}{2 \cdot 2} = \frac{2}{4}$$

We also can begin with $\dfrac{2}{4}$ and get back to $\dfrac{1}{2}$ by dividing both the numerator and denominator by 2.

$$\frac{2}{4} = \frac{2 \div 2}{4 \div 2} = \frac{1}{2}$$

Writing Equivalent Fractions

For any fraction, we can write an equivalent fraction by multiplying or dividing both its numerator and denominator by the same nonzero number.

This rule makes use of the fact that multiplying or dividing both the numerator and denominator by the same nonzero number is equivalent to multiplying or dividing by 1, which does not change a number.

Example 1 Find the missing number that makes the fractions equivalent.

a. $\dfrac{7}{12} = \dfrac{?}{48}$

SOLUTION Because $48 \div 12 = 4, 12 \cdot 4 = 48$. So we multiply the numerator and denominator by 4.

$$\dfrac{7 \cdot 4}{12 \cdot 4} = \dfrac{28}{48} \qquad \text{Multiply the numerator and denominator by 4.}$$

Note: When a fraction is negative, the sign can be placed to the left of the numerator, to the left of the fraction line, or to the left of the denominator.

$$-\dfrac{30}{42} = -\dfrac{5}{7} = \dfrac{-5}{7} = \dfrac{5}{-7}$$

b. $-\dfrac{30}{42} = \dfrac{?}{7}$

SOLUTION Because $42 \div 7 = 6, 42 \div 6 = 7$. So divide the numerator and denominator by 6.

$$-\dfrac{30 \div 6}{42 \div 6} = -\dfrac{5}{7} \qquad \text{Divide the numerator and denominator by 6.}$$

Your Turn 1 Find the missing number that makes the fractions equivalent.

a. $\dfrac{9}{16} = \dfrac{?}{32}$

b. $-\dfrac{54}{72} = \dfrac{3}{?}$

Objective ❷ Write equivalent fractions with the LCD. It is sometimes necessary to write fractions that have different denominators, such as $\dfrac{1}{2}$ and $\dfrac{1}{3}$, as equivalent fractions with a common denominator. A common denominator can be any number that is a **multiple** of the denominators.

Definition *Multiple:* A multiple of a given integer n is the product of n and an integer.

We can generate multiples of a given number by multiplying the given number by the integers. For example, to generate the positive **multiples** of 2 and 3, we multiply 2 and 3 by 1, 2, 3, 4, and so on.

Note: Every multiple of 2 is evenly divisible by 2. Similarly, every multiple of 3 is evenly divisible by 3. In general, every multiple of a number n is evenly divisible by n.

Multiples of 2	Multiples of 3
$2 \cdot 1 = 2$	$3 \cdot 1 = 3$
$2 \cdot 2 = 4$	$3 \cdot 2 = 6$
$2 \cdot 3 = 6$	$3 \cdot 3 = 9$
$2 \cdot 4 = 8$	$3 \cdot 4 = 12$
$2 \cdot 5 = 10$	$3 \cdot 5 = 15$
$2 \cdot 6 = 12$	$3 \cdot 6 = 18$

Answers Your Turn 1

a. $\dfrac{9}{16} = \dfrac{18}{32}$ b. $-\dfrac{54}{72} = -\dfrac{3}{4}$

Notice that some common multiples of 2 and 3 in our lists are 6 and 12. When we rewrite fractions with a common denominator, it is usually preferable to use the **least common multiple** of the denominators. Such a denominator is called the **least common denominator**. From our lists, we see that the least common multiple of 2 and 3 is 6. So 6 is the least common denominator of $\frac{1}{2}$ and $\frac{1}{3}$.

Definitions   *Least common multiple (LCM):* The smallest number that is a multiple of each number in a given set of numbers.

Least common denominator (LCD): The least common multiple of the denominators of a given set of fractions.

We can now write equivalent fractions for $\frac{1}{2}$ and $\frac{1}{3}$ with their LCD, 6.

$$\frac{1}{2} = \frac{1 \cdot 3}{2 \cdot 3} = \frac{3}{6} \quad \text{and} \quad \frac{1}{3} = \frac{1 \cdot 2}{3 \cdot 2} = \frac{2}{6}$$

Example 2   Write each pair as equivalent fractions with the LCD.

a. $\frac{5}{8}$ and $\frac{1}{6}$

SOLUTION   The LCD of 8 and 6 is 24.

$$\frac{5}{8} = \frac{5 \cdot 3}{8 \cdot 3} = \frac{15}{24} \quad \text{and} \quad \frac{1}{6} = \frac{1 \cdot 4}{6 \cdot 4} = \frac{4}{24}$$

b. $\frac{2}{3}$ and $\frac{5}{9}$

SOLUTION   The LCD of 3 and 9 is 9.

$\frac{2}{3} = \frac{2 \cdot 3}{3 \cdot 3} = \frac{6}{9}$, and $\frac{5}{9}$ already has the LCD as its denominator.

Your Turn 2   Write each pair as equivalent fractions with the LCD.

a. $\frac{4}{9}$ and $\frac{5}{6}$ **b.** $-\frac{3}{4}$ and $-\frac{5}{8}$

Objective ③ Write the prime factorization of a number.   We now consider simplifying fractions. When simplifying fractions, we work with **factors**, which are numbers or variables that are multiplied together.

Definition   *Factors:* If $a \cdot b = c$, then a and b are factors of c.

In the multiplication statement $20 = 2 \cdot 10$, the 2 and 10 are factors of 20. Notice that a factor of a given number divides the given number evenly (leaves a remainder of 0). The numbers 1, 2, 4, 5, 10, and 20 are factors of 20 because they all divide 20 evenly. When a number is expressed as a product of its factors, it is said to be in *factored form*.

Answers   Your Turn 2

a. $\frac{8}{18}$ and $\frac{15}{18}$ **b.** $-\frac{6}{8}$ and $-\frac{5}{8}$

Factored form is sometimes referred to as a *factorization*. Here are some factorizations for 20:

$$20 = 1 \cdot 20 \quad \text{or} \quad 20 = 2 \cdot 10 \quad \text{or} \quad 20 = 4 \cdot 5$$

Some natural numbers have only two distinct factors, 1 and the number itself. These numbers are called **prime numbers**.

Definition   *Prime number:* A natural number that has exactly two different factors, 1 and the number itself.

Note: A natural number that has factors other than 1 and itself is called a **composite number**. Some examples are 4, 6, 8, and 9. Notice that 1 is not composite, so it is neither prime nor composite.

Note that 0 is not a prime number because it is not a natural number. (0 is a whole number.) Also, 1 is not a prime number because the only factor for 1 is itself and, by definition, a prime number must have *two different* factors.

The first prime number is 2 because it is the first natural number greater than 1 that has exactly two factors, 1 and 2. The number 4 is not a prime number because it is divisible by a number other than 1 and 4, namely 2. Here is a list of prime numbers:

$$2, 3, 5, 7, 11, 13, 17, 19, 23, 29, 31, 37, 41, 43, \ldots$$

When simplifying a fraction, it is useful to write the **prime factorization** of the numerator and denominator. For example, the prime factorization for 20 is $2 \cdot 2 \cdot 5$.

Definition   *Prime factorization:* A factorization that contains only prime factors.

We can find the prime factorization of a number using any factorization of the number as a starting point. Consider 20 again and suppose we factored it as $4 \cdot 5$. Because 4 is not a prime number, we can factor 4 as $2 \cdot 2$. It looks like this:

$$20 = 4 \cdot 5$$
$$= 2 \cdot 2 \cdot 5$$

We will use factor trees to find prime factorizations. The idea is to draw two branches beneath any composite number and write two factors of that number at the end of the branches. Following are two ways we can use a factor tree to find the prime factorization of 20. Notice that when a branch ends in a prime factor, we circle the prime factor. This is optional, but many people find it helpful in making the prime factors stand out.

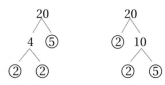

We usually write the prime numbers in a prime factorization in ascending order. So we would write $20 = 2 \cdot 2 \cdot 5$ (instead of $20 = 5 \cdot 2 \cdot 2$ or $20 = 2 \cdot 5 \cdot 2$). Note that no matter how we write the prime factorization, the product of two 2s and one 5 is always 20. No two numbers have the same prime factorization. Consequently, we say that a number's prime factorization is unique.

Example 3   Find the prime factorization of 360.

SOLUTION   Use a factor tree.

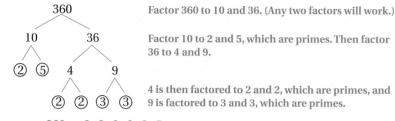

Factor 360 to 10 and 36. (Any two factors will work.)

Factor 10 to 2 and 5, which are primes. Then factor 36 to 4 and 9.

4 is then factored to 2 and 2, which are primes, and 9 is factored to 3 and 3, which are primes.

Answer $360 = 2 \cdot 2 \cdot 2 \cdot 3 \cdot 3 \cdot 5$

Your Turn 3   Find the prime factorization.

   a. 84   **b.** 280

Objective ④   Simplify a fraction to lowest terms.   We can use prime factorizations to simplify fractions to **lowest terms**.

Definition   *Lowest terms:* Given a fraction $\dfrac{a}{b}$ and $b \neq 0$, if the only factor common to both a and b is 1, then the fraction is in lowest terms.

For example, $\dfrac{3}{4}$ is in lowest terms because the only factor common to both 3 and 4 is the number 1, whereas $\dfrac{6}{8}$ is not in lowest terms because 6 and 8 have a common factor of 2.

To develop the method of reducing fractions using primes, we must understand that if the numerator and denominator are identical and not both 0, then the fraction is equivalent to 1.

$$1 = \frac{1}{1} = \frac{2}{2} = \frac{3}{3} = \frac{4}{4} = \frac{5}{5} = \cdots$$

This suggests the following rule:

Rule   **Simplifying a Fraction with the Same Nonzero Numerator and Denominator**

$$\frac{n}{n} = \frac{1}{1} = 1, \text{ when } n \neq 0.$$

This rule can be extended to common factors within a fraction. It allows us to eliminate factors that are identical in the numerator and denominator of a fraction.

Rule   **Eliminating a Common Factor in a Fraction**

$$\frac{an}{bn} = \frac{a \cdot 1}{b \cdot 1} = \frac{a}{b}, \text{ when } b \neq 0 \text{ and } n \neq 0.$$

These rules allow us to write fractions in lowest terms using prime factorizations. The idea is to replace the numerator and denominator with their prime factorizations and then eliminate the prime factors that are common to both the numerator and

Answers   Your Turn 3
a. $84 = 2 \cdot 2 \cdot 3 \cdot 7$
b. $280 = 2 \cdot 2 \cdot 2 \cdot 5 \cdot 7$

denominator. Consider the fraction $\frac{10}{15}$. Replacing the 10 and 15 with their prime factorizations, we have the following:

$$\frac{10}{15} = \frac{2 \cdot 5}{3 \cdot 5}$$

$$= \frac{2 \cdot 1}{3 \cdot 1}$$

$$= \frac{2}{3}$$

Note: The common factor of 5 is replaced with 1. In Example 4, we leave this step out.

This process of eliminating common factors is called *dividing out* the common factors and is sometimes written this way:

$$\frac{10}{15} = \frac{2 \cdot \overset{1}{\cancel{5}}}{3 \cdot \underset{1}{\cancel{5}}} = \frac{2}{3}$$

Although there are different styles of showing which common factors are eliminated, we simply highlight the common factors that are eliminated.

Procedure   **Simplifying a Fraction to Lowest Terms**

To simplify a fraction to lowest terms:

1. Replace the numerator and denominator with their prime factorizations.
2. Eliminate (divide out) all prime factors common to the numerator and denominator.
3. Multiply the remaining factors.

Connection You may remember a method for simplifying fractions in which you divide out the greatest common factor. For example, to simplify $\frac{28}{70}$, we recognize that 14 is the greatest common factor of 28 and 70 and divide both of those numbers by 14 to get lowest terms:
$\frac{28}{70} = \frac{28 \div 14}{70 \div 14} = \frac{2}{5}$. This method is a perfectly good way of simplifying fractions. However, we show the prime method to prepare for simplifying rational expressions in Chapter 8.

Example 4   Simplify to lowest terms.

a. $\dfrac{28}{70}$

SOLUTION $\dfrac{28}{70} = \dfrac{2 \cdot 2 \cdot 7}{2 \cdot 5 \cdot 7} = \dfrac{2}{5}$ Replace the numerator and denominator with their prime factorizations; then eliminate the common prime factors.

b. $\dfrac{156}{210}$

SOLUTION $\dfrac{156}{210} = \dfrac{2 \cdot 2 \cdot 3 \cdot 13}{2 \cdot 3 \cdot 5 \cdot 7} = \dfrac{26}{35}$

Your Turn 4   Simplify to lowest terms.

a. $\dfrac{36}{48}$ b. $\dfrac{126}{315}$ c. $\dfrac{312}{378}$

Example 5   A newspaper reports that over the last year, 160 households per 1000 households were victims of a crime. Write a fraction in simplest form representing the fraction of households that were victims of crime.

Answers   Your Turn 4

a. $\dfrac{3}{4}$ b. $\dfrac{2}{5}$ c. $\dfrac{52}{63}$

SOLUTION If 160 out of 1000 households were victims of crime, the fraction of households that were victims of crime is $\frac{160}{1000}$. Now we can simplify the fraction.

$$\frac{160}{1000} = \frac{2 \cdot 2 \cdot 2 \cdot 2 \cdot 2 \cdot 5}{2 \cdot 2 \cdot 2 \cdot 5 \cdot 5 \cdot 5} = \frac{4}{25}$$

Answer The fraction, in simplest form, of households that were victims of crime is $\frac{4}{25}$, which means that 4 out of 25 households were victims of crime.

Your Turn 5 Researchers gave 560 volunteers a new medication and found that 48 of them developed a headache within 30 minutes of taking the medication. Write a fraction in simplest form representing the portion of the volunteers that developed a headache.

Answer   Your Turn 5

$\frac{3}{35}$

📠 Calculator Tips

On most scientific calculators, fractions are entered using the [a^b/c] key. To reduce a fraction, enter it and then press [=]. For example, to reduce $\frac{156}{210}$, type 156 [a^b/c] 210 [=] or, more specifically,

[1] [5] [6] [a^b/c] [2] [1] [0] [=]. You should now see 26/35.

On most graphing calculators, enter a fraction using the [÷] key. To reduce the fraction, press the

[MATH] key and select 1:Frac; then press [ENTER]. For example, to reduce $\frac{156}{210}$ on a graphing calculator,

type [1] [5] [6] [÷] [2] [1] [0] [MATH] [1] [ENTER].

1.2 Exercises   For Extra Help   *MyMathLab*   Math XL WATCH DOWNLOAD
PRACTICE

1. What are two ways to generate fractions equivalent to a given fraction?

2. Is 4 a multiple of 12? Explain.

3. Is $2 \cdot 2 \cdot 4 \cdot 7$ a prime factorization? Explain.

4. Is it possible for two different numbers to have the same prime factorization? Explain.

5. Is every whole number a composite number?

6. How do you know if a fraction is in lowest terms?

For Exercises 7–10, answer true or false; then explain.

7. Every prime number is a whole number.

8. Prime numbers are rational numbers.

9. Every composite number has at least one factor that is a prime number.

10. The number 1 is neither prime nor composite.

For Exercises 11–14, identify the fraction represented by each shaded region.

11.

12.

13.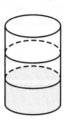

14.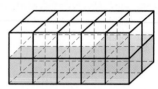

For Exercises 15–20, write the length of each line segment in lowest terms.

15.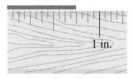

16.

17.

18.

19.

20.

For Exercises 21–28, find the missing number that makes the fractions equivalent.

21. $\dfrac{3}{10} = \dfrac{?}{40}$

22. $\dfrac{5}{8} = \dfrac{?}{16}$

23. $-\dfrac{7}{12} = \dfrac{?}{60}$

24. $\dfrac{2}{5} = \dfrac{6}{?}$

25. $\dfrac{4}{7} = \dfrac{?}{28}$

26. $\dfrac{6}{8} = \dfrac{?}{4}$

27. $\dfrac{8}{60} = \dfrac{2}{?}$

28. $\dfrac{27}{30} = \dfrac{9}{?}$

For Exercises 29–36, write the fractions as equivalent fractions with the LCD. See Example 2.

29. $\dfrac{4}{5}$ and $\dfrac{2}{3}$

30. $\dfrac{5}{7}$ and $\dfrac{3}{11}$

31. $\dfrac{4}{9}$ and $\dfrac{7}{6}$

32. $\dfrac{5}{8}$ and $\dfrac{7}{12}$

33. $-\dfrac{11}{18}$ and $-\dfrac{17}{24}$ **34.** $-\dfrac{9}{20}$ and $-\dfrac{7}{15}$ **35.** $-\dfrac{1}{16}$ and $-\dfrac{15}{36}$ **36.** $-\dfrac{13}{21}$ and $-\dfrac{9}{14}$

For Exercises 37–44, write the prime factorization for each number. See Example 3.

37. 44   **38.** 33   **39.** 36   **40.** 42

41. 64   **42.** 48   **43.** 250   **44.** 810

For Exercises 45–52, simplify to lowest terms. See Example 4.

45. $\dfrac{72}{90}$ **46.** $\dfrac{48}{84}$ **47.** $\dfrac{63}{99}$ **48.** $\dfrac{42}{91}$

49. $-\dfrac{78}{104}$ **50.** $-\dfrac{30}{54}$ **51.** $-\dfrac{48}{90}$ **52.** $-\dfrac{24}{162}$

For Exercises 53–56, determine whether each simplification is correct. If it isn't, explain why. See Example 4.

53. $\dfrac{\overset{1}{\cancel{2}} + 3}{\underset{1}{\cancel{2}}} = \dfrac{1 + 3}{1} = 4$

54. $\dfrac{3 - 4 \cdot \overset{1}{\cancel{2}}}{\underset{1}{\cancel{2}}}$

55. $\dfrac{84}{240} = \dfrac{\overset{1}{\cancel{2}} \cdot \overset{1}{\cancel{2}} \cdot \overset{1}{\cancel{3}} \cdot 7}{2 \cdot \underset{1}{\cancel{2}} \cdot \underset{1}{\cancel{2}} \cdot \underset{1}{\cancel{3}} \cdot 3 \cdot 5} = \dfrac{7}{30}$

56. $\dfrac{300}{108} = \dfrac{\overset{1}{\cancel{2}} \cdot \overset{1}{\cancel{2}} \cdot \overset{1}{\cancel{3}} \cdot 5 \cdot 5}{2 \cdot \underset{1}{\cancel{2}} \cdot \underset{1}{\cancel{2}} \cdot \underset{1}{\cancel{3}} \cdot 3} = \dfrac{25}{3}$

For Exercises 57–60, write each fraction in lowest terms. See Example 5.

57. A student scores 294 points out of a total of 336 points in a college course. What fraction of the total points is the student's score?

58. The nutrition label on a package of frozen fish sticks indicates that a serving of fish sticks has 250 calories, with 130 of those calories coming from fat. What fraction of the calories in a serving of fish sticks comes from fat?

59. At a company, 300 of the 575 employees have optional life insurance coverage as part of their benefits package. What fraction of the employees have optional life insurance coverage?

60. Laura uses 120 square feet of a room in her home as an office for her business. The total living space of her home is 1830 square feet. For tax purposes, she needs to compute the fraction of her home that is used as an office. What is that fraction?

For Exercises 61–64, use the following bar graph, which shows the number of hours that Carla spends each week in each activity. Write all fractions in lowest terms. See Example 5.

61. What fraction of a week does Carla spend working?

62. What fraction of a week does Carla spend sleeping?

63. What fraction of a week does Carla spend in all of the listed activities combined?

64. What fraction of the week does Carla have for free time, which is time away from all of the listed activities?

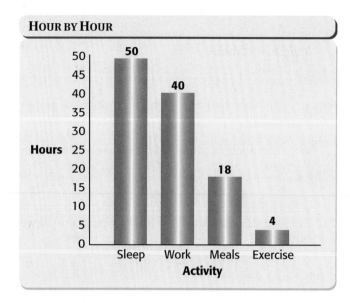

For Exercises 65–72, use the following table, which shows the number of households per 1000 that were victims of property crime in each year. Write all fractions in lowest terms. See Example 5.

National Crime Victimization Survey, Property Crime Trends, 1985–2005

Year	Total Property Crime*	Burglary	Thefts	Motor Vehicle Thefts
1985	385	75	296	14
1986	373	74	284	15
1987	380	75	289	16
1988	378	74	287	18
1989	373	68	287	19
1990	349	65	264	21
1991	354	65	267	22
1992	325	59	248	19
1993	319	58	242	19
1994	310	56	235	19
1995	291	49	224	17
1996	266	47	206	14
1997	248	45	190	14
1998	217	39	168	11
1999	198	34	154	10
2000	178	32	138	9
2001	167	29	129	9
2002	159	28	122	9
2003	163	30	124	9
2004	161	30	123	9
2005	154	30	116	8

*Victimizations per 1000 households

Source: U.S. Department of Justice, Bureau of Crime Statistics.

65. What fraction of households were victims of property crime in 1987?

66. What fraction of households were victims of property crime in 1992?

67. In 1987, what fraction of households were not victims of property crime?

68. In 1994, what fraction of households were not victims of property crime?

69. a. What year had the highest number of property crimes?

  b. What fraction of households were victims in that year?

70. a. What year had the lowest number of burglaries?

  b. What fraction of households were victims of burglary in that year?

71. In 1988, what fraction of the total property crimes were burglaries?

72. In 1985, what fraction of total property crimes were motor vehicle theft?

For Exercises 73–80, write each fraction in lowest terms. See Example 5.

73. What fraction of a year is 30 days? Use 365 days for the number of days in a year.

74. What fraction of an hour is 8 minutes?

75. What fraction of a minute is 40 seconds?

76. What fraction of a foot is 4 inches?

77. In the first session of 2007, the U.S. House of Representatives had 233 Democrats and 202 Republicans. What fraction of the House of Representatives was Republican? (*Source:* Official list of United States House of Representatives.)

78. In 2007, the U.S. Senate had 49 Democrats, 49 Republicans, and 2 Independents. What fraction of the Senate was not Democrat? (*Source:* Official list of United States Senators.)

79. One molecule of lactose contains 12 carbon atoms, 22 hydrogen atoms, and 11 oxygen atoms. What fraction of the atoms that make up the molecule are carbon?

80. One molecule of glucose contains 6 carbon atoms, 12 hydrogen atoms, and 6 oxygen atoms. What fraction of the atoms that make up the molecule are not carbon?

> **Of Interest**
>
> It takes Earth one year to complete one revolution around the Sun, which is actually $365\frac{1}{4}$ days. Because it is impractical to have $\frac{1}{4}$ days on the calendar, every four years those $\frac{1}{4}$ days add up to a full day, which is why we have a leap year with 366 days.

> **Of Interest**
>
> The chemical formula for a molecule of lactose is $C_{12}H_{22}O_{11}$. Notice that the subscripts in a chemical formula indicate the number of atoms of each element that are present in the molecule.

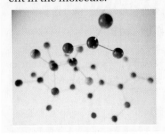

Review Exercises

Exercises 1–3 CONSTANTS AND VARIABLES

[1.1]   **1.** Write a set containing the names of the four planets closest to the Sun.

[1.1]   **2.** Is 0.8 a rational or irrational number? Explain.

[1.1]   **3.** Graph 3.7 on a number line.

Exercises 4–6 EXPRESSIONS

[1.1]   **4.** Is $5x + 2$ an expression or an equation? Explain.

[1.1]   **5.** Simplify: $|27|$

[1.1]   **6.** Use $<$, $>$, or $=$ to make a true statement: $|-6|$ ___ 6

1.3 Adding and Subtracting Real Numbers; Properties of Real Numbers

Objectives
1. Add integers.
2. Add rational numbers.
3. Find the additive inverse of a number.
4. Subtract rational numbers.

Objective 1 Add integers.   We now turn our attention to adding and subtracting numbers. We begin with addition. First, consider the parts of an addition statement. The numbers added are called *addends*, and the answer is called a *sum*.

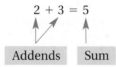

$$2 + 3 = 5$$

Addends   Sum

There are three important properties of addition that are true for all real numbers. First, the sum of a number and 0 is that number. Because adding 0 to a number does not change the identity of the number, 0 is called the *additive identity*.

$$2 + 0 = 2$$

Second, the order of addends can be changed without affecting the sum. This is known as the *commutative property of addition*.

$$2 + 3 = 5 \quad \text{or} \quad 3 + 2 = 5$$

Third, when adding three or more numbers, we can group the addends differently without affecting the sum. This is known as the *associative property of addition*.

$$2 + (3 + 4) \qquad \text{or} \qquad (2 + 3) + 4$$
$$= 2 + 7 \qquad\qquad\qquad = 5 + 4$$
$$= 9 \qquad\qquad\qquad\quad = 9$$

Following is a summary of these properties.

Properties of Addition	Symbolic Form	Word Form
Additive Identity	$a + 0 = a$	The sum of a number and 0 is that number.
Commutative Property of Addition	$a + b = b + a$	Changing the order of addends does not affect the sum.
Associative Property of Addition	$a + (b + c) = (a + b) + c$	Changing the grouping of three or more addends does not affect the sum.

Example 1   Indicate whether each equation illustrates the additive identity, commutative property of addition, or associative property of addition.

a. $(4 + 7) + 2 = 4 + (7 + 2)$

Answer   Associative property of addition because the grouping is changed.

b. $0 + (-7) = -7$

Answer   Additive identity because the sum of a number and 0 is that number.

c. $(-3 + 5) + 2 = 2 + (-3 + 5)$

Answer   Commutative property of addition because the order of the addends is changed.

Your Turn 1   Indicate whether each equation illustrates the additive identity, commutative property of addition, or associative property of addition.

  a. $-6 + (3 + 5) = (-6 + 3) + 5$ **b.** $3 + 0 = 3$ **c.** $-7 + 3 = 3 + (-7)$

Adding Numbers with the Same Sign

Keeping these properties in mind, let's consider how to add signed numbers. First, we learn how to add numbers that have the same sign. It is helpful to relate adding signed numbers to money. For example, adding two positive numbers is like adding deposits, whereas adding two negative numbers is like adding debts.

$$15 + 10 = 25 \qquad\qquad\qquad -20 + (-8) = -28$$

Note: This addition is like having $15 in an account and depositing another $10 into the same account, bringing the total to $25.

Note: This addition is like having a debt balance of $20 on a credit card and charging another $8 on the same card, bringing the balance to a total debt of $28.

We can also illustrate each of these cases using a number line. The first addend is the starting point, and the second addend indicates the direction and distance to travel on the number line so that we finish at the sum.

Answers   Your Turn 1

a. associative property of addition

b. additive identity

c. commutative property of addition

 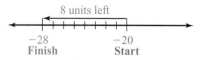

These two cases are similar and suggest the following procedure:

Procedure

Adding Numbers with the Same Sign

To add two numbers that have the same sign, add their absolute values and keep the same sign.

Example 2 Add.

a. $15 + 13$

SOLUTION $15 + 13 = 28$

b. $-14 + (-23)$

SOLUTION $-14 + (-23) = -37$

Note: In terms of money, part a illustrates two deposits, whereas part b illustrates two debts.

Your Turn 2 Add.

a. $32 + 19$

b. $-62 + (-13)$

Adding Numbers with Different Signs

Now consider adding two numbers with different signs, which is like making a payment toward a debt. Consider the following cases:

$$-15 + 3 = -12$$

$$-9 + 20 = 11$$

Note: This addition is like having a debt of \$15 and a payment of \$3 is made toward that debt. This decreases the debt to \$12.

Note: This addition is like having a debt of \$9 and a payment of \$20 is made toward that debt. Because the payment is more than the debt, we now have a credit of \$11.

Using number lines:

Note that the commutative property of addition indicates that in the examples, the addends can be rearranged without affecting the sum.

$$3 + (-15) = -12 \qquad\qquad 20 + (-9) = 11$$

These examples suggest the following procedure:

Procedure

Adding Numbers with Different Signs

To add two numbers that have different signs, subtract the smaller absolute value from the greater absolute value and keep the sign of the number with the greater absolute value.

Answers Your Turn 2

a. 51 **b.** −75

Example 3 Add.

a. 22 + (−14)

SOLUTION 22 + (−14) = 8

b. −13 + 20

> **Note:** In terms of money, parts a and b illustrate situations in which the deposit is greater than the debt. Therefore, the result is a credit, or positive.

SOLUTION −13 + 20 = 7

c. 12 + (−18)

SOLUTION 12 + (−18) = −6

d. −24 + 5

> **Note:** In terms of money, parts c and d illustrate situations in which a payment toward a debt is not enough to pay off the debt. Therefore, the result is still debt, or negative.

SOLUTION −24 + 5 = −19

Your Turn 3 Add.

a. 15 + (−6)   **b.** 27 + (−35)   **c.** −13 + 19   **d.** −29 + 14

Objective ② Add rational numbers.

Adding Fractions with the Same Denominator

Recall that the set of rational numbers contains any number that can be expressed as a ratio of integers. This set includes the integers themselves because every integer can be expressed as a ratio with 1 as the denominator. Now let's turn our attention to the rest of the rational number set, namely fractions and decimals. Consider the intuitive fact that adding two quarters equals a half. (In money, this would be a half-dollar.) The following circle shows this sum.

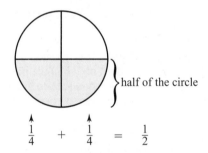

Notice that if the numerators are added and the denominator remains the same, the result simplifies to $\frac{1}{2}$.

$$\frac{1}{4} + \frac{1}{4} = \frac{1+1}{4} = \frac{2}{4} = \frac{1}{2}$$

Procedure To add fractions with the same denominator, add the numerators and keep the same denominator; then simplify.

Example 4 Add.

a. $\frac{5}{8} + \frac{1}{8}$

SOLUTION $\dfrac{5}{8} + \dfrac{1}{8} = \dfrac{5+1}{8} = \dfrac{6}{8}$

Add the numerators and keep the same denominator. Because $\dfrac{6}{8}$ is not in lowest terms, we must simplify.

$= \dfrac{2 \cdot 3}{2 \cdot 2 \cdot 2} = \dfrac{3}{4}$

Replace 6 and 8 with their prime factorizations; divide out the common factor, 2; and then multiply the remaining factors.

b. $-\dfrac{3}{10} + \left(-\dfrac{1}{10}\right)$

SOLUTION $-\dfrac{3}{10} + \left(-\dfrac{1}{10}\right) = \dfrac{-3 + (-1)}{10} = -\dfrac{4}{10}$

Note: Because the fractions have the same sign, we add and keep the same sign.

$= -\dfrac{2 \cdot 2}{2 \cdot 5} = -\dfrac{2}{5}$

Simplify to lowest terms by dividing out the common factor, 2.

c. $\dfrac{7}{9} + \left(-\dfrac{4}{9}\right)$

Note: Because the two addends have different signs, we subtract the smaller absolute value from the greater absolute value and keep the sign of the number with the greater absolute value.

SOLUTION $\dfrac{7}{9} + \left(-\dfrac{4}{9}\right) = \dfrac{7 + (-4)}{9} = \dfrac{3}{9}$

$= \dfrac{3}{3 \cdot 3} = \dfrac{1}{3}$

Simplify to lowest terms by dividing out the common factor, 3.

Your Turn 4 Add.

a. $\dfrac{3}{8} + \dfrac{1}{8}$ **b.** $-\dfrac{5}{9} + \left(-\dfrac{2}{9}\right)$ **c.** $-\dfrac{7}{12} + \dfrac{5}{12}$

Adding Fractions with Different Denominators

If the denominators are different, we must first find a common denominator. Recall that in Section 1.2, we wrote equivalent fractions with the LCD. Consider the addition $\dfrac{3}{4} + \dfrac{1}{6}$. The LCD of 4 and 6 is 12. We now write each fraction as an equivalent fraction with the LCD.

$$\dfrac{3}{4} + \dfrac{1}{6} = \dfrac{3(3)}{4(3)} + \dfrac{1(2)}{6(2)}$$

$$= \dfrac{9}{12} + \dfrac{2}{12}$$

$$= \dfrac{11}{12}$$

Note: To write $\dfrac{3}{4}$ as an equivalent fraction with 12 as the denominator, we multiply the numerator and denominator by 3.

Note: To write $\dfrac{1}{6}$ as an equivalent fraction with 12 as the denominator, we multiply the numerator and denominator by 2.

Procedure **Adding Fractions**

To add fractions with different denominators:

1. Write each fraction as an equivalent fraction with the LCD.
2. Add the numerators and keep the LCD.
3. Simplify.

Example 5 Add.

a. $\dfrac{1}{6} + \dfrac{5}{9}$

SOLUTION The LCD of 6 and 9 is 18.

$$\dfrac{1}{6} + \dfrac{5}{9} = \dfrac{1(3)}{6(3)} + \dfrac{5(2)}{9(2)}$$ Write equivalent fractions with 18 in the denominator.

$$= \dfrac{3}{18} + \dfrac{10}{18}$$ Add numerators and keep the common denominator.

$$= \dfrac{3 + 10}{18}$$ Because the addends have the same sign, we add and keep the same sign.

$$= \dfrac{13}{18}$$

b. $-\dfrac{4}{5} + \dfrac{2}{3}$

SOLUTION The LCD of 5 and 3 is 15.

$$-\dfrac{4}{5} + \dfrac{2}{3} = -\dfrac{4(3)}{5(3)} + \dfrac{2(5)}{3(5)}$$ Write equivalent fractions with 15 in the denominator.

$$= -\dfrac{12}{15} + \dfrac{10}{15}$$ Add numerators and keep the common denominator.

$$= \dfrac{-12 + 10}{15}$$ Because the addends have different signs, we subtract and keep the sign of the number with the greater absolute value.

$$= -\dfrac{2}{15}$$

c. $-\dfrac{7}{12} + \dfrac{7}{30}$

SOLUTION The LCD of 12 and 30 is 60.

$$-\dfrac{7}{12} + \dfrac{7}{30} = -\dfrac{7(5)}{12(5)} + \dfrac{7(2)}{30(2)}$$ Write equivalent fractions with 60 as the denominator.

$$= -\dfrac{35}{60} + \dfrac{14}{60}$$ Add numerators and keep the common denominator.

$$= \dfrac{-35 + 14}{60}$$

$$= -\dfrac{21}{60}$$ Reduce to lowest terms.

$$= -\dfrac{3 \cdot 7}{2 \cdot 2 \cdot 3 \cdot 5}$$ Write the numerator and denominator in terms of prime factors.

$$= -\dfrac{3 \cdot 7}{2 \cdot 2 \cdot 3 \cdot 5}$$ Divide out the common factor of 3.

$$= -\dfrac{7}{20}$$

Your Turn 5 Add.

a. $\dfrac{1}{15} + \dfrac{5}{12}$ **b.** $\dfrac{7}{8} + \left(-\dfrac{1}{5}\right)$ **c.** $-\dfrac{7}{18} + \dfrac{5}{30}$

Answers **Your Turn 5**

a. $\dfrac{29}{60}$ **b.** $\dfrac{27}{40}$ **c.** $-\dfrac{2}{9}$

Adding Decimal Numbers

Now let's review adding decimal numbers, which are also a part of the set of rational numbers. Recall that to add decimal numbers, we add like place values.

Example 6 Angela has a balance of $578.26 and incurs a debt of $92.45. What is Angela's new balance?

Solution A debt of $92.45 is −$92.45. Her balance is

$$578.26 + (-92.45) = \$485.81$$

Your Turn 6 Daryl has a balance of −$452.75 on a credit card, and he makes a payment of $120. What is his new balance?

Objective ③ Find the additive inverse of a number. What happens if we add two numbers that have the same absolute value but different signs, such as $5 + (-5)$? In money terms, this is like making a $5 payment toward a debt of $5. Notice that the payment pays off the debt so that the balance is 0.

$$5 + (-5) = 0$$

Because their sum is 0, we say that 5 and −5 are **additive inverses**, or opposites.

Definition *Additive inverses:* Two numbers whose sum is 0.

Example 7 Find the additive inverse of the given numbers.

a. 9
Answer −9 because $9 + (-9) = 0$
b. −7
Answer 7 because $-7 + 7 = 0$
c. 0
Answer 0 because $0 + 0 = 0$

Answer Your Turn 6
−$332.75

📟 **Calculator Tips**

To enter a negative number or find an additive inverse on a graphing or scientific calculator, use the $\boxed{(-)}$ *key. Note that this key is different from the subtraction key.*

To find the additive inverse of 9, type $\boxed{(-)}$ $\boxed{9}$ $\boxed{\text{ENTER}}$.

To find the additive inverse of −7, type $\boxed{(-)}$ $\boxed{(-)}$ $\boxed{7}$ $\boxed{\text{ENTER}}$.

On older scientific calculators, the $\boxed{+/-}$ *key is used to find the additive inverse of a number. You enter the value first, then press the* $\boxed{+/-}$ *key. Each time you press the* $\boxed{+/-}$ *key the sign will change.*

To find the additive inverse of 9, type $\boxed{9}$ $\boxed{+/-}$.

To find the additive inverse of −7, type $\boxed{7}$ $\boxed{+/-}$ $\boxed{+/-}$.

Your Turn 7 Find the additive inverse of the given numbers.

a. −12 **b.** 0.67 **c.** $-\dfrac{3}{8}$

A minus sign can be used to indicate additive inverse.

Example 8 Simplify.

a. −(−8)

Answer −(−8) indicates that we need to find the additive inverse of −8. So −(−8) = 8.

b. −|6|

Answer −|6| indicates that we need to find the additive inverse of the absolute value of 6. So −|6| = −6.

c. −|−4|

Answer −|−4| indicates that we need to find the additive inverse of the absolute value of −4. So −|−4| = −4.

Your Turn 8 Simplify.

a. −(−5) **b.** −|8| **c.** −|−3|

Objective ④ **Subtract rational numbers.** Now let's turn our attention to subtraction. The parts of subtraction are

$$8 - 5 = 3$$

Minuend Subtrahend Difference

Recall that when we add two numbers that have different signs, we actually subtract. This implies that subtraction and addition are related. In fact, every subtraction statement can be written as an equivalent addition statement. Notice that the subtraction statement 8 − 5 = 3 and the addition statement 8 + (−5) = 3 are equivalent: They are both ways of finding the balance given an $8 deposit with a $5 debt. Note that in writing 8 − 5 as an equivalent addition 8 + (−5), we change the operation sign from a minus sign to a plus sign and change the 5 to its additive inverse, −5.

Any subtraction statement can be written as an equivalent addition statement using this procedure. Consider 8 − (−5), which represents an $8 deposit subtracting a debt of $5. What does it mean to subtract a debt? If a bank records a debt and then realizes it made a mistake, it must reverse that debt. Subtracting, or reversing, a debt means that the bank must record a deposit to correct the mistake. In math terms, this means that 8 − (−5) is equivalent to 8 + 5, which equals 13. Notice that this still follows the process we just discussed.

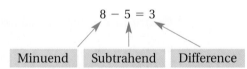

$$8 - (-5)$$

Change the operation from minus to plus. Change the subtrahend to its additive inverse.

$$= 8 + 5$$
$$= 13$$

Rewriting Subtraction

To write a subtraction statement as an equivalent addition statement, change the operation symbol from a minus sign to a plus sign and change the subtrahend to its additive inverse.

Example 9 Subtract.

a. $-15 - (-6)$

SOLUTION Write the subtraction as an equivalent addition.

$$-15 - (-6)$$

Change the operation from minus to plus. ——— Change the subtrahend to its additive inverse.

$$= -15 + 6$$
$$= -9$$

Note: We can think of this as subtracting a debt of \$6 from a debt of \$15. Subtracting a debt from a debt is equivalent to making a deposit or payment against the existing debt balance. So this decreases the amount of debt.

b. $-\dfrac{2}{3} - \dfrac{3}{4}$

SOLUTION Write an equivalent addition; then find a common denominator and continue the process of adding fractions.

$$-\frac{2}{3} - \frac{3}{4}$$

Change the operation from minus to plus. ——— Change the subtrahend to its additive inverse.

$$= -\frac{2}{3} + \left(-\frac{3}{4}\right)$$

$$= -\frac{2(4)}{3(4)} + \left(-\frac{3(3)}{4(3)}\right)$$

Write equivalent fractions with the common denominator, 12.

Note: If we begin with a debt and subtract a positive amount, we go deeper in debt.

$$= -\frac{8}{12} + \left(-\frac{9}{12}\right)$$

$$= -\frac{17}{12}$$

Calculator Tips

To enter the problem in Example 9(b) into a scientific calculator, type

$\boxed{(-)}$ $\boxed{2}$ $\boxed{a^{b/c}}$ $\boxed{3}$ $\boxed{-}$ $\boxed{3}$ $\boxed{a^{b/c}}$ $\boxed{4}$ $\boxed{\text{ENTER}}$.

On a graphing calculator, type

$\boxed{(-)}$ $\boxed{2}$ $\boxed{\div}$ $\boxed{3}$ $\boxed{-}$ $\boxed{3}$ $\boxed{\div}$ $\boxed{4}$ $\boxed{\text{ENTER}}$.

To see the answer in fraction form, press $\boxed{\text{MATH}}$ *and select the 1:Frac option from the menu; then press* $\boxed{\text{ENTER}}$.

c. $5.04 - 8.01$

SOLUTION   Write an equivalent addition statement.

Note: We begin with a positive balance of $5.04 and subtract $8.01. Because we are subtracting more than we have, the result is negative.

$5.04 - \textbf{8.01}$
$= 5.04 + (-\textbf{8.01})$
$= -2.97$

Note: To calculate the sum, the number with the smaller absolute value must be subtracted from the number with the greater absolute value.

$$\begin{array}{r} 8.01 \\ -5.04 \\ \hline 2.97 \end{array}$$

Because the number with the greater absolute value is negative, the result is negative.

Your Turn 9   Subtract.

a. $25 - (-17)$

b. $-\dfrac{3}{8} - \left(-\dfrac{1}{3}\right)$

c. $0.06 - 4.02$

Example 10   On Mars, the daytime high temperature can reach $-15°C$, while at night, the temperature can plummet to $-140°C$. What is the difference between the high and low temperatures?

SOLUTION   *Difference* indicates that we subtract the temperatures. Because the wording in the problem is to find the "difference between the high and low temperatures," we arrange the subtraction with the low temperature subtracted from the high temperature.

$$-15 - (-140) = -15 + 140 \qquad \text{Write as an equivalent addition.}$$
$$= 125 \qquad \text{Add.}$$

Answer   The difference between the high and low temperatures is $125°C$.

Answers   Your Turn 9

a. 42   **b.** $-\dfrac{1}{24}$ **c.** -3.96

Answer   Your Turn 10

$34.4°C$

Your Turn 10   In an experiment, a mixture begins at a temperature of $5.8°C$. The mixture is then cooled to a temperature of $-28.6°C$. Find the difference between the initial and final temperatures.

1.3 Exercises   For Extra Help   *MyMathLab*

Note: Exercises marked with a ★ represent challenging exercises.

1. Explain the difference between the commutative property of addition and the associative property of addition.

2. Explain why 4 and -4 are additive inverses.

3. Why is 0 called the additive identity?

4. In your own words, explain the process of adding or subtracting two fractions with different denominators.

5. In your own words, explain how to add two numbers that have the same sign.

6. In your own words, explain how to add two numbers that have different signs.

7. Explain how to write a subtraction statement as an equivalent addition statement.

8. Explain why $6 - 5$ and $5 - 6$ have different answers, whereas $-6 - 5$ and $-5 - 6$ have the same answer. (*Hint:* Think about the properties of addition.)

For Exercises 9–20, indicate whether each equation illustrates the additive identity, commutative property of addition, associative property of addition, or additive inverse. See Examples 1 and 7.

9. $5 + 6 = 6 + 5$

10. $-3 + 7 = 7 + (-3)$

11. $0 + 8 = 8$

12. $-5 + 0 = -5$

13. $0.8 + (-0.8) = 0$

14. $-\dfrac{4}{9} + \dfrac{4}{9} = 0$

15. $(3 + 5) + 2 = 3 + (5 + 2)$

16. $6 + (2 + 3) = (6 + 2) + 3$

17. $-8 + (7 + 3) = (7 + 3) + (-8)$

18. $(5 + 3) - 4 = -4 + (5 + 3)$

19. $6.3 + (2.1 - 2.1) = 6.3 + 0$

20. $(-4.6 + 4.6) + 9.5 = 0 + 9.5$

For Exercises 21–56, add. See Examples 2–6.

21. $8 + 13$

22. $15 + 7$

23. $-6 + (-12)$

24. $-5 + (-7)$

25. $-4 + 13$

26. $-5 + 16$

27. $-14 + 5$

28. $-17 + 8$

29. $27 + (-13)$

30. $29 + (-7)$

31. $-28 + 12$

32. $-16 + 13$

33. $\dfrac{5}{8} + \dfrac{1}{8}$

34. $\dfrac{9}{16} + \dfrac{5}{16}$

35. $-\dfrac{4}{9} + \left(-\dfrac{2}{9}\right)$

36. $-\dfrac{3}{5} + \left(-\dfrac{1}{5}\right)$

37. $\dfrac{1}{6} + \left(-\dfrac{5}{6}\right)$

38. $-\dfrac{9}{14} + \dfrac{3}{14}$

39. $\dfrac{3}{4} + \dfrac{1}{6}$

40. $\dfrac{1}{4} + \dfrac{7}{8}$

41. $-\dfrac{1}{12} + \left(-\dfrac{2}{3}\right)$

42. $-\dfrac{2}{5} + \left(-\dfrac{3}{20}\right)$

43. $-\dfrac{5}{6} + \dfrac{4}{21}$

44. $-\dfrac{5}{16} + \dfrac{3}{12}$

45. $0.21 + 0.05$

46. $0.06 + 0.17$ **47.** $-0.18 + 6.7$ **48.** $-15.81 + 4.28$ **49.** $-0.28 + (-4.1)$ **50.** $-7.8 + (-9.16)$

★51. $-42 + |-14|$ **★52.** $-31 + |-54|$ **★53.** $|-2.4| + |-0.78|$ **★54.** $|-0.6| + |-9.1|$ **★55.** $\left|-\frac{3}{8}\right| + \left|\frac{5}{6}\right|$

★56. $\left|-\frac{4}{5}\right| + \left|\frac{3}{4}\right|$

For Exercises 57–70, find the additive inverse. See Example 7.

57. 5 **58.** 7 **59.** -12 **60.** -6 **61.** 0 **62.** -9 **63.** $\frac{5}{6}$

64. $-\frac{6}{17}$ **65.** -0.29 **66.** 2.8 **67.** $-x$ **68.** b **69.** $\frac{m}{n}$ **70.** $-\frac{a}{b}$

For Exercises 71–78, simplify. See Example 8.

71. $-(-2)$ **72.** $-(-15)$ **★73.** $-(-(-4))$ **★74.** $-(-(-1))$

75. $-|4|$ **76.** $-|10|$ **77.** $-|-12|$ **78.** $-|-5|$

For Exercises 79–102, subtract. See Example 9.

79. $6 - 15$ **80.** $8 - 20$ **81.** $-4 - 9$ **82.** $-7 - 15$ **83.** $4 - (-3)$

84. $6 - (-7)$ **85.** $-8 - (-2)$ **86.** $-13 - (-6)$ **87.** $-\frac{1}{5} - \left(-\frac{1}{5}\right)$ **88.** $-\frac{3}{4} - \left(-\frac{3}{4}\right)$

89. $\frac{7}{10} - \left(-\frac{3}{5}\right)$ **90.** $\frac{3}{8} - \left(-\frac{5}{6}\right)$ **91.** $-\frac{4}{5} - \left(-\frac{2}{7}\right)$ **92.** $-\frac{1}{2} - \left(-\frac{1}{3}\right)$ **93.** $4.01 - 3.65$

94. $8.1 - 4.76$ **95.** $0.07 - 5.82$ **96.** $0.107 - 5.802$ **97.** $-6.1 - (-4.5)$ **98.** $-7.1 - (-2.3)$

★99. $-|-4| - |-6|$ **★100.** $-|-9| - |-12|$ **★101.** $|6.2| - |-7.1|$ **★102.** $|4.6| - |-7.3|$

103. The Walt Disney Company's third-quarter financial report for 2007 contains the following information. Find the net income.

Third-quarter report for the year ending September 29, 2007

Income (in millions)	Expenditures (in millions)
Revenues = $8930	Costs and expenses = $7359
Equity in the income of investees = $96	Net interest expense = $163
	Income taxes = $521
	Minority interests = $100
	Loss from discontinued operations = $6

104. Following is a balance sheet of income and expenditures for a small business. Find the net profit or loss.

June 2010 balance sheet

Income	Expenditures
Revenue = $24,572.88	Materials = $1545.75
Dividends = $1284.56	Lease = $2700
	Utilities = $865.45
	Employee wages = $21,580.50

105. In engineering, the resultant force on an object is the sum of all forces acting on the object. The diagram shows the forces acting on a steel beam. Find the resultant force. What does the sign of the resultant force indicate?

820.7 N   915.6 N

−2004.5 N

106. A family's taxable income is its income less deductions. Following is a table that shows the Brendel family's income and deductions for the 2010 tax year. What is the Brendel family's taxable income?

Income	Deductions
Mr. Brendel = $31,672.88	Mortgage interest = $6545.75
Mrs. Brendel = $32,284.56	Charitable donations = $1200
Dividends = $124.75	Medical expenses = $165.45
Miscellaneous income = $2400	Exemptions = $10,800

107. On February 13, 2008, the Dow Jones Industrial Average (DJIA) closed at 12,373.41. On February 12, 2008, the DJIA closed at 12,240.01. Find the difference in closing value from February 13 to February 12.

108. On January 28, 2008, the closing price of Dell's stock was $20.35. On January 29, 2008, the closing price was $20.56. Find the difference in closing price from January 29 to January 28.

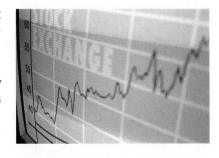

109. On February 11, 2008, the closing price of AT&T's stock was $24.78. If the price had a change of −$0.32 from the previous trading date's closing price, what was the closing price on February 8, 2008?

110. On February 14, 2008, an investor found that the NASDAQ closed at 2,320.04, which was a change of −53.89 from the previous day's closing value. What was the closing value on February 13, 2008?

111. The temperature of liquid nitrogen is −208°C. When an apple is placed in the liquid, the apple raises the temperature of the liquid to its boiling point of −196°C. Write an expression that describes the difference between the boiling point and the initial temperature; then calculate the difference.

112. Absolute zero is the temperature at which molecular motion is at a minimum. This temperature is −273.15°C. A piece of metal is cooled to −256.5°C. Write an expression that describes the difference between the piece of metal's current temperature and absolute zero; then calculate the difference.

Use the following line graph to answer Exercises 113 and 114. The graph shows the mean composite ACT score of entering college freshmen from 1985 to 2006.

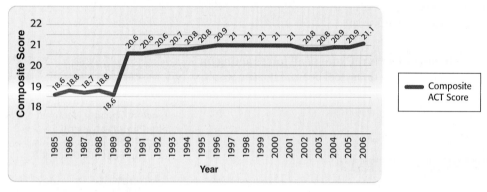

Source: U.S. Department of Education.

113. a. Write an expression that describes the difference between the mean composite scores in 1989 and 1986.

   b. Calculate the difference.

   c. What does the sign of the difference indicate about how the scores changed?

114. a. Write an expression that describes the difference between the mean composite scores in 2006 and 1986.

   b. Calculate the difference.

   c. What does the sign of the difference indicate about how the scores changed?

Use the bar graph to the right to answer Exercises 115 and 116. The graph shows the median annual income by level of education completed and gender from 1995 to 2004.

115. In what year was the difference between salaries for males and females with a bachelor's degree the greatest? How much was that difference?

116. In what year was the difference in salaries greatest between a person with a high school degree and a person with a bachelor's degree? How much was that difference?

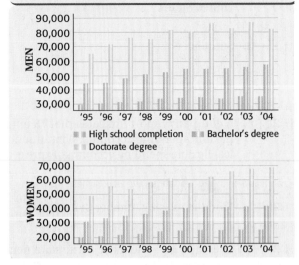

MEDIAN ANNUAL INCOME OF YEAR-ROUND FULL-TIME WORKERS

25 YEARS OLD AND OVER, BY LEVEL OF EDUCATION COMPLETED AND GENDER: 1995 TO 2004

Source: Information Please® Database, © Pearson Education, Inc. All rights reserved.

Puzzle Problem

Fill in each square with a number 1 through 9 so that the sum of the numbers in each row, each column, and the two diagonals is the same.

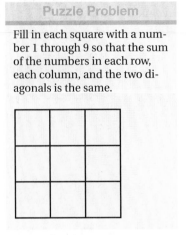

Collaborative Exercises   Dollars and Sense

Your group is to prepare a simple monthly household budget. The budget should be for a "typical" student or for a new employee in your field of study. (However, you should not use any of your personal information.) Remember to be realistic about both income and expenditures.

1. Discuss and agree upon an amount of income you think an average working student makes each month. (Or use a starting salary for a person in your field of study.)
2. Estimate average expense amounts for each of the following. If your group believes that an expense does not apply, explain why.

Taxes	Transportation/Gasoline
Housing	Child care
Electricity	Loan repayments
Water	Hair care
Telephone (home and cell)	Clothing
Cable TV	Magazine/Newspaper subscriptions
Internet service	Allowance
Food	Movies
Household supplies	Other entertainment (renting DVDs, attending concerts,
Medicine	eating out)

3. Using your estimates in Exercises 1 and 2, calculate the monthly net.
4. Discuss and list ways that a person could cut expenses to improve his or her monthly net.

Review Exercises

Exercises 1–4 EXPRESSIONS

[1.1] **1.** Write a set containing the last names of the first four presidents of the United States.

[1.1] **2.** Simplify: $|-7|$

[1.2] **3.** Find the prime factorization of 100.

[1.2] **4.** A company report indicates that 78 employees out of 91 have medical flexible spending accounts. What fraction of the employees have medical flexible spending accounts? Express the fraction in lowest terms.

Exercises 5–6 EQUATIONS AND INEQUALITIES

[1.2] **5.** Find the missing number that makes $\dfrac{40}{48} = \dfrac{?}{6}$ true.

[1.2] **6.** Use $<$, $>$, or $=$ to make a true statement: $-\dfrac{4}{5} \quad \boxed{} \quad -\dfrac{8}{20}$

1.4 Multiplying and Dividing Real Numbers; Properties of Real Numbers

Objectives

1. Multiply integers.
2. Multiply more than two numbers.
3. Multiply rational numbers.
4. Find the multiplicative inverse of a number.
5. Divide rational numbers.

Objective 1 Multiply integers. In a multiplication statement, *factors* are multiplied to equal a *product*. In Section 1.3, we discussed properties of addition. Multiplication has similar properties to addition. The following table shows the properties of multiplication.

$$2 \cdot 3 = 6$$

Factors   Product

Properties of Multiplication	Symbolic Form	Word Form
Multiplicative property of 0	$0 \cdot a = 0$	The product of a number multiplied by 0 is 0.
Multiplicative identity	$1 \cdot a = a$	The product of a number multiplied by 1 is the number.
Commutative property of multiplication	$ab = ba$	Changing the order of factors does not affect the product.
Associative property of multiplication	$a(bc) = (ab)c$	Changing the grouping of three or more factors does not affect the product.
Distributive property of multiplication over addition	$a(b + c) = ab + ac$	A sum multiplied by a factor is equal to the sum of that factor multiplied by each addend.

Example 1 Give the name of the property of multiplication that is illustrated by each equation.

a. $4(-3) = -3 \cdot 4$

Answer Commutative property of multiplication because the order of the factors is different.

b. $6(-2 \cdot 7) = [6(-2)] \cdot 7$

Answer Associative property of multiplication because the grouping of factors is different.

c. $5(8 - 3) = 5 \cdot 8 - 5 \cdot 3$

Answer Distributive property of multiplication over addition.

Your Turn 1 Give the name of the property of multiplication that is illustrated by each equation.

  a. $-8 \cdot 0 = 0$ **b.** $9 \cdot 1 = 9$ **c.** $-2(7 + 4) = -2 \cdot 7 - 2 \cdot 4$

Multiplying Numbers with Different Signs

To determine the rules for multiplying signed numbers, consider the following pattern. Pay attention to the product as we decrease the second factor.

$$2 \cdot 4 = 8$$
$$2 \cdot 3 = 6$$

As we decrease the second factor by 1, the product decreases by 2, from 8 to 6.

$$2 \cdot 2 = 4$$

Notice that the same pattern continues with the product decreasing by 2 each time we decrease the factor by 1.

$$2 \cdot 1 = 2$$
$$2 \cdot 0 = 0$$
$$2 \cdot (-1) = -2$$

Notice that when we decrease the 0 factor to -1, we continue the pattern and decrease the product by 2 so that it is -2.

$$2 \cdot (-2) = -4$$

Decreasing the factor by 1 again decreases the product by 2, resulting in -4.

This pattern suggests that multiplying a positive number by a negative number equals a negative product. Note that by the commutative property of multiplication, we can exchange the factors and conclude that multiplying a negative number times a positive number also equals a negative product.

$$2 \cdot (-3) = (-3) \cdot 2 = -6$$

Rule **Multiplying Two Numbers with Different Signs**

When multiplying two numbers that have different signs, the product is negative.

Example 2 Multiply.

a. $9(-8)$

Warning: Make sure you see the difference between $9(-8)$, which indicates multiplication, and $9 - 8$, which indicates subtraction.

SOLUTION $9(-8) = -72$

b. $(-12)3$

SOLUTION $(-12)3 = -36$

Your Turn 2 Multiply.

  a. $13(-7)$ **b.** $-5 \cdot 9$

Multiplying Numbers with the Same Sign

Now let's examine multiplying numbers with the same sign. Consider the following pattern to determine the rule.

$$(-2) \cdot 4 = -8$$ From the rule we already established, a negative number times a positive number is a negative product.

$$(-2) \cdot 3 = -6$$ ◄── As we decrease the positive factor by 1, the product *increases* by 2, from -8 to -6.

$$(-2) \cdot 2 = -4$$ Notice that the same pattern continues with the product increasing by 2 each time we decrease the factor by 1.

$$(-2) \cdot 1 = -2$$

$$(-2) \cdot 0 = 0$$

$$(-2) \cdot (-1) = 2$$ ◄── To continue the same pattern, when we decrease the 0 factor to -1, we must continue to increase the product by 2. This means that the product must become positive.

$$(-2) \cdot (-2) = 4$$

This pattern indicates that the product of two negative numbers is a positive number. We have already seen that the product of two positive numbers is a positive number. So we can draw the following conclusion:

Rule **Multiplying Two Numbers with the Same Sign**

When multiplying two numbers that have the same sign, the product is positive.

Example 3 Multiply.

a. $-6(-8)$

SOLUTION $-6(-8) = 48$

b. $(-9)(-7)$

SOLUTION $(-9)(-7) = 63$

Your Turn 3 Multiply.

   a. $-6(-9)$ 　　　　　　　　　　　　**b.** $(-3)(-13)$

Objective ② Multiply more than two numbers. Suppose we have to multiply more than two numbers, such as $(-2)(-3)(5)$. The associative property of multiplication allows us to group the factors any way we like.

$$(-2)(-3)(5) = 6(5)$$ Multiply the first two factors.
$$= 30$$

or

$$(-2)(-3)(5) = (-2)(-15)$$ Multiply the second and third factors.
$$= 30$$

Notice that no matter which way we choose to multiply the factors, the result is the same. Also notice that the result is positive because there are two negative factors involved in the multiplication. What if we have three negative factors, as in $(-2)(-3)(-5)$?

$$(-2)(-3)(-5) = 6(-5)$$ Multiply the first two factors.
$$= -30$$

Answers **Your Turn 3**
a. 54　**b.** 39

or

$$(-2)(-3)(-5) = (-2)(15) \quad \text{Multiply the second and third factors.}$$
$$= -30$$

With three negative factors, the result is negative. This suggests the following rule:

Rule **Multiplying with Negative Factors**

The product of an even number of negative factors is positive, whereas the product of an odd number of negative factors is negative.

Example 4 Multiply.

a. $(-1)(-1)(-2)(6)$

SOLUTION Because there are three negative factors (an odd number of negative factors), the result is negative.

$$(-1)(-1)(-2)(6) = -12$$

b. $(-1)(-3)(-7)(2)(-4)$

SOLUTION Because there are four negative factors (an even number of negative factors), the result is positive.

$$(-1)(-3)(-7)(2)(-4) = 168$$

Your Turn 4 Multiply.

a. $(-3)(4)(-5)(2)$ **b.** $(6)(-5)(-1)(2)(-1)$

Objective ③ Multiply rational numbers.

Multiplying Fractions

On a ruler, note that half of a fourth is an eighth. When preceded by a fraction, the word *of* means multiply. So the sentence would translate as follows:

Half of a fourth is an eighth.

$$\frac{1}{2} \cdot \frac{1}{4} = \frac{1}{8}$$

Notice that the numerators multiply to equal the numerator and the denominators multiply to equal the denominator. This suggests the following rule for multiplying fractions:

Rule **Multiplying Fractions**

$$\frac{a}{b} \cdot \frac{c}{d} = \frac{ac}{bd}, \text{ where } b \neq 0 \text{ and } d \neq 0.$$

Multiplying and Simplifying

In mathematics, we are expected to simplify results to lowest terms. Consider $\frac{3}{4} \cdot \frac{5}{6}$. We can multiply first, then simplify.

$$\frac{3}{4} \cdot \frac{5}{6} = \frac{15}{24} = \frac{3 \cdot 5}{2 \cdot 2 \cdot 2 \cdot 3} = \frac{5}{8}$$

Note: After multiplying, we divide out the common factor, 3.

Or as an alternative, we can divide out the common factors before multiplying the fractions. In our example, we can divide out the common factor of 3 and then multiply.

$$\frac{3}{4} \cdot \frac{5}{6} = \frac{3}{2 \cdot 2} \cdot \frac{5}{2 \cdot 3} = \frac{5}{8}$$

Note: Some people prefer to divide out the common factors without breaking down to primes, like this:

$$\frac{\overset{1}{3}}{4} \cdot \frac{5}{\underset{2}{6}} = \frac{5}{8}$$

We can divide out common factors before or after multiplying because every factor in the original numerators becomes a factor in the product's numerator and every factor in the original denominators becomes a factor in the product's denominator. Now let's put this together with the sign rules for multiplication.

Connection Using the slash marks style, Example 5(a) looks like this:

$$-\frac{\overset{1}{2}}{3} \cdot \frac{7}{\underset{4}{8}} = -\frac{7}{12}$$

Notice that we are still dividing out a factor of 2 in the 2 and 8.

 **Calculator Tips**

On most scientific calculators, for Example 5(a), type

Example 5 Multiply.

a. $-\frac{2}{3} \cdot \frac{7}{8}$

SOLUTION $-\frac{2}{3} \cdot \frac{7}{8} = -\frac{2}{3} \cdot \frac{7}{2 \cdot 2 \cdot 2}$ Divide out the common factor, 2.

$$= -\frac{7}{12}$$ Because we are multiplying two numbers that have different signs, the product is negative.

b. $-\frac{4}{15} \cdot \left(-\frac{9}{16}\right) \cdot \frac{14}{15}$

SOLUTION $-\frac{4}{15} \cdot \left(-\frac{9}{16}\right) \cdot \frac{14}{15} = -\frac{2 \cdot 2}{3 \cdot 5} \cdot \left(-\frac{3 \cdot 3}{2 \cdot 2 \cdot 2 \cdot 2}\right) \cdot \frac{2 \cdot 7}{3 \cdot 5}$ Divide out the common factors.

$$= \frac{7}{50}$$ Because there is an even number of negative factors, the product is positive.

Your Turn 5 Multiply.

a. $\frac{5}{8} \cdot \left(-\frac{3}{10}\right)$ **b.** $-\frac{4}{5} \cdot \left(\frac{1}{6}\right) \cdot \left(-\frac{3}{9}\right)$

Multiplying Decimal Numbers

Recall that decimal numbers name fractions with denominators that are powers of 10. In the problem $0.3 \cdot 0.7$, we could multiply by replacing the decimal factors with their fraction equivalents.

$$0.3 \cdot 0.7 = \frac{3}{10} \cdot \frac{7}{10} = \frac{21}{100} = 0.21$$

Answers **Your Turn 5**

a. $-\frac{3}{16}$ **b.** $\frac{2}{45}$

Because we multiply the denominators, if we multiply tenths times tenths, the product will be hundredths. Notice that if we count the number of decimal places in the factors, this corresponds to the total places in the product.

$$
\begin{array}{lll}
0.7 & \longrightarrow & 1 \text{ place} \\
\underline{\times\ 0.3} & \longrightarrow & \underline{+\ 1 \text{ place}} \\
0.21 & \longrightarrow & 2 \text{ places}
\end{array}
$$

Likewise, hundredths (2 places) times tenths (1 place) equals thousandths (3 places).

$$
0.03 \cdot 0.7 = \frac{3}{100} \cdot \frac{7}{10} = \frac{21}{1000} = 0.021 \quad \text{or} \quad
\begin{array}{lll}
0.7 & \longrightarrow & 1 \text{ place} \\
\underline{\times\ 0.03} & \longrightarrow & \underline{+\ 2 \text{ places}} \\
0.021 & \longrightarrow & 3 \text{ places}
\end{array}
$$

Procedure **Multiplying Decimal Numbers**

To multiply decimal numbers:

1. Multiply as if they were whole numbers.
2. Place the decimal in the product so that it has the same number of decimal places as the total number of decimal places in the factors.

Example 6 Multiply.

a. $(-5.6)(0.03)$

SOLUTION First, we calculate the value and disregard the signs for now.

$$
\begin{array}{lll}
0.03 & & 2 \text{ places} \\
\underline{\times\quad 5.6} & & +\ 1 \text{ place} \\
0\,1\,8 & & \\
\underline{+\,0\,1\,5} & & \\
0.1\,6\,8 & \longleftarrow & 3 \text{ places}
\end{array}
$$

Answer -0.168 **Note:** When we multiply two numbers with different signs, the product is negative.

b. $(-2)(4.3)(1.8)(-7.1)$

SOLUTION First, we calculate the value and disregard the signs for now. Multiply from left to right.

$$
\underbrace{(2)(4.3)}\qquad\qquad \underbrace{(8.6)(1.8)}
$$

$$
(2)(4.3)(1.8)(7.1) = (8.6)(1.8)(7.1) = (15.48)(7.1) = 109.908
$$

Answer 109.908 **Note:** The product of an even number of negative factors is positive. The factors have a total of three decimal places, so the product has three decimal places.

Your Turn 6 Multiply.

a. $(-0.07)(-2.65)$ **b.** $(-1)(-0.9)(-24)(0.2)$

Objective ④ Find the multiplicative inverse of a number. Two numbers that multiply to equal 1 are called **multiplicative inverses**.

Definition *Multiplicative inverses:* Two numbers whose product is 1.

For example, $\frac{2}{3}$ and $\frac{3}{2}$ are multiplicative inverses because their product is 1.

$$\frac{2}{3} \cdot \frac{3}{2} = \frac{6}{6} = 1$$

Notice that to write a number's multiplicative inverse, we simply invert the numerator and denominator. Multiplicative inverses are also known as *reciprocals*.

Example 7 Find the multiplicative inverse.

a. $\frac{5}{8}$

Answer The multiplicative inverse of $\frac{5}{8}$ is $\frac{8}{5}$ because $\frac{5}{8} \cdot \frac{8}{5} = 1$.

b. $-\frac{1}{4}$

Answer The multiplicative inverse of $-\frac{1}{4}$ is -4 because $-\frac{1}{4} \cdot (-4) = -\frac{1}{4} \cdot \left(-\frac{4}{1}\right) = 1$.

c. -6

Answer The multiplicative inverse of -6 is $-\frac{1}{6}$ because $-6 \cdot \left(-\frac{1}{6}\right) = -\frac{6}{1} \cdot \left(-\frac{1}{6}\right) = 1$.

Your Turn 7 Find the multiplicative inverse.

a. $\frac{4}{5}$ **b.** $\frac{1}{7}$ **c.** -9

Objective ⑤ Divide rational numbers. Now let's turn our attention to division. The parts of a division statement are shown next.

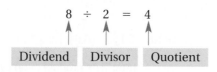

Connection Division can also be expressed in fraction form. For example, the division statement $8 \div 2 = 4$ can also be written as $\frac{8}{2} = 4$.

Sign Rules for Division

The sign rules for division are the same as for multiplication. We can see this by looking at the relationship between division and multiplication. We can use one operation to check the other. For example, if $8 \cdot 2 = 16$, then $16 \div 2 = 8$ or $16 \div 8 = 2$. In checking a multiplication, we divide the product by one of the factors to equal the other factor. If we place signs in the problem, we can say that if $8 \cdot (-2) = -16$, then $-16 \div (-2) = 8$ or $-16 \div 8 = -2$.

Answers Your Turn 7

a. $\frac{5}{4}$ b. 7 c. $-\frac{1}{9}$

Dividing Signed Numbers

When dividing two numbers that have the *same* sign, the quotient is positive.
When dividing two numbers that have *different* signs, the quotient is negative.

Example 8   Divide.

a. $48 \div (-6)$

SOLUTION $48 \div (-6) = -8$

b. $-35 \div (-7)$

SOLUTION $-35 \div (-7) = 5$

Your Turn 8   Divide.

   a. $(-60) \div 12$ **b.** $-42 \div (-6)$

Division Involving 0

What if 0 is involved in division? First, consider when the *dividend* is 0 and the *divisor* is not 0, as in $0 \div 9$. The result should be $0 \div 9 = 0$. Remember that we can check division by multiplying the quotient and the divisor to equal the dividend. So we can verify that $0 \div 9 = 0$ by writing the following multiplication:

$$\text{Quotient} \cdot \text{Divisor} = \text{Dividend}$$
$$0 \quad \cdot \quad 9 \quad = \quad 0$$

Conclusion   If the dividend is 0 and the divisor is not zero, the quotient is 0.

What if the divisor is 0, as in $8 \div 0$? Again, think about the check statement.

$$\text{Quotient} \cdot \text{Divisor} = \text{Dividend}$$
$$? \quad \cdot \quad 0 \quad = \quad 8$$

There is no number that can make $? \cdot 0 = 8$ true because any number multiplied by 0 equals 0, not 8. So there also is no number that can make $8 \div 0 = ?$ true, and we say that dividing a nonzero number by 0 is *undefined*.

Conclusion   If the divisor is 0 with a nonzero dividend, the statement is undefined.

What if both the dividend and divisor are 0, as in $0 \div 0$? Again, think about the check statement.

$$\text{Quotient} \cdot \text{Divisor} = \text{Dividend}$$
$$? \quad \cdot \quad 0 \quad = \quad 0$$

Notice that any number will work for this quotient because any number times 0 equals 0. Because we cannot determine a unique quotient here, we say that $0 \div 0$ is *indeterminate*.

Conclusion   If both the dividend and divisor are 0, the statement is indeterminate.

Following is a summary of these rules.

Connection This rule says that the numerator of a fraction can equal 0 (unless the denominator equals 0) but that the denominator can never equal 0.

Calculator Tips

Try dividing any number by 0 on your calculator. When you press the equal key, the screen will display "error," which is your calculator's way of indicating undefined or indeterminate.

Dividing Fractions

To determine how to divide when the divisor is a fraction, consider this problem: How many quarters are in \$5? A quarter is $\frac{1}{4}$ of a dollar, so this question translates to the following division problem:

$$\text{Number of quarters} = 5 \div \frac{1}{4}$$

Now since there are 4 quarters in each of those 5 dollars, there must be 20 quarters in \$5. Notice in analyzing the situation that we translated the division problem into a multiplication problem. We can say the following:

$$\text{Number of quarters} = 5 \div \frac{1}{4} = 5 \cdot 4 = 20$$

Notice that $\frac{1}{4}$ and 4 are multiplicative inverses. We can write a division problem as an equivalent multiplication problem by changing the divisor to its multiplicative inverse.

Rule **Dividing Fractions**

$$\frac{a}{b} \div \frac{c}{d} = \frac{a}{b} \cdot \frac{d}{c}, \text{ where } b \neq 0, c \neq 0, \text{ and } d \neq 0.$$

Connection Using slash marks, Example 9 looks like this:

$$-\frac{7}{8} \div \frac{5}{6} = -\frac{7}{\overset{}{8}_4} \cdot \frac{\overset{3}{6}}{5} = -\frac{21}{20}$$

Example 9 Divide. $-\frac{7}{8} \div \frac{5}{6}$

SOLUTION $-\frac{7}{8} \div \frac{5}{6} = -\frac{7}{8} \cdot \frac{6}{5}$ Write an equivalent multiplication.

$$= -\frac{7}{2 \cdot 2 \cdot 2} \cdot \frac{2 \cdot 3}{5}$$ Divide out the common factor, 2.

$$= -\frac{21}{20}$$ Because we are dividing two numbers that have different signs, the result is negative.

Your Turn 9 Divide.

a. $\frac{4}{9} \div \left(-\frac{8}{15}\right)$ **b.** $-\frac{8}{21} \div \left(-\frac{6}{7}\right)$

Dividing Decimal Numbers

Division with decimal numbers can be broken into two categories: (1) if the divisor is an integer and (2) if the divisor contains a decimal point. The following procedure summarizes how to divide decimal numbers.

Answers **Your Turn 9**

a. $-\frac{5}{6}$ **b.** $\frac{4}{9}$

Dividing Decimal Numbers

To divide decimal numbers, set up a long division and consider the divisor.

Case 1: If the divisor is an integer, divide as if the dividend were a whole number and place the decimal point in the quotient directly above its position in the dividend.

Case 2: If the divisor is a decimal number,

1. Move the decimal point in the divisor to the right enough places to make the divisor an integer.
2. Move the decimal point in the dividend the same number of places.
3. Divide the divisor into the dividend as if both numbers were whole numbers. Make sure you align the digits in the quotient properly.
4. Write the decimal point in the quotient directly above its new position in the dividend.

In either case, continue the division process until you get a remainder of 0 or a repeating digit (or block of digits) in the quotient.

Example 10 Divide. $-12.8 \div (-0.09)$

SOLUTION Because the divisor is a decimal number, we move the decimal point enough places to the right to create an integer—in this case, two places. Then we move the decimal point two places to the right in the dividend. Because we are dividing two numbers with the same sign, the result is positive.

$$
\begin{array}{r}
142.22 \\
9\overline{)1280.00} \\
-9 \\
\overline{38} \\
-36 \\
\overline{20} \\
-18 \\
\overline{20}
\end{array}
$$

Note: Because the 2 digit repeats without end, we write a repeat bar over the 2 just after the decimal point.

Answer $142.\overline{2}$

Your Turn 10 Divide.

a. $32.04 \div (-12)$ **b.** $-13.12 \div 6.4$ **c.** $-0.88 \div (-0.6)$

Applications

It is often necessary to use signed numbers when solving applications problems.

Example 11 Solve.

a. For a chemistry experiment, Fred used $\frac{3}{5}$ of the contents of a bottle of hydrochloric acid that was $\frac{3}{4}$ full. What fractional part of a full bottle did he use?

SOLUTION To find $\frac{3}{5}$ of $\frac{3}{4}$, multiply $\frac{3}{5}$ and $\frac{3}{4}$.

$$
\frac{3}{5} \cdot \frac{3}{4} = \frac{3 \cdot 3}{5 \cdot 4} = \frac{9}{20}
$$

Answer Fred used $\frac{9}{20}$ of a full bottle of hydrochloric acid.

b. An object's weight is a downward force due to gravity, and this force is calculated by multiplying the object's mass by the acceleration due to gravity (Force = Mass × Acceleration.) Find the force due to gravity of a truck with a mass of 1800 kilograms if the acceleration due to gravity is −9.8 m/sec². The force will be in newtons.

SOLUTION Force = Mass × Acceleration.   **Substitute for mass and acceleration.**

$$Force = (1800)(-9.8)$$
$$= -17,640$$

Answer The force due to gravity of the truck is −17,640 newtons.

Your Turn 11 Solve.

a. At a company picnic, $\frac{2}{5}$ of the people attending were men and of those men, $\frac{3}{4}$ were over the age of 50. What fraction of those attending were men over the age of 50?

b. A truck is carrying a load that has a weight of 4900 newtons, which means that the downward force due to gravity is −4900 newtons. Find the mass of the load in kilograms.

Answers Your Turn 11

a. $\frac{3}{10}$ **b.** 500 kg

1.4 Exercises   For Extra Help   *MyMathLab*

Note: Exercises marked with a ★ represent challenging exercises.

For Exercises 1–4, state whether the result is positive or negative.

1. When multiplying or dividing two numbers that have the same sign, the sign of the result is _____.

2. When multiplying or dividing two numbers that have different signs, the sign of the result is _____.

3. When a multiplication problem contains an odd number of negative factors, the sign of the product is _____.

4. When a multiplication problem contains an even number of negative factors, the sign of the product is _____.

5. Why is $5 \div 0$ undefined?

6. Why is $0 \div 0$ indeterminate?

7. Why are 5 and $\frac{1}{5}$ multiplicative inverses?

8. Explain how to divide fractions.

9. Explain why $\frac{-a}{b} = \frac{a}{-b} = -\frac{a}{b}$, where a and b are any real numbers and $b \neq 0$.

10. Explain the difference between $6(-7)$ and $6 - 7$.

For Exercises 11–22, indicate whether each equation illustrates the multiplicative property of 0, the multiplicative identity, the commutative property of multiplication, the associative property of multiplication, or the distributive property. See Example 1.

11. $6(3 + 4) = 6 \cdot 3 + 6 \cdot 4$

12. $7(2 - 5) = 7 \cdot 2 - 7 \cdot 5$

13. $1 \cdot \left(-\dfrac{5}{6}\right) = -\dfrac{5}{6}$

14. $1 \cdot (-7) = -7$

15. $0 \cdot (-6.3) = 0$

16. $\dfrac{2}{3} \cdot 0 = 0$

17. $\dfrac{3}{5} \cdot \left(-\dfrac{1}{4}\right) = -\dfrac{1}{4} \cdot \dfrac{3}{5}$

18. $-\dfrac{2}{3} \cdot \dfrac{4}{5} = \dfrac{4}{5} \cdot \left(-\dfrac{2}{3}\right)$

19. $[4(-2)] \cdot 3 = 4(-2 \cdot 3)$

20. $-6 \cdot (5 \cdot 4) = (-6 \cdot 5) \cdot 4$

21. $(-4.5)(2 + 0.6) = (2 + 0.6)(-4.5)$

22. $-6.2(-4.3 + 7.1) = (-4.3 + 7.1)(-6.2)$

For Exercises 23–56, multiply. See Examples 2–6.

23. $6(-3)$

24. $4(-7)$

25. $(-10)5$

26. $(-8)(5)$

27. $(9)(-6)$

28. $(12)(-4)$

29. $-4(-5)$

30. $(-4)(-3)$

31. $(-7)(-6)$

32. $(-8)(-12)$

33. $\dfrac{1}{2} \cdot \left(-\dfrac{1}{2}\right)$

34. $-\dfrac{4}{5} \cdot \left(\dfrac{20}{3}\right)$

35. $\left(-\dfrac{2}{3}\right)\left(\dfrac{3}{4}\right)$

36. $\left(-\dfrac{5}{6}\right)\left(-\dfrac{6}{5}\right)$

37. $\left(-\dfrac{5}{6}\right)\left(\dfrac{8}{15}\right)$

38. $\left(\dfrac{2}{9}\right)\left(-\dfrac{21}{26}\right)$

39. $3.8(-10)$

40. $8(-2.5)$

41. $-1.6(-4.2)$

42. $-7.1(-0.5)$

43. $-4.3(1.52)$

44. $8.1(-2.75)$

45. $6(-9)(-1)$

46. $-4(5)(-3)$

47. $6(-2)(4)$

48. $3(7)(-8)$

49. $(-4)(-6)(-7)$

50. $(-5)(-3)(-2)$

51. $12(-6)(-2)(-1)$

52. $-5(3)(-4)(-2)$

53. $(-1)(-6)(-40)(-3)$

54. $(-2)(-4)(-30)(-1)$

55. $(-2)(3)(-4)(-1)(2)$

56. $(-1)(-1)(4)(-5)(-3)$

For Exercises 57–64, find the multiplicative inverse. See Example 7.

57. $\dfrac{2}{3}$

58. $\dfrac{20}{3}$

59. $-\dfrac{5}{2}$

60. $-\dfrac{6}{7}$

61. -5

62. 17

63. 0

64. -1

For Exercises 65–92, divide. See Examples 8–10.

65. $-6 \div 2$

66. $42 \div (-7)$

67. $-56 \div (-4)$

68. $-12 \div (-4)$

69. $\dfrac{-18}{9}$

70. $\dfrac{75}{-3}$

71. $\dfrac{-63}{-7}$

72. $\dfrac{-48}{-6}$

73. $\dfrac{0}{-6}$

74. $\dfrac{0}{5}$

75. $\dfrac{-8}{0}$

76. $-21 \div 0$

77. $\dfrac{0}{0}$

78. $0 \div 0$

79. $6 \div \dfrac{2}{3}$

80. $-8 \div \dfrac{3}{4}$

81. $\dfrac{3}{5} \div \left(-\dfrac{9}{10}\right)$

82. $-\dfrac{4}{5} \div \dfrac{4}{5}$

83. $-\dfrac{2}{7} \div \left(-\dfrac{8}{21}\right)$

84. $-\dfrac{1}{3} \div \left(-\dfrac{3}{2}\right)$

85. $-\dfrac{4}{9} \div \dfrac{10}{21}$

86. $\dfrac{7}{15} \div \left(-\dfrac{35}{24}\right)$

87. $9.03 \div 4.3$

88. $8.1 \div 0.6$

89. $-36.72 \div (-0.4)$

90. $-10.65 \div (-7.1)$

91. $-14 \div 0.3$

92. $19 \div (-0.06)$

For Exercises 93–98, solve. See Example 11a.

93. In a poll where respondents can agree, disagree, or have no opinion, $\dfrac{3}{4}$ of the respondents said that they agreed and $\dfrac{2}{3}$ of those that agreed were women. What fraction of all respondents were women who agreed with the statement in the poll?

94. On a standard-size guitar, the length of the strings between the saddle and nut is $25\dfrac{1}{2}$ inches. Guitar makers must place the 12th fret at exactly half of the length of the string. How far from the saddle or nut should a guitar maker measure to place the 12th fret?

Saddle

Nut

12th fret

95. A financial planner estimates that at the rate one of his clients is increasing debt, she will have $3\dfrac{1}{2}$ times her current debt in five years. If her current debt is represented by $-\$2480$, how can her debt in five years be represented?

Of Interest

Three components affect the pitch of a string on a stringed instrument: diameter, tension, and length. Given two strings of the same diameter and same tension, if one string is half the length of the other, it will have a pitch that is one octave higher than the longer string. On a guitar, placing a finger at the 12th fret cuts the string length in half, thereby creating a tone that is an octave higher than the string's open tone.

96. Mario finds that $\frac{2}{3}$ of his credit card debt is from dining out. If his total credit card debt is represented by $-\$858$, how much of the debt is from dining out?

97. In 2000, the Garret family's only debt was credit card balance represented by $-\$258.75$. In 2010, they have a mortgage, two car loans, and three credit cards with a total debt represented by $-\$158,572.85$. How many times greater is their debt in 2010 than it was in 2000?

98. A company's stock loses value by an amount represented by $-\$\frac{3}{8}$ each day for four days. Write a representation of the total loss in value at the close of the 4th day.

For Exercises 99–102, use the fact that an object's weight is a downward force due to gravity and is calculated by multiplying the object's mass by the acceleration due to gravity, which is a constant (Force = Mass × Acceleration). The following table indicates the units. See Example 11b.

	Force	=	Mass	×	Acceleration
American measurement	pounds (lb)		slugs (s)		-32.2 ft./sec.^{2}
Metric measurement	newtons (N)		kilograms (kg)		-9.8 m/sec.^{2}

99. Find the force due to gravity on an object with a mass of 12.5 slugs. What does the sign of the weight indicate?

100. Find the force due to gravity on a person with a mass of 70.4 kilograms.

101. The blue whale is the largest animal on Earth. The largest blue whale ever caught weighed 1,658,181.8 newtons, which means that the downward force due to gravity was $-1,658,181.8$ newtons. What was the whale's mass in kilograms? (*Source: The Whale Watcher's Guide: Whale Watching Trips in North America*, Corrigan, 1999)

102. The Liberty Bell weighs 2080 pounds, which means the downward force due to gravity is -2080 pounds. What is its mass in slugs? (*Source: Ring in the Jubilee: The Epic of America's Liberty Bell*, Boland, 1973)

For Exercises 103 and 104, use the fact that in an electrical circuit, voltage (V) is equal to the product of the current, measured in amperes (A), and the resistance of the circuit, measured in ohms (Ω) (Voltage = Current × Resistance).

103. Suppose the current in a circuit is -6.4 amperes and the resistance is 8 ohms. Find the voltage.

104. An electrical technician measures the voltage in a circuit to be -15 volts and the current to be -8 amperes. What is the resistance of the circuit?

For Exercises 105 and 106, use the fact that in an electrical circuit, power, which is measured in watts (W), is the product of the voltage and the current (Power = Voltage × Current).

105. An engineer measures the voltage in a circuit to be −120 volts and the current to be −30 amperes. What is the power?

★**106.** An electrical technician determines that the power in a circuit is 400 watts. If the current is measured to be −6.5 amperes, what is the resistance in the circuit?

Review Exercises

Exercises 1–3 CONSTANTS AND VARIABLES

[1.1] **1.** Is π a rational or irrational number?

[1.1] **2.** Graph −7.2 on a number line.

[1.3] **3.** Find the additive inverse of $\frac{2}{3}$.

Exercises 4–6 EXPRESSIONS

[1.1] **4.** Simplify: $|-6.8|$

[1.3] **5.** Add: $-7 + (-15)$

[1.3] **6.** Subtract: $-\frac{5}{8} - \left(-\frac{1}{6}\right)$

1.5 Exponents, Roots, and Order of Operations

Objectives
1. Evaluate numbers in exponential form.
2. Evaluate square roots.
3. Use the order-of-operations agreement to simplify numerical expressions.
4. Find the mean of a set of data.

Objective 1 **Evaluate numbers in exponential form.** Sometimes, problems involve repeatedly multiplying the same number. In such problems, we can use an **exponent** to indicate that a **base** number is repeatedly multiplied.

Definitions ***Exponent:*** A symbol written to the upper right of a base number that indicates how many times to use the base as a factor.

Base: The number that is repeatedly multiplied.

When we write a number with an exponent, we say that the expression is in *exponential form*. The expression 2^4 is in exponential form, where the base is 2 and the exponent is 4. It is read "two to the fourth power," or simply "two to the fourth." To *evaluate* 2^4, write 2 as a factor 4 times; then multiply.

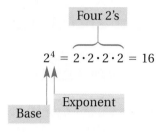

$$2^4 = 2 \cdot 2 \cdot 2 \cdot 2 = 16$$

Four 2's

Exponent

Base

Procedure Evaluating an Exponential Form

To evaluate an exponential form raised to a natural number exponent, write the base as a factor the number of times indicated by the exponent; then multiply.

Example 1   Evaluate.

a. $(-7)^2$

SOLUTION   The exponent 2 indicates that we have two factors of -7. Because we multiply two negative numbers, the result is positive.

$$(-7)^2 = (-7)(-7) = 49$$

> **Note:** When a base is raised to the second power, we say it is **squared**. The expression $(-7)^2$ can be read as "negative seven **squared**."

b. $\left(-\dfrac{2}{3}\right)^3$

SOLUTION   The exponent 3 means that we write the base as a factor three times.

$$\left(-\frac{2}{3}\right)^3 = \left(-\frac{2}{3}\right)\left(-\frac{2}{3}\right)\left(-\frac{2}{3}\right)$$
$$= -\frac{8}{27}$$

> **Note:** When a base is raised to the third power, we can say it is **cubed**. The expression $\left(-\dfrac{2}{3}\right)^3$ can be read as "negative two-thirds **cubed**."

Example 1 suggests the following sign rules:

Rule Evaluating Exponential Forms with Negative Bases

- If the base of an exponential form is a negative number and the exponent is even, the product is positive.

- If the base is a negative number and the exponent is odd, the product is negative.

Your Turn 1   Evaluate.

a. $\left(-\dfrac{1}{2}\right)^5$

b. $(-0.3)^4$

There is a great deal of difference between expressions of the form $(-a)^n$ and $-a^n$. The base of the exponent in $(-a)^n$ is $-a$. So $(-a)^n$ means multiply n factors of $-a$. The base of the exponent in $-a^n$ is a. So $-a^n$ means find the additive inverse of a^n.

Note: This is read as "negative 2 to the sixth power."

Note: This is read as "the additive inverse of two to the sixth power."

Example 2   Evaluate.

a. $(-2)^6$

SOLUTION $(-2)^6 = (-2)(-2)(-2)(-2)(-2)(-2) = 64$

b. -2^6

SOLUTION $-2^6 = -2 \cdot 2 \cdot 2 \cdot 2 \cdot 2 \cdot 2 = -64$

c. $(-3)^3$

SOLUTION $(-3)^3 = (-3)(-3)(-3) = -27$

d. -3^3

SOLUTION $-3^3 = -3 \cdot 3 \cdot 3 = -27$

Your Turn 2   Evaluate.

a. $(-3)^4$ **b.** -3^4 **c.** $(-2)^3$ **d.** -2^3

Objective ❷ Evaluate square roots.   Roots are inverses of exponents. More specifically, a square root is the inverse of a square. So a square root of a given number is a number that, when squared, equals the given number. For example, 5 is a square root of 25 because $5^2 = 25$. Notice that -5 is also a square root of 25 because $(-5)^2 = 25$.

Conclusion   For every real number, there are *two* square roots, a positive root and a negative root.

For convenience, if asked to find all square roots of 25, we can write both the positive and negative answers in a compact expression like this: ± 5.

Answers   Your Turn 2

a. 81   **b.** -81 **c.** -8 **d.** -8

Example 3   Find all square roots of the given number.

a. 36

Answer   ±6

b. −25

Answer   No real-number square roots exist.

> **Note:** We do not say that this situation is undefined or indeterminate because we eventually will define the square root of a negative number. Such roots will be defined using imaginary numbers.

Explanation   Because of the sign rules for multiplication, there is no way to square a real number and get anything other than a positive result.

Our work so far suggests the following rules:

Rule   **Square Roots**

- Every positive number has two square roots, a positive root and a negative root.
- Negative numbers have no real-number square roots.

Your Turn 3   Find all square roots of the given number.

a. 144   **b.** −81

The Principal Square Root

A symbol, $\sqrt{\ }$, called the *radical*, is used to indicate finding only the *positive* (or *principal*) square root of a given number. The given number or expression inside the radical is called the *radicand*.

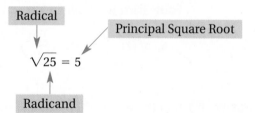

What about the square root of a fraction such as $\sqrt{\dfrac{4}{9}}$? Because $\left(\dfrac{2}{3}\right)^2 = \dfrac{2}{3} \cdot \dfrac{2}{3} = \dfrac{4}{9}$, we can say that $\sqrt{\dfrac{4}{9}} = \dfrac{2}{3}$. Notice that we can evaluate $\sqrt{\dfrac{4}{9}}$ by finding the square roots of the numerator and denominator separately, or $\sqrt{\dfrac{4}{9}} = \dfrac{\sqrt{4}}{\sqrt{9}} = \dfrac{2}{3}$. We now have the following rules for square roots involving the radical sign.

Rule   **Square Roots Involving the Radical Sign**

- The radical symbol $\sqrt{\ }$ denotes only the positive (principal) square root.

- $\sqrt{\dfrac{a}{b}} = \dfrac{\sqrt{a}}{\sqrt{b}}$, where $a \geq 0$ and $b > 0$.

Answers   **Your Turn 3**

a. ±12
b. No real-number square roots exist.

Calculator Tips

To calculate a square root, use the $\sqrt{}$ function, which may require pressing [2nd] *and then the* [x²] *key. To find the square root of 144, type*

[2nd] [x²] [1] [4] [4] [ENTER] .

Some graphing calculators give an open parenthesis when the $\sqrt{}$ function is used; so you may need to type a closed parenthesis before you press [ENTER] .

Some scientific calculators have a [√x] *key. To find the square root of 144 on one of these calculators, type*

[1] [4] [4] [√x] .

Example 4   Evaluate the square roots.

a. $\sqrt{144}$ b. $\sqrt{\dfrac{49}{81}}$

SOLUTION $\sqrt{144} = 12$ **SOLUTION** $\sqrt{\dfrac{49}{81}} = \dfrac{\sqrt{49}}{\sqrt{81}} = \dfrac{7}{9}$

c. $\sqrt{0.81}$ d. $\sqrt{-81}$

SOLUTION $\sqrt{0.81} = 0.9$ **SOLUTION** not a real number

Your Turn 4   Evaluate the square roots.

a. $\sqrt{121}$ b. $\sqrt{\dfrac{81}{169}}$ c. $\sqrt{0.49}$ d. $\sqrt{-0.09}$

Objective ③ Use the order-of-operations agreement to simplify numerical expressions. Some expressions are too complicated to evaluate using only the associative and commutative properties. If an expression contains a mixture of operations, we must be careful about the order in which we perform them. Performing the operations in a different order can change the results. Consider the expression $2 + 5 \cdot 3$.

Adding first: $2 + 5 \cdot 3$ Multiplying first: $2 + 5 \cdot 3$
$= 7 \cdot 3$ $= 2 + 15$
$= 21$ $= 17$

Changing the order of the operations changes the outcome. To ensure that everyone arrives at the same answer, mathematicians developed an order of operations that everyone agrees to follow.

Procedure   Order-of-Operations Agreement

Perform operations in the following order:

1. Within grouping symbols: parentheses ( ), brackets [ ], braces { }, above/below fraction bars, absolute value | |, and radicals $\sqrt{}$.
2. Exponents/Roots from left to right, in order as they occur.
3. Multiplication/Division from left to right, in order as they occur.
4. Addition/Subtraction from left to right, in order as they occur.

If we follow this agreement to simplify $2 + 5 \cdot 3$, the correct order is to multiply first and then add so that the answer is 17.

Embedded Parentheses

When one set of grouping symbols contains another set of grouping symbols, we say that they are *nested* or *embedded*. Work from the innermost set of grouping symbols outward.

Example 5   Simplify.

a. $-34 + 12 \div (-3) \cdot 5$

Note: The division and multiplication must be calculated before the addition. As we read from left to right, we see the division before the multiplication; so we divide first.

Solution $-34 + 12 \div (-3) \cdot 5$

$= -34 + (-4) \cdot 5$ Divide $12 \div (-3) = -4$.

$= -34 + (-20)$ Multiply $(-4) \cdot 5 = -20$.

$= -54$ Add $-34 + (-20) = -54$.

Note: Recall that the expressions $(-a)^n$ and $-a^n$ are different. The expression $(-a)^n$ means "multiply n factors of $-a$" and $-a^n$ means "find the additive inverse of a^n."

b. $-2^4 + 4|18 - 24|$

Solution $-2^4 + 4|18 - 24|$

$= -2^4 + 4|-6|$ Subtract inside the absolute value: $18 - 24 = -6$.

$= -2^4 + 4 \cdot 6$ Simplify the absolute value $|-6| = 6$.

$= -16 + 4 \cdot 6$ Evaluate the exponential form $-2^4 = -16$.

$= -16 + 24$ Multiply $4 \cdot 6 = 24$.

$= 8$ Add $-16 + 24 = 8$.

c. $(-5)^2 - 8[4 - (6 + 3)] - \sqrt{36}$

Solution $(-5)^2 - 8[4 - (6 + 3)] - \sqrt{36}$

$= (-5)^2 - 8(4 - 9) - \sqrt{36}$ Calculate within the innermost parentheses: $6 + 3 = 9$.

$= (-5)^2 - 8(-5) - \sqrt{36}$ Calculate within the brackets.

$= 25 - 8(-5) - 6$ Evaluate the exponential form and square root.

$= 25 - (-40) - 6$ Multiply $8(-5) = -40$.

$= 25 + 40 - 6$ Write $25 - (-40)$ as an equivalent addition.

$= 65 - 6$ Add $25 + 40 = 65$.

$= 59$ Subtract $65 - 6 = 59$.

Your Turn 5   Simplify.

a. $28 - 36 \div 3(-4)$

b. $6|-5 - 4| + 2(-3)^3$

c. $15.8 - 0.2[(-7)^2 - (12 + 8) \div 4]$

Radical Symbols

Radical symbols can be grouping symbols. Consider $\sqrt{36 \cdot 4}$. Following the order of operations, we multiply first, then find the root of the product. However, if a radical contains multiplication, we can find the roots of the factors and multiply those roots to get the same answer.

Multiplying first:		Multiplying the roots:
$\sqrt{36 \cdot 4} = \sqrt{144}$	or	$\sqrt{36 \cdot 4} = \sqrt{36} \cdot \sqrt{4}$
$= 12$		$= 6 \cdot 2$
		$= 12$

Both approaches yield the same result. The same holds true for division under a radical.

$$\sqrt{36 \div 4} = \sqrt{9} \qquad \text{or} \qquad \sqrt{36 \div 4} = \sqrt{36} \div \sqrt{4}$$
$$= 3 \qquad\qquad\qquad\qquad = 6 \div 2$$
$$= 3$$

Rule **Square Root of a Product or Quotient**

If a square root contains multiplication or division, we can multiply or divide first, then find the square root of the result, or we can find the square roots of the individual numbers, then multiply or divide the square roots. In symbols, $\sqrt{\dfrac{a}{b}} = \dfrac{\sqrt{a}}{\sqrt{b}}$ for $a \geq 0$ and $b > 0$ and $\sqrt{ab} = \sqrt{a} \cdot \sqrt{b}$ for $a \geq 0$ and $b \geq 0$.

Note that this rule does *not* hold for addition or subtraction under a radical, such as in $\sqrt{9 + 16}$.

Adding first: Adding the roots:

$$\sqrt{9 + 16} = \sqrt{25} \qquad\qquad \sqrt{9 + 16} = 3 + 4$$
$$= 5 \qquad\qquad\qquad\qquad\quad = 7$$

Because the results are different, we do not have a choice of how to approach the situation when the radical contains addition. The correct approach is to treat the radical as a grouping symbol and add, then find the root of the sum. Thus, the correct answer for $\sqrt{9 + 16}$ is 5. The same procedure applies when a radical contains subtraction.

Rule **Square Root of a Sum or Difference**

When a radical contains addition or subtraction, we must add or subtract first, then find the root of the sum or difference.

Example 6   Simplify.

a. $12.4 \div 5(-3)^2 + 4\sqrt{169 - 25}$

SOLUTION $12.4 \div 5(-3)^2 + 4\sqrt{169 - 25}$

$= 12.4 \div 5(-3)^2 + 4\sqrt{144}$ Subtract within the radical: $169 - 25 = 144$.

$= 12.4 \div 5(9) + 4(12)$ Evaluate the exponential form and root.

$= 2.48(9) + 4(12)$ Divide $12.4 \div 5 = 2.48$.

$= 22.32 + 48$ Multiply $2.48(9) = 22.32$ and $4(12) = 48$.

$= 70.32$ Add $22.32 + 48 = 70.32$.

Note: We divide before we multiply here because the order is to multiply or divide from left to right in the order in which those operations occur.

b. $\left(-\dfrac{1}{4}\right)^3 - \left(\dfrac{1}{2} + \dfrac{3}{8}\right) \div \sqrt{\dfrac{48}{3}}$

SOLUTION $\left(-\dfrac{1}{4}\right)^3 - \left(\dfrac{1(4)}{2(4)} + \dfrac{3}{8}\right) \div \sqrt{\dfrac{48}{3}}$

$= \left(-\dfrac{1}{4}\right)^3 - \left(\dfrac{4}{8} + \dfrac{3}{8}\right) \div \sqrt{\dfrac{48}{3}}$ Write equivalent fractions with a common denominator in order to add within the parentheses.

$= \left(-\dfrac{1}{4}\right)^3 - \dfrac{7}{8} \div \sqrt{16}$ Add within the parentheses and divide within the radical.

$$= -\frac{1}{64} - \frac{7}{8} \div 4 \quad \text{Evaluate the exponential form and square root.}$$

$$= -\frac{1}{64} - \frac{7}{8} \cdot \frac{1}{4} \quad \text{Write an equivalent multiplication using the reciprocal of the divisor.}$$

$$= -\frac{1}{64} - \frac{7(2)}{32(2)} \quad \text{Multiply } \frac{7}{8} \cdot \frac{1}{4} \text{ to get } \frac{7}{32}; \text{ then write it as an equivalent fraction with the common denominator 64 in order to subtract.}$$

$$= -\frac{1}{64} - \frac{14}{64}$$

$$= -\frac{15}{64} \quad \text{Subtract.}$$

Your Turn 6 Simplify.

a. $-\dfrac{3}{5} \div \dfrac{1}{10} \cdot 4 + \sqrt{64 + 36}$ **b.** $5 + (0.2)^3 - 3(8 - 14)$

Fraction Lines

Sometimes fraction lines are used as grouping symbols. When they are, we simplify the numerator and denominator separately, then divide the results.

Connection The expression in Example 7 is equivalent to the expression $[5(-6) - 2^3] \div [3(7) - 2]$.

Example 7 Simplify.

a. $\dfrac{5(-6) - 2^3}{3(7) - 2}$

SOLUTION $\dfrac{5(-6) - 2^3}{3(7) - 2}$

$$= \frac{5(-6) - 8}{21 - 2} \quad \text{Evaluate the exponential form in the numerator and multiply in the denominator.}$$

$$= \frac{-30 - 8}{19} \quad \text{Multiply in the numerator and subtract in the denominator.}$$

$$= \frac{-38}{19} \quad \text{Subtract in the numerator.}$$

$$= -2 \quad \text{Divide.}$$

b. $\dfrac{9(3) + 15}{4^2 + 2(-8)}$

SOLUTION $\dfrac{9(3) + 15}{4^2 + 2(-8)}$

$$= \frac{27 + 15}{16 + 2(-8)} \quad \text{Multiply in the numerator and evaluate the exponential form in the denominator.}$$

$$= \frac{42}{16 + (-16)} \quad \text{Add in the numerator and multiply in the denominator.}$$

$$= \frac{42}{0} \quad \text{Add in the denominator.}$$

Because the denominator or divisor is 0, the answer is undefined.

Answers **Your Turn 6**

a. -14 **b.** 23.008

a. $\dfrac{7(3-7)+1}{-1-2^3}$

b. $\dfrac{2[9-4(3+5)]}{25-(6-1)^2}$

Objective ④ **Find the mean of a set of data.** Fraction line notation can be used to indicate the proper sequence of operations in calculating an arithmetic mean, or average, of a set of data.

Procedure **Finding the Arithmetic Mean**

To find the arithmetic mean, or average, of n numbers, divide the sum of the numbers by n.

> **Note:** The subscripts indicate that each x represents a different given number. So x_1 represents the first given number, x_2 represents the second, and so on, until x_n, which represents the last given number.

$$\text{Arithmetic mean} = \dfrac{x_1 + x_2 + \cdots + x_n}{n}$$

Example 8 Jacky has the following test scores: 86, 95, 78, 82, and 84. Find the average of her test scores.

SOLUTION $\dfrac{86+95+78+82+84}{5} = \dfrac{425}{5}$ Divide the sum of the 5 scores by 5.

$$= 85$$

Your Turn 8 This table shows the daily rainfall accumulations for the month of April in a certain city. Find the average daily accumulation for April. (Remember, April has 30 days.)

Date	April 2	April 7	April 9	April 15	April 21	April 26
Accumulation (in inches)	0.5	1.25	1.0	1.25	0.25	1.0

Answers Your Turn 7

a. 3   b. undefined

Answer Your Turn 8

0.175 in./day

1.5 Exercises   For Extra Help   *MyMathLab*

Note: Exercises marked with a ★ represent challenging exercises.

1. What is another way to say "two to the third power"?

2. What is another way to say "three to the second power"?

3. Explain the difference between squaring a number and finding its square root.

4. If the base of an exponential form is a negative number and the exponent is even, will the sign of the result be positive or negative?

5. When simplifying a square root of a product of two numbers, what is the proper order of operations?

6. When simplifying a square root of a sum of two numbers, what is the proper order of operations?

For Exercises 7–12, identify the base and the exponent; then translate the expression to words. See Objective 1.

7. 7^2

8. 9^4

9. $(-5)^3$

10. $(-8)^2$

11. -2^7

12. -3^8

For Exercises 13–34, evaluate. See Examples 1 and 2.

13. 3^4

14. 2^5

15. $(-8)^2$

16. $(-2)^4$

17. -8^2

18. -2^4

19. $(-5)^3$

20. $(-3)^5$

21. -5^3

22. -3^5

23. $-(-2)^3$

24. $-(-3)^3$

25. $-(-1)^6$

26. $-(-1)^4$

27. $\left(-\dfrac{1}{5}\right)^2$

28. $\left(-\dfrac{2}{7}\right)^2$

29. $\left(-\dfrac{3}{4}\right)^3$

30. $\left(-\dfrac{1}{3}\right)^5$

31. $(0.2)^3$

32. $(0.3)^4$

33. $(-4.1)^2$

34. $(-0.2)^4$

For Exercises 35–42, find all square roots of each number. See Example 3.

35. 121

36. 49

37. -81

38. -36

39. 196

40. 169

41. 256

42. 225

For Exercises 43–54, evaluate the square roots. See Example 4.

43. $\sqrt{16}$

44. $\sqrt{36}$

45. $\sqrt{144}$

46. $\sqrt{289}$

47. $\sqrt{0.49}$

48. $\sqrt{0.01}$

49. $\sqrt{-64}$

50. $\sqrt{-25}$

51. $\sqrt{\dfrac{64}{81}}$

52. $\sqrt{\dfrac{9}{100}}$

53. $\sqrt{\dfrac{50}{2}}$

54. $\sqrt{\dfrac{48}{3}}$

For Exercises 55–102, simplify using the order of operations. See Examples 5–7.

55. $5 \cdot 3 + 4$

56. $4 \cdot 6 - 5$

57. $8 \div 2 - 4$

58. $18 \div 2 + 3$

59. $8 + 4 \div 2$

60. $9 + 6 \div 3$

61. $4 \cdot 6 - 7 \cdot 5$

62. $-3 \cdot 4 - 2 \cdot 7$

63. $12 - 2^4$

64. $8 - 3^2$

65. $8 - 3(-4)^2$

66. $16 - 5(-2)^2$

67. $4^2 - 36 \div 4(12 - 8)$

68. $3^2 - 18 \div 3(6 - 3)$

69. $-4 + 3(-1)^4 + 18 \div 3 \cdot 3$

70. $12 - 2(-2)^3 - 64 \div 4 \cdot 2$

71. $-2^3 + 8 - 7(4 - 3)$

72. $-3^3 - 16 - 5(7 - 2)$

73. $24 \div (-9 + 3)(3 + 4)$

74. $18 \div (-6 + 3)(4 + 1)$

75. $13.02 \div (-3.1) + 6^2 - \sqrt{25}$

76. $-15.54 \div 3.7 + (-2)^4 + \sqrt{49}$

77. $18.2 + 3.4[(5 + 9) \div 7 - 6^2]$

78. $16.3 + 2.8[(8 + 7) \div 5 - 4^2]$

79. $-4^3 - 3^2 + 4|7 - 12|$

80. $-2|9 - 15| + 5^2 - 3^2$

81. $-\dfrac{3}{4} \div \left(\dfrac{1}{8}\right) + \left(-\dfrac{2}{5}\right)(-3)(-4)$

82. $\dfrac{5}{6} \div \left(-\dfrac{2}{3}\right) + \left(-\dfrac{2}{7}\right)(5)(-14)$

83. $-36 \div (-2)(3) + \sqrt{169 - 25} + 12$

84. $\sqrt{100 - 64} + 18 \div (-3)(-2)$

85. $8 - 3[5 - (7 + 4)] - \sqrt{49}$

86. $4 - 8[3 - (9 + 3)] + \sqrt{64}$

87. $\sqrt{71 - 35} - 3^2[6 - (4 - 9)] + 4^3$

88. $\sqrt{83 - 58} - 2^2[9 - (3 - 8)] + 3^4$

89. $\dfrac{9}{8} \cdot \left(-\dfrac{2}{3}\right) + \left(\dfrac{1}{5} - \dfrac{2}{3}\right) \div \sqrt{\dfrac{125}{5}}$

90. $\left(\dfrac{3}{4} - \dfrac{2}{3}\right) \div \sqrt{\dfrac{9}{81}} - \left(\dfrac{16}{27}\right) \div \left(\dfrac{4}{9}\right)$

91. $\dfrac{2}{5} \div \left(-\dfrac{1}{10}\right) \cdot (-3) + \sqrt{64 + 36}$

92. $\dfrac{5}{6}(-18) \div \left(\dfrac{3}{2}\right) - \sqrt{9 + 16}$

93. $-16 \cdot \left(\dfrac{3}{4}\right) \div (-2) + |9 - 3(4 + 1)|$

94. $18 \cdot \left(-\dfrac{5}{6}\right) \div (-3) + 2|4 + 2(7 - 3)|$

95. $\dfrac{7 - |4(5) - 13|}{6^2 - 3(2 - 14)}$

96. $\dfrac{|6(-3) + 7| - 11}{5^3 - 2(6 - 12)}$

97. $\dfrac{4[5 - 8(2 + 1)]}{3 - 6 - (-4)^2}$

98. $\dfrac{3[24 - 4(6 - 2)]}{-3^3 + 4^2 + 3}$

99. $\dfrac{5^3 - 3(4^3 - 41)}{19 - (3 - 10)^2 + 38}$

100. $\dfrac{6^2 - 3(4 + 2^5)}{4 + 20 - (2 + 4)^2}$

101. $\dfrac{4(5^2 - 10) + 3}{\sqrt{25 - 16} - 3}$

102. $\dfrac{5(4 - 9) + 1}{2^3 - \sqrt{100 - 36}}$

In Exercises 103–106, a property of arithmetic was correctly used as an alternative to the order of operations. Determine what property of arithmetic was applied and explain how it is different from the order-of-operations agreement.

103. $14 - 2 \cdot 6 \cdot 3 + 8^2$
$= 14 - 2 \cdot 18 + 64$
$= 14 - 36 + 64$
$= 42$

104. $2(3 + 8) - \sqrt{81}$
$= 6 + 16 - 9$
$= 13$

105. $-6(5 + 3^2) - \sqrt{14 + 11}$
$= -6(5 + 9) - \sqrt{25}$
$= -30 + (-54) - 5$
$= -89$

106. $(-4)^3 + 2[-10 + 8 + (-3)]$
$= -64 + 2(-13 + 8)$
$= -64 + 2(-5)$
$= -64 + (-10)$
$= -74$

Find

Ⓧ *the mistake*   **For Exercises 107–110, explain the mistakes, then simplify.**

107. $24 \div 4 \cdot 2 - 11$
$= 24 \div 8 - 11$
$= 3 - 11$
$= -8$

108. $19 - 6(10 - 8)$
$= 19 - 6(2)$
$= 13(2)$
$= 26$

109. $40 \div 2 + \sqrt{25} - 9$
$= 20 + \sqrt{25} - 9$
$= 20 + 5 - 3$
$= 22$

110. $-3^4 + 20 \div 5 - (16 - 24)$
$= 81 + 4 - (-8)$
$= 85 + 8$
$= 93$

For Exercises 111–118, solve. See Example 8.

111. Tomeka has the following test scores in a history course: 82, 76, 64, 90, and 74. What is the average of her test scores?

112. Will's math instructor gives quizzes worth 10 points each. Will has the following quiz scores: 9, 8, 4, 8, 7, 7, 6, 9, and 8. If his instructor drops the lowest quiz score, what is Will's quiz average?

★ **113.** Michael will not have to take his chemistry final if he has a test average greater than or equal to 90 on the five tests in the course. His current test scores are 96, 88, 86, and 84. Using trial and error, determine the minimum test score on the last test that will give him a test average of 90.

★ **114.** To get an A in her psychology course, Lisa must have a test average greater than or equal to 90 on four out of five tests. (The lowest test score is dropped.) Her test scores on the first four tests are 98, 68, 84, and 86. What is the minimum score on the last test that will give her a test average of 90?

115. The following chart shows the average weekly earnings for workers in the United States in a recent year. Find the average weekly earnings for the year.

AVERAGE WEEKLY EARNINGS			
Month	**Average Earnings**	**Month**	**Average Earnings**
January	$577.98	July	$589.81
February	$578.29	August	$591.50
March	$583.42	September	$592.85
April	$583.05	October	$593.87
May	$585.42	November	$596.23
June	$589.86	December	$598.60

Source: Bureau of Labor Statistics.

116. The following graphic shows the number of people 20 years and older who were unemployed in the United States for each month during a recent year. Find the average number of people unemployed per month during that year.

ALL IN A MONTH'S WORK			YEAR 2007
Month	**Unemployed (in thousands)**	**Month**	**Unemployed (in thousands)**
January	7043	July	7137
February	6837	August	7133
March	6738	September	7246
April	6829	October	7291
May	6863	November	7181
June	6997	December	7655

Source: Bureau of Labor Statistics, *Labor Force Statistics from Current Population Survey.*

117. The following bar graph shows the daily closing price of Walmart stock over a one-week period from February 4 to February 8, 2008. Find the average of the daily prices for that week.

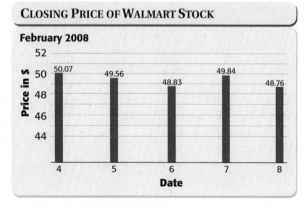

Source: www.marketwatch.com.

118. The following bar graph shows the Dow Jones Industrial Average (DJIA) at the close of each of the days listed. What is the average of the DJIA for that week?

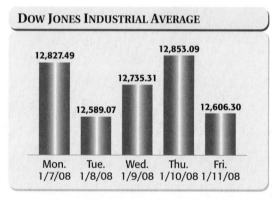

Source: www.measuringworth.com.

Review Exercises

Exercises 1–3 **CONSTANTS AND VARIABLES**

[1.1] **1.** Write a set containing the whole numbers through 10.

[1.1] **2.** Graph $-\dfrac{4}{5}$ on a number line.

[1.4] **3.** Find the multiplicative inverse of -9.

[1.1]   **4.** Is $5x - 4y$ an equation or an expression? Explain.

[1.4]   **5.** Multiply: $-4(3)(-2)$

[1.4]   **6.** Divide: $\dfrac{-15}{-5}$

1.6 Translating Word Phrases to Expressions

Objective ① Translate word phrases to expressions. An important technique in solving problems is to translate words in the problem to math symbols. In this section, we focus on how to translate word phrases to expressions containing variables, but we begin by translating expressions containing variables to word phrases for addition, subtraction, multiplication, and division.

Operation	Variable Expression	Word Phrases
Addition	$x + 3$	Some number plus three Three added to some number The sum of some number and three Three more than some number Some number increased by three
Subtraction	$y - 6$	Some number minus six Six subtracted from some number The difference of some number and six Six less than some number Some number decreased by six Some number less six
	$6 - y$	Six minus some number Some number subtracted from six The difference of six and some number Some number less than six Six decreased by some number Six less some number
Multiplication	$2a$	Twice some number Two times some number The product of two and some number Some number multiplied by two
Division	$\dfrac{x}{5}$	Some number divided by five The quotient of some number and five The ratio of some number and five
	$\dfrac{5}{x}$	Five divided by some number The quotient of five and some number The ratio of five and some number

The first step in translating a word phrase to math symbols is to identify the unknown amount. If a variable is not already given, select a variable to represent that unknown amount. The above translations from variable expressions to word phrases can be used as a guide.

Translating Basic Phrases

The following table contains some basic phrases and their translations.

Addition	Translation
The sum of x and three	$x + 3$
h plus k	$h + k$
t added to seven	$7 + t$
Three more than a number	$n + 3$
y increased by two	$y + 2$

Subtraction	Translation
The difference of three and x	$3 - x$
h minus k	$h - k$
Seven subtracted from t	$t - 7$
Three less than a number	$n - 3$
Two decreased by y	$2 - y$

Note: Since addition is a commutative operation, it doesn't matter in what order we write the translation.

For "the sum of x and three," we can write $x + 3$ or $3 + x$.

Note: Subtraction is not a commutative operation; therefore, the way we write the translation matters. We must translate each key phrase exactly as it was presented above. Notice that when we translate "less than" or "subtracted from," the translation is in reverse order from what we read.

Multiplication	Translation
The product of x and three	$3x$
h times k	hk
Twice a number	$2n$
Triple the number	$3n$
Two-thirds of a number	$\frac{2}{3}n$

Division	Translation
The quotient of x and three	$x \div 3$ or $\frac{x}{3}$
h divided by k	$h \div k$ or $\frac{h}{k}$
h divided into k	$k \div h$ or $\frac{k}{h}$
The ratio of a to b	$a \div b$ or $\frac{a}{b}$

Note: Like addition, multiplication is a commutative operation. This means that we can write the translation order any way we like.

h times k can be hk or kh.

Note: Division is like subtraction in that it is not a commutative operation; therefore, we must translate division phrases exactly as presented above. Notice how "divided into" is translated in reverse order of what we read.

Exponents	Translation
c squared	c^2
The square of b	b^2
k cubed	k^3
The cube of b	b^3
n to the fourth power	n^4
y raised to the fifth power	y^5

Roots	Translation
The square root of x	$\sqrt{x}$

Consider the translated phrases that follow. The key words *sum, difference, products* and *quotient* indicate the answer for their respective operations. Also notice that all of the phrases involve the word *and*. In the translation, the word *and* separates the parts and can therefore be translated to the operation symbol indicated by the key word *sum, difference, product,* or *quotient.*

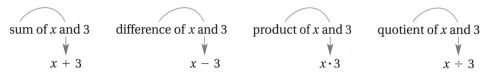

sum of x and 3 $\quad$ difference of x and 3 $\quad$ product of x and 3 $\quad$ quotient of x and 3

$$x + 3 \qquad\qquad x - 3 \qquad\qquad x \cdot 3 \qquad\qquad x \div 3$$

Combinations of Basic Phrases

Now that we have seen the basic phrases, let's translate phrases that involve combinations of the basic phrases.

Example 1 Translate to an algebraic expression.

Note: The commutative property of addition allows us to write the expression in either order.

a. four more than three times a number

Translation $\quad 4 + 3n$ or $3n + 4$

b. six less than the square of a number

Translation $\quad n^2 - 6$

Note: Because subtraction is not commutative, we must translate the subtraction in a specific order. *Less than* indicates that 6 is the subtrahend and the *square of a number* is the minuend.

c. the sum of h raised to the fifth power and fifteen

Translation $\quad h^5 + 15$

Note: When coupled with the word *ratio*, the word *to* translates to the fraction line. The amount to the left of the word *to* goes in the numerator, and the amount to the right of the word *to* goes in the denominator.

d. the ratio of m times n to r cubed

Translation $\quad \dfrac{mn}{r^3}$

e. one-half of v divided by the square root of t

Translation $\quad \dfrac{1}{2}v \div \sqrt{t}$

Note: When the word *of* is preceded by a fraction, it means multiply.

Your Turn 1 Translate to an algebraic expression.

a. two-thirds subtracted from the product of nine and a number

b. -6.2 increased by 9.8 times a number

Translating Phrases Involving Parentheses

Sometimes the word phrases imply an order of operations that would require us to use parentheses in the translation. These situations arise when the phrase indicates that a sum or difference is to be calculated before a higher-order operation such as multiplication, division, exponent, or root is performed.

Note: Without the parentheses, the expression is $5x + y$, which indicates that we are to multiply 5 times x first and then add y to the result. The parentheses indicate that we add x and y first, then multiply the resulting sum by 5.

Example 2 Translate to an algebraic expression.

a. five times the sum of x and y

Translation $\quad 5(x + y)$

b. The sum of five times x and y

Translation $\quad 5x + y$

c. the square root of the difference of the square of x and the square of y

Translation $\quad \sqrt{x^2 - y^2}$

Note: The *square root of the difference* indicates that we are to calculate the difference before we calculate the square root. We indicate this symbolically by placing the entire subtraction under the radical.

Answers Your Turn 1

a. $9n - \dfrac{2}{3}$ **b.** $-6.2 + 9.8n$

d. the product of x and y divided by the sum of x^2 and 5

Translation $xy \div (x^2 + 5)$ or $\dfrac{xy}{x^2 + 5}$

Your Turn 2   Translate to an algebraic expression.

   a. negative two times the sum of n and seven

   b. the sum of eight times a and b

   c. the difference of m and n, all raised to the fourth power

   d. the product of 5 and m divided by the difference of m and 3

Example 3   Isaac Newton developed an expression that describes the gravitational effect between two objects. The relationship is the product of the masses of the two objects divided by the square of the distance between the objects. Translate the relationship to an expression.

Translation $Mm \div d^2$ or $\dfrac{Mm}{d^2}$

Note: If the same letter is to describe two different quantities, uppercase and lowercase can be used to distinguish the quantities. Subscripts can also be used, as in $m_1 m_2 \div d^2$.

Answers   Your Turn 2

a. $-2(n + 7)$ **b.** $8a + b$

c. $(m - n)^4$ **d.** $\dfrac{5m}{m - 3}$

Answer   Your Turn 3

$d \div (v_1 + v_2)$ or $\dfrac{d}{v_1 + v_2}$

Your Turn 3   Two asteroids are on a collision course with each other. We can calculate the time to impact by dividing the distance that separates the asteroids by the sum of their velocities. Translate the relationship to an expression.

1.6 Exercises   For Extra Help   *MyMathLab*

Note: Exercises marked with a ★ represent challenging exercises.

1. List three key words that indicate addition.

2. List three key words that indicate subtraction.

3. List three key words that indicate multiplication.

4. List three key words that indicate division.

5. The phrase *nine subtracted from n* translates to $n - 9$, which is in reverse order of what we read. What other key words for subtraction translate in reverse order?

6. What key words for division translate in reverse order?

7. Translate the algebraic expression $x + 7$ into five different word phrases.

8. Translate the algebraic expression $x - 5$ into six different word phrases.

9. Translate the algebraic expression $\dfrac{7}{b}$ into three different word phrases.

10. Translate the algebraic expression $3n$ into three different word phrases.

For Exercises 11–44, translate each phrase to an algebraic expression. See Examples 1 and 2.

11. four times a number

12. the product of a number and four

13. the sum of four times x and sixteen

14. five more than y

15. the difference of seven times x and eight

16. six less than T

17. the quotient of negative four and the cube of y

18. the ratio of seven to the square of m

19. the product of eight and p decreased by four

20. thirteen subtracted from twice a number

21. fourteen divided into m

22. r divided by six

23. x to the fourth power increased by five

24. the sum of b cubed and seven

25. one-fifth subtracted from the product of seven and a number

26. two-thirds added to the product of four and a number

27. the product of negative three and the difference of a number and two

28. the product of three and the sum of the number and four

29. the sum of four and n, all raised to the fifth power

30. the difference of two and l, all raised to the third power

31. five less than the product of m and n

32. the product of three and a number, increased by five

33. the quotient of four and a number, decreased by two

34. seven added to the quotient of x and y

35. negative twenty-seven decreased by the sum of a and b

36. the difference of m and n subtracted from negative eight

37. six-tenths decreased by the product of four and the difference of y and two

38. eighty-one hundredths increased by the product of eight and the sum of x and three tenths

39. the difference of p and q decreased by the sum of m and n

40. the sum of a and b subtracted from the difference of c and d

41. the product of m and n subtracted from the square root of y

42. the square root of x subtracted from the product of a and b

43. a number minus the product of three and the difference of the number and six

44. the product of five and a number minus the sum of the number and two

For Exercises 45–48, explain each mistake. Then translate the phrase correctly.

45. seventeen less than three times t
Translation: $17 - 3t$

46. four subtracted from the square of m
Translation: $4 - m^2$

47. nine times the sum of x and y
Translation: $9x + y$

48. Nineteen divided into the product of h and k
Translation: $19 \div hk$

For Exercises 49–58, translate each phrase. See Example 3.

49. The length of a rectangle is five more than the width. If the width is represented by w, write an expression that describes the length.

50. The width of a rectangle is four less than the length. If the length is represented by l, write an expression that describes the width.

51. The length of a rectangle is three times the width. If the width is represented by the variable w, write an expression that describes the length.

52. The width of a rectangle is one-fourth of the length. If the length is represented by l, write an expression that describes the width.

53. The radius of a circle is one-half of the diameter. If d represents the diameter, write an expression for the radius.

54. The diameter of a circle is twice the radius. If r represents the radius, write an expression for the diameter.

55. Lindsey has 42 coins in her change purse that are either dimes or quarters. If n represents the number of quarters she has, write an expression in terms of n that describes the number of dimes.

56. Sherice owns a total of 60 shares of stock in two companies. If n represents the number of the higher-priced stock, write an expression in terms of n for the number of shares of the lower-priced stock.

★**57.** Don passes mile marker 51 on the highway. One-fourth of an hour later, a state trooper traveling in the same direction passes the same mile marker. If t represents the amount of time it takes the trooper to catch up to Don, write an expression in terms of t that describes the amount of time Don has traveled since passing marker 51.

★**58.** Barry is jogging along a trail and passes a sign. One-third of an hour later, Dedra, who is traveling in the same direction, passes the same sign. If t represents the amount of time it takes Dedra to catch up to Barry, write an expression in terms of t that describes the amount of time Barry has traveled since passing the sign.

Exercises 59–66 contain word descriptions of expressions from mathematics and physics. Translate the descriptions to symbolic form. See Examples 1–3.

59. The perimeter of a rectangle is twice the width plus twice the length.

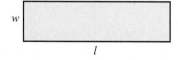

60. The area of a circle can be found by multiplying the square of the radius by π.

61. The volume of a cone is one-third of the product of π, the square of the radius, and the height of the cone.

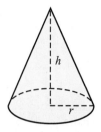

62. The volume of a sphere is four-thirds the product of π and the cube of the radius.

63. Albert Einstein developed an expression that describes the energy of a particle at rest, which is the product of the mass, m, of the particle and the square of the speed of light, which is represented by c.

64. The centripetal acceleration of an object traveling in a circular path is found by an expression that is the ratio of the square of the velocity, v, to the radius, r, of the orbital path.

★**65.** René Descartes developed an expression for the distance between two points in the coordinate plane. The distance between two points is the square root of the sum of the square of the difference of x_2 and x_1 and the square of the difference of y_2 and y_1.

★**66.** Albert Einstein's theory of relativity includes a mathematical expression that is the square root of the difference of one and the ratio of the square of the velocity, v, of an object in motion to the square of the speed of light, c.

For Exercises 67–72, the algebraic expressions have been translated to a word phrase that is ambiguous (unclear). Indicate why the translation is unclear and give a translation that is not ambiguous. There is more than one correct answer.

67. $3x + 4$
Translation: three times x plus four

68. $2a - 7$
Translation: two times a minus seven

69. $5(x - 1)$
Translation: five times x minus one

70. $4(y + 6)$
Translation: four times y plus six

71. $n + 5(n - 6)$
Translation: n plus five times the difference of n and six

72. $m - 3(m + 2)$
Translation: m minus three times the sum of m and two

For Exercises 73–80, translate the expression to a word phrase.

73. The area of a triangle: $\frac{1}{2}bh$

74. The area of a trapezoid: $\frac{1}{2}h(a + b)$

75. The volume of a box: lwh

76. The volume of a cylinder: $\pi r^2 h$

77. The surface area of a box: $2(lw + lh + wh)$

78. The surface area of a cylinder: $2\pi r(r + h)$

79. The slope of a line: $\dfrac{y_2 - y_1}{x_2 - x_1}$

80. A quadratic expression: $ax^2 + bx + c$

Puzzle Problem

Consecutive integers follow one another in a pattern such as 1, 2, 3, . . .
Let n represent the first integer in a set of consecutive integers.

a. Write expressions for the next two consecutive integers.

b. Suppose n is odd. Write expressions for the next two consecutive odd integers.

c. Suppose n is even. Write expressions for the next two consecutive even integers.

Review Exercises

Exercises 1–6  EXPRESSIONS

[1.4] **1.** Rewrite and simplify $2(5 + 2)$ using the distributive property.

[1.3] *For Exercises 2 and 3, simplify.*

    2. $4 - 8 + 6$ **3.** $\dfrac{2}{3} - \dfrac{1}{4}$

[1.5] *For Exercises 4–6, simplify.*

    4. $-6^2 \div 3 + 4(5 - 9)$ **5.** $-4|3 - 8| + 2^5$ **6.** -3^2

1.7 Evaluating and Rewriting Expressions

Objectives

1 Evaluate an expression.

2 Determine all values that cause an expression to be undefined.

3 Rewrite an expression using the distributive property.

4 Rewrite an expression by combining like terms.

To be able to solve equations in Chapter 2, we must learn how to manipulate algebraic expressions. There are two actions we can perform with algebraic expressions. We can *evaluate* them and *rewrite* them.

Objective 1 Evaluate an expression.   First, we learn how to evaluate an expression.

Procedure   **Evaluating an Algebraic Expression**

To evaluate an algebraic expression:

1. Replace the variables with their corresponding given values.
2. Calculate the numerical expression using the order of operations.

Example 1 Evaluate each expression using the given values.

a. $4m - 3(n - 4)$ when $m = 3$ and $n = 2$

SOLUTION $4m - 3(n - 4)$

$4(3) - 3(2 - 4)$ Replace m with 3 and n with 2.

$= 4(3) - 3(-2)$ Simplify inside the parentheses first.

$= 12 + 6$ Multiply.

$= 18$ Add.

b. $y^2 - 0.2xy + 5$ when $x = 6$ and $y = -4$

SOLUTION $y^2 - 0.2xy + 5$

$(-4)^2 - 0.2(6)(-4) + 5$ Replace x with 6 and y with -4.

$= 16 - 0.2(6)(-4) + 5$ Begin calculating by simplifying the exponential form.

$= 16 - (-4.8) + 5$ Multiply.

$= 16 + 4.8 + 5$ Write the subtraction as an equivalent addition.

$= 25.8$ Add from left to right.

c. $\dfrac{5r^2}{d - 3}$ when $r = 2$ and $d = -5$

SOLUTION $\dfrac{5r^2}{d - 3}$

$\dfrac{5(2)^2}{-5 - 3}$ Replace r with 2 and d with -5.

$= \dfrac{5(4)}{-8}$ Calculate the top and bottom expressions separately.

$= \dfrac{20}{-8}$

$= -\dfrac{5}{2}$ or -2.5

Your Turn 1 Evaluate the expression using the given values.

a. $-5a + 3(b - 6)$ when $a = -2$ and $b = 4$

b. $m^3 - 6n^2$ when $m = -2$ and $n = 3$

c. $\dfrac{3a^2 - b}{a + 6}$ when $a = -4$ and $b = 2$

Objective ② **Determine all values that cause an expression to be undefined.** In Example 1c, what if $d = 3$? Notice that the denominator would become 0: $\dfrac{5r^2}{3 - 3} = \dfrac{5r^2}{0}$, which is undefined if $r \neq 0$ and indeterminate if $r = 0$. When asked to evaluate a division expression in which the divisor or denominator contains a variable or variables, we must be careful about what values replace the variable(s). Often we need to know what values replace the variable(s) and cause the expression to be undefined or indeterminate.

Example 2 Determine all values that cause each expression to be undefined.

a. $\dfrac{5}{x + 7}$

Answer If $x = -7$, we have $\dfrac{5}{-7 + 7} = \dfrac{5}{0}$, which is undefined because the denominator is 0.

b. $\dfrac{-y}{(y + 2)(y - 6)}$

SOLUTION Note that because $(y + 2)(y - 6)$ is a product, if either factor is 0, the entire denominator is 0.

Notice that if $y = -2$, we have $\dfrac{-(-2)}{(-2 + 2)(-2 - 6)}$

$$= \dfrac{2}{(0)(-8)}$$

$$= \dfrac{2}{0}, \text{ which is undefined}$$

Also, if $y = 6$, we have $\dfrac{-6}{(6 + 2)(6 - 6)}$

$$= \dfrac{-6}{(8)(0)}$$

$$= \dfrac{-6}{0}, \text{ which is undefined}$$

> **Connection** This principle—that if either factor is 0, then the entire product is 0—is known as the *zero-factor theorem*. We use this theorem extensively in Chapter 6 when solving equations such as $(y + 2)(y - 6) = 0$.

Answer Both $y = -2$ and $y = 6$ cause the expression to be undefined.

Your Turn 2 Determine all values that cause each expression to be undefined.

a. $\dfrac{9}{n + 6}$

b. $\dfrac{2m}{(m - 3)(m + 4)}$

Objective ③ Rewrite an expression using the distributive property. Recall from Section 1.4 the distributive property of multiplication over addition.

Rule **Distributive Property of Multiplication over Addition**
$$a(b + c) = ab + ac$$

This property gives us an alternative to the order of operations. Look at the numerical expression $2(5 + 6)$ and compare using the order of operations versus using the distributive property.

Following order of operations:		Using the distributive property:
$2(5 + 6) = 2(11)$	or	$2(5 + 6) = 2 \cdot 5 + 2 \cdot 6$
$\qquad\quad\;\; = 22$		$\qquad\qquad\;\; = 10 + 12$
		$\qquad\qquad\;\; = 22$

Note: Using the order of operations, we add within the parentheses first, then multiply.

Note: Using the distributive property, we multiply each addend by the factor outside the parentheses, then add the products.

Answers Your Turn 2

a. $n = -6$

b. both $m = 3$ and $m = -4$

The result is the same either way, so the distributive property allows us to say that $2(5 + 6)$ and $2 \cdot 5 + 2 \cdot 6$ are equivalent expressions.

Conclusion   We can use the distributive property to rewrite an expression in another form that is equivalent to the original form.

Example 3   Use the distributive property to write an equivalent expression and simplify.

a. $2(x + 5)$

SOLUTION $2(x + 5) = 2 \cdot x + 2 \cdot 5$
$$= 2x + 10$$

b. $-6(n - 9)$

SOLUTION $-6(n - 9) = -6 \cdot n - (-6) \cdot 9$
$$= -6n - (-54)$$
$$= -6n + 54$$

Connection   We can apply the distributive property to $-6(n - 9)$ even though the parentheses contain subtraction instead of addition because subtraction can be expressed as addition like this:

$$-6(n - 9) = -6[n + (-9)]$$

In fact, it is helpful to think of the expression this way because when we multiply the -6 by the -9, we get positive 54.

c. $\dfrac{3}{8}\left(2m + \dfrac{4}{5}\right)$

SOLUTION $\dfrac{3}{8}\left(2m + \dfrac{4}{5}\right) = \dfrac{3}{8} \cdot 2m + \dfrac{3}{8} \cdot \dfrac{4}{5}$ Use the distributive property.

$$= \dfrac{3}{\underset{4}{8}} \cdot \dfrac{\overset{1}{2}}{1}m + \dfrac{3}{\underset{2}{8}} \cdot \dfrac{\overset{1}{4}}{5}$$ Divide out common factors.

$$= \dfrac{3}{4}m + \dfrac{3}{10}$$

Your Turn 3   Use the distributive property to write an equivalent expression and simplify.

a. $3(n + 2)$ **b.** $-9(7 - 3x)$ **c.** $\dfrac{4}{5}\left(\dfrac{1}{2}y - 10\right)$

Objective 4 Rewrite an expression by combining like terms.   Many expressions are sums of expressions called **terms**. For example, the expression $5x + 3$ is a sum of the terms $5x$ and 3.

Definition   *Terms:* Expressions that are the addends in an expression that is a sum.

If an expression contains subtraction, as in $4x^2 - 6x + 8$, we can identify its terms by writing the subtraction as addition. Because $4x^2 - 6x + 8 = 4x^2 + (-6x) + 8$, its terms are $4x^2, -6x$, and 8. Notice we can see that the second term in $4x^2 - 6x + 8$ is $-6x$ without writing the equivalent addition by recognizing that the sign to the left of the term is its sign.

The numerical factor in a term is called the numerical **coefficient** or, simply, the coefficient of the term.

Definition   *Coefficient:* The numerical factor in a term.

Answers   Your Turn 3
a. $3n + 6$ **b.** $-63 + 27x$
c. $\dfrac{2}{5}y - 8$

The coefficient of $4x^2$ is 4.

The coefficient of $-6x$ is -6.

The coefficient of 8 is 8.

The coefficient of x is 1 because $1 \cdot x = x$.

The coefficient of $-x$ is -1 because $-1 \cdot x = -x$.

The coefficient of $\dfrac{2x}{3}$ is $\dfrac{2}{3}$ because $\dfrac{2}{3}x = \dfrac{2x}{3}$.

Another way to rewrite an expression is to combine **like terms**.

Definition

Like terms: Constant terms or variable terms that have the same variable(s) raised to the same exponents.

Connection We can use the distributive property in reverse to confirm our procedure for combining like terms. Consider the expression $3x + 5x$ again. Notice that there is a common factor of x in both terms. When applying the distributive property in reverse, we write this common factor outside the parentheses and write the remaining factors as addends within the parentheses. It looks like this:

$$3x + 5x = (3 + 5)x$$

Because the parentheses contain only the addition of two numbers, we simplify by adding the numbers.

$$3x + 5x = (3 + 5)x$$
$$= 8x$$

Examples of like terms:	Examples of unlike terms:	
$3x$ and $5x$	$2x$ and $8y$	(different variables)
$4y^2$ and $9y^2$	$4t^2$ and $4t^3$	(different exponents)
$7xy$ and $3xy$	x^2y and xy^2	(different exponents)
6 and 15	12 and $12x$	(different variables)

Combining Like Terms

Now consider an expression that is a sum of like terms, such as $3x + 5x$. We can rewrite this expression in a more compact form. Multiplication of 3 times x means that x is repeatedly added three times. Likewise, 5 times x means that x is repeatedly added five times. We can expand the expression out like this:

$$\underbrace{3x}_{} + \underbrace{5x}_{}$$
$$= \underbrace{x + x + x + x + x + x + x + x}_{}$$
$$= 8x$$

Note: After expanding, we see that there are a total of eight x's repeatedly added. We can write those eight x's as a single term, $8x$.

Notice that when we combine the like terms, we get an expression that is more compact. When we rewrite an expression in a more compact form, we say that we are *simplifying* the expression. Also notice that when combining like terms, we simply add the coefficients and keep the variable the same.

Procedure

Combining Like Terms

To combine like terms, add or subtract the coefficients and keep the variables and their exponents the same.

Example 4   Combine like terms.

a. $9y + 7y$

SOLUTION $9y + 7y = 16y$ **Note:** We think: Nine y's plus seven y's equals sixteen y's.

b. $7x - 2x$

SOLUTION $7x - 2x = 5x$ **Note:** We think: Seven x's minus two x's leaves five x's.

c. $14n^2 - n^2$

Note: The coefficient of n^2 is 1, so we think fourteen n^2's minus one n^2 leaves thirteen n^2's.

SOLUTION $14n^2 - n^2 = 13n^2$

Your Turn 4   Combine like terms.

a. $6r + 8r$

b. $6.5x + 2.3x$

c. $\dfrac{1}{3}n^3 - \dfrac{3}{4}n^3$

Answers   Your Turn 4

a. $14r$ **b.** $8.8x$ **c.** $-\dfrac{5}{12}n^3$

Collecting Like Terms

Sometimes expressions are more complex and contain different sets of like terms. In such cases, we combine the like terms and copy any unlike terms in the final expression. Many people like to use the commutative property of addition and rearrange the expression first so that the like terms are together, then combine them. This type of manipulation is called *collecting* the like terms.

Example 5 Combine like terms in $4y^2 + 5 + 3y^2 - 9$.

SOLUTION $4y^2 + 5 + 3y^2 - 9$

$= \underbrace{4y^2 + 3y^2} + \underbrace{5 - 9}$ Collect the like terms.

$= 7y^2 - 4$ Combine like terms.

Collecting like terms is optional. Alternatively, many people mark through the terms as they combine them as a way of keeping track of what has been combined. Using this technique with the expression from Example 5 looks like this:

$$4y^2 + 5 + 3y^2 - 9$$

$$= 7y^2 - 4$$

Example 6 Combine like terms in $15y + 8x - y - 8x$.

SOLUTION $15y + 8x - y - 8x$ **Note:** The term $-y$ is equivalent to $-1y$.

$= 14y + 0$

$= 14y$

Sometimes collecting the like terms is beneficial, as in Example 7, where there are like terms with fraction coefficients. Some people find it easier to "see" the steps in writing the equivalent fractions when the fractions are closer together.

Example 7 Combine like terms in $\dfrac{1}{3}n - 9m + 2 - m + \dfrac{1}{4}n$.

SOLUTION $\dfrac{1}{3}n - 9m + 2 - m + \dfrac{1}{4}n$

$= \dfrac{1}{3}n + \dfrac{1}{4}n - 9m - m + 2$ Collect the like terms.

$= \dfrac{1(4)}{3(4)}n + \dfrac{1(3)}{4(3)}n - 9m - m + 2$ Write fraction coefficients as equivalent fractions with their LCD, 12.

$= \dfrac{4}{12}n + \dfrac{3}{12}n - 9m - m + 2$

$= \dfrac{7}{12}n - 10m + 2$ Combine like terms.

Answers **Your Turn 7**

a. $10x^3 - 4y$ **b.** $4.2x^2 - 6.7x - 3$

c. $-\dfrac{3}{8}y + 4x + 2$

Your Turn 7 Combine like terms.

a. $6x^3 + 2y + 4x^3 - 6y$ **b.** $4.2x^2 - 6x + 9 - 0.7x - 12$

c. $\dfrac{3}{8}y - x + 2 - \dfrac{3}{4}y + 5x$

Connection One way to verify that two expressions are equivalent is to evaluate both using the same value(s) for the variable(s). Consider the expression in Example 5. If we evaluate the original expression $4y^2 + 5 + 3y^2 - 9$ and the simplified expression $7y^2 - 4$ using the same chosen value for y, we should get the same answer. Let's choose the number 2 for y.

Original Expression	Simplified Expression
$4y^2 + 5 + 3y^2 - 9$	$7y^2 - 4$
$4(2)^2 + 5 + 3(2)^2 - 9$	$7(2)^2 - 4$
$= 4(4) + 5 + 3(4) - 9$	$= 7(4) - 4$
$= 16 + 5 + 12 - 9$	$= 28 - 4$
$= 24$	$= 24$

Warning: Because some expressions that are not equivalent may give the same answer using this method, this "check" does not guarantee that the expressions are equivalent. It is, however, a good sign.

Notice that the simplified expression gave the same answer, yet was easier to work with. This is one of the reasons we simplify expressions. If we can simplify an expression, we have an easier time evaluating it.

1.7 Exercises For Extra Help

Note: Exercises marked with a ★ represent challenging exercises.

1. In your own words, explain how to evaluate an expression.

2. Explain the difference between evaluating and rewriting an expression.

3. What causes an expression to be undefined?

4. What are like terms?

5. What is the coefficient in a term?

6. Explain how to combine like terms.

For Exercises 7–28, evaluate the expressions using the given values. See Example 1.

7. $3(a + 5) - 4b;\ a = 2, b = 3$

8. $8n - 2(m + 1);\ m = 5, n = 3$

9. $4 - 0.2(x + 3);\ x = 1$

10. $6 - 0.4(y - 2);\ y = 5$

11. $2p^2 - 3p - 4;\ p = -3$

12. $n^2 - 8n + 1;\ n = -1$

13. $2y^2 - 4y + 3;\ y = -\dfrac{1}{2}$

14. $3r^2 - 9r + 6;\ r = -\dfrac{1}{3}$

15. $8 - 2(y + 4);\ y = -2.3$

16. $-6 - 2(l - 5);\ l = -0.4$

17. $-|3x| + |4y^3|;\ x = 5, y = -1$

18. $-|2m^2| - |4n|;\ m = 3, n = -2$

19. $|2r^2 - 3q|;\ r = -3, q = 5$

20. $|2m^2 + 2n|;\ m = -4, n = -5$

21. $\sqrt{c} + 4ab^2;\ a = -1, b = -2, c = 16$

22. $-2x^3y + \sqrt{z}$; $x = -2$, $y = -3$, $z = 4$ **23.** $-5\sqrt{a} + 2\sqrt{b}$; $a = 25$, $b = 4$ **24.** $-3\sqrt{h} + 3\sqrt{k}$; $h = 16$, $k = 9$

25. $\dfrac{6x^3}{y - 8}$; $x = 2$, $y = 7$ **26.** $\dfrac{4m^2}{n + 4}$; $m = 2$, $n = 4$ **27.** $\dfrac{15 - b^2}{2\sqrt{c + d}}$; $b = 3$, $c = 25$, $d = 144$

28. $\dfrac{5 - a^2}{3\sqrt{x + y}}$; $a = 1$, $x = 64$, $y = 36$

29. The expression $b^2 - 4ac$ is called the *discriminant* and is used to determine the types of solutions for quadratic equations. Find the value of the discriminant given the following values.

   a. $a = 2$, $b = 1$, $c = -3$ **b.** $a = 1$, $b = -2$, $c = 4$

30. The expression $ad - bc$ is used to calculate the determinant of a matrix. Find the determinant given the following values.

   a. $a = 1$, $b = 0.5$, $c = -4$, $d = 6$ **b.** $a = -3$, $b = \dfrac{4}{5}$, $c = 2$, $d = \dfrac{1}{2}$

> **Connection** You will learn all of the details about each expression presented in Exercises 29–32 in future chapters (and future courses). Notice that even without a full understanding of the context in which the expressions are used, you can still evaluate them.

31. The expression $\dfrac{y_2 - y_1}{x_2 - x_1}$ is used to calculate the slope of a line. Find the slope given each set of the following values.

   a. $x_1 = 2$, $y_1 = 1$, $x_2 = 5$, $y_2 = 7$ **b.** $x_1 = -1$, $y_1 = 2$, $x_2 = -7$, $y_2 = -2$

32. The expression $\sqrt{(x_2 - x_1)^2 + (y_2 - y_1)^2}$ is used to calculate the distance between two points in the coordinate plane. Evaluate the expression using the following values.

   a. $x_1 = 2$, $y_1 = 1$, $x_2 = 5$, $y_2 = 7$ **b.** $x_1 = -1$, $y_1 = 2$, $x_2 = -7$, $y_2 = -2$

For Exercises 33–40, determine all values that cause each expression to be undefined. See Example 2.

33. $\dfrac{-7}{5 + y}$ **34.** $\dfrac{8}{x + 3}$ **35.** $\dfrac{6m}{(m + 1)(m - 3)}$ **36.** $\dfrac{-5a}{(a - 4)(a - 2)}$

37. $\dfrac{6 + x^2}{x}$ **38.** $\dfrac{7 - y}{y}$ ★**39.** $\dfrac{x + 1}{3x - 2}$ ★**40.** $\dfrac{3y}{2y + 1}$

For Exercises 41–48, use the distributive property to write an equivalent expression and simplify. See Example 3.

41. $6(a + 2)$ **42.** $4(b - 5)$ **43.** $-8(4 - 3y)$ **44.** $-7(3 - 2m)$

45. $\dfrac{7}{8}\left(\dfrac{1}{2}c - 16\right)$ **46.** $\dfrac{4}{5}\left(-10h + \dfrac{2}{9}\right)$ **47.** $0.2(3n - 8)$ **48.** $-1.5(6x + 7)$

For Exercises 49–58, identify the coefficient of each term. See Objective 4.

49. $-6x^3$ **50.** $-14y$ **51.** m^9 **52.** y^5 **53.** $-b$

54. $-n$ **55.** $-\dfrac{2}{3}m$ **56.** $\dfrac{5}{8}a$ **57.** $\dfrac{y}{5}$ **58.** $-\dfrac{u}{3}$

For Exercises 59–84, simplify by combining like terms. See Examples 4–7.

59. $3y + 5y$ **60.** $6m + 7m$ **61.** $6a - 11a$ **62.** $5b - 13b$

63. $14x - 3x$ **64.** $-5y + 12y$ **65.** $-3r - 8r$ **66.** $-7m - 6m$

67. $6.3n - 8.2n$ **68.** $-5.1x^4 + 3.4x^4$ **69.** $\dfrac{1}{2}b^2 - \dfrac{5}{6}b^2$ **70.** $\dfrac{3}{4}z - \dfrac{7}{5}z$

71. $-11c - 10c - 5c$ **72.** $-15w - 6w - 11w$ **73.** $6x + 7 - x - 8$ **74.** $5y^2 + 6 + 3y^2 - 8$

75. $-8x + 3y - 7 - 2x + y + 6$ **76.** $-4a + 9b - a + 5 + 2b - 8$

77. $-10m + 7n + 1 + m - 2n - 12 + y$ **78.** $-3h + 7k - 5 - 8h - 7k + 19 + x$

79. $1.5x + y - 2.8x + 0.3 - y - 0.7$ **80.** $0.4t^2 + t - 2.8 - t^2 + 0.9t - 4$

81. $\dfrac{1}{6}c + 3d + \dfrac{2}{3}c + \dfrac{1}{7} - \dfrac{1}{2}d$ **82.** $\dfrac{5}{8}y + 4 - \dfrac{3}{4}x + \dfrac{2}{3} - \dfrac{1}{4}y$

83. $7a + \dfrac{4}{5}b^2 - 5 - \dfrac{3}{4}a - \dfrac{2}{7}b^2 + 9$ **84.** $\dfrac{1}{2}m - 3n + 14 - \dfrac{3}{8}m - \dfrac{9}{10}n - 5$

★**85. a.** Translate to an algebraic expression: fourteen plus the difference of six times a number and eight times the same number.

    b. Simplify the expression.

    c. Evaluate the expression when the number is -3.

★**86. a.** Translate to an algebraic expression: the sum of negative five times a number and eight minus two times the same number.

    b. Simplify the expression.

    c. Evaluate the expression when the number is 0.2.

Puzzle Problem

Each letter in the addition problem to the right represents a different whole number, 0–9. What number does each letter represent?

```
  FORTY
    TEN
+   TEN
------
  SIXTY
```

Review Exercises

Exercise 1 CONSTANTS AND VARIABLES

[1.1] 1. Arrange in order from least to greatest.

$$4.2, |-6|, 4\frac{5}{8}, -2.5, -3$$

Exercises 2–6 EXPRESSIONS

[1.2] 2. Simplify: $\dfrac{72}{420}$

For Exercises 3–6, perform the indicated operation(s).

[1.3] 3. $-8 + (-14)$ 　　　[1.4] 4. $\dfrac{1}{2}(12.5)(8)$ 　　　[1.5] 5. $4(-2) + 9$ 　　　[1.4] 6. $\dfrac{3}{4}\left(5\dfrac{1}{3}\right) - 1$

Chapter 1 Summary

Defined Terms

Section 1.1
Variable (p. 2)
Constant (p. 2)
Expression (p. 2)
Equation (p. 2)
Inequality (p. 2)
Set (p. 3)
Rational number (p. 4)
Irrational number (p. 5)
Real numbers (p. 5)
Absolute value (p. 6)

Section 1.2
Fraction (p. 10)
Multiple (p. 11)
LCM (p. 12)
LCD (p. 12)
Factors (p. 12)
Prime number (p. 13)
Prime factorization
(p. 13)
Lowest terms (p. 14)

Section 1.3
Additive inverses (p. 27)

Section 1.4
Multiplicative inverses
(p. 42)

Section 1.5
Exponent (p. 50)
Base (p. 50)

Section 1.7
Terms (p. 73)
Coefficient (p. 73)
Like terms (p. 74)

The Real Number System

Real Numbers

Rational Numbers: Real numbers that can be expressed in the form $\frac{a}{b}$, where a and b are integers and $b \neq 0$, such as $-4\frac{3}{4}$, $-\frac{2}{3}$, 0.018, $0.\overline{3}$, and $\frac{5}{8}$.

Integers: $\ldots, -3, -2, -1, 0, 1, 2, 3, \ldots$

Whole Numbers: $0, 1, 2, 3, \ldots$

Natural Numbers: $1, 2, 3, \ldots$

Irrational Numbers: Any real number that is not rational, such as $-\sqrt{2}$, $-\sqrt{3}$, $\sqrt{0.8}$, and π.

Arithmetic Summary Diagram

Each operation has an inverse operation. In the diagram, the operations build from the top down. Addition leads to multiplication, which leads to exponents. Subtraction leads to division, which leads to roots.

Properties of Arithmetic

In each of the following, a, b, and c represent real numbers.

Additive Identity
$$a + 0 = a$$

Commutative Property of Addition
$$a + b = b + a$$

Associative Property of Addition
$$(a + b) + c = a + (b + c)$$

Multiplicative Property of 0
$$a \cdot 0 = 0$$

Multiplicative Identity
$$a \cdot 1 = a$$

Commutative Property of Multiplication
$$ab = ba$$

Associative Property of Multiplication
$$(ab)c = a(bc)$$

Distributive Property
$$a(b + c) = ab + ac$$

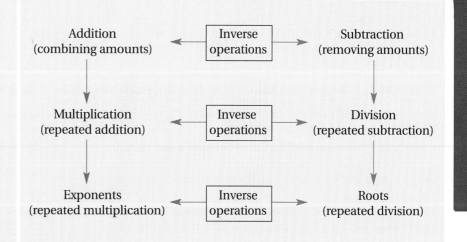

Order of Operations
1. Within grouping symbols
2. Exponents and roots from left to right
3. Multiply or divide from left to right
4. Add or subtract from left to right

Procedures, Rules, and Key Examples

Procedures/Rules	Key Examples
SECTION 1.1 Number Sets and the Structure of Algebra	
To write a set, write the elements or members of the set within braces, { }.	**Example 1:** Write the set containing the natural numbers divisible by 3. Answer: $\{3, 6, 9, 12, \dots\}$
The absolute value of a number is positive or 0.	**Example 2:** Find the absolute value. **a.** $\lvert 8 \rvert = 8$ **b.** $\lvert -15 \rvert = 15$ **c.** $\lvert 0 \rvert = 0$
For any two real numbers a and b, a is greater than b if a is to the right of b on a number line.	**Example 3:** Use $<$ or $>$ to make a true sentence. **a.** $-8 \quad -10$ **b.** $-6 \quad 9$ Answers: **a.** $-8 > -10$ **b.** $-6 < 9$
SECTION 1.2 Fractions	
For any fraction, we can write an equivalent fraction by multiplying or dividing both its numerator and denominator by the same nonzero number.	**Example 1:** Find the missing number in $\dfrac{5}{6} = \dfrac{?}{24}$ so that the fractions are equivalent. Answer: $\dfrac{5}{6} = \dfrac{5 \cdot 4}{6 \cdot 4} = \dfrac{20}{24}$

(continued)

Procedures/Rules	**Key Examples**

SECTION 1.2   Fractions (*continued*)

To simplify a fraction to lowest terms: 1. Replace the numerator and denominator with their prime factorizations. 2. Eliminate (divide out) all prime factors common to the numerator and denominator. 3. Multiply the remaining factors.	**Example 2:** Reduce $\dfrac{54}{60}$ to lowest terms. $\dfrac{54}{60} = \dfrac{2 \cdot 3 \cdot 3 \cdot 3}{2 \cdot 2 \cdot 3 \cdot 5} = \dfrac{9}{10}$

SECTION 1.3   Adding and Subtracting Real Numbers; Properties of Real Numbers

When adding two numbers that have the same sign, add their absolute values and keep the same sign. When adding two numbers that have different signs, subtract the smaller absolute value from the greater absolute value and keep the sign of the number with the greater absolute value.	**Example 1:** Add. **a.** $4 + 9 = 13$ **b.** $-4 + (-9) = -13$ **c.** $-4 + 9 = 5$ **d.** $4 + (-9) = -5$
To add fractions with the same denominator, add the numerators and keep the same denominator; then simplify. To add fractions with different denominators: 1. Write each fraction as an equivalent fraction with the LCD. 2. Add the numerators and keep the LCD. 3. Simplify.	**Example 2:** Add. **a.** $\dfrac{1}{10} + \dfrac{3}{10} = \dfrac{4}{10} = \dfrac{2}{5}$ **b.** $-\dfrac{3}{4} + \dfrac{5}{6} = -\dfrac{3(3)}{4(3)} + \dfrac{5(2)}{6(2)}$ $\quad\quad = -\dfrac{9}{12} + \dfrac{10}{12}$ $\quad\quad = \dfrac{1}{12}$
To write a subtraction statement as an equivalent addition statement, change the operation symbol from a minus sign to a plus sign and change the subtrahend to its additive inverse.	**Example 3:** Subtract. **a.** $4 - 9 = 4 + (-9) = -5$ **b.** $-4 - 9 = -4 + (-9) = -13$ **c.** $4 - (-9) = 4 + 9 = 13$ **d.** $-4 - (-9) = -4 + 9 = 5$

SECTION 1.4   Multiplying and Dividing Real Numbers; Properties of Real Numbers

The product of an even number of negative factors is positive, whereas the product of an odd number of negative factors is negative.	**Example 1:** Multiply. **a.** $(5)(7) = 35$ **b.** $(-5)(-7) = 35$ **c.** $(-1)(-2)(-5)(-7) = 70$ **d.** $(-5)(7) = -35$ **e.** $(5)(-7) = -35$ **f.** $(-2)(-5)(-7) = -70$
Rule for multiplying fractions: $\dfrac{a}{b} \cdot \dfrac{c}{d} = \dfrac{ac}{bd}$, where $b \neq 0$ and $d \neq 0$.	**Example 2:** Multiply. **a.** $\dfrac{3}{4} \cdot \dfrac{5}{7} = \dfrac{15}{28}$ **b.** $\dfrac{5}{6} \cdot \dfrac{4}{9} = \dfrac{5}{2 \cdot 3} \cdot \dfrac{2 \cdot 2}{9} = \dfrac{10}{27}$
To multiply decimal numbers: 1. Multiply as if they were whole numbers. 2. Place the decimal in the product so that it has the same number of decimal places as the total number of decimal places in the factors.	**Example 3:** Multiply $(3.6)(2.4)$. $\begin{array}{r} 2.4 \\ \times\ 3.6 \\ \hline 144 \\ +\ 72\ \\ \hline 8.64 \end{array}$ $\quad\longrightarrow\quad$ 1 place $\quad\longrightarrow\quad$ + 1 place $\longleftarrow$ 2 places

(continued)

Procedures/Rules	Key Examples

SECTION 1.4   Multiplying and Dividing Real Numbers; Properties of Real Numbers (*continued*)

When dividing two numbers that have the same sign, the quotient is positive.

When dividing two numbers that have different signs, the quotient is negative.

Rules for division involving 0: $0 \div n = 0$ when $n \neq 0$
$n \div 0$ is undefined when $n \neq 0$
$0 \div 0$ is indeterminate

Rule for dividing fractions: $\dfrac{a}{b} \div \dfrac{c}{d} = \dfrac{a}{b} \cdot \dfrac{d}{c}$, where $b \neq 0, c \neq 0,$ and $d \neq 0$.

Example 4: Divide.
a. $24 \div 8 = 3$
b. $-24 \div (-8) = 3$
c. $-24 \div 8 = -3$
d. $24 \div (-8) = -3$
e. $0 \div 5 = 0$
f. $14 \div 0$ is undefined.

Example 5: Divide $\dfrac{5}{8} \div \dfrac{3}{4}$.

$$\frac{5}{8} \div \frac{3}{4} = \frac{5}{8} \cdot \frac{4}{3} = \frac{5}{2 \cdot 2 \cdot 2} \cdot \frac{2 \cdot 2}{3} = \frac{5}{6}$$

When dividing decimal numbers, set up a long division and consider the divisor.
Case 1:   If the divisor is an integer, divide as if the dividend were a whole number and place the decimal point in the quotient directly above its position in the dividend.
Case 2:   If the divisor is a decimal number,
1. Move the decimal point in the divisor to the right enough places to make the divisor an integer.
2. Move the decimal point in the dividend the same number of places.
3. Divide the divisor into the dividend as if both numbers were whole numbers. Make sure you align the digits in the quotient properly.
4. Write the decimal point in the quotient directly above its new position in the dividend.
In either case, continue the division process until you get a remainder of 0 or a repeating digit (or block of digits) in the quotient.

Example 6: Divide $12.88 \div 0.06$.

$$\begin{array}{r} 214.66 \\ 6\overline{)1288.00} \\ \underline{-12} \\ 08 \\ \underline{-6} \\ 28 \\ \underline{-24} \\ 40 \\ \underline{-36} \\ 40 \end{array}$$

Answers: $214.\overline{6}$

SECTION 1.5   Exponents, Roots, and Order of Operations

To evaluate an exponential form raised to a natural number exponent, write the base as a factor the number of times indicated by the exponent; then multiply.

If the base of an exponential form is a negative number and the exponent is even, the product is positive.

If the base is a negative number and the exponent is odd, the product is negative.

Every positive number has two square roots, a positive root and a negative root.

Negative numbers have no real-number square roots.

Example 1: Evaluate.
a. $3^4 = 3 \cdot 3 \cdot 3 \cdot 3 = 81$
b. $-3^4 = -3 \cdot 3 \cdot 3 \cdot 3 = -81$

c. $(-2)^4 = (-2)(-2)(-2)(-2) = 16$

d. $(-2)^5 = (-2)(-2)(-2)(-2)(-2)$
$= -32$

Example 2: Find all square roots.
a. 49

Answer: ± 7

b. -36

Answer: No real-number square roots exist.

(*continued*)

Procedures/Rules	Key Examples

SECTION 1.5 Exponents, Roots, and Order of Operations (*continued*)

The radical symbol $\sqrt{}$ indicates to find only the positive (principal) square root.

$$\sqrt{\frac{a}{b}} = \frac{\sqrt{a}}{\sqrt{b}}, \text{ where } a \geq 0 \text{ and } b > 0.$$

Example 3: Simplify.

a. $\sqrt{81} = 9$

b. $\sqrt{-49}$ is not a real number.

c. $\sqrt{\frac{25}{64}} = \frac{\sqrt{25}}{\sqrt{64}} = \frac{5}{8}$

Order-of-Operations Agreement

Perform operations in the following order:
1. Within grouping symbols.
2. Exponents/Roots from left to right, in order as they occur.
3. Multiplication/Division from left to right, in order as they occur.
4. Addition/Subtraction from left to right, in order as they occur.

Example 4: Simplify.

$$35 - [29 - 2(12 + 4)] + 4^3$$
$$= 35 - [29 - 2(16)] + 4^3$$
$$= 35 - (29 - 32) + 4^3$$
$$= 35 - (-3) + 4^3$$
$$= 35 - (-3) + 64$$
$$= 35 + 3 + 64$$
$$= 102$$

To find the arithmetic mean, or average, of n numbers, divide the sum of the numbers by n.

$$\text{Arithmetic mean} = \frac{x_1 + x_2 + \cdots + x_n}{n}$$

Example 5: Carl has the following test scores: 68, 76, 82, and 78. Find the average of his test scores.

$$\frac{68 + 76 + 82 + 78}{4} = \frac{304}{4} = 76$$

SECTION 1.6 Translating Word Phrases to Expressions

To translate a word phrase to an expression, identify the unknowns, constants, and key words; then write the corresponding symbolic form.

Example 1: Translate to an algebraic expression:

a. Five more than seven times a number

Answer: $7n + 5$

b. Six times the difference of a number and nine

Answer: $6(n - 9)$

c. Four divided by a number cubed

Answer: $4 \div n^3$

SECTION 1.7 Evaluating and Rewriting Expressions

To evaluate an algebraic expression:
1. Replace the variables with their corresponding given values.
2. Calculate the numerical expression using the order of operations agreement.

Example 1: Evaluate $4x^2 - 5x$ when $x = -3$.

$$4(-3)^2 - 5(-3)$$
$$= 4(9) - 5(-3)$$
$$= 36 - (-15)$$
$$= 36 + 15$$
$$= 51$$

Distributive Property of Multiplication over Addition

$$a(b + c) = ab + ac$$

Example 2: Use the distributive property to write an equivalent expression.

$$3(x + 7) = 3 \cdot x + 3 \cdot 7$$
$$= 3x + 21$$

To combine like terms, add or subtract the coefficients and keep the variables and their exponents the same.

Example 3: Combine like terms.

$$15x^2 + 8x + 3x^2 + 7 - 9x$$
$$= 15x^2 + 3x^2 + 8x - 9x + 7$$
$$= 18x^2 - x + 7$$

Chapter 1 Review Exercises

For Exercises 1–6, answer true or false.

[1.1] **1.** π is a rational number.

[1.1] **2.** $-\sqrt{2}$ is a rational number.

[1.7] **3.** $-6x$ and $2x^3$ are like terms.

[1.7] **4.** $8xy$ and yx are like terms.

[1.5] **5.** $-6^2 = -36$

[1.5] **6.** $(-6)^2 = 36$

For Exercises 7–10, fill in the blank.

[1.1] **7.** The absolute value of every number is either ———— or 0.

[1.2] **8.** A natural number that has exactly two different factors, 1 and the number itself, is a(n) ———— number.

[1.3] **9.** When adding two addends that have the same sign, add their absolute values and ———— the sign.

[1.4] **10.** If there are a(n) ———— number of negative factors, the product is positive.

[1.1] **For Exercises 11–14, write a set representing each description.**

11. The months beginning with the letter M

12. The letters in the word *algebra*

13. The even natural numbers

14. The even integers

[1.1] **For Exercises 15–18, graph each number on a number line.**

15. $-3\dfrac{2}{5}$

16. $2\dfrac{1}{4}$

17. 8.2

18. -4.6

Exercises 19–22 EXPRESSIONS

[1.1] **For Exercises 19–22, simplify.**

19. $|-6.3|$

20. $|8.46|$

21. $\left|2\dfrac{1}{6}\right|$

22. $\left|-4\dfrac{3}{8}\right|$

Exercises 23–26 EQUATIONS AND INEQUALITIES

[1.1] **For Exercises 23–26, use $=$, $+$, $<$, or $>$ to write a true sentence.**

23. $|-6.4|$ ▨ 6.4

24. -6.9 ▨ 0

25. -14 ▨ -9

26. $\left|5\dfrac{6}{7}\right|$ ▨ $\left|-5\dfrac{6}{7}\right|$

[1.2] *For Exercises 27 and 28, name the fraction represented by each shaded region.*

27.

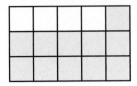

28.

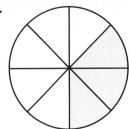

Exercises 29–32 EQUATIONS AND INEQUALITIES

[1.2] *For Exercises 29–32, find the missing number that makes the fractions equivalent.*

29. $-\dfrac{6}{7} = -\dfrac{?}{14}$

30. $\dfrac{2}{5} = \dfrac{10}{?}$

31. $\dfrac{15}{30} = \dfrac{?}{10}$

32. $\dfrac{-5}{2} = \dfrac{?}{4}$

Exercises 33–40 EXPRESSIONS

[1.2] *For Exercises 33–36, write the prime factorization for each number.*

33. 51

34. 100

35. 108

36. 84

[1.2] *For Exercises 37–40, simplify to lowest terms.*

37. $\dfrac{152}{200}$

38. $\dfrac{61}{122}$

39. $\dfrac{250}{360}$

40. $\dfrac{26}{39}$

Exercises 41–46 EQUATIONS AND INQUALITIES

[1.3–1.4] *For Exercises 41–46, indicate whether the expression illustrates the commutative property of addition, the associative property of addition, the commutative property of multiplication, the associative property of multiplication, or the distributive property of multiplication over addition.*

41. $4 + (7 + 3) = (4 + 7) + 3$

42. $3(9 + 1) = 27 + 3$

43. $\dfrac{1}{3} \cdot \dfrac{5}{6} = \dfrac{5}{6} \cdot \dfrac{1}{3}$

44. $5.6 + (11.2 + 4.3) = 5.6 + (4.3 + 11.2)$

45. $(n - 8)(-3) = -3(n - 8)$

46. $(2 \cdot 3) \cdot 7 = 2 \cdot (3 \cdot 7)$

Exercises 47–92 EXPRESSIONS

[1.3–1.5] *For Exercises 47–68, perform the indicated operation(s).*

47. $-8 - (-4.2)$

48. $6 - (-9.1)$

49. $\dfrac{2}{5} + \dfrac{1}{3}$

50. $-\dfrac{1}{4} + \dfrac{2}{3}$

51. $-7(-13)$

52. $6(-8)$

53. $-25 \div 5$

54. $-30 \div (-3)$

55. -6^2

56. $(-3)^2$

57. $7(6 + 2) - 48 \div 12$

58. $6(9 - 4) \div 3 - 1$

59. $(-2)^3(4)(-5)$

60. $3^5 \div 3^2 \div 3^2 \div 3$

61. $-\dfrac{1}{3} - \dfrac{3}{2} \div \left(\dfrac{1}{6}\right)$

62. $-6.8 + (-4.1 + 2.3)$

63. $4|-2| \div (-8)$

64. $10 \div 5 + (-5)(-5)$

65. $6 + \{3(4 - 5) + 2[6 + (-4)]\}$

66. $-1 - 3[4^2 - 3(-6)]$

67. $\dfrac{-6 + 3^2(8 - 10)}{2^3 - 16}$

68. $\dfrac{3(4 - 5) - 4 \cdot 3 + (11 - 20)}{(3 - 4)^3}$

[1.5] *For Exercises 69 and 70, find all square roots of each number.*

69. 49

70. -36

[1.6] *For Exercises 71–74, translate to an algebraic expression.*

71. twice a number subtracted from fourteen

72. the ratio of y to seven

73. a number minus twice the sum of the number and four

74. seven times the difference of m and n

[1.7] *For Exercises 75–80, evaluate the expressions using the given values.*

75. $b^2 - 4ac$; $a = -3, b = -1, c = 5$

76. $b^2 - 4ac$; $a = 2, b = 6, c = -2$

77. $-|4x| + |y^2|$; $x = -3, y = -2$

78. $\sqrt{m} + \sqrt{n}$; $m = 16, n = 9$

79. $\sqrt{m + n}$; $m = 16, n = 9$

80. $\dfrac{3l^2}{4 - n}$; $l = -4, n = 16$

[1.7] *For Exercises 81 and 82, determine all value(s) that cause each expression to be undefined.*

81. $\dfrac{n}{n + 6}$

82. $\dfrac{y}{(y - 4)(y + 3)}$

[1.7] *For Exercises 83–86, use the distributive property to write an equivalent expression and simplify.*

83. $5(x + 6)$

84. $-3(5n - 8)$

85. $\dfrac{1}{4}(8y + 3)$

86. $0.6(4.5m - 2.1)$

87. $6x + 3y - 9x - 6y - 15$

88. $5y^2 - 6y + 4y - y^2 + 9y$

89. $-6xy + 9xy - xy$

90. $8x^3 - 4x - 6x^2 - 10x^3 + 4x$

91. $-4m - 4m - 4n + 4n$

92. $14 - 6x + 4y - 8 - 10y - 8x$

[1.3] **93.** The credit card statement to the right shows when each transaction was posted to the account. Find the new balance.

4-01-09	Beginning balance	−$685.92
4-03-09	Dillard's	−$45.80
4-05-09	Payment	$250.00
4-08-09	Applebee's	−$36.45
4-16-09	CVS Pharmacy	−$12.92
4-24-09	Target credit	$32.68
4-30-09	Finance charge	−$5.18

[1.3] **94.** The bar graph to the right shows the closing price for Merck and Co., Inc. stock each day for the week beginning February 4, 2008.
   a. Find the difference between the closing price at the end of the week and the closing price at the beginning of the week.

   b. If a person bought shares at the close of 2/5 and sold at the close of 2/7, what would that person's net profit or loss be on each share?

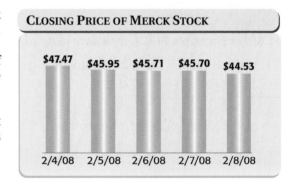

CLOSING PRICE OF MERCK STOCK

$47.47 $45.95 $45.71 $45.70 $44.53

2/4/08   2/5/08   2/6/08   2/7/08   2/8/08

[1.4] **95.** Refer to the graph in Exercise 94.
   a. If Kaye bought 200 shares at the close of 2/4, how much did she spend?

   b. If she sold all 200 shares on 2/8, what was her loss?

[1.4] **96.** The voltage in a circuit can be calculated by multiplying the current and the resistance. Suppose the current in a circuit is −9.5 amperes and the resistance is 16 ohms. Find the voltage.

[1.4] **97.** In 2000, Larry has a loan balance of −$2405.80. If he makes no payment for ten years, his balance will become −$14,896.08. How many times greater is his debt in 2010 than it was in 2000?

[1.5] **98.** A company reports the following revenues for the first six months of 2010. Find the average revenue.

Month	Revenue
January	$45,320
February	$38,250
March	$61,400
April	$42,500
May	$74,680
June	$62,800

Chapter 1 Practice Test

For Extra Help

Step-by-step test solutions are found on the Chapter Test Prep Videos available via the Video Resources on DVD, in *MyMathLab*, and on YouTube (search "CarsonElemAlg" and click on "Channels").

1. Graph $-3\frac{2}{7}$ on a number line.

2. Simplify: $|-3.67|$

3. Write the prime factorization of 100.

4. Simplify: $\dfrac{17}{51}$

5. Find the missing number: $-\dfrac{6}{5} = -\dfrac{?}{10}$

For Exercises 6 and 7, indicate whether the expression illustrates the commutative property of addition, the associative property of addition, the commutative property of multiplication, the associative property of multiplication, or the distributive property of multiplication over addition.

6. $3(2 + 5) = 6 + 15$

7. $4(7 + 1) = 4(1 + 7)$

For Exercises 8–13, calculate.

8. $8 + (-4)$

9. $\dfrac{7}{8} - \left(-\dfrac{5}{6}\right)$

10. $(-1.5)(-0.4)$

11. $-\dfrac{5}{6} \div \dfrac{2}{3}$

12. $(-4)^3$

13. -3^4

For Exercises 14–17, simplify.

14. $-12 \div 3\,(6 - 2^2)$

15. $\dfrac{6^2 + 14}{(2 - 7)^2}$

16. $-8 \div |4 - 2| + 3^2$

17. $\sqrt{25 - 16} + [(14 + 2) - 3^2]$

For Exercises 18 and 19, translate the indicated phrase.

18. twice the sum of m and n

19. five less than three times w

20. After making a payment, Karen's balance on her credit card is $-\$854.80$. If her balance prior to making the payment was $-\$1104.80$, how much did she pay toward the balance?

21. On six days over a three-month period, a city receives 6 inches, 10 inches, 4 inches, 3 inches, 2.5 inches, and 8 inches of snow. Calculate the average daily amount of snowfall over those days.

22. Evaluate $-|3x^2 + 2y|$, when $x = -1$ and $y = 4$.

23. Evaluate $\sqrt{a - b}$, when $a = 64$ and $b = -36$.

24. Use the distributive property to write an equivalent expression: $-5(4y + 9)$

25. Simplify: $3.5x - 8 + 2.1x + 9.3$

Solving Linear Equations and Inequalities

2

In Chapter 1, we reviewed numbers, arithmetic, and expressions, which form a foundation for studying equations. Remember from Chapter 1 that our most basic components are variables and constants, which make up expressions. We saw that we can perform two actions with expressions: (1) evaluate and (2) rewrite. In this chapter, we build on that foundation by creating equations and inequalities out of those expressions; then we *solve* those equations and inequalities. This development is repeated throughout the course. We study and classify new types of expressions, evaluate and rewrite those expressions, and solve equations and inequalities made from the new expressions.

Remember, as we build algebra concepts, they become more complex, just as a building becomes more complex during its construction. Try not to get discouraged if a new concept does not make sense right away. Learning new concepts is like constructing a building: During the early stages, the building is not as solid as it will be upon completion. This chapter is like the frame of the walls of a house: The frame is not as solid as the walls will be after the drywall and roof are put in place. Similarly, as we revisit equations and inequalities throughout the course, we reinforce and solidify the concepts you will learn in this chapter.

2.1 Equations, Formulas, and the Problem-Solving Process

Objectives

1. Verify solutions to equations.
2. Use formulas to solve problems.

We now move to the top tier of the Algebra Pyramid and focus on equations. An **equation** was first defined in Section 1.1. We repeat the definition, but state it a little differently.

Definition *Equation:* Two expressions set equal.

For example, $4x + 5 = 9$ is an equation made from the expressions $4x + 5$ and 9.

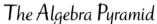

The Algebra Pyramid

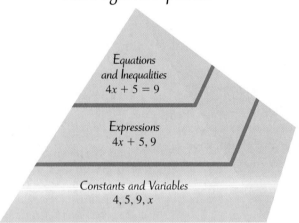

Equations and Inequalities
$4x + 5 = 9$

Expressions
$4x + 5, 9$

Constants and Variables
$4, 5, 9, x$

> **Connection** We can think of an equation as a complete sentence and the equal sign as the verb *is*. We can read $4x + 5 = 9$ as "Four x plus five is equal to nine" or "Four x plus five is nine."
>
> $$4x + 5 = 9$$
> ↕
> "Four x plus five is 9."

Objective 1 Verify solutions to equations. An equation can be true or false. For example, the equation $3 + 1 = 4$ is true, whereas the equation $3 + 1 = 5$ is false. When given an equation that contains a variable for a missing number, such as $x + 2 = 7$, our goal is to solve it, which means to find every number that can replace the variable and make the equation true. Such a number is called a **solution** to the equation.

Definition *Solution:* A number that makes an equation true when it replaces the variable in the equation.

For example, the number 5 is a solution to the equation $x + 2 = 7$ because when 5 replaces x, it makes the equation true.

$$x + 2 = 7$$
↓
$$5 + 2 = 7 \qquad \text{True}$$

By showing that the equation $x + 2 = 7$ is a true statement when x is replaced with 5, we verify, or *check*, that 5 is the solution to the equation.

Procedure **Checking a Possible Solution**

To determine whether a value is a solution to a given equation, replace the variable in the equation with the value. If the resulting equation is true, the value is a solution.

Example 1 Check to see if the given value is a solution to the equation.

a. $4x + 9 = 1; x = -2$

SOLUTION $4x + 9 = 1$

$$4(-2) + 9 \overset{?}{=} 1 \qquad \text{Replace } x \text{ with } -2 \text{ and see if the equation is true.}$$
$$-8 + 9 \overset{?}{=} 1$$
$$1 = 1$$

Note: The symbol $\overset{?}{=}$ indicates that we are asking, "How does the left side compare with the right side? Are they equal or not equal?"

Because -2 makes the equation true, it is a solution to $4x + 9 = 1$.

b. $n^2 + 0.8 = 3n - 0.1; n = 0.4$

SOLUTION $n^2 + 0.8 = 3n - 0.1$

$$(0.4)^2 + 0.8 \overset{?}{=} 3(0.4) - 0.1 \qquad \text{Replace } n \text{ with 0.4 and see if the equation is true.}$$
$$0.16 + 0.8 \overset{?}{=} 1.2 - 0.1$$
$$0.96 \neq 1.1$$

Because 0.4 does not make the equation true, it is not a solution to $n^2 + 0.8 = 3n - 0.1$.

c. $\dfrac{3}{4}y - 1 = \dfrac{y}{2} + \dfrac{1}{3}; y = 5\dfrac{1}{3}$

SOLUTION $\dfrac{3}{4}y - 1 = \dfrac{1}{2}y + \dfrac{1}{3}$

Note: It is convenient to write $\dfrac{y}{2}$ as $\dfrac{1}{2}y$, which we can do because $\dfrac{y}{2} = y \div 2 = y \cdot \dfrac{1}{2} = \dfrac{1}{2}y$.

$$\dfrac{3}{4}\left(5\dfrac{1}{3}\right) - 1 \overset{?}{=} \dfrac{1}{2}\left(5\dfrac{1}{3}\right) + \dfrac{1}{3} \qquad \text{Replace } y \text{ with } 5\dfrac{1}{3} \text{ and see if the equation is true.}$$

$$\dfrac{3}{4}\overset{1}{\underset{1}{\cancel{}}}\left(\dfrac{\overset{4}{\cancel{16}}}{\underset{1}{\cancel{3}}}\right) - 1 \overset{?}{=} \dfrac{1}{2}\left(\dfrac{\overset{8}{\cancel{16}}}{3}\right) + \dfrac{1}{3} \qquad \text{Rewrite } 5\dfrac{1}{3} \text{ as an improper fraction, divide out common factors, and then multiply.}$$

$$4 - 1 \overset{?}{=} \dfrac{8}{3} + \dfrac{1}{3}$$

$$3 = 3 \qquad \text{Therefore, } 5\dfrac{1}{3} \text{ is a solution to } \dfrac{3}{4}y - 1 = \dfrac{y}{2} + \dfrac{1}{3}.$$

Your Turn 1 Check to see if the given value is a solution to the equation.

a. $5n - 8 = 3n + 4; n = 6$ **b.** $4y - 3.5 = 2y + 3(y + 0.2); y = -1.5$

c. $\dfrac{2}{3}x + \dfrac{1}{4} = \dfrac{1}{5} - x^2; x = \dfrac{1}{6}$

Objective 2 **Use formulas to solve problems.** A primary purpose of studying mathematics is to develop and improve problem-solving skills. George Polya proposed the idea that all problem solving follows a four-step outline: (1) Understand the problem, (2) devise a plan for solving the problem, (3) execute the plan, and (4) check the results. Following is an outline for problem solving based on Polya's four stages. We'll see Polya's four stages illustrated throughout the rest of the text in application problems.

Answers **Your Turn 1**
a. yes   **b.** no   **c.** no

Problem-Solving Outline

1. **Understand** the problem.
   a. Read the question(s) (not the whole problem, just the question at the end) and write a note to yourself about what you are to find.
   b. Read the whole problem, underlining the key words.
   c. If possible or useful, draw a picture, make a list or table to organize what is known and unknown, simulate the situation, or search for a related example problem.
2. **Plan** your solution by searching for a formula or using the key words to translate to an equation.
3. **Execute** the plan by solving the equation/formula.
4. **Answer** the question. Look at the note about what you were to find and make sure you answer that question. Include appropriate units.
5. **Check** results.
   a. Try finding the solution in a different way, reversing the process, or estimating the answer and making sure the estimate and actual answer are reasonably close.
   b. Make sure the answer is reasonable.

As the course develops, we will explore the various strategies listed in the outline. In this section, we focus on using **formulas** to solve problems. Later in this chapter, we solve problems by using key words to translate to an equation.

Definition

Formula: An equation that describes a mathematical relationship.

First, let's consider using formulas from geometry. The table on page 95 lists some common geometric formulas. Before we examine the formulas, let's review a few terms.

Definitions

Perimeter: The distance around a figure.
Area: The total number of square units that fill a figure.
Volume: The total number of cubic units that fill a space.
Circumference: The distance around a circle.
Radius: The distance from the center of a circle to any point on the circle.
Diameter: The distance across a circle through its center.

Procedure

Using a Formula

To use a formula,

1. Replace the variables with the corresponding given values.
2. Solve for the missing value.

Geometric Formulas

Plane Figures

Square

$P = 4s$
$A = s^2$

Rectangle

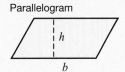

$P = 2l + 2w$
$A = lw$

Parallelogram

$A = bh$

Trapezoid

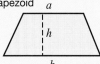

$A = \frac{1}{2}h(a + b)$

> **Note:** A parallelogram is a four-sided figure with two pairs of parallel sides. A trapezoid has only one pair of parallel sides.

Triangle

$A = \frac{1}{2}bh$

Circle

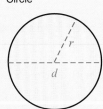

$C = \pi d$ or $C = 2\pi r$
$A = \pi r^2$

> **Note:** A circle's radius is half its diameter. (Or its diameter is twice the radius.)

> **Note:** Recall that π represents an irrational number. The π key on most calculators approximates the value as 3.141592654. Some people prefer to round the value to 3.14, which is simpler but less accurate.

Solids

Box

$V = lwh$
$SA = 2lw + 2lh + 2wh$

Pyramid

$V = \frac{1}{3}lwh$

Cylinder

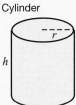

$V = \pi r^2 h$

Cone

$V = \frac{1}{3}\pi r^2 h$

Sphere

$V = \frac{4}{3}\pi r^3$

Connection Notice that the volume formulas for a box and a pyramid are similar. This is because a pyramid has $\frac{1}{3}$ the volume of a box with the same-size base and same height. The same relationship is true for cylinders and cones with the same-size base and same height.

Problems Involving a Single Formula

Example 2   **a.** A mason is building a rectangular foundation wall that is to be 75 feet by 40 feet. What is the total distance around the wall?

Understand   The "total distance around the wall" is the perimeter.

Plan   Because the shape is a rectangle, we can use the formula $P = 2l + 2w$.

Execute   Replace l with **75** feet and w with **40** feet; then calculate.

$$P = 2l + 2w$$
$$P = 2(\textbf{75 ft.}) + 2(\textbf{40 ft.})$$
$$P = 150 \text{ ft.} + 80 \text{ ft.}$$
$$P = 230 \text{ ft.}$$

Answer   The mason must have enough supplies to build a wall that is 230 feet long.

Check   In this case, the solution can be verified using an alternative method. One could add all four side lengths.

$$P = 75 \text{ ft.} + 75 \text{ ft.} + 40 \text{ ft.} + 40 \text{ ft.} = 230 \text{ ft.}$$

b. A tabletop is in the shape of a circle whose radius is 4 feet. If the tabletop is to be covered with Formica, how much Formica is needed?

Understand   The amount of Formica needed to cover the tabletop is the same as the area of the tabletop.

Plan   Because the tabletop is in the shape of a circle, we use the formula $A = \pi r^2$.

Execute   Replace π with **3.14** and r with **4 ft.** and simplify.

$$A = \pi r^2$$
$$A = (\textbf{3.14})\,(\textbf{4 ft.})^2$$
$$A = (3.14)(16 \text{ ft.}^2)$$
$$A = 50.24 \text{ ft.}^2$$

Answer   About 50.24 square feet of Formica is needed to cover the tabletop.

Check   Verify the reasonableness of the answer using estimation. If we round π to 3, the answer is $(3)(4 \text{ ft.})(4 \text{ ft.}) = 48 \text{ ft.}^2$. Because π is a little more than 3 and our answer is a little more than 48 ft.^{2}, our answer is reasonable.

Your Turn 2   A wallpaper hanger is to paper a bedroom wall that is 18 feet long and 12 feet high. How much wallpaper is needed to cover the wall?

Problems Involving Combinations of Formulas

A *composite figure* contains more than one geometric figure and requires a combination of formulas.

Procedure   **Calculating the Area of Composite Figures**

For composite figures:

1. To calculate the area of a figure composed of two or more figures that are next to each other, add the areas of the individual figures.
2. To calculate the area of a region defined by a smaller figure within a larger figure, subtract the area of the smaller figure from the area of the larger figure.

Answer   **Your Turn 2**

216 square feet

Example 3 Following is a drawing of a room that is to be carpeted. Calculate the area of the room.

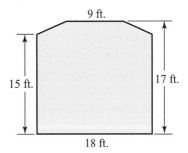

Understand This figure can be viewed as a trapezoid and a rectangle. The height of the trapezoid is found by subtracting 15 feet from 17 feet.

Plan Because the figure consists of a trapezoid and rectangle combined, we can add the areas of these shapes to find the total area of the figure. The area of a rectangle is found using the formula $A = lw$, and the area of a trapezoid is found using the formula $A = \dfrac{1}{2}h(a + b)$. The total area is

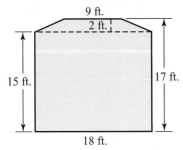

$$A = \text{Area of the rectangle} + \text{Area of the trapezoid}$$

$$A = \qquad\quad lw \qquad\quad + \quad \frac{1}{2}h(a + b)$$

Execute Replace the variables with the corresponding values and calculate.

$$A = lw + \frac{1}{2}h(a + b)$$

$$A = (15\,\text{ft.})(18\,\text{ft.}) + \frac{1}{2}(2\,\text{ft.})(18\,\text{ft.} + 9\,\text{ft.}) \qquad \begin{array}{l} l = 15\,\text{ft.}, w = 18\,\text{ft.}, h = 2\,\text{ft.}, \\ a = 18\,\text{ft., and } b = 9\,\text{ft.} \end{array}$$

$$A = (15\,\text{ft.})(18\,\text{ft.}) + \frac{1}{2}(2\,\text{ft.})(27\,\text{ft.})$$

$$A = 270\,\text{ft.}^2 + 27\,\text{ft.}^2$$

$$A = 297\,\text{ft.}^2$$

Answer The total area is 297 square feet.

Check We can verify the reasonableness of the answer using estimation. Suppose the figure had been a rectangle measuring 18 feet by 17 feet. We would expect the area of this rectangle to be slightly greater than the area of the actual figure. The area of an 18-foot by 17-foot rectangle is $A = (18\,\text{ft.})(17\,\text{ft.}) = 306\,\text{ft.}^2$, which indicates that 297 square feet is reasonable.

Your Turn 3 Calculate the area of the following sign. (Use $\pi \approx 3.14$.)

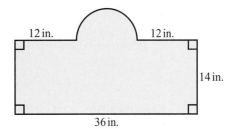

Sometimes, a figure may be placed within another figure and we must calculate the area of the region *between* the outside figure and inside figure.

Example 4   A portion of the front of a house is to be covered with siding (shaded in the figure). The window is 2.8 feet by 4.8 feet. Calculate the area to be covered.

Understand   The shaded area is the entire triangle excluding the rectangular window.

Plan   We can exclude the area of the window by calculating the area of the triangle and subtracting the area of the window.

$$A = \text{Area of the triangle} - \text{Area of the rectangle}$$
$$A = \frac{1}{2}bh - lw$$

Execute $A = \dfrac{1}{2}(12.5\text{ ft.})(8\text{ ft.}) - (2.8\text{ ft.})(4.8\text{ ft.})$ $b = 12.5$ ft., $h = 8$ ft., $l = 2.8$ ft., $w = 4.8$ ft.

$$A = 50\text{ ft.}^2 - 13.44\text{ ft.}^2$$
$$A = 36.56\text{ ft.}^2$$

Answer   The area to be covered is 36.56 square feet.

Check   Estimate the area by rounding the decimal numbers. Rounding the measurements of the triangle to the nearest whole number, we have a base of 13 feet and a height of 8 feet, which means an area of 52 square feet. Rounding the window dimensions to the nearest whole number, we have a 3-foot by 5-foot window, which has an area of 15 square feet. Subtracting these areas, we have a final area of 37 square feet, which indicates that our answer of 36.56 square feet is reasonable. Because all of our estimates are greater than or equal to the actual amount, the estimated answer is larger than the actual answer.

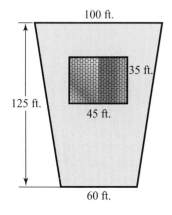

Your Turn 4   A house is to be situated on a lot as shown at the left. Once the house is built, the remaining area will be landscaped. Calculate the area to be landscaped.

Problems Involving Nongeometric Formulas

Following is a list of other formulas you may encounter.

The distance, d, an object travels given its rate, r, and the time of travel, t:	$d = rt$
The average rate of travel, r, given the total distance, d, and total time, t:	$r = \dfrac{d}{t}$
The voltage, V, in a circuit with a current, i, in amperes (A), and a resistance, R, in ohms (Ω):	$V = iR$
The temperature in degrees Celsius given degrees Fahrenheit:	$C = \dfrac{5}{9}(F - 32)$
The temperature in degrees Fahrenheit given degrees Celsius:	$F = \dfrac{9}{5}C + 32$

Example 5   A truck driver begins a delivery at 9 A.M. and travels 150 miles before taking a 30-minute break. He then travels another 128 miles, arriving at his destination at 2 P.M. What was his average driving rate?

Understand   We are given travel distances and times, and we are to find the average driving rate.

Plan To find the average rate, we first need the total distance traveled and the total time spent driving. We can then use the formula $r = \dfrac{d}{t}$.

Execute Total distance $= 150 + 128 = 278$ miles $= d$

Total time driving $= 5 - 0.5 = 4.5$ hours $= t$

Note: From 9 A.M. to 2 P.M. is 5 hours, but we must deduct the 0.5-hour (30-minute) break because he was not traveling during that time.

Now we can calculate the average rate.

$$r = \frac{278 \text{ miles}}{4.5 \text{ hours}}$$

$$r = 61.\overline{7} \text{ mph (miles per hour)}$$

Answer His average driving rate was $61.\overline{7}$ miles per hour.

Check We can use $d = rt$ to verify that if he traveled an average rate of $61.\overline{7}$ miles per hour for 4.5 hours, he would travel 278 miles.

$$d \approx 61.8(4.5)$$

$$d \approx 278.1 \text{ miles}$$

Note: We rounded $61.\overline{7}$, so the distance is an approximation.

Answer Your Turn 5

$3\dfrac{3}{17}$ or ≈ 3.2 mph

Your Turn 5 Hilda hikes 10 miles then rests for 20 minutes, after which she hikes another 8 miles. If she began her hike at 10:30 A.M. and stopped at 4:30 P.M., what was her average rate?

2.1 Exercises   **For Extra Help** *MyMathLab*   PRACTICE WATCH DOWNLOAD

Note: Exercises marked with a ★ represent challenging exercises.

1. What symbol is present in an equation but not in an expression?

2. What is a solution to an equation?

3. Explain how to check a value to see if it is a solution to a given equation.

4. If the dimensions of a figure are given in inches, in what unit will the area be expressed?

For Exercises 5–20, check to see if the given number is a solution to the given equation. See Example 1.

5. $3x - 5 = 41; x = 2$

6. $4a + 7 = 51; a = 11$

7. $7y - 1 = y + 3; y = \dfrac{2}{3}$

8. $-8t - 3 = 2t - 15; t = -\dfrac{6}{5}$

9. $-4(x - 5) + 2 = 5(x - 1) + 3; x = -2$

10. $2(3m + 2) - 2 = 5m - 1; m = -3$

11. $-\dfrac{1}{2} + \dfrac{1}{3}y = \dfrac{3}{2}; y = 6$

12. $\dfrac{1}{2}p - \dfrac{1}{2} = \dfrac{2}{5}p + \dfrac{3}{2}; p = 20$

13. $4.3z - 5.71 = 2.1z + 0.07; z = 2.2$

14. $12.7a + 12.6 = a + 5.4a; a = -2$

★15. $b^2 - 4b = b^3 + 6; b = -1$

★16. $-x^3 + 9 = 2x^2 - 6x; x = -3$

★17. $|x^2 - 9| = -x + 3; x = 3$

★18. $-|2u - 3| = -3u + 8; u = 5$

★19. $\dfrac{4x - 3}{x - 4} = \sqrt{x + 8}; x = 17$

★20. $\dfrac{-y}{10 + y} = \dfrac{\sqrt{4 - y}}{3}; y = -5$

For Exercises 21–44, solve using geometric formulas. (Answers to exercises involving π were calculated using the π key on a calculator, which approximates the value as 3.141592654.) See Examples 2–4.

21. Janet wants to put a wallpaper border around her rectangular kitchen. The room is 13 feet by 20 feet.

   a. What is the total length of wallpaper border she needs?

   b. If the wallpaper border comes in packages of 12 feet, how many packages must she buy?

   c. If the packages are priced at $8.99 each, what will be the total cost of the wallpaper?

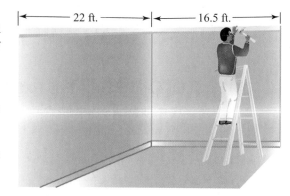

22. Jamal is planning to install crown molding where the walls and ceiling of his rectangular living room meet. The room is 16.5 feet by 22 feet.

   a. What is the total length of crown molding he needs?

   b. If the crown molding comes in strips of 8 feet, how many strips must he purchase?

   c. If the strips cost $9.99 each, what will be the cost of the crown molding?

23. Fermi National Accelerator Laboratory is a circular tunnel that is used to accelerate elementary particles. The radius of the tunnel is about 2 kilometers. If a particle travels one complete revolution around the circumference of the tunnel, what distance does it travel to the nearest hundredth of a kilometer? (*Source:* Cesare Emiliani, *The Scientific Companion*, 2nd ed., Wiley, 1995.)

24. The Chicxulub crater, which is buried partly beneath the Yucatan peninsula and partly beneath the Gulf of Mexico, is circular and is about 180 kilometers in diameter. What is the circumference of this crater to the nearest hundredth of a kilometer?

25. At one time, the Venetian Hotel and Casino in Las Vegas boasted having the largest hotel rooms in the world. If one of the rectangular rooms is 70 feet by 100 feet, what is the total square footage in the room?

26. The world's largest oil production barge was built by Kvaernel Oil and Gas, Inc. of Norway and Single Bouy Moorings, Inc. of Switzerland. The rectangular deck of the barge is 273 meters long and 50 meters wide. What is the area of the deck?

Of Interest

The Chicxulub crater was created when Earth was hit by an asteroid about 65 million years ago. The devastation of the impact is thought to have caused the extinction of about 50 percent of the species on Earth, including the dinosaurs.

27. A large wall in a home is to be painted with a custom accent color. The rectangular wall measures 32.5 feet by 12 feet. A painter charges $2.50 per square foot to paint a wall with a custom color. How much will it cost to have the painter paint the wall?

28. Hardwood floors are to be installed in a rectangular bedroom that measures 14 feet by 15 feet. If the contractor charges $34.50 per square foot to install the floors, what will be the cost of this improvement?

29. Tina needs to buy enough pine straw to cover a rectangular flower bed that measures 6 feet by 13 feet.

   a. What is the total area that she must cover?

   b. A landscape consultant tells Tina that she should use one bale of pine straw for every 10 square feet to be covered. If Tina follows this advice, how many bales must she purchase?

   c. If Tina's local garden center charges $4.50 per bale, what will be the cost of covering her flower bed?

30. Juan is considering installing carpet in his new office space. The office is rectangular and measures 42 feet by 36 feet. Juan has a budget of $3000.

   a. What is the total area that will be carpeted?

   b. How many square yards is this area? (There are 9 square feet in 1 square yard.)

   c. If the contractor quotes Juan a price of $22.50 per square yard to install the carpet, how much will it cost to carpet the space? Is this feasible with Juan's budget?

31. A city planning committee decides to install 20 historic markers around the city. The signs are triangular, measuring 26 inches wide at the base and 20 inches high.

   a. What is the area of each sign?

   b. If a contractor charges $0.75 per square inch to build and install the signs, how much will it cost to have them installed?

32. A parks department is going to install a flower bed in a triangular space created by three trails, as shown. The flower bed measures 32 feet at the base and is 24 feet high.

   a. What is the area of the flower bed?

   b. If a landscaper charges $6.50 per square foot to design and install the bed, what will the total cost be?

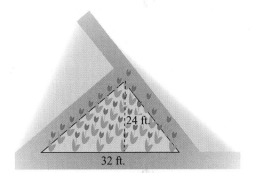

33. Conner and Ellyn purchase a new home. The area around an 8-foot by 9-foot storage building in the rectangular backyard is to be landscaped, as shown in the figure.

   a. What is the total area that is to be landscaped?

   b. If $\frac{2}{3}$ of this area is to be grass, what amount will be covered with grass?

   c. If each pallet of grass sod covers 504 square feet, how many pallets will be needed?

   d. Each pallet cost $106. What is the total cost of the sod?

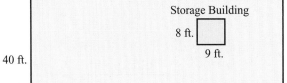

34. A kitchen floor is to be recovered with granite tiles. There is a fixed island in the center of the floor, as shown in the figure.

    a. What is the total area to be tiled?

    b. If each piece of tile covers $\frac{1}{4}$ of a square foot, how many pieces of tile will be needed?

    c. If each tile costs $3.95, what is the cost of the tile?

    d. If the contractor charges $8 per square foot to install the new floor, what will be the installation cost?

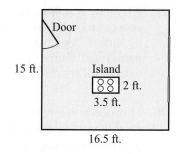

35. A company cuts two circular pieces out of a rectangular sheet of metal, and the rest of the sheet is recycled. The sheet is 4 feet by 8 feet, and the diameter of each circle is 4 feet. Find the area of the sheet that is recycled. Round your answer to the nearest tenth.

36. A compact disc has a diameter of $5\frac{3}{4}$ inches. On a CD, no information is placed within a circle in the center of the CD with a diameter of $1\frac{3}{4}$ inches. Find the area on a CD that can contain information. Round your answer to the nearest tenth.

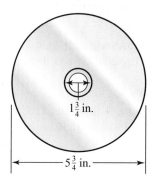

★**37.** The wall shown is to be painted. The door, which will not be painted, is 2.5 feet by 7.5 feet. What area will be painted?

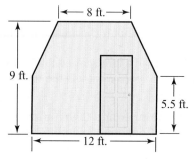

★**38.** The side of the house in the figure needs new siding. The window measures 3 feet by 4.5 feet. Find the area to be covered in siding.

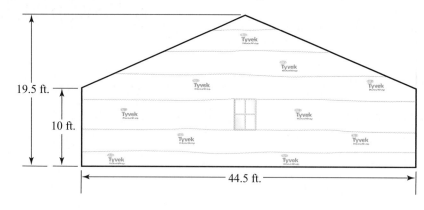

39. A small storage building is shaped like a box that measures 2 meters by 5 meters by 4.75 meters. What is the storage building's volume?

40. A dorm refrigerator measures 2 feet by $1\frac{1}{2}$ feet by 4 feet. What is its volume?

41. A cylindrical bottle has a diameter of 8 centimeters and a height of 20 centimeters. Find the volume of the bottle. Round your answer to the nearest tenth.

42. Earth is roughly a sphere with a radius of approximately 6370 kilometers. What is the approximate volume of Earth?

43. The base of the Great Pyramid at Giza, Egypt, is 745 feet by 745 feet, and its height is 449 feet. Find the volume of the pyramid.

44. If a funnel is approximately a cone, approximate the volume of a funnel with a diameter of $8\frac{1}{2}$ inches and a height of 6 inches. Round your answer to the nearest tenth.

For Exercises 45–50, use the formulas relating distance, rate, and time
$(d = rt, r = \dfrac{d}{t})$. *See Example 5.*

45. A family began a trip of 516 miles at 8 A.M. They arrived at their final destination at 5:30 P.M. If they took two 15-minute breaks and an hour for lunch, what was their average driving rate?

46. A family began a trip at 7:30 A.M. At the beginning of the trip, the car's odometer read 45,362.6. When they arrived at their final destination at 6 P.M., the odometer read 45,785.2. If they took three 15-minute breaks and an hour for lunch, what was their average driving rate?

47. A truck driver is to make a delivery that requires him to drive 285 miles by 4 P.M. If he starts at 10 A.M. and makes three 15-minute stops at weigh stations along the route, what must his average driving rate be to make the delivery on time?

★**48.** In 2005, Lance Armstrong won the Tour de France, which covered a distance of 3606 kilometers, with a time of 82 hours, 34 minutes, and 5 seconds. What was his average rate in kilometers per hour for the race? (*Source: The World Almanac and Book of Facts 2005.*)

★**49.** A flight departs Atlanta at 8:30 A.M. EST to arrive in Las Vegas at 10:00 A.M. PST. If the plane flies at an average rate of $388\frac{1}{3}$ mph, what distance does it travel? (*Hint:* There is a three-hour time difference between EST and PST.)

★**50.** A flight departs at 7:30 A.M. EST from Philadelphia and travels to Dallas–Fort Worth, arriving at 10:10 A.M. CST. If the plane is averaging a speed of 368.2 miles per hour, what distance does it travel? (*Hint:* There is a one-hour time difference between EST and CST. Also express the time using fractions.)

For Exercises 51 and 52, use the formula relating voltage, current, and resistance
($V = iR$). See Example 5.

51. A technician measures the current in a circuit to be −6.5 amperes, and the resistance is 8 ohms. Find the voltage.

52. The current in a circuit is measured to be 4.2 amperes. If the resistance is 16 ohms, find the voltage.

For Exercises 53–58, use the formulas for converting degrees Fahrenheit to degrees Celsius or degrees Celsius to degrees Fahrenheit $\left(C = \dfrac{5}{9}(F - 32), F = \dfrac{9}{5}C + 32 \right)$.
See Example 5.

53. On a drive to work, Tim notices that the temperature reading on a bank's digital thermometer is 34°C. What is this temperature in degrees Fahrenheit?

54. Iron melts at a temperature of 1535°C. What is this temperature in degrees Fahrenheit?

55. Liquid oxygen boils at a temperature of −183°C. What is this temperature in degrees Fahrenheit?

56. The average temperature on Neptune is approximately −360°F. What is this temperature in degrees Celsius?

57. At the South Pole, temperatures can get down to −76°F. What is this temperature in degrees Celsius?

58. The average temperature on Venus is approximately 890°F. What is this temperature in degrees Celsius?

Of Interest

Because of the tilt of Earth, the South Pole experiences daylight for six months, then night for six months. The Sun rises in mid-September and sets in mid-March.

Puzzle Problem

A contest has people guess how many marbles are inside a jar. The jar is a cylinder with a diameter of 9 inches and a height of 12 inches. Suppose each marble has a diameter of 0.5 inch. Calculate the number of marbles that might fit inside the jar. Discuss the accuracy of your calculation.

Review Exercises

Exercises 1–6 **EXPRESSIONS**

[1.5] *For Exercises 1 and 2, simplify.*

1. $6(-3 - 3) + 5(-3)$

2. $6(2 + 3^2) - 4(2 + 3)^2$

[1.7] *For Exercises 3 and 4, multiply using the distributive property.*

3. $\dfrac{1}{3}(x + 6)$

4. $-(4w - 6)$

[1.7] *For Exercises 5 and 6, combine like terms.*

5. $5x - 14 + 3x + 9$

6. $6.2y + 7 - 1.5 - 0.8y$

2.2 The Addition Principle of Equality

Objectives

1. Determine whether a given equation is linear.
2. Solve linear equations in one variable using the addition principle.
3. Solve equations with variables on both sides of the equal sign.
4. Solve identities and contradictions.
5. Solve application problems.

Objective 1 Determine whether a given equation is linear. There are many types of equations. In this section, we introduce and begin solving **linear equations**.

Definition **Linear equation:** An equation in which each variable term contains a single variable raised to an exponent of 1.

Equations that are not linear are *nonlinear equations*.

Connection Every equation has a corresponding graph. The names *linear* and *nonlinear* refer to the graphs of the equations that they describe. The graph of a linear equation is a line; the graph of a nonlinear equation is not.

Example 1 Determine whether the equation is linear or nonlinear.

a. $3x + 5 = 12$

Answer This equation is linear because the variable x has an exponent of 1.

b. $7x^3 + x = 1$

Answer This equation is nonlinear because the variable in the term $7x^3$ has an exponent of 3.

c. $3x + 2y = -10$

Answer This equation is linear because the variables x and y have exponents of 1.

Your Turn 1 Determine whether the given equation is linear.

a. $3x - 4 = 11$ **b.** $-6x + y = 30$ **c.** $n^2 = 5m + 4$ **d.** $y = 3x^2 + 1$

In this chapter, we focus on equations with the same variable throughout the equation. If a linear equation has the same variable throughout, we say it is a **linear equation in one variable**.

Definition **Linear equation in one variable:** An equation that can be written in the form $ax + b = c$, where a, b, and c are real numbers and $a \neq 0$.

$5x + 9 = -6$ is a linear equation in one variable.
$x^2 - 2x + 5 = 8$ is a nonlinear equation in one variable.

In Chapter 4, we consider linear equations in two variables, such as $x + y = 3$.

Objective 2 Solve linear equations in one variable using the addition principle. In Section 2.1, we discussed solutions to equations and how to check an equation's solution. We now consider a technique for solving an equation (which means to find its solution(s)). This technique is called the *balance technique*. Imagine an equation as scales with the equal sign as the fulcrum.

Answers Your Turn 1

a. yes **b.** yes **c.** no **d.** no

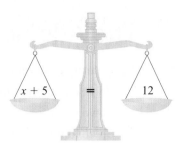

When weight is added or removed on one side of a balanced set of scales, the scales tip out of balance. In this case, we add 4 to the left side.

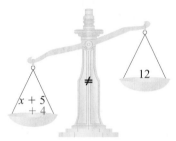

To maintain balance, the same weight must be added or removed on *both* sides of the scale. Adding 4 to the right side balances the scale.

In the language of mathematics, this balance technique is stated in two principles: the *addition principle of equality* and the *multiplication principle of equality*. In this section, we consider the addition principle of equality. In Section 2.3, we consider the multiplication principle of equality. The addition principle of equality says that we can add the same amount to (or subtract the same amount from) both sides of an equation without affecting its solution(s).

Rule **The Addition Principle of Equality**

> If $a = b$, then $a + c = b + c$ is true for all real numbers a, b, and c.

Note: We say that $a = b$ and $a + c = b + c$ are **equivalent equations** because they have the same solution.

In solving an equation, the goal is to write an equivalent equation in a simpler form, $x = c$, where the variable is alone, or isolated, on one side of the equal sign and the solution appears on the other side. For example, to isolate x in $x + 5 = 12$, we need to clear $+5$ from the left-hand side. To clear $+5$, we add -5 (or subtract 5) because $+5$ and -5 are additive inverses, which means that their sum is 0.

Note: To isolate x, we clear $+5$ by adding -5 to (or subtracting 5 from) the left side.

$$
\begin{aligned}
x + 5 &= 12 \\
-5 \quad & -5 \\
\hline
x + 0 &= 7 \\
x &= 7
\end{aligned}
$$

Note: To keep the equation balanced, we add -5 to (or subtract 5 from) the right side as well.

The statement $x = 7$ is the solution statement.

Using the Addition Principle of Equality

To use the addition principle of equality to clear a term in an equation, add the additive inverse of that term to both sides of the equation. (That is, add or subtract appropriately so that the term you want to clear becomes 0.)

Example 2 Solve and check.

a. $x - 13 = -22$

SOLUTION To isolate x, we need to clear -13.

Note: Add 13 to the left-hand side so that $-13 + 13 = 0$.

$$\begin{aligned} x - 13 &= -22 \\ +13 \quad &+13 \\ \hline x + 0 &= -9 \\ x &= -9 \end{aligned}$$

Note: Because we added 13 to the left side, we *must* add 13 to the right side as well.

Check Recall from Section 2.1 that to check, we replace x in the original equation with -9 and verify that the equation is true.

$$\begin{aligned} x - 13 &= -22 \\ -9 - 13 &\overset{?}{=} -22 \qquad \text{Replace } x \text{ with } -9. \\ -22 &= -22 \end{aligned}$$

True; so -9 is the solution.

b. $\dfrac{3}{5} = y + \dfrac{1}{3}$

Note: Here, we illustrate another popular style of writing the addition principle, in which $-\dfrac{1}{3}$ is written beside the expressions on both sides of the equation instead of underneath the corresponding terms.

LEARNING Strategy

VISUAL If you are a visual learner, to become comfortable with the addition principle, consider writing the amount added or subtracted on both sides of an equation in a different color than the rest of the equation. Once you are comfortable seeing the amount added or subtracted, you may not need to use the color any longer.

SOLUTION $\dfrac{3}{5} - \dfrac{1}{3} = y + \dfrac{1}{3} - \dfrac{1}{3}$ Subtract $\dfrac{1}{3}$ from both sides to isolate y.

$\dfrac{9}{15} - \dfrac{5}{15} = y + 0$ Rewrite fractions with their LCD.

$\dfrac{4}{15} = y$ Subtract $\dfrac{5}{15}$ from $\dfrac{9}{15}$.

Check $\dfrac{3}{5} = y + \dfrac{1}{3}$

$\dfrac{3}{5} \overset{?}{=} \dfrac{4}{15} + \dfrac{1}{3}$ Replace y in the original equation with $\dfrac{4}{15}$ and verify that the equation is true.

$\dfrac{3}{5} \overset{?}{=} \dfrac{4}{15} + \dfrac{5}{15}$ Rewrite fractions with their LCD.

$\dfrac{3}{5} \overset{?}{=} \dfrac{9}{15}$ Add $\dfrac{4}{15}$ and $\dfrac{5}{15}$.

$\dfrac{3}{5} = \dfrac{3}{5}$ Simplify to lowest terms.

True; so $\dfrac{4}{15}$ is the solution.

Your Turn 2 Solve and check.

a. $4.7 = x - 9.8$

b. $n + \dfrac{3}{4} = \dfrac{1}{6}$

Simplifying before Isolating the Variable

Some equations have expressions that can be simplified. If like terms are on the same side of the equation, we combine the like terms before isolating the variable. If the equation to be solved contains parentheses, we use the distributive property to clear the parentheses before isolating the variable.

Example 3 Solve and check.

a. $8y - 2.1 - 7y = -3.7 + 9$

SOLUTION Simplify the expressions; then isolate y.

$$8y - 2.1 - 7y = -3.7 + 9$$

Combine $8y$ and $-7y$ to equal y. $\quad y - 2.1 = 5.3 \quad$ Combine -3.7 and 9 to equal 5.3.

$$\underline{+2.1 \quad +2.1} \quad \text{Add 2.1 to both sides to isolate } y.$$
$$y + 0 = 7.4$$
$$y = 7.4$$

Check Replace y in the original equation with 7.4 and verify that the equation is true. We will leave this to the reader.

b. $6(n - 3) - 5n = -25 + 4$

SOLUTION Use the distributive property, simplify, and then isolate n.

$$6(n - 3) - 5n = -25 + 4$$

Distribute 6 to clear parentheses. $\quad 6n - 18 - 5n = -21 \quad$ Combine -25 and 4

Combine $6n$ and $-5n$ to equal n. $\qquad n - 18 = -21 \quad$ to equal -21.

$$\underline{+18 \qquad +18} \quad \text{Add 18 to both sides}$$
$$n + 0 = \quad -3 \quad \text{to isolate } n.$$
$$n = \quad -3$$

Check Replace n with -3 in the original equation and verify that the equation is true. We will leave this to the reader.

Your Turn 3 Solve and check.

a. $9x + 4 - 8x = 1 - 6$

b. $6.2 - 0.4 = 1.5(m - 6) - 0.5m$

Objective ③ **Solve equations with variables on both sides of the equal sign.** Some equations have variable terms on both sides of the equal sign. To isolate the variable, we must first get the variable terms together on the same side of the equal sign. To do this, we use the addition principle.

Answers   Your Turn 2

a. 14.5   b. $-\dfrac{7}{12}$

Answers   Your Turn 3

a. -9 b. 14.8

Example 4 Solve and check. $5x - 8 = 4x - 13$

SOLUTION Use the addition principle to get the variable terms together on the same side of the equal sign. Then isolate the variable.

Note: Remember that x means $1 \cdot x$. It does *not* mean that $x = 1$.

$$
\begin{array}{rcl}
5x - 8 &=& 4x - 13 \\
\underline{-4x} && \underline{-4x} \\
x - 8 &=& 0 - 13 \\
x - 8 &=& -13 \\
\underline{+8} && \underline{+8} \\
x + 0 &=& -5 \\
x &=& -5
\end{array}
$$

Subtract $4x$ from both sides.

Add 8 to both sides to isolate x.

Note: By choosing to subtract $4x$ on both sides, we clear the $4x$ term from the right side and combine it with the $5x$ on the left side. It doesn't matter which term you move first. (See the explanation following the example.)

Check $5x - 8 = 4x - 13$

$$
\begin{array}{rcl}
5(-5) - 8 &\overset{?}{=}& 4(-5) - 13 \\
-25 - 8 &\overset{?}{=}& -20 - 13 \\
-33 &=& -33
\end{array}
$$

Replace x in the original equation with -5 and verify that the equation is true.

True; so -5 is the solution.

When variable terms appear on both sides of the equal sign, it doesn't matter which term you choose to clear first, although some choices make the process easier. Suppose in Example 4 that we clear the $5x$ term first.

$$
\begin{array}{rcl}
5x - 8 &=& 4x - 13 \\
\underline{-5x} && \underline{-5x} \\
0 - 8 &=& -x - 13 \\
-8 &=& -x - 13 \\
\underline{+13} && \underline{+13} \\
5 &=& -x + 0 \\
5 &=& -x \\
-5 &=& x
\end{array}
$$

Note: Clearing the $5x$ term first gives $-x$, whereas when we cleared the $4x$ term, we got x. Most people prefer working with positive coefficients.

Note: $5 = -x$ means that we must find a number whose additive inverse is 5; so x must be equal to -5. In effect, we simply changed the signs of both sides. In Section 2.3, we will see another way to interpret $-x$.

Notice that the term we chose to clear in the first step did not affect the solution, but it did affect our approach to the rest of the problem.

Conclusion In solving an equation that has variable terms on both sides of the equal sign, when we select a variable term to clear, we can avoid negative coefficients by clearing the term with the lesser coefficient.

Your Turn 4 Solve and check.

a. $7y - 3 = 6y + 10$ **b.** $2.6 + 3n = 9.5 + 4n$

Some equations that have variables on both sides require using the distributive property and combining like terms to solve.

Answers **Your Turn 4**
a. 13   **b.** -6.9

Note: When we distribute a minus sign in an expression such as $-(3y + 11)$, we can think of the minus sign as -1; so we have

$$-(3y + 11)$$
$$= -1(3y + 11)$$
$$= -1(3y) + (-1)(11)$$
$$= -3y - 11$$

Example 5   Solve and check. $y - (3y + 11) = 5(y - 2) - 8y$

SOLUTION   Simplify both sides of the equation. Then isolate y.

$$y - (3y + 11) = 5(y - 2) - 8y$$

Distribute the minus sign.

$$y - 3y - 11 = 5y - 10 - 8y \qquad \text{Distribute 5.}$$

Note: The coefficient of $-3y$ is less than the coefficient of $-2y$, so we chose to clear $-3y$ by adding $3y$.

$$-2y - 11 = -3y - 10 \qquad \text{Combine like terms.}$$
$$\underline{+3y \qquad\qquad +3y} \qquad \text{Add } 3y \text{ to both sides.}$$
$$y - 11 = \quad 0 - 10$$
$$y - 11 = -10$$
$$\underline{+11 \qquad +11} \qquad \begin{array}{l}\text{Add 11 to both sides}\\ \text{to isolate } y.\end{array}$$
$$y + 0 = \quad 1$$
$$y = 1$$

Check   Replace y in the original equation with 1 and verify that the equation is true. We will leave this to the reader.

Following is an outline for solving equations based on everything you've learned so far.

Procedure   **Solving Linear Equations**

To solve linear equations requiring the addition principle only,

1. Simplify both sides of the equation as needed.
   a. Distribute to clear parentheses.
   b. Combine like terms.
2. Use the addition principle so that all variable terms are on one side of the equation and all constants are on the other side. Then combine like terms.

Tip: Clear the variable term that has the lesser coefficient to avoid negative coefficients.

Your Turn 5   Solve and check.

   a. $7n - (n - 3) = 4(n + 4) + n$ 　　　 **b.** $11 + 3(5m - 2) = 7(2m - 1) + 9$

Objective ④   Solve identities and contradictions.   In general, a linear equation in one variable has only one real-number solution. However, there are two special cases that we need to consider. First, if a linear equation is an identity, as in Example 6, every real number is a solution. Second, as we'll see in Example 7, some linear equations have no solution.

Equations with an Infinite Number of Solutions

Consider $3x + 9 = 3x + 9$. Because the expressions on each side of the equation are identical, each produces the same result no matter what number replaces the variable x. This means that every real number is a solution. Such an equation is called an **identity**.

Definition   *Identity:* An equation that has every real number as a solution (excluding any numbers that cause an expression in the equation to be undefined).

Answers   Your Turn 5
   a. 13   **b.** -3

The equation $5y - 4 = 7y - 3$ is not an identity because not every real number is a solution. Consider the number 2, for example. If we replace y with 2 in the equation, we see that it does not check.

$$5(2) - 4 \overset{?}{=} 7(2) - 3$$
$$10 - 4 \overset{?}{=} 14 - 3 \qquad \text{The real number we picked, 2, is not a solution for}$$
$$6 \neq 11 \qquad\qquad 5y - 4 = 7y - 3; \text{ so the equation is not an identity.}$$

Sometimes it isn't obvious that an equation is an identity. Simplifying the expressions on each side of the equation can make the identity apparent.

Procedure   **Recognizing an Identity**

When solving a linear equation, if after simplifying each side of the equation the expressions are identical, the equation is an identity and every real number for which the equation is defined is a solution.

Example 6   Solve and check. $5(2t + 1) - 8 = 4t - 3(1 - 2t)$

SOLUTION   Simplify both sides of the equation. Then isolate the variable.

$$5(2t + 1) - 8 = 4t - 3(1 - 2t)$$

Distribute 5. $\qquad 10t + 5 - 8 = 4t - 3 + 6t \qquad$ Distribute -3.

$$10t - 3 = 10t - 3 \qquad \text{Combine like terms.}$$

> **Note:** We can stop here because the equation is obviously an identity. However, if we continue and apply the addition principle, the equation is still an identity.
>
> $$10t - 3 = 10t - 3$$
> $$\underline{-10t \qquad\quad -10t}$$
> $$0 - 3 = \quad 0 - 3$$
> $$-3 = -3$$

Because the linear equation is an identity, every real number is a solution.

Check   Every real number is a solution for an identity, so any number we choose should check in the original equation. We will choose to test the number 1.

$$5(2t + 1) - 8 = 4t - 3(1 - 2t)$$
$$5(2(1) + 1) - 8 \overset{?}{=} 4(1) - 3(1 - 2(1)) \qquad \text{Replace } t \text{ with 1.}$$
$$5(3) - 8 \overset{?}{=} 4 - 3(-1)$$
$$15 - 8 \overset{?}{=} 4 + 3$$
$$7 = 7$$

Because the equation is true, 1 is a solution, which supports but does not prove our conclusion that the equation is an identity. To be more certain, we could choose another number to check in the equation. (Remember, every number should work.) We will leave this to the reader.

Conclusion   If after simplifying, the linear equation is an identity, then every real number is a solution to the equation.

Equations with No Solution

Some equations, called **contradictions**, have no solution.

Definition | ***Contradiction:*** An equation that has no real-number solution.

We recognize linear equations of this type by the fact that after the expressions on each side of the equation have been simplified, the variable terms will match but the constant terms will not. For example, $2x - 5 = 2x - 3$ is a contradiction. In Example 7, watch what happens when we try to solve the equation.

Example 7   Solve and check. $2x - 5 = 2x - 3$

Note: The variable terms are eliminated, and the resulting numeric equation is false, which indicates that the given equation is a contradiction.

SOLUTION

$$
\begin{array}{rcr}
2x - 5 &=& 2x - 3 \\
\underline{-2x} & & \underline{-2x} \\
0 - 5 &=& 0 - 3 \\
-5 &=& -3
\end{array}
$$

Subtract $2x$ from both sides.

Because the equation is a contradiction, it has no solution.

Note: The symbol $\varnothing$, which indicates the empty set, is often used to represent the solution set of an equation that has no solutions.

Check   Because the variable terms, $2x$, are identical on both sides of the equal sign, replacing x with any number will yield an identical product. Because we subtract 5 from that product on the left side and we subtract 3 from that product on the right side, the equation cannot be true. Therefore, it has no solution.

Conclusion   When solving a linear equation in one variable, if applying the addition principle of equality causes the variable terms to be eliminated from the equation and the resulting equation is false, then the equation is a contradiction and has no solution.

Procedure | **Recognizing a Contradiction**

When solving a linear equation, if after simplifying the expressions on each side of the equation the expressions have the same variable term but different constant terms, the equation is a contradiction and has no solution. Equivalently, if the variable terms are eliminated and the resulting equation is false, the equation is a contradiction.

Your Turn 7   Solve and check.

a. $4x - 3(x + 5) = 2(2x - 3) - 3(x + 5)$ **b.** $2(2x + 4) - 11 = 4(x - 2) + 5$

Objective ⑤ Solve application problems.   Now consider some application problems that can be solved using the addition principle.

Example 8   Laura wants to buy a car stereo that costs \$275. She currently has \$142. How much more is needed?

Understand   We are given the total required and the amount she currently has, and we must find how much she needs.

Plan   Let x represent the amount Laura needs. We will write an equation, then solve.

Execute   Current amount + Needed amount = 275

$$
\begin{array}{ccccc}
142 & + & x & = & 275
\end{array}
$$

$$
\begin{array}{rcr}
142 + x &=& 275 \\
\underline{-142} & & \underline{-142} \\
0 + x &=& 133 \\
x &=& 133
\end{array}
$$

Subtract 142 from both sides to isolate x.

Answer   Laura needs $133 to buy the stereo.

Check   Does $142 plus the additional $133 equal $275?

$$142 + 133 \stackrel{?}{=} 275$$
$$275 = 275 \quad \text{It checks.}$$

Your Turn 8   Daryl has a balance of $-\$568$ on a credit card. How much must he pay to bring his balance to $-\$480$?

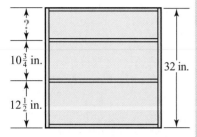

Note: As in Example 2(b), because we are subtracting a fraction from both sides, we are writing the subtraction horizontally instead of vertically. Use whichever approach, horizontal or vertical, you prefer.

Example 9   The figure to the left shows the front-view design of a bookcase. Find the missing length.

Understand   In the drawing, the sum of the dimensions on the left side must be equal to 32 inches.

Plan   Let d represent the unknown distance. We will write an equation and solve for d.

Execute $$12\frac{1}{2} + 10\frac{3}{4} + d = 32$$

$$23\frac{1}{4} + d = 32 \qquad \text{Combine like terms.}$$

$$23\frac{1}{4} - 23\frac{1}{4} + d = 32 - 23\frac{1}{4} \qquad \text{Subtract } 23\frac{1}{4} \text{ from both sides to isolate } d.$$

$$d = 8\frac{3}{4}$$

Answer   The unknown length is $8\frac{3}{4}$ inches.

Check   Verify that the sum of the three lengths on the left side is equal to 32 inches. We will leave this to the reader.

Answer   Your Turn 8

$88

Answer   Your Turn 9

157 students

Your Turn 9   In researching the grades given in a particular course during one semester, a department chair finds that 89 students received an A, 154 received a B, and 245 received a C. If the department does not give D's and 645 students enrolled in the course initially, how many students did not receive a passing grade in the course?

2.2 Exercises   For Extra Help   *MyMathLab*

Note: Exercises marked with a ★ represent challenging exercises.

1. In your own words, explain what the addition principle of equality says.

2. In the equation $5x - 9 = 4x + 1$, how do you use the addition principle to clear the $4x$ term?

3. In the equation $5x - 9 = 4x + 1$, how do you use the addition principle to clear the -9 term?

4. If an equation has like terms on the same side of the equation, what should you do first?

5. If an equation has variable terms on both sides of the equation, what should you do first?

6. How do you handle parentheses in an equation?

For Exercises 7–22, determine whether each equation is linear. See Example 1.

7. $4y + 7 = 2y - 8$

8. $6x - 5 = 3y + 64$

9. $m^2 + 4 = 20$

10. $-1 = w^2 - 5w$

11. $7u^4 + u^2 = 16$

12. $6n^3 - 9n^2 = 4n + 6$

13. $4x - 8 = 2(x + 3)$

14. $5(t - 2) + 7 = 3t - 1$

15. $6x - y = 11$

16. $3x + 2y = 12$

17. $x^2 + y^2 = 1$

18. $3x^2 + 2y^2 = 24$

19. $y = 8$

20. $x = -3$

21. $y = \dfrac{1}{4}x - 3$

22. $y = -0.5x + 2$

For Exercises 23–74, solve and check. See Examples 2–7.

23. $x - 7 = 2$

24. $a - 8 = 30$

25. $m - 8 = -3$

26. $n - 6 = -2$

27. $r - 2 = -9$

28. $a - 3 = -11$

29. $x + 7 = 12$

30. $x + 2 = 8$

31. $n + 12 = 2$

32. $y + 15 = 8$

33. $-16 = y + 9$

34. $-24 = n + 11$

35. $m + \dfrac{7}{8} = \dfrac{4}{5}$

36. $k + \dfrac{5}{9} = -\dfrac{1}{3}$

37. $-\dfrac{5}{8} = y - \dfrac{1}{6}$

38. $\dfrac{3}{4} = c - \dfrac{2}{3}$

39. $15.8 + y = 7.6$

40. $b + 8.8 = 5.4$

41. $-2.1 = n - 7.5 + 0.8$

42. $x + 0.4 - 1.6 = -12.5$

43. $6 - 18 = 7x - 3 - 6x$

44. $2z + 6 - z = 5 - 9$

45. $5m = 4m + 7$

46. $7y = 6y - 8$

47. $7x - 9 = 6x + 4$

48. $12y + 22 = 11y - 3$

49. $-2y - 11 = -3y - 5$

50. $-4t + 9 = -5t + 1$

51. $7x - 2 + x = 10x - 1 - x$

52. $3t + 6 + 4t = 9t - 2 - t$

53. $8n + 7 - 12n = 4n + 5 - 9n$

54. $-10x - 9 + 8x = -4x - 5 + 3x$

55. $2.6 + 7a + 5 = 8a - 5.6$

56. $9c + 4.8 = 7.5 + 4.8 + 8c$

57. $12 - 7(h + 6) + 8h = 14 - 8$

58. $19 - 3(m + 4) + 4m = 42 - 18$

59. $5 - \dfrac{1}{2}(4x + 6) = -3x - 8$

60. $6 - \dfrac{2}{3}(6b - 9) = -5b - 21$

61. $5y - (3y + 7) = y + 9$

62. $-15 - 2x = 16 - (3x - 9)$

63. $3(3x - 5) - 2(4x - 3) = 5 - 10$

64. $5(5x - 3) - 6(4x - 2) = 12 - 15$

65. $1.4(2.5x - 4.5) - (6.2 + 2.5x) = -7.3 + 4.5$

66. $0.5(3.8x - 6.2) - (0.9x - 4) = 2.9 - 4.7$

67. $4 + 3y + y - 12 = 5y + 9 - y - 17$

68. $-9 - 4v - 1 + v = -2v + 5 - v - 15$

69. $7.3 - 0.2x + 1.3 - 0.6x = 12 - 0.8x - 3.6$

70. $2.5y - 3.4 - 1.2y = 6.7 - 9.1 + 1.3y$

★71. $20z + \dfrac{2}{3}(6z - 48) - 9z = 14z + 0.2(50z - 250) - 9z$

★72. $6b - 1.5(8 + 2b) + 4 = 6b - \dfrac{1}{8}(24b + 48)$

★73. $8(2m - 4) + 20(m - 5) = 4m + 5(5m - 20) - 32 + 7m$

★74. $-3(2x + 5) + 8(x + 2) - 7 = 6(x - 5) - 4(x - 6)$

For Exercises 75–86, translate to an equation and then solve. See Examples 8 and 9.

75. Latonia is planning to buy a new car. The down payment is $2373. She has $1947 saved. How much more does she need?

76. Kent owes $12,412 on his Visa card. What payment should he make so he will owe $10,500?

77. Robert knows that the distance from his home to work is 42 miles. Unfortunately, he gets a flat tire 16 miles from work. How far did Robert drive before his tire went flat?

78. Susan is playing Yahtzee. The "chance" score is found by adding the value shown on five dice. If she has a total of 23 on four of the dice, what does the fifth die need to be so that her score will be 28?

79. On May 16, the balance in Nikki's checking account was $1741.62. Afterwards, she writes four checks and doesn't record the amount of the last check. The first check is for $16.82, the second is for $150.88, and the third is for $192.71. On May 21, she finds her balance to be $1286.65. If the checks were the only transactions that could have cleared the bank during that time, what was the amount of the fourth check?

80. A patient must receive 350 cc of a medication in three injections. He has received two injections at 110 cc each. How much should the third injection be?

81. The perimeter of the trapezoid shown is 67.2 centimeters. Find the length of the missing side.

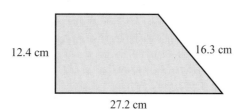

12.4 cm   16.3 cm

27.2 cm

82. The perimeter of the triangle shown is $84\dfrac{1}{2}$ inches. Find the length of the missing side.

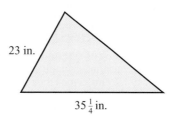

23 in.

$35\frac{1}{4}$ in.

83. In the blueprint shown, what is the distance x?

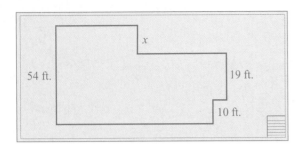

84. What is the distance x in the following figure?

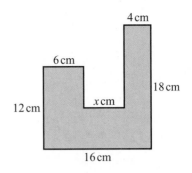

85. John sells scanning devices. The company quota is set at $8500 each month. The spreadsheet shows John's sales as of the end of the second week of April. How much more does John need to sell to make the quota? Do you think John will make the quota? Why or why not?

Date	Item No.	Quantity	Price per Unit
4/2	32072	1	$900
4/9	17032	3	$1500
4/13	48013	2	$1245

86. Tamika works from 8 A.M. until 5 P.M. in the women's department of a major clothing store. She has set a goal for herself of $600 per day in sales. Thus far, she has sold 3 blouses for $25 each, 5 pairs of slacks for $30 each, and 2 pairs of shoes for $85 each. How much more in sales does she need to reach her goal? If it is now 4 P.M., is it likely that she will make her goal? Why or why not?

★**87.** In a survey, respondents were given a statement, and they could agree, disagree, or have no opinion regarding the statement. The results indicate that $\frac{1}{3}$ of the respondents agree with the statement while $\frac{2}{5}$ disagree. What fraction had no opinion?

★**88.** In a *Time*/CNN poll, $\frac{19}{50}$ of the respondents said that they believed that the increase in the number of divorces was due more to changes in women's attitude toward marriage, while $\frac{9}{50}$ said the increase was due more to changes in men's attitude toward marriage. The rest thought that the increase was a result of both men and women equally. What fraction of the respondents believed that the increase was due to men and women equally?

Puzzle Problem

In a study of an experimental medication, half of the participants are given the medication and half are given a placebo. The results indicate that $\frac{1}{3}$ of the participants had improvement in their condition, $\frac{1}{8}$ had no improvement in their condition but experienced known side effects of the medication, and the rest showed no effects at all. What fraction of the group that received the medication showed no discernible effects from it? (Hint: Assume that all of the participants that saw improvement in their condition or experienced known side effects took the medication.)

Review Exercises

Exercises 1–6 ▰▰▰ **EXPRESSIONS**

[1.4]   **1.** What is the multiplicative inverse of $-\dfrac{4}{5}$?

For Exercises 2 and 3, simplify.

[1.4]   **2.** $4(-8.3)$ **[1.4]**   **3.** $27 \div (-3)$

[1.7]   **4.** Simplify: $13 - 6(x + 2)$ **[1.7]**   **5.** Combine like terms: **[1.7]**   **6.** Use the distributive property

$$3x^2 - 6x + 5x^2 + x - 9$$

to rewrite: $12\left(\dfrac{1}{3}x - \dfrac{3}{4}\right)$

2.3 The Multiplication Principle of Equality

Objectives

 1   Solve linear equations using the multiplication principle.

 2   Solve linear equations using both the addition and multiplication principles.

 3   Use the multiplication principle to clear fractions and decimals from equations.

 4   Solve application problems.

Objective 1   **Solve linear equations using the multiplication principle.** In Section 2.2, when we isolated the variable, the coefficient of that variable was always 1. What if we ended up with a coefficient other than 1? Look at the equation $6n - 11 = 4n + 3$. Using the methods of Section 2.2, we have

$$
\begin{array}{rcl}
6n - 11 &=& 4n + 3 \\
\underline{-4n} & & \underline{-4n} \\
2n - 11 &=& 0 + 3 \\
\underline{+11} & & \underline{+11} \\
2n + 0 &=& 14 \\
2n &=& 14
\end{array}
$$

Note: The coefficient is not 1. ▶

How can we solve $2n = 14$? To isolate n, we must clear the coefficient 2 so that we have $1n$, or simply n. Because coefficients multiply variables, we must undo that multiplication using its inverse operation, division. If we are going to divide, we must divide on both sides to keep the equation balanced. We write

$$2n \div 2 = 14 \div 2, \text{ or the more popular form } \dfrac{2n}{2} = \dfrac{14}{2}$$

Notice that the coefficient 2 divides out.

Note: The 2s divide out, leaving $1n$, which can be simplified as n. ▶

$$\dfrac{\cancel{2}n}{\cancel{2}} = \dfrac{14}{2}$$

$$1n = 7$$

$$n = 7$$

You might remember that in Chapter 1, we discussed that multiplicative inverses are numbers whose product is 1. Multiplying the coefficient by its multiplicative inverse is another way to clear the coefficient. In the previous case, if we multiply the coefficient 2 by its multiplicative inverse $\dfrac{1}{2}$, we get the desired $1n$.

$$\frac{1}{2} \cdot 2n = 14 \cdot \frac{1}{2}$$

Note: Throughout the rest of the book, we omit this step of writing the resulting 1 coefficient.

$$\frac{1}{\cancel{2}} \cdot \frac{\overset{1}{\cancel{2}}n}{1} = \frac{\overset{7}{\cancel{14}}}{1} \cdot \frac{1}{\cancel{2}}$$

$$1n = 7$$

$$n = 7$$

Connection Remember that when we divide by a fraction, we write an equivalent multiplication using the multiplicative inverse. Multiplying by the multiplicative inverse is the same as dividing.

If we manipulate an equation using the addition principle and we still have a coefficient other than 1, we can clear that coefficient by dividing by the coefficient (or multiplying by its multiplicative inverse). This is the main purpose of the *multiplication principle*. The multiplication principle says that we can multiply (or divide) both sides of an equation by the same amount without affecting its solution(s).

Rule **The Multiplication Principle of Equality**

If $a = b$, then $ac = bc$ is true for all real numbers a, b, and c, where $c \neq 0$.

Procedure **Using the Multiplication Principle of Equality**

To use the multiplication principle of equality to clear a coefficient in an equation, multiply both sides of the equation by the multiplicative inverse of that coefficient or divide both sides by the coefficient.

Example 1 Solve and check.

a. $-\frac{4}{5}m = \frac{6}{7}$

SOLUTION $-\frac{4}{5}m = \frac{6}{7}$

Note: If the equation contains fractions, the answer will be left as a fraction or an integer. If the equation contains decimals, the answer will be left as a decimal or an integer. If the equation contains integers only, the answer will be expressed as an integer, a fraction, or a decimal, as appropriate.

$$-\frac{\overset{1}{\cancel{5}}}{\cancel{4}} \cdot -\frac{\overset{1}{\cancel{4}}}{\cancel{5}}m = \frac{\overset{3}{\cancel{6}}}{7} \cdot -\frac{5}{\cancel{4}}$$

Clear the coefficient $-\frac{4}{5}$ by multiplying both sides by its multiplicative inverse, $-\frac{5}{4}$.

$$m = -\frac{15}{14}$$

Check $\quad -\frac{4}{5}m = \frac{6}{7}$

$$-\frac{\overset{2}{\cancel{4}}}{\cancel{5}}\left(-\frac{\overset{3}{\cancel{15}}}{\cancel{14}}\right) \overset{?}{=} \frac{6}{7}$$

Replace m in the original equation with $-\frac{15}{14}$ and verify that the equation is true.

$$\frac{6}{7} = \frac{6}{7}$$

True; therefore, $-\frac{15}{14}$ is correct.

b. $-1.9x = -4.56$

SOLUTION $-1.9x = -4.56$

$$\frac{\cancel{-1.9}x}{\cancel{-1.9}} = \frac{-4.56}{-1.9}$$

Clear the coefficient -1.9 by dividing both sides by -1.9.

$$x = 2.4$$

Check $-1.9x = -4.56$

$-1.9(2.4) = -4.56$ Replace x with 2.4 and verify that the equation is true.

$-4.56 = -4.56$

True; therefore, 2.4 is correct.

Your Turn 1   Solve and check.

a. $-56 = -7t$ **b.** $-\dfrac{3}{8}y = 9$ **c.** $0.4x = -0.92$

Objective ② **Solve linear equations using both the addition and multiplication principles.**   Now let's put the multiplication principle together with the addition principle. We follow the same outline as in Section 2.2, with one new step: using the multiplication principle to clear any remaining coefficient at the end.

Procedure   **Solving Linear Equations**

To solve linear equations in one variable,

1. Simplify both sides of the equation as needed.
   a. Distribute to clear parentheses.
   b. Combine like terms.
2. Use the addition principle so that all variable terms are on one side of the equation and all constants are on the other side. (Clear the variable term with the lesser coefficient to avoid negative coefficients.) Then combine like terms.
3. Use the multiplication principle to clear any remaining coefficient.

Example 2   Solve and check. $-3x - 11 = 7$

SOLUTION   Use the addition principle to separate the variable term and constant terms; then use the multiplication principle to clear any remaining coefficient.

$$-3x - 11 = 7$$ Add 11 to both sides to isolate the $-3x$ term.

$$\underline{+11 \quad +11}$$

$$-3x + 0 = 18$$

$$\frac{-3x}{-3} = \frac{18}{-3}$$ Divide both sides by -3 to clear the -3 coefficient.

$$x = -6$$

Check $-3x - 11 = 7$

$-3(-6) - 11 \overset{?}{=} 7$ Replace x in the original equation with -6 and verify that the equation is true.

$18 - 11 \overset{?}{=} 7$

$7 = 7$

True; therefore, -6 is correct.

Note: Multiplying both sides by $-\dfrac{1}{3}$ would also work:

$$\left(-\frac{1}{\underset{1}{3}}\right)\left(-\frac{\overset{1}{3}x}{1}\right) = \left(\frac{\overset{6}{18}}{1}\right)\left(-\frac{1}{\underset{1}{3}}\right)$$

$$x = -6$$

Your Turn 2   Solve and check.

a. $6y - 19 = -22$ **b.** $-12 = 20 - 8t$

Answers   Your Turn 1

a. 8   **b.** -24 **c.** -2.3

Answers   Your Turn 2

a. $-\dfrac{1}{2}$ **b.** 4

Solving Equations with Variable Terms on Both Sides

Recall that when variable terms appear on both sides of the equal sign, we use the addition principle to get the variable terms on one side of the equal sign and the constant terms on the other side.

Example 3 Solve and check. $8y - 5 = 2y - 29$

SOLUTION Use the addition principle to get the variable terms on one side of the equation and the constant terms on the other side; then use the multiplication principle to clear the remaining coefficient.

$$8y - 5 = 2y - 29 \quad \text{Subtract } 2y \text{ from both sides. } (2y \text{ has the lesser coefficient.})$$
$$\underline{-2y \qquad -2y}$$
$$6y - 5 = 0 - 29$$
$$6y - 5 = -29$$
$$\underline{+5 \qquad +5} \quad \text{Add 5 to both sides to isolate the } 6y \text{ term.}$$
$$6y + 0 = -24$$
$$\frac{6y}{6} = \frac{-24}{6} \quad \text{Divide both sides by 6 to clear the 6 coefficient.}$$
$$y = -4$$

Check
$$8y - 5 = 2y - 29$$
$$8(-4) - 5 \stackrel{?}{=} 2(-4) - 29 \qquad \text{Replace } y \text{ in the original equation with } -4$$
$$-32 - 5 \stackrel{?}{=} -8 - 29 \qquad \text{and verify that the equation is true.}$$
$$-37 = -37$$

True; therefore, -4 is correct.

Simplifying First

Recall that when an equation contains parentheses or like terms that appear on the same side of the equal sign, we first clear the parentheses and combine like terms. Then we use the addition principle to separate the variable terms and constant terms.

Example 4 Solve and check. $17 - (4n - 5) = 9n - 3(n + 7)$

SOLUTION $17 - (4n - 5) = 9n - 3(n + 7)$
$$17 - 4n + 5 = 9n - 3n - 21 \qquad \text{Distribute to clear parentheses.}$$
$$22 - 4n = 6n - 21 \qquad \text{Combine like terms.}$$
$$22 - 4n = 6n - 21$$
$$\underline{+4n \quad +4n} \qquad \text{Add } 4n \text{ to both sides.}$$
$$\qquad\qquad\qquad (-4n \text{ has the lesser coefficient.})$$
$$22 + 0 = 10n - 21$$
$$22 = 10n - 21$$
$$\underline{+21 = \qquad +21} \qquad \text{Add 21 to both sides to isolate } 10n.$$
$$43 = 10n + 0$$
$$\frac{43}{10} = \frac{10n}{10} \qquad \text{Divide both sides by 10 to clear the 10 coefficient.}$$
$$\frac{43}{10} \text{ or } 4.3 = n$$

Check

$$17 - (4n - 5) = 9n - 3(n + 7)$$

$$17 - (4(4.3) - 5) \stackrel{?}{=} 9(4.3) - 3(4.3 + 7)$$

$$17 - (17.2 - 5) \stackrel{?}{=} 38.7 - 3(11.3)$$

$$17 - 12.2 \stackrel{?}{=} 38.7 - 33.9$$

$$4.8 = 4.8$$

Replace n in the original equation with 4.3 and verify that the equation is true.

True; therefore, 4.3 is correct.

Your Turn 4 Solve and check.

a. $3t - 10 + 9t = 17 + 4t - 3$ **b.** $13 - 6(x + 2) = 3x - (11x + 5)$

Objective ③ Use the multiplication principle to clear fractions and decimals from equations.
To clear fractions and decimals from an equation, we can use the multiplication principle of equality. Although equations can be solved without clearing the fractions or decimals, most people find equations that contain only integers easier to solve.

Clearing Fractions in an Equation

If the equation contains fractions, we multiply both sides by a number that will clear all of the denominators. We could multiply both sides by any multiple of the denominators; however, using the LCD (least common denominator) results in the simplest equations.

Example 5 Solve and check. $\dfrac{1}{3}x - \dfrac{3}{4} = \dfrac{5}{6}x + 2$

SOLUTION Clear the fractions by multiplying through by the LCD. The LCD of 3, 4, and 6 is 12.

Note: We have chosen to clear the fractions in our first step. Remember, you can clear the fractions at any step in solving the equation or not clear them at all.

$$12\left(\frac{1}{3}x - \frac{3}{4}\right) = \left(\frac{5}{6}x + 2\right)12$$

Clear the fractions by multiplying both sides by the LCD, 12.

$$\frac{\overset{4}{\cancel{12}}}{1} \cdot \frac{1}{\underset{1}{\cancel{3}}}x - \frac{\overset{3}{\cancel{12}}}{1} \cdot \frac{3}{\underset{1}{\cancel{4}}} = \frac{\overset{2}{\cancel{12}}}{1} \cdot \frac{5}{\underset{1}{\cancel{6}}}x + 12 \cdot 2$$

Distribute 12; then divide out the denominators.

$$4x - 9 = 10x + 24$$

$$4x - 9 = 10x + 24$$

$$\underline{-4x \qquad\quad -4x}$$

Subtract $4x$ from both sides.

$$0 - 9 = 6x + 24$$

$$-9 = 6x + 24$$

$$\underline{-24 \qquad\quad -24}$$

Subtract 24 from both sides.

$$-33 = 6x + 0$$

$$\frac{-33}{6} = \frac{6x}{6}$$

Divide both sides by 6 to clear the coefficient.

$$-\frac{11}{2} \text{ or } -5.5 = x$$

Simplify.

Answers **Your Turn 4**
a. 3 **b.** −3

The Multiplication Principle of Equality **121**

Check

$$\frac{1}{3}x - \frac{3}{4} = \frac{5}{6}x + 2$$

$$\frac{1}{3}\left(-\frac{11}{2}\right) - \frac{3}{4} \stackrel{?}{=} \frac{5}{6}\left(-\frac{11}{2}\right) + 2$$

Replace x in the original equation with $-\frac{11}{2}$ and verify that the equation is true.

$$-\frac{11}{6} - \frac{3}{4} \stackrel{?}{=} -\frac{55}{12} + 2$$

$$-\frac{22}{12} - \frac{9}{12} \stackrel{?}{=} -\frac{55}{12} + \frac{24}{12}$$

Write equivalent fractions with their LCD.

$$-\frac{31}{12} = -\frac{31}{12}$$

True; therefore, $-\frac{11}{2}$ is correct.

Clearing Decimals in an Equation

The multiplication principle can also be used to clear decimals in an equation. Because multiplying by a power of 10 will cause the decimal point to move to the right, we can clear decimals by multiplying both sides of the equation by the appropriate power of 10. The power of 10 we use depends on the decimal number with the most decimal places.

Example 6   Solve and check. $0.2(y - 6) = 0.48y + 3$

SOLUTION   Decimals can be cleared at any step in the process of solving an equation. In this case, we will distribute to clear the parentheses first, then clear the decimals.

Connection Clearing decimals and fractions uses the same process. Decimal numbers represent fractions with denominators that are powers of 10. For example, $0.48 = \frac{48}{100}$ and $0.2 = \frac{2}{10}$. The LCD for these fractions is 100.

$$0.2(y - 6) = 0.48y + 3$$

$$0.2y - 1.2 = 0.48y + 3$$ Distribute to clear parentheses.

$$100(0.2y - 1.2) = (0.48y + 3)100$$

Note: The number 0.48 has two decimal places, which is more decimal places than any of the other decimal numbers; so we multiply both sides by 100 to clear the decimals.

$$20y - 120 = 48y + 300$$

$$20y - 120 = 48y + 300$$

$$\underline{-20y \qquad\qquad -20y}$$

$$0 - 120 = 28y + 300$$

$$-120 = 28y + 300$$

$$\underline{-300 \qquad\qquad -300}$$

$$-420 = 28y + 0$$

$$\frac{-420}{28} = \frac{28y}{28}$$

$$-15 = y$$

Check $$0.2(y - 6) = 0.48y + 3$$

$$0.2(-15 - 6) \stackrel{?}{=} 0.48(-15) + 3$$

$$0.2(-21) \stackrel{?}{=} -7.2 + 3$$

Replace y in the original equation with -15 and verify that the equation is true.

$$-4.2 = -4.2$$

True; therefore, -15 is correct.

We can amend the outline for solving equations to include clearing fractions and decimals. Remember that this process is optional.

Procedure   **Solving Linear Equations**

To solve linear equations in one variable,

1. Simplify both sides of the equation as needed.
   a. Distribute to clear parentheses.
   b. Clear fractions or decimals by multiplying through by the LCD. In the case of decimals, the LCD is the power of 10 with the same number of zero digits as decimal places in the number with the most decimal places. (Clearing fractions and decimals is optional.)
   c. Combine like terms.
2. Use the addition principle so that all variable terms are on one side of the equation and all constants are on the other side. (Clear the variable term with the lesser coefficient to avoid negative coefficients.) Then combine like terms.
3. Use the multiplication principle to clear any remaining coefficient.

Your Turn 6   Solve and check.

a. $\dfrac{1}{3}(y - 2) = \dfrac{3}{5}y + \dfrac{1}{6}$

b. $2.5n - 1.04 = 0.15n - 0.1$

Objective ④ Solve application problems.   In Section 2.1, we began solving application problems using formulas. Recall that to use a formula, we replace the variables with the corresponding given numbers, then solve for the unknown amount.

Example 7   Solve.

a. The perimeter of the figure shown is 188 feet. Find the width and length.

Understand   The width is represented by w, and the length is represented by $w + 7$. We are given the perimeter, so we can use the formula $P = 2l + 2w$.

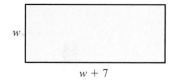

Plan   In the perimeter formula, replace P with **188** and l with $w + 7$ and solve for w.

Execute

$$P = 2l + 2w$$
$$188 = 2(w + 7) + 2w$$
$$188 = 2w + 14 + 2w \qquad \text{Distribute.}$$
$$188 = 4w + 14 \qquad \text{Combine like terms.}$$
$$188 = 4w + 14$$
$$\underline{-14 \qquad\qquad -14} \qquad \text{Subtract 14 from both sides.}$$
$$174 = 4w + 0$$
$$\dfrac{174}{4} = \dfrac{4w}{4} \qquad \text{Divide both sides by 4.}$$
$$\dfrac{87}{2} \text{ or } 43.5 = w$$

Answer   The width is 43.5 feet. To find the length, we evaluate the expression that represents the length, $w + 7$, with $w = 43.5$.

$$\text{Length} = 43.5 + 7 = 50.5 \text{ ft.}$$

Answers   **Your Turn 6**

a. $-\dfrac{25}{8}$ b. 0.4

Check   First, verify that the length is 7 more than the width: $50.5 = 7 + 43.5$. Second, verify that a rectangle with a length of 50.5 feet and a width of 43.5 feet has a perimeter of 188 feet.

$$P = 2l + 2w$$
$$P = 2(50.5) + 2(43.5)$$
$$P = 101 + 87$$
$$P = 188$$

It checks.

b. The total material allotted for the construction of a metal box is 4754 square inches. The length is to be 30 inches, and the width is to be 28 inches. Find the height.

Understand   The material will be used to create the outer shell that is the box, which means that 4754 square inches is the surface area of the box. The formula for the surface area of a box is $SA = 2lw + 2lh + 2wh$.

Plan   Replace SA with **4754**, l with **30**, and w with **28** and solve for h.

Execute

$$SA = 2lw + 2lh + 2wh$$
$$4754 = 2(\mathbf{30})(\mathbf{28}) + 2(\mathbf{30})h + 2(\mathbf{28})h$$

$4754 = 1680 + 60h + 56h$	Simplify.
$4754 = 1680 + 116h$	Combine like terms.

$$4754 = 1680 + 116h$$

$\underline{-1680 \quad -1680}$	Subtract 1680 from both sides.
$3074 = \quad 0 + 116h$	
$\dfrac{3074}{116} = \dfrac{116h}{116}$	Divide both sides by 116.
$26.5 = h$	

Answer   The height is 26.5 inches.

Check   Does a 30-inch by 28-inch by 26.5-inch box have a surface area of 4754 square inches?

$$4754 \overset{?}{=} 2(30)(28) + 2(30)(26.5) + 2(28)(26.5)$$
$$4754 \overset{?}{=} \quad 1680 \quad + \quad 1590 \quad + \quad 1484$$
$$4754 = 4754$$

It checks.

Your Turn 7   Solve.

a. The desired surface area for a box is 98 square feet. If the length of the box is to be 4 feet and the height is to be 6 feet, what will the width be?

b. The circumference of Venus at the equator is 24,170 miles. What is the diameter of Venus at the equator? (Use $C = \pi d$ with $\pi \approx 3.14$.)

Answers   Your Turn 7
a. 2.5 ft.   **b.** 7697 mi.

2.3 Exercises   For Extra Help   *MyMathLab*

Note: Exercises marked with a ★ represent challenging exercises.

1. In your own words, explain what the multiplication principle of equality says.

2. How do you use the multiplication principle to clear a coefficient that is an integer?

3. How do you use the multiplication principle to clear a coefficient that is a fraction?

4. To clear a negative coefficient, do you multiply/divide by a positive or negative number? Why?

5. If an equation contains fractions, how do you transform the equation so that it contains only integers?

6. If an equation contains decimal numbers, how do you transform the equation so that it contains only integers?

For Exercises 7–18, solve and check. See Example 1.

7. $3x = 12$

8. $8x = -24$

9. $-6x = -18$

10. $-5y = 20$

11. $\dfrac{n}{2} = 8$

12. $\dfrac{t}{3} = -4$

13. $\dfrac{3}{4}y = -15$

14. $\dfrac{5}{6}x = 20$

15. $-\dfrac{4}{7}t = -\dfrac{2}{3}$

16. $-\dfrac{3}{8}a = \dfrac{5}{6}$

17. $-\dfrac{2}{5}t = \dfrac{8}{15}$

18. $-\dfrac{7}{9}t = -\dfrac{5}{12}$

For Exercises 19–40, solve and check. See Example 2.

19. $5n - 2n = 21$

20. $7t - 3t = 20$

21. $4x + 1 = 21$

22. $3x + 5 = 11$

23. $4a - 3 = 17$

24. $3x - 8 = 10$

25. $5x + 7 = -8$

26. $3x + 9 = -6$

27. $9 - 2n = 12$

28. $1 - 7y = -8$

29. $\dfrac{5}{8}x - 2 = 8$

30. $7 = \dfrac{3}{4}x + 13$

31. $2(x - 3) = -6$

32. $4(5x + 7) = 28$

33. $3(n + 7) = -6$

34. $4(n - 3) = -8$

35. $3a - 2a + 7 + 6a = -28$

36. $c + 2c + 3 + 4c = 24$

37. $12b - 8(2b - 3) = 16$

38. $2x - 6(x + 8) = -12$

39. $2(y - 33) + 3(y - 2) = 8$

40. $4(r - 8) + 2(r + 3) = -8$

For Exercises 41–62, solve and check. See Examples 3 and 4.

41. $8x + 5 = 2x + 17$

42. $10t + 1 = 6t + 13$

43. $5y - 7 = 2y + 13$

44. $9m + 1 = 3m - 14$

45. $9 - 4k = 15 - k$

46. $6 - 12m = -20m + 22$

47. $4k + 5 = 7k - 7 + 9k$

48. $-11b - 5 = -5b + 23 - 10$

49. $3a - 4a + 9 = -12a + 43 - 6a$

50. $17b - 11b - 17 = -4b + 13 - 5b$

51. $9x + 12 - 3x - 2 = x + 8 + 5x$

52. $12 - 6r - 14 = 9 - 4r - 7 - 2r$

53. $14(w - 2) + 13 = 4w + 5$

54. $2x + 2(3x - 4) = -23 + 5x$

55. $2n - (6n + 5) = 9 - (2n - 1)$

56. $2 - (17 - 5m) = 9m - (m + 7)$

57. $4 - (6x + 5) = 7 - 2(3x + 4)$

58. $-6 - (3z - 2) = 5z - 4(2z + 1)$

59. $5(a - 1) - 9(a - 2) = -3(2a + 1) - 2$

60. $-4(k + 4) + 13(k - 1) = -3(k - 2) + 13$

61. $3(2x - 5) - 4x = 2(x - 3) - 12$

62. $4(2x - 1) - 3(x + 5) = 5(x - 2) + 7$

For Exercises 63–82, use the multiplication principle of equality to eliminate the fractions or decimals; then solve and check. See Examples 5 and 6.

63. $\dfrac{2}{3}n - 1 = \dfrac{1}{4}$

64. $\dfrac{2}{5}n - 1 = \dfrac{3}{2}$

65. $\dfrac{3}{4}n - 2 = -\dfrac{7}{8}n + 4$

66. $-\dfrac{2}{5}t + 1 = \dfrac{3}{10}t - 3$

67. $\dfrac{3}{4}x + \dfrac{5}{6} = \dfrac{1}{4} + \dfrac{4}{3}x$

68. $\dfrac{7}{9}w - \dfrac{13}{6} = \dfrac{7}{2}w + \dfrac{5}{9}$

69. $\dfrac{1}{2}(y + 5) = \dfrac{3}{2} - y$

70. $\dfrac{2}{3}(x - 4) = \dfrac{4}{3} + 2x$

71. $\dfrac{1}{6}(m - 4) = \dfrac{2}{3}(m + 1) - \dfrac{1}{3}m$

72. $\dfrac{1}{5}(y - 3) = \dfrac{3}{10}(y + 5) - \dfrac{2}{5}y$

73. $-2.9u + 3.6u = 6.3$

74. $-4.6z + 2.2z = 4.8$

75. $0.2 - 1.3t - 0.8t = 1 - 2.1t - 0.8$

76. $4.2y - 8.2 + 2.3y = 0.9 + 6.5y - 9.1$

77. $0.3(a - 18) = 0.9a$

78. $0.6(w - 12) = 0.2w$

79. $0.08(15) + 0.4x = 0.05(36 + 7x)$

80. $0.06(25) + 0.27x = 0.3(4 + x)$

81. $9 - 0.2(v - 8) = 0.3(6v - 8)$

82. $0.5 - (t - 6) = 10.4 - 0.4(1 - t)$

Find
ⓧ *the mistake*

For Exercises 83–86, the check indicates that a mistake was made. Find and correct the mistake. See Examples 2–4.

83.

$$6x - 11 = 8x - 15$$
$$\underline{-6x \qquad\quad -6x}$$
$$11 = 2x - 15$$
$$\underline{+15 \qquad +15}$$
$$26 = 2x$$
$$\dfrac{26}{2} = \dfrac{2x}{2}$$
$$13 = x$$

Check $6x - 11 = 8x - 15$
$$6(13) - 11 \overset{?}{=} 8(13) - 15$$
$$78 - 11 \overset{?}{=} 104 - 15$$
$$67 \neq 89$$

84. $6n + 1 = -2n + 21$

$$\underline{{-1} \qquad\qquad {-1}}$$
$$6n + 0 = -2n + 20$$
$$6n = -2n + 20$$
$$\underline{+2n \qquad +2n}$$
$$8n = 0 + 20$$
$$\frac{8n}{8} = \frac{20}{8}$$
$$n = \frac{5}{2}$$

Check $6n + 1 = -2n + 21$

$$\overset{3}{\underset{1}{\cancel{6}}}\left(\frac{5}{\underset{1}{\cancel{2}}}\right) + 1 \overset{?}{=} -\overset{1}{\underset{1}{\cancel{2}}}\left(\frac{5}{\underset{1}{\cancel{2}}}\right) + 21$$
$$3 + 1 \overset{?}{=} -5 + 21$$
$$4 \neq 16$$

85. $4 - (x + 2) = 2x + 3(x - 8)$

$$4 - x + 2 = 2x + 3x - 24$$
$$6 - x = 5x - 24$$
$$\underline{{+x} \qquad {+x}}$$
$$6 + 0 = 6x - 24$$
$$6 = 6x - 24$$
$$\underline{+24 \qquad\qquad +24}$$
$$30 = 6x + 0$$
$$\frac{30}{6} = \frac{6x}{6}$$
$$5 = x$$

Check $4 - (x + 2) = 2x + 3(x - 8)$

$$4 - (5 + 2) \overset{?}{=} 2(5) + 3(5 - 8)$$
$$4 - 7 \overset{?}{=} 10 + 3(-3)$$
$$-3 \overset{?}{=} 10 + (-9)$$
$$-3 \neq 1$$

86. $\dfrac{1}{2}n - 3 = \dfrac{2}{3}n + \dfrac{1}{4}$

$$\frac{\overset{6}{\cancel{12}}}{1}\cdot\frac{1}{\underset{1}{\cancel{2}}}n - 3 = \frac{\overset{4}{\cancel{12}}}{1}\cdot\frac{2}{\underset{1}{\cancel{3}}}n + \frac{\overset{3}{\cancel{12}}}{1}\cdot\frac{1}{\underset{1}{\cancel{4}}}$$
$$6n - 3 = 8n + 3$$
$$\underline{-6n \qquad\quad -6n}$$
$$0 - 3 = 2n + 3$$
$$\underline{-3 \qquad\qquad -3}$$
$$-6 = 2n + 0$$
$$\frac{-6}{2} = \frac{2n}{2}$$
$$-3 = n$$

Check $\dfrac{1}{2}n - 3 = \dfrac{2}{3}n + \dfrac{1}{4}$

$$\frac{1}{2}\left(-\frac{3}{1}\right) - 3 \overset{?}{=} \frac{2}{3}\left(-\frac{\overset{1}{\cancel{3}}}{1}\right) + \frac{1}{4}$$
$$-\frac{3}{2} - 3 \overset{?}{=} -2 + \frac{1}{4}$$
$$-\frac{3}{2} - \frac{6}{2} \overset{?}{=} -\frac{8}{4} + \frac{1}{4}$$
$$-\frac{9}{2} \neq -\frac{7}{4}$$

For Exercises 87–100, solve for the unknown amount. See Example 7.

87. The area of a rectangular deck is known to be 140 square feet. Find the length if the width must be 10 feet. (Use $A = lw$.)

88. A crate manufacturer receives an order for a crate with a volume of 128 cubic feet. The crate must be 5 feet 4 inches by 6 feet. What must the crate's height be? (Use $V = lwh$.)

89. The area of the metal plate shown in the figure is 45.6 square centimeters. What is the length of the side labeled h?

(Use $A = \dfrac{1}{2}bh$.)

h

4.8 cm

90. The volume of a rectangular pyramid is 76.8 cubic inches, and the area of the base is 6.4 square inches. Find the height. (Use $V = \dfrac{1}{3}Ah$.)

91. A fence is to be installed around a rectangular field. The field's perimeter is 212 feet. Find the dimensions of the field if the length of the field is 2 feet more than the width, as shown in the figure. (Use $P = 2l + 2w$.)

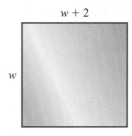

$w + 2$

w

92. The base of a building has a perimeter of 241 feet. The length is 16 feet more than the width, as shown in the figure. Find the dimensions of the base of the building. (Use $P = 2l + 2w$.)

w

$w + 16$

93. The area of the trapezoidal plot of land shown in the figure is 7500 square feet. What is the length of the side labeled b?

(Use $A = \dfrac{1}{2}h(a + b)$.)

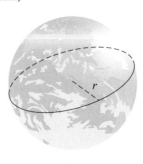

90 ft.

75 ft.

b

94. A tabletop is in the shape of a trapezoid and has an area of 1008 square inches. What is the length of the side labeled a?

(Use $A = \dfrac{1}{2}h(a + b)$.)

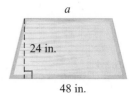

a

24 in.

48 in.

95. A small storage box is designed so that its surface area is 106 square inches. If the length is 7.5 inches and the width is 4 inches, find the height. (Use $SA = 2lw + 2lh + 2wh$.)

96. Jerry has 1992 square inches of wood to build a chest for storing musical equipment. He wants the length to be 22 inches and the width to be 18 inches. What will the height be? (Use $SA = 2lw + 2lh + 2wh$.)

97. The circumference of Earth along the equator is approximately 40,053.84 kilometers. What is the equatorial radius? (Use $C = 2\pi r$ and round your answer to the nearest tenth.)

r

98. A giant sequoia is measured to have a circumference of 26.4 meters. What is the diameter? (Use $C = \pi d$ and round your answer to the nearest tenth.)

Of Interest

The giant sequoia tree is found along the western slopes of the Sierra Nevada range. These trees reach heights of nearly 100 meters, and some are over 4000 years old.

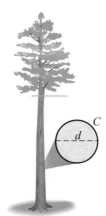

C

d

99. An engineer is designing a chemical storage tank in the shape of a cylinder. The volume of the tank is to be 60 cubic meters, and the radius is to be 2 meters. Find the height. (Use $V = \pi r^2 h$ and round your answer to the nearest tenth.)

2 m

100. The liquid in a standard 12-ounce soda can occupies a volume of 355 cubic centimeters. If the diameter is 6 centimeters, find the height of the liquid inside the can. (Use $V = \pi r^2 h$ and round your answer to the nearest tenth.)

6 cm

12 ounces (355 mL)

★**101.** The area of the L-shaped room shown is 321 square feet. Find all of the missing dimensions.

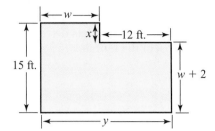

★**102.** The area of the figure below is 192 square inches. Find all of the missing dimensions.

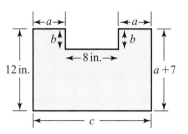

The formula V = iR describes the voltage, V, in a circuit when the current, i, passes through a resistance, R. Voltage is measured in volts (V), current is measured in amps (A), and resistance is measured in ohms (Ω). Use the formula for voltage to solve Exercises 103 and 104. See Example 7.

103. The voltage in a circuit is measured to be −76 V. If the resistance is 8 Ω, find the current.

104. A technician applies 15 V to a circuit. The current is measured to be 2.5 A. What is the resistance?

The formula F = ma describes the force that an object with mass m experiences when accelerated an amount a. If acceleration is measured in meters per second squared (m/sec.²) and the mass is measured in kilograms (kg), the force is in units of newtons (N). Use the force formula to solve Exercises 105–108. See Example 7.

Of Interest

Earth's gravity accelerates all objects at the same rate, −9.8 m/sec.² (or −32.2 ft./sec.²). The accelerations are negative because the direction is downward.

105. A car with a mass of 1200 kilograms exerts a force of 4800 newtons. Find the acceleration.

106. An object weighs 44.1 newtons; that is, it exerts a downward force of −44.1 newtons. The object is dropped so that it accelerates at −9.8 meters per second squared. Find the mass of the object.

Of Interest

The Liberty Bell was cast in 1752 and arrived in Philadelphia in August of the same year. In September 1752, the bell cracked during a test while hanging in the yard of the Pennsylvania state house.

107. The Liberty Bell weighs 9265.9 newtons, so its downward force is −9265.9 newtons. If the acceleration due to gravity is about −9.8 meters per second squared, what is the mass of the bell?

108. Jacob weighs 160 pounds (a downward force of −160 pounds). If the acceleration due to gravity is about −32.2 feet per second squared, what is his mass? (The mass will be in terms of the unit *slugs*.)

109. The formula $w = \dfrac{s}{f}$ describes the wavelength w of a musical note if f is the frequency and s is the speed of sound. A note is sounded with a wavelength of 0.94 meters/cycle and a frequency of 440 cycles/second. Find the speed of sound.

110. The formula $t = \dfrac{V}{g}$ gives the time that a free-falling object falls, where V is the velocity and g is the acceleration due to gravity. Find the velocity if the object has been falling for four seconds and the acceleration due to gravity is 9.8 meters per second squared.

111. The formula $C = 0.3m + 25$ describes the total cost in dollars of using a digital phone for 100 minutes, where m represents the number of minutes beyond 100 minutes. Pam's bill shows that she used the phone for 187 minutes. What is the total cost of her use of the phone?

112. A rental car agency rents a van for a weekend using the formula $C = 0.27m + 200$, where m represents the number of miles and C represents the cost in dollars. Laura rents a van from the company for a weekend when the company is offering 100 free miles. If she returns the van and has driven 421 miles, what is the total cost of the rental?

113. The formula $d = v_i t + \frac{1}{2}at^2$ describes the distance, d, an object travels if the object has an initial velocity (speed) v_i and an acceleration of a for a time t. An object travels 135 feet after being accelerated for three seconds with an acceleration of 20 feet per second squared. Find the object's initial velocity.

114. The formula $F = \frac{9}{5}C + 32$ gives a relationship between the temperature in Celsius and Fahrenheit. If the temperature is 82.4°F, find the temperature in Celsius.

Review Exercises

Exercises 1–3 EXPRESSIONS

[1.5] **1.** Simplify: $-6 + 3(5^2 + 2) - (6 - 1)$

[1.7] **2.** Combine like terms:
$$-3.2x^2 + 0.4x + 7x^2 - 8.03x + 1.5$$

[1.7] **3.** Use the distributive property to rewrite:
$$-\frac{3}{4}(6x + 8)$$

Exercises 4–6 EQUATIONS AND INEQUALITIES

For Exercises 4–6, solve and check.

[2.2] **4.** $6 + x = -14 + 2x$

[2.3] **5.** $-\frac{2}{5}x = 10$

[2.3] **6.** $2x + 3(x - 3) = 10x - (3x - 11)$

2.4 Applying the Principles to Formulas

Objective ❶ **Isolate a variable in a formula using the addition and multiplication principles.** So far when using formulas, we have substituted given numbers for all of the variables except one. We then used the addition and multiplication principles of equality to isolate that variable. However, we can actually isolate a particular variable in a formula without being given numbers for the other variables. We follow the same steps as if we had been given numbers.

Consider the area formula, $A = lw$. If we were given values for A and l and asked to isolate w, or solve for w, we would divide both sides by the value of l. We can perform this manipulation symbolically.

$$\frac{A}{l} = \frac{lw}{l} \qquad \text{To isolate } w, \text{we clear } l \text{ by dividing both sides by } l.$$

$$\frac{A}{l} = w$$

This development suggests the following procedure:

Isolating a Variable in a Formula

To isolate a particular variable in a formula, treat all other variables like constants and isolate the desired variable using the outline for solving equations.

Isolate a Variable Using the Addition Principle

Recall that the addition principle is used to clear terms that are added or subtracted.

Example 1   Isolate R in the formula $P = R - C$.

SOLUTION

$$P = R - C$$

To isolate R, we must clear C. Because C is subtracted from R, we add C to both sides.

$$\underline{+\, C \qquad +\, C}$$

$$P + C = R + 0$$

$$P + C = R$$

Note: Because P and C are not like terms, they cannot be combined.

Of Interest

The formula $P = R - C$ is used to calculate profit, given revenue, R, and cost, C.

Your Turn 1   Isolate v_i in the formula that describes the velocity of an object after being accelerated from an initial velocity (v_i), $v = v_i + at$.

Isolate a Variable Using the Multiplication Principle

Now let's use the multiplication principle to isolate a variable in a formula.

Example 2

a. Isolate w in the formula for the volume of a box, $V = lwh$.

SOLUTION

$$\frac{V}{lh} = \frac{l\!w\!h}{l\!h}$$

To isolate w, we must clear l and h. Because l and h are multiplying w, we divide both sides by l and h.

$$\frac{V}{lh} = w$$

b. Isolate P in the formula $B = P(1 + rt)$.

SOLUTION

$$\frac{B}{1 + rt} = \frac{P(1 + rt)}{1 + rt}$$

To isolate P, we must clear $(1 + rt)$. Because $(1 + rt)$ is multiplying P, we divide both sides by $1 + rt$.

$$\frac{B}{1 + rt} = P$$

Of Interest

The formula $B = P(1 + rt)$ is used in finance to calculate the balance for a simple-interest investment or loan. The variable B represents the final balance, P represents the principal, r represents the interest rate, and t represents time.

Your Turn 2

a. Isolate b in the formula for the area of a parallelogram, $A = bh$.

b. Isolate i in the voltage formula for a series circuit with two resistors, $V = i(R + r)$.

Answer   Your Turn 1

$v - at = v_i$

Answers   Your Turn 2

a. $\dfrac{A}{h} = b$ **b.** $\dfrac{V}{R + r} = i$

Isolate a Variable Using Both Principles

Now consider formulas that require the use of both principles to isolate the desired variable.

Example 3

a. Isolate l in the perimeter formula for rectangles, $P = 2l + 2w$.

SOLUTION First, use the addition principle to isolate the term that contains l. Then use the multiplication principle to clear the coefficient.

$$P = 2l + 2w$$

$$\underline{-2w \qquad\qquad -2w} \qquad \text{Isolate } 2l \text{ by subtracting } 2w \text{ from both sides.}$$

$$P - 2w = 2l + \quad 0$$

$$\frac{P - 2w}{2} = \frac{2l}{2} \qquad \text{Isolate } l \text{ by dividing both sides by 2.}$$

$$\frac{P - 2w}{2} = l$$

b. Isolate d in the formula from physics, $v^2 = v_0^2 + \dfrac{2Fd}{m}$.

SOLUTION First, use the addition principle to isolate the term that contains d. Then use the multiplication principle to clear the coefficient.

$$v^2 = v_0^2 + \frac{2Fd}{m} \qquad \text{Isolate } \frac{2Fd}{m} \text{ by subtracting } v_0^2 \text{ on both sides.}$$

$$\underline{-v_0^2 \qquad\quad -v_0^2}$$

$$v^2 - v_0^2 = 0 + \frac{2Fd}{m}$$

$$v^2 - v_0^2 = \frac{2Fd}{m}$$

$$\frac{m}{2F} \cdot \frac{v^2 - v_0^2}{1} = \frac{\overset{1}{\cancel{m}}}{\underset{1}{\cancel{2F}}} \cdot \frac{\overset{1}{\cancel{2F}} d}{\underset{1}{\cancel{m}}} \qquad \text{Clear the coefficient } \frac{2F}{m} \text{ by multiplying both sides}$$

$$\text{by its reciprocal.}$$

$$\frac{m(v^2 - v_0^2)}{2F} = d$$

Of Interest

In this example, the 0 subscript is used to indicate that v_0 is an initial velocity. We have also seen v_i used to indicate initial velocity. Both forms are used in physics and engineering.

Your Turn 3

a. Isolate t in the formula that describes the velocity of an object after being accelerated from an initial velocity, $v = v_i + at$.

b. Isolate n in the formula $T = \dfrac{nt}{r} - R$.

Answers Your Turn 3

a. $\dfrac{v - v_i}{a} = t$

b. $\dfrac{r}{t}(T + R) = n$ or $\dfrac{r(T + R)}{t} = n$

2.4 Exercises
For Extra Help *MyMathLab*

Note: Exercises marked with a ★ represent challenging exercises.

1. What would you do to isolate w in the formula $l = w - 2$? Why?

2. What would you do to isolate p in the formula $C = np$? Why?

3. To isolate y in $2x + 3y = 8$, what would you do first? Why?

4. How can you verify that your answer is correct after isolating a variable in a formula?

For Exercises 5–24, solve for the indicated variable. See Examples 1–3.

5. $t - 4u = v; t$

6. $x = a + 3y; a$

7. $-5y = x; y$

8. $2n = a; n$

9. $2x - 3 = b; x$

10. $3m + b = y; m$

11. $y = mx + b; m$

12. $ab + c = d; b$

13. $3x + 4y = 8; y$

14. $19 = 2l + 2w; w$

15. $\dfrac{mn}{4} - Y = f; Y$

16. $q = \dfrac{rs}{2} - p; p$

17. $6(c + 2d) = m - np; c$

18. $5(2n + a) = bn - c; a$

19. $\dfrac{x}{3} + \dfrac{y}{5} = 1; y$

20. $\dfrac{a}{4} + \dfrac{b}{6} = 3; a$

21. $\dfrac{3}{4} + 5a = \dfrac{n}{c}; n$

22. $\dfrac{x}{6} + 5y = \dfrac{m}{n}; m$

23. $t = \dfrac{A - P}{pr}; p$

24. $S = \dfrac{C}{1 - M}; M$

For Exercises 25–40, solve the geometric formula for the indicated variable. See Examples 1–3.

25. $P = a + b + c; a$

26. $a + b + c = 180; b$

27. $A = lw; l$

28. $C = 2\pi r; r$

29. $\dfrac{C}{d} = \pi; d$

30. $\dfrac{A}{b} = h; b$

31. $V = \dfrac{1}{3}\pi r^2 h; r^2$

32. $V = \dfrac{4}{3}\pi r^3; r^3$

33. $A = \dfrac{1}{2}bh; b$

34. $S = \dfrac{1}{2}gt^2; t^2$

35. $P = 2l + 2w; w$

36. $S = 2\pi r^2 - 2\pi rh; h$

37. $S = \dfrac{n}{2}(a + l); l$

38. $A = \dfrac{h}{2}(a + b); b$

★**39.** $V = h(\pi r^2 - lw); w$

★**40.** $V = h(\pi r^2 + lw); r^2$

For Exercises 41–48, solve the financial formula for the indicated variable.
See Examples 1–3.

41. $P = R - C; C$ **42.** $P = R - C; R$ **43.** $I = Prt; r$ **44.** $I = Prt; t$

45. $P = \dfrac{C}{n}; C$ **46.** $P = \dfrac{C}{n}; n$ **47.** $A = P + Prt; r$ **48.** $A = P + Prt; t$

For Exercises 49–60, solve the physics formula for the indicated variable. See Examples 1–3.

49. $d = rt; t$ **50.** $F = ma; m$ **51.** $W = Fd; d$ **52.** $V = kT; k$

53. $P = \dfrac{W}{t}; t$ **54.** $D = \dfrac{M}{V}; V$ **55.** $v = -32t + v_0; t$ **56.** $x = x_0 + vt; v$

57. $F = \dfrac{9}{5}C + 32; C$ **58.** $C = \dfrac{5}{9}(F - 32); F$ **59.** $F = G\dfrac{Mm}{R^2}; m$ **60.** $F = \dfrac{kMn}{d^2}; k$

Find
(X) the mistake

For Exercises 61–64, find and explain the mistake in each
solution and then solve correctly. See Examples 1–3.

61. $3n + 7t = 54$; isolate t

$$3n + 7t = 54$$
$$\underline{-3n \qquad\qquad -3n}$$
$$7t = 54 - 3n$$
$$7t = 54 - 3n$$
$$\underline{-7 \quad -7}$$
$$t = 47 - 3n$$

62. $\dfrac{1}{4}kt = r$; isolate t

$$\dfrac{1}{4}kt = r$$
$$\dfrac{4}{1} \cdot \dfrac{1}{4}kt = 4r$$
$$\dfrac{1}{k} \cdot kt = 4r\dfrac{k}{1}$$
$$t = 4kr$$

63. $\dfrac{3m - 2}{5} = nk$; isolate m

$$\dfrac{3m - 2}{5} = nk$$
$$\dfrac{5}{1} \cdot \dfrac{3m - 2}{5} = nk \cdot 5$$
$$3m - 10 = 5nk$$
$$3m - 10 = 5nk$$
$$\underline{+10 \quad +10}$$
$$3m = 5nk + 10$$
$$\dfrac{3m}{3} = \dfrac{5nk + 10}{3}$$
$$m = \dfrac{5nk + 10}{3}$$

64. $7(y - 5) = xv$; isolate y

$$7(y - 5) = xv$$
$$7y - 5 = xv$$
$$7y - 5 = xv$$
$$\underline{+5 \quad\quad +5}$$
$$\dfrac{7y}{7} = \dfrac{xv + 5}{7}$$
$$y = \dfrac{xv + 5}{7}$$

Collaborative Exercises Where Does Speeding Get You?

1. The formula that relates distance, rate, and time is $d = rt$. Solve this formula for t.
2. Use the formula you found in Problem 1 to determine how long it will take you to travel 20 miles (an average daily commute) at a rate of 55 miles per hour. Note that your result will be in hours. Convert the time to minutes by multiplying by 60 and round to the nearest tenth of a minute.
3. Now use the formula from Problem 1 to determine how long it will take you to travel 20 miles at 65 miles per hour. Again, convert the time to minutes and round to the nearest tenth of a minute.
4. How much time does speeding save you? To find out, subtract the time you found in Problem 3 from the time you found in Problem 2.
5. Discuss in your group whether the additional time is worth the additional risk that comes with increasing the speed by 10 miles per hour.
6. Complete the table, rounding all values to the nearest tenth. (You've already calculated the values for the 20-mile trip.)

Trip Distance	Time at 55 mph	Time at 65 mph	Time Saved
20 miles			
30 miles			
40 miles			

7. Is the time saved on longer trips worth the additional risk of increasing your speed by 10 miles per hour? Why or why not? Does the longer distance also increase risk? Why or why not?

Review Exercises

Exercises 1 and 2  EXPRESSIONS

[1.6] *For Exercises 1 and 2, translate each phrase to an expression.*

1. four more than seven times a number

2. nine less than three times the sum of a number and two

Exercises 3–6 EQUATIONS AND INEQUALITIES

[2.3] *For Exercises 3–6, solve.*

3. $4x - 7 = -5$

4. $\dfrac{3}{4}x = -\dfrac{5}{8}$

5. $3(n - 5) = 7n - 12$

6. $\dfrac{1}{4}y + 6 = \dfrac{1}{5}(y + 10)$

2.5 Translating Word Sentences to Equations

Objective ① **Translate sentences to equations using key words, then solve.**   In previous sections, we translated phrases to expressions (Section 1.6) and solved problems using formulas (Section 2.1). In this section, we further develop problem-solving strategies by translating sentences to equations. Because we have not learned how to solve equations containing exponents and roots, we will explore translations that involve only addition, subtraction, multiplication, and division. As in Section 1.6, we begin by translating algebraic equations to word sentences. Several key words or phrases indicate an equal sign.

Key words for an equal sign:	is equal to	is	yields
	is the same as	produces	results in

Translating Algebraic Equations to English Sentences

Equation	Translation
Equations Involving Addition	
$x + 3 = 6$	The sum of x and three is six.
	Three more than x is equal to six.
	x plus three results is six.
	Three added to x produces six.
	x increased by three is the same as six.
Equations Involving Subtraction	
$y - 2 = 7$	The difference of y and two is seven.
	y minus two equals seven.
	Two subtracted from y yields seven.
	Two less than a number results in seven.
	y minus two produces seven.
	A number decreased by two equals seven.
Equations Involving Multiplication	
$\frac{3}{4}m = 9$	The product of three-fourths and m is nine.
	Three-fourths times m is the same as nine.
	Three-fourths of a number equals nine.
Equations Involving Division	
$\frac{x}{4} = -2$	The quotient of x and four results in negative two.
	x divided by four produces negative two.
	Four divided into a number is equal to negative two.
	The ratio of x and four is the same as negative two.
Equations That Involve More Than One Operation	
$4n + 2 = 8$	Two more than the product of four and n is eight.
	Four times n increased by two is equal to eight.
	The sum of four times n and two is the same as eight.
	Two more than the product of four and a number equals eight.
	Two added to four times a number results in eight.
Equations Involving Parentheses	
$4(m - 3) = 6$	Four times the difference of m and three results in six.
	The product of four and the difference of m and three is six.

You also may want to review or reference the key words and their translations in Section 1.6 on pages 63–65.

Procedure | **Translating Word Sentences**

To translate a word sentence to an equation, identify the unknown(s), constants, and key words; then write the corresponding symbolic form.

Problems Involving Addition or Subtraction

Example 1   Translate to an equation and then solve.

a. The sum of twenty-five and a number is equal to fourteen.

Understand   The key word *sum* indicates addition, *is equal to* indicates an equal sign, and *a number* indicates a variable.

Plan   Use the key words to translate to an equation, and then solve the equation. We'll use n as the variable.

Execute   Translate: The sum of twenty-five and a number is equal to fourteen.

$$25 \quad + \quad n \quad = \quad 14$$

Solve: $25 + n = 14$

$$\underline{-25 -25} \qquad \text{Subtract 25 from both sides to isolate } .$$

$$0 + n = -11$$

Answer $n = -11$

Check   In the original sentence, replace the unknown amount with -11 and verify that it makes the sentence true. Verify that the sum of twenty-five and a negative eleven is equal to fourteen.

$$25 + (-11) \stackrel{?}{=} 14$$
$$14 = 14$$

Yes, the sum of 25 and -11 is equal to 14.

b. Nineteen less than a number is seven.

Understand   The key words *less than* indicate subtraction. The key word *is* means an equal sign.

Plan   Use the key words to translate to an equation and then solve. We'll use n for the variable.

Execute   Translate: Nineteen less than a number is seven.

$$n - 19 \qquad = \quad 7$$

> **Note:** The key words *less than* require careful translation. In the sentence, the word *nineteen* comes before the words *less than* and the words *a number* come after. In the translation, this order is reversed.

Solve: $n - 19 = 7$

$$\underline{+19 +19} \qquad \text{Add 19 to both sides.}$$

$$n + 0 = 26$$

Answer $n = 26$

Check   Verify that 19 less than 26 is 7.

$$26 - 19 \stackrel{?}{=} 7$$
$$7 = 7$$

Yes, 19 less than 26 is 7.

Your Turn 1   Translate to an equation and then solve.

a. Fifteen more than a number is negative seven.

b. The difference of x and thirty-five is negative nine.

Problems Involving Multiplication

Example 2   Three-fourths of a number is negative five-eighths. Translate to an equation and then solve.

Understand   When *of* is preceded by a fraction, it means multiply. The word *is* means an equal sign.

Plan   Use the key words to translate to an equation and then solve.

Execute   Translate: Three-fourths of a number is negative five-eighths.

$$\frac{3}{4} \cdot n = -\frac{5}{8}$$

Solve: $\frac{\overset{1}{\cancel{4}}}{\underset{1}{\cancel{3}}} \cdot \frac{\overset{1}{\cancel{3}}}{\underset{1}{\cancel{4}}} n = -\frac{5}{\underset{2}{\cancel{8}}} \cdot \frac{\overset{1}{\cancel{4}}}{3}$ Clear the coefficient $\frac{3}{4}$ by multiplying both sides by its reciprocal, $\frac{4}{3}$.

Answer $n = -\frac{5}{6}$

Check   Verify that $\frac{3}{4}$ of $-\frac{5}{6}$ is $-\frac{5}{8}$.

$$\frac{\overset{1}{\cancel{3}}}{4} \cdot -\frac{5}{\underset{2}{\cancel{6}}} \overset{?}{=} -\frac{5}{8}$$

$$-\frac{5}{8} = -\frac{5}{8}$$

Yes, $\frac{3}{4}$ of $-\frac{5}{6}$ is $-\frac{5}{8}$, so $-\frac{5}{6}$ is correct.

Your Turn 2   The product of 0.6 and a number is −25.08. Translate to an equation and then solve.

Problems Involving More Than One Operation

Example 3   Seven less than the product of four and a number is equal to negative five. Translate to an equation and then solve.

Understand   *Less than* indicates subtraction in reverse order, *product* indicates multiplication, and *is equal to* indicates an equal sign.

Plan   Translate to an equation using the key words and then solve the equation.

Execute   Translate:

Seven less than the product of four and a number is equal to negative five.

$$4n - 7 \qquad\qquad = \qquad -5$$

Solve: $4n - 7 = -5$

$$\underline{+7 \quad +7}$$ Add 7 to both sides to isolate $4n$.

$$4n + 0 = 2$$

$$\frac{4n}{4} = \frac{2}{4}$$ Divide both sides by 4 to clear the coefficient.

Answer $n = \dfrac{1}{2}$ or 0.5

Check Verify that 7 less than the product of 4 and $\frac{1}{2}$ is equal to -5.

$$4\left(\frac{1}{2}\right) - 7 \stackrel{?}{=} -5$$

$$2 - 7 \stackrel{?}{=} -5$$

$$-5 = -5$$

Yes, 7 less than the product of 4 and $\frac{1}{2}$ is equal to -5.

Your Turn 3 Thirteen more than the product of four and a number yields negative seven. Translate to an equation and then solve.

Translations Requiring Parentheses

Sometimes a translation requires parentheses. This occurs when a sum or difference is multiplied or divided. When we multiply or divide a sum or difference, the addition or subtraction is calculated first. Because addition and subtraction follow multiplication and division in the order of operations, parentheses must be inserted around the addition or subtraction so that these operations are performed first.

Example 4 Two-thirds of the difference of six and a number is the same as the number divided by four. Translate to an equation and then solve.

Understand *Of* means to multiply, and the word *difference* indicates subtraction. Because the difference is being multiplied, we write the difference expression in parentheses. *Divided by* indicates division.

Plan Use the key words to translate to an equation and then solve.

Execute Translate:

Two-thirds of the difference of six and a number is the same as the number divided by four.

$$\frac{2}{3} \quad \cdot \quad (6 - n) \qquad\qquad = \qquad \frac{n}{4}$$

Solve:
$$\frac{2}{3}(6 - n) = \frac{n}{4}$$

$$\frac{2}{\overset{}{\underset{1}{3}}} \cdot \frac{\overset{2}{6}}{1} - \frac{2}{3}n = \frac{n}{4} \qquad \text{Distribute to clear the parentheses.}$$

$$4 - \frac{2}{3}n = \frac{n}{4} \qquad \text{Simplify.}$$

$$12\left(4 - \frac{2}{3}n\right) = \left(\frac{n}{4}\right)12 \qquad \begin{array}{l}\text{Multiply both sides by the LCD, 12,}\\\text{to clear the fractions.}\end{array}$$

$$12 \cdot 4 - \frac{\overset{4}{\cancel{12}}}{1} \cdot \frac{2}{\underset{1}{3}}n = \frac{n}{\underset{1}{4}} \cdot \frac{\overset{3}{\cancel{12}}}{1} \qquad \text{Distribute 12 and simplify.}$$

$$48 - 8n = 3n \qquad \begin{array}{l}\text{Add } 8n \text{ to both sides to separate the}\\\text{constant term and variable terms.}\end{array}$$

$$\underline{+8n \quad +8n}$$

$$48 + 0 = 11n$$

$$\frac{48}{11} = \frac{11n}{11} \qquad \begin{array}{l}\text{Clear the coefficient by dividing both}\\\text{sides by 11.}\end{array}$$

Answer Your Turn 3

$4n + 13 = -7; n = -5$

Answer $\dfrac{48}{11} = n$

Check   Verify that $\dfrac{2}{3}$ of the difference of 6 and $\dfrac{48}{11}$ is the same as $\dfrac{48}{11}$ divided by 4.

$$\frac{2}{3}\left(6 - \frac{48}{11}\right) \overset{?}{=} \frac{1}{4}\left(\frac{48}{11}\right)$$

Write 6 as a fraction with a denominator of 11. $\dfrac{2}{3}\left(\dfrac{66}{11} - \dfrac{48}{11}\right) \overset{?}{=} \dfrac{1}{\overset{}{\underset{1}{\cancel{4}}}}\left(\dfrac{\overset{12}{\cancel{48}}}{11}\right)$ Divide out 4.

Subtract. $\dfrac{2}{3}\left(\dfrac{18}{11}\right) \overset{?}{=} \dfrac{12}{11}$ Multiply.

Divide out 3. $\dfrac{2}{\underset{1}{\cancel{3}}}\left(\dfrac{\overset{6}{\cancel{18}}}{11}\right) \overset{?}{=} \dfrac{12}{11}$

Multiply. $\dfrac{12}{11} = \dfrac{12}{11}$

Yes, $\dfrac{2}{3}$ of the difference of 6 and $\dfrac{48}{11}$ is the same as $\dfrac{48}{11}$ divided by 4.

Answers Your Turn 4

a. $3(n-5) = 7n - 12$;

$n = -\dfrac{3}{4}$

b. $\dfrac{1}{4}n + 6 = \dfrac{1}{5}(n+10)$;

$n = -80$

Your Turn 4   Translate to an equation and then solve.

a. Three times the difference of a number and five is equal to seven times the number minus twelve.

b. The sum of one-fourth of a number and six is equal to one-fifth of the sum of the number and ten.

2.5 Exercises   For Extra Help   *MyMathLab*

1. List three key words that indicate addition.

2. List three key words that indicate subtraction.

3. List three key words that indicate multiplication.

4. List three key words that indicate division.

5. Why are "four minus n" and "four less than n" different?

6. Write a sentence that requires parentheses when translated and explain why your sentence requires parentheses.

For Exercises 7–44, translate each sentence to an equation and then solve.
See Examples 1–4.

7. Three more than y is negative 8.

8. Six added to p is negative two.

9. Six decreased by the number x is equal to negative three.

10. Eighteen subtracted from a number is equal to negative three.

11. Eleven times m is negative ninety-nine.

12. The product of negative three and z is -18.

13. Four divided into m is 1.6.

14. The quotient of a number and -6.5 is 4.2.

15. Four-fifths of a number is five-eighths.

16. Three-sevenths of a number is negative nine-eighths.

17. Seven more than three times a number is equal to thirty-four.

18. Nineteen more than triple a number is one hundred.

19. Nine less than five times a number is seventy-six.

20. Twenty-five less than four times a number is eleven.

21. Eight times the sum of eight and a number is equal to one hundred sixty.

22. Four times the sum of a number and three is equal to sixteen.

23. Negative three times the difference of x and two is twelve.

24. Tripling the difference of x and five produces negative fifteen.

25. One-third of the sum of a number and two is one.

26. Half of the difference of a number and two results in four.

27. Eleven less than three times a number is equal to that number added to five.

28. Five more than four times a number is equal to seven subtracted from that number.

29. Ten is the result when one is subtracted from the ratio of a number to four.

30. Two less than the quotient of a number and five is six.

31. The product of three and the sum of a number and four subtracted from twice the number yields negative three.

32. Seven times the sum of a number and six subtracted from four times the number results in negative twenty-one.

33. The difference of a number and twelve is added to the difference of twice the number and eight so that the result is thirteen.

34. The sum of a number and six is added to the difference of the number tripled and four so that the result is fourteen.

35. Five times the sum of a number and two-thirds is equal to three times the number decreased by two-thirds.

36. The product of three and the sum of a number and two-fifths is four-fifths less than negative two times the number.

37. Two times a number decreased by four is equal to triple the number added to sixteen.

38. Six less than three times a number is equal to the difference of fourteen and twice the number.

39. Two-fifths of a number is equal to two less than one-half the number.

40. One-third of a number is the same as half of the number decreased by one.

quotient of three less than a number and three is the same as the quotient of one less than the number and four.

42. The quotient of one more than a number and two is the same as the quotient of six more than twice the number and eight.

43. The product of negative four and the difference of two and a number is six less than the product of two and the difference of four and three times the number.

44. Negative two times the difference of one and three times a number is equal to four added to five times the difference of one and the number.

For Exercises 45–54, translate the equation to a word sentence. There is more than one correct translation.

45. $4x + 3 = 7$

46. $-3y + 8 = 10$

47. $6(y + 4) = -10y$

48. $8(w - 2) = 3w$

49. $\dfrac{1}{2}(x - 3) = \dfrac{2}{3}(x - 8)$

50. $\dfrac{1}{2}(t + 1) = \dfrac{1}{3}(t - 5)$

51. $0.05m + 0.06(m - 11) = 22$

52. $0.05v + 0.03(v - 4.5) = 0.465$

53. $\dfrac{2}{3}x + \dfrac{3}{4}x + \dfrac{1}{2}x = 10$

54. $\dfrac{1}{2}a + \dfrac{1}{3}a + \dfrac{1}{6}a = 5$

Find the mistake   *For Exercises 55–60, explain the mistake in each translation; then translate correctly. See Examples 1–4.*

55. Ten less than a number is forty.
   Translation: $10 - n = 40$

56. Twelve divided into a number is negative eight.
   Translation: $12 \div n = -8$

57. Five times the difference of a number and six is equal to negative two.
   Translation: $5x - 6 = -2$

58. Nine times the sum of a number and eight is the same as three subtracted from the number.
   Translation: $9(y + 8) = 3 - y$

59. Twice a number minus the sum of the number and three is equal to negative six.
   Translation: $2t - t + 3 = -6$

60. Four times the sum of a number and two is equal to the number minus the difference of the number and seven.
   Translation: $4(y + 2) = y - (7 - y)$

For Exercises 61–68, translate to a formula; then use the formula to solve the problem.

61. The perimeter of a rectangle is equal to twice the sum of its length and width. Find the perimeter with the following length and width.

Length	Width
a. 24 ft.	18 ft.
b. 12.5 cm	18 cm
c. $8\frac{1}{4}$ in.	$10\frac{3}{8}$ in.

Width

Length

62. The surface area of a box is equal to twice the sum of its length times its width, its length times its height, and its width times its height. Find the surface area of a box with the following length, width, and height.

Length	Width	Height
a. 15 in.	6 in.	4 in.
b. 9.2 cm	12 cm	6.5 cm
c. 8 ft.	$2\frac{1}{2}$ ft.	3 ft.

Height

Length

Width

63. An isosceles triangle has two sides that are equal in length. The perimeter of an isosceles triangle is the sum of its base and twice the length of one of the other sides. Find the perimeter of each listed isosceles triangle.

Base	Sides
a. 5 in.	11 in.
b. 1.5 m	0.6 m
c. $15\frac{1}{4}$ in.	$9\frac{1}{2}$ in.

Side   Side

Base

64. The perimeter of a semicircle is the sum of the diameter and π times the radius. Find the perimeter of each semicircle listed. Where appropriate, round your answer to the nearest tenth.

a. Diameter = 18 in. (Use $\pi \approx 3.14$.)
b. Radius = 4.5 m (Use $\pi \approx 3.14$.)
c. Diameter = $2\frac{3}{4}$ ft. (Use $\pi \approx \frac{22}{7}$.)

Diameter

65. The simple interest earned after investing an amount of money, called principal, is equal to the product of the principal, the interest rate, and the time in years that the money remains invested. Use the formula to calculate the interest of each of the following investments.

Principal	Rate	Time
a. $4000	0.05	2 years
b. $500	0.03	$\frac{1}{2}$ year
c. $2000	0.06	$\frac{1}{4}$ year

66. Investors often consider a stock's price-to-earnings ratio to help them decide whether a stock is worth investing in. The price-to-earnings ratio is equal to the stock's price divided by its company's earnings per share over the last 12 months. Use your formula to calculate each company's price-to-earnings ratio. Round your answers to the nearest tenth. (*Source:* David and Tom Gardner, *The Motley Fool Investment Guide*, Simon & Schuster, 1996.)

Company	Price	Earnings per Share
a. Microsoft	$47.98	$1.41
b. Intel	$18.79	$0.29
c. General Motors	$46.55	$4.01

67. Skydivers often use a formula to estimate free-fall time. The amount of free-fall time in seconds is approximately equal to the difference of the exit altitude and the parachute deployment altitude divided by 153.8. Use this formula to calculate free-fall times if a skydiver deploys at each of the following altitudes after falling from an exit altitude of 12,500 feet. Round your answers to the nearest second.

   a. 5000 feet
   b. 4500 feet
   c. 3000 feet

68. A person's body mass index (BMI) is a measure of the amount of body fat based on height, in inches, and weight, in pounds. A person's BMI is equal to the ratio of the product of his or her weight and 704.5 to his or her height in inches multiplied by itself. Use your formula to find the BMI of each person listed below. Round your answers to the nearest tenth. (*Source:* National Institutes of Health.)

	Height	Weight
a.	5′4″	138 lb
b.	5′9″	155 lb
c.	6′2″	220 lb

Of Interest

After exiting the plane, a skydiver accelerates from 0 to approximately 110 miles per hour (terminal velocity) in about 10 seconds and falls about 1000 feet. Terminal velocity means that air resistance balances out gravitational acceleration so that the skydiver falls at a constant 110 miles per hour until he or she deploys the parachute. At terminal velocity, the skydiver falls at a rate of about 1000 feet every 6 seconds. Of course, the speed of the fall can be changed by varying body position to increase or decrease air resistance.

Of Interest

The National Institutes of Health use the following categories when analyzing BMI:

Underweight: BMI < 18.5
Normal Weight: BMI = 18.5–24.9
Overweight: BMI = 25–29.9
Obesity: BMI ≥ 30

Note that this way of calculating BMI is a rough estimate and is not accurate for extremely fit people such as athletes and bodybuilders because their "extra" weight is muscle rather than fat.

Puzzle Problem

Fill in each blank with a whole number so that the resulting rational number has the following properties:

   The rational number is between 300 and 500.
   The tenths placeholder is less than the tens placeholder.
   The sum of the tens place and the tenths place is 8.
   The product of all four placeholders is 42.

   _____ _____ _____ . _____

Review Exercises

Exercises 1–6 EQUATIONS AND INEQUALITIES

[1.1]   **1.** Is $19 \geq 19$ true or false? Explain.

[1.1, 1.5] *For Exercises 2 and 3, use* <, >, *or* = *to make a true sentence.*

2. $-|-6|$ $-2(3)$

3. $5(4) - 23$ $2(3 - 8)$

[2.3] *For Exercises 4–6, solve and check.*

4. $7x + 9 = 4x - 6$

5. $\dfrac{5}{6} = -\dfrac{3}{4}x$

6. $\dfrac{1}{3}(y - 2) = \dfrac{3}{5}y + 6$

2.6 Solving Linear Inequalities

Objectives

① Represent solutions to inequalities graphically and using set notation.

② Solve linear inequalities.

③ Solve problems involving linear inequalities.

Objective ① **Represent solutions to inequalities graphically and using set notation.** Not all problems translate to equations. Sometimes a problem can have a range of values as solutions. In mathematics, we can write inequalities to describe situations where a range of solutions is possible. Following are the inequality symbols and their meanings.

Symbol	Meaning
$<$	is less than
$>$	is greater than
$\leq$	is less than or equal to
$\geq$	is greater than or equal to

Throughout the chapter, we have explored techniques for solving linear equations. In this section, we explore how to solve **linear inequalities**.

Definition   *Linear inequality:* An inequality containing expressions in which each variable term contains a single variable with an exponent of 1.

Following are some examples of linear inequalities:

$$x > 5 \qquad n + 2 < 6 \qquad 2(y - 3) \leq 5y - 9 \qquad 2x + 3y \geq 6$$

The solutions of an inequality are any numbers that can replace the variable(s) in the inequality and make it true. Because inequalities have a range of solutions, we often write those solutions in set-builder notation, which we introduced in Section 1.1. For example, the solution set for the inequality $x \geq 5$ contains 5 and every real number greater than 5, which we write as $\{x | x \geq 5\}$. We read this set-builder notation as shown here.

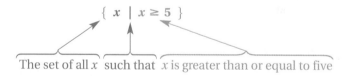

The set of all x such that x is greater than or equal to five

We can graph solution sets for inequalities on a number line. Because the solution set for $x \geq 5$ contains 5 and every real number to the right of 5, we draw a dot (or solid circle) at 5 and shade to the right of 5.

$x \geq 5$ is represented by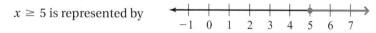

The solution set for $x < 2$ contains every real number to the left of 2, but not 2 itself; so it is written $\{x | x < 2\}$. To graph this solution set, we draw an open circle at 2, indicating that 2 is not included, and shade to the left of 2.

$x < 2$ is represented by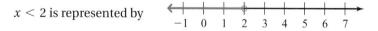

Another popular notation used to indicate ranges of values is *interval notation*, which uses parentheses and brackets to indicate whether end values are included. Parentheses are used for end values that are not included in the interval, and brackets are used for end values that are included. To write interval notation, we imagine traveling from left to right on the number line.

For example, as we travel from left to right on the graph for $x \geq 5$, we encounter 5 and then every real number to the right of 5; so the interval notation is $[5, \infty)$. The symbol ∞ means infinity and is used to indicate the positive extreme. Since ∞ can never be reached, it will always have a parenthesis indicating that it is not included as an end value. As we travel from left to right on the graph for $x < 2$, we go from the negative extreme up to 2, but not including 2; so the interval notation is $(-\infty, 2)$. Notice that $-\infty$ represents the leftmost extreme.

The parentheses and brackets from interval notation, instead of open and solid circles, can be used on the graphs.

$x \geq 5$ is represented by

$x < 2$ is represented by

Your instructor may have a style preference. Because interval notation is a preferred notation in college algebra courses, we draw our graphs with the parentheses and brackets. Notice that the parentheses or brackets "open" in the direction that is shaded.

Procedure   **Graphing Inequalities**

To graph an inequality on a number line,

1. If the symbol is $\leq$ or $\geq$, draw a bracket (or solid circle) on the number line at the indicated number open to the left for $\leq$ and to the right for $\geq$. If the symbol is $<$ or $>$, draw a parenthesis (or open circle) on the number line at the indicated number open to the left for $<$ and to the right for $>$.
2. If the variable is greater than the indicated number, shade to the right of the indicated number. If the variable is less than the indicated number, shade to the left of the indicated number.

Example 1   Write the solution set in set-builder notation and interval notation; then graph the solution set.

a. $x \leq -6$

SOLUTION   Set-builder notation: $\{x | x \leq -6\}$
Interval notation: $(-\infty, -6]$

Note: When writing interval notation, imagine traveling from left to right so this range of values is from $-\infty$ to -6, including -6.

Graph:

b. $n > 0$

SOLUTION   Set-builder notation: $\{n | n > 0\}$
Interval notation: $(0, \infty)$

Connection $\{n | n > 0\}$ is a way of expressing the set of all positive real numbers.

Graph:

Answers   Your Turn 1

a. Set-builder notation:
$\{x | x < 2\}$
Interval notation: $(-\infty, 2)$

Graph: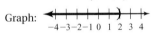

b. Set-builder notation:
$\{t | t \geq -1\}$
Interval notation: $[-1, \infty)$

Graph:

Your Turn 1   Write the solution set in set-builder notation and interval notation; then graph the solution set.

a. $x < 2$ b. $t \geq -1$

Compound Inequalities

Inequalities containing two inequality symbols are called *compound inequalities*. Compound inequalities are useful in writing a range of values between two numbers. For example, $1 < x < 6$ is a compound inequality meaning that x can be any number that is greater than 1 and less than 6. Its solution set contains every real number between 1 and 6, but not 1 and 6.

Set-builder notation: $\{x \mid 1 < x < 6\}$

Interval notation: $(1, 6)$

Graph:

Note: Because the end values 1 and 6 are not included in the solution set, we use parentheses in the interval notation and on the graph.

Example 2 Write the solution set for $-5 \le x < 1$ in set-builder notation and interval notation; then graph the solution set.

SOLUTION Set-builder notation: $\{x \mid -5 \le x < 1\}$

Interval notation: $[-5, 1)$

Graph:

Note: Because the end value -5 is included in the solution set, we use a bracket in the interval notation and on the graph. Because 1 is not included, we use a parenthesis.

Your Turn 2 Write the solution set for $-4 < y \le -1$ in set-builder notation and interval notation; then graph the solution set.

Objective 2 Solve linear inequalities. To solve inequalities such as $n + 2 < 6$ and $2(x - 3) \le 5x - 9$, we follow essentially the same process as for solving equations. The addition and multiplication principles of inequalities are similar to the principles of equality. First, let's examine the addition principle of inequality.

Consider the inequality $3 < 7$. According to the addition principle, if we add or subtract the same amount on both sides, the inequality should still be true. Let's test this by choosing some numbers to add and subtract on both sides.

Add 2 to both sides:

$$\begin{array}{r} 3 < 7 \\ +2 \ +2 \\ \hline 5 < 9 \end{array}$$ Still true

Subtract 9 from both sides:

$$\begin{array}{r} 3 < 7 \\ -9 \ -9 \\ \hline -6 < -2 \end{array}$$ Still true

Rule   The Addition Principle of Inequality

If $a < b$, then $a + c < b + c$ is true for all real numbers a, b, and c. The principle also holds true when $<$ is replaced with $>$, $\le$, or $\ge$.

Note: This principle indicates that we can add (or subtract) the same amount on both sides of an inequality without affecting its solution(s). Although the principle is written in terms of the $<$ symbol, the principle is true for any inequality symbol.

Example 3 Solve $x + 2 \le -3$ and write the solution set in set-builder notation and interval notation; then graph the solution set.

SOLUTION

Note: Subtracting 2 from both sides does not affect the inequality.

$$\begin{array}{r} x + 2 \le -3 \\ -2 \quad -2 \\ \hline x \le -5 \end{array}$$

Isolate x by subtracting 2 from both sides.

Set-builder notation: $\{x | x \le -5\}$

Interval notation: $(-\infty, -5]$

Graph:

Your Turn 3 Solve $n - 6 > -5$ and write the solution set in set-builder notation and interval notation; then graph the solution set.

The multiplication principle, on the other hand, does not work as neatly as it did for equations. As we shall see, sometimes multiplying or dividing both sides of an inequality by the same number can turn a true inequality into a false inequality. Let's multiply both sides of $3 < 7$ by a positive number and then by a negative number.

Multiply both sides by 4:	**Multiply both sides by -2:**
$4(3) < 4(7)$	$-2(3) < -2(7)$
$12 < 28$ Still true	$-6 < -14$ Not true!

Conclusion An inequality remains true when we multiply both sides by a positive number, but not when we multiply both sides by a negative number.

Rule **The Multiplication Principle of Inequality**

If a and b are real numbers, where $a < b$, then $ac < bc$ is true if c is a positive real number.

If a and b are real numbers, where $a < b$, then $ac > bc$ is true if c is a negative real number.

The principle also holds true for $>$, $\le$, and $\ge$.

Note: We can multiply (or divide) both sides of an inequality by the same positive amount without affecting its solution(s). However, if we multiply (or divide) both sides of an inequality by the same negative amount, we must reverse the direction of the inequality symbol to maintain the truth of the inequality.

Example 4 Solve and write the solution set in set-builder notation and interval notation; then graph the solution set.

a. $3x > -7$

SOLUTION

Note: Dividing both sides by a positive number does not affect the inequality.

$$\frac{3x}{3} > \frac{-7}{3}$$

To isolate x, we clear the 3 coefficient by dividing by 3 on both sides.

$$x > -\frac{7}{3}$$

Set-builder notation: $\left\{ x \middle| x > -\frac{7}{3} \right\}$

Interval notation: $\left(-\frac{7}{3}, \infty \right)$

Graph:

b. $-8x \ge 24$

SOLUTION

Note: Because we divided both sides by a negative number, we reversed the direction of the inequality symbol.

$$\frac{-8x}{-8} \le \frac{24}{-8}$$

To isolate x, we clear the -8 coefficient by dividing by -8 on both sides.

$$x \le -3$$

Set-builder notation: $\{x\,|\,x \le -3\}$

Interval notation: $(-\infty, -3]$

Graph:

$-6\ \ -5\ \ -4\ \ -3\ \ -2\ \ -1\ \ \ 0\ \ \ 1\ \ \ 2$

c. $\dfrac{5}{6} \le -\dfrac{3}{4}n$

SOLUTION

> **Note:** You may find it helpful to rewrite the inequality so that the variable is on the left side. The inequality $-\dfrac{10}{9} \ge n$ is the same as $n \le -\dfrac{10}{9}$.

$$-\overset{2}{\underset{3}{\cancel{\dfrac{4}{3}}}}\left(\dfrac{5}{\underset{3}{\cancel{6}}}\right) \ge -\overset{1}{\underset{3}{\cancel{\dfrac{4}{3}}}}\left(-\dfrac{\overset{1}{\cancel{3}}}{\underset{1}{\cancel{4}}}n\right)$$

$$-\dfrac{10}{9} \ge n$$

To isolate n, we clear the $-\dfrac{3}{4}$ coefficient by multiplying by its reciprocal $-\dfrac{4}{3}$ on both sides. Multiplying by a negative number means that we must reverse the direction of the inequality symbol.

Set-builder notation: $\left\{ n\,\middle|\, n \le -\dfrac{10}{9} \right\}$

Interval notation: $\left(-\infty, -\dfrac{10}{9} \right]$

Graph:

$-2\qquad\quad -1\qquad\quad 0\qquad\quad 1$

Your Turn 4 Solve $-\dfrac{2}{5}t < \dfrac{9}{10}$ and write the solution set in set-builder notation and in interval notation; then graph the solution set.

We can solve more complex linear inequalities using an outline similar to that for solving linear equations.

Procedure **Solving Linear Inequalities**

To solve linear inequalities,

1. Simplify both sides of the inequality as needed.
   a. Distribute to clear parentheses.
   b. Clear fractions or decimals by multiplying through by the LCD just as we did for equations. (Clearing fractions and decimals is optional.)
   c. Combine like terms.
2. Use the addition principle so that all variable terms are on one side of the inequality and all constants are on the other side. Then combine like terms.
3. Use the multiplication principle to clear any remaining coefficient. If you multiply (or divide) both sides by a negative number, reverse the direction of the inequality symbol.

Answer   Your Turn 4

$t > -\dfrac{9}{4}$ or $-2\dfrac{1}{4}$

Set-builder notation: $\left\{ t\,\middle|\, t > -\dfrac{9}{4} \right\}$

Interval notation: $\left(-\dfrac{9}{4}, \infty \right)$

Graph:
$-4\qquad -3\qquad\ -2\qquad -1$
$\qquad\qquad\qquad -\dfrac{9}{4}$

Notice that clearing the term with the smaller coefficient in step 2 results in a positive coefficient, which means that you won't have to reverse the direction of the inequality in step 3.

Example 5   Solve $7x + 9 > 4x - 6$ and write the solution set in set-builder notation and interval notation; then graph the solution set.

SOLUTION   Use the addition principle to separate the variable terms and constant terms; then clear the remaining coefficient.

$$7x + 9 > 4x - 6$$

Note: Clearing the $4x$ term results in a positive coefficient after combining like terms; so we won't have to reverse the inequality when we clear the 3 coefficient.

$$\underline{-4x \qquad\qquad -4x}$$ Subtract $4x$ from both sides.

$$3x + 9 > 0 - 6$$
$$3x + 9 > -6$$
$$\underline{-9 \qquad -9}$$ Subtract 9 from both sides.
$$3x + 0 > -15$$
$$\frac{3x}{3} > \frac{-15}{3}$$ Divide both sides by 3 to isolate x.
$$x > -5$$

Set-builder notation: $\{x | x > -5\}$

Interval notation: $(-5, \infty)$

Graph:

```
<----+--+--[--+--+--+--+--+--+---->
    -7 -6 -5 -4 -3 -2 -1  0  1
```

Your Turn 5   Solve $7(x - 3) \le 4x + 3$ and write the solution set in set-builder notation and interval notation; then graph the solution set.

Objective ③ Solve problems involving linear inequalities.   Problems requiring inequalities can be translated using key words much like those we used to translate sentences to equations. The following table lists common key words that indicate inequalities.

Less Than:	
A number is less than seven.	$n < 7$
A number must be smaller than five.	$n < 5$

Greater Than:	
A number is greater than two.	$n > 2$
A number must be greater than three.	$n > 3$
A number must be more than negative six.	$n > -6$

Less Than or Equal To:	
A number is at most nine.	$n \le 9$
The maximum is fourteen.	$n \le 14$

Greater Than or Equal To:	
A number is at least two.	$n \ge 2$
The minimum is eighteen.	$n \ge 18$

Example 6   Three-fourths of a number is at least eighteen. Translate to an inequality; then solve.

Understand   Because the word *of* is preceded by a fraction, it means multiplication. The key words *at least* indicate a greater-than or equal-to symbol.

Plan   Translate the key words; then solve. We'll use n for the variable.

Execute   Translate.

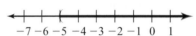

Three-fourths of a number is at least eighteen.

$$\frac{3}{4} \quad \cdot \quad n \qquad \ge \qquad 18$$

Solve: $\frac{3}{4} \cdot n \geq 18$

$$\frac{\overset{1}{\cancel{4}}}{\underset{1}{\cancel{3}}} \cdot \frac{\overset{1}{\cancel{3}}}{\underset{1}{\cancel{4}}} n \geq \frac{4}{\underset{1}{\cancel{3}}} \cdot \frac{\overset{6}{\cancel{18}}}{1}$$ **Multiply both sides by $\frac{4}{3}$ to isolate n.**

Answer $n \geq 24$

Check Verify that 24 and any number greater than 24 satisfy the original sentence.

Test 24:	**Test a number greater than 24, such as 25:**

Is $\frac{3}{4}$ of 24 at least 18?   Is $\frac{3}{4}$ of 25 at least 18?

$$\frac{3}{\underset{1}{\cancel{4}}} \cdot \frac{\overset{6}{\cancel{24}}}{1} \geq 18 \qquad\qquad \frac{3}{4} \cdot \frac{25}{1} \geq 18$$

$$18 \geq 18 \quad \text{True} \qquad\qquad \frac{75}{4} \geq 18$$

$$18\frac{3}{4} \geq 18 \quad \text{True}$$

Your Turn 6 0.2 times the sum of a number and eight is less than the difference of the number and five. Translate to an inequality; then solve.

Example 7 Darwin is planning a garden area, which he will enclose with a fence. Cost restricts him to a total of 240 feet of fencing materials. He wants the garden to span the entire width of his lot, which is 80 feet. What is the maximum length of the garden?

Understand The fence surrounds the garden; so 240 feet is the maximum perimeter for the garden. The width of the garden is to be 80 feet, and we must find the length.

Plan The formula for the perimeter of a rectangle is $P = 2l + 2w$. Because 240 feet is a *maximum* perimeter, we write the perimeter formula as an inequality so that the expression used to calculate perimeter ($2l + 2w$) is less than or equal to 240. Because the width is to be 80 feet, we replace w in the formula with 80.

Execute Translate:

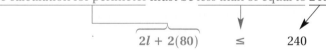

The calculation for perimeter **must be** less than or equal to 240.

$$2l + 2(80) \qquad \leq \qquad 240$$

Solve: $2l + 160 \leq 240$ **Subtract 160 from both sides.**

$$\frac{-160 \quad -160}{2l + 0 \leq 80}$$

$$\frac{2l}{2} \leq \frac{80}{2}$$ **Divide both sides by 2 to isolate l.**

$$l \leq 40$$

Answer The length must be less than or equal to 40 feet, which means that the maximum length is 40 feet.

Check Verify that a garden with a length of 40 feet or less and a width of 80 feet has a perimeter that is 240 feet or less. This will be left to the reader.

Answer   Your Turn 6

$0.2(n + 8) < n - 5$;

$n > 8.25$

Your Turn 7 The surface area of a box is to be at least 678 square centimeters. If the length is to be 12 centimeters and the width is to be 9 centimeters, find the range of values for the height that satisfies the minimum surface area of 678 square centimeters. ($SA = 2lw + 2lh + 2wh$)

2.6 Exercises For Extra Help *MyMathLab*

PRACTICE WATCH DOWNLOAD

Note: Exercises marked with a ★ represent challenging exercises.

1. What is a solution for an inequality?

2. Explain the difference between $x < 8$ and $x \leq 8$.

3. In your own words, explain the meaning of $x \geq -5$.

4. What action causes the direction of an inequality symbol to change?

For Exercises 5–16: *a. Write the solution set in set-builder notation.*
b. Write the solution set in interval notation. *See Examples 1 and 2.*
c. Graph the solution set.

5. $x \geq -3$

6. $n \leq -5$

7. $h < 6$

8. $x > 0$

9. $n < -\dfrac{2}{3}$

10. $a \geq \dfrac{4}{5}$

11. $t \geq 2.4$

12. $p \leq -0.6$

13. $-3 < x < 6$

14. $4 < c \leq 10$

15. $0 \leq n \leq 5$

16. $-1 < a < 5$

For Exercises 17–22, for each graph, write the inequality in set-builder notation and interval notation. See Objective 1.

17.
-4 -3 -2 -1 0 1 2 3 4 5

18.
-6 -5 -4 -3 -2 -1 0 1 2 3

19.
-4 -3 -2 -1 0 1 2 3 4 5

20.
-3 -2 -1 0 1 2 3 4 5 6

21.
-5 -4 -3 -2 -1 0 1 2 3 4

22.
-5 -4 -3 -2 -1 0 1 2 3 4

For Exercises 23–56: **a. Solve.**
b. Write the solution set in set-builder notation.
c. Write the solution set in interval notation. } **See Examples 3–5.**
d. Graph the solution set.

23. $n - 3 > 2$ **24.** $m - 5 < 20$ **25.** $z + 2 \le -4$ **26.** $x + 8 \ge -12$

27. $16y \ge 32$ **28.** $4z \le -20$ **29.** $-3x \ge -12$ **30.** $-4x < -16$

31. $\dfrac{2}{3}x \ge 4$ **32.** $\dfrac{3}{5}a < -6$ **33.** $-\dfrac{3}{4}m < -6$ **34.** $-\dfrac{5}{3}p > -10$

35. $5y + 1 > 16$ **36.** $3y - 2 > 10$ **37.** $1 - 6x < 25$ **38.** $3 - 2c \ge 17$

39. $\dfrac{a}{2} + 1 < \dfrac{3}{2}$ **40.** $\dfrac{k}{3} - \dfrac{2}{3} > -2$ **41.** $10 + 2f \le 2 - 2f$ **42.** $9 + 2n < 3 - 4n$

43. $3 - 6u \ge -5 - 2u$ **44.** $4 - 9k < -4k + 19$ **45.** $2(c - 3) < 3(c + 2)$ **46.** $4(d + 4) \ge 5(d + 2)$

47. $10(9w + 13) - 5(3w - 1) \le 8(11w + 12)$ **48.** $6(n + 13) - 2(4n - 2) > 3(5n - 1)$

49. $\dfrac{1}{2}(3x + 1) \le \dfrac{1}{3}(4x - 5)$ **50.** $\dfrac{1}{5}(6m - 7) > \dfrac{1}{2}(3m - 1)$

51. $\dfrac{1}{6}(2n + 4) - \dfrac{1}{3}(3n - 15) > 1$ **52.** $-\dfrac{1}{3}(2y - 5) + \dfrac{1}{6}(5y - 2) \ge 1$

53. $0.4l - 0.37 < 0.3(l + 2.1)$

54. $0.05 + 0.03(v - 4.5) \geq 0.465$

55. $0.6t + 0.2(t - 8) \leq 0.5t - 0.6$

56. $2.1s + 3(1.4s - 1.5) \geq 0.09s + 20.34$

For Exercises 57–68, translate to an inequality and then solve. See Example 6.

57. Four subtracted from a number is greater than twenty-four.

58. Six less than a number is less than or equal to negative four.

59. Four-ninths of a number is less than or equal to negative eight.

60. Three-fifths of a number is greater than negative twenty-one.

61. Eight times a number less thirty-six is at least sixty.

62. Three subtracted from a number times eleven is at least forty-one.

63. Five added to half of a number is at most two.

64. The sum of two-fifths of a number and two is at most eight.

65. The difference of four times a number and eight is less than two times the number.

66. Triple a number is less than the number decreased by eight.

67. Twenty-five is greater than or equal to seven more than six times a number.

68. Negative five is greater than fifteen less than five times a number.

For Exercises 69–82, solve. See Example 7. (Formulas are listed in the Summary on page 156.)

69. A school is planning a playground. The area of a rectangular playground may not exceed 170 square feet. If the width is 17 feet, what range of values can the length have?

70. Scotty needs to make a rectangular pool cover that is at least 128 square feet. If the pool length is 16 feet, what range of values must the width be?

71. The design of a storage box calls for a width of 27 inches and a length of 41 inches. If the surface area must be at least 4254 square inches, find the range of values for the height.

72. A company builds truck containers in the shape of rectangular solids. One such container has a length of 24 feet and width of 8 feet. What is the range of values for the height if the volume is to be at least 576 cubic feet?

73. Andre builds tables. One particular circular table requires him to glue a strip of veneer around the circumference of the top. He is designing a new circular table and has a strip of veneer that is 150 inches long that he wants to use. What range of values can the radius have? (Use $\pi \approx 3.14$.) Round your answer to the nearest hundredth.

74. The city parks department is putting in a circular flower garden that is to be enclosed using an edging material. If 125 feet of the edging material is available, what is the range of values for the diameter of the flower garden? Round your answer to the nearest hundredth.

75. Jon never exceeds 65 miles per hour when driving on the highway. If he is traveling 410 miles, what range of values can his time spent driving have?

76. A truck driver must make a delivery 298 miles away in five hours or less. What range of values can his average rate be so that he meets the schedule?

77. The fourth-quarter profit goal for a software company is at least $850,000. If it is known that the cost for the quarter will be $625,000, find the revenue that must be generated to achieve the goal. (Use $P = R - C$.)

78. Silver is a solid up to a temperature of 960.8°C. Write an inequality to describe this range of temperatures in degrees Fahrenheit. (Use $F = \dfrac{9}{5}C + 32$.)

79. To get an A in her math course, Tina needs the average of five tests to be at least 90. Her current test scores are 82, 91, 95, and 84. What range of scores on the fifth test would get her an A in the course?

80. In Aaron's English course, the final grade is determined by the average of five papers. The department requires any student whose average falls below 75 to repeat the course. Aaron's scores on the first four papers are 68, 78, 80, and 72. What range of scores on the fifth paper would result in him having to repeat the course?

81. The design of a circuit specifies that the voltage cannot exceed 12 volts. If the resistance of the circuit is 8 ohms, find the range of values that the current can have. (Use $V = iR$.)

82. A label in an elevator indicates that the maximum load is 9000 newtons; that is, the downward force should not exceed -9000 newtons. If the acceleration due to gravity is -9.8 meters per second squared, find the range of values of mass that can be loaded onto the elevator. Round your answer to the nearest hundredth. (Use $F = ma$.)

Puzzle Problem

Write the set of all values such that $|x| = -x$.

Review Exercises

Exercise 1 CONSTANTS AND VARIABLES

[1.1] **1.** Write a set representing the natural numbers up to 10.

Exercises 2–4 EXPRESSIONS

[1.4] **2.** Simplify: $\dfrac{2450}{140}$

[1.6] **3.** Translate to an expression: six more than twice the difference of a number and five.

[1.7] **4.** Evaluate $b^2 - 4ac$ when $a = 1$, $b = -2$, and $c = -3$.

Exercises 5 and 6 EQUATIONS AND INEQUALITIES

[2.3] *For Exercises 5 and 6, solve.*

5. $3\dfrac{1}{5}n = -108$

6. $5n = 7131.04$

Chapter 2 Summary

Defined Terms

Section 2.1
Equation (p. 92)
Solution (p. 92)
Formula (p. 94)
Perimeter (p. 94)
Area (p. 94)

Volume (p. 94)
Circumference (p. 94)
Radius (p. 94)
Diameter (p. 94)

Section 2.2
Linear equation (p. 105)
Linear equation in one
variable (p. 105)
Identity (p. 110)
Contradiction (p. 112)

Section 2.6
Linear inequality (p. 145)

Formulas

Perimeter of a rectangle: $P = 2l + 2w$ (p. 95)

Circumference of a circle: $C = \pi d$ or $C = 2\pi r$ (p. 95)

Area of a parallelogram: $A = bh$ (p. 95)

Area of a rectangle: $A = lw$ (p. 95)

Area of a triangle: $A = \dfrac{1}{2}bh$ (p. 95)

Area of a trapezoid: $A = \dfrac{1}{2}h(a + b)$ (p. 95)

Area of a circle: $A = \pi r^2$ (p. 95)

Surface area of a box: $SA = 2lw + 2lh + 2wh$ (p. 95)

Volume of a box: $V = lwh$ (p. 95)

Volume of a pyramid: $V = \dfrac{1}{3}lwh$ (p. 95)

Volume of a cylinder: $V = \pi r^2 h$ (p. 95)

Volume of a cone: $V = \dfrac{1}{3}\pi r^2 h$ (p. 95)

Volume of a sphere: $V = \dfrac{4}{3}\pi r^3$ (p. 95)

Distance, d, an object travels given its rate, r, and the time of travel, t: $d = rt$ (p. 98)

The average rate of travel, r, given the total time, t:
$r = \dfrac{d}{t}$ (p. 98)

Voltage, V, in a circuit with current, i, in amperes (A), and a resistance, R, in ohms (Ω): $V = iR$ (p. 98)

The temperature in degrees Celsius given degrees Fahrenheit: $C = \dfrac{5}{9}(F - 32)$ (p. 98)

The temperature in degrees Fahrenheit given degrees Celsius: $F = \dfrac{9}{5}C + 32$ (p. 98)

The profit, P, after cost, C, is deducted from revenue, R: $P = R - C$ (p. 131)

Procedure

LEARNING Strategy

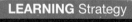 The summaries in the textbook are like the study sheet suggested in the To the Student section that precedes Chapter 1. Remember that your study sheet is a list of the rules, procedures, and formulas that you need to know. If you are a tactile or visual learner, spend a lot of time reviewing and writing the rules or procedures. Try to get to the point where you can write the essence of each rule and procedure from memory. If you are an audio learner, record yourself saying each rule and procedure; then listen to the recording over and over. Consider developing a clever rhyme or song for each rule or procedure to help you remember it.

Problem-Solving Outline

1. **Understand** the problem.
   a. Read the question(s) (not the whole problem, just the question at the end) and write a note to yourself about what you are to find.
   b. Read the whole problem, underlining the key words.
   c. If possible or useful, draw a picture, make a list or table to organize what is known and unknown, simulate the situation, or search for a related example problem.

2. **Plan** your solution by searching for a formula or using the key words to translate to an equation.

3. **Execute** the plan by solving the equation/formula.

4. **Answer** the question. Look at the note about what you were to find and make sure you answer that question. Include appropriate units.

5. **Check** results.
   a. Try finding the solution in a different way, reversing the process, or estimating the answer and making sure the estimate and actual answer are reasonably close.
   b. Make sure the answer is reasonable.

Procedures, Rules, and Key Examples

Procedures/Rules	Key Examples

SECTION 2.1 Equations, Formulas, and the Problem-Solving Process

To determine whether a value is a solution to a given equation, replace the variable in the equation with the value. If the resulting equation is true, the value is a solution.

Example 1: Check to see if the given value is a solution to the given equation.

a. $7n - 13 = 3n + 11; n = 6$

$$7(6) - 13 \stackrel{?}{=} 3(6) + 11$$
$$42 - 13 \stackrel{?}{=} 18 + 11$$
$$29 = 29 \quad \text{Yes, 6 is a solution.}$$

b. $2(x - 3) = 5x + 4; x = 1$

$$2(1 - 3) \stackrel{?}{=} 5(1) + 4$$
$$2(-2) \stackrel{?}{=} 5 + 4$$
$$-4 \neq 9 \quad \text{No, 1 is not a solution.}$$

To use a formula:
1. Replace the variables with the corresponding given values.
2. Solve for the missing value.

Example 2: Find the volume of a box with a length of 4 centimeters, width of 5 centimeters, and height of 6 centimeters.

$$V = lwh$$
$$V = (4\,\text{cm})(5\,\text{cm})(6\,\text{cm})$$
$$V = 120\,\text{cm}^3$$

For composite figures:
1. To calculate the area of a figure composed of two or more figures that are next to each other, add the areas of the individual figures.
2. To calculate the area of a region defined by a smaller figure within a larger figure, subtract the area of the smaller figure from the area of the larger figure.

Example 3: Find the shaded area in the figure shown.

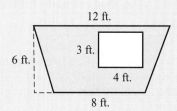

$$A = \text{Area of trapezoid} - \text{Area of rectangle}$$
$$A = \frac{1}{2}h(a + b) - lw$$
$$A = \frac{1}{2}(6)(12 + 8) - (3)(4)$$
$$A = \frac{1}{2}(6)(20) - 12$$
$$A = 60 - 12$$
$$A = 48\,\text{ft.}^2$$

SECTION 2.2 The Addition Principle of Equality

If $a = b$, then $a + c = b + c$ is true for all real numbers a, b, and c.

To use the addition principle of equality to clear a term in an equation, add the additive inverse of that term to both sides of the equation. (That is, add or subtract appropriately so that the term you want to clear becomes 0.)

Example 1: Solve.

a.
$$x + 9 = 15$$
$$\underline{-9 \quad -9}$$
$$x + 0 = 6$$
$$x = 6$$

b.
$$y - 5 = 7$$
$$\underline{+5 \quad +5}$$
$$y + 0 = 12$$
$$y = 12$$

(continued)

Procedures/Rules	Key Examples

SECTION 2.2   The Addition Principle of Equality (*continued*)

To solve linear equations requiring the addition principle only,
1. Simplify both sides of the equation as needed.
   a. Distribute to clear parentheses.
   b. Combine like terms.
2. Use the addition principle so that all variable terms are on one side of the equation and all constants are on the other side. Then combine like terms.

Tip: Clear the variable term that has the lesser coefficient to avoid negative coefficients.

c. $5(n - 2) + 3n = 7n - 4$

$$5n - 10 + 3n = 7n - 4$$
$$8n - 10 = 7n - 4$$
$$\underline{-7n \qquad\quad -7n}$$
$$n - 10 = 0 - 4$$
$$n - 10 = -4$$
$$\underline{+10 \quad +10}$$
$$n + 0 = \quad 6$$
$$n = 6$$

When solving a linear equation, if after simplifying each side of the equation the expressions are identical, the equation is an identity and every real number for which the equation is defined is a solution.

d. $5x - 9 - x = 4(x - 2) - 1$

$$4x - 9 = 4x - 8 - 1$$
$$4x - 9 = 4x - 9 \qquad \text{The equation is an identity, so every real number is a solution.}$$

When solving a linear equation, if after simplifying the expressions on each side of the equation the expressions have the same variable term but different constant terms, the equation is a contradiction and has no solution.

e. $4(2n - 3) - 5n + 2 = 9n - 2(3n - 6)$

$$\text{Distribute.}$$
$$8n - 12 - 5n + 2 = 9n - 6n + 12$$
$$\text{Combine like terms.}$$
$$3n - 10 = 3n + 12$$
$$\text{The equation is a contradiction, so it has no solution.}$$

SECTION 2.3   The Multiplication Principle of Equality

If $a = b$, then $ac = bc$ is true for all real numbers a, b, and c, where $c \neq 0$.

To use the multiplication principle of equality to clear a coefficient in an equation, multiply both sides of the equation by the multiplicative inverse of that coefficient or divide both sides by the coefficient.

Example 1: Solve.

a. $-5x = 15$

$$\frac{-5x}{-5} = \frac{15}{-5}$$
$$x = -3$$

b. $\dfrac{3}{4}n = \dfrac{5}{6}$

$$\frac{\overset{1}{\cancel{4}}}{\underset{1}{\cancel{3}}} \cdot \frac{\overset{1}{\cancel{3}}}{\underset{1}{\cancel{4}}} n = \frac{\overset{2}{\cancel{5}}}{\underset{3}{\cancel{6}}} \cdot \frac{\overset{2}{\cancel{4}}}{3}$$
$$n = \frac{10}{9}$$

To solve linear equations in one variable,
1. Simplify both sides of the equation as needed.
   a. Distribute to clear parentheses.
   b. Clear fractions or decimals by multiplying by the LCD. In the case of decimals, the LCD is the power of 10 with the same number of zero digits as decimal places in the number with the most decimal places. (Clearing fractions and decimals is optional.)
   c. Combine like terms.
2. Use the addition principle so that all variable terms are on one side of the equation and all constants are on the other side. (Clear the variable term with the lesser coefficient to avoid negative coefficients.) Then combine like terms.
3. Use the multiplication principle to clear any remaining coefficient.

c. $7n - (n - 3) = 2n - 5$

$$7n - n + 3 = 2n - 5$$
$$6n + 3 = 2n - 5$$
$$\underline{-2n \qquad\quad -2n}$$
$$4n + 3 = 0 - 5$$
$$4n + 3 = -5$$
$$\underline{-3 \quad -3}$$
$$4n + 0 = -8$$
$$\frac{4n}{4} = \frac{-8}{4}$$
$$n = -2$$

(continued)

Procedures/Rules	**Key Examples**

SECTION 2.4 Applying the Principles to Formulas

To isolate a particular variable in a formula, treat all other variables like constants and isolate the desired variable using the outline for solving equations.

Example 1: Isolate w in the formula $P = 2l + 2w$.

$$P = 2l + 2w$$
$$\underline{-2l \quad -2l}$$
$$P - 2l = 0 + 2w$$
$$\frac{P - 2l}{2} = \frac{2w}{2}$$
$$\frac{P - 2l}{2} = w$$

SECTION 2.5 Translating Word Sentences to Equations

To translate a word sentence to an equation, identify the unknown(s), constants, and key words; then write the corresponding symbolic form.

Example 1: Translate to an equation.
a. Five more than seven times a number is twelve.
Answer: $7n + 5 = 12$
b. Six times the difference of a number and nine is equal to four times the sum of a number and five.
Answer: $6(n - 9) = 4(n + 5)$

SECTION 2.6 Solving Linear Inequalities

To graph an inequality on a number line,
1. If the symbol is $\le$ or $\ge$, draw a bracket (or solid circle) on the number line at the indicated number open to the left for $\le$ and to the right for $\ge$. If the symbol is $<$ or $>$, draw a parenthesis (or open circle) on the number line at the indicated number open to the left for $<$ and to the right for $>$.
2. If the variable is greater than the indicated number, shade to the right of the indicated number. If the variable is less than the indicated number, shade to the left of the indicated number.

Example 1: Write the solution set in set-builder notation and interval notation; then graph the solution set.
a. $x \le 6$
Set-builder notation: $\{x \mid x \le 6\}$
Interval notation: $(-\infty, 6]$
Graph:

b. $x > -2$
Set-builder notation: $\{x \mid x > -2\}$
Interval notation: $(-2, \infty)$
Graph:

c. $-3 \le x < 1$
Set-builder notation: $\{x \mid -3 \le x < 1\}$
Interval notation: $[-3, 1)$
Graph:

To solve linear inequalities:
1. Simplify both sides of the inequality as needed.
   a. Distribute to clear parentheses.
   b. Clear fractions or decimals by multiplying through by the LCD just as we did for equations. (Clearing fractions and decimals is optional.)
   c. Combine like terms.
2. Use the addition principle so that all variable terms are on one side of the inequality and all constants are on the other side. Then combine like terms. (Remember, clearing the term with the lesser coefficient results in a positive coefficient.)
3. Use the multiplication principle to clear any remaining coefficient. If you multiply (or divide) both sides by a negative number, reverse the direction of the inequality symbol.

Example 2: Solve and write the solution set in set-builder notation and interval notation. Then graph the solution set.
a. $-6x > 24$

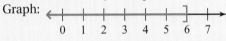

$$\frac{-6x}{-6} < \frac{24}{-6} \qquad \text{Reverse the direction of the inequality symbol.}$$

$$x < -4$$

Set-builder notation: $\{x \mid x < -4\}$
Interval notation: $(-\infty, -4)$
Graph:

(continued)

Procedures/Rules	Key Examples

SECTION 2.6   Solving Linear Inequalities (*continued*)

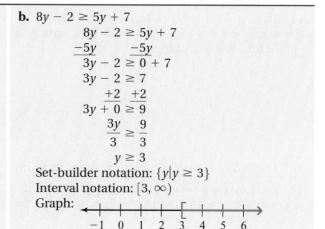

b. $8y - 2 \geq 5y + 7$

$$8y - 2 \geq 5y + 7$$
$$\underline{-5y \qquad -5y}$$
$$3y - 2 \geq 0 + 7$$
$$3y - 2 \geq 7$$
$$\underline{+2 \quad +2}$$
$$3y + 0 \geq 9$$
$$\frac{3y}{3} \geq \frac{9}{3}$$
$$y \geq 3$$

Set-builder notation: $\{y | y \geq 3\}$
Interval notation: $[3, \infty)$
Graph:

Chapter 2 Review Exercises

For Exercises 1–6, answer true or false.

[2.1]   **1.** $6x + 3y = 7x^2 - 4y$ is an equation.

[2.2]   **2.** $5(x + 2) = 10 - 3x + 8x$ is an identity.

[2.1]   **3.** 1 is a solution for $b^2 - 4b = 4$.

[2.5]   **4.** "Five less than a number is fourteen" translates to $5 - n = 14$.

[2.2]   **5.** $5x - 3y = 2$ is a linear equation.

[2.6]   **6.** $3 \leq x < 5$ is a compound inequality.

For Exercises 7–9, fill in the blank.

[2.2]   **7.** If an equation is a(n) _____, then every real number for which its expressions are defined is a solution.

[2.2]   **8.** An equation that is made of expressions in which each variable term contains a single variable raised to an exponent equal to 1 is a(n) _____ equation.

[2.3]   **9.** To solve equations,

    a. _____ both sides of the equation as needed.

       1. _____ to clear parentheses.

       2. Clear fractions or decimals by _____ by the _____.

       3. _____ like terms.

    b. Use the _____ principle so that all variable terms are on one side of the equation and all constants are on the other side.

    c. Use the _____ principle to clear any remaining coefficient.

[2.1]   **10.** In your own words, explain how to check a potential solution.

[2.1] *For Exercises 11–16, check to see if the given number is a solution to the equation.*

11. $y + 6 + 3y - 4 = 14$; $y = 3$

12. $\frac{1}{4}(n - 1) = \frac{1}{2}(n + 1)$; $n = 5$

13. $x^2 + 6.1 = 3x - 0.1$; $x = -0.1$

14. $\frac{5}{6}m - 1 = \frac{m}{3} + \frac{1}{2}$; $m = 3$

15. $\frac{3}{5}(a + 10) = -2$; $a = -20$

16. $6.1(7.2 + b) = 43.92$; $b = 0$

[2.2–2.3] *For Exercises 17–30, solve and check.*

17. $-6x = 30$

18. $-\frac{2}{3}y = -12$

19. $4n - 9 = 15$

20. $5(2a - 3) - 6(a + 2) = -7$

21. $12t + 9 = 4t - 7$

22. $2(c + 4) - 5c = 17 - 8c$

23. $1 - 2(3c - 5) = 10 - 6c$

24. $5(b - 1) - 3b = 7b - 5(1 + b)$

25. $5(u - 1) - (u - 2) = 10 - (2u + 1)$

26. $\frac{3}{4}m - \frac{1}{2} = \frac{2}{3}m + 1$

27. $\frac{1}{12}(4x + 9) = \frac{2}{3}x$

28. $\frac{5}{3} + 7x - 2x = \frac{1}{3}(9x + 2)$

29. $2 - \frac{2}{3}m = 6\left(m - \frac{1}{3}\right) - 4m + 2$

30. $1.4x - 0.5(9 - 6x) = 6 + 2.4x$

[2.4] *For Exercises 31–40, solve for the indicated variable.*

31. $x + y = 1$; for x

32. $a^2 + b^2 = c^2$; for a^2

33. $C = \pi d$; for d

34. $E = mc^2$; for m

35. $A = \frac{1}{2}bh$; for h

36. $F = \frac{mv^2}{r}$; for m

37. $V = \dfrac{1}{3}\pi r^2 h$; for h

38. $y = mx + b$; for m

39. $P = 2l + 2w$; for w

40. $A = \dfrac{1}{2}h(a + b)$; for h

[2.5] *For Exercises 41–44, translate to an equation and solve.*

41. Six times a number is the same as negative eighteen.

42. Three subtracted from half of a number is equal to nine less than one-fourth of the number.

43. One less than two times the sum of a number and four results in one.

44. Twice the difference of a number and one added to triple the number yields twenty less than six times the number.

[2.6] *For Exercises 45–50:* **a. Solve.**
b. Write the solution set in set-builder notation.
c. Write the solution set in interval notation.
d. Graph the solution set.

45. $-6x > 18$

46. $-3x + 2 \le 5$

47. $1 - 2z + 4 \le -3(z + 1) + 7$

48. $\dfrac{1}{4}v < 3 + v$

49. $\dfrac{1}{2}(m - 1) > 5 - \dfrac{7}{2}$

50. $2c - (5c - 7) + 4c \ge 8 - 10$

[2.6] *For Exercises 51–54, translate to an inequality and solve.*

51. Thirteen is greater than three minus ten times a number.

52. Three increased by twice a number is less than three times the difference of a number and five.

53. Negative one-half of a number is greater than or equal to four.

54. Negative two is less than or equal to one minus one-fourth of a number.

For Exercises 55–64, solve.

[2.1] **55.** Best Buy is selling an oak entertainment center. Elijah is concerned it will take up too much room in his den. Using the figure shown, determine the total area.

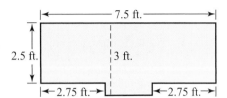

[2.1] **56.** Find the area of the figure shown.

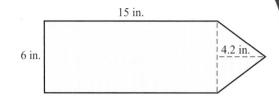

[2.1] **57.** A long-distance company uses the formula $C = 3 + 0.05m$ to describe a calling plan, where C is the total cost of a call lasting m minutes. If a customer uses 415 minutes, what is the total bill?

[2.3] **58.** Bonnie is making a shower curtain. The pattern says that the curtain will use 5760 square inches. If the length is to be 72 inches, how wide will the curtain be?

[2.3] **59.** The area of the trapezoid shown is 58 square inches. Find the missing length.

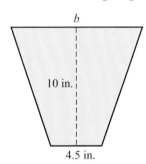

[2.3] **60.** Randy is snowskiing at Lake Tahoe. He knows that the trail is 1.8 miles long, and he is one-third of the way down. How far does he have left to ski?

[2.3] **61.** The surface area of a storage trunk is 2610 square inches. If the height of the trunk is 15 inches and the width is 36 inches, what is the length?

[2.6] **62.** The cost of border material limits the circumference of a circular flower garden to 10 feet. What is the range of values for the diameter of the garden? (Use $\pi \approx 3.14$.)

[2.6] **63.** Terrel has to travel at least 350 miles each day to stay on his schedule. If he averages 60 miles per hour, what is the range of values for his travel time?

[2.6] **64.** A company has a first-quarter profit goal of at least $350,000. If the projected cost is $475,000, what range of values can revenue have to meet the profit goal?

Chapter 2 Practice Test

For Extra Help CHAPTER Test Prep VIDEOS

Step-by-step test solutions are found on the Chapter Test Prep Videos available via the Video Resources on DVD, in *MyMathLab* , and on You Tube (search "CarsonElemAlg" and click on "Channels").

1. Is $5m - 14 = 21$ an expression or an equation? Why?

2. Is $5y + 14 = y^2 - 9$ linear or nonlinear? Why?

3. Check to see if 3 is a solution for $3x^2 + x = 10x$.

4. Check to see if $\frac{2}{3}$ is a solution for $4 - 3(x + 2) = 0$.

> **LEARNING Strategy**
>
> AUDITORY TACTILE VISUAL Because your actual test most likely will not have the same problems as the Practice Test, it is important that you understand how to solve each type of problem. Before taking the Practice Test, write the definition, rule, procedure, or formula that applies to each problem. This will help you focus on how to solve each type. Once you've done this, take a break and then take the test. Repeat this process until every problem on the Practice Test seems easy to solve.

For Exercises 5–10, solve and check.

5. $6x + 9 = -9$

6. $-5.6 + 8a = 7a + 7.6$

7. $8 + 5(x - 3) = 9x - (6x + 1)$

8. $1 - 3(c + 4) + 8c = 1 - 42 + 5(c + 7)$

9. $\frac{1}{2}x + \frac{5}{2} = \frac{3}{2} - x$

10. $-\frac{1}{8}(6m - 32) = \frac{3}{4}m - 2$

For Exercises 11–14, solve for the indicated variable.

11. $2x + y = 7; x$

12. $I = Prt; r$

13. $A = \frac{1}{2}bh; h$

14. $C = \frac{5}{9}(F - 32); F$

For Exercises 15 and 16, write the solution set in set-builder notation and interval notation; then graph the solution set.

15. $x \geq 3$

16. $-1 \leq x < 4$

For Exercises 17–20: a. Solve.
b. Write the solution set in set-builder notation.
c. Write the solution set in interval notation.
d. Graph the solution set.

17. $-5m + 3 > -12$

18. $-2(2 - x) \geq -20$

19. $6(p - 2) - 5 > 3 - 4(p + 6)$

20. $\dfrac{1}{5}(l + 10) \geq \dfrac{1}{2}(l + 5)$

For Exercises 21–24, translate and then solve.

21. Two-thirds of a number minus one-sixth is the same as twice the number.

22. The product of negative seven and a number added to twelve yields five.

23. Three subtracted from five times the difference of a number and two is equal to ten minus four times the difference of the number and one.

24. One minus a number is greater than twice the number.

For Exercises 25–30, solve.

25. A carpenter needs to determine the amount of wood required to trim a window. If the window is 4 feet by 6 feet, what is the perimeter of the window?

26. Find the area of the figure shown.

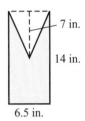

7 in.

14 in.

6.5 in.

27. The perimeter of the trapezoid shown is 63 centimeters. Find the length of the missing side.

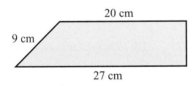

20 cm

9 cm

27 cm

28. A rent-to-own store advertises a television for $10 down with $10 payments each week. The formula $C = 10 + 10w$ represents the cost, C, after making payments for w weeks. If the customer signs a contract for one year, what is the total cost of the television?

29. The area of a sandbox cannot exceed 9 square feet. If the width of the sandbox is to be 3 feet, what range of values can the length have?

30. In Dan's psychology class, if his average on five tests is at least 92, he can be exempt from the final exam. His current scores are 90, 84, 89, and 96. What range of scores on the last test will allow him to be exempt from the final exam?

Chapters 1 and 2 Cumulative Review Exercises

For Exercises 1–4, answer true or false.

[1.3]   **1.** $(2 + 3) + 7 = 2 + (3 + 7)$

[1.1]   **2.** $-16 < -19$

[1.2]   **3.** 51 is a prime number.

[1.5]   **4.** $-2^4 = -16$

For Exercises 5–8, answer.

[2.1]   **5.** List the five steps for probem solving.

[1.5]   **6.** List the order in which we perform operations.

[1.7]   **7.** In your own words, explain how to evaluate an expression.

[1.1]   **8.** Graph $-2\frac{3}{5}$ on a number line.

[1.5] **For Exercises 9–14, simplify**

9. $62 - (-40)$

10. $-7(9)$

11. $|6 - 3(-2)|$

12. $5 - 3^2$

13. $2(-4) - 15 \div (-5)$

14. $6^2 + [19 - 3(4)][6 + 5(2)]$

[1.7]   **15.** Evaluate $3x - 4\sqrt{x + y}$ when $x = 9$ and $y = 16$.

[1.7]   **16.** Combine like terms: $6x^2 - 4x - 9x + 7x^2 - 16$

[2.3] **For Exercises 17–19, solve and check.**

17. $3b - 20 = 11$

18. $2(5 - n) = 7n - 17$

19. $2p + 3(p - 3) = 10p - (3p - 11)$

[2.4] **For Exercises 20 and 21, solve for the indicated variable.**

20. $B = P + Prt; r$

21. $V = \frac{1}{3} lwh; w$

[2.6] **For Exercises 22 and 23: a. Solve.**
b. Write the solution set in set-builder notation
c. Write the solution set in interval notation.
d. Graph the solution set.

22. $-3h - 8 \leq 19$

23. $9m - (2m + 3) < 4(m - 5) + 2$

Problem Solving

3

> " Difficulty, my brethren, is the nurse of greatness—a harsh nurse, who roughly rocks her foster-children into strength and athletic proportion. "
>
> —William Cullen Bryant (1794–1878)
> U.S. poet, editor

> " It is one of man's curious idiosyncrasies to create difficulties for the pleasure of resolving them. "
>
> —Joseph de Maistre (1753–1821)
> French diplomat, philosopher

In Chapter 2, we learned to solve linear equations and developed some techniques for solving problems, namely, using formulas and translating key words. In this chapter, we expand the problem-solving techniques introduced in Chapter 2 and solve problems that are a little more complex. These problems involve proportions, percents, and two unknown amounts.

Some of the exercises you encounter in this chapter will target a specific strategy such as translating key words or using a table to organize information. In the same way that an athlete uses exercise to target specific muscles, we use rather contrived problems to exercise specific problem-solving strategies and techniques. Most of the exercises an athlete performs in weight training are not sport-specific, meaning they are generic exercises designed to strengthen raw muscle tissue. Similarly, many of the problems in the chapter are not what you encounter in your job or everyday life. Rather, they are designed to exercise the strategies and techniques of problem solving.

3.1 Ratios and Proportions

Objectives

1. Solve problems involving ratios.
2. Solve for a missing number in a proportion.
3. Solve proportion problems.
4. Use proportions to solve for missing lengths in figures that are similar.

Objective 1   **Solve problems involving ratios.**   In this section, we solve problems involving **ratios**.

Definition   **Ratio:** A comparison of two quantities using a quotient.

Ratios are usually expressed in fraction form or with a colon. When a ratio is written in English, the word *to* separates the numerator and denominator quantities.

The ratio of **12** to **17** translates to $\dfrac{12}{17}$.

Numerator   Denominator

Note: The quantity preceding the word *to* is written in the numerator, and the quantity following *to* is written in the denominator.

Example 1   A small college has 2450 students and 140 faculty. Write the ratio of students to faculty in simplest form. Interpret the answer.

SOLUTION   The ratio of students to faculty $= \dfrac{2450}{140} = \dfrac{35}{2}$.

This means that there are 35 students for every 2 faculty members.

Your Turn 1   In a local referendum, 6400 people voted for increased school taxes and 4800 people voted against the increase. In simplest form, write the ratio of those voting for higher taxes to those voting against. Interpret the answer.

As noted, the ratio in Example 1 indicates that there are 35 students for every 2 faculty members. Expressing a ratio as a **unit ratio** often makes the ratio easier to interpret.

Definition   **Unit ratio:** A ratio with a denominator of 1.

Example 2   Express the ratio from Example 1 as a unit ratio. Interpret the answer.

SOLUTION $\dfrac{2450}{140} = \dfrac{35}{2} = \dfrac{17.5}{1}$ Divide and write the denominator as 1.

The unit ratio $\dfrac{17.5}{1}$ indicates that there are 17.5 students for every faculty member at the college.

Answer   **Your Turn 1**

$\dfrac{6400}{4800} = \dfrac{4}{3}$; for every four people in favor of higher taxes, three are against them.

Your Turn 2 The price of a 12.5-ounce box of cereal is $2.85. Write the unit ratio of price to weight. Interpret the answer.

Objective ② Solve for a missing number in a proportion. Some problems involve two equal ratios in a **proportion**.

Definition **Proportion:** An equation in the form $\dfrac{a}{b} = \dfrac{c}{d}$, where $b \neq 0$ and $d \neq 0$.

In Section 1.2, we discussed equivalent fractions and saw that fractions such as $\dfrac{3}{8}$ and $\dfrac{6}{16}$ are equivalent because $\dfrac{3}{8}$ can be rewritten as $\dfrac{6}{16}$ and $\dfrac{6}{16}$ can be reduced to $\dfrac{3}{8}$. Note that when fractions are equivalent, the cross products are equal.

$$16 \cdot 3 = 48 \qquad 8 \cdot 6 = 48$$

$$\frac{3}{8} \diagup\!\!\!\!\diagdown \frac{6}{16}$$

Let's see why all proportions have equal cross products. We begin with the general form of a proportion from the definition $\dfrac{a}{b} = \dfrac{c}{d}$, where $b \neq 0$ and $d \neq 0$. We can use the multiplication principle of equality to clear the fractions by multiplying both sides by the LCD. In this case, the LCD is bd.

$$\frac{\overset{1}{\cancel{b}d}}{1} \cdot \frac{a}{\underset{1}{\cancel{b}}} = \frac{\overset{1}{b\cancel{d}}}{1} \cdot \frac{c}{\underset{1}{\cancel{d}}}$$

$$ad = bc$$

Note that ad and bc are the cross products.

$$ad \qquad\qquad bc$$

$$\frac{a}{b} \diagup\!\!\!\!\diagdown \frac{c}{d}$$

Conclusion If two ratios are equal, their cross products are equal.

Rule **Proportions and Their Cross Products**

If $\dfrac{a}{b} = \dfrac{c}{d}$, where $b \neq 0$ and $d \neq 0$, then $ad = bc$.

We can use cross products to solve for a missing number in a proportion.

Procedure **Solving a Proportion**

To solve a proportion using cross products,
1. Calculate the cross products.
2. Set the cross products equal to each other.
3. Use the multiplication principle of equality to isolate the variable.

Answer **Your Turn 2**

$\dfrac{0.228}{1}$; the cereal costs $0.228 per ounce.

Example 3 Solve.

a. $\dfrac{7}{16} = \dfrac{x}{10}$

SOLUTION $10 \cdot 7 = 70$ $16 \cdot x = 16x$ Calculate the cross products.

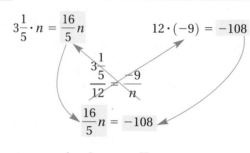

$$\dfrac{7}{16} = \dfrac{x}{10}$$

$$70 = 16x$$ Set the cross products equal to each other.

$$\dfrac{70}{16} = \dfrac{16x}{16}$$ Divide both sides by 16 to isolate x.

Note: We could have expressed the answer as 4375.

$$\dfrac{35}{8} = x$$

b. $\dfrac{3\frac{1}{5}}{12} = \dfrac{-9}{n}$

SOLUTION

$$3\dfrac{1}{5} \cdot n = \dfrac{16}{5}n \qquad 12 \cdot (-9) = -108 \qquad \text{Calculate the cross products.}$$

$$\dfrac{3\frac{1}{5}}{12} = \dfrac{-9}{n}$$

$$\dfrac{16}{5}n = -108 \qquad \text{Set the cross products equal to each other.}$$

$$\dfrac{\overset{1}{\cancel{5}}}{\underset{1}{\cancel{16}}} \cdot \dfrac{\overset{1}{\cancel{16}}}{\underset{1}{\cancel{5}}}n = \dfrac{-\overset{27}{\cancel{108}}}{1} \cdot \dfrac{5}{\underset{4}{\cancel{16}}} \qquad \text{Multiply both sides by } \dfrac{5}{16} \text{ to isolate } n.$$

$$n = -\dfrac{135}{4} \text{ or } -33\dfrac{3}{4}$$

Your Turn 3 Solve.

a. $\dfrac{9}{y} = -\dfrac{5}{6}$ **b.** $\dfrac{\frac{2}{3}}{n} = \dfrac{-5}{4\frac{1}{2}}$ **c.** $-\dfrac{2.4}{8.1} = -\dfrac{t}{12}$

Objective ③ Solve proportion problems. In a typical proportion problem, you are given a ratio and then asked to find an equivalent ratio. The key to translating problems to a proportion is to write the proportion so that the numerators and denominators correspond to each other in a logical way.

Example 4 A company determines that it spends $254.68 every 5 business days on equipment failure. At this rate, how much would the company spend over 28 business days?

Understand We are to find the amount spent on equipment failure over 28 days if the company spends $254.68 every 5 days. Translating $254.68 every 5 days to a ratio, we have $\dfrac{\$254.68}{5 \text{ days}}$.

Plan   *At this rate* indicates that $\dfrac{\$254.68}{5 \text{ days}}$ stays the same for the 28 days; so we can use a proportion. If we let n represent the unknown amount spent over the 28 days, we have a ratio of $\dfrac{\$n}{28 \text{ days}}$. We create the proportion by setting those ratios equal to each other.

Execute $\dfrac{\$254.68}{5 \text{ days}} = \dfrac{\$n}{28 \text{ days}}$

> **Note:** The units in the numerators and denominators correspond.
> **Tip:** One approach to setting up a proportion is to write the ratios so that the numerator units match and the denominator units match.

$$28 \cdot 254.68 = 7131.04 \qquad\qquad 5 \cdot n = 5n$$

$$\dfrac{254.68}{5} \diagup\!\!\!\!\!\diagdown \dfrac{n}{28}$$

$$7131.04 = 5n \qquad \text{Set the cross products equal.}$$

$$\dfrac{7131.04}{5} = \dfrac{5n}{5} \qquad \text{Divide both sides by 5 to isolate } n.$$

$$1426.208 = n$$

Answer   Because the answer is in dollars, we round to the nearest hundredth (or cent). The company spends $1426.21 over 28 business days.

Check   Do a quick estimate to verify the reasonableness of the answer. The company spends $254.68 every 5 business days. Since 28 days is a little less than 6 times the number of days, the company must spend a little less than 6 times the $254.68 spent in the 28 business days. To estimate the calculation, we round $254.68 to $250.00. We then have $6 \cdot \$250 = \$1500$, which is reasonably close to $1426.21.

Checking Proportion Problems

One of the nice things about proportion problems is that if you set up the proportion incorrectly, you usually get an unreasonable answer. A quick estimate check can be a fast way to recognize when you have set up a proportion incorrectly. For example, suppose we had incorrectly set up Example 4 this way:

$$\dfrac{\$254.68}{5 \text{ days}} = \dfrac{28 \text{ days}}{\$n}$$

> **Warning:** This setup is incorrect because the numerators and denominators do not correspond.

$$254.68 \cdot n = 254.68n \qquad\qquad 5 \cdot 28 = 140 \qquad \text{Cross multiply.}$$

$$\dfrac{254.68}{5} \diagup\!\!\!\!\!\diagdown \dfrac{28}{n}$$

$$254.68n = 140 \qquad\qquad \text{Set the cross products equal.}$$

$$\dfrac{254.68n}{254.68} = \dfrac{140}{254.68} \qquad\qquad \text{Divide both sides by 254.68 to isolate } n.$$

$$n \approx 0.55$$

> **Warning:** This answer indicates that the company spends only $0.55 over 28 business days based on spending $254.68 every 5 business days, which is clearly too low.

There are many correct ways to set up a proportion for a particular problem. For example, following are some other correct ways we could have set up the proportion for Example 4. Note that the proportions involve rates that correspond with each other.

$$\frac{5 \text{ days}}{\$254.68} = \frac{28 \text{ days}}{\$n}$$

Note: Think:

In 5 days, In 28 days,
$254.68 $n is spent.
is spent.

$$\frac{5 \text{ days}}{28 \text{ days}} = \frac{\$254.68}{\$n}$$

Note: Think of this version as an analogy:

If 5 days means $254.68, then 28 days means $n.

$$\frac{28 \text{ days}}{5 \text{ days}} = \frac{\$n}{\$254.68}$$

Note: This version is a rearrangement of the previous version.

Procedure   **Solving Proportion Application Problems**

To solve proportion problems,

1. Set up a proportion in which the numerators and denominators of the ratios correspond in a logical manner.
2. Solve using cross products.

Your Turn 4   Pam drove 351.6 miles using 16.4 gallons of gasoline. At this rate, how much gasoline would she need to drive 1000 miles?

Objective ④ **Use proportions to solve for missing lengths in figures that are similar.**   Consider triangles *ABC* and *DEF*.

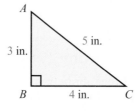
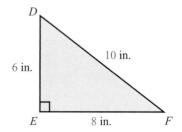

Although the triangles are different in size, their corresponding angle measurements are **congruent**.

Definition   ***Congruent angles:*** Angles that have the same measure. The symbol for congruent is ≅.

In symbols, we write $\angle B \cong \angle E$, $\angle A \cong \angle D$, and $\angle C \cong \angle F$.

Besides having all angles congruent, notice that the sides of triangle *DEF* are twice as long as the corresponding sides of triangle *ABC*. Or we could say that all of the sides of triangle *ABC* are half as long as the corresponding sides of triangle *DEF*. In other words, the corresponding side lengths are proportional.

Ratio of Triangle *DEF* to Triangle *ABC* $\longrightarrow \dfrac{6}{3} = \dfrac{8}{4} = \dfrac{10}{5} = \dfrac{2}{1}$

or

Ratio of Triangle *ABC* to Triangle *DEF* $\longrightarrow \dfrac{3}{6} = \dfrac{4}{8} = \dfrac{5}{10} = \dfrac{1}{2}$

Because the corresponding angles in the two triangles are congruent and the side lengths are proportional, we say that these triangles are **similar figures**.

Answer   Your Turn 4

≈ 46.6 gal.

Definition   ***Similar figures:*** Figures with congruent angles and proportional side lengths.

Example 5   The two trapezoids shown are similar. Find the missing lengths.

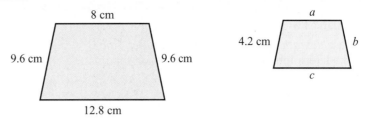

Understand   Because the figures are similar, the lengths of the corresponding sides are proportional. The corresponding sides can be shown as follows:

$$\text{Ratio of larger trapezoid} \atop \text{to} \atop \text{smaller trapezoid} \longrightarrow \frac{9.6}{4.2} = \frac{8}{a} = \frac{9.6}{b} = \frac{12.8}{c}$$

Plan   Write a proportion to find each missing side length.

Execute   It is not necessary to write a proportion to find b. Because two sides in the larger trapezoid measure 9.6 cm, the corresponding sides in the smaller trapezoid must be equal in length; so we can conclude that $b = 4.2$ cm.

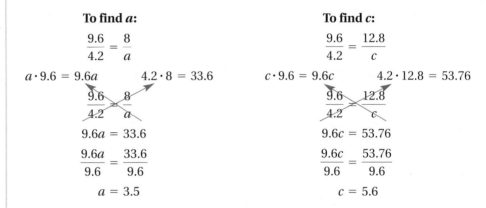

To find a:

$$\frac{9.6}{4.2} = \frac{8}{a}$$

$$a \cdot 9.6 = 9.6a \qquad 4.2 \cdot 8 = 33.6$$

$$\frac{9.6}{4.2} \diagup\!\!\!\!\diagdown \frac{8}{a}$$

$$9.6a = 33.6$$

$$\frac{9.6a}{9.6} = \frac{33.6}{9.6}$$

$$a = 3.5$$

To find c:

$$\frac{9.6}{4.2} = \frac{12.8}{c}$$

$$c \cdot 9.6 = 9.6c \qquad 4.2 \cdot 12.8 = 53.76$$

$$\frac{9.6}{4.2} \diagup\!\!\!\!\diagdown \frac{12.8}{c}$$

$$9.6c = 53.76$$

$$\frac{9.6c}{9.6} = \frac{53.76}{9.6}$$

$$c = 5.6$$

Answer   The missing lengths are $a = 3.5$ cm, $b = 4.2$ cm, and $c = 5.6$ cm.

Check   We can verify the reasonableness of the answers by doing quick estimates of the missing side lengths. Comparing the smaller trapezoid to the larger trapezoid, we see that the 4.2-centimeter side is a little less than half of the corresponding 9.6-centimeter side in the larger trapezoid. Each of the corresponding sides should follow this same rule. The 8-centimeter side of the larger trapezoid should correspond to a length of a little less than 4 for side a of the smaller trapezoid, and the 12.8-centimeter side should correspond to a length of a little less than 6.4 for side c on the smaller trapezoid. These do check out.

Your Turn 5   Find the missing lengths in the similar figures.

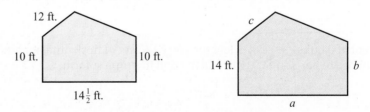

Answers   Your Turn 5

$a = 20\dfrac{3}{10}$ ft. or 20.3 ft.

$b = 14$ ft.

$c = 16\dfrac{4}{5}$ ft. or 16.8 ft.

3.1 Exercises For Extra Help MyMathLab

Note: Exercises marked with a ★ represent challenging exercises.

1. In the phrase *the ratio of the longest side to shortest side,* how do you determine the numerator and denominator of the ratio?

2. In your own words, define a proportion.

3. If $\dfrac{a}{b} = \dfrac{c}{d}$, where $b \neq 0$ and $d \neq 0$, is true, what can we say about the cross products ad and bc?

4. Explain how to solve for a missing number in a proportion.

5. How do you set up and solve a word problem that translates to a proportion?

6. Why are $\dfrac{5}{8} \cdot \dfrac{x}{10}$ and $\dfrac{5}{8} = \dfrac{x}{10}$ different?

For Exercises 7–12, write the ratio in simplest form. See Example 1.

7. One molecule of lactose contains 12 carbon atoms, 22 hydrogen atoms, and 11 oxygen atoms.
   a. Write the ratio of carbon atoms to total atoms in the molecule.

   b. Write the ratio of carbon atoms to hydrogen atoms.

   c. Write the ratio of hydrogen atoms to oxygen atoms.

> **Of Interest**
>
> Lactose is a molecule found in milk. Its chemical formula is $C_{12}H_{22}O_{11}$. Notice that the subscripts indicate the number of atoms of each element in the molecule.

8. One molecule of hematcin contains 16 carbon atoms, 12 hydrogen atoms, and 6 oxygen atoms.
   a. What is the ratio of hydrogen atoms to the total number of atoms in the molecule?

   b. What is the ratio of carbon atoms to hydrogen atoms?

   c. What is the ratio of carbon atoms to oxygen atoms?

9. The roof of a house rises 0.75 feet for every 1.5 feet of horizontal length. Write the ratio of rise to horizontal length as a fraction in simplest form.

10. A steep mountain road has a grade of 8%, which means that the road rises 8 feet vertically for every 100 feet it travels horizontally. Write the grade as a ratio in simplest form.

11. In a certain gear on a bicycle, the back wheel rotates $1\frac{1}{3}$ times for every 2 rotations of the pedals. Write the ratio of rotations of the back wheel to pedal rotations as a fraction in simplest form.

12. A recipe calls for $2\frac{1}{2}$ cups of flour and $1\frac{1}{3}$ cups of milk. Write the ratio of milk to flour as a fraction in simplest form.

For Exercises 13–20, write each as a unit ratio and interpret the answer. See Example 2.

13. Jamial drove 254 miles in 4 hours. What is the unit ratio of miles to hours? Interpret the answer.

14. Janine drove 199.2 miles and used 6 gallons of gas. What is the unit ratio of miles to gallons? Interpret the answer.

15. Six bananas cost 75 cents. What is the unit ratio of cost to bananas? Interpret the answer.

16. An 8-pound bag of ice costs $1.96. What is the unit ratio of cost to pounds of ice. Interpret the answer.

17. In 2009, Omega Corporation's stock earned $0.92 per share. The current price of the stock is $15.15. Write a unit ratio that expresses the price to earnings. What does this ratio indicate?

18. The price of a 10.5-ounce can of soup is $1.68. Write the unit ratio that expresses the price to weight. What does this ratio indicate?

★ **19.** An 8-ounce container of sour cream costs $1.42. A 24-ounce container of the same brand of sour cream costs $3.59. Which is the better buy? Why?

★ **20.** The price of a 12-ounce box of corn flakes is $3.59, and an 18-ounce box of the same cereal is $4.59. Which is the better buy? Why?

For Exercises 21 and 22, use the graph to the right. See Example 2.

21. a. Write the unit ratio of convictions to total cases in 1999.

   b. Write the unit ratio of convictions to total cases in 2004.

   c. Compare the ratios for 1999 to 2004. What do these ratios indicate?

22. a. Write the unit ratio of incarcerated to convictions in 1999.

   b. Write the unit ratio of incarcerated to convictions in 2004.

   c. Compare the ratios for 1999 to 2004. What do these ratios indicate?

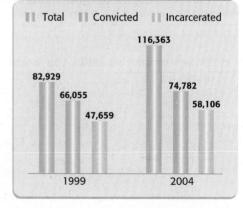

Source: Bureau of Federal Justice Statistics

For Exercises 23 and 24, use the following table. See Example 2.

Mean Earnings in the United States, 2005
Year-Round, Full-Time Workers by Educational Attainment and Age

	Not High School Graduate	High School Graduate or GED	Some College, No Degree	Associate Degree	Bachelor's Degree	Master's Degree	Professional Degree	Doctorate Degree
Male								
25 to 34 Years	$24,533	$33,509	$40,417	$42,200	$53,272	$69,601	$89,474	$72,164
35 to 44 Years	$29,059	$40,885	$49,499	$50,831	$75,812	$88,352	$151,687	$113,799
45 to 54 Years	$30,633	$43,638	$52,314	$53,700	$78,430	$87,248	$168,333	$108,212
55 to 64 Years	$32,225	$44,619	$54,629	$55,887	$79,764	$93,680	$171,879	$117,769
Female								
25 to 34 Years	$20,574	$26,132	$27,927	$32,144	$47,358	$49,694	$71,932	$67,582
35 to 44 Years	$21,980	$28,974	$33,814	$38,079	$51,028	$58,089	$105,850	$97,756
45 to 54 Years	$21,994	$29,729	$36,008	$40,785	$50,770	$62,874	$76,770	$78,396
55 to 64 Years	$21,803	$30,591	$36,032	$37,467	$47,571	$58,913	$78,242	$95,664

Source: Census 2005a

23. **a.** What is the unit ratio of the income for women in the 25–34 year age group with a bachelor's degree to men in the corresponding group? What does this ratio indicate?

b. Calculate the unit ratio for women in the 55–64 age group with a bachelor's degree with men in the corresponding group.

c. Compare the ratios you found for the two age groups in parts a and b. What do these ratios indicate?

24. **a.** What is the unit ratio of the income of men in the 45–54 age group with a bachelor's degree to those in the same age group that are not a high school graduate? What does this ratio indicate?

b. What is the unit ratio of the income of men in the 45–54 age group with a doctorate degree to those in the same age group that are not a high school graduate?

c. Compare the two ratios you found in parts a and b. What do these ratios indicate?

For Exercises 25–34, determine whether the ratios are equal. See Objective 2.

25. $\dfrac{15}{25} \stackrel{?}{=} \dfrac{3}{5}$

26. $\dfrac{6}{8} \stackrel{?}{=} \dfrac{15}{20}$

27. $\dfrac{20}{25} \stackrel{?}{=} \dfrac{25}{30}$

28. $\dfrac{12}{20} \stackrel{?}{=} \dfrac{18}{30}$

29. $\dfrac{7}{4} \stackrel{?}{=} \dfrac{15}{6.2}$

30. $\dfrac{15}{21} \stackrel{?}{=} \dfrac{130.2}{93}$

31. $\dfrac{10}{14} \stackrel{?}{=} \dfrac{2.5}{3.5}$

32. $\dfrac{6.5}{14} \stackrel{?}{=} \dfrac{10.5}{42}$

33. $\dfrac{\frac{2}{5}}{\frac{3}{10}} \stackrel{?}{=} \dfrac{15}{10\frac{1}{2}}$

34. $\dfrac{3\frac{1}{4}}{5\frac{1}{2}} \stackrel{?}{=} \dfrac{4\frac{2}{3}}{6\frac{1}{2}}$

For Exercises 35–50, solve for the missing number. See Example 3.

35. $\dfrac{2}{3} = \dfrac{7}{x}$

36. $\dfrac{7}{8} = \dfrac{y}{40}$

37. $\dfrac{-9}{2} = \dfrac{63}{n}$

38. $\dfrac{11}{m} = \dfrac{132}{-24}$

39. $\dfrac{21}{5} = \dfrac{h}{2.5}$

40. $\dfrac{29}{7} = \dfrac{k}{1.75}$

41. $\dfrac{-17}{b} = \dfrac{8.5}{4}$

42. $\dfrac{4.6}{3} = \dfrac{-23}{c}$

43. $\dfrac{-1.5}{60} = \dfrac{-2\frac{7}{8}}{t}$

44. $\dfrac{u}{-4\frac{7}{16}} = \dfrac{-60}{1.5}$

45. $\dfrac{2\frac{1}{4}}{6\frac{2}{3}} = \dfrac{d}{10\frac{2}{3}}$

46. $\dfrac{\frac{2}{5}}{4\frac{1}{2}} = \dfrac{7\frac{2}{3}}{j}$

★ **47.** $\dfrac{2}{3} = \dfrac{8}{x+8}$

★ **48.** $\dfrac{x+5}{6} = \dfrac{7}{3}$

★ **49.** $\dfrac{4x-4}{8} = \dfrac{2x+1}{5}$

★ **50.** $\dfrac{3}{4} = \dfrac{x+1}{x+3}$

For Exercises 51–74, solve. See Example 4.

51. LaTonia drives 110 miles in three hours. At this rate, how far will she go in five hours?

52. If 90 feet of wire weighs 18 pounds, what will 110 feet of the same type of wire weigh?

53. Jodi pays taxes of $1040 on her house, which is valued at $154,000. She is planning to buy a house valued at $200,000. At the same rate, how much will her taxes be on the new house?

54. Gary notices that his water bill was $24.80 for 600 cubic feet of water. At that rate, what would the charges be for 940 cubic feet of water?

55. A company determines that it spends $1575 every 20 business days (one month) for equipment maintenance. How much does the company spend in 120 business days (six months)?

56. A quality-control manager samples 1200 units of her company's product and finds seven defects. If the company produces about 50,000 units each year, how many of those units would she expect to have a defect?

57. Carson is planning a trip to London. He knows that it takes 1.8946 U.S. dollars to equal a British pound. If he has $800 to take with him, how much will he have in British pounds?

58. Sydney is planning a trip to Mexico. It takes 15.3827 Mexican pesos to equal 1 U.S. dollar. If she has $400 to take with her, how many pesos will she have?

59. Fletcher is building a model of an F-16 jet. If the scale is $1\frac{1}{2}$ inches to 1 foot, what length on the model represents 3.25 feet?

60. The legend on a map indicates that 1 inch = 500 feet. If a road measures $3\frac{3}{4}$ inches on the map, how long is the road in feet?

61. A recipe for pancakes calls for $1\frac{1}{2}$ teaspoons of sugar to make 12 pancakes. How many teaspoons of sugar should be used to make 20 pancakes?

62. A recipe for biscuits calls for $3\frac{1}{2}$ cups of flour to make 8 biscuits. How many cups of flour would be needed to make 12 biscuits?

63. Ford estimates that its 2008 Expedition will travel 504 highway miles on one tank of gas. If the gas tank of the Expedition holds 28 gallons, how far can a driver expect to travel on 20 gallons?

64. Chevrolet estimates that its 2008 Tahoe will travel 520 highway miles on one tank of gas. If the gas tank of the Tahoe holds 26 gallons, how far can a driver expect to travel on 20 gallons?

65. Gregory Arakelian of the United States holds the world record for fastest typing. In 1991, he typed 158 words per minute in the Key Tronic World Invitational Type-Off. If he could maintain that rate, how long would it take him to type a term paper with 2500 words? (*Source: Guinness Book of World Records*)

66. Sean Shannon of Canada holds the world record for the fastest recital of Hamlet's soliloquy "To be or not to be." On August 30, 1995, he recited the 260-word soliloquy in 23.8 seconds. If he could maintain that rate, how long would it take him to recite Shakespeare's entire script of Hamlet, which contains 32,241 words? (*Sources: Guinness Book of World Records*; http://william-shakespeare.info)

67. A consultant recommends that about 9 pounds of rye grass seed be used for every 1000 square feet. If the client's lawn is about 25,000 square feet, how many pounds of seed should he buy?

68. Mark used 3 gallons of paint to cover 1200 square feet of wall space in his home. Mark's neighbor wants to paint 2000 square feet. Based on Mark's rate, how many gallons should the neighbor buy?

★69. A carpet-cleaning company charges $49.95 to clean a rectangular living room that is 20 feet by 15 feet. The client asks the company also to clean the rectangular dining room, which measures 18 feet by 12 feet. If the company charges the same rate, how much should it cost to clean both the living room and dining room?

★70. A painter charged $320 to paint two walls that measured 12 feet by 9 feet and two walls that measured 10 feet by 9 feet. The client asks him to return to paint two walls that measure 15 feet by 12 feet and two walls that measure 18.5 feet by 12 feet. If the painter charges the same rate, what should the price be?

★ **71.** A state wildlife department wants to estimate the number of deer in a state forest. In one month, it captures 25 deer, tags them, and releases them. Later, it captures 45 deer and finds that 17 are tagged. Assuming that the ratio of tagged deer to total deer in the forest remains constant, estimate the number of deer in the preserve.

★ **72.** A marine biologist is studying the population of trout in a mountain lake. During one month, she catches 225 trout, tags them, and releases them. Later she captures 300 trout and finds that 48 are tagged. Assuming that the ratio of tagged trout to total trout in the lake remains constant, estimate the population of trout in the lake.

73. The Hoover Dam has 17 generators and can supply all of the electricity needed by a city of 750,000 people. If only 14 generators are running, how many people will be supplied with electricity?

74. Until the early 1950s, the erosion rate for Niagara Falls was an average of 3 feet per year. The suggested rate of erosion may now be as low as 1 foot per 10 years for Niagara Falls. Given this rate, how much erosion may be expected over the next 50 years?

For Exercises 75–78, find the missing lengths in the similar figures. See Example 5.

75.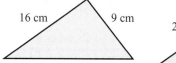

16 cm   9 cm

20 cm   x

76.

$12\frac{1}{2}$ ft.   15 ft.

$6\frac{1}{3}$ ft.   w

77.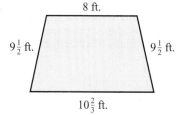

8 ft.

$9\frac{1}{2}$ ft. $9\frac{1}{2}$ ft.

$10\frac{2}{3}$ ft.

$5\frac{1}{4}$ ft.   a   b   c

78.

10.5 km   6.5 km

8 km

z

x   y

12.8 km

18.4 km

79. To estimate the height of the Eiffel Tower, a tourist uses the concept of similar triangles. The Eiffel Tower is found to have a shadow measuring 200 meters in length. The tourist has his own shadow measured at the same time. His shadow measures 1.2 meters. If he is 1.8 meters tall, how tall is the Eiffel Tower?

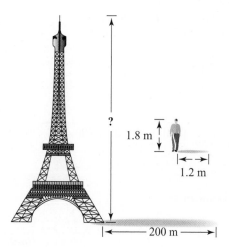

?

1.8 m

1.2 m

200 m

80. To estimate the height of the Great Pyramid in Egypt, the Greek mathematician Thales performed a procedure similar to that in Exercise 79, except that he supposedly used a staff instead of his body. Suppose the shadow from the staff was measured to be 6 feet and at the same time the shadow of the pyramid was measured to be $524\frac{2}{3}$ feet. If the staff was $5\frac{1}{2}$ feet tall, what was the height of the pyramid?

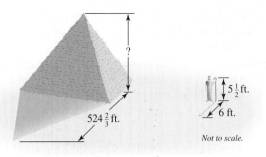

?

$524\frac{2}{3}$ ft.

$5\frac{1}{2}$ ft.

6 ft.

Not to scale.

Of Interest

The Greek mathematician Thales lived from 640 B.C. to 546 B.C. He is the first to have proposed the importance of proof and is considered to be the father of science and mathematics in the Greek culture. (*Source:* D. E. Smith, *History of Mathematics,* Dover, 1951)

81. To estimate the distance across a river, an engineer creates similar triangles. The idea is to create two similar right triangles as shown. Calculate the width of the river.

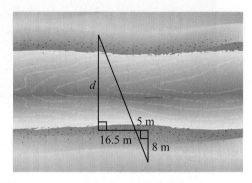

d

5 m

16.5 m

8 m

82. An engineer wants to estimate the distance across the Amazon River at his current location. The engineer creates two similar right triangles as shown. Calculate the distance across the river.

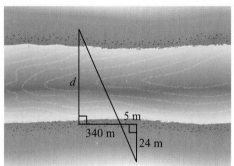

d

5 m

340 m

24 m

Of Interest

At 6275 km long, the Amazon River is the second-longest river in the world. Although the Nile River is slightly longer, the Amazon is the largest river in terms of watershed area, number of tributaries, and volume of water discharged.

Review Exercises

Exercise 1 CONSTANTS AND VARIABLES

[1.1] **1.** Is 0.58 a rational number? Why or why not?

Exercises 2–4 EXPRESSIONS

For Exercises 2–4, simplify.

[1.2] **2.** $\dfrac{42}{350}$

[1.4] **3.** $(0.15)(20.4)$

[1.4] **4.** $45.2 \div 100$

Exercises 5 and 6 EQUATIONS AND INEQUALITIES

For Exercises 5 and 6, solve.

[2.3] **5.** $0.75n = 240$

[2.3] **6.** $60p = 24$

3.2 Percents

Objectives

① Write a percent as a fraction or decimal number.

② Write a fraction or decimal number as a percent.

③ Translate and solve percent sentences.

④ Solve problems involving percents.

Objective ① **Write a percent as a fraction or decimal number.** The word *percent* is a compound word made from the prefix *per* and the suffix *cent*. The prefix *per* means "for each." The suffix *cent* comes from the Latin word *centum*, which means "100." Therefore, **percent** means "for each 100." In other words, percent is a ratio with the denominator 100.

Definition   *Percent:* A ratio representing some part out of 100.

The symbol for percent is %. For example, 20 percent, which means 20 out of 100, is written 20%. Note that the definition can be used to write percents as fractions or decimal numbers.

$$20\% = 20 \text{ out of } 100$$
$$= \frac{20}{100}$$
$$= \frac{1}{5} \text{ or } 0.2$$

Procedure   **Rewriting a Percent**

To write a percent as a fraction or decimal,

1. Write the percent as a ratio with 100 in the denominator.
2. Simplify to the desired form.

Note: When simplifying, remember that dividing a decimal number by 100 moves the decimal point two places to the left.

Example 1   Write each percent as a decimal and as a fraction in simplest form.

a. 20.5%

SOLUTION   Write 20.5 over 100; then simplify. In this case, it might be helpful to write the decimal number first and then write the equivalent fraction from that decimal number.

$$20.5\% = \frac{20.5}{100} \qquad \text{Write as a ratio with 100 in the denominator.}$$
$$= 0.205 \qquad \text{Write the decimal form.}$$
$$= \frac{205}{1000} \qquad \text{Write the fraction form.}$$
$$= \frac{41}{200} \qquad \text{Simplify to lowest terms.}$$

b. $30\dfrac{1}{4}\%$

SOLUTION Write $30\dfrac{1}{4}$ over 100; then simplify. In contrast to part a, you could find the fraction form first and then find the decimal form.

$$30\dfrac{1}{4}\% = \dfrac{30\dfrac{1}{4}}{100} \qquad \text{Write as a ratio with 100 in the denominator.}$$

$$= 30\dfrac{1}{4} \div 100 \qquad \text{Rewrite the division.}$$

$$= \dfrac{121}{4} \cdot \dfrac{1}{100} \qquad \text{Write an equivalent multiplication.}$$

$$= \dfrac{121}{400} \text{ or } 0.3025 \qquad \text{Multiply.}$$

Your Turn 1 Write each percent as a fraction in simplest form and as a decimal.

a. 26% **b.** 9.25% **c.** $30\dfrac{2}{3}\%$

Objective ② **Write a fraction or decimal number as a percent.** We have seen that $20\% = \dfrac{20}{100} = \dfrac{1}{5}$ or 0.2. To write a percent as a fraction or a decimal, we replace the % sign with division by 100. To write a fraction or decimal as a percent, we should multiply by 100% and simplify.

Procedure **Writing a Fraction or Decimal as a Percent**

To write a fraction or decimal number as a percent,

1. Multiply by 100%.
2. Simplify.

Note: Multiplying a decimal number by 100 moves the decimal point two places to the right.

Example 2 Write as a percent. If necessary, round your answer to the nearest tenth of a percent.

a. $\dfrac{5}{8}$

SOLUTION $\dfrac{5}{8} = \dfrac{5}{\overset{8}{\underset{2}{\cancel{8}}}} \cdot \dfrac{\overset{25}{\cancel{100}}}{1}\%$ Multiply by 100%.

$$= \dfrac{125}{2}\% \qquad \text{Simplify.}$$

$$= 62\dfrac{1}{2}\% \text{ or } 62.5\%$$

b. 0.267

SOLUTION $0.267 = 0.267 \cdot 100\%$ Multiply by 100%.

$$= 26.7\%$$

Connection Multiplying an amount by 100% is a way of rewriting the amount because multiplying by 100% is equivalent to multiplying by 1. The percent symbol means $\dfrac{1}{100}$; so 100% literally means

$$100 \cdot \% = 100 \cdot \dfrac{1}{100} = 1.$$

a. $\dfrac{5}{9}$

b. 2.4

Objective ③ Translate and solve percent sentences. To solve problems involving percents, it is often helpful to reduce the problem to the following simple percent sentence:

A **percent** of a **whole** is a **part** of the whole.

Note that there are three pieces in the simple sentence: the *percent*, the *whole*, and the *part*. Any one of those pieces could be unknown in a problem.

	A **percent** of	a **whole**	is a **part**.
Unknown part:	42% of	68	is what amount?
Unknown whole:	76% of	what number	is 63.84?
Unknown percent:	What percent of	72	is 63?

> **Note:** The whole follows the word *of* because we are finding a percentage of the whole.

To solve for an unknown piece of a simple percent sentence, we can translate the sentence to an equation. We will look at two methods of translation: translating word for word and translating to a proportion.

Procedure

Translating Simple Percent Sentences

Method 1. Translate the sentence word for word.

1. Select a variable for the unknown.
2. Translate the word *is* to an equal sign.
3. If *of* is preceded by the percent, translate it to multiplication.
   If *of* is preceded by a whole number, translate it to division.

Method 2. Translate to a proportion by writing the following form:

$$\text{Percent} = \frac{\text{Part}}{\text{Whole}},$$

where the percent is expressed as a fraction with a denominator of 100.

First, let's consider the case where the part is unknown.

Simple Percent Sentence with an Unknown Part

Example 3 42% of 68 is what number?

> **Note:** This question can also be worded this way: "What is 42% of 68?"

SOLUTION Notice that the part is the unknown.

42%   of   68   is   what number?

Percent   of   the whole   is   the part.

Method 1. Word-for-word translation: Because *of* is preceded by the percent, it means multiply, and *is* means equals.

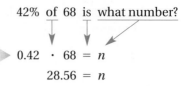

42% of 68 is what number?

Note: To calculate, we write the percent as a decimal (or fraction).

$0.42 \cdot 68 = n$

$28.56 = n$

Method 2. Proportion: The percent, which is a ratio, is equivalent to the ratio of the part to the whole; so we can write a proportion in the following form:

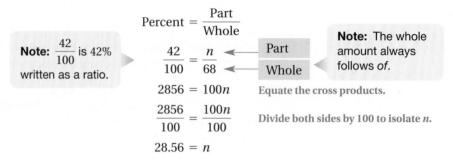

$$\text{Percent} = \frac{\text{Part}}{\text{Whole}}$$

Note: $\frac{42}{100}$ is 42% written as a ratio.

$\dfrac{42}{100} = \dfrac{n}{68}$ ← Part ← Whole

Note: The whole amount always follows *of*.

$2856 = 100n$ Equate the cross products.

$\dfrac{2856}{100} = \dfrac{100n}{100}$ Divide both sides by 100 to isolate n.

$28.56 = n$

Your Turn 3 What number is 15% of 36?

Simple Percent Sentence with an Unknown Whole

Now consider the situation in which the whole amount is unknown.

Example 4 76% of what number is 63.84?

Note: This question can be worded "63.84 is 76% of what number?"

SOLUTION Note the three pieces:

76%	of	what number	is	63.84?
Percent	of	the whole	is	the part.

Method 1. Word-for-word translation:

76% of what number is 63.84?

$0.76 \cdot n = 63.84$

$0.76n = 63.84$

$\dfrac{0.76n}{0.76} = \dfrac{63.84}{0.76}$

$n = 84$

Method 2. Proportion:

$$\text{Percent} = \frac{\text{Part}}{\text{Whole}}$$

Note: This is 76% expressed as a ratio.

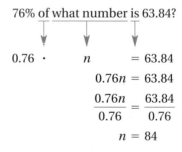

$\dfrac{76}{100} = \dfrac{63.84}{n}$ ← Part ← Whole

$76n = 6384$ Equate the cross products.

$\dfrac{76n}{76} = \dfrac{6384}{76}$ Divide both sides by 76 to isolate n.

$n = 84$

Answer   Your Turn 3

5.4

Your Turn 4   48 is 120% of what number?

Simple Percent Sentence with an Unknown Percent

Finally, consider the case in which the percent is the unknown.

Example 5   What percent of 72 is 63?

SOLUTION   Note the pieces.

What percent   of   72   is   63?

Percent   of   the whole   is   the part.

> **Note:** This question can be worded
> "63 is what percent of 72?"
> or
> "What percent is 63 out of 72?"
> or
> "What percent is 63 of 72?"
> Note how 72, the whole amount,
> always follows *of*.

Method 1. Word-for-word translation:

What percent of 72 is 63?

$$p \cdot 72 = 63$$
$$72p = 63$$
$$\frac{72p}{72} = \frac{63}{72}$$
$$p = 0.875$$

> **Note:** The division
> yields a decimal
> number that *must* be
> written as a percent.

To write 0.875 as a percent, we multiply by 100%.

Answer $0.875 \cdot 100\% = 87.5\%$

Method 2. Proportion:

> **Note:** This is the unknown
> percent written as a ratio. We use
> an uppercase P here because the
> 100 in the proportion causes the
> solution to be slightly different
> from the solution to the equation
> in the word-for-word translation.

$$\frac{P}{100} = \frac{63}{72} \quad \begin{array}{l} \leftarrow \text{Part} \\ \leftarrow \text{Whole} \end{array}$$
$$72P = 6300 \qquad \text{Equate the cross products.}$$
$$\frac{72P}{72} = \frac{6300}{72} \qquad \text{Divide both sides by 72 to isolate } p.$$
$$P = 87.5$$

> **Note:** In calculating the cross
> products, we multiplied by
> 100; so the solution is the
> percent value, and we need to
> add only the percent sign.

Answer   87.5%

As Example 5 shows, when the percent is unknown, the proportion method gives the answer as a percent; so you may want to consider using proportions in this case.

Your Turn 5   21.6 is what percent of 90?

Objective ④ **Solve problems involving percents.**   If a problem involves a percent, it can be helpful to try to write the information in the form of a simple percent sentence.

Answer   Your Turn 4

40

Answer   Your Turn 5

24%

Solving Percent Application Problems

To solve problems involving percent,

1. Determine whether the percent, the whole, or the part is unknown.
2. Write the problem as a simple percent sentence (if needed).
3. Translate to an equation (word for word or proportion).
4. Solve for the unknown.

Example 6   In an English class, 12 of the 28 students received an A. What percent of the class received an A?

Understand   The unknown is the percent of the class that received an A. The number of students in the class is 28, which is the whole. The number of students that received an A, which is the part, is 12.

Plan   Write a simple percent sentence; then solve. We'll let P represent the unknown percent, and we'll translate to a proportion because the proportion's solution will be the percent value.

Execute   What percent of 28 is 12?

$$\frac{P}{100} = \frac{12}{28} \quad \leftarrow \text{Part}$$
$$\phantom{\frac{P}{100} = \frac{12}{28}} \quad \leftarrow \text{Whole}$$

$$28P = 1200 \qquad \text{Equate the cross products.}$$

$$\frac{28P}{28} = \frac{1200}{28} \qquad \text{Divide both sides by 28 to isolate } P.$$

$$P \approx 42.9 \qquad \text{\textbf{Note:} We rounded to the nearest tenth. We need to add the \% sign.}$$

Answer   About 42.9% of the students in the class received an A on their first paper.

Check   Verify that 42.9% of the 28 students is about 12 students.

$$0.429 \cdot 28 = 12.012$$

Because 12.012 is reasonably close to 12, we can conclude that 42.9% is correct.

Your Turn 6   On Monday, January 18, 2010, the *Morning Sentinel* reported that it planned to cut 100 full-time jobs in a bid to return to profitability. This is 6% of the newspaper's workforce. How many employees does the *Sentinel* have prior to this layoff?

Percent Problems Involving Increase or Decrease

Sometimes, a percent problem involves an increase or decrease of some initial amount. Sales tax, interest, and discount are situations in which an initial amount of money is increased or decreased by an amount of money that is a percent of the initial amount.

Example 7   If the sales tax rate in a certain city is 6%, what is the total cost of the following items?

Raisin Bran	$3.79
Milk	$1.29
Orange Juice	$1.49
Yogurt	$2.59

Understand   The total cost is the purchase price plus the sales tax, which is a percent of the purchase price.

Answer   Your Turn 6

1667

Plan   We must first calculate the purchase price. We can then write a simple percent sentence and solve for the sales tax. Finally, we add the tax to the purchase price.

Execute   Purchase price $= 3.79 + 1.29 + 1.49 + 2.59 = \$9.16$

Simple percent sentence: The sales tax is 6% of $9.16. We will translate word for word, letting x represent the unknown sales tax.

$$x = 0.06 \cdot 9.16$$
$$x = 0.5496$$

Sales tax is always rounded up to the next cent, so the sales tax is $0.55.

Answer   The total cost is $9.16 + \$0.55 = \$9.71$.

Check   Determine whether the total cost of $9.71 represents a 6% increase in the total sale of $9.16. To determine the percent of increase, we need to calculate the amount of the increase by subtracting 9.16 from 9.71.

$$9.71 - 9.16 = 0.55 \qquad \text{This verifies that the final addition was correct.}$$

Now to determine the percent of the increase, we can set up a proportion.

$$\frac{P}{100} = \frac{0.55}{9.16} \quad \longleftarrow \begin{array}{l} \text{Part} \\ \text{Whole} \end{array}$$

$$9.16P = 55 \qquad \textbf{Equate the cross products.}$$

$$\frac{9.16P}{9.16} = \frac{55}{9.16} \qquad \textbf{Divide both sides by 9.16 to isolate } P.$$

$$P \approx 6.004 \qquad \blacktriangleleft \quad \textbf{Note:} \text{ This does not match 6\% exactly because we used the rounded sales tax amount, 0.55, instead of the exact value, 0.5496.}$$

Connection The steps in this check help us understand how to approach a type of problem in which the percent of increase or decrease is unknown. We will see this type of problem in Example 8.

Your Turn 7   An advertisement indicates that a coat is on sale at 20% off the marked price. If the marked price is $79.95, what will be the price after the discount?

Connection A clever shortcut for Example 7 is to recognize that we pay 100% of the initial price *plus* 6% in sales tax, so the total is 106% of the initial price.

Total cost $= 1.06 \cdot 9.16 = 9.7096$, which rounds to $9.71.

A similar shortcut exists for the Your Turn problem involving a discount. If we take 30% off the marked price, we pay 70% of the marked price.

Price after discount $= 0.7 \cdot 79.95 = 55.965$, which rounds to $55.97.

The check in Example 7 is a common type of percent problem in which the percent of increase or decrease is unknown.

Example 8   On May 1, 2006, the median price of a house sold in Jacksonville, Florida, reached an all-time high of $322,990. On April 7, 2008, the median price was $253,990. What was the percent of decrease? (*Source:* www.housingtracker.net)

Understand   We are to determine the percent of decrease given the initial price and the price after the decrease.

Plan   To determine the percent of decrease, we need the amount of the decrease, which is found by subtracting the final amount from the initial amount.

Answer   Your Turn 7

$63.96

Execute   Amount of decrease = 322,990 − 253,990 = 69,000

Simple percent sentence: What percent of 322,990 is 69,000?

> **Note:** The whole is the original amount (before the decrease).

Because the percent is unknown, we translate to a proportion, letting P represent the number of the unknown percent.

$$\frac{P}{100} = \frac{69,000}{322,990} \quad \leftarrow \text{Part} \\ \leftarrow \text{Whole}$$

$322,990P = 6,900,000$ Equate the cross products.

$\dfrac{322,990P}{322,990} = \dfrac{6,900,000}{322,990}$ Divide both sides by 322,990 to isolate P.

$P \approx 21.36$

Answer   The percent of decrease in the median price of a house in Jacksonville was about 21.36%.

Check   Determine whether decreasing the initial median price of $322,990 by 21.36% results in a price of $253,990.

$0.2136(322,990) = n$

$68,990.66 = n$

$69,000 \approx n$

> **Note:** When rounded to the nearest thousand, this amount of decrease corresponds to our earlier calculation.

Now subtract the amount of the decrease: 322,990 − 69,000 = 253,990.

In general, when solving for a percent of increase or decrease, the *part* is the amount of increase or decrease and the *whole* is the initial amount.

$$\frac{P}{100} = \frac{\text{Amount of increase/decrease}}{\text{Initial amount}}$$

Your Turn 8   The Air Transportation Association reports that the per-gallon price for jet fuel jumped from $2.10 in 2007 to $2.85 in 2008. What is the percent of increase?

Answer   **Your Turn 8**

35.7%

3.2 Exercises   For Extra Help   *MyMathLab*

Note: Exercises marked with a ★ represent challenging exercises.

1. Explain how to write a percent as a decimal or fraction.

2. Explain how to write a decimal or fraction as a percent.

3. When preceded by a percent, the word *of* indicates what operation?

4. Explain how to translate a simple percent sentence to a proportion.

5. When using word-for-word translation to find a percent, the calculation will not be a percent, whereas if you use the proportion method, the result will be a percent. Why?

6. When given an initial amount and a final amount after an increase, how do you find the percent of the increase?

For Exercises 7–18, write each percent as a decimal and as a fraction in simplest form. See Example 1.

7. 20%

8. 30%

9. 15%

10. 85%

11. 14.8%

12. 18.6%

13. 3.75%

14. 6.25%

15. $45\frac{1}{2}\%$

16. $65\frac{1}{4}\%$

17. $33\frac{1}{3}\%$

18. $18\frac{1}{6}\%$

For Exercises 19–38, write as a percent rounded to the nearest tenth if necessary. See Example 2.

19. $\frac{3}{5}$

20. $\frac{1}{5}$

21. $\frac{3}{8}$

22. $\frac{5}{8}$

23. $\frac{5}{6}$

24. $\frac{4}{9}$

25. $\frac{2}{3}$

26. $\frac{5}{11}$

27. 0.96

28. 0.42

29. 0.8

30. 0.7

31. 0.09

32. 0.01

33. 1.2

34. 3.58

35. 0.028

36. 0.065

37. 4.051

38. 0.007

For Exercises 39–60, translate word for word or translate to a proportion; then solve. See Examples 3–5.

39. 80% of 35 is what number?

40. What number is 40% of 90?

41. 2.5% of 124 is what number?

42. 13.5% of 940 is what number?

43. What number is $9\frac{1}{4}\%$ of 64?

44. $3\frac{3}{4}\%$ of 24 is what number?

45. 120% of 86 is what number?

46. What number is 250% of 62.8?

47. 30% of what number is 15?

48. 45 is 5% of what number?

49. 16.4 is 20.5% of what number?

50. 30.2% of what number is 18.12?

51. 7.8 is $12\frac{1}{2}\%$ of what number?

52. $5\frac{1}{4}\%$ of what number is 1.26?

53. What percent of 45 is 9?

54. 15 is what percent of 40?

55. What percent of 38 is 39.9?

56. 73.44 is what percent of 68?

★**57.** What percent is 18 of 27?

★**58.** 50 of 60 is what percent?

★**59.** What percent is 12 of 15?

★**60.** 18 of 24 is what percent?

For Exercises 61–66, solve. See Example 6.

61. Angela answers 86% of the questions on a psychology multiple choice test correctly. If there were 50 questions on the test, how many did she answer correctly?

62. Terra is a server at a restaurant. She serves a table of seven people, for which the restaurant automatically adds a service charge of 15% of the cost of the meal. If the meal costs $127.40, how much is the service charge?

63. A county's annual property taxes are 4% of the market value of the property. How much is the tax on a property valued at $120,000? If the tax is paid in monthly installments, what amount must be paid each month?

64. How much is the tax on a house valued at $230,000 in the same county as that in Exercise 63? If the tax is paid in monthly installments, what amount must be paid each month?

65. In general, lenders do not want home buyers to spend more than 28% of their gross monthly income on a house payment. If a buyer has a monthly gross income of $3210, what is the most a lender will allow for a mortgage payment?

66. In general, lenders do not want home buyers to spend more than 36% of gross income a month on debts. Using the buyer in Exercise 65, what is the most a lender will accept in total debt payments per month? If she makes the highest mortgage payment allowed each month, what does she have left to spend on other debt and still be within the lender's guidelines?

Of Interest
In finance, the ratio of the house payment to the gross monthly income is known as the *front-end ratio*. The ratio of the monthly debts to the gross monthly income is known as the *back-end ratio*. The front-end and back-end ratios, along with a person's credit history, are primary factors in determining loan qualification.

For Exercises 67 and 68, solve using the circle graph. See Example 6.

67. How many teragrams of carbon dioxide were emitted by the United States?

68. How many teragrams of methane were emitted by the United States?

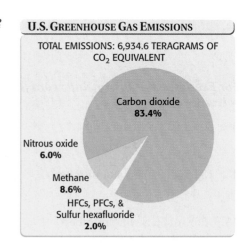

U.S. GREENHOUSE GAS EMISSIONS

TOTAL EMISSIONS: 6,934.6 TERAGRAMS OF CO_2 EQUIVALENT

Carbon dioxide
83.4%

Nitrous oxide
6.0%

Methane
8.6%

HFCs, PFCs, &
Sulfur hexafluoride
2.0%

Source: U.S. Environmental Protection Agency

For Exercises 69–104, solve. See Examples 6–8.

69. Terra, the server from Exercise 62, receives a tip of $8.50. If this was 20% of the cost of the meal, what was the cost of the meal?

70. When a camera is sold on eBay for $50.00 or less, the seller must pay eBay 6% of the selling price. If eBay charges $2.70 for the sale of a camera, what was the selling price of the camera?

71. A theater charges 2.5% of the ticket price as a handling fee to hold tickets at the will-call booth. If the handling fee for four tickets is $7.50, what is the total cost of the tickets?

72. Carol's checking account earns 0.025% of her average daily balance as a dividend every month. If she receives a dividend of 15 cents one month, what was her average daily balance?

73. During his career, Detroit Pistons player Isiah Thomas made 7194 field goals out of 15,904 attempts. What percent of his field goal attempts did he score?

74. In 2008, Chipper Jones of the Atlanta Braves had the number 1 batting average in the National League with 160 hits out of 439 at-bats. What percent of his at bats were hits?

75. According to the 2000 census, the total population of the United States was 281,421,906. If there were 196,899,193 people 21 years of age and older, what percent of the total population was 21 years of age and older? (*Source:* Bureau of the Census, Department of Commerce, 2000 Census)

76. In 2006, 55,521,868 households out of 111,617,402 households in the United States were headed by married couples. What percent of U.S. households was headed by married couples in 2006? (*Source:* Bureau of the Census, Department of Commerce)

For Exercises 77–80, use the following table, which shows the energy use in some of the countries that use the most energy. The unit used is the amount of energy produced by one metric ton of oil. (Source: Firefly's World of Facts, 2007)

Country	Oil	Gas	Coal	Nuclear	Hydroelectric	Total
United States	944.6	570.1	575.4	185.9	60.6	2336.6
China	327.3	42.3	1081.9	11.8	90.8	1554.0
Russia	130.0	364.6	111.6	33.9	39.6	679.6

77. What percent of the energy use in the United States comes from nuclear energy?

78. What percent of the energy use in China comes from coal?

79. For each country, find the percent of energy use that comes from gas. Which country has the greatest percentage?

80. For each country, find the percent of energy use that comes from the least polluting form of energy, hydroelectric. Which country has the smallest percentage?

81. A small company has the following costs in one quarter.

Category	Cost (in thousands)
Plant operations	124.2
Employee wages	248.6
Research and development	115.7

What percent of the company's total cost is spent on research and development?

82. A pharmaceutical company conducts a study on the side effects of a new drug. Following is a report of the data from the study.

Side Effect	Number of Subjects
No side effects	1462
Headache	35
Rash	24
Fever	65

What percent of the subjects in the study experienced fever?

★ **83.** Margaret mixes 50 milliliters of a solution that is 10% HCl with 100 milliliters of a solution that is 25% HCl. What percent of the resulting solution is HCl?

84. Leon invests $3400 in stock that returns 12% the first ★ year and $2200 in stock that returns 15% during the same year. What is his total return as a percent of his total investment?

85. Carl purchases $124.45 in groceries. If the sales tax rate is 5%, what is the amount of the tax and the cost after tax is added?

86. Juanita purchases several books over the Internet that have a total price of $85.79. The online store charges a handling fee of 2.5% of the total cost. What is the handling fee and the price after the fee is added?

87. An appliance store has all refrigerators discounted 10% off the regular price. If the regular price of a refrigerator is $949, what is the amount of the discount and the price after the discount?

88. A furniture store has all sofas discounted 40% off. If the original price of a particular sofa is $785.99, what is the amount of the discount and the price after the discount?

★ **89.** Johanna paid a total of $2675 for a sofa, which included 7% sales tax. What was the price of the sofa before the sales tax was added?

90. Paul paid a total of $1908 for a refrigerator, which in- ★ cluded 6% sales tax. What was the price of the refrigerator before the sales tax was added?

91. Video game sales in a department store increased from $2600 in November to $4550 in December. What was the amount of increase? What was the percent of increase?

92. Fifteen years ago James inherited a marble-top table. It was appraised at $420. The insurance company asked that he have it reappraised. It is now worth $1200. What was the amount of the increase? What was the percent of increase?

93. From 1989–1990, about 580,000 bachelor's degrees were awarded to women in the United States. From 2005–2006, about 845,000 bachelor's degrees were awarded to women. What was the percent of increase in degrees awarded to women? (*Source:* National Center of Education Statistics, U.S. Department of Education)

94. On May 14, 2009, the Dow Jones Industrial Average (DJIA) closed at 8331.72, and on May 5, it closed at 8268.64. What was the percent of decrease?

95. The price of a TI-84 calculator drops from $120 to $79. What is the percent of decrease?

96. After the Vallina family purchased a smaller car, their monthly gasoline bill dropped from $225 per month to $170 per month. What was the percent of decrease in the monthly gasoline bill?

97. In 1990, the carbon monoxide concentration in Baltimore was 7.2 parts per million, and in 2006, the concentration was 2.1 parts per million. What was the percent of decrease in the carbon monoxide concentration? (*Source:* U.S. Environmental Protection Agency, Office of Air Quality Planning and Standards)

98. In 1990, the population of Decatur, Illinois, was 83,900. In 2006, the population was 77,047. What was the percent of decrease in population? (*Source:* U.S. Bureau of the Census, Department of Commerce)

99. In 1938, the federal hourly minimum wage was $0.25. In 2009, the hourly minimum wage was $7.25. What was the percent of increase in the minimum wage from 1938 to 2009? (*Source:* U.S. Department of Labor)

100. A small business owner has an audit performed and finds that the net worth of the company is $58,000. After two years of poor business, an audit shows the net worth is $-20,000$. What is the percent of the decrease in the net worth of the business?

101. The U.S. Bureau of the Census reported a 475% increase in the number of single fathers from 1970 to 2006. If the number of single fathers in 2006 was 2,300,000, how many single fathers were there in 1970? (*Source:* U.S. Census Report, June 18, 2006)

102. The projected population of Florida for the year 2030 is 30.1 million, which is a 986.25% increase from the population in 1950. What was the population in 1950? (*Source:* Floridians for a Sustainable Population, 2008)

103. On "Black Monday," October 19, 1987, the stock market closed at 1739, which was a 22.6% decrease from the previous day's closing price. What was the closing price on Friday, October 16?

104. On March 28, 2008, the closing price of Disney stock was at $31, which was a 24.3% decrease from March 28, 2000. What was the closing price on March 28, 2000?

Collaborative Exercises Occupation Growth

Complete the following table by calculating the amount of change and the percent of the increase for each occupation; then answer the questions. Round to the nearest tenth of a percent.

The 10 Fastest-Growing Occupations, 2002–2012

Occupation	Employment		Change	
	2002	2012	Amount	Percent
Medical assistants	364,600	579,400		
Network systems and data communications analysts	186,000	292,000		
Physician assistants	63,000	93,800		
Social and human service assistants	305,200	453,900		
Home health aides	579,700	858,700		
Medical records and health information technicians	146,900	215,600		
Physical therapist aides	37,000	54,100		
Computer software engineers, applications	394,100	573,400		
Computer software engineers, systems software	281,100	408,900		
Physical therapist assistants	50,200	72,600		

Source: Bureau of Labor Statistics, Office of Occupational Statistics and Employment Projections

1. In 2002, in which occupation were the greatest number of people employed?

2. By 2012, which occupation is projected to have the greatest number of people employed?

3. Explain why the two occupations you listed in Exercises 1 and 2 are not at the top of the list.

4. Which is a better indicator of the demand for people in a particular occupation: the number of people employed in a particular year, the amount of change projected from 2002 to 2012, or the percent of increase in employment from 2002 to 2012?

5. Based on your conclusions in Exercise 4, what occupation will have the greatest demand? What college majors might have the greatest potential for employment in that occupation? Write your conclusions and present them to the class.

Review Exercises

Exercises 1–3 **EXPRESSIONS**

[1.5] **1.** Simplify: $-3°$

[1.7] **2.** Multiply: $-3(x + 7)$

[1.7] **3.** Combine like terms: $7xy + 4x - 8xy + 3y$

Exercises 4–6 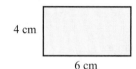 **EQUATIONS AND INEQUALITIES**

For Exercise 4, translate to an equation and then solve.

[2.5] **4.** Three times the sum of a number and nine is equal to eleven less than the number.

[3.1] **5.** Solve $\dfrac{1}{5} = \dfrac{n}{10.5}$.

[3.1] **6.** The figures shown are similar. Find the missing side length.

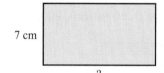

3.3 Problems with Two or More Unknowns

Objectives

1 Solve problems involving two unknowns.

2 Use a table in solving problems with two unknowns.

Objective 1 Solve problems involving two unknowns. Sometimes problems have two or more unknown amounts. In general, if there are two unknowns, there will be two relationships. Our approach to solving these problems is to use the relationships to write an equation that we can solve. The following outline gives a more specific approach.

Procedure **Solving Problems with Two or More Unknowns**

To solve problems with two or more unknowns,

1. Determine which unknown will be represented by a variable.
   Tip: Let the unknown that is acted on be represented by the variable.
2. Use one of the relationships to describe the other unknown(s) in terms of the variable.
3. Use the other relationship to write an equation.
4. Solve the equation.

First, let's look at problems in which we simply translate key words.

Example 1 The greater of two numbers is twice the lesser number. If the sum of the two numbers is 57, find the numbers.

Understand There are two unknowns in the problem: We are to find the greater and lesser numbers. We are given two relationships.

Relationship 1: The greater number is twice the lesser number.

Relationship 2: The sum of the two numbers is 57.

Plan   Translate the relationships to an equation and then solve.

Execute   Use the first relationship to determine which unknown will be represented by a variable and represent the other unknown in terms of that variable.

Relationship 1: The greater number is twice the lesser number.

$$\text{Greater number} = 2 \cdot \text{lesser number}$$

$$\text{Greater number} = 2n$$

Note: We usually choose the unknown amount that is acted on to be the variable. Because the greater number is twice the lesser number, we let n represent the lesser number.

Now use the second relationship to write an equation.

Relationship 2: The sum of the two numbers is 57.

$$\text{Greater number} + \text{lesser number} = 57$$

$$2n \qquad + \qquad n \qquad = 57$$

$$3n = 57 \qquad \text{Combine like terms.}$$

$$\frac{3n}{3} = \frac{57}{3} \qquad \text{Divide both sides by 3.}$$

$$n = 19$$

Answer   Because n represents the lesser number, the lesser number is 19. The greater number is twice the lesser number, so the greater number is $2 \cdot 19 = 38$.

Check   Verify that the sum of the two numbers is 57. $19 + 38 = 57$.

Your Turn 1   The greater of two numbers is 3 times the lesser number. If the sum of the two numbers is 36, find the numbers.

Example 2   Lindsey eats a snack consisting of 1 ounce of almonds and one oatmeal fruit bar. She is surprised to see that the ounce of almonds has 40 more calories than the fruit bar. If her snack has a total of 300 calories, how many calories are in the almonds and how many are in the fruit bar?

Understand   There are two unknowns in the problem: We must find the number of calories in 1 ounce of almonds and the number of calories in an oatmeal fruit bar. We are given two relationships.

Relationship 1: The number of calories in the almonds is 40 more than the number of calories in the fruit bar.

Relationship 2: The total calories in the snack are 300.

Plan   Translate the relationships to an equation and then solve.

Execute   Use the first relationship to determine which unknown will be represented by a variable.

Relationship 1: The number of calories in the almonds is 40 more than the number of calories in the fruit bar.

$$\text{Calories in almonds} = 40 \text{ calories} + \text{Calories in the fruit bar}$$

$$\text{Calories in almonds} = \qquad 40 \qquad + \qquad n$$

Note: We choose the unknown amount that is acted on to be the variable. Because 40 is added to the calories in the fruit bar, we let n represent the calories in a fruit bar.

Now use the second relationship to write an equation that we can solve for n.

Relationship 2: The total calories are 300.

$$\text{Calories in the fruit bar} + \text{Calories in the almonds} = 300$$

$$n \qquad + \qquad 40 + n \qquad = 300$$

$2n + 40 = 300$	Combine like terms.
$2n + 40 = 300$	
$\dfrac{-40 \quad\;\; -40}{2n + 0 = 260}$	Subtract 40 from both sides.
$\dfrac{2n}{2} = \dfrac{260}{2}$	Divide both sides by 2.
$n = 130$	

Answer Because n represents the number of calories in the fruit bar, 1 fruit bar has 130 calories. The almonds contain 40 more calories than the fruit bar, so the almonds contain $40 + 130 = 170$ calories.

Check Verify that the total calories is 300, which it is because $170 + 130 = 300$.

Your Turn 2 In a year, the average person eats 12 pounds more of chicken than pork. If the average person eats a total of 108 pounds of chicken and pork, how many pounds of each does the average person eat in a year?

Geometry Problems

Sometimes, problems do not give all of the needed relationships in an obvious manner. This often occurs in geometry problems in which the definition of a geometry term provides the needed relationship. For example, we must know that *perimeter* means "the sum of the lengths of the sides" in order to solve a problem that involves perimeter.

Example 3 A rectangular frame is to be built so that the length is 4 inches more than the width. The perimeter of the frame is to be 72 inches. Find the dimensions of the frame.

Understand Draw a picture and list the relationships.

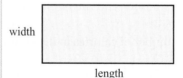
width

length

Relationship 1: The length is 4 inches more than the width.

Relationship 2: The perimeter must be 72 inches.

Plan Translate to an equation using the key words and then solve the equation.

Execute Relationship 1: The length is 4 inches more than the width.

$$\text{Length} = \text{Width} + 4$$
$$\text{Length} = w + 4$$

Note: Because 4 is added to the width and the length is isolated, we let w represent width.

Now use the other given relationship to write an equation to solve.

Relationship 2: The perimeter is 72 inches. Recall that the formula for the perimeter of a rectangle is $P = 2l + 2w$, where l represents the length and w represents the width.

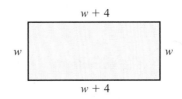

$$\text{Perimeter} = 72$$
$$2l + 2w = P$$

$2(w + 4) + 2w = 72$	Substitute $w + 4$ for l and 72 for P.
$2w + 8 + 2w = 72$	Distribute.
$4w + 8 = 72$	Combine like terms.
$\underline{ -8 \quad -8}$	Subtract 8 from both sides to isolate the
$4w + 0 = 64$	variable term.
$\dfrac{4w}{4} = \dfrac{64}{4}$	Divide both sides by 4 to isolate w.
$w = 16$	

Answer   The width, w, is 16 inches. To determine the length, use Relationship 1. If the length $= w + 4$, then the length $= 16 + 4 = 20$ inches.

Check   Verify that both conditions in the problem are satisfied. First, the length must be 4 inches more than the width, and 20 inches is 4 inches more than 16 inches. Second, the perimeter must be 72 inches, which is true because $2 \cdot 16 + 2 \cdot 20 = 32 + 40 = 72$ inches.

Your Turn 3   Karen is constructing a rectangular flower garden with a wooden border. She wants the length to be twice the width and has 30 feet of border material. Find the dimensions of the garden.

In Example 3, the word *perimeter* was a key word that helped us write the equation that we solved. Other terms in geometry, by their definitions, give information to help us translate a problem into an equation. For example, a problem might involve angles that are **complementary** or **supplementary**.

Definition   ***Complementary angles:*** Two angles are complementary if the sum of their measures is 90°.

In the following figure, $\angle ABD$ and $\angle DBC$ are complementary because $32° + 58° = 90°$.

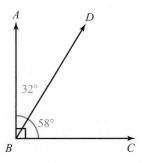

Note: The small square indicates a 90° angle.

Definition   ***Supplementary angles:*** Two angles are supplementary if the sum of their measures is 180°.

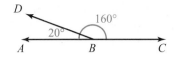

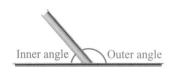

Note: The second relationship comes from the sketch. It shows that the inner and outer angles are supplementary, which means that the sum of their measures is 180°.

A straight line forms an angle that measures 180°. Any line that intersects a straight line will divide the 180° angle into two angles that are supplementary. In the figure in the margin, line segment *DB* joins line segment *AC*, forming two angles, $\angle ABD$ and $\angle DBC$. These two angles are supplementary angles because 20° + 160° = 180°.

Now look at a problem containing one of these terms.

Example 4   A steel beam is to be welded to a crossbeam, creating two angles, an inner angle and an outer angle. If the outer angle is to be 15° less than twice the inner angle, what are the angle measurements?

Understand   We must find the inner and outer angle measurements. The sketch for this situation is shown in the margin. (Notice that it is like the preceding figure.)

Next, list the relationships.

Relationship 1: The outer angle is 15° less than twice the inner angle.

Relationship 2: Inner angle + Outer angle = 180

Plan   Translate the relationships to an equation and then solve.

Execute   Relationship 1: The outer angle is 15° less than twice the inner angle.

$$\text{Outer angle} = 2 \cdot \text{Inner angle} - 15$$
$$\text{Outer angle} = 2a - 15$$

Note: Because the measure of the inner angle is multiplied by 2, we should let the variable represent the inner angle. Let's use the letter a.

Now use relationship 2 to write an equation that we can solve.

$$\text{Inner angle} + \text{Outer angle} = 180$$
$$a + 2a - 15 = 180$$

$3a - 15 = 180$	Combine like terms.
$\underline{+15 \quad +15}$	Add 15 to both sides.
$3a + 0 = 195$	
$\dfrac{3a}{3} = \dfrac{195}{3}$	Divide both sides by 3 to isolate a.
$a = 65$	

Answer   The inner angle, represented by a, is 65°. According to relationship 1, the outer angle = $2a - 15$; so the outer angle = $2(65) - 15 = 115°$.

Check   Verify that all of the conditions in the problem are satisfied. The outer angle measuring 115° is 15° less than twice the inner angle measuring 65°. The sum of the inner and outer angles is 180°, which verifies that they are supplementary.

Your Turn 4   The wooden joists supporting a roof are connected as shown. The measure of $\angle DBC$ is 6° less than three times the measure of $\angle ABD$. Find the measures of $\angle DBC$ and $\angle ABD$.

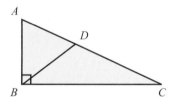

Problems Involving Consecutive Integers

Another group of problems in which the relationships are not obvious involve *consecutive integers*. The word *consecutive* means that the numbers are in sequence. For example, beginning with −3, consecutive integers are −3, −2, −1, 0, 1, 2, ...

Answer   Your Turn 4

$\angle DBC = 66°$
$\angle ABD = 24°$

What relationship exists in the pattern of consecutive integers? Notice that to get from one integer in the sequence to the next integer, we add 1. For the sake of finding the general relationship, suppose the first integer in a sequence of consecutive integers is 1. By noting how we get from the number 1 to its consecutive integer, we can find a general pattern when the first integer is an unknown number, n, as the following table illustrates.

	With 1 as the First Integer	With n as the First Integer
First Integer	1	n
Next Integer	$1 + 1 = 2$	$n + 1$
Next Integer	$1 + 2 = 3$	$n + 2$

From the table, we see that the pattern for consecutive integers is $n, n + 1$, $n + 2, \ldots$.

Now let's determine a general relationship for a sequence of consecutive odd integers, such as 1, 3, 5, 7, ... As we see in the following table, adding 2 to an odd integer gives the next consecutive odd integer.

	With 1 as the First Odd Integer	With n as the First Odd Integer
First Odd Integer	1	n
Next Odd Integer	$1 + 2 = 3$	$n + 2$
Next Odd Integer	$1 + 4 = 5$	$n + 4$

Warning: The pattern $n, n + 1, n + 3, \ldots$ is *not correct* for consecutive odd integers.

So the pattern for consecutive odd integers is $n, n + 2, n + 4, \ldots$.

As we see in the following table, the pattern for consecutive even integers is the same as for consecutive odd integers. Adding 2 to an even integer gives the next consecutive even integer.

	With 2 as the First Even Integer	With n as the First Even integer
First Even Integer	2	n
Next Even Integer	$2 + 2 = 4$	$n + 2$
Next Even Integer	$2 + 4 = 6$	$n + 4$

The pattern for consecutive even integers is $n, n + 2, n + 4, \ldots$ which is the same pattern as that for consecutive odd integers.

Example 5 The sum of three consecutive even integers is 78. Find the integers.

Understand The key word *sum* means to "add." If we let n represent the smallest of the even integers, the pattern for three unknown consecutive even integers is as follows:

Smallest unknown even integer: n

Next even integer: $n + 2$

Third even integer: $n + 4$

Plan Translate to an equation and then solve.

Execute Translate: The sum of three consecutive even integers is 78.

$$\boxed{\begin{array}{c}\text{Smallest even}\\\text{integer}\end{array}} + \boxed{\begin{array}{c}\text{Next even}\\\text{integer}\end{array}} + \boxed{\begin{array}{c}\text{Third even}\\\text{integer}\end{array}} = 78$$

$$n \quad + \quad n + 2 \quad + \quad n + 4 \quad = 78$$

$$3n + 6 = 78 \qquad \text{Combine like terms.}$$

$$\underline{\quad -6 \quad -6 \quad} \qquad \text{Subtract 6 from both sides.}$$

$$3n + 0 = 72$$

$$3n = 72$$

$$\frac{3n}{3} = \frac{72}{3} \qquad \begin{array}{l}\text{Divide both sides by 3 to}\\\text{isolate } n.\end{array}$$

$$n = 24$$

Answer The smallest of the three unknown integers, represented by n, is 24.

$$\text{Next even integer: } n + 2 = 24 + 2 = 26$$
$$\text{Third even integer: } n + 4 = 24 + 4 = 28$$

Check The numbers 24, 26, and 28 are three consecutive even integers, and $24 + 26 + 28 = 78$.

Your Turn 5 The sum of three consecutive integers is 96. Find the integers.

Objective ❷ Use a table in solving problems with two unknowns. We have seen how using key words, drawing pictures, making lists, and looking for patterns have helped us understand problems. Now we consider a group of problems in which a table is helpful in organizing information.

Example 6 A home-improvement store sells two sizes of cans of wood stain. A small can sells for $8.95, and a large can sells for $15.95. One day, the store sold twice as many of the large cans as the small cans. If the total revenue that day for these cans of wood stain was $694.45, how many cans of each size were sold?

Understand We know that the total revenue is $694.45. From this we can say

$$\boxed{\begin{array}{c}\text{Revenue from}\\\text{the small cans}\end{array}} + \boxed{\begin{array}{c}\text{Revenue from}\\\text{the large cans}\end{array}} = 694.45$$

We can also describe the revenue from the sale of each size can. We know the price of each size. If we knew the number of cans sold, we could multiply that number by the price per can to calculate the revenue from that particular size.

$$\boxed{\begin{array}{c}\text{Price}\\\text{per can}\end{array}} \cdot \boxed{\begin{array}{c}\text{Number of}\\\text{cans sold}\end{array}} = \boxed{\begin{array}{c}\text{Revenue for}\\\text{that size can}\end{array}}$$

With so much information, it is helpful to use a table to list it all. We use a four-column table with the labels *Categories*, *Price per Can*, *Number of Cans*, and *Revenue*. The categories are the two sizes of wood stain.

Categories	Price per Can	Number of Cans	Revenue
Small	8.95	n	$8.95n$
Large	15.95	$2n$	$15.95(2n)$

Note: We were given the price.

Note: We selected n to represent the number of small cans, then translated "twice as many of the large cans as the small cans."

Note: We multiplied straight across because

$$\boxed{\text{Price per can}} \cdot \boxed{\text{Number of cans}} = \boxed{\text{Revenue from each size}}$$

The expressions in the last column in the table give the revenue from the sale of each size can.

Plan Translate the information to an equation and then solve.

Execute Now we can use our initial relationship.

$$\boxed{\text{Revenue from the small cans}} + \boxed{\text{Revenue from the large cans}} = 694.45$$

$$8.95n + 15.95(2n) = 694.45$$

$8.95n + 31.9n = 694.45$ **Multiply $2n$ by 15.95.**

$40.85n = 694.45$ **Combine like terms.**

$$\dfrac{40.85n}{40.85} = \dfrac{694.45}{40.85}$$ **Divide both sides by 40.85 to isolate n.**

$$n = 17$$

Answer The number of small cans sold, n, is 17. The number of large cans sold, $2n$, is 34 (found by multiplying $2 \cdot 17$).

Check First, verify that twice the number of small cans sold, 17, is 34, which is true. Second, verify that the total revenue from the sale of wood stain is $694.45.

$$8.95(17) + 15.95(34) = 152.15 + 542.30 = 694.45$$ **It checks.**

The table used in Example 6 is a common type of table. In general, when a problem involves two or more categories, we can construct a table with a column labeled *Categories* and a row for each category in the problem. We then create a column for each parameter described in the problem. For example, parameters in a problem might be the *value* of each item, the *number* of items, and the total *amount* of money in each item. The table would look like this:

Categories	Value	Number	Amount

Because *value · number = amount*, the column labeled *Amount* contains the products of the expressions in the Value and Number columns.

Your Turn 6 Complete a table, write an equation, and then solve.

A computer store sells two different ink cartridges for a particular printer. The black ink cartridge sells for $12.75, and the color cartridge sells for $24.95. A company orders five more black ink cartridges than color cartridges. If the total cost is $629.25, how many of each cartridge were purchased?

Now consider a problem in which a total number of items is given and you must determine how they are split up.

Example 7   A company sells two versions of the same software. The regular version sells for $29.95. The deluxe version sells for $39.95. Due to an inventory error, the number sold of each version was lost. However, the company knows that there were 458 units sold of the two versions combined and that the total revenue was $14,667.10. How many of each version were sold?

Understand   We are given the combined income, $14,667.10, which indicates the following relationship:

$$\boxed{\begin{array}{c}\text{Revenue from the} \\ \text{regular version}\end{array}} + \boxed{\begin{array}{c}\text{Revenue from the} \\ \text{deluxe version}\end{array}} = 14{,}667.10$$

Because the problem involves two different items, their value, and the number of each item, we can use a four-column table to organize the information. We begin filling in the table with the categories and the given value of each version.

Categories	Selling Price (Value)	Number	Revenue (Amount)
Regular	29.95		
Deluxe	39.95		

Filling in the Number column is tricky. We are told that a total of 458 units were sold, so we can say

$$\boxed{\begin{array}{c}\text{Number of the} \\ \text{regular version}\end{array}} + \boxed{\begin{array}{c}\text{Number of the} \\ \text{deluxe version}\end{array}} = 458$$

Notice that we can isolate one of the two unknowns using related subtraction statements. We will isolate the number of the regular version and let n represent the number of the deluxe version so that we have

$$\boxed{\begin{array}{c}\text{Number of the} \\ \text{regular version}\end{array}} = 458 - \boxed{\begin{array}{c}\text{Number of the} \\ \text{deluxe version}\end{array}}$$

$$= 458 - n$$

Note: By letting the variable represent the number of the larger-valued item, which is the deluxe version, we will avoid negative coefficients in solving the equation, as you will see in the Execute step of the solution.

Now we can complete the table.

Categories	Selling Price (Value)	Number	Revenue (Amount)
Regular	29.95	$458 - n$	$29.95(458 - n)$
Deluxe	39.95	n	$39.95n$

Note: In general, if given a total number of items, select a variable to represent the number of one of the items. The number of the other item will be

Total number − Variable.

Note: Because selling price · number = revenue, we multiply straight across the columns to get the expressions in the revenue column.

204 **Chapter 3**   Problem Solving

Plan Translate the information to an equation and then solve.

Execute Use the initial relationship.

| Revenue from the regular version | + | Revenue from the deluxe version | = 14,667.10 |

$$29.95(458 - n) \quad + \quad 39.95n \quad = 14,667.10$$

$$13,717.10 - 29.95n + 39.95n = 14,667.10 \quad \text{Distribute.}$$

$$13,717.10 + 10n = 14,667.10 \quad \text{Combine like terms.}$$

$$\underline{-13,717.10 \qquad\qquad -13,717.10} \quad \begin{array}{l}\text{Subtract 13,717.10}\\\text{from both sides.}\end{array}$$

$$0 + 10n = 950.00$$

$$\frac{10n}{10} = \frac{950}{10} \quad \begin{array}{l}\text{Divide both sides}\\\text{by 10 to isolate } n.\end{array}$$

$$n = 95$$

Answer The number of the deluxe versions, n, is 95. The number of the regular versions, $458 - n$, is 363 (found by subtracting $458 - 95$).

Check First, verify that the sum of 95 deluxe versions and 363 regular versions gives 458 copies of the software sold, which is true. Second, check the total revenue.

$$29.95(363) + 39.95(95) = 10,871.85 + 3795.25 = 14,667.10$$

It checks.

Your Turn 7 Complete a table, write an equation, and then solve.

A company offers two dental plans for its employees: individual coverage at $22.75 each month or family coverage at $87.50 each month. The company has 26 employees with dental coverage who pay a total of $1757 each month in premiums. How many employees have individual coverage, and how many have family coverage?

Answer Your Turn 7

8 have individual coverage, 18 have family coverage

3.3 Exercises   For Extra Help   *MyMathLab*

PRACTICE   WATCH   DOWNLOAD

Note: Exercises marked with a ★ represent challenging exercises.

1. Suppose a problem gives the following information: "One number is five times a second number." Which is easier: letting the *one number* be represented by n or letting the *second number* be represented by n? What expression would describe the other unknown?

2. Suppose you are given that a rectangle has a width that is 5 centimeters less than the length. Which is easier: representing the length or the width with a variable? What expression would describe the other unknown?

3. In describing consecutive even integers or consecutive odd integers, we use the same pattern:

Consecutive Odd	Consecutive Even
first odd integer $= n$	first even integer $= n$
second odd integer $= n + 2$	second even integer $= n + 2$
third odd integer $= n + 4$	third even integer $= n + 4$

Explain why the same pattern can be used to describe both odd and even consecutive integers.

4. Consider the following problem: Grant has a jar containing only quarters and half-dollars. If there are a total of 42 coins and the total value is $13.75, how many of each type of coin are there?

    a. Although we could let either unknown be represented by n, what is advantageous about letting n represent the number of half-dollars, which is the greater-valued coin?

    b. If n represents the number of half-dollars, what expression describes the number of quarters?

For Exercises 5–16, represent each item in terms of the given variable.

5. The price of a rosebush is $3 more than the price of an azalea. If x represents the price of an azalea,
    a. Represent the price of a rosebush in terms of x.

    b. Write and simplify an expression that represents the price of 6 rosebushes and 8 azaleas.

6. The weight of a truck is 800 pounds more than the weight of a car. If the weight of the car is represented by x pounds,
    a. Represent the weight of the truck in terms of x.

    b. Write and simplify an expression that represents the sum of the weights of the two vehicles.

7. The length of a rectangle is 4 more than three times the width. If w represents the width,
    a. Represent the length in terms of w.

    b. Write and simplify the expression for the perimeter in terms of w.

8. The width of a rectangle is 8 less than twice the length. If l represents the length,
    a. Represent the width in terms of l.

    b. Write and simplify an expression for the perimeter in terms of l.

9. Two angles are complementary, and the measure of one angle is 10° less than 4 times its complement. If $x°$ represents the measure of the complement,
    a. Represent the measure of the other angle in terms of x.

    b. Write the equation that would be used to find the value of x.

10. Two angles are supplementary, and the measure of one angle is 10° more than twice the measure of its supplement. If $d°$ represents the measure of the supplement,
    a. Represent the measure of the other angle in terms of d.

    b. Write the equation that would be used to find the value of d.

11. a. Given the integer -12, what are the next two greater consecutive even integers? What is the sum of the three integers?

b. Given the even integer y, what are the next two greater consecutive even integers?

c. Write and simplify an expression for the sum of the three integers in part b.

12. a. Given the integer 17, what are the next two greater consecutive odd integers? What is the sum of the three integers?

b. Given the odd integer z, what are the next two greater consecutive odd integers?

c. Write and simplify an expression for the sum of the three integers in part b.

13. A construction supply store sells maple flooring for $7.99 per square foot and pecan flooring for $11.99 per square foot. In one day, the store sold three times as much maple flooring as pecan. If n represents the number of square feet of pecan flooring sold,
a. Represent the number of square feet of maple flooring sold.

b. Write an expression for the revenue for each type of flooring.

c. Write and simplify an expression for the total revenue for both types of flooring.

14. A convenience store sells soft drinks in 12-ounce and 16-ounce bottles. The 12-ounce drinks sell for $1.29 each, and the 16-ounce drink sells for $1.69. One day the store sold 28 more 12-ounce drinks than 16-ounce drinks. If n represents the number of 16-ounce drinks sold,
a. Represent the number of 12-ounce drinks sold.

b. Write an expression for the revenue for each type of drink.

c. Write and simplify an expression for the total revenue for both drink sizes.

15. A paint store sells one brand of interior paint for $12.99 per gallon and the same brand of exterior paint for $16.99 per gallon. A large construction company purchases a total 372 gallons of both types of paint. If x represents the number of gallons of exterior paint purchased,
a. Represent the number of gallons of interior paint purchased.

b. Represent the amount paid for each type of paint.

c. Write and simplify an expression for the total amount paid for both types of paint.

16. A nursery sells nandina shrubs for $7.95 each and mums for $4.95 each. In one day, the nursery sold a total of 83 of both types of plants. If x represents the number of nandina shrubs sold,
a. Represent the number of mums sold.

b. Represent the revenue for each type of plant.

c. Write and simplify an expression for the total revenue for both types of plants.

For Exercises 17–68, translate to an equation and then solve. See Examples 1–5.

17. One number is three times another number. The sum of the numbers is 52. what are the numbers?

18. One number is four times another. The sum of the numbers is 35. What are the numbers?

19. One number is seven more than another. The sum of the numbers is 27. What are the numbers?

20. One number is nine more than another. The sum of the numbers is 55. What are the numbers?

21. The greater of two numbers is 5 less than twice the lesser number. The sum of the two numbers is 28. What are the numbers?

22. The greater of two numbers is 4 more than three times the lesser number. If the sum of the two numbers is 32, find the numbers.

23. The lesser of two numbers is 2 less than the greater number. If 10 is subtracted from six times the lesser number, the result is the same as 6 less than four times the greater number. Find the numbers.

24. The lesser of two numbers is 6 less than the greater number. If three times the greater number is subtracted from 52, the answer is the same as when 56 is subtracted from seven times the lesser number. Find the numbers.

25. The greater of two numbers is twice the lesser number. Three times the sum of the lesser number and 8 is the same as 36 more than the greater number. Find the numbers.

26. The greater of two numbers is three times the lesser number. Twice the sum of 15 and the lesser number is equal to 15 plus the greater number. What are the numbers?

27. A positive number is one-fifth of another positive number. The greater number less the lesser number is 8. What are the numbers?

28. One positive number is one-third of another number. The sum of the two numbers is 24. What are the numbers?

29. One cup of canned unsweetened grapefruit juice has 5 fewer grams of carbohydrates than 1 cup of frozen diluted orange juice. If a person drinking a mixture of 1 cup of each consumes 49 total grams of carbohydrates, how many carbohydrates do 1 cup of orange juice and 1 cup of grapefruit juice contain?

30. One slice of white bread has 9 fewer milligrams of sodium than one slice of wheat bread. If a person ate one slice of each, he or she would consume a total of 267 milligrams of sodium. How many milligrams of sodium are in a slice of each type of bread?

31. A McDonald's Big Mac contains 60 more calories than a large order of french fries. If a person eats both, he or she consumes a total of 1100 calories. Find the number of calories in a Big Mac and the number of calories in a large order of french fries. (*Source:* McDonald's USA)

32. In one workout, Jennifer walked on a treadmill for 30 minutes, then spent 30 minutes on a bicycle. She burned 75 more calories on the bicycle than on the treadmill. If she burned a total of 527 calories during that hour of exercise, how many calories did she burn on each piece of equipment?

Of Interest

Adding a 32-ounce chocolate triple-thick shake to the meal discussed in Exercise 31 brings the total calories for the meal to 2260 calories. A general recommendation for calorie intake is around 1950 for an adult female and 2500 for an adult male *per day*.

---

33. The brightness of light is measured in a unit called lumens. A 60-watt bulb produces 20 more than four times the number of lumens produced by a 25-watt bulb. Combined they produce 1095 lumens. Find the number of lumens produced by each.

34. In 2006, the number one tourist destination was France, with Spain number two. The number of tourists visiting France was 37.9 million less than twice the number visiting Spain. If the total number of tourists visiting both countries was 137.6 million, find the number of tourists visiting each country.

35. The length of a rectangular garden is 20 feet more than its width, and the perimeter is 240 feet. What are the length and width?

36. The length of a rectangular room is 8 feet more than the width, and the perimeter is 76 feet. What are the length and width?

37. The length of a rectangular hallway is three times its width, and the perimeter is 32 feet. Find the length and width.

38. The length of a rectangular picture is twice the width, and the perimeter is 108 inches. Find the length and width.

39. The height of a rectangular wall is 10 feet less than the length. If the perimeter of the wall is 72 feet, find the length and width.

40. The width of a rectangular field is 12 meters less than its length. If the perimeter is 400 meters, find the length and width.

41. The perimeter of a rectangular frame is 36 centimeters. If the length is 3 less than twice the width, find the length and width.

42. The perimeter of a rectangular tabletop is 22 feet. If the length is 1 foot less than three times the width, find the length and width.

43. The width of a rectangular bathroom is $\frac{2}{3}$ of its length. If the perimeter is 25 feet, what are its length and width?

44. An artist creates a rectangular sculpture with a width that is $\frac{3}{5}$ of the length. If the perimeter of the sculpture is 32 feet, find the length and width.

45. Two angles are supplementary. If the measure of one angle is 26° more than the other angle, find the measure of the two angles.

46. Two angles are supplementary. If the measure of one angle is 44° more than the other angle, find the measures of the two angles.

47. Two angles are supplementary. One of the angles is 30° more than twice the other angle. What are the measures of the two angles?

48. Two angles are supplementary. One of the angles is 12° more than three times the other angle. What are the measures of the two angles?

49. Two angles are complementary. If the measure of one angle is 38° less than the other angle, find the measure of the two angles.

50. Two angles are complementary. If the measure of one angle is 6° less than the measure of the other angle, find the measure of the two angles.

51. Two angles are complementary. The smaller angle is $\frac{1}{5}$ of the measure of the larger angle. What are the measures of the two angles?

52. Two angles are complementary. The smaller angle is $\frac{1}{4}$ of the measure of the larger angle. What are the measures of the two angles?

53. The sum of two consecutive integers is 147. What are the integers?

54. The sum of two consecutive integers is 123. What are the integers?

55. Find two consecutive even integers such that three times the lesser integer is eight more than twice the greater integer.

56. Find two consecutive even integers such that four times the lesser integer is ten more than three times the greater integer.

57. The sum of three consecutive integers is 279. What are the integers?

58. The sum of three odd consecutive integers is 237. What are the integers.

59. The sum of three consecutive even integers is 234. What are the integers?

60. The sum of three consecutive even integers is 42. What are the integers?

61. The Smith, Lopez, and Jones families live in houses whose numbers are consecutive odd integers and whose sum is 375. What are the house numbers?

62. The Windmueller, Wu, and Blindert families live in houses whose numbers are consecutive even integers and whose sum is 264. What are the house numbers?

63. Jenny, Bryan, and Tommy live in the same apartment building with apartment numbers that are consecutive integers such that the sum of the smallest and largest integers is 24 more than the middle integer. Find the apartment numbers.

64. Carole, Sue, and Karen attended a concert and had front row seats whose numbers were consecutive integers such that the sum of the smallest and middle integers was 13 more than the largest integer. Find the seat numbers.

★ 65. There are three consecutive integers such that the sum of the first and second integers decreased by the third integer will result in 68. What are the integers?

66. Find three consecutive integers such that the sum of the second and third integers decreased by the first integer is 44.
★

★ **67.** Find three consecutive odd integers such that twice the smallest integer plus three times the largest integer will result in the middle integer being increased by 70.

★ **68.** Find three consecutive even integers such that the sum of three times the smallest integer and twice the largest integer is eight more than four times the middle integer.

For Exercises 69–76, complete a table, write an equation, then solve. See Examples 6 and 7.

69. Ember is playing Monopoly and has three times as many $5 bills as $50 bills. If the total value of the two types of bills is $520, find the number of $5 and $50 bills that she has.

70. Kelly purchased some plants from a garden center where the prices are based on the size of the pot. Plants in 1-gallon pots cost $9.00 each, and plants in half-gallon pots cost $5.50 each. She purchases twice as many plants in 1-gallon pots as in half-gallon pots. If she paid a total of $141 for the plants, how many of each sized pots did she buy?

71. Dirk has some stock worth $18 per share and other stock worth $32 per share. The number of shares at $32 is one more than twice the number at $18. If the total value of the stocks is $1836, find the number of shares of each type.

72. The college bookstore sells CDs for $15.50 and study guides for $21.50. One day, the number of study guides sold was four more than three times the number of CDs. If the total sales were $566, how many of each did the bookstore sell?

73. The local high school is performing *Oklahoma* as its spring play. The tickets cost $5.00 for students and $8.50 for the general public. Although the organizers lost track of the ticket count, they know they sold 310 more general public tickets than student tickets. The total sales were $5605. How many of each type of ticket were sold?

74. A vending machine has 12-ounce and 16-ounce drinks. The 12-ounce drinks are $0.50, and the 16-ounce drinks are $1.00. If 3600 total drinks were sold in one month and the total sales were $2225, how many of each size drink were sold?

75. Darryl has some $5 bills and $10 bills in his wallet. If he has a total of 16 bills worth a total of $110, how many of each bill are in the wallet?

76. Tax preparation software is sold in the standard version for $19.95 and the deluxe version for $10 more. If a store sold 174 copies of the software for total sales of $4011.30, how many of each type were sold?

Puzzle Problem

The product of three integers is 84. Two of the integers are prime numbers, and the third is not. If the sum of the composite number and one of the two prime numbers is 17, what are the three numbers?

Review Exercises

Exercises 1–6 EQUATIONS AND INEQUALITIES

[2.1]   **1.** Check to see if $x = -2$ is a solution for $5x + 7 = x - 1$.

[2.2]   **2.** Is $3x - 8 = 12$ a linear equation?

[2.3] *For Exercises 3–5, solve and check.*

3. $-\dfrac{3}{4}x = 12$ **4.** $4x + 5 = -27$ **5.** $9x - 7(x + 2) = -3x + 1$

[2.1]   **6.** Carla knows from running on a treadmill that her average running speed is about 4.5 miles per hour. If she maintains this rate running on roads in her neighborhood for 1.5 hours, what distance will she travel? (Use $d = rt$.)

3.4 Rates

Objectives
1. Solve problems involving two objects traveling in opposite directions.
2. Solve problems involving two objects traveling in the same direction.

In Section 3.3, we developed the use of tables to organize information. In this section, we use a similar table to organize information in problems that involve distance, rate, and time. Recall that the general formula relating distance, rate, and time is $d = rt$.

Objective 1 Solve problems involving two objects traveling in opposite directions. First, consider the situation in which two people or objects are traveling in opposite directions. Whether the two people or objects are moving in opposite directions toward or away from each other doesn't matter because the math involved will be the same.

Example 1 In practicing maneuvers, two fighter jets fly toward each other. One flies east at 582 miles per hour, and the other flies west at 625 miles per hour. If the two are 22 miles apart, how much time will it take for them to meet?

Understand Draw a picture of the situation.

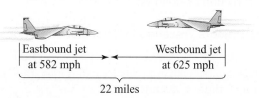

> **Note:** The westbound jet will travel the greater distance because it is going faster.

From the diagram, we see that the sum of the individual distances traveled will be the total distance separating the two jets, which is 22 miles.

$$\boxed{\begin{array}{c}\text{Distance traveled by}\\\text{the eastbound jet}\end{array}} + \boxed{\begin{array}{c}\text{Distance traveled by}\\\text{the westbound jet}\end{array}} = 22$$

Use a table to find expressions for the individual distances.

Categories	Rate	Time	Distance
Eastbound	582	t	$582t$
Westbound	625	t	$625t$

Note: Both jets start at the same time and meet at the same moment in time, so they traveled the same amount of time, t.

Note: Multiplying the rate value by the time value gives the distance value.

Plan   Use the information to write an equation and then solve.

Execute

$$\boxed{\begin{array}{c}\text{Distance traveled by}\\\text{the eastbound jet}\end{array}} + \boxed{\begin{array}{c}\text{Distance traveled by}\\\text{the westbound jet}\end{array}} = 22$$

$$582t \quad + \quad 625t \quad = 22$$
$$1207t = 22$$
$$\frac{1207t}{1207} = \frac{22}{1207}$$
$$t \approx 0.018$$

Connection When we combined the coefficients in the like terms, we added the individual speeds of the jets, yielding the relative closing speed of the jets.

$$582 \text{ mph} + 625 \text{ mph} = 1207 \text{ mph}$$

The closing speed for these jets is 1207 miles per hour, which is the same as if one of them remained still and the other traveled 1207 miles per hour for the entire 22 miles.

The same reasoning applies if the two jets were going away from each other. They would separate at a rate of 1207 miles per hour.

Note: In solving for time, the time units will match the unit of time in the rate. In this problem, the rate is in miles per hour, so the time unit is hours.

Answer   The jets will meet in 0.018 hours, which is approximately 1.09 minutes, or about 1 minute and 6 seconds.

Check   Verify that in 0.018 hours, the jets will travel a combined distance of 22 miles.

Eastbound jet: $d = 582(0.018) = 10.476$ miles

Westbound jet: $d = 625(0.018) = 11.25$ miles

Note: The westbound jet goes farther because it is traveling faster.

The combined distance, in miles, is $10.476 + 11.25 = 21.726 \approx 22$. Because we rounded the time, 21.726 is reasonable.

We can summarize how to solve problems involving two objects traveling in opposite directions as follows:

Procedure   **Two Objects Traveling in Opposite Directions**

To solve for time when two objects are moving in opposite directions,

1. Use a table with columns for categories, rate, time, and distance. Use the fact that rate · time = distance.
2. Write an equation that is the sum of the individual distances equal to the total distance of separation.
3. Solve the equation.

Your Turn 1 Kari is running west at 5 miles per hour along a trail. Vernon is walking east on the same trail at $3\frac{1}{4}$ miles per hour. If they are $\frac{1}{2}$ mile apart, how long will it be until they meet?

Objective ② Solve problems involving two objects traveling in the same direction. If two objects are traveling in the same direction, we can use the same type of four-column table to organize the information. However, we will see that the equation is different.

Example 2 Carol and Richard are traveling north in separate cars on the same highway. Carol is traveling at 65 miles per hour; Richard, at 70 miles per hour. Carol passes Exit 102 at 1:30 P.M. Richard passes the same exit at 1:45 P.M. At what time will Richard catch up to Carol?

Understand To determine the time of day at which Richard catches up to Carol, we must calculate the amount of time it will take him to catch up to her. We can then add that amount of time to 1:45 P.M.

Richard passed the same exit 15 minutes after Carol did. So that we can use $d = rt$, the time units must match the time units in the rate. Because 65 and 70 were in miles per hour (mph), the time units must be hours. What portion of an hour is 15 minutes? Since there are 60 minutes in an hour, we can write

$$\frac{15}{60} = \frac{1}{4} = 0.25 \text{ hr.}$$

Now we can begin completing the table. We let t represent the amount of time it takes Richard to catch up. Carol's travel time is a bit tricky. Suppose Richard catches up at 2:30 P.M. Notice that this is 1 hour after Carol passed Exit 102 and 45 minutes after Richard passed Exit 102. We see that at the time Richard catches up, Carol's travel time from Exit 102 is 15 minutes (0.25 hours) more than Richard's travel time from the same exit. Because this will be the case no matter what time he catches up, Carol's travel time is $t + 0.25$.

Categories	Rate	Time	Distance
Carol	65	$t + 0.25$	$65(t + 0.25)$
Richard	70	t	$70t$

Note: Because $d = rt$, we multiply the rate and time values to equal the distance values.

What can we conclude about the distances each person will have traveled when Richard catches up? Notice that Exit 102 is the common reference point from which we describe the time for Richard to catch up. When Richard catches up, he and Carol will be the same distance from that exit; so their distances are equal.

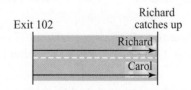

Plan Set the expressions of their individual distances equal and solve for t.

Execute   Richard's distance = Carol's distance (at the time he catches up)

$$70t = 65(t + 0.25)$$
$$70t = 65t + 16.25$$
$$\underline{-65t \quad -65t}$$
$$5t = 0 + 16.25$$
$$\frac{5t}{5} = \frac{16.25}{5}$$
$$t = 3.25$$

Connection The coefficients of the like terms are the individual rates; so $70 - 65 = 5$ means that Richard's rate relative to Carol's is 5 miles per hour. This means that he must make up the 16.25 miles that separates them at a rate of 5 miles per hour.

Connection Because 65 is the rate and 0.25 is the additional amount of time, their product, 16.25 miles, is the distance that Carol is ahead of Richard. Richard must make up this distance to catch her.

Connection Dividing 16.25 miles by 5 miles per hour calculates the time it takes Richard to make up the distance between Carol and him.

Answer   It will take Richard 3.25 hours, which is 3 hours and 15 minutes, after passing Exit 102 to catch up to Carol. Because he passed Exit 102 at 1:45 P.M., he catches up to Carol at 5:00 P.M.

Check   Verify that Richard and Carol are equal distances from Exit 102 after traveling for their respective times. If Richard travels for 3.25 hours after Exit 102, then Carol travels for $3.25 + 0.25 = 3.5$ hours after Exit 102. Using $d = rt$

Richard: $d = (70)(3.25)$ Carol: $d = (65)(3.5)$
$d = 227.5$ miles $d = 227.5$ miles

Following is a summary of the process for problems involving two objects traveling in the same direction in which we are to calculate the time for one of the objects to catch up to the other.

Procedure   **Two Objects Traveling in the Same Direction**

To solve problems involving two objects traveling in the same direction in which the objective is to determine the time for one object to catch up to the other,

1. Use a table to organize the rates and times. Let t represent the time for the object to catch up, and add the time difference to t to represent the other object's time.
2. Set the expressions for the individual distances equal.
3. Solve the equation.

Your Turn 2   Juan and Angela are bicycling along the same trail. Juan passes a marker at 9:00 A.M., and Angela passes the same marker at 9:05 A.M. Juan is traveling at 8 miles per hour, while Angela is traveling at 10 miles per hour. What time will Angela catch up to Juan?

Answer   Your Turn 2

9:25 A.M.

3.4 Exercises   For Extra Help *MyMathLab*

Note: Exercises marked with a ★ represent challenging exercises.

1. Suppose two objects begin moving toward each other on a collision course. What can we conclude about the amount of time it takes them to meet?

2. Suppose two objects start back-to-back and travel in opposite directions. Object 1 travels $24t$ miles, while object 2 travels $45t$ miles. In the following figure, which arrow represents which object? Explain.

3. Suppose the two objects described in Exercise 2 are separated by 10 miles after traveling t hours. How do you write the equation?

4. A car passes under a bridge on a highway. Forty-five minutes later, a second car passes under the same bridge traveling in the same direction at a speed greater than the first car. Suppose the second car catches up to the first car in t hours. Write an expression for the first car's travel time (in hours) since passing under the bridge.

5. Draw a picture of the situation described in Exercise 4.

6. Regarding the cars described in Exercise 4, what can you say about their distances from the bridge?

For Exercises 7–12, answer the questions and complete the charts.

7. Use $d = rt$ to find the following.
   a. Find the distance a car travels in 3 hours at an average rate of 55 miles per hour.

   b. Represent the distance traveled in t hours at 48 miles per hour.

   c. Represent the distance traveled in $t + 2$ hours at 63 miles per hour.

8. Use $d = rt$ to find the following.
   a. Find the distance a bus travels in 6 hours at an average rate of 58 miles per hour.

   b. Represent the distance traveled in t hours at 53 miles per hour.

   c. Represent the distance traveled in $t - 1$ hours at 47 miles per hour.

9. Complete the table, which contains a description of two unknowns and a variable for one of the unknowns. Use the description to describe the second unknown in terms of the variable.

	Sal's time	Jorge's time
Sal's time equals Jorge's time.	t	
Sal's time is 2 hours more than Jorge's time.		t
Jorge's time is 3 hours less than Sal's time.	t	
Sal's time is twice Jorge's time.		t

10. Complete the table, which contains a description of two unknowns and a variable for one of the unknowns. Use the description to describe the second unknown in terms of the variable.

	Eve's time	Caleb's time
Eve's time equals Caleb's time.		t
Caleb's time is one-half of Eve's time.	t	
Eve's time is 1 hour less than Caleb's time.		t
Caleb's time is 2 hours more than Eve's time.	t	

11. Atlanta, Georgia, and Charleston, West Virginia, are 500 miles apart. A truck leaves Atlanta traveling toward Charleston at an average rate of 58 miles per hour. At the same time, a bus leaves Charleston traveling toward Atlanta at an average rate of 62 miles per hour. Assume that they are traveling on the same route. Answer the following questions, which will lead to an equation to find the number of hours until they meet.

   a. At the moment the truck and bus meet, how are their times related?

   b. Draw a figure describing the situation.

   c. Complete the table.

Categories	Rate	Time	Distance
Truck	58		
Bus	62		

   d. What is the total distance traveled by the two vehicles?

   e. Write the equation illustrated by the following.

$$\boxed{\text{Distance the truck traveled}} + \boxed{\text{Distance the bus traveled}} = \boxed{\text{Distance between Atlanta and Charleston}}$$

12. Tameka leaves Kansas City traveling west on I-70 at an average rate of 56 miles per hour. At the same time, Akiva leaves Kansas City traveling east on I-70 at an average rate of 48 miles per hour.

   a. At the moment they are 300 miles apart, how are their times related?

   b. Draw a figure describing the situation.

   c. Complete the table.

Categories	Rate	Time	Distance
Tameka	56		
Akiva	48		

   d. The unknown is how many hours until they are 300 miles apart. Write the equation illustrated by the following.

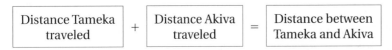

$$\boxed{\text{Distance Tameka traveled}} + \boxed{\text{Distance Akiva traveled}} = \boxed{\text{Distance between Tameka and Akiva}}$$

13. Polly leaves Boston traveling north on I-93 at an average rate of 55 miles per hour. Two hours later Erick also leaves Boston traveling north on I-93 at an average rate of 68 miles per hour. Erick catches up to Polly.

   a. How does Polly's time compare with Erick's?

   b. Draw a figure describing the situation.

c. How do Polly's and Erick's distances compare?

d. Complete the table.

Categories	Rate	Time	Distance
Polly	55		
Erick	68	t	

e. Write the equation illustrated by the following.

Polly's distance	=	Erick's distance

14. Bill leaves Houston traveling west on I-10 at an average rate of 43 miles per hour. An hour later, Bryan also leaves Houston traveling west on I-10 at an average rate of 57 miles per hour. Bryan catches up to Bill.
a. How does Bill's time compare with Bryan's time?

b. Draw a figure describing the situation.

c. How do Bill's and Bryan's distances compare?

d. Complete the table.

Categories	Rate	Time	Distance
Bill	43		
Bryan	57	t	

e. Write the equation illustrated by the following.

Bill's distance	=	Bryan's distance

Exercises 15–26 involve people or objects moving in opposite directions. Complete a four-column table, write an equation, and then solve. See Example 1.

15. An Expedition and a Mustang leave a gas station and travel in opposite directions. The Mustang travels at 50 miles per hour, and the Expedition travels at 45 miles per hour. In how many hours will the two cars be 380 miles apart?

16. Monica and Chandler pass each other along a straight road traveling in opposite directions. Monica is driving at 40 miles per hour; Chandler, at 50 miles per hour. In how many hours will the cars be 225 miles apart?

17. If a plane flying from Cincinnati travels west at 900 kilometers per hour and another leaving at the same time travels east at 1050 kilometers per hour, in how many hours will the planes be 6825 kilometers apart?

18. Two planes leave Denver at the same time, one flying north and the other flying south. If the plane flying south is traveling at 653 miles per hour and the plane flying north is traveling 560 miles per hour, after how many hours will the planes be 2426 miles apart?

19. Two cars leave a restaurant at the same time with one heading north and the other heading south. If the first car travels at 55 miles per hour and the second car travels at 60, how long will it take them to be 230 miles apart?

20. If two planes leave the Dallas airport at the same time with one flying west at 530 miles per hour and the other flying east at 620 miles per hour, how long will it take them to be 1725 miles apart?

21. Two trains started at the same time from Grand Central Station and traveled in opposite directions. One traveled at a rate of 45 miles per hour; the other, at the rate of 60 miles per hour. In how many hours were they 210 miles apart?

22. Two trains start at the same time from Victoria Station and travel in opposite directions. One travels at a rate of 40 miles per hour; the other, at the rate of 55 miles per hour. After how many hours will they be 190 miles apart?

23. Kim and Karmen begin walking toward each other from opposite directions at 8 A.M. If Kim is walking at 3.5 miles per hour and Karmen is walking at 4 miles per hour, what time will they meet if they are 12 miles apart?

24. Two trains traveling on parallel tracks are going toward each other from a distance of 378 miles. If the freight train is moving at 22 miles per hour and the passenger train is moving at 50 miles per hour, how long will it take for them to pass each other?

★25. Two cars pass each other traveling in opposite directions on a highway. One car is going $1\frac{1}{2}$ times as fast as the other car. At the end of $3\frac{1}{2}$ hours, they are $192\frac{1}{2}$ miles apart. Find each car's rate.

★26. Dustin drives 5 miles per hour faster than Kelly. If he leaves town 1 hour after Kelly and they travel in opposite directions, they will have driven equal distances after Dustin has traveled 7 hours. How far apart are they?

Exercises 27–38 involve people or objects moving in the same direction. Complete a table, write an equation, and then solve. See Example 2.

27. At 9 A.M., a freight train leaves Seattle traveling at 45 miles per hour. At 10 A.M., a passenger train leaves the same station traveling in the same direction at 55 miles per hour. How long will it take the passenger train to overtake the freight train? How far will they be from the station at this time?

28. At 6 A.M., a freight train leaves Washington, D.C., traveling at 50 miles per hour. At 9 A.M., a passenger train leaves the same station traveling in the same direction at 75 miles per hour. How long will it take the passenger train to overtake the freight train? How far will they be from the station at this time?

29. Dan and Kathy are both traveling south for spring break. Dan is taking a bus traveling 60 miles per hour, and Kathy is driving her car at 70 miles per hour. If Dan's bus passes Jacksonville at 2:00 P.M. and Kathy passes through Jacksonville at 2:30 P.M., what time will Kathy catch up to Dan?

30. Two truckers are delivering materials from the same site. One leaves the site at 8 A.M., and the other leaves at 8:15 A.M. If the first trucker travels at 60 miles per hour and the second at 65 miles per hour, at what time will the second trucker catch up to the first?

31. Daphne and Niles are both driving west on Route 66. At 4 P.M., Daphne is 5 miles west of Niles. A little later, Niles passes Daphne. If Niles is driving at 55 miles per hour and Daphne at 50 miles per hour, at what time does Niles pass Daphne?

32. Janet and Paul were both traveling west from Charleston, West Virginia, to Rupp Arena in Lexington, Kentucky, for a concert. At 6 P.M., Janet, traveling at 45 miles per hour, was 10 miles west of Paul. A little later, Paul, traveling at 50 miles per hour, passed Janet. What time did Paul pass Janet?

33. Jane leaves Oklahoma City at 9 A.M. heading for Kansas City. Harrison leaves at 10 A.M. on the same highway, heading toward Kansas City. By driving 15 miles per hour faster, Harrison overtakes Jane at 1 P.M.

   a. How fast is Jane driving?

   b. How far have they traveled when Harrison catches up with Jane?

34. Morrison leaves Cincinnati at 4:00 P.M., heading for New York City. Barbara leaves at 4:30 P.M. on the same highway, heading toward New York City. By driving 10 miles per hour faster, Barbara overtakes Morrison at 7:30 P.M.

   a. How fast is Barbara driving?

   b. How far have they traveled when Barbara catches up with Morrison?

★ **35.** If two cars start driving in the same direction from the same place while one is traveling 45 miles per hour and the other 50 miles per hour, how long will it take them to be 30 miles apart?

★ **36.** A car traveling 70 miles per hour passes a truck traveling 62 miles per hour in the same direction on the highway. If they maintain their speeds, how long will it take them to be 1 mile apart?

★ **37.** If two joggers leave the same point and run in the same direction while one jogs at 1.5 miles per hour and the other jogs at 2.5 miles per hour, how long will it take them to be 4 miles apart?

★ **38.** Two brothers start a bicycling race at the same time. If one brother rides at a rate of 15 miles per hour and the other brother rides at a rate of 12 miles per hour, how long will it take them to be 0.75 miles apart?

For Exercises 39–42, use a table and the techniques discussed in this section to solve.

★ **39.** On the way to a ski weekend in Vail, Carlita drove 2 hours in a snowstorm. When it finally stopped snowing, she was able to increase her speed by 35 miles per hour and drove another 3 hours. If the entire trip was 325 miles, how fast did she drive when it finally stopped snowing?

★ **40.** While driving from Los Angeles, Tori had a visibility problem due to the smog. After $1\frac{1}{2}$ hours, she hit the open highway outside the city and was able to increase her speed by 40 miles per hour, driving another $5\frac{1}{2}$ hours. If the entire trip was 514 miles, how fast was she traveling in the smog?

★ **41.** A trucker traveling along a highway encounters road construction so that traffic is down to a single lane. After driving 30 minutes, the road construction ends and the trucker is able to increase his speed by 64 miles per hour for 4.5 hours. If his entire trip is 303 miles, what was his speed in the road construction? What was his speed after the road construction?

★ **42.** An athlete is training for a triathlon. She is working on the running and cycling legs of the event. She runs for 45 minutes, then bikes for 1.5 hours. If her combined distance running and cycling is 31.5 miles and her cycling speed is three times her speed running, what is her running speed?

Review Exercises

Exercises 1–4 EXPRESSIONS

[1.4] **1.** What property of arithmetic is illustrated by $3 \cdot (5 \cdot 7) = 3 \cdot (7 \cdot 5)$?

[1.5] **2.** Simplify $|14 - 9 \cdot 3| + (4 - 9)^2$.

[1.7] **3.** Evaluate $-3x^2 - 4y^3$, where $x = 5$ and $y = -2$.

[1.7] **4.** Simplify $9x - 5y - 7 + 12y + 13 - x$.

Exercises 5 and 6 EQUATIONS AND INEQUALITIES

[2.3] **5.** Solve $0.15n + 0.4n = 0.25(200)$.

[3.2] **6.** 12% of what number is 84?

3.5 Investment and Mixture

Objectives

1. Use a table to solve problems involving two investments.
2. Use a table to solve problems involving mixtures.

In Sections 3.3 and 3.4, we began using tables to organize information in problems with two unknowns. In this section, we continue the use of tables in problems involving investments and mixtures.

Objective 1 **Use a table to solve problems involving two investments.** When money is invested in an account that earns a percentage of the money invested, the amount earned is called *interest* and is represented by *I*. The amount of money invested is called the *principal* and is represented by *P*. With regard to interest, the *percentage rate, r,* is usually an annual percentage rate (APR), which is used to calculate the interest if the principal is invested for one year. The relationship for the interest after one year is $I = Pr$. For example, if a person invests $500 at 8% for one year, the interest would be calculated as follows:

$$I = 0.08(500)$$
$$I = \$40$$

> **Connection** The formula $I = Pr$ comes from the basic percent sentence
>
> | Percent | of a | whole | is a | part of the whole. |
> |---------|------|-------|------|--------------------|
> | Percent | of the | principal | is the | interest. |
> | r | $\cdot$ | P | $=$ | I |

Let's explore problems in which an investor invests money in two different accounts.

Example 1   Han invests a total of $8000 in two different accounts. The first account earns a 5% APR, while the second account earns 8%. If the total interest earned after one year is $565, what principal was invested in each account?

Understand   Since Han invests a total of $8000, we can say that

$$\boxed{\begin{array}{c}\text{Principal invested}\\\text{in first account}\end{array}} + \boxed{\begin{array}{c}\text{Principal invested}\\\text{in second account}\end{array}} = 8000$$

Note that we can isolate one of the unknown amounts by writing a related subtraction statement. We will isolate the principal invested in the first account and let *P* represent the principal invested in the second account.

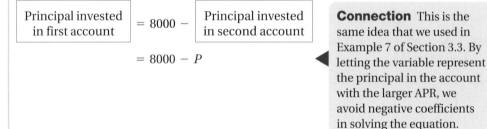

$$\boxed{\begin{array}{c}\text{Principal invested}\\\text{in first account}\end{array}} = 8000 - \boxed{\begin{array}{c}\text{Principal invested}\\\text{in second account}\end{array}}$$

$$= 8000 - P$$

> **Connection** This is the same idea that we used in Example 7 of Section 3.3. By letting the variable represent the principal in the account with the larger APR, we avoid negative coefficients in solving the equation.

We can use a table similar to those used in prior sections. Because $I = Pr$, we need columns for rate, principal, and interest, in addition to a column for the account names.

Accounts	Rate	Principal	Interest
First account	0.05	$8000 - P$	$0.05(8000 - P)$
Second account	0.08	P	$0.08P$

Note: These rates were given. Don't forget to write the percents as decimals.

Note: From our analysis, if P represents the principal invested in the second account, then the principal invested in the first account is $8000 - P$.

Note: Multiply each principal by the corresponding rate to get the expressions for interest.

Rate $\cdot$ Principal $=$ Interest

Plan Write an equation. Then solve for P.

Execute Because the total interest after one year is $565, we can say that

$$\boxed{\begin{array}{c}\text{Interest from}\\\text{first account}\end{array}} + \boxed{\begin{array}{c}\text{Interest from}\\\text{second account}\end{array}} = \boxed{\text{Total interest}}$$

$$
\begin{aligned}
0.05(8000 - P) + \qquad 0.08P &= 565 \\
400 - 0.05P + 0.08P &= 565 &&\text{Distribute.}\\
400 + 0.03P &= 565 &&\text{Combine like terms.}\\
\underline{-400\qquad\qquad -400} && &&\text{Subtract 400 from both sides.}\\
0 + 0.03P &= 165 \\
\frac{0.03P}{0.03} &= \frac{165}{0.03} &&\text{Divide both sides by 0.03.}\\
P &= 5500
\end{aligned}
$$

Answer The principal Han invested in the second account, P, is $5500. In the first account, he invested $8000 - P$, or $2500 (found by subtracting $8000 - 5500$).

Check Verify that investing $5500 at 8% and $2500 at 5% results in a total of $565 in interest.

$$5500(0.08) + 2500(0.05) = 440 + 125 = 565 \qquad \text{It checks.}$$

The following procedure can be used as a general guideline for solving the type of problems illustrated in Example 1.

Procedure **Solving Problems Involving Investment**

To solve problems involving two interest rates in which the total interest is given,

1. Use a table to organize the interest rates and principals. Multiply the individual rates and principals to get expressions for the individual interests.
2. Write an equation that is the sum of the expressions for interest set equal to the given total interest.
3. Solve the equation.

Your Turn 1   Marvin invests $12,000 in two plans. Plan 1 is at an APR of 6%, and plan 2 is at an APR of 9%. If the total interest after one year is $828, what principal was invested in each plan?

Objective ② **Use a table to solve problems involving mixtures.**   Chemicals are often mixed to achieve a solution that has a particular concentration. *Concentration* refers to the portion of a solution that is pure. For example, an 80-milliliter solution may have a 5% concentration of hydrochloric acid (HCl), which means that 5% of that 80 milliliters is pure HCl and the rest is water. To calculate the actual number of milliliters that is pure HCl, we use the simple percent sentence.

$$\begin{array}{cccccc} \text{Percent} & \text{of} & \text{a whole} & \text{is} & \text{a part of the whole.} \\ 0.05 & \cdot & 80 & = & 4 \end{array}$$

This means that 4 milliliters out of the 80 milliliters is pure HCl. The general relationship is as follows:

$$\text{Concentration} \cdot \text{Whole solution volume} = \text{Volume of the particular chemical}$$

Let's consider a problem in which two solutions are mixed to obtain a solution with a particular concentration.

Example 2   Margaret has a bottle containing 50 milliliters of 10% HCl solution and a bottle of 25% HCl solution. She wants a 20% HCl solution. How much of the 25% solution must be added to the 10% solution so that a 20% concentration is created?

Understand   Because more than one solution is involved, a table is helpful in organizing the information. There are three solutions in this problem: the 10% solution, the 25% solution, and the new 20% solution that is created by mixing the 10% and 25% solutions. Therefore, the table has a row for each solution. We need columns for the concentration of HCl, the solution volume, and the volume of HCl in each solution. Because we are to find the volume of the 25% solution, we choose a variable, n, to represent this volume.

Solutions	Concentration of HCl	Volume of Solution	Volume of HCl
10% solution	0.10	50	0.10(50)
25% solution	0.25	n	0.25n
20% solution	0.20	50 + n	0.20(50 + n)

Note: These are the concentration percents written in decimal form.

Note: Because the 10% solution and 25% solution are combined to form the 20% solution, we add their volumes to get the volume of the 20% solution.

Note: Multiply straight across the columns to generate the expressions for the volume of HCl in each solution.

$$\text{Concentration} \cdot \dfrac{\text{Volume}}{\text{of solution}} = \dfrac{\text{Volume}}{\text{of HCl}}$$

Plan   Write an equation that describes the mixture. Then solve for n.

Execute The 10% solution and 25% solution combine to equal the 20% solution. We can say that

Volume of HCl in the 10% solution	+	Volume of HCl in the 25% solution	=	Volume of HCl in the 20% solution
$0.10(50)$	+	$0.25n$	=	$0.20(50 + n)$

$$5 + 0.25n = 10 + 0.20n \qquad \text{Multiply } 0.10(50) \text{ and distribute the } 0.20.$$

$$
\begin{array}{ll}
5 + 0.25n = 10 + 0.20n & \\
\underline{\quad\quad -0.20n \quad\quad -0.20n} & \text{Subtract } 0.20n \text{ from both sides.} \\
5 + 0.05n = 10 + 0 & \\
\underline{-5 \quad\quad\quad\quad -5} & \text{Subtract 5 from both sides.} \\
0 + 0.05n = 5 & \\
\dfrac{0.05n}{0.05} = \dfrac{5}{0.05} & \text{Divide both sides by } 0.05. \\
n = 100 &
\end{array}
$$

Answer 100 milliliters of the 25% solution must be added to 50 milliliters of the 10% solution to create a solution that is 20% HCl.

Check Verify that the volume of HCl in the two original solutions combined is equal to the volume of HCl in the combined solution.

$$0.10(50) + 0.25(100) \stackrel{?}{=} 0.20\,(150)$$
$$5 + 25 \stackrel{?}{=} 30$$
$$30 = 30 \qquad \text{True}$$

Following is a procedure that can be used as a general guideline for solving problems of the type illustrated in Example 2.

Procedure **Solving Mixture Problems**

To solve problems involving mixing solutions,

1. Use a table to organize the concentrations and volumes. Multiply the individual concentrations and solution volumes to get expressions for the volume of the particular chemical.
2. Write an equation that the sum of the volumes of the chemical in each solution set is equal to the volume in the combined solution.
3. Solve the equation.

Your Turn 2 Andre has 75 milliliters of a 5% sulfuric acid solution. How much of a 20% sulfuric acid solution must be added to create a solution that is 15% sulfuric acid?

Answer **Your Turn 2**

150 ml

Note: Exercises marked with a ★ represent challenging exercises.

1. Suppose an investor has $10,000 to invest in two accounts. The table lists some ways to split the principal that he is considering.
   a. Complete the table.
   b. How did you determine the unknown principals in the table?

Account 1	Account 2
$9000	$1000
$8500	
	$6400
$3200	

   c. In general, if he invests *P* in one of the accounts, how do you describe the amount invested in the other account?

2. An investor always invests in two accounts according to the pattern shown in the table.
   a. Complete the table.
   b. In general, if he invests *P* in account 1, how would you describe the amount invested in account 2?

Account 1	Account 2
$500	$1000
$700	$1400
$900	
	$2400

   c. If he invests *P* in account 2, how would you describe the amount invested in account 1?

3. Complete the following table.

Principal	Interest Rate	Annual Interest
$5500	4%	
$10,450	3.75%	
$P	5%	
$(15,000 − P)	6.5%	

4. Complete the following table.

Principal	Interest Rate	Annual Interest
$6800	6%	
$12,650	$4\frac{1}{2}\%$	
$P	6.5%	
$(50,000 − P)	4.3%	

5. Complete the table, which contains a description of two unknowns and a variable for one of the unknowns. Use the description to describe the second unknown in terms of the variable.

Description	Account 1	Account 2
Carol invests twice as much in account 2 as in account 1.	*P*	
Jennifer invests $1200 more in account 1 as in account 2.		*P*
Tim invests $6000 in the two accounts.	*P*	
Johnny invests $4000 in the two accounts.		*P*

6. Complete the table, which contains a description of two unknowns and a variable for one of the unknowns. Use the description to describe the second unknown in terms of the variable.

Description	Account 1	Account 2
Ahmad invests two-thirds as much in account 2 as in account 1.	*P*	
Johanna invests $3500 less in account 1 than in account 2.		*P*
Franz invests $5000 more in account 2 than in account 1.	*P*	
Murlene invests a total of $10,000 in the two accounts.		*P*

7. Given the total interest from an investment in two plans, once you have the expressions for the two interest amounts, how do you write the equation?

8. When completing a table for mixture problems, what are the three categories?

9. The following table shows some possible combinations of two solutions. Look for a pattern; then complete the table.

Volume of Solution 1	Volume of Solution 2	Combined Solution
25 ml	50 ml	75 ml
40 ml	80 ml	
100 ml		
n ml		

10. The following table shows possible combinations of two solutions. Look for a pattern and complete the table.

Volume of Solution 1	Volume of Solution 2	Combined Solution
10 oz.	30 oz.	40 oz.
40 oz.	120 oz.	
75 oz.		
x oz.		

11. Complete the following table.

Volume of Solution	Percent Pure Alcohol	Volume of Alcohol in Solution
130 oz.	5%	
65 oz.	16.5%	
x oz.	$12\frac{1}{2}\%$	
$(50 - x)$ oz.	8.2%	

12. Complete the following table.

Volume of Solution	Percent Pure Sulfuric Acid	Volume of Sulfuric Acid in Solution
30 ml	45%	
150 ml	$5\frac{3}{4}\%$	
y ml	3.5%	
$(125 - y)$ ml	27.6%	

Exercises 13–28 involve investment. Complete a table, write an equation, and then solve. See Example 1.

13. Janice invests in a plan that has an APR of 2%. She invests three times as much in a plan that has an APR of 4%. If the total interest after one year from the investments is $644, how much was invested in each plan?

14. Gayle invests in a plan that has an APR of 3%. She invests twice as much in a plan that has an APR of 5%. If the total interest after one year from the investments is $637, how much was invested in each plan?

15. Tory invests money in two plans. She invests two-fifths of the money at an annual return rate of 2% in a money market account. She invests the remainder of the money in a short-term mutual fund account with a return rate of 3%. If the total annual interest from the investments is $104, how much was invested in each plan?

16. Dwight invests money in two plans. He invests two-fifths of the money at an APR of 8% and the rest at an APR of 7.6%. If the total interest after one year from the investments is $620.80, how much was invested in each plan?

17. Boyd invests in a plan that has an APR of 8%. He invests in a 12% APR account $650 more than what he invested in the 8% account. If the total interest after one year from the investments is $328, how much was invested in each plan?

18. Deon invests in an account that has an APR of 2.5%. He invests in a 3.5% APR account $400 more than what he invests in the 2.5% account. If the total annual interest from the investments is $242, how much was invested in each account?

19. Agnes invests $4500 in two plans. Plan 1 is at an APR of 4%, and plan 2 is at an APR of 6%. If the total interest after one year is $234, what principal was invested in each plan?

20. Juan invests $12,800 in two plans. Plan 1 is at an APR of 7%, and plan 2 is at an APR of 5%. If the total interest after one year is $706, what principal was invested in each plan?

21. Kennon invested $1600. She was able to put some of this money into a CD at a 5% APR and the rest into a CD at only 4%. If she achieved an annual interest income of $73.40 on these investments, how much was invested in each plan?

22. Shanequa has $4000. She put some of the money into savings that pays 6% annually and the rest in an account that pays 7%. If her total interest for the year is $264, how much did she invest at each rate?

23. Annette invested $5000. She put some into an investment earning 5.5% annually and the rest earning 7%. If her interest income during this year was $335, how much did she invest at each rate?

24. Buddy inherited $12,000. He invested part of the money at a 4% APR and the rest at 5.5%. If his annual interest income from these investments is $532.50, how much did he invest at each rate?

★ **25.** Deloris invests in a plan that has an APR of 6%. She invests twice as much in a plan that ends up as a loss of 2% (that is, a −2% APR). If the net gain from the investments is $84, how much was invested in each plan?

★ **26.** Dominique invests a total of $10,000 in two plans. Plan 1 is at an APR of 5%, while plan 2 ends up returning a loss of 9% (that is, a −9% APR). If the return on the investments after one year is a net loss of $312, what principal was invested in each plan?

★ **27.** Cody has two investments totaling $10,000. Plan A is at an APR of 4%, and plan B is at an APR of 8%. After one year, the interest earned from plan B is $320 more than the interest from plan A. How much did Cody invest in each plan?

★ **28.** Curtis has two investments totaling $12,000. Plan A is at an APR of 5%, and plan B is at an APR of 7%. After one year, the interest earned from plan B is $360 more than the interest earned from plan A. How much did Curtis invest in each plan?

Exercises 29–46 involve mixtures. Complete a table, write an equation, and then solve. See Example 2.

29. How many liters of a 40% solution of HCl are added to 2000 liters of a 20% solution to obtain a 35% solution?

30. Charles has 80 milliliters of a 15% acid solution. How much of a 20% acid solution must be added to create a solution that is 18% acid?

31. Wilson has a bottle containing 35 milliliters of 15% saline solution and a bottle of 40% saline solution. He wants a 30% solution. How much of the 40% solution must be added to the 15% solution so that a 30% concentration is created?

32. Sharon has a bottle containing 45 milliliters of 15% HCl solution and a bottle of 35% HCl solution. She wants a 25% solution. How much of the 35% solution must be added to the 15% solution so that a 25% concentration is created?

33. A dairy is making a 30% buttermilk cream. If it mixes a 26% buttermilk cream with a 35% buttermilk cream, how much of each does it need to use to produce 300 pounds of the 30% buttermilk cream?

34. A pharmacist has a 45% acid solution and a 35% acid solution. How many liters of each must be mixed to form 80 liters of a 40% acid solution?

35. To form a 10% copper alloy weighing 75 grams, a 6% copper alloy is combined with an 18% copper alloy. How much of each type should be used?

36. A jeweler is mixing two silver alloys. How many ounces of an alloy containing 25% silver must be mixed with an alloy containing 20% silver to obtain 50 ounces of a 22% silver alloy?

37. A cough medicine contains 25% alcohol. How much nonalcoholic liquid should a pharmacist add to 120 milliliters of the cough medicine so that it contains only 20% alcohol?

38. A solution of gasoline and oil used for a gas-powered weed trimmer is 8% oil. How much gasoline must be added to 3 gallons of the solution to obtain a new solution that is 5% oil?

39. How many quarts of pure antifreeze must be added to 6 quarts of a 40% antifreeze solution to obtain a 50% antifreeze solution?

40. A 30-milliliter solution of alcohol and water is 10% alcohol. How much pure alcohol must be added to yield a 40% solution?

41. A candy store is selling two types of candy, one at $1.20 per pound and the other at $0.90 per pound. To sell a mixture at $1.11 a pound, how many pounds of each must the store mix to make 80 pounds?

42. Daly's Farm Feed sells two kinds of horse feed. One is 20% oats, and the other is 28% oats. How many pounds of each should be mixed to form 50 pounds of a feed that is 25% oats?

43. The Market Street Sweets sells two types of pralines: a caramel version for $1.65 a pound and a chocolate version for $1.25 a pound. A mixture of these pralines can be sold for $1.49 a pound. How many pounds of each kind should be used to make 5 pounds of the $1.49 mixture?

44. Cromer's sells salted peanuts for $2.50 per pound and cashews for $6.75 per pound. How many pounds of each are needed to obtain a 10-pound mixture costing $4.20 per pound?

45. The Candy Shoppe wants to mix 115 pounds of candy to sell for $0.80 per pound. How many pounds of $0.60 candy must be mixed with a candy costing $1.20 per pound to make the desired mix?

46. A coffee shop wants to mix 20 pounds of coffee that will sell for $7 per pound. How many pounds of coffee at $5 per pound must be mixed with coffee costing $8 per pound to make the mix?

Puzzle Problem

Herschal has $18,000 to invest. He invests $\frac{2}{3}$ in one account and the rest in a second account.

The first account returned 2% more than the second account. The total return on his investment was $1050. What were the percentage rates of the two accounts?

Review Exercises

Exercises 1–3 **EXPRESSIONS**

[1.5] *For Exercises 1 and 2 , simplify.*

1. $\dfrac{3}{4} - \dfrac{5}{6} \cdot \left(\dfrac{2}{3}\right)^2$

2. $(-4)^3$

$\begin{bmatrix} \mathbf{1.6} \\ \mathbf{1.7} \end{bmatrix}$ **3.** Write an expression in simplest form that describes the perimeter of the rectangle shown.

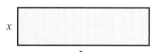

[1.1] **4.** Use $<$, $>$, or $=$ to make a true statement:

$-|-14|$ ___ 14

Exercises 4–6 **EQUATIONS AND INEQUALITIES**

[2.3] *For Exercises 5 and 6, solve.*

5. $4(y - 3) + y = 7y - 8$

6. $\dfrac{3}{4}x - 5 = \dfrac{2}{3}x + 1$

Chapter 3 Summary

Defined Terms

Section 3.1
Ratio (p. 170)
Unit ratio (p. 170)
Proportion (p. 171)
Congruent angles (p. 174)
Similar figures (p. 174)

Section 3.2
Percent (p. 183)

Section 3.3
Complementary angles (p. 199)
Supplementary angles (p. 199)

Formulas

Simple percent sentence: A percent of a whole is a part of the whole.
(p. 185)

$\text{Percent} = \dfrac{\text{Part}}{\text{Whole}}$ (p. 185)

Perimeter of a rectangle: $P = 2l + 2w$ (p. 199)

Distance, rate, and time: $d = rt$ (p. 211)

Simple interest: $I = Pr$ (p. 220)

Procedures, Rules, and Key Examples

Procedures/Rules	Key Examples

SECTION 3.1   Ratios and Proportions

If $\dfrac{a}{b} = \dfrac{c}{d}$, where $b \neq 0$ and $d \neq 0$, then $ad = bc$.	**Example 1:** Determine whether the ratios are equal. $$\frac{5}{6} \overset{?}{=} \frac{15}{18}$$ $18 \cdot 5 = 90 \qquad 6 \cdot 15 = 90$ $$\frac{5}{6} \underset{}{\overset{}{\times}} \frac{15}{18}$$ Because the cross products are equal, the ratios are equal.
To solve a proportion using cross products, 1. Calculate the cross products. 2. Set the cross products equal to each other. 3. Use the multiplication principle of equality to isolate the variable.	**Example 2:** Solve for the missing number in the proportion. $$\frac{5}{12} = \frac{x}{18}$$ $90 = 12x$ Cross products $\dfrac{90}{12} = \dfrac{12x}{12}$ Divide by 12. $7.5 = x$ *(continued)*

Procedures/Rules	Key Examples

SECTION 3.1 Ratios and Proportions (*continued*)

To solve proportion problems,
1. Set up a proportion in which the numerators and denominators of the ratios correspond in a logical manner.
2. Solve using cross products.

Example 3: A car can travel about 350 miles on 16 gallons of gasoline. How many gallons would be needed to travel 600 miles?

$$\frac{350}{16} = \frac{600}{x}$$
$$350x = 9600 \qquad \text{Cross products}$$
$$\frac{350x}{350} = \frac{9600}{350} \qquad \text{Divide by 350.}$$
$$x \approx 27.4 \text{ gal.}$$

SECTION 3.2 Percents

To write a percent as a fraction or decimal,
1. Write the percent as a ratio with 100 in the denominator.
2. Simplify to the desired form.

Note: When simplifying, remember that dividing a decimal number by 100 moves the decimal point two places to the left.

Example 1: Write each percent as a fraction in lowest terms and as a decimal.

a. $42\% = \dfrac{42}{100} = \dfrac{21}{50} = 0.42$

b. $8.5\% = \dfrac{8.5}{100} = \dfrac{85}{1000} = \dfrac{17}{200} = 0.085$

c. $20\dfrac{1}{2}\% = \dfrac{20\frac{1}{2}}{100} = \dfrac{41}{200} = 0.205$

To write a fraction or decimal number as a percent,
1. Multiply by 100%.
2. Simplify.

Note: Multiplying a decimal number by 100 moves the decimal point two places to the right.

Example 2: Write as a percent.
a. $0.453 = 0.453 \cdot 100\% = 45.3\%$

b. $\dfrac{5}{8} = \dfrac{5}{\underset{2}{\cancel{8}}} \cdot \dfrac{\overset{25}{\cancel{100}}}{1}\% = \dfrac{125}{2}\% = 62\dfrac{1}{2}\%$

To translate simple percent sentences:

Method 1. Translate the sentence word for word.
1. Select a variable for the unknown.
2. Translate the word *is* to an equal sign.
3. If *of* is preceded by the percent, translate it to multiplication. If *of* is preceded by a whole number, translate it to division.

Method 2. Translate to a proportion by writing the following form:

$$\text{Percent} = \frac{\text{Part}}{\text{Whole}},$$

where the percent is expressed as a fraction with a denominator of 100.

Example 3: 15% of what number is 33?

Method 1: Direct translation

$$0.15n = 33$$
$$\frac{0.15n}{0.15} = \frac{33}{0.15}$$
$$n = 220$$

Method 2: Proportion

$$\frac{15}{100} = \frac{33}{n}$$
$$15n = 3300$$
$$\frac{15n}{15} = \frac{3300}{15}$$
$$n = 220$$

(*continued*)

Procedures/Rules	Key Examples

SECTION 3.2   Percents (*continued*)

To solve problems involving percent,
1. Determine whether the percent, the whole, or the part is unknown.
2. Write the problem as a simple percent sentence (if needed).
3. Translate to an equation (word for word or proportion).
4. Solve for the unknown.

Example 4: A student answered 24 questions out of 30 correctly. What percent of the questions did the student answer correctly?

Simple percent sentence: What percent of 30 is 24?

Direct Translation:

$$p \cdot 30 = 24$$
$$\frac{30p}{30} = \frac{24}{30}$$
$$p = 0.8$$

Answer:
$$0.8 \cdot 100\% = 80\%$$

Proportion:

$$\frac{P}{100} = \frac{24}{30}$$
$$30P = 2400$$
$$\frac{30p}{30} = \frac{2400}{30}$$
$$P = 80$$

Answer: 80%

SECTION 3.3   Problems with Two or More Unknowns

To solve problems with two or more unknowns,
1. Determine which unknown will be represented by a variable. *Tip:* Let the unknown that is acted on be represented by the variable.
2. Use one of the relationships to describe the other unknown(s) in terms of the variable.
3. Use the other relationship to write an equation.
4. Solve the equation.

Example 1: One number is 15 more than twice another. The sum of the numbers is 39. Find the numbers.

Relationship 1: "One number is fifteen more than twice another."

Translation: $2x + 15$, where x represents the smaller number.

Relationship 2: "The sum of the numbers is 39."

Translation: $x + 2x + 15 = 39$
$$3x + 15 = 39$$
$$\underline{-15 \quad -15}$$
$$3x + 0 = 24$$
$$\frac{3x}{3} = \frac{24}{3}$$
$$x = 8$$

The smaller number is 8, and the other number is $2(8) + 15 = 31$.

SECTION 3.4   Rates

To solve for time when two objects are moving in opposite directions,
1. Use a table with columns for categories, rate, time, and distance. Use the fact that rate · time = distance.
2. Write an equation that is the sum of the individual distances equal to the total distance of separation.
3. Solve the equation.

Example 1: Mark and Latonia are 17 miles apart traveling toward each other. Mark is traveling at a rate of 45 miles per hour, and Latonia is traveling at a rate of 40 miles per hour. How long will it be until they meet?

Categories	Rate	Time	Distance
Mark	45	t	$45t$
Latonia	40	t	$40t$

Equation: $45t + 40t = 17$
$$85t = 17$$
$$\frac{85t}{85} = \frac{17}{85}$$
$$t = 0.2 \text{ hr. or } 12 \text{ min.}$$

(*continued*)

Procedures/Rules	Key Examples

SECTION 3.4 Rates (*continued*)

To solve problems involving two objects traveling in the same direction in which the objective is to determine the time for one object to catch up to the other,
1. Use a table to organize the rates and times. Let t represent the time for the object to catch up and add the time difference to t to represent the other object's time.
2. Set the expressions for the individual distances equal.
3. Solve the equation.

Example 2: Frank is traveling at a rate of 64 miles per hour and passes Exit 43 at 4:35 P.M. Juanita is traveling at a rate of 72 miles per hour in the same direction and passes the same exit at 4:50 P.M. At what time does she catch up with Frank?

Categories	Rate	Time	Distance
Frank	64	$t + 0.25$	$64(t + 0.25)$
Juanita	72	t	$72t$

Equation:
$$64(t + 0.25) = 72t$$
$$64t + 16 = 72t$$
$$\underline{-64t \qquad\qquad -64t}$$
$$0 + 16 = 8t$$
$$\frac{16}{8} = \frac{8t}{8}$$
$$2 = t$$

Answer: Juanita catches up at 6:50 P.M.

SECTION 3.5 Investment and Mixture

To solve problems involving two interest rates in which the total interest is given,
1. Use a table to organize the interest rates and principals. Multiply the individual rates and principals to get expressions for the individual interests.
2. Write an equation that is the sum of the expressions for interest set equal to the given total interest.
3. Solve the equation.

Example 1: Dan invests a total of $4000 in two different plans. The first plan returns 9% annually, while the second plan returns 6%. If the total interest earned after one year is $285, what principal was invested in each plan?

Categories	Rate	Principal	Interest
Plan 1	0.09	p	$0.09p$
Plan 2	0.06	$4000 - p$	$0.06(4000 - p)$

Equation:
$$0.09p + 0.06(4000 - p) = 285$$
$$0.09p + 240 - 0.06p = 285$$
$$0.03p + 240 = 285$$
$$\underline{\qquad -240 \quad -240}$$
$$0.03p + 0 = 45$$
$$\frac{0.03p}{0.03} = \frac{45}{0.03}$$
$$p = 1500$$

Answer: $1500 in plan 1 and
$4000 - $1500 = $2500 in
plan 2

(continued)

SECTION 3.5 Investment and Mixture (*continued*)

To solve problems involving mixing solutions,

1. Use a table to organize the concentrations and volumes. Multiply the individual concentrations and solution volumes to get expressions for the volume of the particular chemical.
2. Write an equation that is the sum of the volumes of the chemical in each solution set equal to the volume in the combined solution.
3. Solve the equation.

Example 2: 200 milliliters of a 15% alcohol solution is mixed with a 30% alcohol solution to make a solution that is 20% alcohol. How much of the 30% solution is needed?

Categories	Concentration	Volume of Solution	Volume of Alcohol
15% solution	0.15	200	$0.15(200)$
30% solution	0.30	n	$0.30n$
20% solution	0.20	$200 + n$	$0.20(200 + n)$

Equation:

$$0.15(200) + 0.30n = 0.20(200 + n)$$
$$30 + 0.30n = 40 + 0.20n$$
$$\underline{-0.20n \qquad\quad -0.20n}$$
$$30 + 0.10n = 40 + 0$$
$$30 + 0.10n = 40$$
$$\underline{-30 \qquad\qquad -30}$$
$$0 + 0.10n = 10$$
$$\frac{0.10n}{0.10} = \frac{10}{0.10}$$
$$n = 100$$

Answer: 100 ml of the 30% solution is needed.

Chapter 3 Review Exercises

For Exercises 1–6, answer true or false.

[3.1] **1.** The cross products of a proportion are equal.

[3.2] **2.** A percent is a ratio out of 100.

[3.2] **3.** $7.8\% = 0.78$

[3.3] **4.** Consecutive odd integers can be represented as $x, x + 1, x + 3, \ldots$.

[3.3] **5.** Complementary angles have a sum of $180°$.

[3.3] **6.** "One less than a number" can be represented as $1 - x$.

For Exercises 7–10, fill in the blank.

[3.2] **7.** To write a percent as a fraction or decimal,
1. Write the numeral over _____.
2. Simplify to the desired form.

[3.1] **8.** Angles that have the same measurement are _____.

[3.3] 9. Angles whose sum is 90° are _____.

[3.1] 10. An equation in the form $\dfrac{a}{b} = \dfrac{c}{d}$, where $b \neq 0$ and $d \neq 0$, is a _____.

[3.1] 11. A recipe calls for $\dfrac{3}{4}$ of a cup of sugar and 2 cups of flour. Write the ratio of sugar to flour as a fraction in simplest form.

[3.1] 12. The price of a 10.5-ounce can of vegetables is $0.89. Write the unit ratio of price to weight rounded to the nearest thousandth. Interpret the answer.

Exercises 13–32 EQUATIONS AND INEQUALITIES

[3.1] *For Exercises 13–16, determine whether the ratios are equal.*

13. $\dfrac{2}{5} \overset{?}{=} \dfrac{10}{20}$

14. $\dfrac{1}{3} \overset{?}{=} \dfrac{13}{39}$

15. $-\dfrac{2}{8} \overset{?}{=} -\dfrac{1}{4}$

16. $\dfrac{15}{21} \overset{?}{=} \dfrac{130.2}{93}$

[3.1] *For Exercises 17–22, solve for the missing number.*

17. $\dfrac{3}{5} = \dfrac{15}{x}$

18. $\dfrac{-11}{12} = \dfrac{n}{-24}$

19. $\dfrac{\frac{2}{5}}{4\frac{1}{2}} = \dfrac{7}{p}$

20. $\dfrac{k}{15.3} = \dfrac{5}{8}$

21. $\dfrac{-60}{w} = \dfrac{20}{40}$

22. $\dfrac{c}{12.7} = \dfrac{5}{3}$

[3.1] *For Exercises 23–28, translate to a proportion and solve.*

23. Jason has driven 225 miles in 3 hours. If he averages the same speed after lunch, how far can he expect to travel in 4 hours?

24. Gutzon Borglum used a simple ratio of 1 to 12 (1 inch to 1 foot) when he transferred plaster models to granite for Mount Rushmore. Each sculpture is 60 feet tall. How tall was a plaster model?

25. It costs $5.85 to mail a 4.5-pound package. How much will it cost to mail a 7-pound package?

26. Tamika pays property taxes of $1848 on her house, which is assessed at $125,000. She is planning to buy a house assessed at $175,000. At the same rate, how much will her taxes be on the new home?

27. Ferrari estimates that the 2007 612 Scaglietti two-door coupe will travel 209 city miles on one tank of gas. If the gas tank of the automobile holds 23.7 gallons, how far can a driver expect to travel on 25 gallons?

28. Dan notes on his natural gas bill that he paid $195.16 for 159 Therms in November. If he knows he usually uses around 220 Therms in January, how much can he expect his bill to be in January?

[3.1] *For Exercises 29–32, find the missing lengths in the similar shapes.*

29.

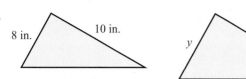

30.

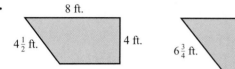

31.

32.

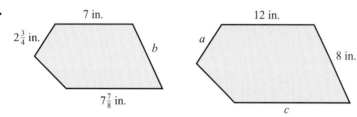

Exercises 33–44 EXPRESSIONS

[3.2] *For Exercises 33–38, write each percent as a decimal and as a fraction in simplest form.*

33. 15%   **34.** 82.5%   **35.** $12\frac{1}{2}\%$ **36.** 2.45%   **37.** 10%   **38.** $33\frac{1}{3}\%$

[3.2] *For Exercises 39–44, write as a percent.*

39. $\dfrac{2}{5}$ **40.** $\dfrac{1}{3}$ **41.** $\dfrac{4}{11}$ **42.** 0.35   **43.** 1.2   **44.** 2.016

Exercises 45–68 EQUATIONS AND INEQUALITIES

[3.2] *For Exercises 45–48, translate the percent sentence to an equation; then solve.*

45. 16% of 91 is what number?   **46.** 14.5% of what number is 42?

47. 6.5 is what percent of 20?   **48.** What percent is 18 out of 32?

For Exercises 49–68, solve.

[3.2] 49. In a Gallup poll, 47% of those polled knew someone who lost a job during the past year. If 1016 Americans were polled, how many knew someone who lost a job?

[3.2] 50. Karen buys a new kitchen table for $759.99. If the tax rate in her area is 6%, what is the cost after tax is added?

[3.2] 51. In 2007, the Boston Red Sox won 96 out of 162 games. What percent of its games did the Red Sox win?

[3.2] 52. In 1995, cotton farmers received an average price for upland cotton of 75.4 cents per pound. In 2006, the average price of upland cotton was 48.2. What was the percent of the decrease in average price? (*Source:* National Agricultural Statistics Services, U.S. Department of Agriculture)

[3.3] 53. One number is four times another. The sum of the numbers is 55. What are the numbers?

[3.3] 54. The greater of two numbers is five more than three times the lesser number. The sum of the two numbers is 15. What are the numbers?

[3.3] 55. The page numbers of a book are two consecutive integers whose sum is 83. What are the page numbers?

[3.3] 56. Jack and Jill are neighbors whose house numbers are consecutive even integers with Jack's house number less than Jill's. Two times Jack's house number is 22 more than Jill's house number. What are their house numbers?

[3.3] 57. The sum of three consecutive odd integers is 63. What are the integers?

[3.3] 58. Find three consecutive integers such that twice the sum of the first two is six less than triple the third number.

[3.3] 59. The length of a rectangle is 6 inches more than the width, and the perimeter is 48 inches. What are the length and width?

[3.3] 60. The length of a rectangle is 15 feet more than twice its width, and the perimeter is 120 feet. What are the length and width?

[3.3] 61. Two angles are supplementary. One of the angles is 32° less than the measure of the other angle. What are the measures of the two angles?

[3.3] 62. Two angles are complementary. One of the angles is three less than twice the other. What are the measures of the two angles?

[3.3] 63. While buying pantyhose, Sophie notices that regular fit costs $6.50 and control top costs $9.50. If she buys five pairs of hose and spends $38.50, how many of each type did she buy?

[3.3] 64. A paint contractor bought paint for the interior and exterior of a house. He bought 4 gallons more of exterior paint than interior paint. The interior paint cost $12 per gallon, the exterior paint cost $18 per gallon, and the total cost of the paint was $372. How many gallons of interior and exterior paint did he buy?

[3.4] 65. In Cincinnati, a Land Rover and an Escort enter I-71 at Exit 2 at the same time and travel in opposite directions. The Land Rover traveling south is caught in traffic and travels at 45 miles per hour. The Escort traveling north encounters light traffic and is able to travel at 65 miles per hour. In how many hours will the cars be 385 miles apart?

[3.4] 66. If a plane flying from O'Hare travels west at 550 miles per hour and another leaving at the same time travels east at 450 miles per hour, how many hours until the planes are 2600 miles apart?

[3.4] **67.** Wayne and Libby are both driving south. At 3 P.M., Wayne was 5 miles south of Libby. A little later Libby passed Wayne. If Libby is driving at 55 miles per hour and Wayne at 50 miles per hour, what time did Libby pass Wayne?

[3.4] **68.** Two trains started from Victoria Station and traveled in the same direction on parallel tracks. The freight train leaves at 10 A.M. The passenger train leaves at 11 A.M., but traveling at 25 miles per hour faster, it overtakes the freight train at 2 P.M. How fast is the freight train traveling?

[3.5] **69.** Terrell invests in a retirement plan that has an APR of 8%. He invests twice as much in a second plan that has an APR of 10%. If the total annual interest from the investments is $5600, how much was invested in each plan?

[3.5] **70.** Brian invests in a retirement plan that has an APR of 7%. He invests three times as much in a second plan that ended up as a loss of 2% (that is, at −2% APR). If the net annual gain from the investments is $102, how much was invested in each plan?

[3.5] **71.** A solution that is 90% alcohol is to be mixed with another solution that is 40% alcohol. The chemist is preparing 100 liters of a 60% alcohol solution. How many liters of each must he use?

[3.5] **72.** How many ounces of a 20% saline solution must be added to 40 ounces of a 10% solution to produce a 16% solution?

Chapter 3 Practice Test

For Extra Help

CHAPTER
Test Prep
VIDEOS

Step-by-step test solutions are found on the Chapter Test Prep Videos available via the Video Resources on DVD, in *MyMathLab* , and on YouTube (search "CarsonElemAlg" and click on "Channels").

1. The roof of a building rises 8 inches for every 20 inches of horizontal length. Write the ratio of rise to horizontal length as a fraction in simplest form.

2. A 14.5-ounce box of cereal costs $3.77. Write the unit ratio of price to weight. Interpret the answer.

For Exercises 3 and 4, solve for the missing number.

3. $\dfrac{m}{8} = \dfrac{5}{12}$

4. $\dfrac{-32}{-9} = \dfrac{8}{b}$

5. On a trip, Michael drives 857.5 miles using 35 gallons of fuel. If he has another trip planned that will cover about 1200 miles, how many gallons of fuel should he expect to use?

6. The following figures are similar. Find the missing side length.

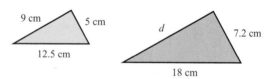

For Exercises 7 and 8, write each percent as a decimal and as a fraction in simplest form.

7. 22%

8. $3\dfrac{1}{3}\%$

For Exercises 9 and 10, write as a percent.

9. 3.2

10. $\frac{2}{5}$

For Exercises 11–20, solve.

11. 12 is what percent of 60?

12. 20.5% of 70 is what number?

13. 70 is 40% of what number?

14. According to a survey conducted by the National Center for Health, 37.1% of Americans say that they are in excellent health. Based on these findings, in a group of 500 Americans, how many would say that they are in excellent health? (*Source:* National Center for Health Statistics)

15. In 2007, Phoenix was the top seed of the Western Conference in the WNBA with 23 wins out of 34 games. What percent of its games did Phoenix win? (*Source: The World Almanac and Book of Facts*, 2008)

16. In March, Andrea notes that her electric bill is $148.40. In April, her bill is $166.95. What is the percent of the increase in her bill?

17. The ostrich lays the largest egg of any bird. An average ostrich egg weighs 14.4 ounces more than twice the average weight of an albatross egg. If the combined average weight of an ostrich egg and an albatross egg is 77.4 ounces, find the average weight of each type of egg.

18. The sum of two consecutive even integers is 110. Find the integers.

19. The length of a rectangle is one more than double the width. If the perimeter of the rectangle is 32 kilometers, what are the length and width?

20. Two angles are supplementary. If the larger angle is 24° more than the smaller angle, find the measure of the two angles.

21. Christie owns stock in both Walmart and Target. The number of shares of Target stock is 7 more than the number of shares of Walmart stock. On January 4, 2008, the Target stock was worth $48 per share, the Walmart stock was worth $46 per share, and the total value of the stock was $1746. How many shares of each stock does Christie have?

22. Two runners begin running toward each other along the same path. One person runs at a speed of 4 miles per hour, and the other person runs at 4.5 miles per hour. If they begin $4\frac{1}{4}$ miles apart, how long will it be until they meet?

23. An F-22 jet passes a landmark at a speed of 450 miles per hour. A second F-22 traveling in the same direction at 570 miles per hour passes the same landmark ten minutes later. How long will it take the second jet to catch up with the first jet?

24. Kari invests a total of $4500 in two plans. After one year, she found that plan A earned 8% interest and plan B earned 6% interest. If she earned a total of $330 that year, how much did she invest in each plan?

25. How many ounces of a 15% saline solution must be added to 30 ounces of a 25% solution to produce a 20% solution?

Chapters 1–3 Cumulative Review Exercises

For Exercises 1–4, answer true or false.

[1.1] **1.** $\sqrt{2}$ is an irrational number.

[1.5] **2.** When evaluating an expression using the order of operations, you must always do multiplication before division.

[1.6] **3.** "Four less than a number" can be translated to $4 - n$.

[2.2] **4.** The equation $2(x - 1) - 3x = 5 - x - 7$ is an identity.

For Exercises 5 and 6, fill in the blank.

[1.2] **5.** A prime number is a natural number that has exactly two different factors, 1 and _____.

[2.3] **6.** To clear fractions from an equation, we can use the multiplication principle of equality and multiply both sides by the _____.

[1.2] **7.** Find the prime factorization of 120.

[1.7] **8.** Determine the values that make $\dfrac{4x}{x - 3}$ undefined.

[1.7] **9.** Which property of arithmetic is illustrated by $4(x + 3) = 4x + 12$?

[3.1] **10.** Explain in your own words how to solve a proportion.

Exercises 11–17 ◢ **EXPRESSIONS**

[1.5] *For Exercises 11–16, simplify.*

11. $\left(-\dfrac{2}{3}\right)^2$

12. $\sqrt{16} + \sqrt{9}$

13. $\dfrac{5 - 2(-1)^2}{3\sqrt{64} + 36}$

14. $-|3 \cdot (-5)| + |(-1)^3|$

15. $2\left(-\dfrac{1}{2}\right)^2 - 4\left(\dfrac{3}{5}\right) + 3$

16. $-6 - 2(3 + 0.2)^2$

[1.7] **17.** Use the distributive property to simplify $-3\left(\dfrac{1}{6}x - 9\right)$.

Exercises 18–30 ◢ **EQUATIONS AND INEQUALITIES**

[2.2 and 2.3] *For Exercises 18–21, solve and check.*

18. $3.23c - 8.75 = 1.41c + 7.63$

19. $2m + \dfrac{1}{2} = 4m - 3\dfrac{1}{2} + 8m$

20. $6x - 2 = 3\left(2x - \dfrac{2}{3}\right)$

21. $-7 = -6(p + 2) + 5(2p - 3)$

[2.4] *For Exercises 22 and 23, solve for the indicated variable.*

22. $A = \dfrac{1}{2}bh$ for b

23. $P = a + b + c$ for c

For Exercises 24–30, solve.

[3.2] **24.** What is 15% of 42.3?

[3.3] **25.** The sum of three consecutive odd integers is -3. What are the integers?

[2.6] **26.** Jon scored an 82 on his first test and a 70 on his second test. What range of scores does he need on his third test to have an average of at least 80?

[3.5] **27.** How many liters of a 60% solution of boric acid should be added to 10 liters of a 30% solution to obtain a 50% solution?

[3.4] **28.** Archie and Veronica are both driving east on Riverdale Drive. At 3 P.M., Archie, driving at 50 miles per hour, was 5 miles east of Veronica. A little later Veronica, driving at 55 miles per hour, passed Archie. At what time did Veronica pass Archie?

[3.5] **29.** Cromers sells cashews for $7.00 per pound and peanuts for $3.00 per pound. If a 10-pound bag is mixed selling at $4.50 per pound, how many pounds of each type nut should be used?

[3.1] **30.** If a car travels 120 miles on 4 gallons of gasoline, how far can it travel on 10 gallons?

Graphing Linear Equations and Inequalities

4

"There's a compelling reason to master information and news. Clearly there will be better job and financial opportunities. Other high stakes will be missed by people if they don't master and connect information."

—Everette Dennis,
Media foundation executive and professor of media and entertainment

"Where is the wisdom we have lost in knowledge?
Where is the knowledge we have lost in information?"

—T. S. Eliot (1888–1965)
American-born British critic and writer

All equations and inequalities have corresponding pictures, or graphs. The graph of an equation or inequality offers information about the solutions to the equation or inequality. Graphs can be helpful in solving problems. For example, we may be given an equation or inequality that describes a particular problem. The graph of that equation may offer additional insight about possible solutions to the problem. Or a problem may have some data that we can graph. We can often use such a graph to help us write an equation or inequality that describes the situation. Both approaches can be quite useful in solving problems.

4.1 The Rectangular Coordinate System

Objectives

1. Determine the coordinates of a given point.
2. Plot points in the coordinate plane.
3. Determine the quadrant for a given ordered pair.
4. Determine whether the graph of a set of data points is linear.

Objective 1 Determine the coordinates of a given point. In 1619, René Descartes, the French philosopher and mathematician, recognized that positions of points in a plane could be described using two number lines that intersect at a right angle. Each number line is called an *axis*.

Two perpendicular axes form the *rectangular*, or Cartesian, *coordinate system*, named in honor of René Descartes. Usually, we call the horizontal axis the *x-axis* and the vertical axis the *y-axis*.

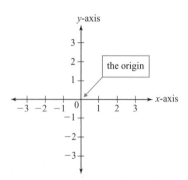

The point where the axes intersect is 0 for both the *x*-axis and *y*-axis. This position is called the *origin*. The positive numbers are to the right and up from the origin, whereas negative numbers are to the left and down from the origin.

Any point in the plane can be described using two numbers, one number from each axis. To avoid confusion, these two numbers are written in a specific order. The number representing a point's *horizontal distance* from the origin is given *first*, and the number representing a point's *vertical distance* from the origin is given *second*. Because the order in which we say or write these two numbers matters, we say that they form an *ordered pair*. Each number in an ordered pair is called a *coordinate* of the ordered pair. The notation for writing ordered pairs is (horizontal coordinate, vertical coordinate).

Consider the point labeled *A* in the coordinate plane shown. The point is drawn at the intersection of the 3rd line to the right of the origin and the 4th line up from the origin. The ordered pair that describes point *A* is $(3, 4)$.

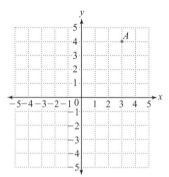

Connection Think of the intersecting lines in the grid as avenues and streets in a city such as New York, where avenues are north–south and streets are east–west. To describe an intersection, a person would say the avenue first, then the street. Point *A* is at the intersection of 3rd Avenue and 4th Street, or, as a New Yorker would say, "3rd and 4th."

Procedure | **Identifying the Coordinates of a Point**

To determine the coordinates of a given point in the rectangular system,

1. Follow a vertical line from the point to the *x*-axis (horizontal axis). The number at this position on the *x*-axis is the first coordinate.
2. Follow a horizontal line from the point to the *y*-axis (vertical axis). The number at this position on the *y*-axis is the second coordinate.

Example 1 Write the coordinates for each point shown.

Answers

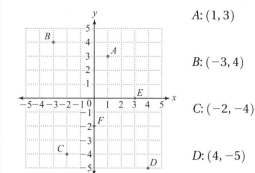

$A: (1, 3)$ — From point A, a vertical line intersects the x-axis at 1 and a horizontal line intersects the y-axis at 3.

$B: (-3, 4)$ — From point B, a vertical line intersects the x-axis at -3 and a horizontal line intersects the y-axis at 4.

$C: (-2, -4)$ — From point C, a vertical line intersects the x-axis at -2 and a horizontal line intersects the y-axis at -4.

$D: (4, -5)$ — From D, a vertical line intersects the x-axis at 4 and a horizontal line intersects the y-axis at -5.

$E: (3, 0)$ — From E, a vertical line intersects the x-axis at 3 and a horizontal line intersects the y-axis at 0.

$F: (0, -2)$ — From F, a vertical line intersects the x-axis at 0 and a horizontal line intersects the y-axis at -2.

Your Turn 1 Write the coordinates for each point shown.

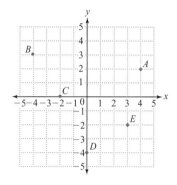

Objective 2 **Plot points in the coordinate plane.** Any point can be plotted in the coordinate plane using its coordinates. Remember that the first coordinate in an ordered pair indicates the distance to move right or left and that the second coordinate indicates the distance to move up or down. To plot the ordered pair $(3, 4)$, we begin at the origin, move to the right 3, move up 4, and draw a dot to indicate the point.

Procedure **Plotting a Point**

To graph or plot a point given its coordinates,

1. Beginning at the origin, $(0, 0)$, move to the right or left along the x-axis the amount indicated by the first coordinate.
2. From that position on the x-axis, move up or down the amount indicated by the second coordinate.
3. Draw a dot to represent the point described by the coordinates.

Example 2 Plot the point described by the coordinates.

a. $(3, -4)$ **b.** $(-4, 2)$ **c.** $(0, 5)$

Answers Your Turn 1

$A: (4, 2)$ $D: (0, -4)$
$B: (-4, 3)$ $E: (3, -2)$
$C: (-2, 0)$

SOLUTION

Note: For $(-4, 2)$, we begin at the origin and move to the left 4, then up 2.

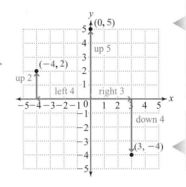

Note: For $(0, 5)$, because the first coordinate is 0, we do not move right or left. Because the second coordinate is 5, we move straight up 5, drawing the point on the *y*-axis.

Note: For $(3, -4)$, we begin at the origin and move to the right 3, then down 4.

Your Turn 2 Plot the point described by the coordinates.

$A: (-3, 2)$ $\qquad$ $B: (1, 4)$ $\qquad$ $C: (-2, -4)$ $\qquad$ $D: (3, 0)$

Objective ③ Determine the quadrant for a given ordered pair. The two perpendicular axes divide the coordinate plane into four regions called *quadrants*.

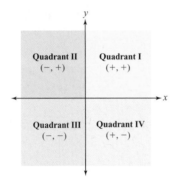

The quadrants are numbered using Roman numerals, as shown to the left. Note that the signs of the coordinates determine the quadrant in which a point lies. For points in quadrant I, both coordinates are positive. Every point in quadrant II has a negative first coordinate (horizontal) and a positive second coordinate (vertical). Points in quadrant III have both coordinates negative. Every point in quadrant IV has a positive first coordinate and a negative second coordinate. Points on axes are not in any quadrant.

Procedure **Identifying Quadrants**

To determine the quadrant for a given ordered pair, consider the signs of the coordinates.

$(+, +)$ means the point is in quadrant I.

$(-, +)$ means the point is in quadrant II.

$(-, -)$ means the point is in quadrant III.

$(+, -)$ means the point is in quadrant IV.

Note: If either coordinate is 0, the point is on an axis and not in a quadrant.

Answer **Your Turn 2**

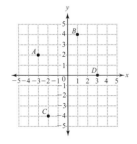

Example 3 State the quadrant in which each point is located. If the point is on an axis, state which axis.

a. $(-61, 23)$

Answer Quadrant II because the first coordinate is negative and the second coordinate is positive.

b. $\left(14, -37\dfrac{2}{3}\right)$

Answer Quadrant IV because the first coordinate is positive and the second coordinate is negative.

c. $(0, -12)$

Answer Because the *x*-coordinate is 0, this point is on the *y*-axis and is not in a quadrant.

Your Turn 3 State the quadrant in which each point is located. If the point is on an axis, state which axis.

a. $(-42, -109)$ **b.** $(37, -15.9)$ **c.** $(4.7, 0)$ **d.** $\left(-16\frac{3}{4}, 124\right)$ **e.** $\left(4.3, 1\frac{7}{8}\right)$

Objective ④ Determine whether the graph of a set of data points is linear. In many problems, data are listed as ordered pairs. For example, in a report on stock prices, we might have the listing in the margin of the closing price of a particular stock for each day over four days.

Day	Closing Price
1	$24.50
2	$25.75
3	$27.00
4	$28.25

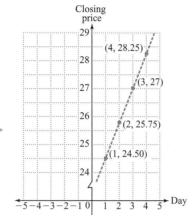

If we plot each pair of data as an ordered pair in the form (day, closing price), we see that the points lie on a straight line.

Because the points can be connected to form a straight line, they are said to be *linear*. Points that do not form a straight line are *nonlinear*.

Note: Because the vertical axis has dollar values that involve fractions, there are two marks for every whole dollar. This allows halves to be plotted accurately and quarters to be estimated well.

Note: The broken line in the vertical axis indicates that values between 0 and 24 were skipped to conserve space.

Time (in seconds)	Height (in feet)
0	6.00
0.5	7.50
1.0	8.25
1.5	7.50
2.0	6.00
2.5	2.00

Example 4 The data points in the margin track the trajectory, or path, of an object over a period of time. Plot the points with the time along the horizontal axis and the height along the vertical axis. Then state whether the trajectory is linear or nonlinear.

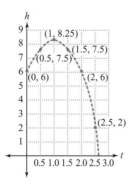

SOLUTION Because the data points do not form a straight line when connected, the trajectory is nonlinear.

Note: The graph for Example 4 does not have the origin at the center of the graph. It is often necessary to adjust the location of the origin, and consequently the axes, to fit the data. This frequently happens when the values plotted on the vertical and/or horizontal axes are nonnegative.

Note: In general, when a graph involves time, the time values are plotted along the horizontal axis.

Answers Your Turn 3

a. III   **b.** IV   **c.** on the *x*-axis

d. II   **e.** I

Your Turn 4 The following data points show the velocity of an object as time passes. Plot the points with the time along the horizontal axis and the velocity along the vertical axis. Then state whether the points are linear or nonlinear.

Time (in seconds)	Height (in feet)
0	2.0
1	2.5
2	3.0
3	3.5

Answer   Your Turn 4

The points are linear.

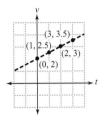

4.1 Exercises   For Extra Help   *MyMathLab*   Math XL PRACTICE   WATCH   DOWNLOAD

Note: Exercises marked with a ★ represent challenging exercises.

1. When writing an ordered pair, which is written first, the horizontal-axis coordinate or the vertical-axis coordinate?

2. Describe how to use the coordinates of an ordered pair to plot a point in the rectangular coordinate system.

3. Draw the axes in the rectangular coordinate system and label the quadrants. Include the signs for coordinates in each quadrant.

4. Describe how to determine whether a set of data points is linear.

For Exercises 5–8, write the coordinates for each point. See Example 1.

5.
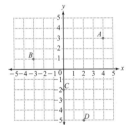

6.
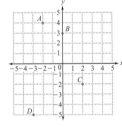

7.
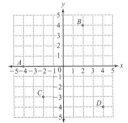

8.
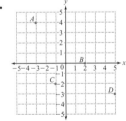

For Exercises 9–12, on a piece of graph paper, draw and label x- and y-axes. Then plot and label the points indicated by the coordinate pairs. See Example 2.

9. $(5, 4), (2, -3), (-1, -3), (2, 0)$

10. $(-3, 0), (0, 1), (-2, -2), (4, -1)$

11. $(-2, 0), (3, -3), (-1, -5), (2, 2)$

12. $(-1, -4), (0, -2), (4, -2), (-4, 2)$

For Exercises 13–24, state the quadrant in which the point is located. If the point lies on an axis, state which axis. See Example 3.

13. $(-6.5, 1050)$

14. $\left(-42\frac{1}{3}, -500\right)$

15. $(620, 50)$

16. $(-42, 50)$

17. $(57, -82.71)$

18. $(16, 27)$

19. $\left(-41\frac{1}{2}, -82\right)$

20. $\left(47, -51\frac{5}{8}\right)$

21. $(0, -9)$

22. $(0, 5.3)$

23. $(-0.6, 0)$

24. $(8.2, 0)$

For Exercises 25–34, determine whether the set of points is linear or nonlinear. See Example 4.

25. $(-2, -9), (0, -5), (2, -1), (4, 3), (5, 5)$

26. $(-5, 7), (-1, 3), (2, 0), (4, -2), (6, -4)$

27. $(-4, 8), (-2, 2), (0, 0), (1, 0.5), (3, 4.5)$

28. $(-1, -3), (0, 0), (2, 0), (3, -3)$

29. (pounds of chicken, cost): $(1, 2.5), (2, 5), (3, 7.5), (4, 10)$

30. (time, distance): $(0, 2), (1, 5), (2, 8), (3, 11), (4, 14)$

31. (time, height): $(0, 8.0), (0.5, 9.5), (1.0, 11.0), (1.5, 9.5),$ $(2.0, 8.0), (2.5, 6.5)$

32. (days, stock closing price): $(1, 23), (2, 21), (3, 19), (4, 17)$

★ **33.** (copy costs, number of copies): $(15, 300), (18, 400), (21, 500), (24, 600)$

★ **34.** (number of hours, plumbing cost): $(1, 120), (2, 160),$ $(3, 200), (4, 240)$

35. List the coordinates of three points on the line shown.

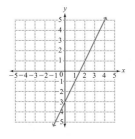

36. List the coordinates of three points on the line shown.

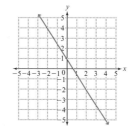

A vertex on a figure is a point where two line segments form a corner or joint on the figure. For Exercises 37 and 38, find the coordinates of each labeled vertex. See Example 1.

37.

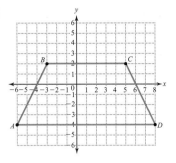

38.

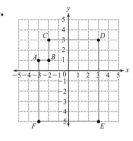

★ **39.** A parallelogram has been moved from its original position, shown with the dashed lines, to a new position, shown with solid lines.

    a. List the coordinates of each vertex of the parallelogram in its original position.

    b. List the coordinates of each vertex of the parallelogram in its new position.

    c. Write a rule that describes in mathematical terms how to move each point on the parallelogram in its original position to each point on the parallelogram in its new position.

40. A trapezoid has been moved from its original position, shown with the dashed lines, to a new position, shown with solid lines.

    a. List the coordinates of each vertex of the trapezoid in its original position.

    b. List the coordinates of each vertex of the trapezoid in its new position.

    c. Write a rule that describes in mathematical terms how to move each point on the trapezoid in its original position to each point on the trapezoid in its new position.

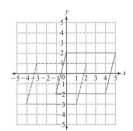

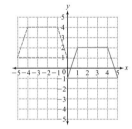

41. The ordered pairs $(-3, 1), (-1, 1), (-1, 3), (2, 3), (2, -2),$ and $(-3, -2)$ form the vertices of a figure.

    a. Plot the points and connect them to form the figure.

    b. Find the perimeter of the figure.

    c. Find the area of the figure.

42. The ordered pairs $(-4, 2), (-2, 2), (-2, 4), (1, 4), (1, 3), (3, 3), (3, -4),$ and $(-4, -4)$ form the vertices of a figure.

    a. Plot the points and connect them to form the figure.

    b. Find the perimeter of the figure.

    c. Find the area of the figure.

Review Exercises

Exercise 1 **CONSTANTS AND VARIABLES**

[1.1]   **1.** Graph -3 on a number line.

Exercises 2–4 **EXPRESSIONS**

[1.7]   **2.** Evaluate $2x - 3y$ when $x = 4$ and $y = 2$.   **[1.7]**   **3.** Evaluate $-4x - 5y$ when $x = -3$ and $y = 0$.

[1.7]   **4.** Evaluate $-\dfrac{1}{3}x + 2$ when $x = 6$.

Exercises 5 and 6 **EQUATIONS AND INEQUALITIES**

[2.3] *For Exercises 5 and 6, solve.*

5. $2x - 5 = 10$

6. $-\dfrac{5}{6}x = \dfrac{3}{4}$

4.2 Graphing Linear Equations

Objectives

1 Determine whether a given pair of coordinates is a solution to a given equation with two unknowns.

2 Find solutions for an equation with two unknowns.

3 Graph linear equations.

Objective 1 **Determine whether a given pair of coordinates is a solution to a given equation with two unknowns.**   In Chapter 2, we considered linear equations with one variable. We now consider linear equations that have two variables, such as $x + y = 4$ and $y = 2x - 3$.

A solution for an equation with two variables is a pair of numbers, one number for each variable, that can replace the corresponding variables and make the equation true. These solutions are ordered pairs, so we can write them using coordinates. The ordered pair $(1, 3)$ is a solution for $x + y = 4$ because replacing x with 1 and y with 3 makes the equation true.

Note: Unless otherwise specified, the coordinates are written in alphabetical order; so $(1, 3)$ means that $x = 1$ and $y = 3$.

$$x + y = 4$$
$$\downarrow \quad \downarrow$$
$$1 + 3 = 4 \qquad \text{The equation is true, so the ordered pair is a solution.}$$

Procedure   **Checking a Potential Solution for an Equation with Two Variables**

To determine whether a given ordered pair is a solution for an equation with two variables,

1. Replace the variables in the equation with the corresponding coordinates.
2. Verify that the equation is true.

Example 1   Determine whether the ordered pair is a solution for the equation.

a. $(-4, -10)$; $y = 2x - 3$

SOLUTION $-10 \overset{?}{=} 2(-4) - 3$ Replace x with -4 and y with -10 and see if the equation is true.

$-10 \overset{?}{=} -8 - 3$

$-10 \neq -11$ Because the equation is not true, $(-4, -10)$ is not a solution for $y = 2x - 3$.

b. $(4, -1)$; $2m - n = 9$

SOLUTION $2(4) - (-1) \overset{?}{=} 9$ Unless otherwise specified, ordered pairs are expressed in alphabetical order; therefore, replace m with 4 and n with -1 and see if the equation is true.

$8 + 1 \overset{?}{=} 9$

$9 = 9$ The equation is true, so $(4, -1)$ is a solution for $2m - n = 9$.

c. $\left(-2, -\dfrac{1}{3}\right)$; $y = \dfrac{2}{3}x + 1$

SOLUTION $-\dfrac{1}{3} \overset{?}{=} \dfrac{2}{3}(-2) + 1$ Replace x with -2 and y with $-\dfrac{1}{3}$ and see if the equation is true.

$-\dfrac{1}{3} \overset{?}{=} -\dfrac{4}{3} + \dfrac{3}{3}$

$-\dfrac{1}{3} = -\dfrac{1}{3}$ The equation is true, so $\left(-2, -\dfrac{1}{3}\right)$ is a solution for $y = \dfrac{2}{3}x + 1$.

Your Turn 1   Determine whether the ordered pair is a solution for the equation.

a. $(-8, -2)$; $x - 3y = -2$ **b.** $(1, -6)$; $q = -5p + 1$ **c.** $\left(\dfrac{3}{4}, \dfrac{9}{4}\right)$; $y = \dfrac{x}{3} + 2$

Objective ❷ Find solutions for an equation with two unknowns.   Consider the equation $x + y = 4$. We have already seen that $(1, 3)$ is a solution. But there are other solutions. We list some of those solutions in the following table.

x	y	Ordered Pair
0	4	$(0, 4)$
1	3	$(1, 3)$
2	2	$(2, 2)$
3	1	$(3, 1)$
4	0	$(4, 0)$
5	−1	$(5, -1)$

Note: Every solution for $x + y = 4$ is an ordered pair of numbers whose sum is 4.

In fact, $x + y = 4$, and every other linear equation, has an infinite number of solutions. For every x-value, there is a corresponding y-value that will add to the x-value to equal 4 and vice versa, which gives a clue about how to find solutions. We can simply choose a value for x or y and solve for the corresponding value of the other variable. (In this case, our equation was easy enough that we could solve for y mentally.)

Finding Solutions to Linear Equations with Two Variables

To find a solution to a linear equation with two variables,

1. Choose a value for one of the variables (any value).
2. Replace the corresponding variable with your chosen value.
3. Solve the equation for the value of the other variable.

Example 2 Find three solutions for the equation.

a. $3x + y = 2$

SOLUTION To find a solution, we replace one of the variables with a chosen value and then solve for the value of the other variable.

For the first solution, we will choose x to be 0.

$$3x + y = 2$$
$$3(0) + y = 2$$
$$y = 2$$

Solution $(0, 2)$

Note: Choosing x (or y) to equal 0 usually makes the equation very easy to solve.

For the second solution, we will choose x to be 1.

$$3x + y = 2$$
$$3(1) + y = 2$$
$$3 + y = 2$$
$$y = -1$$

Subtract 3 from both sides.

Solution $(1, -1)$

For the third solution, we will choose y to be -4.

$$3x + y = 2$$
$$3x + (-4) = 2$$
$$3x = 6$$
$$x = 2$$

Divide both sides by 3.

Solution $(2, -4)$

We can summarize the solutions in a table.

If we choose x to be 0, then y is equal to 2.
If we choose x to be 1, then y is equal to -1.
If we choose y to be -4, then x is equal to 2.

x	y	Ordered Pair
0	2	$(0, 2)$
1	-1	$(1, -1)$
2	-4	$(2, -4)$

Keep in mind that there are an infinite number of correct solutions for a given equation in two variables; so your solutions may be different from the solutions someone else finds.

b. $y = -\dfrac{1}{3}x + 2$

SOLUTION Notice that this equation has y isolated. If we select values for x, we won't have to isolate y as we did in part a. We simply calculate the y value. Also notice that the coefficient for x is a fraction. Because we can choose any value for x, let's choose values such as 3 and 6 that will divide out nicely with the denominator of 3.

For the first solution, we will choose x to be 0.

$$y = -\frac{1}{3}x + 2$$
$$y = -\frac{1}{3}(0) + 2$$
$$y = 2$$

For the second solution, we will choose x to be 3.

$$y = -\frac{1}{3}x + 2$$
$$y = -\frac{1}{3}(3) + 2$$
$$y = -\frac{1}{\cancel{3}_1}\left(\frac{\cancel{3}^1}{1}\right) + 2$$

For the third solution, we will choose x to be 6.

$$y = -\frac{1}{3}x + 2$$
$$y = -\frac{1}{3}(6) + 2$$
$$y = -\frac{1}{\cancel{3}_1}\left(\frac{\cancel{6}^2}{1}\right) + 2$$

Solution $(0, 2)$

$y = -1 + 2$
$y = 1$

Solution $(3, 1)$

$y = -2 + 2$
$y = 0$

Solution $(6, 0)$

x	y	Ordered Pair
0	2	$(0, 2)$
3	1	$(3, 1)$
6	0	$(6, 0)$

SOLUTION In summary,

If we choose x to be 0, then y is equal to 2.

If we choose x to be 3, then y is equal to 1.

If we choose x to be 6, then y is equal to 0.

Your Turn 2 Find three solutions for each equation. (Answers may vary.)

a. $x + 4y = 8$

b. $3x - 2y = 6$

c. $y = \frac{1}{2}x + 5$

Objective ③ Graph linear equations. We have learned that equations in two variables have an infinite number of solutions. Because of this fact, there is no way that all solutions to an equation can be found, much less listed. However, all of the solutions can be represented using a graph. Consider some of the solutions for $x + y = 4$, which we listed in the table on page 250: $(0, 4), (1, 3), (2, 2), (3, 1), (4, 0),$ and $(5, -1)$.

Note: The arrows on either end of the line indicate that the solutions continue beyond our "window" in both directions.

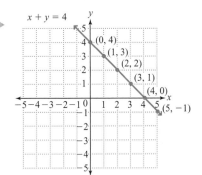

Notice that when we plot our ordered pair solutions for $x + y = 4$, the corresponding points lie in a straight line. In fact, all solutions for $x + y = 4$ are on that same line, which is why we say it is a *linear* equation. By connecting the points, we are graphically representing every possible solution for $x + y = 4$.

The graph of the solutions of every linear equation will be a straight line. Because two points determine a line in a plane, we need a minimum of two ordered pairs. This means that we must find at least two solutions in order to graph the line. However, it is wise to find three solutions, using the third solution as a check. If we plot all three points and they cannot be connected with a straight line, then we know something is wrong.

Procedure **Graphing Linear Equations**

To graph a linear equation,

1. Find at least two solutions to the equation.
2. Plot the solutions as points in the rectangular coordinate system.
3. Connect the points to form a straight line.

Answers Your Turn 2

Remember, your solutions may be different.

a. $(0, 2), (4, 1), (8, 0)$

b. $(0, -3), (2, 0), \left(1, -\frac{3}{2}\right)$

c. $(0, 5), (2, 6), (4, 7)$

Example 3 Graph each equation.

a. $3x + y = 2$

SOLUTION We found three solutions to this equation in Example 2(a). Those solutions are listed in the following table. We plot each solution as a point in the rectangular coordinate system and then connect the points to form a straight line.

x	y	Ordered Pair
0	2	$(0, 2)$
1	-1	$(1, -1)$
2	-4	$(2, -4)$

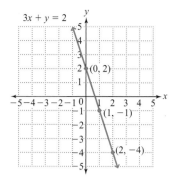

b. $y = -\dfrac{1}{3}x + 2$

SOLUTION We found three solutions to this equation in Example 2(b). Those solutions are listed in the following table. We plot each solution as a point in the rectangular coordinate system and then connect the points to form a straight line.

x	y	Ordered Pair
0	2	$(0, 2)$
3	1	$(3, 1)$
6	0	$(6, 0)$

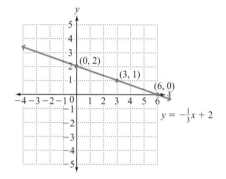

Your Turn 3   Graph each equation.

a. $x + 2y = 4$

b. $y = \dfrac{1}{2}x - 1$

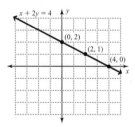
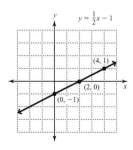
The graph of an equation is a visual representation of the entire solution set of the equation. The graph of every ordered pair that solves the equation lies on the line, and the coordinates of every point on the line is a solution of the equation. Looking at the graph of Example 3(b), we see that $(-3, 3)$ are the coordinates of a point on the line; so $(-3, 3)$ is a solution of the equation $y = -\dfrac{1}{3}x + 2$. We verify by substituting and simplifying.

$$y = -\frac{1}{3}x + 2$$

$$3 = -\frac{1}{3}(-3) + 2$$

$$3 = 1 + 2$$

$$3 = 3$$

We have graphed linear equations in two variables. Now let's consider the graphs of equations in one variable, such as $y = 3$ or $x = 4$, in which the variable is equal to a constant.

Example 4   Graph $y = 3$.

SOLUTION   The equation $y = 3$ indicates that y is equal to a constant, 3. To establish ordered pairs, we can rewrite the equation $y = 3$ as $0x + y = 3$. Because the coefficient of x is 0, y is always 3 no matter what we choose for x. We could complete a table of solutions like this:

If we choose x to be 0, then y equals 3.

If we choose x to be 2, then y is 3.

If we choose x to be 4, then y is still 3.

x	y	Ordered Pair
0	3	$(0, 3)$
2	3	$(2, 3)$
4	3	$(4, 3)$

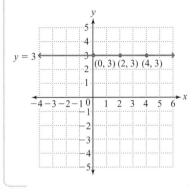

Note: The graph of $y = 3$ is a horizontal line parallel to the x-axis that passes through the y-axis at a point with coordinates $(0, 3)$.

Rule   **Horizontal Lines**

The graph of $y = c$, where c is a real-number constant, is a horizontal line parallel to the x-axis that passes through the y-axis at a point with coordinates $(0, c)$.

Example 5   Graph $x = 4$.

SOLUTION   The equation $x = 4$ indicates that x is equal to a constant, 4. If the y-variable is missing from an equation, it means that its coefficient is 0; so we can write $x = 4$ as $x + 0y = 4$. Notice that x is always 4 no matter what we choose for y. We could complete a table of solutions like this:

If we choose y to be 0, then x equals 4.

If we choose y to be 1, then x is 4.

If we choose y to be 2, then x is still 4.

x	y	Ordered Pair
4	0	$(4, 0)$
4	1	$(4, 1)$
4	2	$(4, 2)$

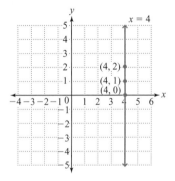

Note: The graph of $x = 4$ is a vertical line parallel to the y-axis that passes through the x-axis at a point with coordinates $(4, 0)$.

Rule Vertical Lines

The graph of $x = c$, where c is a real-number constant, is a vertical line parallel to the y-axis that passes through the x-axis at a point with coordinates $(c, 0)$.

Your Turn 5 Graph.

a. $y = -4$

b. $x = -3$

Answers Your Turn 5

a.
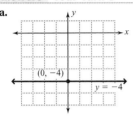

b. $x = -3$
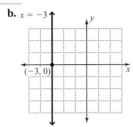

4.2 Exercises   For Extra Help   *MyMathLab*

Note: Exercises marked with a ★ represent challenging exercises.

1. How do you determine whether an ordered pair is a solution for a given equation?

2. If a given ordered pair is not a solution for a linear equation with two variables, does that mean the equation has no solutions? Explain.

3. How do you find a solution for a given equation with two unknowns?

4. What does the graph of an equation represent?

5. What is the minimum number of points needed to draw a straight line? Explain.

6. In your own words, explain the process of graphing a linear equation.

For Exercises 7–22, determine whether the given ordered pair is a solution for the equation. See Example 1.

7. $(2, 3); x + 2y = 8$

8. $(3, 1); 2x - y = 5$

9. $(4, 9); 3x - 2y = -6$

10. $(-5, 10); 5x + 3y = 5$

11. $(-5, 2); y - 4x = 3$

12. $(4, -6); y - 3x = -1$

13. $(9, 0); y = -2x + 18$

14. $(0, -1); y = 2x + 1$

15. $(6, -2); y = -\dfrac{2}{3}x$

16. $(-10, -4); y = \dfrac{2}{5}x$ **17.** $(-8, -8); y = \dfrac{3}{4}x - 2$ **18.** $(-5, 0); y = -\dfrac{3}{5}x + 3$

19. $\left(-1\dfrac{2}{5}, 0\right); y - 3x = 5$ **20.** $\left(0, -2\dfrac{1}{3}\right); 4x - 3y = 7$ **21.** $(2.2, -11.2); y + 6x = -2$

22. $(-1.5, -1.3); y = 0.2x - 1$

For Exercises 23–58, find three solutions for the given equation. Then graph. (Answers may vary for the three solutions.) See Examples 2–5.

23. $y = x$ **24.** $y = -x$ **25.** $y = 2x$

26. $y = 3x$ **27.** $y = -5x$ **28.** $y = -2x$

29. $y = x - 3$ **30.** $y = x + 4$ **31.** $y = -x - 2$

32. $y = -x + 5$ **33.** $y = 2x - 5$ **34.** $y = 3x + 2$

35. $y = -2x + 4$ **36.** $y = -5x - 1$ **37.** $y = \dfrac{1}{2}x$

38. $y = \dfrac{1}{3}x$ **39.** $y = -\dfrac{2}{3}x$ **40.** $y = -\dfrac{3}{4}x$

41. $y = -\dfrac{2}{3}x + 4$

42. $y = \dfrac{4}{5}x - 1$

43. $x - y = 8$

44. $x + y = -5$

45. $2x + y = 6$

46. $3x - y = 9$

47. $2x + 3y = 12$

48. $3x + 5y = 15$

49. $4x - 3y = -12$

50. $3x - 2y = -6$

51. $x - \dfrac{1}{4}y = 2$

52. $\dfrac{1}{3}x + y = -1$

53. $y = -5$

54. $y = 4$

55. $x = 7$

56. $x = -6$

★57. $y = 0.4x - 2.5$

★58. $1.2x + 0.5y = 6$

59. Compare the graphs of $y = x$, $y = 2x$, and $y = 3x$ from Exercises 23, 25, and 26. For an equation in the form $y = ax$, where $a > 0$, what effect does increasing a seem to have on the graph?

60. Compare the graphs of $y = x$ and $y = -x$ from Exercises 23 and 24; then compare the graphs of $y = 2x$ and $y = -2x$ from Exercises 25 and 28. For an equation in the form $y = ax$, what can you conclude about the graph when a is positive versus when a is negative?

61. Compare the graphs of $y = 3x$ and $y = 3x + 2$ from Exercises 26 and 34; then compare the graphs of $y = -2x$ and $y = -2x + 4$ from Exercises 28 and 35. For an equation in the form $y = ax + b$, where $b > 0$, what effect does adding b to $y = ax$ seem to have on the graph?

62. Compare the graphs of $y = x$ and $y = x - 3$ from Exercises 23 and 29; then compare the graphs of $y = -5x$ and $y = -5x - 1$ from Exercises 27 and 36. For an equation in the form $y = ax - b$, where $b > 0$, what effect does subtracting b from $y = ax$ seem to have on the graph?

For Exercises 63–70, solve.

63. A plumber charges \$80 plus \$40 per hour of labor. The equation $c = 40n + 80$ describes the total cost that she would charge for a service visit, where n represents the number of hours of labor.
  a. Find the total cost if labor is 2 hours.

  b. If a client's total charges are \$240, for how many hours of labor was the client charged?

  c. Graph the equation with n along the horizontal axis and c along the vertical axis. (*Hint:* Let each grid mark on the c-axis represent \$20.) Because n and c are nonnegative, the graph is restricted to the first quadrant only.
  d. What does the point where the graph intersects the c-axis represent?

64. An academic tutor charges \$20 for supplies and \$25 per hour of tutoring. The equation $c = 25n + 20$ describes the total cost that she would charge for tutoring, where n represents the number of hours of tutoring.
  a. Find the total cost if the tutor works 3 hours.

  b. If a client's total charges are \$145, for how many hours of tutoring was the client charged?

  c. Graph the equation with n along the horizontal axis and c along the vertical axis. Because n and c are nonnegative, the graph is restricted to the first quadrant only.
  d. What does the point where the graph intersects the c-axis represent?

65. A copy center charges \$15.00 for the first 300 copies plus \$0.03 per copy after that. The equation $c = 0.03n + 15$ describes the total cost that would be charged for a copy service, where n represents the number of copies above 300.
  a. Find the total cost for 400 copies.

  b. If a client's total charges are \$21, for how many copies was the client charged?

  c. Graph the equation with n along the horizontal axis and c along the vertical axis. Because n and c are nonnegative, the graph is restricted to the first quadrant only. (*Hint:* Let each grid mark on the n-axis represent 50 copies and each grid mark on the c-axis represent \$3.)

66. A wireless company charges $20.00 per month for 300 anytime minutes plus $0.10 per minute above 300. The equation $c = 0.10n + 20$ describes the total cost that the company would charge for a monthly bill, where n represents the number of minutes used above 300.

   a. Find the total cost if a customer uses 400 anytime minutes.

   b. If a client's total charges are $22, for how many additional minutes was the client charged?

   c. Graph the equation with n along the horizontal axis and c along the vertical axis. Because n and c are nonnegative, the graph is restricted to the first quadrant only.

67. It is recommended that a hot tub be emptied and cleaned regularly. The hot tub Benjamin owns holds approximately 500 gallons of water. He can drain 3.5 gallons of water per minute using a garden hose. The equation $g = -3.5m + 500$ describes the amount of water in the hot tub in gallons, where g represents the number of gallons remaining after pumping out water for m minutes.

   a. After 25 minutes, what is the amount of water in the hot tub?

   b. How long does it take to pump half of the water out of the hot tub?

   c. How long does it take to pump all of the water out of the hot tub?

   d. Graph the equation. Because m and g are nonnegative, the graph is restricted to the first quadrant only.

   e. What does the point where the graph intersects the g-axis represent?

   f. What does the point where the graph intersects the m-axis represent?

68. The battery in Jenny's laptop is fully charged (100%). While using the laptop continuously on a flight from Kansas to California, she depletes the battery at a rate of 0.75% per minute. The equation $P = -0.75m + 100$ describes the battery's power, where P represents the percent of the battery's power that remains after using the laptop for m minutes.

   a. After 1 hour, what percent of the battery's power remains?

   b. How long does it take for the battery to be at one-fourth of its full power?

   c. How long does it take for the battery to reach 0%?

   d. Graph the equation. Because m and P are nonnegative, the graph is restricted to the first quadrant only.

   e. What does the point where the graph intersects the P-axis represent?

   f. What does the point where the graph intersects the m-axis represent?

69. A small business owner buys a new computer for $2400. Each year the computer is in use, she can deduct its depreciated value when calculating her taxable income. The equation $c = -300n + 2400$ describes the value of the computer, where c is the value after n years of use.

   a. Find the value of the computer after 4 years.
   b. In how many years will the computer be worth half of its initial value?
   c. After how many years will the computer have a value of $0?
   d. Graph the equation with n along the horizontal axis and c along the vertical axis. Because n and c are nonnegative, the graph is restricted to the first quadrant only.

70. As of 2005, it is estimated that Earth is being deforested at a rate of about 40,000 square miles per year. The equation $A = -40,000n + 2,000,000$ describes the area of forest remaining, where A represents the area remaining n years after 2005. (*Source:* www.wikipedia.org)

    a. Find the area of forest remaining in 2015.

    b. If the rate does not change, in what year will Earth have half of the area of the forest that was present in 2005?

    c. If the rate does not change, in what year will forest be gone from Earth?

    d. Graph the equation with n along the horizontal axis and A along the vertical axis. Because n and A are nonnegative, the graph is restricted to the first quadrant only.

Puzzle Problem

Without lifting your pencil from the paper, connect the dots using only four straight lines:

Review Exercises

Exercises 1–3 EXPRESSIONS

[1.7]   **1.** Evaluate $\frac{2}{3}x + 2$ when $x = 3$.

[2.1]   **2.** Calculate the perimeter of a rectangle with a length of 8.5 inches and a width of 7.25 inches.

[2.1]   **3.** Calculate the area of a triangle with a base of 4.2 inches and a height of 2.7 inches.

Exercises 4–6 EQUATIONS AND INEQUALITIES

[2.3]   **4.** Solve $0 = 3x - 5$.

[2.4]   **5.** Isolate x in the formula $v = mx + k$.

[4.2]   **6.** Graph the line that passes through the points $(3, 0)$ and $(0, -2)$.

4.3 Graphing Using Intercepts

Objectives

1. Given an equation, find the coordinates of the x- and y-intercepts.
2. Graph linear equations using intercepts.

Objective 1 **Given an equation, find the coordinates of the x- and y-intercepts.** Look at the following graph.

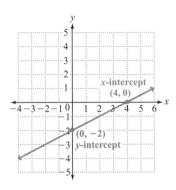

Notice that the line intersects the x-axis at $(4, 0)$ and the y-axis at $(0, -2)$. These points are called *intercepts*. A point where a graph intersects the x-axis is called an **x-intercept**. A point where a graph intersects the y-axis is called a **y-intercept**.

Definitions **x-intercept:** A point where a graph intersects the x-axis.

y-intercept: A point where a graph intersects the y-axis.

Note that the y-coordinate of any x-intercept is always 0. Similarly, the x-coordinate of any y-intercept is always 0. These facts about intercepts suggest the following procedures.

Procedure **Finding the x- and y-intercepts**

To find an x-intercept,	To find a y-intercept,
1. Replace y with 0 in the given equation.	1. Replace x with 0 in the given equation.
2. Solve for x.	2. Solve for y.

Example 1 For the equation $5x - 2y = 10$, find the x- and y-intercepts.

SOLUTION

For the x-intercept, replace y with 0 and solve for x.

$5x - 2(0) = 10$

$5x = 10$

$x = 2$ Divide both sides by 5.

x-intercept: $(2, 0)$

For the y-intercept, replace x with 0 and solve for y.

$5(0) - 2y = 10$

$-2y = 10$

$y = -5$ Divide both sides by -2.

y-intercept: $(0, -5)$

Note: We no longer show the steps containing the addition or multiplication principles as we did in Chapters 2 and 3.

Your Turn 1 For the equation $3x - 4y = 8$, find the x- and y-intercepts.

Answers **Your Turn 1**

x-intercept: $\left(\dfrac{8}{3}, 0\right)$

y-intercept: $(0, -2)$

The equation in Example 1 is in the form $Ax + By = C$, where A, B, and C are real-number constants with both A and B not equal to 0. A linear equation written in this form is said to be in *standard form*. One of the advantages of having a linear equation in standard form is that it is easy to find the intercepts. Throughout this section, we introduce other forms for linear equations and explore some of the information that can be discerned from equations written in various forms.

Example 2   For the equation $y = \dfrac{2}{3}x$, find the x- and y-intercepts.

SOLUTION   Replace x with 0 and solve for y. Then replace y with 0 and solve for x.

y-intercept: $y = \dfrac{2}{3}(0)$ $\qquad\qquad$ x-intercept: $0 = \dfrac{2}{3}x$

$\qquad\qquad\qquad y = 0$ $\qquad\qquad\qquad\qquad\qquad$ $0 = x$ $\qquad$ **Multiply both sides by** $\dfrac{3}{2}$.

x- and y-intercept: $(0, 0)$

The equation in Example 2 is a linear equation in the form $y = mx$, where m is any real number. The graph of any equation in the form $y = mx$ is always a line that passes through the origin, which means that the x- and y-intercepts are both at $(0, 0)$. So to graph an equation of the form $y = mx$, at least one other point must be found.

Rule   **Intercepts for $y = mx$**

If an equation can be written in the form $y = mx$, where m is a real number other than 0, then the x- and y-intercepts are at the origin, $(0, 0)$.

Example 3   For the equation $y = 2x + 1$, find the x- and y-intercepts.

SOLUTION   Replace y with 0 and solve for x. Then replace x with 0 and solve for y.

x-intercept: $0 = 2x + 1$ $\qquad\qquad\qquad\qquad$ y-intercept: $y = 2(0) + 1$

$\qquad\qquad -1 = 2x$ $\qquad$ **Subtract 1 from both sides.** $\qquad\qquad\qquad y = 1$

$\qquad\qquad -\dfrac{1}{2} = x$ $\qquad$ **Divide both sides by 2.** $\qquad$ y-intercept: $(0, 1)$

x-intercept: $\left(-\dfrac{1}{2}, 0\right)$

Note: When we replace x with 0, we are left with the constant 1 for the y-intercept.

The equation in Example 3 is in the form $y = mx + b$, where m and b can be any real number. In this form, when we replace x with 0 to find the y-intercept, we are left with the constant b.

Rule   **The y-intercept for $y = mx + b$**

If an equation is in the form $y = mx + b$, where m and b are nonzero real numbers, then the y-intercept will be $(0, b)$.

Your Turn 3   Find the x- and y-intercepts for the following equations.

a. $y = -\dfrac{3}{4}x$ $\qquad\qquad\qquad\qquad\qquad$ **b.** $y = 3x - 5$

Answers   Your Turn 3

a. x- and y-intercept: $(0, 0)$

b. x-intercept: $\left(\dfrac{5}{3}, 0\right)$

$\quad$ y-intercept: $(0, -5)$

Example 4   For the equation $y = 3$, find the x- and y-intercepts.

SOLUTION   Remember from Section 4.2, Example 4, that the graph of $y = 3$ is a horizontal line parallel to the x-axis that passes through the y-axis at the point $(0, 3)$. Notice that this point is the y-intercept. Because the line is parallel to the x-axis, it will never intersect the x-axis. Therefore, there is no x-intercept.

Example 4 suggests the following rule.

Rule   Intercepts for $y = c$

The graph of an equation in the form $y = c$, where c is a nonzero real number, has no x-intercept and the y-intercept is $(0, c)$. The graph of $y = 0$ is the x-axis.

We can draw a similar conclusion about the graph of an equation in the form $x = c$.

Rule   Intercepts for $x = c$

The graph of an equation in the form $x = c$, where c is a nonzero real number, has no y-intercept and the x-intercept is $(c, 0)$. The graph of $x = 0$ is the y-axis.

Your Turn 4   Find the x- and y-intercepts for the following equations.

a. $y = -5$

b. $x = -2$

Objective ② Graph linear equations using intercepts.   Because two points determine a line in a plane, we can use the x- and y-intercepts to draw the graph of a linear equation.

Example 5   Graph using the x- and y-intercepts: $5x - 2y = 10$

SOLUTION   In Example 1, the intercepts were found to be $(2, 0)$ and $(0, -5)$. We plot these intercepts and connect them to graph the line. It is helpful to find a third solution and verify that all three points can be connected to form a straight line. For this third point, we will choose $x = 3$ and solve for y.

Third solution: (check) If $x = 3$, then

$$5(3) - 2y = 10$$
$$15 - 2y = 10$$
$$-2y = -5 \quad \text{Subtract 15 from both sides.}$$
$$y = \frac{5}{2} \quad \text{Divide both sides by } -2. \quad \textbf{Third solution: } \left(3, \frac{5}{2}\right)$$

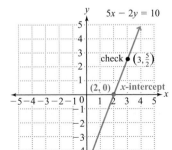

Table of Solutions

	x	y
x-intercept:	2	0
y-intercept:	0	-5
Third solution: (check)	3	$\dfrac{5}{2}$

Your Turn 5 Graph using the x- and y-intercepts: $2x - y = 4$

As previously mentioned, if the x- and y-intercepts are both at the origin, we must find an additional point in order to graph the linear equation.

Example 6 Graph using the x- and y-intercepts.

a. $y = \dfrac{2}{3}x$

SOLUTION In Example 2 we found $(0, 0)$ to be both the x- and the y-intercept. Because this single point is not enough to determine the line in the plane, we must find at least one more solution. We will also find a third solution as a check.

Second solution:
Choose x to be 3.

$$y = \frac{2}{3}(3)$$

$$y = \frac{2}{\underset{1}{3}}\left(\frac{\overset{1}{3}}{1}\right)$$

$$y = 2$$

Second solution: $(3, 2)$

Third solution (check):
Choose x to be -3.

$$y = \frac{2}{3}(-3)$$

$$y = \frac{2}{\underset{1}{3}}\left(-\frac{\overset{1}{3}}{1}\right)$$

$$y = -2$$

Third solution: $(-3, -2)$

Now plot the solutions and connect the points to form the line.

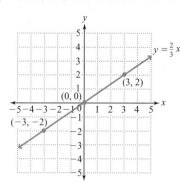

Table of Solutions

	x	y
x- and y-intercept:	0	0
Second solution:	3	2
Third solution: (check)	-3	-2

b. $y = \dfrac{2}{3}x + 2$

SOLUTION Notice that this equation is in the form $y = mx + b$. Because $b = 2$ in this case, we can conclude that the y-intercept is $(0, 2)$. For the x-intercept, let $y = 0$ and solve for x. Also, as a check, we will choose $x = 3$ to generate a third point.

x-intercept: $0 = \dfrac{2}{3}x + 2$

$\qquad -2 = \dfrac{2}{3}x$ $\quad$ Subtract 2 from both sides.

$\dfrac{3}{\underset{1}{2}} \cdot \dfrac{-\overset{1}{2}}{1} = \dfrac{\overset{1}{3}}{2} \cdot \dfrac{\overset{1}{2}}{\underset{1}{3}}x$ $\quad$ Multiply both sides by $\dfrac{3}{2}$.

$\qquad -3 = x$

x-intercept: $(-3, 0)$

Third solution: (check) $y = \dfrac{2}{3}(3) + 2$

$$y = \frac{2}{\underset{1}{3}} \cdot \frac{\overset{1}{3}}{1} + 2$$

$$y = 2 + 2$$

$$y = 4$$

Third solution: $(3, 4)$

Answer **Your Turn 5**

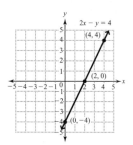

Now plot the solutions and connect the points to form the line.

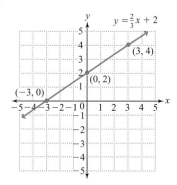

$y = \frac{2}{3}x + 2$

Table of Solutions

	x	y
x-intercept:	−3	0
y-intercept:	0	2
Third solution: (check)	3	4

Your Turn 6 Graph using the x- and y-intercepts.

   a. $y = \frac{1}{2}x$
                                          b. $y = \frac{2}{3}x - 2$

Compare the graphs in Examples 6(a) and 6(b). If we place both lines on the same grid (shown here), we see that they are parallel, which means that they do not intersect at any point. Further, each point on the graph of $y = \frac{2}{3}x + 2$ is 2 units higher than each point with the same x-coordinate on the graph of $y = \frac{2}{3}x$.

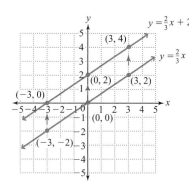

$y = \frac{2}{3}x + 2$

Table of Solutions

x	$y = \frac{2}{3}x$	$y = \frac{2}{3}x + 2$
−3	−2	0
0	0	2
3	2	4

Answers Your Turn 6

a.

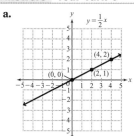

$y = \frac{1}{2}x$

b.

$y = \frac{2}{3}x - 2$

From the table, we see that the difference in an equation in the form $y = mx$ and an equation in the form $y = mx + b$ is the value of the y-intercept. In other words, the line $y = mx$ shifts upward or downward by b units. The graph of $y = \frac{2}{3}x - 2$ is a line parallel to $y = \frac{2}{3}x$, only shifted down 2 units so that its y-intercept is $(0, -2)$. We will explore this further in Section 4.5.

Note: Exercises marked with a ★ represent challenging exercises.

1. Explain what an *x*-intercept is.

2. Explain what a *y*-intercept is.

3. How do you find an *x*-intercept?

4. How do you find a *y*-intercept?

5. If an equation is in the form $y = mx$, where *m* is a real-number constant, what are the coordinates of the *x*- and *y*-intercepts? Explain.

6. Compare the graphs of $y = 2x$ and $y = 2x + 5$.

For Exercises 7–14, find the x- and y-intercepts of each graph. See Objective 1.

7.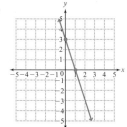

8.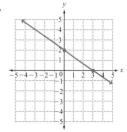

9.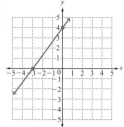

10.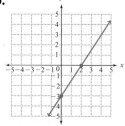

11.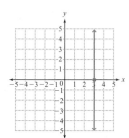

12.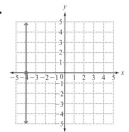

13.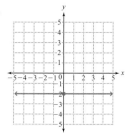

14.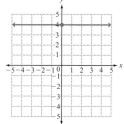

For Exercises 15–32, find the x- and y-intercepts. See Examples 1–4.

15. $x - y = 4$

16. $x + y = 7$

17. $2x + 3y = 6$

18. $2x + 3y = 12$

19. $3x - 4y = -10$

20. $3x - 4y = 2$

21. $y = 3x - 6$

22. $y = 2x + 8$

23. $y = 2x + 5$

24. $y = 3x - 1$

25. $y = 2x$

26. $y = \dfrac{2}{5}x$

27. $2x + 3y = 0$

28. $x + 3y = 0$

29. $x = 5$

30. $x = -2$

31. $y = -2$

32. $y = 7$

For Exercises 33–64, graph using the x- and y-intercepts. See Examples 5 and 6.

33. $x + y = 7$

34. $x + y = 10$

35. $3x - y = 6$

36. $x - 5y = 15$

37. $2x + 3y = 6$

38. $3x + 4y = 12$

39. $5x + 2y = 10$

40. $3x - 2y = 18$

41. $4x + 3y = -12$

42. $3x + 2y = -6$

43. $5x - 3y = -15$

44. $2x - 5y = -10$

45. $y = 2x$

46. $y = -3x$

47. $y = \dfrac{2}{5}x$

48. $-\dfrac{1}{4}x = y$

49. $y = x - 4$

50. $y = x + 2$

51. $y = 2x - 6$

52. $y = 3x + 9$

53. $6x + 3y = 9$

54. $2x - y = -1$

55. $4x + 2y = 5$

56. $3x + 2y = 7$

57. $x = -4$ **58.** $x = 5$ **59.** $y = -2$ **60.** $y = 5$

61. $y - 2 = 0$ **62.** $y + 4 = 0$ **63.** $x - 2 = 0$ **64.** $x + 3 = 0$

★**65.** Which of the following could be the graph of $1.4x - 0.9y = 2.7$? Explain.

a.   b.   c.   d.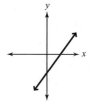

★**66.** Which of the following could be the graph of $0.65x - 2.9y = -3.5$? Explain.

a.   b. 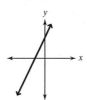   c. 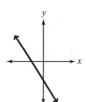   d.

★**67.** Which of the following could be the graph of $\dfrac{2}{3}x - \dfrac{3}{4}y = \dfrac{10}{3}$? Explain.

a.   b.   c.   d.

★ **68.** Which of the following could be the graph of $\frac{3}{2}x + \frac{6}{5}y = -\frac{8}{5}$? Explain.

a.

b.

c.

d.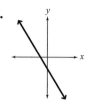

69. If a circle with a radius of 2.7 units was to be drawn in the coordinate plane with its center at the origin, what would be the coordinates of its x- and y-intercepts?

70. If a circle with a diameter of 6.8 units was to be drawn in the coordinate plane with its center at the origin, what would be the coordinates of its x- and y-intercepts?

★ **71.** A salesperson sells two different sizes of facial cleansing lotion. She sells the small size for $15 and the large size for $20.
   a. If x represents the number of units of the small size sold and y represents the number of units of the large size sold, write an expression that describes her total sales.

   b. If her supervisor sets a goal of selling $2000 worth of the cleansing lotion, use the expression from part a to write an equation describing the sales goal.

   c. Find the x- and y-intercepts.

   d. What do the x- and y-intercepts mean in terms of sales?

   e. Find a third combination of sales numbers that would equal the sales goal.

   f. Graph the equation. Because x and y are nonnegative, the graph is restricted to the first quadrant only.

★ **72.** A comic book collector will pay $50 for any Superman comic published prior to 1970 if it is in mint condition and $30 if it is not in mint condition.
   a. If x represents the number of comic books in mint condition and y represents the number of comic books not in mint condition, write an expression that describes the amount the collector would pay for a combination of comic books in mint and nonmint condition.

   b. Suppose the collector visits a convention for collectors and has a budget of $1800 to spend. Use the expression from part a to write an equation describing his expenditure on a combination of the comics if he spends all of his budget.

   c. Find the x- and y-intercepts.

   d. What do the x- and y-intercepts mean in terms of purchases?

   e. Find a third combination of comic books he could purchase to equal his budget amount.

   f. Graph the equation. Because x and y are nonnegative, the graph is restricted to the first quadrant only.

Review Exercises

Exercises 1 and 2 EXPRESSIONS

[1.5] *For Exercises 1 and 2, evaluate.*

    1. $\dfrac{4 - (-2)}{-2 - 1}$

    2. $\dfrac{-3 - 4}{-2 - (-2)}$

Exercises 3–6 EQUATIONS AND INEQUALITIES

[2.4]   **3.** Solve for w in the equation $P = 2l + 2w$.

[2.5]   **4.** Translate to an equation and then solve. Four more than three times a number is equal to five times the sum of the number and six.

[4.1]   **5.** Where is the point located whose coordinates are $(0, 4)$?

[4.2]   **6.** Determine whether $(-8, -2)$ is a solution for $x + 3y = -5$.

4.4 Slope–Intercept Form

Objectives

  1 Compare lines with different slopes.

  2 Graph equations in slope–intercept form.

  3 Use slope–intercept form to write the equation of a line.

  4 Find the slope of a line given two points on the line.

In Section 4.3, we discussed various forms of linear equations, one of which was $y = mx + b$, where m and b are real-number constants. Further, we discovered that the graph of an equation in this form is a line with its y-intercept at $(0, b)$. In this section, we explore how the coefficient m affects the graph.

Objective 1 Compare lines with different slopes. First, we consider graphs of equations in which $b = 0$ so that the equation is $y = mx$ and the graphs all have the origin $(0, 0)$ as the x- and y-intercepts.

> **Example 1** Graph each of the following equations on the same grid.
>
> $$y = x \qquad y = 2x \qquad y = 3x$$
>
> **SOLUTION** In the following table, the same choice for x has been substituted into each equation and the corresponding y-coordinate has been found. We plot the ordered pairs and connect the points to graph the lines.

If x is	and $y = x$, then y is	and $y = 2x$, then y is	and $y = 3x$, then y is
0	0	0	0
1	1	2	3
2	2	4	6

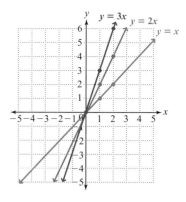

From Example 1, we see that for equations of the form $y = mx$, as the coefficient m increases, the graphs get steeper. Because the coefficient m affects how steep a line is, m is called the *slope* of the line.

For $y = x$, the slope is 1 ($m = 1$).

For $y = 2x$, the slope is 2 ($m = 2$).

For $y = 3x$, the slope is 3 ($m = 3$).

We have seen that if the slope is greater than 1, the corresponding line has a steeper incline than the graph of $y = x$. What values for m would cause a line to be less inclined than $y = x$? The pattern in the graphs suggests that we should explore values less than 1.

First, consider values less than 1 but greater than 0, such as $\frac{1}{2}$ and $\frac{1}{3}$.

Example 2 Graph each of the following equations on the same grid.

$$y = x \qquad y = \frac{1}{2}x \qquad y = \frac{1}{3}x$$

SOLUTION For consistency, we use the same three choices for x that we used in Example 1. After finding the corresponding y-coordinates for each equation, we plot the ordered pairs and connect the points to graph the lines.

If x is	and $y = x$, then y is	and $y = \frac{1}{2}x$, then y is	and $y = \frac{1}{3}x$, then y is
0	0	0	0
1	1	$\frac{1}{2}$	$\frac{1}{3}$
2	2	1	$\frac{2}{3}$

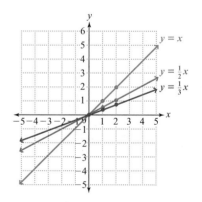

Notice that if the slope is a fraction between 0 and 1, the smaller the fraction, the less inclined or flatter the lines. So far, we have considered only positive slope values. If we "travel" from left to right along lines with positive slopes, we travel uphill. What would a line with a negative slope look like?

Example 3 Graph each of the following equations on the same grid.

$$y = -x \qquad y = -2x \qquad y = -\frac{1}{2}x$$

SOLUTION Again, we make a table of ordered pairs, then plot those ordered pairs and connect the points to graph the lines.

If x is	and $y = -x$, then y is	and $y = -2x$, then y is	and $y = -\frac{1}{2}x$, then y is
0	0	0	0
1	-1	-2	$-\frac{1}{2}$
2	-2	-4	-1

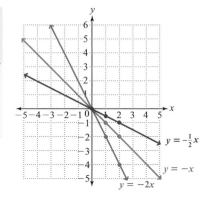

Note: When we speak of traveling along a graph, the convention is to travel from left to right, just as we do when we read English.

Notice that if we "travel" from left to right along lines with negative slopes, we travel downhill. As the slope becomes more negative, the downhill incline becomes steeper.

Rule **Graphs of $y = mx$**

Given an equation of the form $y = mx$, the graph of the equation is a line passing through the origin and having the following characteristics:

If $m > 0$ (slope is positive), the graph is a line that slants uphill from left to right.

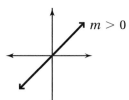

If $m < 0$ (slope is negative), the graph is a line that slants downhill from left to right.

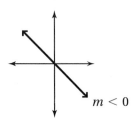

The greater the absolute value of m, the steeper the line.

Your Turn 3

a. Is $y = 5x$ steeper or less inclined than $y = x$?

b. Is the graph of $y = -\frac{2}{3}x$ uphill or downhill from left to right?

Answers   Your Turn 3

a. $y = 5x$ is steeper than $y = x$.

b. downhill

Graph equations in slope–intercept form. We have seen that slope determines the incline of a line. What more can we discover about slope? Notice that if we isolate m

in $y = mx$, we get $m = \dfrac{y}{x}$, which suggests that **slope** is a ratio of the amount of vertical change to the amount of horizontal change. A change up or to the right is a positive change, and a change down or to the left is a negative change.

Definition   *Slope:* The ratio of the vertical change between any two points on a line to the horizontal change between these points.

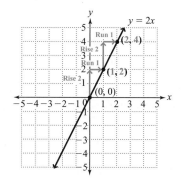

Note: If m is an integer, think of it as a fraction whose denominator is 1, as in $2 = \dfrac{2}{1}$.

For example, the slope of $y = 2x$ is 2, which can be written as $\dfrac{2}{1}$. This slope value means that rising vertically 2 units and then running horizontally 1 unit from any point on the line locates a second point on the line. For example, if we rise 2 units and run 1 unit from $(0, 0)$, we end up at $(1, 2)$. Similarly, rising 2 units and running 1 unit from $(1, 2)$ puts us at $(2, 4)$.

We have learned that in an equation of the form $y = mx + b$, the slope is m and the y-intercept is $(0, b)$. For this reason, we say that equations of the form $y = mx + b$ are in *slope–intercept* form. We can graph equations in slope–intercept form using the y-intercept as a starting point and then using the slope to locate other points on the line.

Procedure   **Graphing Equations in Slope–Intercept Form**

To graph an equation in slope–intercept form, $y = mx + b$,

Note: The "rise" will be a "fall" if the slope is negative and the run is positive.

1. Plot the y-intercept, $(0, b)$.
2. Plot a second point by rising the number of units indicated by the numerator of the slope, m, then running the number of units indicated by the denominator of the slope, m.
3. Draw a straight line through the two points.

Note: You can check by locating additional points using the slope. Every point you locate using the slope should be on the line.

Example 4   For the equation $y = -\dfrac{1}{3}x + 2$, determine the slope and the y-intercept. Then graph the equation.

SOLUTION   Because the equation is in the form $y = mx + b$, where $m = -\dfrac{1}{3}$ and $b = 2$, the slope of the line is $-\dfrac{1}{3}$ and the y-intercept is $(0, 2)$. To graph the line, we can plot the y-intercept and then use the slope to find other points. To interpret $-\dfrac{1}{3}$ in terms of "rise" and "run," we place the negative sign in either the numerator or denominator. We can "rise" -1 and "run" 3, which gives the point $(3, 1)$. Or we can "rise" 1 and "run" -3, which gives the point $(-3, 3)$. Either interpretation gives another point on the same line.

The points we found using the slope can be verified in the equation. We can check the ordered pair $(-3, 3)$ in the equation.

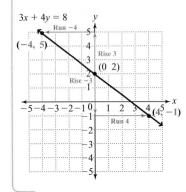

$$y = -\frac{1}{3}x + 2$$

$$3 \overset{?}{=} -\frac{1}{3}(-3) + 2$$

$$3 \overset{?}{=} -\frac{1}{\underset{1}{3}} \cdot -\frac{\overset{1}{3}}{1} + 2$$

$$3 \overset{?}{=} 1 + 2$$

$$3 = 3 \qquad \text{This checks.}$$

We will leave the check of $(3, 1)$ to the reader.

Your Turn 4 For the equation $y = \frac{2}{5}x - 2$, determine the slope and the y-intercept. Then graph the equation.

If an equation is not in slope–intercept form, as in $3x + 4y = 8$, and we need to determine the slope and y-intercept, we can write the equation in slope–intercept form by isolating y.

Example 5 For the equation $3x + 4y = 8$, determine the slope and the y-intercept. Then graph the equation.

Solution Write the equation in slope–intercept form by isolating y.

$$3x + 4y = 8$$

$$4y = -3x + 8 \qquad \text{Subtract } 3x \text{ from both sides to isolate } 4y.$$

$$\frac{4y}{4} = \frac{-3x + 8}{4} \qquad \text{Divide both sides by 4 to isolate } y.$$

$$y = -\frac{3}{4}x + \frac{8}{4} \qquad \text{Simplify.}$$

$$y = -\frac{3}{4}x + 2$$

The slope is $-\frac{3}{4}$, and the y-intercept is $(0, 2)$. To graph the line, we begin at $(0, 2)$ and rise -3 and run 4, which gives the point $(4, -1)$. Or we can begin at $(0, 2)$ and rise 3 and run -4, which gives the point $(-4, 5)$.

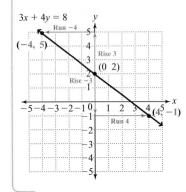

We can check the ordered pairs in the equation.

For $(-4, 5)$:

$$3x + 4y = 8$$
$$3(-4) + 4(5) \overset{?}{=} 8$$
$$-12 + 20 \overset{?}{=} 8$$
$$8 = 8$$

For $(4, -1)$:

$$3x + 4y = 8$$
$$3(4) + 4(-1) \overset{?}{=} 8$$
$$12 + (-4) \overset{?}{=} 8$$
$$8 = 8$$

Both ordered pairs check.

Answer   Your Turn 4

$m = \frac{2}{5}$; y-intercept: $(0, -2)$

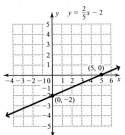

Your Turn 5   For the equation $3x - 2y = 6$, determine the slope and the y-intercept. Then graph the equation.

Objective ③   **Use slope–intercept form to write the equation of a line.**   If you are given the slope, m, and the y-intercept, $(0, b)$, of a line, use the slope–intercept form of the equation to write the equation of the line.

Procedure   **Equation of a Line Given Its Slope and y-Intercept**

To write the equation of a line given its slope, m, and its y-intercept, $(0, b)$, use the slope–intercept form of the equation, $y = mx + b$.

Example 6   A line with a slope of 5 crosses the y-axis at the point $(0, 3)$. Write the equation of the line.

Solution   We use $y = mx + b$, the slope–intercept form of the equation, replacing m with the slope, 5, and b with the y-coordinate of the y-intercept, 3.

$$y = 5x + 3$$

Your Turn 6   A line has a slope of -3. If the y-intercept is $(0, -4)$, write an equation of the line in slope–intercept form.

Objective ④   **Find the slope of a line given two points on the line.**   Given two points on a line, we can determine the slope of the line. Consider the points $(-3, 0)$ and $(3, 4)$ on the graph of $y = \frac{2}{3}x + 2$, which we graphed in Example 6(b) of Section 4.3. In the equation, we see that the slope is $\frac{2}{3}$. Recall that slope is the ratio of the vertical change between any two points on a line to the horizontal change between those points. Therefore, the ratio of the vertical distance between $(-3, 0)$ and $(3, 4)$ to the horizontal distance between those points should be $\frac{2}{3}$. Let's verify.

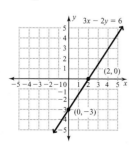
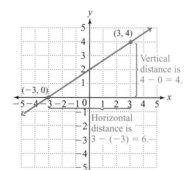

To find the vertical distance between the two points, we can subtract their y-coordinates: $4 - 0 = 4$. Similarly, the horizontal distance between the points is the difference of the x-coordinates: $3 - (-3) = 6$. Now we can write the ratio.

$$\frac{\text{Vertical distance}}{\text{Horizontal distance}} = \frac{4 - 0}{3 - (-3)} = \frac{4}{6} = \frac{2}{3}$$

Our example suggests the following formula:

Procedure **The Slope Formula**

Given two points (x_1, y_1) and (x_2, y_2), where $x_2 \neq x_1$, the slope of the line connecting the two points is given by the formula $m = \dfrac{y_2 - y_1}{x_2 - x_1}$.

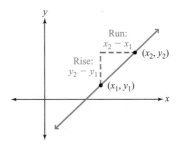

Example 7   Find the slope of the line connecting the given points.

a. $(3, 5)$ and $(-1, 7)$

SOLUTION   Using $m = \dfrac{y_2 - y_1}{x_2 - x_1}$, replace the variables with their corresponding values and then simplify. We'll let $(3, 5)$ be (x_1, y_1) and $(-1, 7)$ be (x_2, y_2).

$$m = \frac{7 - 5}{-1 - 3} = \frac{2}{-4} = -\frac{1}{2}$$

It doesn't matter which ordered pair is (x_1, y_1) and which is (x_2, y_2). If we let $(-1, 7)$ be (x_1, y_1) and $(3, 5)$ be (x_2, y_2), we get the same slope.

$$m = \frac{5 - 7}{3 - (-1)} = \frac{-2}{4} = -\frac{1}{2}$$

b. $(-3, -4)$ and $(1, -4)$

SOLUTION   We'll let $(-3, -4)$ be (x_1, y_1) and $(1, -4)$ be (x_2, y_2).

$$m = \frac{-4 - (-4)}{1 - (-3)} = \frac{0}{4} = 0$$

LEARNING Strategy

VISUAL Many people, especially visual learners, find it helpful to write the coordinate labels beneath their corresponding value.

$$(3, 5) \qquad (-1, 7)$$
$$(x_1, y_1) \qquad (x_2, y_2)$$

Also, some people prefer to stack the ordered pairs.

$$(-1, 7)$$
$$(3, 5)$$

Your Turn 7   Find the slope of the line connecting the given points.

a. $(2, 5)$ and $(6, 1)$ **b.** $(-5, -2)$ and $(7, -5)$

In Example 7(b), we found the slope of the line connecting $(-3, -4)$ and $(1, -4)$ to be 0. What does a line with a slope of 0 look like? Let's plot the points $(-3, -4)$ and $(1, -4)$ and see.

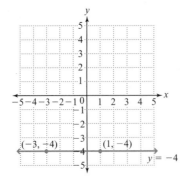

Because the two points have the same y-coordinate, they form a horizontal line parallel to the x-axis. The equation of this line is $y = -4$.

Answers Your Turn 7

a. -1 **b.** $-\dfrac{1}{4}$

Rule **Zero Slope**

Two points with different x-coordinates and the same y-coordinates, (x_1, c) and (x_2, c), will form a line with a slope of 0 (a horizontal line) and the equation $y = c$.

What is the slope of a vertical line? Two points on a vertical line have the same x-coordinates but different y-coordinates, as in $(2, 4)$ and $(2, -3)$.

Example 8 Find the slope of the line connecting the points $(2, 4)$ and $(2, -3)$.

SOLUTION Note that $(2, 4)$ and $(2, -3)$ connect to form a vertical line whose equation is $x = 2$. The slope of this line is found using the slope formula.

$$m = \frac{-3 - 4}{2 - 2}$$ Replace the variables in $m = \frac{y_2 - y_1}{x_2 - x_1}$ with their corresponding values.

$$m = \frac{-7}{0}$$ Simplify.

Because we never divide by 0, the slope is *undefined*.

Rule **Undefined Slope**

Two points with the same x-coordinates and different y-coordinates, (c, y_1) and (c, y_2), will form a line with a slope that is undefined (a vertical line) and the equation $x = c$.

Your Turn 8 Find the slope of the line connecting the given points.

a. $(4, -1)$ and $(-3, -1)$ **b.** $(3, 2)$ and $(3, -4)$

🖩 Calculator Tips

To graph an equation on a graphing calculator, you first enter the equation using the [Y =] *key. When you press the* [Y =] *key, you will see a list of Y's: Y1, Y2, Y3, and so on. Each of the Y's can be used to enter an equation, which means you can graph multiple equations on the same grid if you like.*

To enter the equation $y = 3x + 2$, *press* [Y =]; *at Y1, type* [3] [X, T, θ, n] [+] [2].

Notice that the [X, T, θ, n] *key is used to type the variable x. To have the calculator draw the graph, press* [GRAPH].

Sometimes you may need to adjust or change your viewing window. To change the range of tick marks on your axes, press [WINDOW]. *For example, if you want the x-axis to range from* -10 *to* 10, *press* [WINDOW]; *then at Xmin, enter* [(−)] [1] [0] *and at Xmax, enter* [1] [0]. *Xscl sets the frequency of the marks in the range. For example, setting the Xscl to 1 places a mark at each integer in the range; so if the range is* -10 *to 10, a mark will appear at* $-10, -9, -8, \ldots, 9, 10$. *If the Xscl is 2 and the range is* -10 *to 10, a mark will appear at* $-10, -8, -6, \ldots, 8, 10$. *Similarly, Ymax and Ymin change the range on the y-axis and Yscl affects how many marks are shown in that range.*

1. In your own words, what is the slope of a line?

2. What is the formula for finding the slope of a line connecting two points (x_1, y_1) and (x_2, y_2)?

3. Does the graph of $y = -2x + 1$ go uphill from left to right or downhill? Why?

4. Does the graph of $3x - 5y = 10$ go uphill from left to right or downhill? Why?

5. Describe a line with a slope of 0. Explain why a slope of 0 produces this type of line.

6. Describe a line with a slope that is undefined. Explain why an undefined slope produces this type of line.

For Exercises 7–14, graph each set of equations on the same grid. For each set of equations, compare the slopes, y-intercepts, and their effects on the graphs. See Examples 1–3.

7. $y = \dfrac{1}{2}x$

   $y = x$

   $y = 2x$

8. $y = \dfrac{1}{3}x$

   $y = x$

   $y = 3x$

9. $y = x$

   $y = \dfrac{1}{2}x$

   $y = \dfrac{1}{5}x$

10. $y = x$

    $y = \dfrac{1}{3}x$

    $y = \dfrac{1}{6}x$

11. $y = -\dfrac{1}{2}x$

    $y = -x$

    $y = -2x$

12. $y = -\dfrac{1}{3}x$

    $y = -x$

    $y = -3x$

13. $y = x$

    $y = x + 2$

    $y = x - 2$

14. $y = 2x$

    $y = 2x - 1$

    $y = 2x - 3$

For Exercises 15–38, determine the slope and the y-intercept. Then graph the equation. See Examples 4 and 5.

15. $y = 2x + 3$

16. $y = 4x - 3$

17. $y = -2x - 1$

18. $y = -3x + 5$

19. $y = \dfrac{1}{3}x + 2$

20. $y = \dfrac{1}{2}x - 4$

21. $y = \dfrac{3}{4}x - 2$

22. $y = \dfrac{2}{5}x - 5$

23. $y = -\dfrac{2}{3}x + 8$

24. $y = -\dfrac{4}{3}x + 7$

25. $2x + y = 4$

26. $3x + y = 6$

27. $3x - y = -1$

28. $2x - y = 5$

29. $2x + 3y = 6$

30. $4x - 3y = 12$

31. $x + 2y = -4$

32. $2x + 3y = -6$

33. $3x - 2y + 7 = 0$

34. $2x - 3y + 8 = 0$

35. $2y = -3x + 5$

36. $3y = 2x - 4$

37. $0.6x - 0.2y = 1$

38. $0.3x + 1.5y = -6$

For Exercises 39–50, write the equation of the line in slope–intercept form given the slope and the coordinates of the y-intercept. See Example 6.

39. $m = 2; (0, 3)$

40. $m = 3; (0, -2)$

41. $m = -3; (0, -9)$

42. $m = -4; (0, -5)$

43. $m = \frac{3}{4}; (0, 5)$

44. $m = \frac{3}{7}; (0, -1)$

45. $m = -\frac{2}{5}; \left(0, \frac{7}{8}\right)$

46. $m = \frac{5}{2}; \left(0, \frac{3}{8}\right)$

47. $m = 0.8; (0, -5.1)$

48. $m = -0.75; (0, 2.5)$

49. $m = -3; (0, 0)$

50. $m = 1; (0, 0)$

For Exercises 51–66, find the slope of the line through the given points. See Examples 7 and 8.

51. $(4, 1), (8, 11)$

52. $(1, 5), (3, 8)$

53. $(2, 4), (5, 2)$

54. $(6, 2), (3, 7)$

55. $(-3, 5), (4, 7)$

56. $(3, -2), (5, 5)$

57. $(10, -12), (4, -4)$

58. $(-8, 14), (4, 6)$

59. $(3, -3), (-15, -15)$

60. $(8, -12), (-4, -18)$

61. $(0, 5), (4, 0)$

62. $(-3, 0), (0, 5)$

63. $(-3, 5), (4, 5)$

64. $(-2, -8), (4, -8)$

65. $(6, 1), (6, -8)$

66. $(-3, 2), (-3, 5)$

For Exercises 67–76, match the equation with the appropriate graph. See Examples 4 and 5.

67. $y = 2x + 4$

68. $3x + 2y = 6$

69. $x = y$

70. $y = 3$

71. $x - 2 = 0$

72. $x - 2y = 8$

73. $y = \frac{2}{3}x - 2$

74. $y = -\frac{3}{4}x$

75. $x = -3$

76. $y + 4 = 0$

a.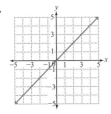

b.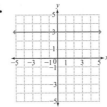

c.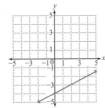

d.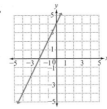

e.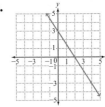

f.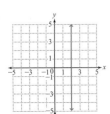

g.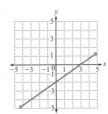

h.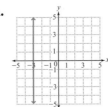

i.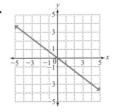

j.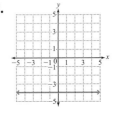

For Exercises 77–88, find the slope of the line and the y-intercept and write the equation of the line in slope–intercept form. See Examples 4 and 8.

77.

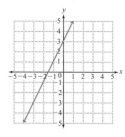

78.

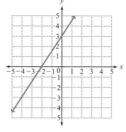

79.

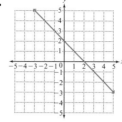

80.

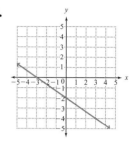

81.

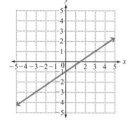

82.

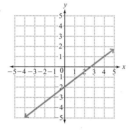

83.

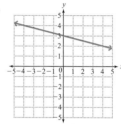

84.

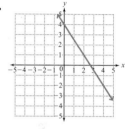

85.

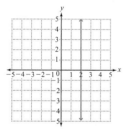

86.

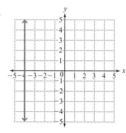

87.

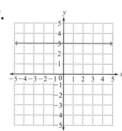

88.

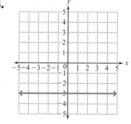

89. What is the equation of a horizontal line through $(0, -2)$? A vertical line through $(3, 0)$?

90. What is the equation of a vertical line through $(-3.5, 0)$? A horizontal line through $(0, 2.6)$?

91. What equation describes the *x*-axis?

92. What equation describes the *y*-axis?

93. A parallelogram has vertices at $(1, -2)$, $(3, 4)$, $(-3, -2)$, and $(-1, 4)$.
  a. Plot the vertices in a coordinate plane; then connect them to form the parallelogram.
  b. Find the slope of each side of the parallelogram.
  c. What do you notice about the slopes of the parallel sides?

94. A right triangle has vertices at $(-2, 3)$, $(2, 1)$, and $(3, 3)$.

   a. Plot the vertices in a coordinate plane; then connect them to form the right triangle with the right angle at $(2, 1)$.

   b. Find the slope of each side of the triangle.

   c. What do you notice about the slopes of the perpendicular sides?

95. In architecture, the slope of a roof is called its *pitch*. The roof of a house rises 3.5 feet for every 6 feet of horizontal distance. Find the pitch of the roof. Write the pitch as a decimal number and a fraction in simplest form.

96. The slope of a hill is referred to as the *grade* of the hill. Heartbreak Hill, which is part of the Boston Marathon, rises from an elevation of 150 feet to an elevation of 225 feet over a horizontal distance of about 5000 feet. Find the grade of Heartbreak Hill. (*Source:* Boston Athletic Association)

97. The Great Pyramid at Giza rises 29.3 meters for every 23 meters of horizontal distance. Find the slope of the pyramid.

98. The steps of the Pyramid of the Sun at Teotihuacan, Mexico, rise 65 meters over a horizontal distance of 112.5 meters. Find the slope of the steps of the Pyramid of the Sun.

Collaborative Exercises Population Growth

The following data show the total U.S. population by decade.

Decade	1900	1910	1920	1930	1940	1950	1960	1970	1980	1990	2000
Population (in millions)	76.2	92.2	106.0	123.2	132.2	151.3	179.3	203.3	226.5	248.7	281.4

1. Plot each ordered pair as a point on a graph with the decades along the horizontal axis and the population along the vertical axis.
2. Do the points have a linear relationship?
3. Draw a straight line that connects the point at 1900 to the point at 1950. Find the slope of this line.
4. Draw a straight line that connects the point at 1950 to the point at 2000. Find the slope of this line.
5. Compare the slopes of the two lines you have drawn. Which has the greater slope? What does this indicate about population growth in the United States during the 20th century?

6. Use the line you drew from 1950 to 2000 to predict the population at 2110.

Review Exercises

Exercise 1 CONSTANTS AND VARIABLES

[1.4] **1.** Find the multiplicative inverse of $-\dfrac{2}{3}$.

Exercises 2 and 3 EXPRESSIONS

[1.7] **2.** Simplify $12x - 9y + 7 - x + 9y$.   **[1.7]** **3.** Multiply $-2(x + 7)$.

Exercises 4–6 EQUATIONS AND INEQUALITIES

[2.4] **4.** Isolate x in the equation $Ax + By = C$.   **[4.2]** **5.** Graph $2x + 3y = 8$.

[4.3] **6.** Find the x- and y-intercepts for $3x - 5y = 7$.

4.5 Point–Slope Form

Objectives

1. **Use point–slope form to write the equation of a line.**
2. **Write the equation of a line parallel to a given line.**
3. **Write the equation of a line perpendicular to a given line.**

In this section, we explore how to write an equation of a line given information about the line, such as its slope and the coordinates of a point on the line.

Objective 1 **Use point–slope form to write the equation of a line.** In the previous section, we used the slope–intercept form to write the equation of a line because we were given the slope and y-intercept. However, we need a more general approach so that we can write the equation of a line given *any* two points on the line. For this more general approach, we can use the slope formula to derive a new form of the equation of a line called the *point–slope* form. Recall the slope formula.

> **Connection** We also can view the formula for slope as a proportion and cross multiply to get the point–slope form.
>
> $$\frac{m}{1} \diagup \frac{y_2 - y_1}{x_2 - x_1}$$
>
> $(x_2 - x_1)m = y_2 - y_1$

$$m = \frac{y_2 - y_1}{x_2 - x_1}$$

$$(x_2 - x_1) \cdot m = \left(\frac{y_2 - y_1}{x_2 - x_1}\right) \cdot \left(\frac{x_2 - x_1}{1}\right) \qquad \text{Multiply both sides by } x_2 - x_1 \text{ to isolate the } y\text{'s.}$$

$$(x_2 - x_1)m = y_2 - y_1 \qquad \text{Simplify.}$$

$$y_2 - y_1 = m(x_2 - x_1) \qquad \text{Rewrite with } y\text{'s on the left side to resemble slope–intercept form.}$$

To write the equation of a line with this formula, we replace m with the slope and substitute one of the given ordered pairs for x_1 and y_1, leaving x_2 and y_2 as variables. To indicate that x_2 and y_2 remain variables, we remove their subscripts so that we have $y - y_1 = m(x - x_1)$, which is called the point–slope form of the equation of a line.

Procedure | **Using the Point–Slope Form of the Equation of a Line**

To write the equation of a line given its slope and any point, (x_1, y_1), on the line, use the point–slope form of the equation of a line, $y - y_1 = m(x - x_1)$. If given a second point, (x_2, y_2), and not the slope, we first calculate the slope using $m = \dfrac{y_2 - y_1}{x_2 - x_1}$; then we use $y - y_1 = m(x - x_1)$.

Example 1 Write the equation of a line with a slope of 2 passing through the point $(3, 5)$. Write the equation in slope–intercept form.

SOLUTION Because we are given the coordinates of a point and the slope of a line passing through the point, we begin with the point–slope form. After replacing m, x_1, and y_1 with their corresponding values, we isolate y to get slope–intercept form.

$$y - y_1 = m(x - x_1)$$
$$y - 5 = 2(x - 3) \qquad \text{Replace } m \text{ with 2, } x_1 \text{ with 3, and } y_1 \text{ with 5.}$$
$$y - 5 = 2x - 6 \qquad \text{Simplify.}$$
$$y = 2x - 1 \qquad \text{Add 5 to both sides to isolate } y.$$
$$\text{This is now slope–intercept form.}$$

Your Turn 1 Write the equation of the line in slope–intercept form with the given slope passing through the given point.

$$m = 0.6; (-4, 2)$$

Example 2 Write the equation of a line passing through the points $(3, 2)$ and $(-3, 6)$. Write the equation in slope–intercept form.

SOLUTION Because we do not have the slope, we calculate it.

$$m = \frac{6 - 2}{-3 - 3} = \frac{4}{-6} = -\frac{2}{3} \qquad \text{Use } m = \frac{y_2 - y_1}{x_2 - x_1}.$$

Now we can use the point–slope form, then isolate y to write the slope–intercept form. Because we are given two points, we can use either point for (x_1, y_1) in the point–slope equation. We select $(3, 2)$ to be (x_1, y_1).

Note: If one of the given points is the y-intercept, after finding the slope, you can use the slope–intercept form to write the equation of the line.

$$y - y_1 = m(x - x_1)$$
$$y - 2 = -\frac{2}{3}(x - 3) \qquad \text{Replace } m \text{ with } -\frac{2}{3}, x_1 \text{ with 3, and } y_1 \text{ with 2.}$$
$$y - 2 = -\frac{2}{3}x + 2 \qquad \text{Simplify.}$$
$$y = -\frac{2}{3}x + 4 \qquad \text{Add 2 to both sides to isolate } y.$$

Answer   Your Turn 1

$y = 0.6x + 4.4$

Your Turn 2 Write the equation of the line passing through the given points. Write the equation in slope–intercept form.

$$(-2, -4), (4, -1)$$

Writing Linear Equations in Standard Form

Equations can also be written in standard form. Recall from Section 4.3 that standard form is $Ax + By = C$, where A, B, and C are real numbers. We will manipulate the point–slope form of the equation so that the x and y terms are on the same side of the equation and a constant appears on the other side of the equation.

Example 3 The following data points relate the velocity of an object as time passes. (You may recall this data from Your Turn 4 in Section 4.1.) The graph shows that the points are in a line. Write the equation of the line in standard form.

Time (x) (in seconds)	Velocity (y) (in ft./sec.)
0	2.0
1	2.5
2	3.0
3	3.5

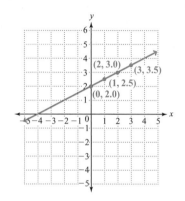

SOLUTION First, we find the slope of the line using any two ordered pairs in the slope formula. We will use $(0, 2)$ and $(1, 2.5)$.

$$m = \frac{2.5 - 2}{1 - 0} = 0.5 \text{ or } \frac{1}{2} \qquad \text{Use } m = \frac{y_2 - y_1}{x_2 - x_1}.$$

Because $(0, 2)$ is the y-intercept, we can write the equation in slope–intercept form, then manipulate the equation so that x and y appear on the same side of the equal sign to get standard form.

$$y = \frac{1}{2}x + 2 \qquad \text{Slope–intercept form}$$

$$y - \frac{1}{2}x = 2 \qquad \text{Subtract } \frac{1}{2}x \text{ from both sides to get } x \text{ and } y \text{ together.}$$

When we write an equation in standard form ($Ax + By = C$), it is customary to write the equation so that the x term is first with a positive coefficient and, if possible, so that A, B, and C are all integers. Let's manipulate $y - \frac{1}{2}x = 2$ to this more polished form.

$$-\frac{1}{2}x + y = 2 \qquad \text{Use the commutative property of addition to write the } x \text{ term first.}$$

$$(-2)\left(-\frac{1}{2}x + y\right) = (-2)(2) \qquad \text{Multiply both sides by } -2 \text{ so that the coefficient of } x \text{ is a positive integer.}$$

$$x - 2y = -4 \qquad \text{Standard form with } A, B, \text{ and } C \text{ integers and } A > 0$$

Answer   Your Turn 2

$$y = \frac{1}{2}x - 3$$

Example 4   A line connects the points $(1, 5)$ and $(-3, 2)$. Write the equation of the line in the form $Ax + By = C$, where A, B, and C are integers and $A > 0$.

SOLUTION   Find the slope; then write the equation of the line using point–slope form. To finish, rewrite the equation in standard form.

$$\text{Find the slope:}\quad m = \frac{2 - 5}{-3 - 1} = \frac{-3}{-4} = \frac{3}{4}\qquad \text{Use } m = \frac{y_2 - y_1}{x_2 - x_1}.$$

Because we were not given the y-intercept, we use the point–slope form of the linear equation.

$$y - 5 = \frac{3}{4}(x - 1)\qquad \text{Use } y - y_1 = m(x - x_1) \text{ and } (1, 5) \text{ for } (x_1, y_1).$$

Now manipulate the equation to get the form $Ax + By = C$, where A, B, and C are integers and $A > 0$.

$$y - 5 = \frac{3}{4}x - \frac{3}{4}\qquad \text{Distribute } \frac{3}{4} \text{ to clear parentheses.}$$

$$4(y - 5) = 4\left(\frac{3}{4}x - \frac{3}{4}\right)\qquad \text{Multiply both sides by the LCD, 4.}$$

$$4y - 20 = 3x - 3$$

$$4y - 3x - 20 = -3\qquad \text{Subtract } 3x \text{ from both sides to get } x \text{ and } y \text{ together.}$$

$$4y - 3x = 17\qquad \text{Add 20 to both sides to get the constant terms on the right-hand side of the equation.}$$

$$-3x + 4y = 17\qquad \text{Use the commutative property of addition to write the } x \text{ term first.}$$

$$(-1)(-3x + 4y) = (-1)(17)\qquad \text{Multiply both sides by } -1 \text{ so that the coefficient of } x \text{ is positive.}$$

$$3x - 4y = -17\qquad \text{Standard form with } A, B, \text{ and } C \text{ integers and } A > 0$$

Your Turn 4   Write the equation of the line through the given points in the form $Ax + By = C$, where A, B, and C are integers and $A > 0$.

$$(-3, 6), (1, -4)$$

Objective ② Write the equation of a line parallel to a given line.   Consider the graphs of $y = 2x - 3$ and $y = 2x + 1$.

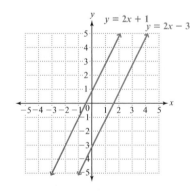

These two lines are parallel, which means that they will never intersect. Notice in the equations that their slopes are equal, which is why they are parallel.

Answer   Your Turn 4

$5x + 2y = -3$

Parallel Lines

Nonvertical parallel lines have equal slopes and different y-intercepts. Vertical lines are parallel.

Example 5   Write the equation of a line that passes through $(1, -4)$ and is parallel to $y = -3x + 5$. Write the equation in slope–intercept form.

SOLUTION   In the given equation, we see that the slope of the line is -3; so the slope of the parallel line will also be -3. We can now use the point–slope form to write the equation of the parallel line.

$$y - y_1 = m(x - x_1)$$
$$y - (-4) = -3(x - 1)$$ Replace m with -3, x_1 with 1, and y_1 with -4.
$$y + 4 = -3x + 3$$ Simplify.
$$y = -3x - 1$$ Subtract 4 from both sides to isolate y.

Your Turn 5   Write the equation of the line that passes through the given point and is parallel to the given line. Write the equation in slope–intercept form.

$$(-4, -2); y = \frac{3}{4}x - 7$$

Objective ③ Write the equation of a line perpendicular to a given line.   Consider the graphs of $y = \frac{2}{3}x - 4$ and $y = -\frac{3}{2}x + 1$.

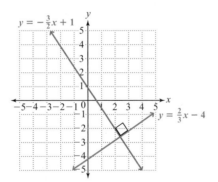

These two lines are perpendicular, which means that they intersect at a 90° angle. Notice in the equations that their slopes are negative reciprocals. In other words, $-\frac{3}{2}$ is the negative reciprocal of $\frac{2}{3}$.

Perpendicular Lines

The slope of a line perpendicular to a line with a slope of $\frac{a}{b}$ will be $-\frac{b}{a}$. Horizontal and vertical lines are perpendicular.

Another way of stating the rule is that the slopes of perpendicular lines are opposites *and* reciprocals. Also note that $\left(\frac{a}{b}\right)\left(-\frac{b}{a}\right) = -1$; so two lines are perpendicular if and only if the product of their slopes is -1.

Answer   Your Turn 5

$$y = \frac{3}{4}x + 1$$

Example 6 Write the equation of a line that passes through $(3, -1)$ and is perpendicular to $3x + 4y = 8$. Write the equation in slope–intercept form.

SOLUTION To write the equation of the perpendicular line, we need its slope. To determine its slope, we need the slope of the given line. Because the equation of the given line is in standard form, to determine its slope, we rewrite it in slope–intercept form.

$$3x + 4y = 8$$
$$4y = -3x + 8 \qquad \text{Subtract } 3x \text{ from both sides.}$$
$$y = -\frac{3}{4}x + 2 \qquad \text{Divide both sides by 4.}$$

We see that the slope of the given line is $-\frac{3}{4}$; so the slope of any perpendicular line will be $\frac{4}{3}$. We can now use point–slope form to write the equation of the perpendicular line through $(3, -1)$.

$$y - y_1 = m(x - x_1)$$
$$y - (-1) = \frac{4}{3}(x - 3) \qquad \text{Replace } m \text{ with } \frac{4}{3}, x_1 \text{ with 3, and } y_1 \text{ with } -1.$$
$$y + 1 = \frac{4}{3}x - 4 \qquad \text{Simplify.}$$
$$y = \frac{4}{3}x - 5 \qquad \text{Subtract 1 from both sides to isolate } y.$$

Your Turn 6 Write the equation of a line that passes through $(-10, 3)$ and is perpendicular to $5x - 2y = 6$. Write the equation in slope–intercept form.

By using slopes, it is possible to determine whether the graphs of two lines are parallel, perpendicular, or neither parallel nor perpendicular.

Example 7 Determine whether the given lines are parallel, perpendicular, or neither.

a. $y = \frac{2}{3}x - 4$ and $y = -\frac{3}{2}x + 2$

SOLUTION The slope of $y = \frac{2}{3}x - 4$ is $\frac{2}{3}$, and the slope of $y = -\frac{3}{2}x + 2$ is $-\frac{3}{2}$. Because the slopes are opposites (negatives) and reciprocals, the lines are perpendicular.

b. $2x + 4y = 7$ and $3x + 6y = 8$

SOLUTION To determine the slopes of the lines, write the equations in slope–intercept form.

$$2x + 4y = 7 \qquad\qquad 3x + 6y = 8$$
$$4y = -2x + 7 \qquad\qquad 6y = -3x + 8$$
$$y = -\frac{1}{2}x + \frac{7}{4} \qquad\qquad y = -\frac{1}{2}x + \frac{4}{3}$$

Because the slope of both lines is $-\frac{1}{2}$ and the y-intercepts differ, the lines are parallel.

Your Turn 7 Determine whether the given lines are parallel, perpendicular, or neither.

a. $y = 2x + 4$ and $y = -2x - 1$ **b.** $x - 2y = 6$ and $3x - 6y = 12$

Answer **Your Turn 7**
a. neither **b.** parallel

4.5 Exercises   For Extra Help   *MyMathLab*

Note: Exercises marked with a ★ represent challenging exercises.

1. If given the slope and the *y*-intercept and asked to write the equation of the line, which form of the equation would be easiest to use? Why?

2. If given two ordered pairs, neither of which is the *y*-intercept, and asked to write the equation of the line, which form of the equation would be easiest to use? Why?

3. What is the first step when you are given two ordered pairs and asked to write the equation of the line connecting the two points?

4. When given two ordered pairs and asked to write an equation of a line, does it matter which ordered pair is used in the point–slope equation? Explain.

5. How do you rewrite an equation that is in standard form in slope–intercept form?

6. How do you rewrite an equation that is in point–slope form in standard form?

For Exercises 7–18, write the equation of the line in slope–intercept form with the given slope passing through the given point. See Example 1.

7. $m = 3; (5, 2)$

8. $m = 2; (3, -5)$

9. $m = -1; (-1, -1)$

10. $m = -5; (2, 0)$

11. $m = \dfrac{2}{3}; (6, 3)$

12. $m = \dfrac{3}{2}; (-2, 3)$

13. $m = -\dfrac{3}{4}; (-1, -5)$

14. $m = -\dfrac{2}{5}; (-2, -1)$

15. $m = -\dfrac{4}{3}; (-2, 4)$

16. $m = \dfrac{2}{9}; (-1, -7)$

17. $m = 2; (0, 0)$

18. $m = -3; (0, 0)$

For Exercises 19–22, write the equation of a line. Leave your answer in standard form. (Hint: Find the slope and a point or two points on the line.) See Examples 1 and 2.

19.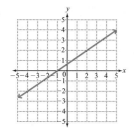

20.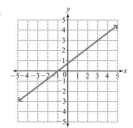

21.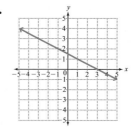

22.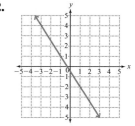

For Exercises 23–34, write the equation of a line passing through the given points. Write the equation in slope–intercept form. See Example 2.

23. $(4, -3), (-1, 7)$

24. $(-1, -5), (-4, 1)$

25. $(3, 1), (5, 6)$

26. $(-5, -2), (7, 5)$

27. $(0, 0), (-1, -7)$

28. $(0, 0), (-5, 1)$

29. $(8, -1), (8, -10)$

30. $(-3, 2), (-3, 5)$

31. $(-9, -1), (2, -1)$

32. $(-2, 5), (6, 5)$

★ **33.** $(1.4, 5), (3, 9.8)$

★ **34.** $(3.9, -2.1), (4.1, -1)$

For Exercises 35–38, write the equation of the line in slope–intercept form given the y-intercept and one other point. See Objective 1.

35. $(3, 4), (0, -2)$

36. $(0, 2), (4, 10)$

37. $(0, 3), (5, -3)$

38. $(-3, -1), (0, -5)$

For Exercises 39–50, write the equation of a line through the given points in the form $Ax + By = C$, where A, B, and C are integers and $A > 0$. See Example 4.

39. $(2, 3), (4, 8)$

40. $(4, 1), (8, 4)$

41. $(2, 0), (0, -3)$

42. $(-6, 0), (0, 6)$

43. $(-4, -1), (7, -5)$

44. $(5, -2), (-3, -3)$

45. $(-3, 2), (-6, 4)$

46. $(4, 8), (-4, -8)$

47. $(-1, 0), (-1, 8)$

48. $(-2, 5), (-2, -6)$

49. $(3, -1), (4, -1)$

50. $(2, -4), (3, -4)$

For Exercises 51–62, write the equation of a line that passes through the given point and is parallel to the given line. See Example 5.
a. Write the equation in slope–intercept form.
b. Write the equation in the form $Ax + By = C$, where A, B, and C are integers and $A > 0$.

51. $(4, 2); y = 4x + 1$

52. $(0, -1); y = 3x + 2$

53. $(-2, 1); y = -3x + 1$

54. $(0, 3); y = -2x + 4$

55. $(5, 2); y = \dfrac{1}{3}x + 4$

56. $(-2, 4); y = \dfrac{3}{4}x - 3$

57. $(-1, -3); y = -\dfrac{3}{4}x + 1$

58. $(-1, -1); y = -\dfrac{2}{3}x + 6$

59. $(2, -3); 2x + y = 5$

60. $(5, 1); 6x + 3y = 8$

61. $(-4, 2); 3x - 4y = 7$

62. $(2, -6); 3x - 2y = 5$

For Exercises 63–74, write the equation of a line that passes through the given point and is perpendicular to the given line. See Example 6.
a. **Write the equation in slope–intercept form.**
b. **Write the equation in the form $Ax + By = C$, where A, B, and C are integers and $A > 0$.**

63. $(-4, -1); y = 2x + 3$ **64.** $(3, -3); y = 3x - 1$ **65.** $(3, -4); y = \frac{1}{4}x + 2$ **66.** $(2, 3); y = \frac{1}{5}x - 4$

67. $(2, -4); y = -2x + 7$ **68.** $(-1, -1); y = -3x + 4$ **69.** $(-2, 3); y = \frac{3}{2}x - \frac{5}{2}$ **70.** $(4, 2); y = \frac{3}{4}x - 11$

71. $(-3, -1); 2x - 5y = 10$ **72.** $(-2, -4); 3x - 4y = 8$ **73.** $(3, -4); 2x + 3y = 12$ **74.** $(3, 7); 5x + 2y = 3$

For Exercises 75–90, determine whether the given lines are parallel, perpendicular, or neither. See Example 7.

75. $y = 4x - 7$
$y = 4x + 2$

76. $y = 3x + 2$
$y = 3x - 2$

77. $y = \frac{1}{2}x$
$y = -2x + 4$

78. $y = -x + 3$
$y = x - 2$

79. $y = \frac{1}{2}x + 3$
$y = 2x - 1$

80. $y = 3x + 4$
$y = \frac{1}{3}x - 3$

81. $4x + 6y = 5$
$8x + 12y = 9$

82. $3x + 6y = 10$
$2x + 4y = 9$

83. $5x + 10y = 12$
$4x - 2y = 5$

84. $5x - 3y = 11$
$3x + 5y = 8$

85. $2x + 3y = -6$
$3x + 2y = 4$

86. $4x - 2y = 5$
$2x - 4y = 3$

87. $y = 1$
$x = -3$

88. $x = 2$
$y = -4$

89. $x = 4$
$x = -2$

90. $y = 4$
$y = -3$

91. From Logan Airport in Boston, a taxi charges a $7.75 initial fee plus $0.30 for each one-eighth of a mile.

    a. Let n represent the number of eighths of a mile and c represent the total cost of the taxi. Write an equation in slope–intercept form that describes how much it costs to hire a taxi from Logan Airport.

    b. What will be the total cost of traveling 5 miles from the airport?

    c. Graph the equation with n plotted along the horizontal axis and c along the vertical axis. Because n and c are nonnegative, restrict the graph to the first quadrant only.

92. A company manager is examining profits. He sees that the profit in May was $52,000. He then discovers that his company's profits have been declining $1000 each month thereafter.

    a. Let n represent the number of months and p represent the profit. Write an equation in slope–intercept form that describes the profit n months after May.

b. If the decline continues at the same rate, what will be the profit in nine months?

c. Graph the equation with n plotted along the horizontal axis and p along the vertical axis. Because n and p are nonnegative, restrict the graph to the first quadrant only.

93. In the 1996–1997 school year, about 2.47 million females were participating in high school athletics. Over the next ten years, participation increased in an approximately linear pattern so that in 2006–2007, 3.02 million females were participating in high school athletics. (*Source:* 2007 High School Participation Survey)

a. Consider 1996–1997 to be year 0 so that 2006–2007 is year 10. Plot the two ordered pairs with the number of years along the horizontal axis and the number of females participating in athletics along the vertical axis. Then draw a line through the two points. Because the number of athletes and years are both nonnegative, restrict the graph to the first quadrant.

b. Find the slope of the line.

c. Let n represent the number of years after the 1996–1997 school year and p represent the number of female participants in high school athletics. Write an equation of the line in slope–intercept form.

d. Using your equation, how many females participated in high school athletics in the 2001–2002 school year?

e. If the linear trend continues, how many females will participate in high school athletics in the 2012–2013 school year?

94. The population of the United States was 274 million in 1997 and continued to increase in a linear pattern until 2007, when the population was 303 million.

a. Consider 1997 to be year 0 so that 2007 is year 10. Plot the two ordered pairs with the year along the horizontal axis and the population (in millions) along the vertical axis. Then draw a line through the two points. Because the years and the population are both nonnegative, restrict the graph to the first quadrant.

b. Find the slope of the line.

c. Let n represent the number of years after 1997 and P represent the population in millions. Write an equation of the line in slope–intercept form.

d. Using your equation, what was the population of the United States in 2003?

e. If the linear trend continues, what will be the population in 2015?

95. A stock begins a decline in price following a linear pattern of depreciation. The stock's initial price before the decline began was $36. On the 5th day, the closing price was $28.50.

a. Graph the line with the number of days after the decline begins plotted along the horizontal axis and the price along the vertical axis. Because the number of days and price are nonnegative, the graph is restricted to the first quadrant only.

b. Find the slope of the line.

c. Write an equation of the line in slope–intercept form with n representing the number of days of decline in price and p representing the price.

d. Find the p-intercept. Explain what the p-intercept represents in the problem.

e. Find the *n*-intercept. Explain what the *n*-intercept represents in the problem.

f. If the decline in price were to continue at the same rate, what would be the price of the stock after eight days of decline?

96. Keith deducts the value of his computer on his taxes. The amount of the deduction is based on the value of the computer, which depreciates each year. The initial value of the computer was $2100. After two years in service, the computer's value is $1500.

a. Graph the line with the number of years plotted along the horizontal axis and the value along the vertical axis. Because the number of years and the value are nonnegative, the graph is restricted to the first quadrant only.

b. Find the slope of the line.

c. Write an equation of the line in slope–intercept form with *n* representing the number of years the computer is in service and *v* representing the value.

d. Find the *v*-intercept. Explain what the *v*-intercept represents in the problem.

e. Find the *n*-intercept. Explain what the *n*-intercept represents in the problem.

f. If the decline in value were to continue at the same rate, what would be the computer's value after five years in service?

Review Exercises

Exercises 1–6 EQUATIONS AND INEQUALITIES

[1.1] **1.** Use $<$, $>$, or $=$ to write a true statement:

$0.6 \; ? \; \dfrac{3}{4}$

[2.3] **2.** Solve: $\dfrac{3}{4}x - \dfrac{4}{5} = \dfrac{1}{2}x + 1$

[2.6] *For Exercises 3 and 4, solve and graph the solution set on the number line.*

3. $4x - 9 \geq 6x + 5$

4. $2(x + 1) - 5x \geq 3x - 6$

[4.4] **5.** What are the slope and *y*-intercept for $y = -\dfrac{3}{4}x + 2$?

[4.3] **6.** Find the *x*- and *y*-intercepts for $2x - 7y = 14$.

4.6 Graphing Linear Inequalities

Objectives

1 Determine whether an ordered pair is a solution for a linear inequality with two variables.

2 Graph linear inequalities.

Objective 1 **Determine whether an ordered pair is a solution for a linear inequality with two variables.** Now that we have learned how to graph linear equations, let's turn our attention to linear inequalities. Linear inequalities have the same form as linear equations except that they contain an inequality symbol instead of an equal sign. Examples of linear inequalities are as follows:

$$2x + 3y > 6$$

$$y \leq x - 3$$

Remember that a solution for a linear equation in two variables is an ordered pair that makes the equation true. Similarly, a solution to a linear inequality in two variables is an ordered pair that makes the inequality true. For example, the ordered pair $(2, 1)$ is a solution for $2x + 3y > 6$.

$$2x + 3y > 6$$

$$2(2) + 3(1) \overset{?}{>} 6 \qquad \text{Replace } x \text{ with 2 and } y \text{ with 1.}$$

$$4 + 3 \overset{?}{>} 6$$

$$7 > 6 \qquad \text{This is true, so } (2, 1) \text{ is a solution.}$$

Procedure | **Checking an Ordered Pair**

To determine whether an ordered pair is a solution for an inequality, replace the variables with the corresponding coordinates and see if the resulting inequality is true. If it is, the ordered pair is a solution.

Example 1 Determine whether $(-4, 7)$ is a solution for $y \geq -2x + 3$.

SOLUTION $y \geq -2x + 3$

$$7 \overset{?}{\geq} -2(-4) + 3 \qquad \begin{array}{l} \text{Replace } x \text{ with } -4 \text{ and } y \text{ with 7; then} \\ \text{determine whether the inequality is true.} \end{array}$$

$$7 \overset{?}{\geq} 8 + 3$$

$$7 \geq 11$$

This statement is false, so $(-4, 7)$ is not a solution.

Your Turn 1 Determine whether the given ordered pair is a solution for the given inequality.

   a. $3x - 5y < 6; (-1, -2)$ **b.** $y \leq -4x + 1; (2, -7)$

Objective ② Graph linear inequalities. Graphing linear inequalities is very much like graphing linear equations. Consider $2x + 3y \geq 6$. The greater than or equal to sign indicates that solutions to the inequality $2x + 3y \geq 6$ will be ordered pairs that satisfy $2x + 3y > 6$ and $2x + 3y = 6$. The graph of $2x + 3y = 6$ is a line. Let's graph this line using intercepts.

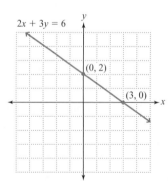

Find the x- and y-intercepts.

x-intercept:	y-intercept:
$2x + 3y = 6$	$2x + 3y = 6$
$2x + 3(0) = 6$	$2(0) + 3y = 6$
$2x = 6$	$3y = 6$
$x = 3$	$y = 2$

x-intercept: $(3, 0)$
y-intercept: $(0, 2)$

The "greater than" portion of the inequality consists of ordered pairs that satisfy $2x + 3y > 6$. To find one of these pairs, we select a point on one side of the line and see if its coordinates satisfy $2x + 3y > 6$. Let's select the point with coordinates $(1, 4)$.

$$2x + 3y > 6$$
$$2(1) + 3(4) \overset{?}{>} 6 \qquad \text{Replace } x \text{ with 1 and } y \text{ with 4.}$$
$$2 + 12 \overset{?}{>} 6$$
$$14 > 6 \qquad \text{This statement is true; therefore, } (1, 4) \text{ is a solution.}$$

It can be shown that every ordered pair in the region containing the point $(1, 4)$, which is above the line, will satisfy $2x + 3y > 6$. It also can be shown that no ordered pair on the other side of the line will satisfy $2x + 3y > 6$. For example, consider the origin $(0, 0)$.

$$2x + 3y > 6 \qquad \text{Replace } x \text{ with 0 and } y \text{ with 0.}$$
$$2(0) + 3(0) \overset{?}{>} 6$$
$$0 > 6 \qquad \text{This statement is false, so } (0, 0) \text{ is not a solution.}$$

The line that is the graph of $2x + 3y = 6$ is called a *boundary*. The region on one side of this boundary line contains all of the ordered pair solutions satisfying $2x + 3y > 6$. The region on the other side contains all ordered pairs that satisfy $2x + 3y < 6$. To indicate that every ordered pair above the boundary line is a solution to the inequality, we shade the region above the line.

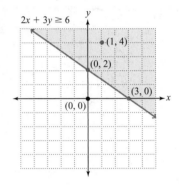

If the inequality had been $2x + 3y > 6$, the ordered pairs that satisfy $2x + 3y = 6$ would not be solutions. Because these ordered pairs form the boundary line, we draw a dashed line instead of a solid line to indicate that those ordered pairs are not solutions. The graph of $2x + 3y > 6$ is shown here.

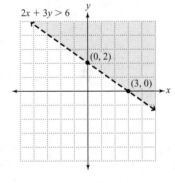

Connection When graphing inequalities with the symbols $\leq$ or $\geq$ on a number line, we used a bracket (or closed circle) to indicate that equality is included. When the inequalities have $<$ or $>$, we used a parenthesis (or open circle) to indicate that equality is excluded. When linear inequalities are graphed in the rectangular coordinate system, a solid line is like a bracket and a dashed line is like a parenthesis.

For the inequalities $2x + 3y \leq 6$ and $2x + 3y < 6$, we shade the region on the other side of the boundary line.

The graph of $2x + 3y \leq 6$:

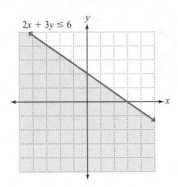

The graph of $2x + 3y < 6$:

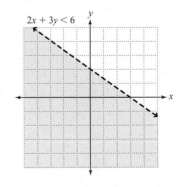

Procedure Graphing Linear Inequalities

To graph a linear inequality in two variables,

1. Graph the related equation (the boundary line). The related equation has an equal sign in place of the inequality symbol. If the inequality symbol is $\leq$ or $\geq$, draw a solid line. If the inequality symbol is $<$ or $>$, draw a dashed line.

2. Choose an ordered pair on one side of the boundary line and test this ordered pair in the inequality. If the ordered pair satisfies the inequality, shade the region that contains it. If the ordered pair does not satisfy the inequality, shade the region on the other side of the boundary line.

Example 2 Graph the linear inequalities.

a. $y > 3x + 1$

SOLUTION First, graph the related equation $y = 3x + 1$. Two ordered pairs that satisfy $y = 3x + 1$ are $(0, 1)$ and $(1, 4)$.

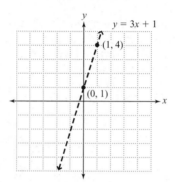

Note: The equation $y = 3x + 1$ could also be graphed using the slope and y-intercept.

Because the inequality is strictly greater than, we draw a dashed line to indicate that ordered pairs on this boundary line are not solutions.

Now we choose an ordered pair on one side of the line and test this ordered pair in the inequality. We can choose any point, but for easier calculations, let's choose $(0, 0)$.

Note: You cannot choose the origin if it lies on the boundary line.

$$y > 3x + 1$$
$$0 \overset{?}{>} 3(0) + 1 \qquad \text{Replace } x \text{ with 0 and } y \text{ with 0.}$$
$$0 > 1 \qquad \text{This statement is false, so } (0, 0) \text{ is not a solution for the inequality.}$$

Because $(0, 0)$ did not satisfy the inequality, we shade the region on the other side of the boundary line.

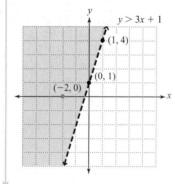

We can confirm that our shading is correct by choosing an ordered pair in this region. We will choose the ordered pair $(-2, 0)$.

$$y > 3x + 1$$
$$0 \overset{?}{>} 3(-2) + 1 \qquad \text{Replace } x \text{ with } -2 \text{ and } y \text{ with 0.}$$
$$0 > -5 \qquad \text{This statement is true, which confirms that we have shaded the correct side of the boundary line.}$$

Shading can be added to a graph on a graphing calcula-tor. Using the Y = *key, enter the related equation for the given inequality. On the same line you have entered the equation, use the arrow keys to move the cursor to the left of the variable* Y. *Normally, in this position, you see a slash, which indicates that the graph will be a solid line. Pressing enter changes this slash to various other options for the graph. There are two options for shading:*

◥ *means that the graph will have the region above the line shaded.*
◣ *means that the graph will have the region below the line shaded.*

Notice that it is up to you to determine which region should be shaded.

b. $x - 2y \le 4$

SOLUTION First, graph the related equation $x - 2y = 4$. Two ordered pairs that satisfy $x - 2y = 4$ are $(0, -2)$ and $(4, 0)$.

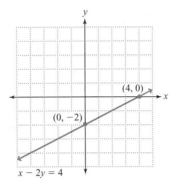

Because the inequality symbol is less than or equal to, we draw a solid line to indicate that ordered pairs on the boundary line are solutions.

Now we choose an ordered pair on one side of the line and test this ordered pair in the inequality. We will choose the origin $(0, 0)$.

$$x - 2y \le 4$$
$$0 - 2(0) \stackrel{?}{\le} 4 \quad \text{Replace } x \text{ with 0 and } y \text{ with 0.}$$
$$0 \le 4 \quad \text{This statement is true; therefore, } (0, 0) \text{ is a solution for the inequality.}$$

Because $(0, 0)$ satisfies the inequality, we shade the region that contains it.

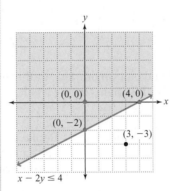

We can confirm that the region on the other side of the line should not be shaded by choosing a point in that region, such as $(3, -3)$.

$$x - 2y \le 4$$
$$3 - 2(-3) \stackrel{?}{\le} 4 \quad \text{Replace } x \text{ with 3 and } y \text{ with } -3.$$
$$3 + 6 \stackrel{?}{\le} 4$$
$$9 \le 4$$

This statement is false, which confirms that the region containing the point with coordinates $(3, -3)$ should not be shaded.

c. $x > 2$

SOLUTION First, graph the related line $x = 2$, which is a vertical line intersecting the x-axis at $(2, 0)$.

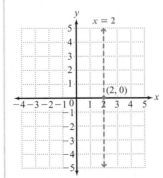

Because the inequality symbol is greater than, we draw a dashed line to indicate that the ordered pairs on the boundary line are not solutions.

Now we choose an ordered pair on one side of the line and test this ordered pair in the inequality. We will choose the origin $(0, 0)$.

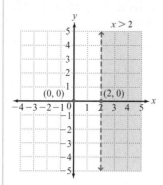

$x > 0$

$0 > 0$ Replace x with 0, which results in a false statement. So shade the side of the line opposite $(0, 0)$.

Note: It is possible to determine the solutions without using a test point. We are looking for all ordered pairs whose x-values are greater than 2. The ordered pairs whose x-values are greater than 2 are to the right of 2, much like graphing on a number line.

d. $y \le -3$

SOLUTION First, graph the related line $y = -3$, which is a horizontal line intersecting the y-axis at $(0, -3)$.

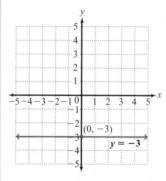

Because the inequality symbol is less than or equal to, we draw a solid line to indicate that the ordered pairs on the boundary line are solutions.

Now we choose an ordered pair on one side of the line and test this ordered pair in the inequality. We will choose the origin $(0, 0)$.

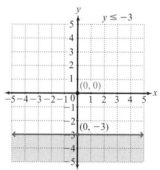

$y \le -3$

$0 \le -3$ Replace y with 0, which results in a false statement. So shade the side of the line on the opposite side from $(0, 0)$.

Note: Again, it is possible to determine the solutions without using a test point. We are looking for all ordered pairs whose y-values are less than or equal to -3. The ordered pairs whose y-values are less than or equal to -3 are either on or below the line.

Your Turn 2 Graph the linear inequalities.

a. $y < -\dfrac{1}{3}x + 2$ **b.** $2x + 3y \ge 6$ **c.** $x \le -3$ **d.** $y > 2$

Answers   Your Turn 2

a. $y < -\frac{1}{3}x + 2$

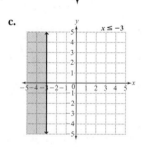

b. $2x + 3y \ge 6$

c. $x \le -3$

d. $y > 2$

1. How do you determine whether an ordered pair is a solution for an inequality?

2. How do you determine the related equation for a given inequality?

3. In graphing a linear inequality, when do you draw a dashed line and when do you draw a solid line?

4. When graphing a linear inequality, how do you determine which side of the boundary line to shade?

For Exercises 5–12, determine whether the ordered pair is a solution for the linear inequality. See Example 1.

5. $(5, 5)$; $y > -x + 2$

6. $(0, 1)$; $y > -x + 2$

7. $(2, 0)$; $y \geq \dfrac{1}{2}x + 1$

8. $(0, 0)$; $y < -\dfrac{2}{3}x + 2$

9. $(-1, 2)$; $3x - y \leq -8$

10. $(3, -1)$; $x - y < -2$

11. $(-1, -2)$; $x - 2y > -7$

12. $(0, -2)$; $x + 4y \geq 8$

For Exercises 13–40, graph the linear inequality. See Example 2.

13. $y \leq -x + 1$

14. $y \leq x - 8$

15. $y > -2x + 5$

16. $y < -3x + 2$

17. $y > 2x$

18. $y < x$

19. $y > \dfrac{1}{3}x$

20. $y < \dfrac{2}{5}x$

21. $y \geq \dfrac{3}{4}x + 2$

22. $y \geq -\dfrac{2}{3}x - 3$

23. $3x + y \leq 9$

24. $2x + y > 6$

25. $5x - 2y < 10$

26. $3x - 2y \geq 6$

27. $3x + y \geq 8$

28. $5x + y \geq -10$

29. $2x + 3y < 9$

30. $4x + 2y \leq 3$

31. $3x + y \geq 0$

32. $-3x - 2y < 0$

33. $x > -4$

34. $x \leq 6$

35. $y \leq 2$

36. $y > -1$

37. $x + 2 \geq 0$

38. $x - 4 < 0$

39. $y - 3 < 0$

40. $y + 2 \geq 0$

41. A company sells two sizes of bottles of hand lotion. The revenue for each unit of the large bottle is $2.00. The revenue for each unit of the small bottle is $1.50. For the company to break even during the first quarter, the company must generate $280,000 in revenue. The inequality $2x + 1.5y \geq 280,000$ describes the amount of revenue that must be generated from the sale of both size bottles for the company to break even or turn a profit.

   a. What do x and y represent?

   b. Graph the inequality. Because x and y are both nonnegative, restrict the graph to the first quadrant.

   c. What does the boundary line represent? What does the shaded region represent?

   d. List three combinations of sales numbers of each size that allow the company to break even.

   e. List three combinations of sales numbers of each size that allow the company to turn a profit.

   f. In reality, is every combination of units sold represented by the line and shaded region a possibility? Explain.

42. A company produces two versions of tax preparation software. The regular version costs $12.50 per unit to produce, and the deluxe version costs $15.00 per unit to produce. Management plans for the cost of production to be a maximum of $300,000 for the first quarter. The inequality $12.50x + 15y \le 300,000$ describes the total cost as prescribed by management.
   a. What do x and y represent?

   b. Graph the inequality. Because x and y are both nonnegative, restrict the graph to the first quadrant.
   c. What does the line represent?

   d. What does the shaded region represent?

   e. List three combinations of numbers of units produced of each version of software that yield a total cost of $300,000.

   f. List three combinations of numbers of units produced of each version of software that yield a total cost less than $300,000.

   g. In reality, is every combination of units produced represented by the line and shaded region a possibility? Explain.

43. Andrea visits a garden center to purchase new plants for her garden. She has a $100 gift certificate. The plants that she is considering come in two sizes and are priced according to size. Plants in 1-gallon containers are priced at $5.50 each, and plants in 5-gallon containers are priced at $12.50 each.
   a. Select a variable for the number of each plant size and write an inequality that has Andrea's total cost within the amount of the gift certificate.

   b. Graph the inequality. Because both quantities are nonnegative, restrict the graph to the first quadrant.
   c. What does the line represent?

   d. What does the shaded region represent?

   e. What are some possible combinations of containers she could purchase?

   f. In reality, could she purchase any combinations that would yield a total in the exact amount of the gift certificate? Explain.

44. Michelle is coordinating a fund-raiser that will sell two different cookbooks. One cookbook sells for $10; the other, for $12. The goal is to raise at least $30,000.
   a. Select variables for each type of cookbook and write an inequality in which total sales is at least $30,000.

   b. Graph the inequality. Because both quantities are nonnegative, restrict the graph to the first quadrant.
   c. What does the line represent?

   d. What does the shaded region represent?

e. Find two combinations of book sales that yield exactly $30,000 in total sales.

f. Find two combinations of units sold that yield more than $30,000.

45. A building with a rectangular base is to be designed so that the maximum perimeter is 180 feet.

    a. Select variables for the length and width and write an inequality in which the maximum perimeter is 180 feet.

    b. Graph the inequality. Because the length and width are both positive, restrict the graph to the first quadrant.

    c. What does the line represent?

    d. What does the shaded region represent?

    e. Find three combinations of length and width that yield a perimeter of exactly 180 feet.

    f. Find three combinations of length and width that yield a perimeter of less than 180 feet.

46. Steel beams are to be welded together to form a frame that is an isosceles triangle (two sides of equal length). The perimeter of the frame cannot exceed 60 feet.

    a. Select a variable for the length of the base and a second variable for the sides of equal length. Then write an inequality in which the maximum perimeter is 60 feet.

    b. Graph the inequality. Because the lengths are positive, restrict the graph to the first quadrant.

    c. What does the line represent?

    d. What does the shaded region represent?

    e. Find sets of dimensions that yield a perimeter of exactly 60 feet.

    f. Find sets of dimensions that yield a perimeter of less than 60 feet.

> **Puzzle Problem**
>
> The sum of two different whole numbers is equal to 8, while the difference of the two numbers is greater than or equal to 1. Find all possible ordered pairs that satisfy both constraints.

Review Exercises

Exercise 1   CONSTANTS AND VARIABLES

[1.1]   **1.** Write a set containing all vowels in the English alphabet.

Exercises 2–6   EQUATIONS AND INEQUALITIES

[1.7]   **2.** Evaluate $\frac{3}{4}x - 5$ when $x = 8$.   **[2.3]**   **3.** Solve and check: $3x - 4(x + 7) = -15$

[3.3]   **4.** The length of a rectangle is four more than the width. If the perimeter is 88 feet, find the dimensions.

[4.3]   **5.** Find the x- and y-intercepts for $3x - 2y = 8$.   **[4.3]**   **6.** Graph $-x + y = 7$.

4.7 Introduction to Functions and Function Notation

Objectives

1. Identify the domain and range of a relation.
2. Identify functions and their domains and ranges.
3. Find the value of a function.
4. Graph linear functions.

In this section, we study relations and functions, which develop naturally from our work with graphing because they involve ordered pairs.

Objective 1 Identify the domain and range of a relation. To formally define a function, we first discuss **relations**.

Definition   *Relation:* A set of ordered pairs.

A linear equation such as $y = 2x$ is a relation because it pairs a given x-value with a corresponding y-value. You might recall that with an equation in this form, we thought of the x-values as input values and the y-values as output values. In a relation, the set of all of the input values (x-values) is called the **domain** and the set of all of the output values (y-values) is called the **range**.

Definitions   *Domain:* The set of all input values (x-values) for a relation.

Range: The set of all output values (y-values) for a relation.

The following table lists some values in the domain and range for $y = 2x$.

Domain (x-values)	Range (y-values)
0	0
1	2
2	4
3	6

Example 1   Determine the domain and range of the relation $\{(2, -1), (3, 0), (4, 5), (5, 8)\}$.

SOLUTION   The domain is the set containing all of the x-values $\{2, 3, 4, 5\}$, and the range is the set containing all of the y-values $\{-1, 0, 5, 8\}$.

Domain and Range with a Graph

We can determine the domain and range of a relation from its graph. Look at the graph of $y = 2x$.

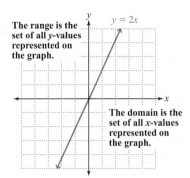

The range is the set of all *y*-values represented on the graph.

The domain is the set of all *x*-values represented on the graph.

It is convenient to imagine traveling along a graph from left to right. As we travel along $y = 2x$ from left to right, every *x*-value along the *x*-axis has a corresponding *y*-value. Thus, the domain is a set containing all real numbers. Similarly, every *y*-value is paired with an *x*-value; so the range also contains all real numbers.

Procedure **Determining Domain and Range**

To determine the domain of a relation given its graph, answer this question: What are all of the *x*-values that have a corresponding *y*-value? To determine the range, answer this question: What are all of the *y*-values that have a corresponding *x*-value?

Example 2 Determine the domain and range of the relation.

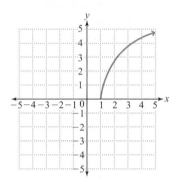

SOLUTION From left to right, this graph "begins" at the point $(1, 0)$. Values along the *x*-axis begin at 1 and continue infinitely; so the domain is $\{x \mid x \geq 1\}$ in set-builder notation and $[1, \infty)$ in interval notation. For the range, the *y*-values begin at 0 and continue infinitely; so the range is $\{y \mid y \geq 0\}$ in set-builder notation and $[0, \infty)$ in interval notation.

Your Turn 2 Determine the domain and range of the relation.

a. $\{(-2, -4), (0, -1), (3, 6), (7, 5)\}$

b.

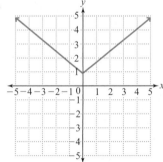

Answers Your Turn 2

a. Domain: $\{-2, 0, 3, 7\}$
   Range: $\{-4, -1, 6, 5\}$

b. Domain: All real numbers or $(-\infty, \infty)$
   Range: $\{y \mid y \geq 1\}$ or $[1, \infty)$

Objective ② Identify functions and their domains and ranges. Now we turn to a special type of relation called a *function*. Let's reconsider the trajectory data from Example 4 of Section 4.1 (p. 245), which we have listed here as a relation in the following table.

Time in Seconds (domain)	Height in Feet (range)
0	6.00
0.5	7.50
1.0	8.25
1.5	7.50
2.0	6.00
2.5	2.00

Note that each value for time is paired with one and only one height value. This is the key feature that makes this relation a **function**.

Definition   *Function:* A relation in which every value in the domain is paired with (or assigned to) exactly one value in the range.

From the data in the preceeding table, we see that a function can have different values in the domain assigned to the same value in the range. For example, domain values 0.5 seconds and 1.5 seconds each correspond to the same range value 7.50 feet. Also, the domain values 0 seconds and 2.0 seconds each correspond to 6.00 feet in the range. Using arrows to match the values in the domain with their corresponding values in the range can be a helpful technique in determining whether a relation is a function. Such a map might look like this:

Domain: {0,   0.5,   1.0,   1.5,   2.0,   2.5}

Note: Each element in the domain has a single arrow pointing to an element in the range.

Range: {2.00,   6.00,   7.50,   8.25}

Note that every function is a relation but not every relation is a function. For example, consider the following relation that assigns baseball outfield positions (the domain) to the 2008 Atlanta Braves outfielders (the range).

Note: Elements in the domain have more than one arrow pointing to an element in the range.

Domain: {left field, center field, right field}

Range: {Diaz, Anderson, Kotsay, Anderson, Francoeur, Anderson}

Notice that all three positions are assigned to more than one player. Since an element in the domain is assigned to more than one element in the range, this relation is not a function.

Conclusion   If any value in the domain is assigned to more than one value in the range, the relation is not a function.

Example 3   Determine whether the relation is a function.

a. The table to the right indicates the actual and projected national health expenditures, in billions of dollars, for selected years. (*Source:* www.healthcare.org)

Year (domain)	Expenditure (range)
2000	1310.0
2001	1424.5
2002	1547.6
2003	1660.5
2008	2354.6
2010	3079.8

Solution This relation is a function because each year value in the domain has only one expenditure value in the range.

b. The following relation assigns a birth date to the corresponding person.

Domain: {May 1, June 4, July 8, August 2}

Range: {Danielle, Gerard, Juan, René, Candice}

Solution This relation is not a function because an element in the domain, July 8, is assigned to two people in the range.

c. $\{(1, 4), (3, -2), (-5, 0)\ (3, 4)\}$

> **Note:** The domain of Example 3(c) is $\{1, 3, -5\}$. Even though 3 occurs twice, it is listed only once.

Solution This relation is not a function because an element in the domain, 3, is paired with two elements, -2 and 4, in the range.

Your Turn 3 Determine whether the relation is a function.

a.

Index Topic (domain)	Page Number(s) (range)
Celestial sphere	56
Celsius	36
Centigrade	36
Centimeter	6
Centripetal force	239, 305

b. The table shows the top five finishers in the 2009 Daytona 500 race.

Final Position (domain)	Car Number (range)
1st	17 (Matt Kenseth)
2nd	29 (Kevin Harvick)
3rd	44 (A. J. Allmendinger)
4th	33 (Clint Bowyer)
5th	19 (Elliott Sadler)

c. $\{(-2, -1), (5, -2), (0, 0), (3, -2)\}$

We can determine whether a relation is a function by its graph. Remember that in Example 4 of Section 4.1 (p. 245), we created a graph of the trajectory data (that data also appears in this section on p. 305) by plotting the ordered pairs as points in the coordinate plane.

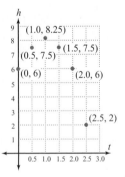

Notice that the domain (time in this case) is plotted along the horizontal axis and the range (height) is plotted along the vertical axis.

To determine whether a relation is a function from its graph, we can perform a test called the *vertical line test*. The nature of the test is to draw or imagine a vertical line through every point in the domain.

Answers Your Turn 3

a. not a function

b. function

c. function

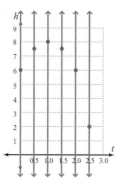

Here, we've drawn lines in blue through each value in the domain. Note that each vertical line intersects the graph at one and only one point, which means that each value in the domain corresponds to exactly one value in the range. This relation is a function.

Procedure

The Vertical Line Test

To determine whether a relation is a function from its graph, perform a vertical line test.

1. Draw or imagine vertical lines through each point in the domain.
2. If each vertical line intersects the graph in at most one point, the graph is the graph of a function.
3. If any vertical line intersects the graph at two or more different points, the graph is not the graph of a function.

Example 4 For each graph, determine the domain and range. Then state whether the relation is a function.

a.

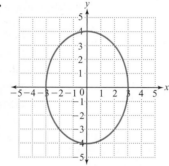
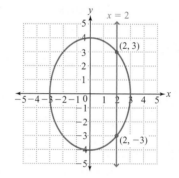

SOLUTION Domain: $\{x \mid -3 \leq x \leq 3\}$ or $[-3, 3]$

Range: $\{y \mid -4 \leq y \leq 4\}$ or $[-4, 4]$

This relation is not a function because a vertical line can be drawn that intersects the graph at two different points.

For example, the vertical line $x = 2$ (shown in blue) passes through two different points on the graph, $(2, 3)$ and $(2, -3)$.

b.

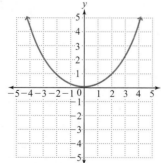
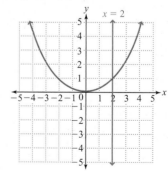

SOLUTION Domain: all real numbers

Range: $\{y \mid y \geq 0\}$ or $[0, \infty)$

This relation is a function. A vertical line through any value along the horizontal axis will intersect the graph at only one point. We have shown $x = 2$ to illustrate one such line.

c.

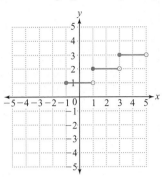

SOLUTION Domain: $\{x \mid -1 \le x < 5\}$ or $[-1, 5)$

Range: $\{y \mid y = 1, 2, 3\}$

This relation is a function. It may seem that the vertical lines $x = 1$ and $x = 3$ (shown in blue) intersect the graph at two different points. However, the open circles at $(1, 1)$, $(3, 2)$, and $(5, 3)$ indicate that those points are not part of the graph. The closed circles at $(-1, 1)$, $(1, 2)$, and $(3, 3)$ are part of the graph. The vertical line $x = 1$ "passes through" the open circle at $(1, 1)$ and intersects the graph at only one point, the closed circle at $(1, 2)$. Similarly, $x = 3$ passes through the open circle at $(3, 2)$ and intersects the graph at only one point, $(3, 3)$.

SOLUTION

Domain:

$\{t \mid 9:30 \text{ A.M.} \le t \le 4:00 \text{ P.M.}\}$ or $[9:30 \text{ A.M.}, 4:00 \text{ P.M.}]$

Range:

$\{N \mid 2507 \le N \le 2569\}$ or $[2507, 2569]$

This relation is a function. A vertical line through any value along the horizontal axis will intersect the graph at only one point. In the terminology from the graph, at each time during this business day, the Nasdaq had only one value.

d.

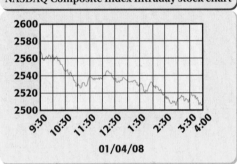

Your Turn 4 For each graph, determine the domain and range. Then state whether the relation is a function.

a.

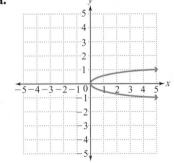

b.

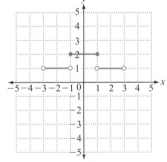

Answers Your Turn 4

a. Domain: $\{x \mid x \ge 0\}$ or $[0, \infty)$
Range: all real numbers
not a function

b. Domain: $\{x \mid -3 < x < 3\}$
or $(-3, 3)$
Range: $\{y \mid y = 1, 2\}$
function

Function Notation

When written as an equation, the notation for a function is a modification of an equation in two variables. Instead of $y = 2x + 5$, in function notation, we would write $f(x) = 2x + 5$ to mean the same equation. Notice that the notation $f(x)$, which is read "a function in terms of x," or "f of x," replaces the variable y. The following information shows how equations in two variables can be written using function notation.

Equation in two variables: Function notation:

$$y = 3x \qquad\qquad f(x) = 3x$$
$$y = -4x + 1 \qquad\qquad f(x) = -4x + 1$$
$$y = \frac{2}{3}x - 5 \qquad\qquad f(x) = \frac{2}{3}x - 5$$

Warning: The notation $f(x)$ does *not* mean multiply f and x.

Finding the Value of a Function

Remember that when we found solutions to equations with y isolated, we often spoke of the x-value as an input value and the corresponding y-value as the output value. Finding the value of a function is the same as finding a solution to an equation in two variables: We input a value for x and calculate the output value, $f(x)$. The following tables list the same ordered pairs. The table on the left is for $y = 2x + 5$, and the table on the right is for $f(x) = 2x + 5$. Notice that the only difference is in labeling the output values.

(input) x	(output) y
0	$2(0) + 5 = 5$
1	$2(1) + 5 = 7$
2	$2(2) + 5 = 9$

(input) x	(output) $f(x)$
0	$2(0) + 5 = 5$
1	$2(1) + 5 = 7$
2	$2(2) + 5 = 9$

Function notation offers a clever way to indicate that a specific x-value is to be used. For example, given a function $f(x) = 3x$, the notation $f(2)$ means to find the value of the function where $x = 2$ so that $f(2) = 3(2) = 6$, which is another way of representing the ordered pair $(2, 6)$.

Procedure   **Finding the Value of a Function**

Given a function $f(x)$, to find $f(a)$, where a is a real number in the domain of f, replace x in the function with a and calculate the value.

Example 5   For the function $f(x) = \dfrac{4}{x - 3}$, find the following.

a. $f(2)$

SOLUTION $f(2) = \dfrac{4}{2 - 3}$ Replace x in $\dfrac{4}{x-3}$ with 2; then calculate.

$$= \frac{4}{-1}$$

$$= -4$$

b. $f(-5)$

SOLUTION $f(-5) = \dfrac{4}{-5-3}$ Replace x in $\dfrac{4}{x-3}$ with -5; then calculate.

$= \dfrac{4}{-8}$

$= -\dfrac{1}{2}$

c. $f(3)$

SOLUTION $f(3) = \dfrac{4}{3-3}$ Replace x in $\dfrac{4}{x-3}$ with 3; then calculate.

$= \dfrac{4}{0}$ Therefore, $f(3)$ is undefined.

Your Turn 5   For the function $\sqrt{x+4}$, find the following.

   a. $f(5)$ **b.** $f(3)$ **c.** $f(-6)$

Finding the Value of a Function Given Its Graph

We can also find the value of a function given its graph. For a given value in the domain (x-value), we look on the graph for the corresponding value in the range (y-value).

Example 6   Using the graph, find the value of the function.

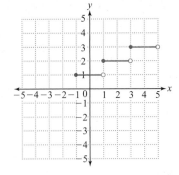

a. $f(0)$

SOLUTION   The notation $f(0)$ means to find the value of the function (y-value), where $x = 0$. On the graph, we see that when $x = 0$, the corresponding y-value is 1; so we say that $f(0) = 1$.

b. $f(1)$

SOLUTION   When $x = 1$, we see that $y = 2$; so $f(1) = 2$.

c. $f(-4)$

SOLUTION   Remember from Example 4(c) that the domain is $\{x \mid -1 \le x < 5\}$ or $[-1, 5)$. Because the value -4 is not in the domain, we say that $f(-4)$ is undefined.

Your Turn 6   Use the graph in the margin to find the value of the function.

   a. $f(-1)$ **b.** $f(0)$ **c.** $f(5)$ **d.** $f(-3)$

Objective ④   **Graph linear functions.**   We create the graph of a function the same way we create the graph of an equation in two variables. In this chapter, we graph only linear functions. Linear functions have a form similar to linear equations in slope–intercept form.

Slope–intercept form: $y = mx + b$

Linear function: $f(x) = mx + b$

Example 7 Graph: $f(x) = -2x + 3$

SOLUTION Think of the function as the equation $y = -2x + 3$. We can make a table of ordered pairs or use the fact that the slope is -2 and the y-intercept is 3.

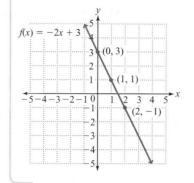

$f(x) = -2x + 3$

$(0, 3)$

$(1, 1)$

$(2, -1)$

Table of ordered pairs

x	$f(x)$
0	3
1	1
2	-1

Answer   Your Turn 7

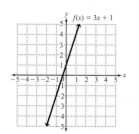

$f(x) = 3x + 1$

Your Turn 7 Graph $f(x) = 3x + 1$.

4.7 Exercises For Extra Help *MyMathLab*

Note: Exercises marked with a ★ represent challenging exercises.

1. Explain domain of a relation in your own words.

2. Explain range of a relation in your own words.

3. Given a graph of a relation, how do you determine the domain?

4. Given a graph of a relation, how do you determine the range?

5. When using the vertical line test on a graph, how do you determine whether the graph is the graph of a function?

6. True or false: All relations are functions. Explain.

For Exercises 7–12, determine the domain and range of the relation. See Example 1.

7. $\{(2, 1), (3, -2), (1, 4), (-1, -1)\}$

8. $\{(-2, 1), (4, 3), (2, 5), (-3, -2)\}$

9. $\{(4, 1), (2, 1), (3, -5), (-4, 0)\}$

10. $\{(-2, 1), (8, 11), (-3, 5), (4, -4)\}$

11. $\{(0, 0), (-3, -3), (2, 9), (8, -16)\}$

12. $\{(5, 4), (-3, 8), (10, -2), (0, -5)\}$

For Exercises 13–20, determine whether the relation is a function. In each table, the domain is in the left column and the range is in the right column. See Example 3.

13. Average college tuition in the United States for 2005–2006:

Sector	Cost ($)
Two-year public	1935
Two-year private	11,180
Four-year public	5351
Four-year private	19,292

Source: National Center for Education Statistics, U.S. Department of Education

14. The number of U.S. radio stations by format in 2007:

Format	Radio Stations
Country	2034
News/Talk	1370
Spanish	777
Oldies	711
Adult Contemporary	661

Source: The World Almanac and Book of Facts, 2008

15. Results of the 2007 Masters golf tournament:

Position	Name
1st	Zach Johnson
2nd	Rory Sabbatini; Retief Goosen, Tiger Woods
5th	Jerry Kelly, Justin Rose
7th	Padrig Harrington, Stuart Appleby

(*Source:* www.masters.org)

16. Top-five cable TV networks in 2006:

Rank	Channel
1	Discovery Channel
2	ESPN and CNN
3	TNT, LIFE, and USA
4	Weather
5	Nickelodeon and History

Source: The World Almanac and Book of Facts, 2007

17. Highest average second-quarter home prices by metropolitan area in 2007:

Metropolitan Area	Average Price ($)
San Jose-Sunnyvale-Santa Clara	865,000
San Francisco-Oakland-Fremont	846,800
Anaheim-Santa Ana	727,000
Honolulu	665,000
San Diego-Carlsbad-San Marcos	614,000
Los Angeles-Long Beach-Santa Ana	593,000
New York-Wayne-White Plains	557,500

Source: National Association of Realtors

18. Top brands of ice cream by total sales:

Brand	Sales ($)
Private Label	846,152,600
Breyers	618,797,200
Dreyer's Edy's Grand	426,366,100
Dreyer's Edy's Slow-Churned	342,514,200
Haagen-Dazs	290,367,400

Source: Information Resources Inc.

19. Average high temperature for Orlando, Florida:

Average High (°F)	Month
92	July, August
91	June
89	September
88	May
84	October
83	April
77	March, November
73	February, December
72	January

20. Average rainfall for Seattle:

Avg. Rainfall (inches)	Month
6	January, December
4	February, March
3	October, November
2	April, May, September
1	June, July, August

For Exercises 21–36, determine whether the relation is a function and explain your answer. See Examples 3 and 4.

21. $\{(2, 7), (3, 4), (-1, 4), (0, 5)\}$

22. $\{(1, 2), (4, -7), (9, 5), (6, 14)\}$

23. $\{(8, 2), (4, -5), (3, 0), (8, -1)\}$

24. $\{(-1, 5), (2, 6), (2, 9), (-3, -2)\}$

25. The grades on a recent calculus test:

Student	Score
Jonathon	54
Nathan	79
Rebecca	84
Terri	97
Allen	97

26. The sources of distraction among inattentive drivers:

Distraction	Percentage of Drivers
Outside person, object, or event	29.4
Adjusting radio/CD	11.4
Other occupant	10.9
Using/dialing cell phone	1.5
Smoking related	0.9

Source: AAA Foundation for Traffic Safety/UNC Safety Research Center

27. Ten most home runs hit in a single season:

Player	Home Runs
Barry Bonds	73
Mark McGwire	70
Sammy Sosa	66
Mark McGwire	65
Sammy Sosa	64
Sammy Sosa	63
Roger Maris	61
Babe Ruth	60
Babe Ruth	59
Mark McGwire	58

28. Number of NASCAR championships (3 or more):

Number	Driver
7	Richard Perry
	Dale Earnhardt
4	Jeff Gordon
3	David Pearson
	Darrell Waltrip
	Cale Yarborough
	Lee Petty
	Jimmie Johnson

29.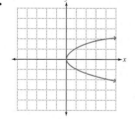

30.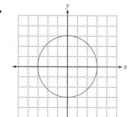

31.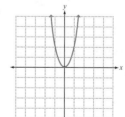

32.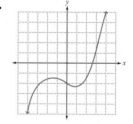

33.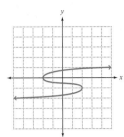

34.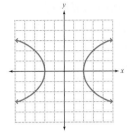

35.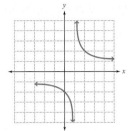

36.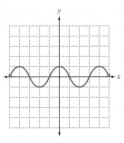

For Exercises 37–46, determine the domain and range. Then state whether the relation is a function. See Examples 2 and 4.

37. This graph shows the number of Catholics in the United States from 1900 to 2005.

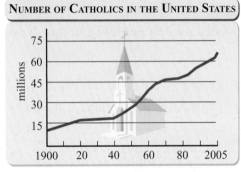

NUMBER OF CATHOLICS IN THE UNITED STATES

Source: BBC, 2005

38. The line graph represents postage costs for first-class letters from 1991 to 2009.

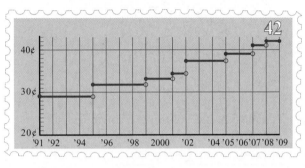

Source: Postal Rate Commission

39. This graph shows the percentage of the population that was foreign-born at each census.

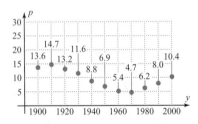

Source: Bureau of the Census, U.S. Department of Commerce

40. This graph shows the total number of people enrolled in a program each month after the program began.

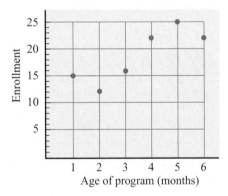

41.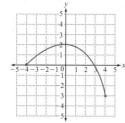

42.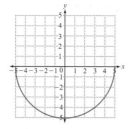

43.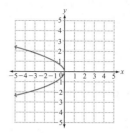

44.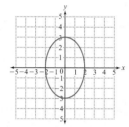

45.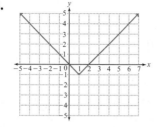

46.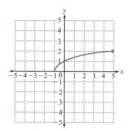

For Exercises 47–70, find the value of the function. See Example 5.

47. $f(x) = 2x - 5$
  a. $f(0)$ **b.** $f(1)$

  c. $f(-2)$ **d.** $f\left(\dfrac{1}{2}\right)$

48. $f(x) = 5x + 3$
  a. $f(1)$ **b.** $f(0)$

  c. $f(-1)$ **d.** $f(t)$

49. $f(x) = x^2 - 3x + 7$
  a. $f(0)$ **b.** $f(1)$

  c. $f(-1)$ **d.** $f(-2)$

50. $f(x) = 3x^2 + 2x - 1$
  a. $f(-2)$ **b.** $f(0.2)$

  c. $f(a)$ **d.** $f(3)$

51. $f(x) = \sqrt{x + 3}$
  a. $f(0)$ **b.** $f(-4)$

  c. $f(-1)$ **d.** $f(a)$

52. $f(x) = \sqrt{x + 5}$
  a. $f(-2)$ **b.** $f(6)$

  c. $f(-5)$ **d.** $f(a)$

53. $f(x) = \sqrt{x^2 - 4}$
  a. $f(2)$ **b.** $f(1)$

  c. $f(-2)$ **d.** $f(3)$

54. $f(x) = \sqrt{x^2 + 2x}$
  a. $f(-2)$ **b.** $f(-1)$

  c. $f(0)$ **d.** $f(1)$

55. $f(x) = \dfrac{1}{x - 1}$
  a. $f(0)$ **b.** $f(1)$

  c. $f(-1)$ **d.** $f(a)$

56. $f(x) = \dfrac{2}{x + 2}$
  a. $f(-2)$ **b.** $f(-1)$

  c. $f(3)$ **d.** $f(a)$

57. $f(x) = x^2 - 2x - 3$
  a. $f(0)$ **b.** $f(0.1)$

  c. $f(a)$ **d.** $f(a - 3)$

58. $f(x) = 2x^2 + x - 3$
  a. $f\left(\dfrac{3}{4}\right)$ **b.** $f(-1)$

  c. $f(a)$ **d.** $f(x + 2)$

59. $f(x) = \dfrac{1}{3}x - 2$

   a. $f(0)$ **b.** $f(1)$

   c. $f(-1)$ **d.** $f(a)$

60. $f(x) = \dfrac{2}{3}x + 1$

   a. $f(-3)$ **b.** $f(-1)$

   c. $f(3)$ **d.** $f(t)$

61. $f(x) = \dfrac{x + 3}{x - 4}$

   a. $f(0)$ **b.** $f(4)$

   c. $f(-2)$ **d.** $f(3)$

62. $f(x) = \dfrac{9 - x}{x + 2}$

   a. $f(0)$ **b.** $f(-2)$

   c. $f(3)$ **d.** $f(-1)$

63. $f(x) = \dfrac{1}{\sqrt{x + 2}}$

   a. $f(-2)$ **b.** $f(1)$

   c. $f(-3)$ **d.** $f(a)$

64. $f(x) = \dfrac{2}{\sqrt{2 - x}}$

   a. $f(-2)$ **b.** $f(-1)$

   c. $f(3)$ **d.** $f(a)$

65. $f(x) = |4x - 2|$

   a. $f(0)$ **b.** $f(1.5)$

   c. $f(-1)$ **d.** $f(-2)$

66. $f(x) = |3x - 5|$

   a. $f(-2)$ **b.** $f(-1)$

   c. $f(2)$ **d.** $f(3)$

67. $f(x) = x^2 - 3x + 2,\ g(x) = \dfrac{x - 4}{x + 2}$

   a. $f(2)$ **b.** $g(2)$

   c. $f(-3)$ **d.** $g(-3)$

68. $f(x) = \sqrt{x^2 - 3x},\ g(x) = \dfrac{2}{x + 4}$

   a. $g(-3)$ **b.** $f(-1)$

   c. $g(6)$ **d.** $f(4)$

★69. $f(x) = 2x^2 + 3x + 1$

   a. $f(2a)$ **b.** $f(a^2)$

   c. $f(a + 2)$ **d.** $f(2a - 1)$

★70. $f(x) = 3x^2 - 2x - 4$

   a. $f(3a)$ **b.** $f(a^3)$

   c. $f(a - 1)$ **d.** $f(2a + 2)$

For Exercises 71–74, use the given graph to find the value of the function. See Example 6.

71.

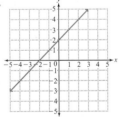

   a. $f(1)$

   b. $f(2)$

   c. $f(0)$

72.

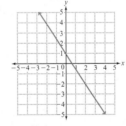

   a. $f(4)$

   b. $f(2)$

   c. $f(0)$

73.

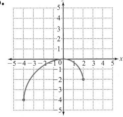

   a. $f(0)$

   b. $f(2)$

   c. $f(3)$

74.

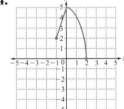

   a. $f(0)$

   b. $f(2)$

   c. $f(-2)$

For Exercises 75–84, graph. See Example 7.

75. $f(x) = 2x + 1$

76. $f(x) = 3x - 5$

77. $f(x) = \frac{1}{3}x + 3$

78. $f(x) = \frac{1}{2}x - 3$

79. $f(x) = -4x + 1$

80. $f(x) = -x - 1$

81. $f(x) = -\frac{2}{3}x$

82. $f(x) = \frac{2}{5}x$

83. $f(x) = -\frac{1}{4}x - 2$

84. $f(x) = -\frac{1}{3}x + 4$

85. The cost, $C(x)$, as a function of the number of items produced, x, is given by $C(x) = 20x + 150$.
  a. Find the cost of producing ten items.

  b. Find the cost of producing fifteen items.

  c. What does $C(30) = 750$ mean?

86. For a rental car, the cost, $C(x)$, as a function of the number of miles driven, x, is given by $C(x) = 0.12x + 22$.
  a. Find the cost of driving 100 miles.

  b. Find the cost of driving 350 miles.

  c. What does $C(250) = 52$ mean?

87. The value, $V(x)$, of a car as a function of the number of years after it was purchased, x, is given by $V(x) = -1500x + 18,000$.
  a. Find the value of the car after three years.

  b. Find the value of the car after six years.

  c. What does $V(10) = 3000$ mean?

88. The monthly salary, $S(x)$, of a salesperson as a function of her sales, x, is given by $S(x) = 0.10x + 175$.
  a. Find her salary for sales of $15,000.

  b. Find her salary for sales of $10,000.

  c. What does $S(20,000) = 2175$ mean?

Review Exercises

Exercises 1 and 2 **EXPRESSIONS**

For Exercises 1 and 2, simplify.

[1.5]   **1.** $\dfrac{2}{5} - \dfrac{5}{8} \div \left(\dfrac{3}{4}\right)$

[1.7]   **2.** $x + 3y - 5x + 7 - 3y - 9$

Exercises 3–6 **EQUATIONS AND INEQUALITIES**

[1.3]   **3.** Which property is illustrated in the equation $4(x + 9) = 4(9 + x)$?

[4.3]   **4.** Graph $2x - y = 4$ using the intercepts.

[4.4]   **5.** Solve $3x - 2y = 6$ for y.

[4.5]   **6.** Are the graphs of $2x + 3y = 9$ and $4x + 6y = 12$ parallel, perpendicular, or neither?

Chapter 4 Summary

Defined Terms

Section 4.3
x-intercept (p. 261)
y-intercept (p. 261)

Section 4.4
Slope (p. 273)

Section 4.7
Relation (p. 303)
Domain (p. 303)
Range (p. 303)
Function (p. 305)

Forms of a Linear Equation

Slope–intercept form: $y = mx + b$ (p. 273)
Point–slope form: $y - y_1 = m(x - x_1)$ (p. 284)
Standard form: $Ax + By = C$, where A, B, and C are integers and $A > 0$ (p. 286)

Formula

Slope: $m = \dfrac{y_2 - y_1}{x_2 - x_1}$, $x_2 \neq x_1$ (p. 276)

Procedures, Rules, and Key Examples

Procedures/Rules	Key Examples

SECTION 4.1   The Rectangular Coordinate System

To determine the coordinates of a given point in the rectangular system,
1. Follow a vertical line from the point to the x-axis (horizontal axis). The number at this position on the x-axis is the first coordinate.
2. Follow a horizontal line from the point to the y-axis (vertical axis). The number at this position on the y-axis is the second coordinate.

To graph or plot a point given its coordinates,
1. Beginning at the origin, $(0, 0)$, move to the right or left along the x-axis the amount indicated by the first coordinate.
2. From that position on the x-axis, move up or down the amount indicated by the second coordinate.
3. Draw a dot to represent the point described by the coordinates.

To determine the quadrant for a given ordered pair, consider the signs of the coordinates.

$(+, +)$ means the point is in quadrant I.
$(-, +)$ means the point is in quadrant II.
$(-, -)$ means the point is in quadrant III.
$(+, -)$ means the point is in quadrant IV.

Note: If either coordinate is 0, the point is on an axis and not in a quadrant.

Example 1: Plot each of the following:
$(-2, 4), (5, 0), (0, -2), (-4, -3), (3, -4), (2, 3)$

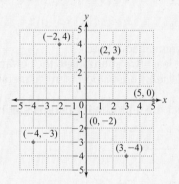

Example 2: State the quadrant in which each point is located. If the point lies on an axis, state which axis.

$(19, 78)$: quadrant I
$(-67, 45)$: quadrant II
$(-107, -36)$: quadrant III
$(58, -92)$: quadrant IV
$(0, -16)$: This point is on the y-axis.

(*continued*)

Procedures/Rules	Key Examples

SECTION 4.2 Graphing Linear Equations

To determine whether a given ordered pair is a solution for an equation with two variables,
1. Replace the variables in the equation with the corresponding coordinates.
2. Verify that the equation is true.

Example 1: Determine whether $(3, 5)$ is a solution for $4x - y = 7$.

Solution: $4(3) - 5 \overset{?}{=} 7$ Replace x with 3 and y with 5.

$$12 - 5 \overset{?}{=} 7$$
$$7 = 7$$
$$(3, 5) \text{ is a solution.}$$

Example 2: Determine whether $(4, -9)$ is a solution for $4x - y = 7$.

Solution: $4(4) - (-9) \overset{?}{=} 7$ Replace x with 4 and y with -9.

$$16 + 9 \overset{?}{=} 7$$
$$25 \neq 7$$
$$(4, -9) \text{ is not a solution.}$$

To find a solution to a linear equation with two variables,
1. Choose a value for one of the variables (any value).
2. Replace the corresponding variable with your chosen value.
3. Solve the equation for the value of the other variable.

Example 3: Find two solutions for the equation $y = 2x - 3$.

First solution:
Let $x = 0$

$$y = 2(0) - 3$$
$$y = -3$$

Second solution:
Let $x = 1$

$$y = 2(1) - 3$$
$$y = 2 - 3$$
$$y = -1$$

Solution: $(0, -3)$ Solution: $(1, -1)$

To graph a linear equation,
1. Find at least two solutions to the equation.
2. Plot the solutions as points in the rectangular coordinate system.
3. Connect the points to form a straight line.

Example 4: Graph $y = 2x - 3$.
We found two solutions above: $(0, -3)$ and $(1, -1)$.

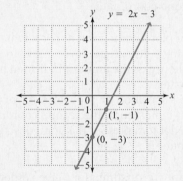

The graph of $y = c$, where c is a real-number constant, is a horizontal line parallel to the x-axis that passes through the y-axis at a point with coordinates $(0, c)$.

The graph of $x = c$, where c is a real-number constant, is a vertical line parallel to the y-axis that passes through the x-axis at a point with coordinates $(c, 0)$.

Example 5: Graph   **a.** $y = 4$
            b. $x = -3$

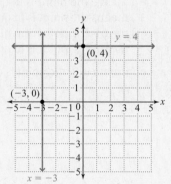

(continued)

Procedures/Rules	Key Examples

SECTION 4.3 Graphing Using Intercepts

To find an x-intercept,
1. Replace y with 0 in the given equation.
2. Solve for x.

To find a y-intercept,
1. Replace x with 0 in the given equation.
2. Solve for y.

Example 1: Find the x- and y-intercepts for $2x - 3y = 18$.

Solution:

x-intercept:

$$2x - 3(0) = 18$$
$$2x = 18$$
$$x = 9$$

y-intercept:

$$2(0) - 3y = 18$$
$$-3y = 18$$
$$y = -6$$

x-intercept: $(9, 0)$ y-intercept: $(0, -6)$

If an equation can be written in the form $y = mx$, where m is a real number other than 0, then the x- and y-intercepts are at the origin, $(0, 0)$.

Example 2: The graph of the equation $y = 3x$ has $(0, 0)$ as both the x- and y-intercept.

If an equation is in the form $y = mx + b$, where m and b are nonzero real numbers, then the y-intercept will be $(0, b)$.

Example 3: The y-intercept of the graph of $y = 3x - 2$ is $(0, -2)$.

The graph of an equation in the form $y = c$, where c is a nonzero real number, has no x-intercept and the y-intercept is $(0, c)$. The graph of $y = 0$ is the x-axis.

Example 4: The graph of the equation $y = 4$ has no x-intercept, and its y-intercept is $(0, 4)$.

The graph of an equation in the form $x = c$, where c is a nonzero real number, has no y-intercept and the x-intercept is $(c, 0)$. The graph of $x = 0$ is the y-axis.

Example 5: The graph of the equation $x = -3$ has no y-intercept, and its x-intercept is $(-3, 0)$.

SECTION 4.4 Slope–Intercept Form

Given an equation of the form $y = mx + b$,

If $m > 0$ (slope is positive) the graph is a line that slants uphill from left to right.

If $m < 0$ (slope is negative) the graph is a line that slants downhill from left to right.

To graph an equation in slope–intercept form, $y = mx + b$,
1. Plot the y-intercept, $(0, b)$.
2. Plot a second point by rising the number of units indicated by the numerator of the slope, m, then running the number of units indicated by the denominator of the slope, m.
3. Draw a straight line through the two points.

Note: You can check by locating additional points using the slope. Every point you locate using the slope should be on the line.

Example 1: Graph.
a. $y = 2x + 1$ **b.** $y = -\dfrac{3}{4}x - 2$

Solution: For $y = 2x + 1$, the slope is 2 and the y-intercept is $(0, 1)$. The positive slope indicates an uphill line from left to right. To graph, we can rise 2, then run 1 from the y-intercept, $(0, 1)$, to get a second point $(1, 3)$.

For $y = -\dfrac{3}{4}x - 2$, the slope is $-\dfrac{3}{4}$ and the y-intercept is $(0, -2)$. The negative slope indicates a downhill line from left to right. To graph, we rise -3, then run 4 from the y-intercept, $(0, -2)$, to get a second point $(4, -5)$.

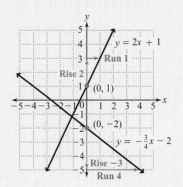

(*continued*)

Procedures/Rules	Key Examples

SECTION 4.4 Slope–Intercept Form (*continued*)

To write the equation of a line given its slope, m, and its y-intercept, $(0, b)$, use the slope–intercept form of the equation, $y = mx + b$.

Example 2: Write the slope–intercept form of the equation of a line with slope $\frac{3}{4}$ and y-intercept $(0, -1)$.

Solution: $y = \frac{3}{4}x - 1$

Given two points (x_1, y_1) and (x_2, y_2) where $x_2 \neq x_1$, the slope of the line connecting the two points is given by the formula

$$m = \frac{y_2 - y_1}{x_2 - x_1}$$

Example 3: Find the slope of a line connecting the points $(2, 7)$ and $(-8, 3)$.

Solution: $m = \frac{3 - 7}{-8 - 2} = \frac{-4}{-10} = \frac{2}{5}$

Two points with different x-coordinates and the same y-coordinates, (x_1, c) and (x_2, c), will form a line with a slope of 0 (a horizontal line) and the equation $y = c$.

Example 4: The slope of a line connecting the points $(2, 4)$ and $(1, 4)$ is 0, and the equation of the line is $y = 4$.

Two points with the same x-coordinates and different y-coordinates, (c, y_1) and (c, y_2), will form a line with a slope that is undefined (a vertical line) and the equation $x = c$.

Example 5: The slope of a line connecting the points $(-3, 7)$ and $(-3, -2)$ is undefined and the equation of the line is $x = -3$.

SECTION 4.5 Point–Slope Form

To write the equation of a line given its slope and any point, (x_1, y_1), on the line, use the point–slope form of the equation of a line, $y - y_1 = m(x - x_1)$. If given a second point, (x_2, y_2), and not the slope, we first calculate the slope using $m = \frac{y_2 - y_1}{x_2 - x_1}$; then we use $y - y_1 = m(x - x_1)$.

Example 1: Write the slope–intercept form of the equation of a line with slope -3 that passes through $(2, 5)$.

Solution:
$$\begin{aligned}
y - y_1 &= m(x - x_1) \\
y - 5 &= -3(x - 2) \qquad x_1 = 2, y_1 = 5 \\
y - 5 &= -3x + 6 \\
y &= -3x + 11
\end{aligned}$$

Example 2: Write the equation of a line connecting $(2, 3)$ and $(-3, -1)$ in the form $Ax + By = C$, where A, B, and C are integers and $A > 0$.

Solution: Find the slope:

$$m = \frac{-1 - 3}{-3 - 2} = \frac{-4}{-5} = \frac{4}{5}$$

Write the equation: $y - 3 = \frac{4}{5}(x - 2)$

$5(y - 3) = 5 \cdot \frac{4}{5}(x - 2)$ $\left\{\begin{array}{l}\text{Multiply by 5 to} \\ \text{clear the fraction.}\end{array}\right.$

$5y - 15 = 4(x - 2)$

$5y - 15 = 4x - 8$ Distribute 4.

$5y - 4x - 15 = -8$ Subtract $4x$ from both sides.

$5y - 4x = 7$ Add 15 to both sides.

$-4x + 5y = 7$ Rearrange so that the $-4x$ term is first.

$-1(-4x + 5y) = -1 \cdot 7$ $\left\{\begin{array}{l}\text{Multiply by } -1 \text{ so that the} \\ x \text{ term is positive.}\end{array}\right.$

$4x - 5y = -7$ Simplify.

(continued)

Procedures/Rules	Key Examples

SECTION 4.5 Point–Slope Form (*continued*)

Nonvertical parallel lines have equal slopes and different *y*-intercepts. Vertical lines are parallel.

The slope of a line perpendicular to a line with a slope of $\dfrac{a}{b}$ will be $-\dfrac{b}{a}$. Horizontal and vertical lines are perpendicular.

Example 3: Find the slope of a line parallel and perpendicular to $3x + 2y = 4$.

Solution: Write $3x + 2y = 4$ in slope–intercept form.

$$2y = -3x + 4 \qquad \text{Subtract } 3x \text{ from both sides.}$$
$$y = -\frac{3}{2}x + 2 \qquad \text{Divide both sides by 2.}$$

The slope of $3x + 2y = 4$ is $-\dfrac{3}{2}$.

The slope of a line parallel to $3x + 2y = 4$ is $-\dfrac{3}{2}$. The slope of a line perpendicular to $3x + 2y = 4$ is $\dfrac{2}{3}$.

SECTION 4.6 Graphing Linear Inequalities

To determine whether an ordered pair is a solution for an inequality, replace the variables with the corresponding coordinates and see if the resulting inequality is true. If it is, the ordered pair is a solution.

Example 1: Determine whether $(4, 3)$ is a solution for $2x - 4y \le 8$.

$$2x - 4y \le 8$$
$$2(4) - 4(3) \stackrel{?}{\le} 8 \qquad x = 4, y = 3$$
$$8 - 12 \stackrel{?}{\le} 8$$
$$-4 \le 8 \qquad \text{This is true, so } (4, 3) \text{ is a solution.}$$

Example 2: Determine whether $(3, -1)$ is a solution for $2x - 4y \le 8$.

$$2x - 4y \le 8$$
$$2(3) - 4(-1) \stackrel{?}{\le} 8 \qquad x = 3, y = -1$$
$$6 + 4 \stackrel{?}{\le} 8$$
$$10 \le 8 \qquad \text{This is false, so } (3, -1) \text{ is not a solution.}$$

To graph a linear inequality in two variables,
1. Graph the related equation (the boundary line). The related equation has an equal sign in place of the inequality symbol. If the inequality symbol is $\le$ or $\ge$, draw a solid line. If the inequality symbol is $<$ or $>$, draw a dashed line.
2. Choose an ordered pair on one side of the boundary line and test this ordered pair in the inequality. If the ordered pair satisfies the inequality, shade the region that contains it. If the ordered pair does not satisfy the inequality, shade the region on the other side of the boundary line.

Example 3: Graph $2x - 4y \le 8$.

Solution: Graph $2x - 4y = 8$ as a solid line. We found $(0, 0)$ to be a solution; therefore, shade the region containing $(0, 0)$.

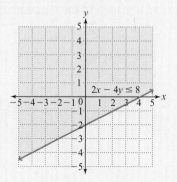

(*continued*)

SECTION 4.7 **Introduction to Functions and Function Notation**

To determine the domain of a relation given its graph, answer this question: What are all of the *x*-values that have a corresponding *y*-value?

To determine the range, answer this question: What are all of the *y*-values that have a corresponding *x*-value?

Example 1: Give the domain and range of the relation shown.

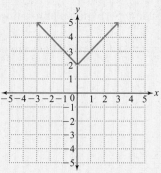

Domain: all real numbers

Range: $\{\, y \mid y \geq 2 \,\}$ or $[\,2, \infty\,)$

To determine whether a graph of a relation is a graph of a function, perform a vertical line test.
1. Draw or imagine vertical lines through each point in the domain.
2. If each vertical line intersects the graph at only one point, the graph is the graph of a function.
3. If any vertical line intersects the graph at two or more different points, the graph is not the graph of a function.

Example 2: Determine whether the graph of the relation is the graph of a function.

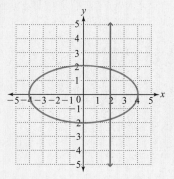

Solution: The relation in the graph is not a function because a vertical line can be drawn that intersects the graph at two points. This means that there is a domain value that corresponds to two values in the range.

Given a function $f(x)$, to find $f(a)$, where *a* is a real number in the domain of *f*, replace *x* in the function with *a* and calculate the value.

Example 3: For the function $f(x) = x^2 - 3x + 1$, find $f(-2)$.

$$f(-2) = (-2)^2 - 3(-2) + 1$$
$$= 4 + 6 + 1$$
$$= 11$$

Chapter 4 Review Exercises

For Exercises 1–6, answer true or false.

[4.1] **1.** When writing an ordered pair, the horizontal coordinate is written first.

[4.3] **2.** An x-intercept is a point where a graph intersects the horizontal axis.

[4.4] **3.** The slope formula is $m = \dfrac{x_2 - x_1}{y_2 - y_1}$.

[4.6] **4.** A linear inequality has just one solution.

[4.5] **5.** Given any two points on a line, the equation of the line may be found.

[4.5] **6.** Given the slope and any point on a line, the equation of the line may be found.

For Exercises 7–10, complete the rule.

[4.3] **7.** To find an x-intercept,
    a. Replace _____ with 0 in the given equation.
    b. Solve for _____.

[4.3] **8.** To find a y-intercept,
    a. Replace _____ with 0 in the given equation.
    b. Solve for _____.

[4.4] **9.** If an equation is in the form $y = mx + b$, where m and b are nonzero real numbers, the _____ will be $(0, b)$.

[4.4] **10.** Given an equation of the form $y = mx + b$,
    if $m > 0$, the graph will be a line that slants _____ from left to right.
    if $m < 0$, the graph will be a line that slants _____ from left to right.

[4.1] **For Exercises 11 and 12, write the coordinates for each point.**

11.

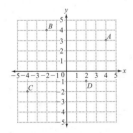

12.

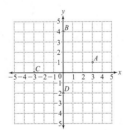

[4.1] **For Exercises 13 and 14, plot and label the points indicated by the coordinate pairs.**

13. $(3, 4), (-2, 0), (0, 1), (-1, -5)$

14. $(-2, 2), (0, -3), (4, 1), (5, 0)$

[4.1] *For Exercises 15–18, state the quadrant in which the point is located. If the point lies on an axis, state which axis.*

15. $(-2.8, 1203)$ **16.** $\left(-\dfrac{1}{8}, -16\right)$ **17.** $(421, 5300)$ **18.** $\left(52\dfrac{2}{3}, -0.36\right)$

Exercises 19–85   EQUATIONS AND INEQUALITIES

[4.2] *For Exercises 19–24, determine whether the given ordered pair is a solution for the equation.*

19. $(-1, 3); x + 3y = 8$ **20.** $(2, 1); 2x - y = 3$ **21.** $(0, 0); 6x = 2y + 1$

22. $(-2, -2); y = \dfrac{2}{3}x$ **23.** $\left(\dfrac{3}{2}, -\dfrac{1}{2}\right); x + y = 2$ **24.** $(2.1, 3.2); y - 3x = 5$

[4.2] *For Exercises 25–30, find three solutions for the given equation. Then graph.*

25. $y = -2x$ **26.** $x = y$ **27.** $y = -x + 7$

28. $x + 4y = 5$ **29.** $2x + 3y = 6$ **30.** $y + \dfrac{1}{3}x = 3$

[4.3] *For Exercises 31–36, find the x- and y-intercepts.*

31. $x + 2y = 12$ **32.** $y = -3x$ **33.** $2x - 3.1y = 6.2$

34. $y = \dfrac{2}{3}x + 5$ **35.** $x = -2$ **36.** $y = 7$

[4.3] *For Exercises 37–42, graph using the x- and y-intercepts.*

37. $3x + 2y = 6$ **38.** $y = -2x + 1$ **39.** $y = -\dfrac{1}{5}x + 3$

40. $2x = 5y + 10$ **41.** $7 = y$ **42.** $x = -1$

[4.4] *For Exercises 43 and 44, graph each equation on the same grid.*

43. $y = x$
$y = 2x$
$y = 4x$

44. $y = -x$
$y = -2x$
$y = -3x$

[4.4] *For Exercises 45–50, determine the slope and the y-intercept and then graph.*

45. $y = -2x + 3$ **46.** $y = -\dfrac{1}{4}x + 2$ **47.** $y = 3x$

48. $x + y = 5$ **49.** $x - 3y = 7$ **50.** $2x - 3y = 8$

[4.4] *For Exercises 51–56, find the slope of the line connecting the given points.*

51. $(2, 7), (-1, -2)$ **52.** $(0, -1), (3, 2)$ **53.** $(-3, 1), (3, -1)$

54. $(-6, 2), (-1, -1)$ **55.** $(2, 8), (-1, 8)$ **56.** $(-7, 3), (-7, -5)$

[4.4] *For Exercises 57–60, write the equation of the line in slope–intercept form given the slope and the y-intercept.*

57. $m = -1; (0, 7)$ **58.** $m = -\dfrac{1}{5}; (0, -8)$ **59.** $m = 0.2; (0, 6)$ **60.** $m = 1; (0, 0)$

[4.5] *For Exercises 61–66, write the equation of a line in slope–intercept form with the given slope that passes through the given point.*

61. $m = -2; (1, 7)$

62. $m = 1; (3, -6)$

63. $m = \dfrac{1}{3}; (-5, 0)$

64. $m = -\dfrac{2}{5}; (3, 3)$

65. $m = 6.2; (-2, -3)$

66. $m = -0.4; (-1, 2)$

[4.5] *For Exercises 67–70, write the equation of a line connecting the given points in slope–intercept form and in standard form.*

67. $(7, -3), (2, 2)$

68. $(5, -3), (9, 2)$

69. $(3, 2), (5, 1)$

70. $(-3, -2), (-8, -1)$

[4.5] *For Exercises 71–74, write the equations in slope–intercept form.*

71. Find the equation of a line that passes through $(0, -2)$ and is parallel to the line $y = 2x + 7$.

72. Find the equation of a line that passes through $(1, 5)$ and is parallel to the line $2x + 3y = 6$.

73. Find the equation of a line that passes through $(-1, -2)$ and is perpendicular to the line $y = \dfrac{3}{5}x - 1$.

74. Find the equation of a line that passes through $(-2, 4)$ and is perpendicular to the line $y = -\dfrac{2}{5}x - 6$.

[4.2] **75.** A salesperson receives $1000 per month plus a commission of 5% of his total sales. The equation $p = 0.05s + 1000$ describes the salesperson's gross pay each month, where p represents the gross pay and s represents the total sales in dollars.

    a. Find the gross pay if the person sells $24,000 worth of merchandise.

    b. Find the point where the graph intersects the p-axis. Explain what it indicates.

    c. Graph the equation.

[4.5] **76.** Andrew purchased a photocopier for his business in 2004 for $18,000. In 2008, he considers selling the copier on eBay and notes that others like it are selling for $12,000.

    a. Assuming that the depreciation is linear, plot the two given data points with 2004 being year 0 so that 2008 is year 4. Draw a line connecting the two points.

    b. What is the slope of the line?

    c. Let n represent the number of years the copier is in service and v represent the value of the copier. Write the equation of the line in slope–intercept form.

    d. If the depreciation continues at the same rate, in what year was the copier worth half of its original value?

    e. In what year will the copier be worth $0?

[4.6] *For Exercises 77–80, determine whether the ordered pair is a solution for the linear inequality.*

77. $(3, 2); x + 2y > 5$

78. $(-1, 0); y < x + 2$

79. $(-5, 8); 2x - 6y \leq 17$

80. $(0, 0); x + y \geq 5$

[4.6] *For Exercises 81–85, graph the linear inequality.*

81. $y < 3x - 5$

82. $y \geq -\dfrac{2}{3}x$

83. $x - y \geq 3$

84. $-3x - 5y < -15$

85. $x \geq -1$

[4.6] **86.** A company produces two versions of its product. The lower-priced package costs $6 to make, whereas the higher-priced package costs $8 to make.
   a. Let a represent the number of the lower-priced packages produced and b represent the number of the higher-priced packages produced. Write an inequality in which the total cost is at most $12,000.

   b. Graph the inequality.
   c. Give a combination of the number of each package that the company could produce that has a cost equal to $12,000.

   d. Give a combination of the number of each package that the company could produce that has a cost less than $12,000.

[4.7] *For Exercises 87 and 88, find the domain and range of the relation.*

87. $\{(2, 3), (-1, 5), (3, 6), (-3, -4)\}$

88. Cars holding their value for resale:

Automobile	Percentage of Original Value
Volkswagen	52.2
Honda	49.7
Toyota	49.0
Subaru	47.8
Nissan	45.8

Source: Automotive Lease Guide

[4.7] *For Exercises 89 and 90, determine whether the relation is a function.*

89. The following relation shows the courses taught by each instructor during one semester.

Domain (instructor name)	Range (courses taught)
Hames	Math 100
Carson	Math 100, Math 102
Pritchard	Math 035
Webb	Math 100, Math 110

90. The following relation shows the price of a particular brand of dog food based on the size of the bag.

Domain (size of bag)	Range (price)
5 lb.	$3.95
10 lb.	$7.90
20 lb.	$15.00
40 lb.	$29.95

[4.7] *For Exercises 91–93, give the domain and range. Then determine whether the graph is the graph of a function.*

91.

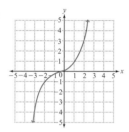

92.

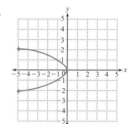

93.

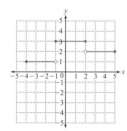

[4.7] **94.** Use the graph in Exercise 93 to find the value of the function at the indicated values.

    a. $f(-3)$ **b.** $f(-1)$

    c. $f(0)$ **d.** $f(3)$

[4.7] *For Exercises 95 and 96, find the value of the function.*

95. $f(x) = x^3 - 5x + 2$

    a. $f(2)$ **b.** $f(0)$ **c.** $f(-3)$

96. $f(x) = \dfrac{x}{x-3}$

    a. $f(6)$ **b.** $f(3)$ **c.** $f(-5)$

Chapter 4 Practice Test

For Extra Help

CHAPTER Test Prep VIDEOS

Step-by-step test solutions are found on the Chapter Test Prep Videos available via the Video Resources on DVD, in *MyMathLab*, and on You Tube (search "CarsonElemAlg" and click on "Channels").

1. Determine the coordinates for each point.

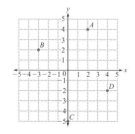

2. Plot and label the points indicated by the coordinate pairs.

$(0, -3), (4, -1), (-3, 2), (2, 0), (-1, -5)$

3. State the quadrant in which $\left(4\frac{2}{3}, 5115\right)$ is located. If the point lies on an axis, state which axis.

4. Determine whether $(1, 7)$ is a solution for $y = -\dfrac{2}{5}x - 8.$

For Exercises 5–8, find the x- and y-intercepts and then graph.

5. $x - 2y = 4$

6. $2x + 5y = 10$

7. $y = -2x$

8. $y = \dfrac{3}{4}x - 1$

For Exercises 9 and 10, determine the slope and the y-intercept.

9. $y = \dfrac{3}{4}x + 11$

10. $5x - 3y = 8$

For Exercises 11 and 12, determine the slope of the line connecting the given points.

11. $(-5, 6), (-2, -4)$

12. $(6, 2), (3, 2)$

For Exercises 13 and 14, write the equation of the line in slope–intercept form.

13. $m = \dfrac{3}{5}$; y-intercept $(0, 4)$

14. Passing through the points $(8, -1), (-7, 5)$

15. Write the equation of a line in the form $Ax + By = C$ through the points $(2, 1)$ and $(5, 3)$.

16. Write the equation of a line in the form $Ax + By = C$ through the point $(4, -2)$ and perpendicular to the line $y = -3x + 4.$

17. From 1999 to 2006, the number of people traveling more than 50 miles for leisure increased at a linear rate. In 1999, the number of people traveling 50 miles or more for leisure was 848.6 million. In 2006, that number was 990.2 million people. (*Source:* Travel Industry Association of America, TravelScope)
    a. Let 1999 be year 0 so that 2006 is year 7. Let x represent the number of years after 1999 and y represent the number of people traveling 50 miles or more. Plot the two data points in the coordinate plane; then draw a line connecting them.
    b. Find the slope of the line.

c. Write the equation of the line in slope–intercept form.

d. If the trend continues, predict the number of people traveling more than 50 miles for leisure in 2010. Round to the nearest tenth.

18. Determine whether $(2, 7)$ is a solution for the linear inequality $y \geq 2x - 9$.

For Exercises 19 and 20, graph the linear inequality.

19. $y \geq \dfrac{2}{5}x - 1$

20. $x - 3y < 8$

21. Alex works two part-time jobs. He receives $10 per hour when working as a cook in a restaurant and $8 per hour when working at a music store. To pay all of his monthly expenses, he needs to make at least $360 per week.

  a. Let x represent the number of hours he works at the restaurant and y represent the number of hours he works at the music store. Write an inequality in which his total weekly income is at least $360.

  b. Graph the inequality. Because he can work only a positive number of hours, restrict the graph to the first quadrant.

  c. Give a combination of hours that provides him with an income of exactly $360.

  d. Give a combination of hours that provides him with an income of more than $360.

22. The following relation shows U.S. per capita income for each year. Is the relation a function?

Year	Income ($)
2000	29,469
2001	30,413
2002	30,906
2003	31,632
2004	33,050
2005	34,586
2006	36,276

Source: U.S. Department of Commerce

23. Determine whether the graph is the graph of a function.

24. Find the indicated value the function $f(x) = \dfrac{x^2}{x - 4}$.

  a. $f(2)$ **b.** $f(4)$ **c.** $f(-3)$

25. **a.** Give the domain and range of the function graphed below.

  b. Find $f(1)$.

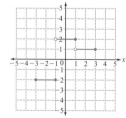

Chapters 1–4 Cumulative Review Exercises

For Exercises 1–4, answer true or false.

[1.1] **1.** π is a rational number.

[4.4] **2.** $y = mx + b$ is the equation of a line in slope–intercept form.

[1.5] **3.** $6° = 6$

[1.5] **4.** $\sqrt{-36} = -6$

For Exercises 5 and 6, fill in the blank.

[2.2] **5.** An equation that has every real number for a solution is a(n) _____.

[3.2] **6.** To write a percent as a fraction or decimal,
    a. Write the percent as a ratio with _____ in the denominator.
    b. Simplify to the desired form.

[3.2] **7.** Write 16.5% as a fraction.

Exercises 8–14 **EXPRESSIONS**

[2.3] **8.** Explain in your own words how to clear decimals from an equation.

[1.5] For Exercises 9–14, simplify.

9. $(-2)^3 - 3\sqrt{9 + 16}$

10. $\{4 - 2[6 + (-10)]\} + 3$

11. $-|-12 - (-2)(-3)|$

12. -3^4

13. $-5\sqrt{25} + 20 \div (-4) \cdot 2 - 6$

14. $4^2 - 7 \cdot 3 + \sqrt{144} - 25 \div (-5)$

Exercises 15–30 **EQUATIONS AND INEQUALITIES**

[4.3] **15.** Find the x- and y-intercepts for $2x - 3y = 6$.

[4.2] **16.** Find three solutions for $y = -3x + 4$.

[4.3] **17.** Graph $2y - 3x = 9$.

[4.4] **18.** Determine the slope of the line $x + 3y = 8$.

[4.5] **19.** Write the equation of a line through the points $(5, 3)$ and $(-3, -1)$.

[4.6] **20.** Graph $2x + 5y > 10$.

For Exercises 21–23, solve and check:

[2.3] **21.** $\dfrac{3}{5}a - 8 = \dfrac{1}{2}$

[3.1] **22.** $\dfrac{u}{14} = -\dfrac{6}{7}$

[2.3] **23.** $1.6x - 14 = 8 + 2.4(x - 5)$

[2.4] **24.** Solve $C = 2\pi r$ for r.

For Exercises 25–30, solve.

[3.2] **25.** Using the pie chart, if 1000 Internet users were polled, how many would check their e-mail weekly?

[3.2] **26.** If a married couple makes a combined income of $48,241 per year, what percent does the couple spend on holiday gifts?

HOW OFTEN WE CHECK E-MAIL

23% Weekly

76% Daily

1% Less than once a week

Source: UCLA Center for Communication Policy

HOLIDAY GIFT SPENDING

Married $924

Parent $867

Single $658

Non-parent $631

Source: The Gallup Organization

[3.3] **27.** Two angles are complementary. The smaller angle is $\dfrac{2}{3}$ the measure of the larger angle. What are the measures of the two angles?

[3.2] **28.** The ticket prices for the 2008 Super Bowl in Glendale, Arizona, ranged from $3270 for end zone tickets to $7725 for lower-level 30–50 yard line tickets. What is the percent of increase from an end zone ticket to a lower-level 30–50 yard line ticket?

[3.3] **29.** Angela has been saving her quarters and dimes for several months. If she has 190 coins totaling $30.70, how many of each coin does she have?

[3.5] **30.** Juanita invests in a plan that has an APR of 8%. She invests twice as much in a plan that has an APR of 10%. If the total interest from the investments is $5600, how much did she invest in each plan?

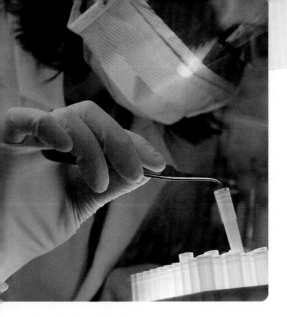

Systems of Linear Equations and Inequalities

5

In Chapter 3, we solved problems involving two unknown amounts, such as finding two consecutive integers whose sum is 63. At that time, we only knew how to solve equations with a single variable; so we had to let one of the unknown amounts be represented by a variable, then describe the other unknown amount in terms of that variable. For example, in the above consecutive integer problem, we would let x represent the first integer and then describe the consecutive integer as $x + 1$.

In this chapter, we again solve problems involving two or more unknowns; however, we assign a different variable to each unknown and translate the problem to several equations, called a *system of equations*. As we shall see, it is easier to translate a problem involving two or more unknowns to a system of equations than to try to translate the problem to a single equation in terms of a single variable. The challenge with systems of equations is in solving them. We will learn several techniques for solving a system of linear equations: graphing (5.1), substitution (5.2), and elimination (5.3). In Section 5.4, we revisit some of the same types of problems involving two or more unknowns as in Chapter 3 and solve them using systems of a equations. Finally, in Section 5.5, we solve systems of linear inequalities.

5.1 Solving Systems of Linear Equations Graphically

Objectives

1. Determine whether an ordered pair is a solution for a system of equations.
2. Solve a system of linear equations graphically.
3. Classify systems of linear equations in two unknowns.

Objective 1 Determine whether an ordered pair is a solution for a system of equations. Problems involving multiple unknowns can be represented by a group of equations called a **system of equations**.

Definition *System of equations:* A group of two or more equations.

For example, look at the following problem: The sum of two numbers is 3. Twice the first number plus three times the second number is 8. What are the two numbers? If x represents the first number and y represents the second number, we can translate each sentence to an equation.

Sentence in the problem:	Translation:	
The sum of two numbers is 3.	$x + y = 3$	(Equation 1)
Twice the first number plus three times the second number is 8.	$2x + 3y = 8$	(Equation 2)

The two equations together form a system of equations that describes the problem.

$$\text{System of equations} \quad \begin{cases} x + y = 3 & \text{(Equation 1)} \\ 2x + 3y = 8 & \text{(Equation 2)} \end{cases}$$

A **solution for a system of equations** is an ordered set of numbers that satisfies *all* equations in the system.

Definition *Solution for a system of equations:* An ordered set of numbers that makes all equations in the system true.

For example, in the preceding system of equations, $x = 1$ and $y = 2$ make both equations true; so they form a solution to the system of equations. We can check by substituting the values in place of the corresponding variables.

Equation 1:	Equation 2:
$x + y = 3$	$2x + 3y = 8$
$1 + 2 \overset{?}{=} 3$	$2(1) + 3(2) \overset{?}{=} 8$
$3 = 3$ True	$2 + 6 = 8$ True

We can write a solution to a system of equations in two variables as an ordered pair, in this case, (1, 2). This suggests the following procedure for checking solutions.

Procedure **Checking a Solution to a System of Equations**

To verify or check a solution to a system of equations,

1. Replace each variable in each equation with its corresponding value.
2. Verify that each equation is true.

Example 1 Determine whether each ordered pair is a solution to the system of equations.

$$\begin{cases} x + y = 5 & \text{(Equation 1)} \\ y = 2x - 4 & \text{(Equation 2)} \end{cases}$$

a. $(-1, 6)$

SOLUTION

$x + y = 5$ (Equation 1)	$y = 2x - 4$ (Equation 2)	
$-1 + 6 \overset{?}{=} 5$	$6 \overset{?}{=} 2(-1) - 4$	**In both equations,**
$5 = 5$ True	$6 \overset{?}{=} -2 - 4$	**replace x with -1**
	$6 = -6$ False	**and y with 6.**

Because $(-1, 6)$ does not satisfy both equations, it is not a solution for the system.

b. $(3, 2)$

SOLUTION

$x + y = 5$ (Equation 1)	$y = 2x - 4$ (Equation 2)	
$3 + 2 \overset{?}{=} 5$	$2 \overset{?}{=} 2(3) - 4$	**In both equations,**
$5 = 5$ True	$2 \overset{?}{=} 6 - 4$	**replace x with 3 and y**
	$2 = 2$ True	**with 2.**

Because $(3, 2)$ satisfies both equations, it is a solution for the system.

Your Turn 1 Determine whether each ordered pair is a solution to the system of equations.

$$\begin{cases} 2x + y = 5 & \text{(Equation 1)} \\ y = 1 - x & \text{(Equation 2)} \end{cases}$$

a. $(3, -1)$ **b.** $(4, -3)$

Objective ❷ Solve a system of linear equations graphically. A system of two linear equations in two variables can have one solution, no solution, or an infinite number of solutions. To see why, let's look at the graphs of the equations in three different systems. First, let's graph the equations in the system from Example 1.

$$\begin{cases} x + y = 5 \\ y = 2x - 4 \end{cases}$$

Note: In Chapter 4, we learned that the graph of an equation represents every ordered pair solution for the equation; so when two graphs intersect, the point of intersection is a solution for both equations.

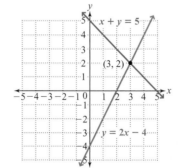

Notice that the graphs of $x + y = 5$ and $y = 2x - 4$ intersect at the point $(3, 2)$, which is the solution for the system. If two linear graphs intersect at a single point, the system has a *single solution* at the point of intersection.

Note: These lines also have different slopes, which is always the case when a system of two linear equations has a single solution.

Now let's look at a system with no solution. Look at the graphs of the equations in the system $\begin{cases} y = 3x + 1 \\ y = 3x - 2 \end{cases}$.

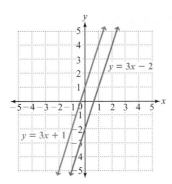

Notice that the equations $y = 3x + 1$ and $y = 3x - 2$ have the same slope, 3, and different y-intercepts; so their graphs are parallel lines. Because the graphs of the equations in this system have no point of intersection, the system has no solution.

Finally, let's examine a system with an infinite number of solutions, such as

$$\begin{cases} x + 2y = 4 & \text{(Equation 1)} \\ 2x + 4y = 8 & \text{(Equation 2)} \end{cases}$$

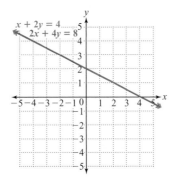

Notice that the graphs of the two equations are identical. Consequently, an infinite number of solutions to this system of equations lie on that line.

Note: Having an infinite number of solutions is not the same as having all real numbers as solutions, as in the solutions of identities.

By writing the equations in slope–intercept form, we can see without graphing that the two equations describe the same line.

$x + 2y = 4$ (Equation 1)

$2y = -x + 4$	Subtract x from both sides.
$y = -\dfrac{1}{2}x + \dfrac{4}{2}$	Divide both sides by 2 to isolate y.
$y = -\dfrac{1}{2}x + 2$	Simplify the fraction.

$2x + 4y = 8$ (Equation 2)

$4y = -2x + 8$	Subtract $2x$ from both sides.
$y = -\dfrac{2}{4}x + \dfrac{8}{4}$	Divide both sides by 4 to isolate y.
$y = -\dfrac{1}{2}x + 2$	Simplify the fraction.

Our work suggests the following graphical method for solving systems of equations.

Procedure   **Solving Systems of Linear Equations Graphically**

To solve a system of linear equations graphically,

1. Graph each equation.
    a. If the lines intersect at a single point, the coordinates of that point form the solution.
    b. If the lines are parallel, there is no solution.
    c. If the lines are identical, there are an infinite number of solutions, which are the coordinates of all points on that line.
2. Check your solution.

Example 2   Solve the system of equations graphically.

a. $\begin{cases} y = -3x + 1 & \text{(Equation 1)} \\ 2x - y = 4 & \text{(Equation 2)} \end{cases}$

SOLUTION   Graph each equation.

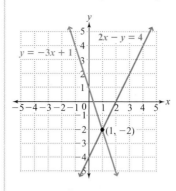

The lines intersect at a single point, which appears to be $(1, -2)$. We can verify that it is the solution by substituting into the equations.

$$y = -3x + 1 \qquad\qquad 2x - y = 4$$
$$-2 \overset{?}{=} -3(1) + 1 \qquad 2(1) - (-2) \overset{?}{=} 4$$
$$-2 \overset{?}{=} -3 + 1 \qquad\qquad 2 + 2 \overset{?}{=} 4$$
$$-2 = -2 \quad \text{True} \qquad\qquad 4 = 4 \qquad \text{True}$$

Warning: Graphing by hand can be imprecise, and not all solutions have integer coordinates. So you should always check your solutions by substituting them into the original equations.

Answer   Because $(1, -2)$ makes both equations true, it is the solution.

b. $\begin{cases} 2x - 3y = 6 & \text{(Equation 1)} \\ y = \dfrac{2}{3}x + 1 & \text{(Equation 2)} \end{cases}$

SOLUTION   Graph each equation.

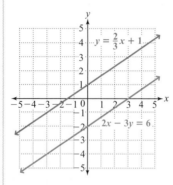

The lines appear to be parallel, which we can verify by comparing the slopes and y-intercepts. The slope of equation 2 is $\dfrac{2}{3}$, and the y-intercept is $(0, 1)$. To determine the slope and y-intercept of equation 1, we rewrite it in slope–intercept form.

$$2x - 3y = 6$$
$$-3y = -2x + 6 \qquad \text{Subtract } 2x \text{ from both sides.}$$
$$y = \frac{2}{3}x - 2 \qquad \text{Divide both sides by } -3 \text{ to isolate } y.$$

The slope of equation 1 also is $\dfrac{2}{3}$, and the y-intercept is $(0, -2)$. Because the slopes are the same and the y-intercepts are different, the lines are indeed parallel.

Answer   The system has no solution.

c. $\begin{cases} 6x - 2y = 2 & \text{(Equation 1)} \\ y = 3x - 1 & \text{(Equation 2)} \end{cases}$

SOLUTION   Graph each equation.

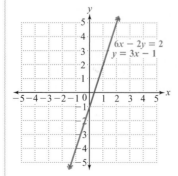

The lines appear to be identical, which we can verify by rewriting $6x - 2y = 2$ in slope–intercept form.

$$6x - 2y = 2$$
$$-2y = -6x + 2 \qquad \text{Subtract } 6x \text{ from both sides.}$$
$$y = 3x - 1 \qquad \text{Divide both sides by } -2 \text{ to isolate } y.$$

The equations are identical.

Answer   All ordered pairs along the line $6x - 2y = 2$ (or $y = 3x - 1$).

Your Turn 2 Solve the system of equations graphically.

a. $\begin{cases} y = \dfrac{3}{4}x + 1 \\ 2x - y = -6 \end{cases}$
　　　　　　　　b. $\begin{cases} x + y = -3 \\ y = -x + 2 \end{cases}$

Calculator Tips

To solve a system of equations using a graphing calculator, begin by graphing each equation. Remember that you must write the equations in slope–intercept form in order to enter them in the calculator. Using the [Y=] *key, enter one equation in Y1 and the second equation in Y2; then press* [GRAPH] *to see both graphs.*

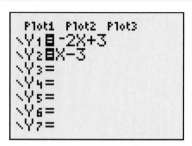

You can find the coordinates of the point of intersection by using the CALC feature, which you find by pressing [2nd] [TRACE]. *Select INTERSECT from the menu. The calculator will then prompt you to verify which lines you want it to find the intersection of. So press* [ENTER] *for each line. The calculator will then prompt you to indicate which point of intersection you want. Because these straight lines will have only one point of intersection, you can simply press* [ENTER]. *The coordinates of the point of intersection will appear at the bottom of the screen.*

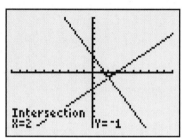

Objective ③ Classify systems of linear equations in two unknowns.
As we have seen, a system of two equations with two variables can have one solution, an infinite number of solutions along a common line, or no solution. Special terms are used to indicate each of these three classifications. The first terms describe whether or not a system has a solution.

Definitions

Consistent system of equations: A system of equations that has at least one solution.

Inconsistent system of equations: A system of equations that has no solution.

Answers   Your Turn 2
a. The solution is $(-4, -2)$.

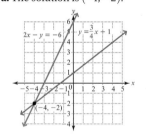

b. no solution

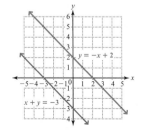

We have seen two types of consistent systems of equations. One type has a single solution because the two equations produce different lines that intersect at a single point. The other type has an infinite number of solutions because the two equations are identical and therefore produce the same line. To distinguish the two consistent cases, we use additional terms. If the equations in a consistent system produce different lines so that the system has one solution, we say that the equations are *independent*. If the equations in a consistent system produce the same line so that this system has an infinite number of solutions along that line, we say that the equations are *dependent*.

We have learned that linear equations whose graphs are different have different slopes or, in the case of parallel lines, have the same slopes and different *y*-intercepts. Linear equations whose graphs are identical have the same slope and same *y*-intercept. Therefore, by observing the slope and *y*-intercept of the equations in a system of equations, we can determine which of the three cases we are dealing with, which suggests the following procedure.

Procedure   **Classifying Systems of Equations**

To classify a system of two linear equations in two unknowns, write the equations in slope–intercept form and compare the slopes and y-intercepts.

Consistent system with independent equations: The system has a single solution at the point of intersection. The graphs are different. They have different slopes.	**Consistent system with dependent equations:** The system has an infinite number of solutions. The graphs are identical. They have the same slope and same y-intercept.	**Inconsistent system:** The system has no solution. The graphs are parallel lines. They have the same slope, but different y-intercepts.

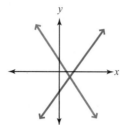 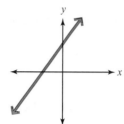 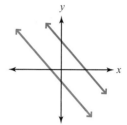

Example 3   For each of the systems of equations in Example 2, determine whether the system is consistent with independent equations, consistent with dependent equations, or inconsistent. How many solutions does the system have?

SOLUTION   When we solved the system $\begin{cases} y = -3x + 1 \\ 2x - y = 4 \end{cases}$ in Example 2(a), we found that the graphs intersected at a single point. Therefore, the system is consistent (it has a solution) with independent equations (different graphs) and has one solution: $(1, -2)$.

When we solved the system $\begin{cases} 2x - 3y = 6 \\ y = \dfrac{2}{3}x + 1 \end{cases}$ in Example 2(b), we found the graphs to be parallel lines. Therefore, the system is inconsistent and has no solution:

When we solved the system $\begin{cases} 6x - 2y = 2 \\ y = 3x - 1 \end{cases}$ in Example 2(c), we found the graphs to coincide. Therefore, the system is consistent (it has a solution) with dependent equations (same graph) and has an infinite number of solutions.

Your Turn 3

a. Determine whether the following system is consistent with independent equations, consistent with dependent equations, or inconsistent.

b. How many solutions does the system have?

$$\begin{cases} 5x + y = 1 \\ y = -5x - 2 \end{cases}$$

Answers   **Your Turn 3**

a. inconsistent   **b.** no solution

1. How do you check a solution for a system of equations?

2. If a system of linear equations has at least one solution (consistent), what does this indicate about the graphs of the equations?

3. If a system of linear equations has no solution (inconsistent), what does this indicate about the graphs of the equations?

4. If two equations are dependent, what does this mean about their graphs?

5. Can a system of linear equations have more than one solution? Explain.

6. What are some weaknesses in using graphing to find a solution to a system of equations?

For Exercises 7–16, determine whether the given ordered pair is a solution to the given system of equations. See Example 1.

7. $(1, 1)$; $\begin{cases} x = y \\ 9y - 13 = -4x \end{cases}$

8. $(3, -2)$; $\begin{cases} 2x = -3y \\ x - 2y = 7 \end{cases}$

9. $(2, -3)$; $\begin{cases} 4x + 3y = -1 \\ 2x - 5y = -11 \end{cases}$

10. $(4, 3)$; $\begin{cases} 3x + 4y = 24 \\ 3x + 2y = 18 \end{cases}$

11. $\left(\dfrac{2}{3}, \dfrac{4}{3}\right)$; $\begin{cases} x + y = 2 \\ 2x - y = 0 \end{cases}$

12. $\left(\dfrac{3}{2}, -\dfrac{3}{2}\right)$; $\begin{cases} x - y = 3 \\ x + y = 0 \end{cases}$

13. $(2, -4)$; $\begin{cases} 0.5x + 1.25y = 4 \\ x + y = -2 \end{cases}$

14. $(1, 5)$; $\begin{cases} 0.25x + 0.75y = 4 \\ x + y = -7 \end{cases}$

15. $(5, -2)$; $\begin{cases} x + 0.5y = 4 \\ \dfrac{1}{5}x - \dfrac{1}{2}y = 2 \end{cases}$

16. $(-1, -2)$; $\begin{cases} \dfrac{2}{3}x - y = 7 \\ 0.25x + y = -1 \end{cases}$

For Exercises 17–42, solve the system of linear equations graphically. See Example 2.

17. $\begin{cases} x + y = 5 \\ 2x - y = 7 \end{cases}$

18. $\begin{cases} x - y = 4 \\ x + y = 8 \end{cases}$

19. $\begin{cases} x + y = 1 \\ y = 7 + x \end{cases}$

20. $\begin{cases} x + y = 4 \\ y - x = 6 \end{cases}$

21. $\begin{cases} y = x - 5 \\ 2x - y = 8 \end{cases}$

22. $\begin{cases} y = x - 1 \\ 3x + y = 11 \end{cases}$

23. $\begin{cases} x = y - 4 \\ x + y = 2 \end{cases}$

24. $\begin{cases} x = y + 4 \\ x + y = 2 \end{cases}$

25. $\begin{cases} 4x + y = 8 \\ 2x - 3y = 18 \end{cases}$

26. $\begin{cases} 2x + 3y = 6 \\ x + 4 = 2y \end{cases}$

27. $\begin{cases} 2x + y = 5 \\ x + 2y = -2 \end{cases}$

28. $\begin{cases} 2x + y = 1 \\ 2x + 3y = 3 \end{cases}$

29. $\begin{cases} 2x + 2y = -2 \\ 3x - 2y = 12 \end{cases}$

30. $\begin{cases} 2x - 4y = 2 \\ 6x - 2y = -4 \end{cases}$

31. $\begin{cases} 3y = -2x + 6 \\ 4x + 6y = 18 \end{cases}$

32. $\begin{cases} 2x + y = 0 \\ 2y = 3 - 4x \end{cases}$

33. $\begin{cases} 2x + 3y = 2 \\ 6x + 9y = 6 \end{cases}$

34. $\begin{cases} x + y = 4 \\ -2x - 2y = -8 \end{cases}$

35. $\begin{cases} 3x - 2y = -6 \\ x = 2 \end{cases}$

36. $\begin{cases} 5x + 2y = -10 \\ x = -4 \end{cases}$

37. $\begin{cases} x - 2y = -6 \\ y = 2 \end{cases}$

38. $\begin{cases} 3x - y = 6 \\ y = -3 \end{cases}$

39. $\begin{cases} y = 3 \\ x = -2 \end{cases}$

40. $\begin{cases} x = 1 \\ y = -4 \end{cases}$

41. $\begin{cases} y = -\dfrac{2}{5}x \\ x - y = 7 \end{cases}$

42. $\begin{cases} y = \dfrac{1}{4}x \\ x - y = 3 \end{cases}$

For Exercises 43–48, (a) determine whether the graph shows a consistent system with independent equations, a consistent system with dependent equations, or an inconsistent system and (b) determine how many solutions the system has. See Example 3.

43.
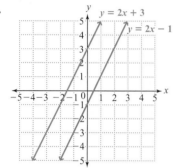

44.
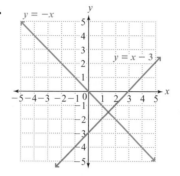

45.
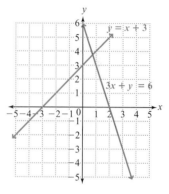

46.
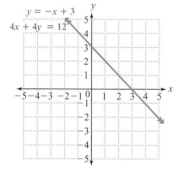

47.
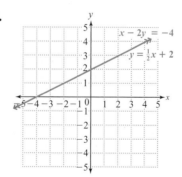

48.
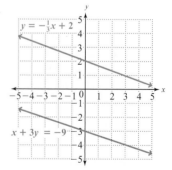

For Exercises 49–56, (a) determine whether the system of equations is consistent with independent equations, consistent with dependent equations, or inconsistent and (b) determine how many solutions the system has. See Example 3.

49. $\begin{cases} x - 4y = 9 \\ 2x - 3y = 8 \end{cases}$

50. $\begin{cases} -4y = 2x - 12 \\ x - 2y = -4 \end{cases}$

51. $\begin{cases} 3x + 2y = 5 \\ -6x - 4y = 1 \end{cases}$

52. $\begin{cases} 3x + 4y = 15 \\ 9x + 12y = 8 \end{cases}$

53. $\begin{cases} x - y = 1 \\ 2x - 2y = 2 \end{cases}$

54. $\begin{cases} 2x - y = 1 \\ -4x + 2y = -2 \end{cases}$

55. $\begin{cases} 2x - y = 1 \\ x + y = 4 \end{cases}$

56. $\begin{cases} x - y = 5 \\ x + 5y = 11 \end{cases}$

57. A business breaks even when its costs and revenue are equal. To the right is a graph of the cost of a product based on the number of units produced and the revenue based on the number of units sold.

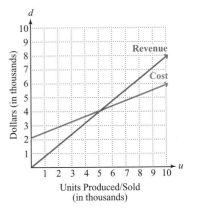

   a. How many units must be produced and sold if the business is to break even?

   b. What amount of revenue is needed for the business to break even?

   c. Write an inequality that describes the number of units that must be sold for the business to make a profit.

58. John's job requires that he drive between two cities using a toll road. He has two choices for toll plans. The first plan is to stop and pay cash at each booth, which amounts to $15 per trip. The second plan is to purchase a toll pass for $12 per trip, which allows him to drive through the toll booths without stopping. However, to use the pass he also must purchase a transponder for a one-time fee of $24. To the right, the cost of using each plan has been graphed.

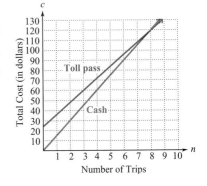

   a. Which plan costs less if John makes 6 trips?

   b. How many trips must he make for the costs to be equal?

   c. What other factors should he consider when choosing a plan?

59. The graph shows two long-distance plans. The Unlimited plan allows you to call anywhere in the United States at any time for $30 per month. The One-Rate plan costs $3.75 per month plus $0.07 per minute for each long-distance call.

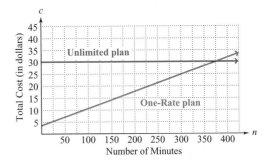

   a. Which plan costs less if a person does not use more than 350 minutes of long-distance service each month?

   b. How many minutes would a person have to talk in a month for the plans to cost the same amount?

   c. How many minutes would a person need to talk per month for the Unlimited plan to cost less?

60. Deciding whether to purchase or rent a home is a difficult decision based on many factors. The graph shows the monthly mortgage payment ($850 per month) compared to the monthly rent payment (initially $600 per month and raised $50 every six months) for a 1200-square-foot house.

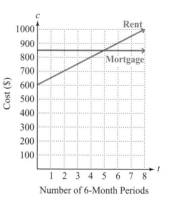

a. If a customer plans to stay only two years, ignoring any tax benefits, which costs less per month, renting or buying?

b. After how many months will the monthly rent be the same as the monthly mortgage payment?

c. In the fourth year, which costs less per month?

d. What other factors should a person consider when deciding whether to purchase or rent a home?

Review Exercises

Exercise 1  EXPRESSIONS

[1.7]   **1.** Simplify: $2y - 5(y + 7)$

Exercises 2–6 EQUATIONS AND INEQUALITIES

[2.4]   **2.** Solve for y: $x + 3y = 10$ [2.4]   **3.** Solve for x: $\frac{1}{3}x - 2y = 10$

[2.1]   **4.** A cone made of marble is used as a decorative accent piece. It is 10 inches tall, and the radius of its base is 2 inches. Find its volume. (Use $V = \frac{1}{3}\pi r^2 h$.)

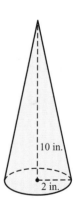

10 in.

2 in.

[3.3]   **5.** One number exceeds another number by 62. If five times the smaller number is subtracted from two times the larger number, the difference is 91. What are the numbers?

[4.7]   **6.** Given $f(x) = 2x + 9$, find $f(x + 4)$.

5.2 Solving Systems of Linear Equations by Substitution

Objectives

① Solve systems of linear equations using substitution.

② Solve applications involving two unknowns using a system of equations.

In Section 5.1, we solved systems of equations by graphing. However, if the solution to a system of equations contains fractions or decimal numbers, it could be difficult to determine those values using the graphing method. For example, look at the following system of equations:

$$\begin{cases} x + 3y = 10 & \text{(Equation 1)} \\ y = x + 4 & \text{(Equation 2)} \end{cases}$$

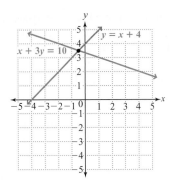

In the graph, notice that the point of intersection is between grid lines, which means the solution contains fractions. We could make a guess at the solution and check our guess in the equations, but it would be better to use a method that does not require guessing. One such method is the *substitution* method.

Objective 1 Solve systems of linear equations using substitution. Remember that in a system's solution, the same x and y values satisfy both equations. In the preceding system, notice that Equation 2 indicates that y is equal to $x + 4$; so y must be equal to $x + 4$ in Equation 1 as well. Therefore, we can substitute $x + 4$ for y in Equation 1.

$$x + 3y = 10$$
$$x + 3\overbrace{(x + 4)} = 10$$

Now we have an equation in terms of a single variable, x, which allows us to solve for x.

$$
\begin{array}{ll}
x + 3(x + 4) = 10 & \\
x + 3x + 12 = 10 & \text{Distribute 3.} \\
4x + 12 = 10 & \text{Combine like terms.} \\
4x = -2 & \text{Subtract 12 from both sides.} \\
x = -\dfrac{1}{2} & \text{Divide both sides by 4 and simplify.}
\end{array}
$$

We can now find the y-value by substituting $-\dfrac{1}{2}$ for x in either of the original equations. Equation 2 is easier because y is isolated.

$$
\begin{array}{ll}
y = x + 4 & \\
y = -\dfrac{1}{2} + 4 & \text{Substitute } -\dfrac{1}{2} \text{ for } x. \\
y = 3\dfrac{1}{2} &
\end{array}
$$

The solution to the system of equations is $\left(-\dfrac{1}{2}, 3\dfrac{1}{2}\right)$. Notice in our graph that the lines do appear to intersect at $\left(-\dfrac{1}{2}, 3\dfrac{1}{2}\right)$, which supports our solution graphically.

Procedure **Solving Systems of Two Linear Equations Using Substitution**

To find the solution of a system of two linear equations using the substitution method,

1. Isolate one of the variables in one of the equations.
2. In the other equation, substitute the expression you found in step 1 for that variable.
3. Solve this new equation. (It will now have only one variable.)
4. Using one of the equations containing both variables, substitute the value you found in step 3 for that variable and solve for the value of the other variable.
5. Check the solution in the original equations.

Example 1   Solve the system of equations using substitution.

$$\begin{cases} x = 3 - y \\ 5x + 3y = 5 \end{cases}$$

SOLUTION

Step 1: We isolate a variable in one of the equations. Because x is already isolated in the first equation, we proceed to step 2.

Step 2: Substitute the expression $3 - y$ for x in the second equation.

$$5x + 3y = 5$$
$$5(3 - y) + 3y = 5 \qquad \text{Substitute } 3 - y \text{ for } x.$$

Step 3: We now have a linear equation in terms of a single variable, y; so we solve for y.

$$15 - 5y + 3y = 5 \qquad \text{Distribute 5.}$$
$$15 - 2y = 5 \qquad \text{Combine like terms.}$$
$$-2y = -10 \qquad \text{Subtract 15 from both sides.}$$
$$y = 5 \qquad \text{Divide both sides by } -2 \text{ to isolate } y.$$

Step 4: Now we solve for x by substituting 5 for y in one of the equations containing both variables. We will use $x = 3 - y$.

$$x = 3 - y$$
$$x = 3 - 5 \qquad \text{Substitute 5 for } y.$$
$$x = -2$$

> **Note:** We chose $x = 3 - y$ because x is isolated.

Connection The graphs of $x = 3 - y$ and $5x + 3y = 5$ are lines that intersect at the point whose coordinates are $(-2, 5)$.

The solution is $(-2, 5)$. We will let the reader check that $(-2, 5)$ makes both equations true.

Your Turn 1   Solve the system of equations using substitution.

$$\begin{cases} x - 3y = 11 \\ y = 2x + 3 \end{cases}$$

In step 1, if no variable is isolated in any of the given equations, we select an equation and isolate one of its variables. It is easiest to isolate a variable that has a coefficient of 1, as this will avoid fractions.

Example 2   Solve the system of equations using substitution.

$$\begin{cases} x + y = 6 \\ 5x + 2y = 8 \end{cases}$$

SOLUTION

Step 1: Isolate a variable in one of the equations. Because both x and y have a coefficient of 1 in $x + y = 6$, isolating either variable is easy. We will isolate y.

$$x + y = 6$$
$$y = 6 - x \qquad \text{Subtract } x \text{ from both sides to isolate } y.$$

Step 2: Substitute $6 - x$ for y in the second equation.

$$5x + 2y = 8$$

$$5x + 2\overbrace{(6 - x)} = 8 \qquad \text{Substitute } 6 - x \text{ for } y.$$

Step 3: Solve for x using the equation we found in step 2.

$$5x + 12 - 2x = 8 \qquad \text{Distribute 2.}$$
$$3x + 12 = 8 \qquad \text{Combine like terms.}$$
$$3x = -4 \qquad \text{Subtract 12 from both sides.}$$
$$x = -\frac{4}{3} \qquad \text{Divide both sides by 3 to isolate } x.$$

Step 4: Find the value of y by substituting $-\dfrac{4}{3}$ for x in one of the equations containing both variables. Because we isolated y in $y = 6 - x$, we will use this equation.

$$y = 6 - x$$
$$y = 6 - \left(-\frac{4}{3}\right) \qquad \text{Substitute } -\frac{4}{3} \text{ for } x.$$
$$y = \frac{18}{3} + \frac{4}{3} \qquad \begin{array}{l}\text{Rewrite as addition of equivalent fractions}\\ \text{with a common denominator.}\end{array}$$
$$y = \frac{22}{3}$$

The solution is $\left(-\dfrac{4}{3}, \dfrac{22}{3}\right)$.

Step 5: We will leave the check to the reader.

Your Turn 2 Solve the system of equations by substitution.

$$\begin{cases} x - y = 1 \\ 3x + y = 11 \end{cases}$$

If none of the variables in the system of equations has a coefficient of 1 or -1, for step 1, select the equation that seems easiest to work with and the variable in that equation that seems easiest to isolate. When selecting that variable, recognize that you'll eventually divide out its coefficient; so choose the variable whose coefficient will divide evenly into most, if not all, of the other numbers in its equation. In Section 5.3, we will discuss another method of solving systems that reduces the amount of work with fractions.

Example 3 Solve the system of equations using substitution.

$$\begin{cases} 3x + 4y = 6 \\ 2x + 6y = -1 \end{cases}$$

SOLUTION

Step 1: We will isolate x in $3x + 4y = 6$.

$$3x + 4y = 6$$
$$3x = 6 - 4y \qquad \text{Subtract } 4y \text{ from both sides.}$$
$$x = 2 - \frac{4}{3}y \qquad \text{Divide both sides by 3 to isolate } x.$$

Note: We chose to isolate x because its coefficient, 3, divides evenly into one of the other numbers in the equation, 6, whereas y's coefficient, 4, does not divide evenly into any of the other numbers.

Step 2: Substitute $2 - \dfrac{4}{3}y$ for x in $2x + 6y = -1$.

$$2x + 6y = -1$$

$$2\left(2 - \dfrac{4}{3}y\right) + 6y = -1 \qquad \text{Substitute } 2 - \dfrac{4}{3}y \text{ for } x.$$

Step 3: Solve for y using the equation we found in step 2.

$$4 - \dfrac{8}{3}y + 6y = -1 \qquad \text{Distribute to clear the parentheses.}$$

$$3 \cdot 4 - 3 \cdot \dfrac{8}{3}y + 3 \cdot 6y = 3 \cdot (-1) \qquad \text{Multiply both sides by 3 to clear the fraction.}$$

$$12 - 8y + 18y = -3$$

$$12 + 10y = -3 \qquad \text{Combine like terms.}$$

$$10y = -15 \qquad \text{Subtract 12 from both sides.}$$

$$y = -\dfrac{15}{10} \qquad \text{Divide both sides by 10 to isolate } y.$$

$$y = -\dfrac{3}{2} \qquad \text{Simplify.}$$

Step 4: Find the value of x by substituting $-\dfrac{3}{2}$ for y in one of the equations containing both variables. Because x is isolated in $x = 2 - \dfrac{4}{3}y$, we will use that equation.

$$x = 2 - \dfrac{4}{3}y$$

$$x = 2 - \dfrac{\overset{2}{\cancel{4}}}{\underset{1}{\cancel{3}}} \cdot \left(-\dfrac{\overset{1}{\cancel{3}}}{\underset{1}{\cancel{2}}}\right) \qquad \text{Substitute } -\dfrac{3}{2} \text{ for } y \text{ and simplify.}$$

$$x = 2 + 2$$

$$x = 4$$

The solution is $\left(4, -\dfrac{3}{2}\right)$.

Step 5: We will leave the check to the reader.

Your Turn 3   Solve the system of equations using substitution.

$$\begin{cases} 5x - 2y = 10 \\ 3x - 6y = 2 \end{cases}$$

Inconsistent Systems of Equations

In Section 5.1, we learned that inconsistent systems of linear equations have no solution because the graphs of the equations are parallel lines. Let's see what happens when we solve an inconsistent system of equations using the substitution method.

Answer   **Your Turn 3**

$\left(\dfrac{7}{3}, \dfrac{5}{6}\right)$

Example 4 Solve the system of equations using substitution.

$$\begin{cases} x + 2y = 4 \\ y = -\dfrac{1}{2}x + 1 \end{cases}$$

SOLUTION Because y is isolated in the second equation, we substitute $-\dfrac{1}{2}x + 1$ in place of y in the first equation.

$$x + 2y = 4$$

$$x + 2\left(-\dfrac{1}{2}x + 1\right) = 4 \qquad \text{Substitute } -\dfrac{1}{2}x + 1 \text{ for } y.$$

Now solve for x.

$$x - x + 2 = 4 \qquad \text{Distribute to clear the parentheses.}$$
$$2 = 4 \qquad \text{Combine like terms.}$$

Notice that $2 = 4$ no longer has a variable and is false. This false equation with no variable indicates that the system is inconsistent and has no solution.

Connection If we were to solve the system in Example 4 using graphing, we would see that the lines are parallel.

Consistent Systems with Dependent Equations

Recall from Section 5.1 that consistent systems with dependent equations have an infinite number of solutions. Let's see what happens when we use the substitution method to solve a system of dependent equations.

Example 5 Solve the system of equations using substitution.

$$\begin{cases} x = 3y + 4 \\ 2x - 6y = 8 \end{cases}$$

SOLUTION Because x is isolated in the first equation, we substitute $3y + 4$ in place of x in the second equation.

$$2x - 6y = 8$$

$$2(3y + 4) - 6y = 8 \qquad \text{Substitute } 3y + 4 \text{ for } x.$$

Now solve for y.

$$6y + 8 - 6y = 8 \qquad \text{Distribute to clear the parentheses.}$$
$$8 = 8 \qquad \text{Combine like terms.}$$

Notice that $8 = 8$ no longer has a variable and is true. This true equation with no variable indicates that the equations in the system are dependent; so there are an infinite number of solutions that are all of the ordered pairs along $x = 3y + 4$ (or $2x - 6y = 8$).

Connection If we were to solve the system in Example 5 using graphing, we would see that the lines are identical.

Your Turn 5 Solve the systems of equations using substitution.

a. $\begin{cases} y = 2x - 7 \\ 4x - 2y = -3 \end{cases}$

b. $\begin{cases} 3x - y = 4 \\ 2y - 6x = -8 \end{cases}$

Answers Your Turn 5
a. no solution
b. all ordered pairs along
$3x - y = 4$ (or $2y - 6x = -8$)

Objective ❷ Solve applications involving two unknowns using a system of equations. To solve a problem involving two unknowns, we must be given two relationships about those unknowns. In Chapter 3, we learned to solve these problems by selecting a variable for one of the unknowns and using one relationship to write the other unknown in terms of that variable. We then translated the second relationship to a single equation in terms of the variable.

We now solve the same types of problems using systems of equations. Using a system of equations is easier because we select two variables to represent the two unknowns, then translate each of the two given relationships to an equation in terms of those variables. We can then solve the system of equations by substitution (or graphing).

Procedure

Solving Applications Using a System of Equations

To solve a problem with two unknowns using a system of equations,

1. Select a variable for each of the unknowns.
2. Translate each relationship to an equation.
3. Solve the system.

We repeat Example 3 of Section 3.3 using a system of equations.

Example 6 A rectangular frame is to be built so that the length is 4 inches more than the width. The perimeter of the frame is to be 72 inches. Find the dimensions of the frame.

Understand We are given two relationships about the frame, and we are to find the length and width.

Plan Select a variable for the length and another variable for the width, translate the relationships to a system of equations, and then solve the system.

Execute Let l represent the length and w represent the width.

Relationship 1: The length is 4 more than the width.

Translation: $l = w + 4$

Relationship 2: The perimeter is 72 inches.

Translation: $2l + 2w = 72$

$$\text{Our system: } \begin{cases} l = w + 4 \\ 2l + 2w = 72 \end{cases}$$

Because the first equation has l isolated, we substitute $w + 4$ for l in the second equation.

$$2l + 2w = 72$$
$$2(w + 4) + 2w = 72 \quad \text{Substitute } w + 4 \text{ for } l.$$
$$2w + 8 + 2w = 72$$
$$4w + 8 = 72 \quad \text{Combine like terms.}$$
$$4w = 64 \quad \text{Subtract 8 from both sides of the equation.}$$
$$w = 16 \quad \text{Divide both sides by 4.}$$

Now we can find the value of l using $l = w + 4$.

$$l = 16 + 4 \quad \text{Substitute 16 for } w.$$
$$l = 20$$

Answer The length should be 20 inches, and the width should be 16 inches.

Check Verify that the length is 4 more than the width: $20 = 16 + 4$. Also verify that the perimeter is 72 inches: $2(20) + 2(16) = 40 + 32 = 72$.

Connection Prior to learning about systems of equations, when we solved problems involving two unknowns, we actually used the concepts of substitution. For example, in Section 3.3, we set up this example as follows. We selected a variable, *w*, for the width and translated the first relationship like this:

$$\text{Length} = w + 4$$

Then we translated the perimeter relationship like this:

$$2 \cdot \text{Length} + 2 \cdot \text{Width} = 72$$
$$2(w + 4) + 2w = 72$$

Notice that we substituted $w + 4$ in place of the word *length* in the perimeter equation.

Your Turn 6 One number is four times a second number. The larger number minus the smaller number equals 45. Find the two numbers.

Answer Your Turn 6

15, 60

5.2 Exercises   For Extra Help   *MyMathLab*

PRACTICE   WATCH   DOWNLOAD

1. What advantages does the method of substitution have over graphing for solving systems of equations?

2. Suppose you are given the system of equations
$$\begin{cases} y = x - 6 \\ 4x + 3y = 1 \end{cases}.$$
   a. Which variable would you replace in the substitution process?

   b. What expression would replace that variable?

3. Suppose you are given the system of equations
$$\begin{cases} 5x + y = 4 \\ 3x - 2y = 9 \end{cases}.$$ Which variable in which equation would you isolate first? Why?

4. Suppose you are given the system of equations
$$\begin{cases} 3x + 4y = 1 \\ x - 5y = 12 \end{cases}.$$ Which variable would you isolate first? Why?

5. How do you know that a system has no solution (inconsistent system) when using substitution?

6. How do you know that a system has an infinite number of solutions (consistent with dependent equations) when using substitution?

For Exercises 7–34, solve the system of equations using substitution. Note that some systems may be inconsistent or consistent with dependent equations. See Examples 1–5.

7. $\begin{cases} x + y = 6 \\ x = 2y \end{cases}$

8. $\begin{cases} x - y = 8 \\ y = 2x \end{cases}$

9. $\begin{cases} x = -3y \\ 2x - 5y = 44 \end{cases}$

10. $\begin{cases} x + 2y = 9 \\ y = -2x \end{cases}$

11. $\begin{cases} 2x - y = 4 \\ y = 3x + 6 \end{cases}$

12. $\begin{cases} x = 3y - 2 \\ 2x - 3y = 2 \end{cases}$

13. $\begin{cases} y = -2x \\ 4x + 2y = 0 \end{cases}$

14. $\begin{cases} x = 2y - 6 \\ -2x + 4y = 12 \end{cases}$

15. $\begin{cases} 2x + y = -4 \\ 3x + 2y = -5 \end{cases}$　　**16.** $\begin{cases} 5x + y = -4 \\ 3x + 2y = -1 \end{cases}$　　**17.** $\begin{cases} 5x - y = 1 \\ 3x + 2y = 24 \end{cases}$　　**18.** $\begin{cases} x + 3y = 5 \\ 2y - x = 10 \end{cases}$

19. $\begin{cases} 5x - 3y = 4 \\ y - x = -2 \end{cases}$　　**20.** $\begin{cases} 2x - y = 39 \\ 14 + 3x = 74y \end{cases}$　　**21.** $\begin{cases} 2x = 10 - y \\ 3x - y = 5 \end{cases}$　　**22.** $\begin{cases} 3y = 7 - x \\ 2x - y = 0 \end{cases}$

23. $\begin{cases} -x + 2y = 1 \\ 3x - 2y = 1 \end{cases}$　　**24.** $\begin{cases} 2x + 3y = -1 \\ 5x - y = 6 \end{cases}$　　**25.** $\begin{cases} 3x + 2y = 8 \\ x + 4y = 1 \end{cases}$　　**26.** $\begin{cases} 4x + y = -2 \\ 8x - y = 11 \end{cases}$

27. $\begin{cases} x - 3y = -4 \\ -5x + 15y = 6 \end{cases}$　　**28.** $\begin{cases} 2x - y = -1 \\ 6x - 3y = 3 \end{cases}$　　**29.** $\begin{cases} 4x + 3y = -2 \\ 8x - 2y = 12 \end{cases}$　　**30.** $\begin{cases} 2x + 3y = -1 \\ 6x + 3y = -9 \end{cases}$

31. $\begin{cases} 5x + 4y = 12 \\ 7x - 6y = 40 \end{cases}$　　**32.** $\begin{cases} 4x - 3y = 2 \\ 3x + 5y = 16 \end{cases}$　　**33.** $\begin{cases} 5x + 6y = 2 \\ 10x + 3y = -2 \end{cases}$　　**34.** $\begin{cases} 7x - 6y = 1 \\ 4x - 6y = 3 \end{cases}$

For Exercises 35–38, solve using substitution; then verify the solution by graphing the equations on a graphing utility. See Examples 1–5.

35. $\begin{cases} y = 2x + 1 \\ 3x + y = 11 \end{cases}$　　**36.** $\begin{cases} y = -4x + 3 \\ y - 2x = -3 \end{cases}$　　**37.** $\begin{cases} 2x + y = 4 \\ x + 2y = 5 \end{cases}$　　**38.** $\begin{cases} 4x - y = 12 \\ x - 3y = 14 \end{cases}$

Find

ⓧ the mistake　　**For Exercises 39 and 40, find the mistake.**

39. $\begin{cases} 3x + 2y = 5 \\ x - 3y = 6 \end{cases}$

Rewrite: $x = 6 + 3y$

Substitute: $3(6 + 3y) + 2y = 5$

$18 + 3y + 2y = 5$

$18 + 5y = 5$

$5y = -13$

$y = -\dfrac{13}{5}$

$x = 6 + 3y$

$x = 6 + 3\left(-\dfrac{13}{5}\right)$

$x = 6 - \dfrac{39}{5}$

$x = -\dfrac{9}{5}$

Solution: $\left(-\dfrac{9}{5}, -\dfrac{13}{5}\right)$

40. $\begin{cases} 2x + 3y = 8 \\ x + 2y = 6 \end{cases}$

Rewrite: $x = 6 - 2y$

Substitute: $x + 2y = 6$

$6 - 2y + 2y = 6$

$6 - 6 - 2y + 2y = 6 - 6$

$0 = 0$

Solution: Infinite solutions along $x + 2y = 6$

For Exercises 41–56, translate the problem to a system of equations and then solve. See Example 6.

41. Find two numbers whose sum is 40 and whose difference is 6.

42. Find two numbers whose sum is 18 and whose difference is 4.

43. Three times a number added to twice a smaller number is 4. Twice the smaller number less than twice the larger number is 6. Find the numbers.

44. Find two numbers whose sum is 14 if one number is three times as large as the other number.

45. The greater of two integers is 10 more than twice the smaller integer. The sum of the two integers is -8. Find the integers.

46. One integer is six less than three times another integer. The sum of the integers is -42. Find the integers.

47. Jon is 13 years older than Tony. The sum of their ages is 27. How old are they?

48. Patrick is twice as old as his sister Sherry. The sum of their ages is 12. How old are they?

49. At a local community college, the difference between a professor's salary and an associate professor's salary is $14,800. The sum of the salaries is $132,000. Find the salary of an associate professor at the college.

50. The sum of the semester hour fees for both out-of-state and in-state students at Austin Community College is $345. The difference between the two tuitions is $237. What is the semester hour fee for an out-of-state student at ACC? (*Source:* Austin Community College)

51. The length of a rectangle is 1 meter more than twice the width. What are the dimensions of the rectangle if the perimeter is 32 meters?

52. The length of a rectangle is 1 meter more than triple the width. The length is also five less than five times the width. Find the length and width of the rectangle.

53. The width of a rectangular garden plot is 4 feet less than the length. The perimeter of the plot is 48 feet. Find the length and width.

54. A rectangular garden is 6 feet longer than it is wide. If the perimeter is 132 feet, what are the dimensions of the garden?

55. The length of a volleyball court is twice the width. If the perimeter of the court along the doubles lines is 54 meters, what are the dimensions? (*Source: USA Volleyball Rule Book*)

56. The length of the Reflecting Pool at the National Mall in Washington, D.C., is 800 feet more than ten times the width. What are the dimensions if the perimeter is 4900 feet? (*Source:* www.answers.com)

Review Exercises

Exercises 1–4 EXPRESSIONS

[1.7] *For Exercises 1 and 2, multiply.*

1. $\frac{1}{4}(20x - 16y)$

2. $-3(x - 7y)$

[1.7] *For Exercises 3 and 4, add or subtract.*

3. $6x + 2y + 3x - 2y$

4. $5x + 3y - 5x + 10y$

Exercises 5 and 6 EQUATIONS AND INEQUALITIES

[2.3] *For Exercises 5 and 6, solve.*

5. $\frac{1}{4}x = 20$

6. $-8y = 12$

5.3 Solving Systems of Linear Equations by Elimination

Objectives

   1 Solve systems of linear equations using elimination.

   2 Solve applications using elimination.

We have seen two methods for solving a system of equations: graphing and substitution. Each method has advantages and disadvantages. The substitution method is advantageous over the graphing method when solutions involve fractions. Also, substitution is easy when a variable's coefficient is 1. However, substitution can be tedious when no coefficients are 1. For this reason, we turn to yet a third method, the *elimination* method.

Objective 1 **Solve systems of linear equations using elimination.** The elimination method uses the addition principle of equality to add equations so that a new equation emerges with one of the variables eliminated. We will work through some examples to get a sense of the method before stating it formally.

Example 1 Solve the system of equations using elimination.

$$\begin{cases} x + y = 5 & \text{(Equation 1)} \\ 2x - y = 7 & \text{(Equation 2)} \end{cases}$$

> **Note:** This method is also called the elimination by addition method, or simply the addition method.

SOLUTION In Chapter 2, we learned that the addition principle of equality says that adding the same amount to both sides of an equation will not affect its solution(s). Because Equation 1 indicates that $x + y$ and 5 are the same amount, if we add $x + y$ to the left side of Equation 2 and 5 to the right side of Equation 2, we are applying the addition principle of equality. Most people prefer to stack the equations and combine like terms vertically like this:

$$\begin{array}{ll} x + y = 5 & \text{(Equation 1)} \\ \underline{2x - y = 7} & \text{(Equation 2)} \\ 3x + 0 = 12 & \text{Add Equation 1 to Equation 2.} \end{array}$$

> **Connection** The terms y and $-y$ are additive inverses, or opposites, which means that their sum is 0.

Notice that y is eliminated in this new equation; so we can easily solve for the value of x.

$$3x = 12$$
$$x = 4 \qquad \text{Divide both sides by 3 to isolate } x.$$

Now that we have the value of x, we can find y by substituting 4 for x in one of the original equations. We will use $x + y = 5$:

$$x + y = 5$$
$$4 + y = 5 \qquad \text{Substitute 4 for } x.$$
$$y = 1 \qquad \text{Subtract 4 from both sides to isolate } y.$$

The solution is $(4, 1)$. We can check by verifying that $(4, 1)$ makes both of the original equations true. We will leave this to the reader.

Solve the system of equations using elimination.

$$\begin{cases} 4x + 3y = 8 \\ x - 3y = 7 \end{cases}$$

Multiplying One Equation by a Number to Create Additive Inverses

Connection In Chapter 2, we learned that the multiplication principle of equality says that we can multiply both sides of an equation by the same amount without affecting its solution(s).

You may have noted that the expressions $x + y$ and $2x - y$ conveniently contained the additive inverses y and $-y$, which eliminated the y's when the expressions were added. If no such pairs of additive inverses appear in a system of equations, we will use the multiplication principle of equality to multiply both sides of one equation by a number in order to create additive inverse pairs.

Example 2 Solve the system of equations using elimination.

$$\begin{cases} x + y = 6 & \text{(Equation 1)} \\ 2x - 5y = -16 & \text{(Equation 2)} \end{cases}$$

Solution Because no variables are eliminated when $x + y = 6$ and $2x - 5y = -16$ are added, we will rewrite one of the equations so that it has a term that is the additive inverse of one of the terms in the other equation. We will multiply both sides of Equation 1 by 5 so that its y term becomes $5y$, which is the opposite of the $-5y$ term in Equation 2.

$$x + y = 6$$
$$5 \cdot x + 5 \cdot y = 5 \cdot 6 \qquad \text{Multiply both sides by 5.}$$
$$5x + 5y = 30 \qquad \text{(Equation 1 rewritten)}$$

Note: We could have multiplied both sides of Equation 1 by -2. We would get a $-2x$ term in Equation 1 that is the opposite of the $2x$ term in Equation 2.

Now we add the rewritten Equation 1 to Equation 2 to eliminate a variable, just as we did in Example 1.

Note: Multiplying Equation 1 by 5 made the elimination of the y terms possible.

$$\begin{array}{ll} 5x + 5y = 30 & \text{(Equation 1 rewritten)} \\ \underline{2x - 5y = -16} & \text{(Equation 2)} \\ 7x + 0 \ = 14 & \text{Add rewritten Equation 1 to} \\ & \text{Equation 2 to eliminate } y. \end{array}$$

We can now solve for x.

$$7x = 14$$
$$x = 2 \qquad \text{Divide both sides by 7 to isolate } x.$$

To finish, we substitute 2 for x in one of the equations containing both variables. We will use $x + y = 6$.

$$x + y = 6$$
$$2 + y = 6 \qquad \text{Substitute 2 for } x.$$
$$y = 4 \qquad \text{Subtract 2 from both sides to isolate } y.$$

The solution is $(2, 4)$. We will leave the check to the reader.

Your Turn 2 Solve the system of equations using elimination.

$$\begin{cases} -3x - 5y = 6 \\ x + 2y = -1 \end{cases}$$

Multiplying Each Equation by a Number to Create Additive Inverses

If every coefficient in a system of equations is other than 1, we may have to multiply each equation by a number to generate a pair of additive inverses.

Example 3 Solve the system of equations using elimination.

$$\begin{cases} 4x - 3y = -2 & \text{(Equation 1)} \\ 6x - 7y = 7 & \text{(Equation 2)} \end{cases}$$

SOLUTION We will choose to eliminate x; so we must multiply both equations by numbers that make the x terms additive inverses. This is like finding the least common multiple (or denominator) of the coefficients of x, but with opposite signs. We will multiply Equation 1 by 3 and Equation 2 by -2.

$$4x - 3y = -2 \xrightarrow{\text{Multiply by 3.}} 12x - 9y = -6$$
$$6x - 7y = 7 \xrightarrow{\text{Multiply by } -2.} -12x + 14y = -14$$

Note: The x terms are now additive inverses, $12x$ and $-12x$.

Now we can add the rewritten equations to eliminate the x term.

$$\begin{array}{rl} 12x - 9y = -6 & \\ \underline{-12x + 14y = -14} & \\ 0 + 5y = -20 & \text{Add the rewritten equations to eliminate } x. \\ y = -4 & \text{Divide both sides by 5 to isolate } y. \end{array}$$

Note: It doesn't matter which equation we use because x should be the same value in any equation in the system.

To finish, we substitute -4 for y in one of the equations and solve for x. We will use $4x - 3y = -2$.

$$\begin{array}{rl} 4x - 3y = -2 & \\ 4x - 3(-4) = -2 & \text{Substitute } -4 \text{ for } y. \\ 4x + 12 = -2 & \\ 4x = -14 & \text{Subtract 12 from both sides.} \\ x = -\dfrac{14}{4} & \text{Divide both sides by 4 to isolate } y. \\ x = -\dfrac{7}{2} & \text{Simplify.} \end{array}$$

The solution is $\left(-\dfrac{7}{2}, -4 \right)$. We will leave the check to the reader.

Your Turn 3 Solve the system of equations using elimination.

$$\begin{cases} 3x - 2y = 7 \\ 4x - 3y = 10 \end{cases}$$

Answer Your Turn 2

$(-7, 3)$

Answer Your Turn 3

$(1, -2)$

Fractions or Decimals in a System

If any of the equations in a system of equations contains fractions or decimals, it is helpful to use the multiplication principle of equality to clear those fractions or decimals so that the equations contain only integers.

Note: Since this system contains both fractions and decimals, the answers can be left as either a fraction or a decimal.

Example 4   Solve the system of equations using elimination.

$$\begin{cases} \dfrac{1}{2}x - y = \dfrac{3}{4} & \text{(Equation 1)} \\ 0.4x - 0.3y = 1 & \text{(Equation 2)} \end{cases}$$

SOLUTION   To clear the fractions in Equation 1, we can multiply both sides by the LCD, which is 4. To clear the decimals in Equation 2, we can multiply both sides by 10.

$$\frac{1}{2}x - y = \frac{3}{4} \qquad \xrightarrow{\text{Multiply by 4.}} \qquad 2x - 4y = 3$$

$$0.4x - 0.3y = 1 \qquad \xrightarrow{\text{Multiply by 10.}} \qquad 4x - 3y = 10$$

Now that both equations contain only integers, it will be easier to solve the system. We will choose to eliminate x. So we multiply the first equation by -2, then combine the equations.

$$2x - 4y = 3 \qquad \xrightarrow{\text{Multiply by } -2.}$$

$$4x - 3y = 10$$

$$\begin{aligned} -4x + 8y &= -6 \\ \underline{4x - 3y} &= \underline{10} \\ 0 + 5y &= 4 \end{aligned} \qquad \text{Add the rewritten equations to eliminate } x.$$

$$y = \frac{4}{5} \text{ or } 0.8 \qquad \text{Divide both sides by 5 to isolate } y.$$

To finish, we substitute $\dfrac{4}{5}$ for y in one of the equations and solve for x. We will use $\dfrac{1}{2}x - y = \dfrac{3}{4}$.

Note: We could have multiplied both sides of the equation by 20 to clear the fraction.

$$\frac{1}{2}x - \frac{4}{5} = \frac{3}{4} \qquad \text{Substitute } \frac{4}{5} \text{ for } y.$$

$$\frac{1}{2}x = \frac{3}{4} + \frac{4}{5} \qquad \text{Add } \frac{4}{5} \text{ to both sides.}$$

$$\frac{1}{2}x = \frac{15}{20} + \frac{16}{20} \qquad \text{Write the fractions with their LCD, 20.}$$

$$\frac{1}{2}x = \frac{31}{20} \qquad \text{Add the fractions.}$$

$$x = \frac{31}{\underset{10}{20}} \cdot \frac{\overset{1}{2}}{1} \qquad \text{Multiply both sides by 2 to isolate } x.$$

$$x = \frac{31}{10} \text{ or } 3.1$$

The solution is $\left(\dfrac{31}{10}, \dfrac{4}{5}\right)$ or $(3.1, 0.8)$. We will leave the check to the reader.

Solve the system of equations using elimination.

$$\begin{cases} 0.6x + y = 3 \\ \dfrac{2}{5}x + \dfrac{1}{4}y = 1 \end{cases}$$

Rewriting the Equations in the Form $Ax + By = C$

Notice in Examples 1 through 4 that all of the equations are in standard form, which we learned in Chapter 4 to be $Ax + By = C$. When the elimination method is used, the equations need to be written in standard form. For example, in the system that we solved in Example 1, the equations could have been given in a nonstandard form, as follows:

$$\begin{cases} y = 5 - x & \text{(Equation 1)} \\ 2x - y = 7 & \text{(Equation 2)} \end{cases}$$

Adding x to both sides of Equation 1 puts it in standard form so that we can use elimination as we did in Example 1.

$$\begin{cases} x + y = 5 & \text{(Equation 1)} \\ 2x - y = 7 & \text{(Equation 2)} \end{cases}$$

We can now summarize the elimination method with the following procedure.

Procedure

Solving Systems of Two Linear Equations Using Elimination

To solve a system of two linear equations using the elimination method,

1. Write the equations in standard form ($Ax + By = C$).
2. Use the multiplication principle to clear fractions or decimals (optional).
3. If necessary, multiply one or both equations by a number (or numbers) so that they have a pair of terms that are additive inverses.
4. Add the equations. The result should be an equation in terms of one variable.
5. Solve the equation from step 4 for the value of that variable.
6. Using an equation containing both variables, substitute the value you found in step 5 for the corresponding variable and solve for the value of the other variable.
7. Check your solution in the original equations.

Inconsistent Systems and Dependent Equations

How would we recognize an inconsistent system or a consistent system with dependent equations using the elimination method?

Example 5 Solve the system of equations.

a. $\begin{cases} 2x - y = 1 \\ 2x - y = -3 \end{cases}$

SOLUTION Notice that the left sides of the equations match. Multiplying one of the equations by -1 and then adding the equations will eliminate both variables.

$$\begin{array}{l} 2x - y = 1 \\ 2x - y = -3 \end{array} \xrightarrow{\text{Multiply by } -1.} \begin{array}{l} -2x + y = -1 \\ \underline{2x - y = -3} \\ 0 = -4 \quad \text{Add the rewritten} \\ \text{equations.} \end{array}$$

Answer Your Turn 4

$\left(1, \dfrac{12}{5}\right)$ or $(1, 2.4)$

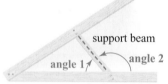

Connection If we graph the equations in Example 5a, the lines will be parallel, indicating there is no solution.

Both variables have been eliminated, and the resulting equation, $0 = -4$, is false. Therefore, there is no solution. This system of equations is inconsistent.

b. $\begin{cases} 3x + 4y = 5 \\ 9x + 12y = 15 \end{cases}$

SOLUTION To eliminate x, we could multiply the first equation by -3, then combine the equations.

$$\begin{array}{l} 3x + 4y = 5 \\ 9x + 12y = 15 \end{array} \quad \xrightarrow{\text{Multiply by } -3.} \quad \begin{array}{l} -9x - 12y = -15 \\ \underline{9x + 12y = 15} \\ 0 = 0 \end{array}$$

Add the rewritten equations.

Connection If graphed, both equations in Example 5(b) generate the same line, indicating that the equations are dependent.

Both variables have been eliminated, and the resulting equation, $0 = 0$, is true. This means that the equations are dependent. So there are an infinite number of solutions, which are all of the ordered pairs along the line $3x + 4y = 5$ (or $9x + 12y = 15$).

Your Turn 5 Solve the system of equations.

a. $\begin{cases} x - 4y = 2 \\ 5x - 20y = 10 \end{cases}$ **b.** $\begin{cases} x + 2y = 3 \\ x + 2y = 1 \end{cases}$

Objective ② Solve applications using elimination. In Section 5.2, we solved problems in which the substitution method was advantageous because after translating, one of the equations had an isolated variable. In this section, we focus on problems in which elimination is advantageous. These problems translate to a system of linear equations in which every equation is in standard form ($Ax + By = C$).

Example 6 In the construction of a roof frame, a support beam is connected to a horizontal joist forming two angles as shown in the margin. If the larger angle minus twice the smaller angle equals 6°, what are the measures of the two angles?

Understand We are to find two angle measurements, but it appears that we are given only one relationship. Because the joist is a straight line, it represents an angle of 180°; therefore, the two angles are supplementary, which is our second relationship.

Plan Translate the relationships to a system of equations and then solve the system.

Execute We will let x represent the measure of angle 1 and y represent the measure of angle 2. Now we translate the two relationships to a system of two equations.

Relationship 1: The larger angle minus twice the smaller angle equals 6°.

Translation: $y - 2x = 6$

Relationship 2: From the picture, we see that the angles are supplementary.

Translation: $x + y = 180$

We can now solve the system using elimination. We will eliminate y.

Note: We rearranged the terms to align the like terms vertically.

$$\begin{cases} -2x + y = 6 \\ x + y = 180 \end{cases} \quad \xrightarrow{\text{Multiply by } -1.} \quad \begin{array}{l} 2x - y = -6 \\ \underline{x + y = 180} \\ 3x + 0 = 174 \end{array}$$

Add the rewritten equations to eliminate y.

$$x = 58$$

Divide both sides by 3 to isolate x.

Answers **Your Turn 5**

a. all ordered pairs along $x - 4y = 2$ or $5x - 20y = 10$ (dependent)

b. no solution (inconsistent)

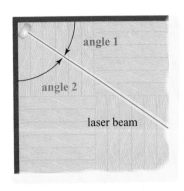

angle 1

angle 2

laser beam

Now we can find the value of y using $x + y = 180$.

$$58 + y = 180 \qquad \text{Substitute 58 for } x.$$
$$y = 122 \qquad \text{Subtract 58 from both sides to isolate } y.$$

Answer Angle 1 measures 58°, and angle 2 measures 122°.

Check Verify that the measure of the larger angle minus twice the measure of the smaller angle equals 6°: $122 - 2(58) = 122 - 116 = 6$. Also verify that the angles are supplementary: $58 + 122 = 180$.

Your Turn 6 A museum's security system uses a laser placed in the corner just above the floor, with its beam projecting across the room parallel to the floor as shown in the figure in the margin. The beam makes two angles in the corner of the room. The difference of the measures of the two angles is 22°. Find the angle measurements.

Answer   Your Turn 6

34°, 56°

5.3 Exercises   For Extra Help   *MyMathLab*

PRACTICE   WATCH   DOWNLOAD

1. What advantages does the method of elimination have over graphing and substitution?

2. Suppose you are given the system of equations
$$\begin{cases} x - y = 2 \\ 3x + y = 5 \end{cases}$$
Which variable would you eliminate? Why?

3. Suppose you are given the system of equations
$$\begin{cases} 6x - 2y = 1 \\ 3x + 5y = 7 \end{cases}$$
   a. Which variable is easier to eliminate? Why?

4. Suppose you are given the system of equations
$$\begin{cases} 5x + 4y = 8 \\ 3x - 6y = 2 \end{cases}$$
   a. Which variable is easier to eliminate? Why?

   b. How would you eliminate the variable you chose?

   b. How would you eliminate the variable you chose?

5. When using elimination, how do you know if a system of equations has no solution (inconsistent system)?

6. When using elimination, how do you know if a system of equations has an infinite number of solutions (dependent equations)?

For Exercises 7–42, solve the system of equations using the elimination method. Note that some systems may be inconsistent or consistent with dependent equations. See Examples 1–5.

7. $\begin{cases} x - y = 14 \\ x + y = -2 \end{cases}$

8. $\begin{cases} x + y = 9 \\ 5x - y = 3 \end{cases}$

9. $\begin{cases} 3x + y = 9 \\ 2x - y = 1 \end{cases}$

10. $\begin{cases} 3x + y = 10 \\ -2x - y = -7 \end{cases}$

11. $\begin{cases} 3x + 2y = -7 \\ 5x - 2y = -1 \end{cases}$

12. $\begin{cases} 2x - 3y = 8 \\ 4x + 3y = 16 \end{cases}$

13. $\begin{cases} 4x + 3y = 17 \\ 2x + 3y = 13 \end{cases}$

14. $\begin{cases} x - 3y = 7 \\ x - 5y = 13 \end{cases}$

15. $\begin{cases} 5x + y = 14 \\ 2x + y = 5 \end{cases}$

16. $\begin{cases} 3x - y = 7 \\ 5x - y = 15 \end{cases}$

17. $\begin{cases} 3x - 5y = -17 \\ 4x + y = -15 \end{cases}$

18. $\begin{cases} 7x - 4y = 4 \\ 5x + y = 26 \end{cases}$

19. $\begin{cases} 12x - 2y = -54 \\ 13x + 4y = -40 \end{cases}$

20. $\begin{cases} 5x + 2y = 3 \\ 7x - 6y = 13 \end{cases}$

21. $\begin{cases} 2x + y = -2 \\ 8x + 4y = -8 \end{cases}$

22. $\begin{cases} 4x - 2y = 3 \\ -8x + 4y = -6 \end{cases}$

23. $\begin{cases} 2x - 5y = 4 \\ -8x + 20y = -20 \end{cases}$

24. $\begin{cases} 5x + 5y = 50 \\ x + y = 2.5 \end{cases}$

25. $\begin{cases} 5x + 6y = 11 \\ 2x - 4y = -2 \end{cases}$

26. $\begin{cases} -6x + 7y = -2 \\ 9x - 5y = -8 \end{cases}$

27. $\begin{cases} 4x + 5y = -4 \\ 3x + 8y = -20 \end{cases}$

28. $\begin{cases} 3x - 4y = -14 \\ 5x + 7y = 45 \end{cases}$

29. $\begin{cases} 5x + 6y = 2 \\ 10x + 3y = -2 \end{cases}$

30. $\begin{cases} 8x + 6y = 5 \\ 2x + y = 1 \end{cases}$

31. $\begin{cases} 10x + 5y = 3.5 \\ 3x - 4y = -0.6 \end{cases}$

32. $\begin{cases} 2x - y = -0.8 \\ 3x - 2y = -1 \end{cases}$

33. $\begin{cases} \dfrac{3}{2}x - \dfrac{4}{3}y = \dfrac{11}{3} \\ \dfrac{1}{4}x - \dfrac{2}{3}y = -\dfrac{7}{6} \end{cases}$

34. $\begin{cases} \dfrac{3}{4}x - \dfrac{2}{7}y = -5 \\ \dfrac{5}{8}x + \dfrac{1}{4}y = -\dfrac{3}{4} \end{cases}$

35. $\begin{cases} 0.7x - \dfrac{1}{2}y = 9.4 \\ 0.9x + \dfrac{7}{10}y = 0 \end{cases}$

36. $\begin{cases} \dfrac{1}{3}x + 0.2y = -0.2 \\ \dfrac{2}{3}x - 0.75y = -5 \end{cases}$

37. $\begin{cases} y = 2x - 5 \\ x - y = 9 \end{cases}$

38. $\begin{cases} x + 3y = 8 \\ 2y = x + 6 \end{cases}$

39. $\begin{cases} y = \dfrac{3}{4}x + 1 \\ 4x = 2y + 3 \end{cases}$

40. $\begin{cases} y = -\dfrac{2}{3}x - 5 \\ 3y = 2x - 27 \end{cases}$

41. $\begin{cases} 0.2x - y = -3.2 \\ y = \dfrac{3}{4}x + 1 \end{cases}$

42. $\begin{cases} y = \dfrac{1}{5}x - 2 \\ 6y = 0.3x - 1.2 \end{cases}$

Find

(X) *the mistake*   **For Exercises 43 and 44, find the mistake.**

43. $\begin{cases} x - y = 1 \\ 9x + 8y = 77 \end{cases}$

$\begin{array}{l} -9(x - y) = 1 \\ 9x + 8y = 77 \end{array} \longrightarrow \begin{array}{r} -9x + 9y = 1 \\ 9x + 8y = 77 \\ \hline 17y = 78 \\ y = \dfrac{78}{17} \end{array}$

$x - y = 1$

$x - \left(\dfrac{78}{17}\right) = 1$

$x = \dfrac{95}{17}$

Solution: $\left(\dfrac{95}{17}, \dfrac{78}{17}\right)$

44. $\begin{cases} x + y = 1 \\ x + 2y = 2 \end{cases}$

$\begin{array}{l} x + y = 1 \\ -1(x + 2y = 2) \end{array} \longrightarrow \begin{array}{r} x + y = 1 \\ -x - 2y = 2 \\ \hline -y = 3 \\ y = -3 \end{array}$

$x + y = 1$

$x + (-3) = 1$

$x = 4$

Solution: $(4, -3)$

For Exercises 45–62, translate the problem to a system of equations; then solve using the elimination method. See Example 6.

45. One number is 62 more than another number. If twice the greater number is subtracted from five times the lesser number, the result is 155. What are the two numbers?

46. The difference of two numbers is 4. Three times the greater number is two more than five times the lesser number. Find the numbers.

47. One number is 7 less than another number. The lesser number subtracted from three-fourths of the greater number is 3. Find the numbers.

48. One number exceeds another number by 9. If two-thirds of the lesser number is increased by 22, the result equals the greater number. What are the two numbers?

49. Janet is four years younger than her brother. If the sum of their ages is 15, what are their ages?

50. The sum of Morgan and Angela's ages is 56, and their age difference is 14. If Morgan is older than Angela, find their ages.

51. Two angles are supplementary. One of the angles is 20° more than twice the other angle. Find the measure of the two angles.

52. Two angles are supplementary. One angle is 10° more than four times the other angle. Find the measure of the two angles.

53. Two angles are complementary. One of the angles is twice the other angle. Find the measure of the two angles.

54. Two angles are complementary. The larger angle is 6° less than twice the smaller angle. Find the measure of the two angles.

55. What are the length and width of a rectangle if the length exceeds the width by 14 inches and the perimeter is 60 inches?

56. What are the length and width of a rectangle if the width is 27 feet less than its length and the perimeter is 398 feet?

57. During the 2006 Winter Olympic Games in Turin, Italy, Germany won the most medals, followed closely by the United States. The total medals that both countries won was 54, and the difference between the numbers of medals that each country won was 4. Find the total number of medals that each country won. (*Source:* www.wikipedia.org)

58. During the same Olympics, the United States won three times as many gold medals as France. If the total gold medals awarded to both countries was 12, find the number of gold medals awarded to each country. (*Source:* www.wikipedia.org)

59. In Super Bowl XLI, the Indianapolis Colts scored 12 points more than the Chicago Bears. If the teams scored a total of 46 points during the game, find the number of points that each team scored.

60. In Super Bowl XL, the Pittsburgh Steelers scored one point more than twice the number of points scored by the Seattle Seahawks. If the teams scored a total of 31 points during the game, find the number of points that each team scored.

61. Jack Nicklaus has the most victories ever won by a golfer in the Masters Tournament. He exceeds the number of Arnold Palmer's victories by 2 wins. If they have won a total of 10 tournaments, how many times has each man won?

62. During the 2007 season, Alex Rodriguez of the New York Yankees hit four more home runs than Prince Fielder of the Milwaukee Brewers. If their combined number of home runs was 104, how many did each player hit? (*Source:* www.mlb.com)

Puzzle Problem

Use algebraic methods to prove that $0.999\ldots$ is equal to 1. (Hint: Let $x = 0.\overline{9}$. Then write a second equation using the multiplication principle so that when the two equations are added, the repeated 9 is eliminated.)

Review Exercises

Exercises 1–6 EQUATIONS AND INEQUALITIES

[3.3] **1.** Monique has $4.05 in her purse. If she has 24 coins, all dimes and quarters, how many of each does she have?

[3.4] **2.** Jay and Lisa are traveling west in separate cars on the same highway. Jay is traveling at 60 miles per hour; Lisa, at 70 miles per hour. Jay passes Exit 54 at 2:30 P.M. Lisa passes the same exit at 2:36 P.M. At what time will Lisa catch up with Jay?

[3.5] **3.** The total value of two bank accounts is $1100. The annual interest earned on the checking account is 5%, while the CD pays 6%. The total interest earned is $61. How much was deposited in each account?

[3.5] **4.** A chemist has 80 milliliters of 10% HCl solution in a container. He has a large amount of 25% HCl solution in another container. How much of the 25% solution should he add to the 10% solution to make a solution that is 20% HCl?

[5.3] **5.** Use the elimination method to solve the system
$$\begin{cases} x + y = 68 \\ 2x + 3y = 160 \end{cases}.$$

[5.2] **6.** Use substitution to solve the system
$$\begin{cases} y = 2x \\ 1800x + 2500y = 27{,}200 \end{cases}.$$

5.4 More Applications with Systems of Equations

Objective **1** **Solve problems using systems of equations.** In this section, we use systems of equations to solve a variety of problems involving two unknown amounts. We learn how to determine which of the three methods already presented is best for a particular problem.

In general, graphing is the least desirable method because it can be time-consuming and inaccurate. Substitution is simpler than elimination if one of the equations has an isolated variable. If all equations in the system are in standard form, elimination is better.

Currency Problems

As in Section 3.3, we will use a table to organize the information in problems. But we will select a variable for each unknown and write a system of equations using the relationships in the problem. We repeat Example 6 of Section 3.3 using a system of equations.

Example 1 A home improvement store sells two sizes of cans of wood stain. A small can sells for $8.95, and a large can sells for $15.95. One day the store sold twice as many of the large cans as the small cans. If the total revenue that day for these cans of stain was $694.45, how many cans of each size were sold?

Understand The two unknowns are the number of each size of cans of stain sold. There are also two relationships, one involving the number sold and the other involving the total revenue.

Plan and Execute Let x represent the number of small cans of stain sold and let y represent the number of large cans. We can use a table to organize the information.

Category	Price	Number	Revenue
Small can	8.95	x	$8.95x$
Large can	15.95	y	$15.95y$

Note: Sales revenue is found by multiplying an item's price by the number sold.

Relationship 1: The store sold twice as many of the small cans as the large.

  Translation: $y = 2x$

Relationship 2: The total revenue was $694.45.

  Translation: $8.95x + 15.95y = 694.45$

$$\text{Our system: } \begin{cases} y = 2x \\ 8.95x + 15.95y = 694.45 \end{cases}$$

To solve this system of equations, substitution is the best method because y is isolated in the first equation. We replace y in the second equation with $2x$ from the first equation.

$$8.95x + 15.95y = 694.45$$
$$8.95x + 15.95(2x) = 694.45 \qquad \text{Substitute } 2x \text{ for } y.$$
$$8.95x + 31.9x = 694.45$$
$$40.85x = 694.45 \qquad \text{Combine like terms.}$$
$$x = 17 \qquad \text{Divide both sides by 40.85 to isolate } x.$$

Now we can find the value of y using $y = 2x$ by substituting 17 for x.

$$y = 2(17) = 34$$

Answer The home improvement store sold 17 small cans of stain and 34 large cans.

Check Verify both given relationships. The number of large cans of stain sold, 34, is twice the number of small cans sold, 17. The total revenue, $8.95(17) + 15.95(34) = 694.45$, is also correct.

Your Turn 1 A cookware consultant sells two sizes of a rectangular baking dish. The small size sells for $24, and the large size sells for $35. In one month, she sold 6 more of the small-size dish than the large-size dish. If she made a total of $439 from the sale of both dishes, how many of each size baking dish did she sell?

Answer   Your Turn 1

11 small, 5 large

Example 2   Aaron sells concessions at a sporting event. In one hour, he sells 68 drinks. The drink sizes are small, which sell for $2 each, and large, which sell for $3 each. If his total sales revenue was $160, how many of each size did he sell?

Understand   The two unknowns are the number of each size drink. One relationship involves the number of drinks (68 total), and the other relationship involves the total sales in dollars ($160).

Plan and Execute   Let x represent the number of small drinks sold and y represent the number of large drinks sold. We can use a table to organize the information.

Category	Price	Number	Revenue
Small	2	x	$2x$
Large	3	y	$3y$

Relationship 1:   The total number of drinks sold is 68.

Translation: $x + y = 68$

Relationship 2:   The total sales revenue is $160.

Translation: $2x + 3y = 160$

Our system: $\begin{cases} x + y = 68 \\ 2x + 3y = 160 \end{cases}$

Because the equations are in standard form ($Ax + By = C$), we will use the elimination method to solve this system. We will eliminate x.

$$x + y = 68 \qquad \xrightarrow{\text{Multiply by } -2.} \qquad -2x - 2y = -136$$
$$2x + 3y = 160 \qquad\qquad\qquad \underline{2x + 3y = 160}$$
$$y = 24 \qquad \text{Add the equations to eliminate } x.$$

Now we can find the value of x by substituting 24 for y in one of the equations. We will use $x + y = 68$.

$$x + y = 68$$
$$x + 24 = 68 \qquad \text{Substitute 24 for } y.$$
$$x = 44 \qquad \text{Subtract 24 from both sides to isolate } x.$$

Answer   Aaron sold 44 small drinks and 24 large drinks.

Check   Verify both given relationships. The total number of drinks sold: $24 + 44 = 68$. The total sales: $2(44) + 3(24) = 88 + 72 = \$160$.

Your Turn 2   A college bookstore receives a shipment of 480 copies of a particular book. Some of the books have tutorial software shrinkwrapped with the book, while others do not. The bookstore pays $54 for each book with the tutorial software and $48 for each copy without the software. If the total invoice for the shipment is $24,000, how many books in the shipment have software and how many do not?

Rate Problems

We first solved rate problems in Section 3.4 using a table similar to the one we used with currency problems.

Answer   Your Turn 2

160 with software, 320 without software

Example 3   Jasmine and Darius are traveling east in separate cars on the same highway. Jasmine is traveling at 60 miles per hour; Darius, at 70 miles per hour. Jasmine passes Exit 82 at 3:15 P.M. Darius passes the same exit at 3:30 P.M. At what time will Darius catch up with Jasmine?

Understand   To determine the time at which Darius catches up with Jasmine, we must calculate the amount of time it will take him to catch up with her. We can then add that amount of time to 3:30 P.M.

Plan and Execute   We will let x represent Jasmine's travel time after passing Exit 82 and y represent Darius's travel time after passing Exit 82.

Category	Rate	Time	Distance
Jasmine	60	x	$60x$
Darius	70	y	$70y$

Note: Distance is the product of rate and time.

Relationship 1:   Darius passes the exit 15 minutes after Jasmine; so when he catches up with her, she will have traveled 15 minutes longer than he has.

Translation: $x = y + \dfrac{1}{4}$

Note: We use $\dfrac{1}{4}$ of an hour instead of 15 minutes because the rates are in miles per hour and the time units must be consistent.

Relationship 2:   When Darius catches up, they will have traveled the same distance from Exit 82; so we set the expressions for their individual distances equal to each other.

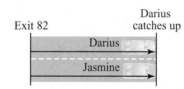

Translation: $60x = 70y$

Our system: $\begin{cases} x = y + \dfrac{1}{4} \\ 60x = 70y \end{cases}$

Because we have an isolated variable, x, we will solve the system using the substitution method.

$$60x = 70y$$
$$60\left(y + \frac{1}{4}\right) = 70y \quad \text{Substitute } y + \frac{1}{4} \text{ for } x.$$
$$60y + 15 = 70y \quad \text{Distribute 60.}$$
$$15 = 10y \quad \text{Subtract 60}y \text{ from both sides.}$$
$$1.5 = y \quad \text{Divide both sides by 10 to isolate } y.$$

Answer   It will take Darius 1.5 hours to catch up with Jasmine. So he catches up at 5 P.M.

Check   At 60 miles per hour, in $\dfrac{1}{4}$ of an hour, Jasmine travels 15 miles. Because Darius is traveling 70 miles per hour, he is gaining on her at a rate of 10 miles per hour (the difference of their rates). So he must make up the 15 miles that separate them at a rate of 10 miles per hour, which will indeed take 1.5 hours.

Your Turn 3   At 10:30 A.M., Celeste begins walking along a trail at an average rate of 3 miles per hour. At 10:40 A.M., Chris begins bicycling along the same trail at 8 miles per hour. At what time will Chris catch up with Celeste?

Los Angeles

Atlanta

2000 miles

Example 4   Airplanes traveling west fly against the jet stream, which decreases their average speed. When flying east, the jet stream increases average speed. Suppose a large jet flies from Atlanta to Los Angeles, a distance of approximately 2000 miles, in 5 hours. On the return trip, the plane takes only 4 hours. Find the plane's speed in still air. How much does the jet stream change the speed of the plane?

Understand   We assume that without the jet stream, the plane would average the same speed going west as going east. We also assume that the amount the jet stream adds to the plane's speed is the same as the amount subtracted depending on which way the plane is traveling.

Plan and Execute   Let x represent the plane's speed in still air. Let y represent the amount the jet stream increases or decreases the plane's speed depending on which way it is traveling. We can use a table to organize the information.

Category	Rate	Time	Distance
West	$x - y$	5	$5(x - y)$
East	$x + y$	4	$4(x + y)$

Relationship 1:   The distance from Atlanta to Los Angeles is 2000 miles.

Translation: $5(x - y) = 2000$

$$\frac{5(x - y)}{5} = \frac{2000}{5}$$
Divide both sides by 5 to streamline the equation.

$$x - y = 400$$

Relationship 2:   The distance from Los Angeles to Atlanta is 2000 miles.

Translation: $4(x + y) = 2000$

$$\frac{4(x + y)}{4} = \frac{2000}{4}$$
Divide both sides by 4 to streamline the equation.

$$x + y = 500$$

Our system: $\begin{cases} x - y = 400 \\ x + y = 500 \end{cases}$

Because the equations are in standard form, we will use the elimination method.

$$\begin{array}{l} x - y = 400 \\ \underline{x + y = 500} \\ 2x + 0 = 900 \qquad \text{Add the equations to eliminate } y. \\ 2x = 900 \\ x = 450 \qquad \text{Divide both sides by 2 to isolate } x. \end{array}$$

Now we can find the value of y by substituting 450 for x in one of the equations. We will use $x + y = 500$.

$$\begin{array}{l} x + y = 500 \\ 450 + y = 500 \qquad \text{Substitute 450 for } x. \\ y = 50 \qquad \text{Subtract 450 from both sides to isolate } y. \end{array}$$

Answer   The plane travels 450 miles per hour in still air. The jet stream changes the plane's speed by 50 miles per hour depending whether it is traveling east or west.

Check   Verify the time of travel for each part of the round trip.

Going west: $450 - 50 = 400$ miles per hour   Going east: $450 + 50 = 500$ miles per hour

Time: $t = \dfrac{d}{r} = \dfrac{2000}{400} = 5$ hours   Time: $t = \dfrac{d}{r} = \dfrac{2000}{500} = 4$ hours

Your Turn 4   To train for the swimming portion of a triathlon, an athlete swims in a river to gauge how the current will affect him. When swimming against the current, he can swim a mile in 1.25 hours. When swimming with the current, he can swim a mile in 0.5 hours. What is his rate in still water? How much did the current increase or decrease his rate?

Investment Problems

We first solved investment problems in Section 3.5.

Example 5   Jeff invests $5000 in two different accounts. The first account has an annual percentage rate (APR) of 4%. The second account has an APR of 6%. If the total interest earned after one year is $252, what principal was invested in each account?

Understand   The two unknowns are the principals invested in each account. One relationship involves the total principal ($5000), and the other relationship involves the total interest earned ($252).

Plan and Execute   Let x and y represent the two amounts invested. We can use a table to organize the information.

> **Note:** Recall from Chapter 3 that we write percentage rates as decimal numbers. Also, interest is the product of the APR and the principal.

Category	APR	Principal	Interest
Account 1	0.04	x	$0.04x$
Account 2	0.06	y	$0.06y$

Relationship 1:   The total principal is $5000.

Translation: $x + y = 5000$

Relationship 2:   The total interest is $252.

Translation: $0.04x + 0.06y = 252$

Our system: $\begin{cases} x + y = 5000 \\ 0.04x + 0.06y = 252 \end{cases}$

Because the equations are in standard form, elimination is the better method. We will eliminate x by multiplying both sides of the first equation by -0.04.

> **Note:** We could have first cleared the decimal numbers, but we chose not to because it is easy to eliminate x or y by the elimination method since both coefficients in $x + y = 5000$ are 1.

$x + y = 5000$ $\xrightarrow{\text{Multiply by } -0.04.}$

$0.04x + 0.06y = 252$

$\begin{aligned} -0.04x - 0.04y &= -200 \\ \underline{0.04x + 0.06y} &= \underline{252} \\ 0.02y &= 52 \\ y &= 2600 \end{aligned}$

Add the equations to eliminate x.

Divide both sides by 0.02 to isolate y.

Now we can find x by substituting 2600 for y in one of the equations. We will use $x + y = 5000$.

$\begin{aligned} x + y &= 5000 \\ x + 2600 &= 5000 \qquad \text{Substitute 2600 for } y. \\ x &= 2400 \qquad \text{Subtract 2600 from both sides to isolate } x. \end{aligned}$

> **Answer**   **Your Turn 4**
> His rate in still water is 1.4 miles per hour. The current changes his rate by 0.6 miles per hour.

Answer   Jeff invested $2400 in the 4% account and $2600 in the 6% account.

Check   The total investment is $2400 + $2600 = $5000, and the total interest is $0.04(2400) + 0.06(2600) = $96 + $156 = $252.

Your Turn 5   Monique invests a total of $8000 in two different accounts. The first account returns 5%, while the second account returns 8%. If the total interest earned after one year is $532, what principal was invested in each account?

Mixture Problems

We first solved mixture problems in Section 3.5.

Example 6   How much 10% HCl solution and 30% HCl solution must be mixed together to make 500 milliliters of 15% HCl solution?

Understand   The two unknowns are the volumes of 10% solution and 30% solution that are mixed. One relationship involves concentrations of each solution in the mixture, and the other relationship involves the total volume of the final mixture (500 ml).

Plan and Execute   Let x and y represent the two amounts to be mixed.

Solution	Concentration	Volume	Amount of HCl
10% solution	0.10	x	$0.10x$
30% solution	0.30	y	$0.30y$
15% solution	0.15	500	$0.15(500)$

Relationship 1:   The total volume is 500 ml.

Translation: $x + y = 500$

> **Note:** The amount of HCl is the product of the concentration and the volume. Notice that we were given the volume of the 15% solution.

Relationship 2:   The combined amount of HCl in the two mixed solutions is to be 15% of the total amount of the mixture.

Translation: $0.10x + 0.30y = 0.15(500)$

Our system: $\begin{cases} x + y = 500 \\ 0.10x + 0.30y = 0.15(500) \end{cases}$

Because the equations are in standard form, elimination is the better method. We will eliminate x by multiplying both sides of the first equation by -0.10.

$$x + y = 500 \xrightarrow{\text{Multiply by } -0.10.} -0.10x - 0.10y = -50$$

$$0.10x + 0.30y = 75 \qquad \underline{\quad 0.10x + 0.30y = 75 \quad}$$

Add the equations to eliminate x.

$$0.20y = 25$$

$$y = 125 \qquad \text{Divide both sides by 0.20 to isolate } y.$$

Now we can find x by substituting 125 for y in one of the equations. We will use $x + y = 500$.

$$x + y = 500$$

$$x + 125 = 500 \qquad \text{Substitute 125 for } y.$$

$$x = 375 \qquad \text{Subtract 125 on both sides to isolate } x.$$

Answer   Your Turn 5

$3600 at 5%, $4400 at 8%

Answer Mixing 375 ml of 10% solution with 125 ml of 30% solution gives 500 ml of 15% solution.

Check The mixture volume is $125 + 375 = 500$. The amount of HCl is $0.10(375) + 0.30(125) = 37.5 + 37.5 = 75$, which means that 75 ml is HCl out of the 500 ml mixture. We can verify that 75 ml is 15% of 500 ml by multiplying: $0.15(500) = 75$.

Your Turn 6 How much 20% HCl solution and 40% HCl solution must be mixed together to make 600 milliliters of 35% HCl solution?

In summary, we can write the following approach to solving problems involving two unknown amounts using a system of equations.

Procedure **Solving Problems Using Systems of Equations**

To solve problems involving two unknowns using a system of equations,

1. Select two variables to represent the two unknowns.
2. Using the two relationships in the problem, write a system of two equations in terms of the two variables.
3. Solve the system.

Note: If there is an isolated variable in one of the equations, the substitution method is the better method. Otherwise, the elimination method is best.

Answer Your Turn 6
450 ml of 40%, 150 ml of 20%

5.4 Exercises   For Extra Help   *MyMathLab*

Note: Exercises marked with a ★ represent challenging exercises.

1. Suppose there are three times as many $5 bills as $10 bills in a wallet. If x represents the number of $5 bills and y represents the number of $10 bills, write an equation relating the number of bills.

2. Suppose $3000 is invested in two accounts. Select variables to represent the amount invested in each account and write an equation describing the total amount invested.

3. If a problem translates to a system of equations with every equation in the form $Ax + By = C$, which method is better for solving the system?

4. If a problem translates to a system of equations and one of the equations has a variable isolated, which method is better for solving the system?

5. Clara went to a music store and bought some CDs and DVDs. She paid $14 for each CD and $20 for each DVD. Let x represent the number of CDs that she purchased and y represent the number of DVDs.
   a. Write an equation that states that she bought 3 more CDs than DVDs.

   b. Represent the cost of the CDs and DVDs.

   c. Write an equation that states that the total cost of the CDs and DVDs was $212.

   d. Solve the system of equations from parts a and c and find the number of CDs and DVDs purchased.

6. For a cookout, Fred bought beef for $7 per pound and pork for $5 per pound. Let x represent the number of pounds of beef and y represent the number of pounds of pork.
   a. Write an equation that states that Fred bought a total of 20 pounds of meat.

   b. Represent the cost of the beef and the pork.

   c. Write an equation that states that the total cost of the meat was $125.

   d. Solve the system of equations from parts a and c and find the amount of beef and pork purchased.

7. John passes through Savannah going north on I-95 at an average rate of 60 miles per hour. Thirty minutes later Susan also passes through Savannah headed north on I-95, but at an average rate of 75 miles per hour. Let x represent the number of hours after John passes through Savannah until Susan catches up with him and let y represent Susan's time after passing through Savannah.
   a. Write an equation that shows the relationship between John's and Susan's times.

   b. Represent the distances that John and Susan each travel since passing through Savannah.

   c. Write an equation that states that the distance John travels is equal to the distance that Susan travels.

   d. Solve the system of equations from parts a and c and give the number of hours after John passes through Savannah until Susan catches up with him.

8. In 4 hours, a boat can go 152 kilometers downstream and 104 kilometers upstream. Let x represent the speed of the boat in still water and y represent the speed of the current.
   a. Represent the speed of the boat upstream and the speed of the boat downstream in terms of x and y.

   b. Using the formula $d = rt$, write an equation indicating that the distance traveled downstream is 152 kilometers.

   c. Using the formula $d = rt$, write an equation indicating that the distance traveled upstream is 104 kilometers.

   d. Solve the system of equations from parts b and c and give the speed of the boat in still water and the speed of the current.

9. The president of a major corporation receives a huge bonus. She invests part of the money in bonds that pay 7% annual interest and the remainder in mutual funds that earn 9% the first year. The amount she invested in mutual funds was $250,000 more than twice the amount she invested in bonds. She earned $63,500 on her two investments. Let x represent the amount invested in bonds and y the amount invested in mutual funds.
   a. Complete the table.

Category	APR	Principal	Interest
Bonds	0.07		
Mutual Funds	0.09		

   b. Write an equation showing the relationship between the amount of money invested in bonds and in mutual funds.

   c. Write an equation showing that she earned a total interest of $63,500 on the two investments.

   d. Solve the system of equations from parts b and c and give the amount of money in each investment.

10. How much of a solution that is 30% alcohol must be added to a solution that is 60% alcohol to obtain 165 milliliters of a solution that is 40% alcohol? Let x represent the amount of 30% solution and y the amount of 60% solution to be mixed.
    a. Complete the table.

Solution	Concentration	Volume	Amount of Alcohol
30% solution	0.30		
60% solution	0.60		
40% solution	0.40		

    b. Write an equation that involves the volumes.

    c. Write an equation that states that the amount of alcohol in the 30% solution and in the 60% solution is the same as the amount of alcohol in the 40% mixture.

    d. Solve the system of equations from parts b and c and give the amounts of the 30% and the 60% solutions used.

For Exercises 11–20, solve the currency problems. See Examples 1 and 2.

11. If David has 19 bills in his wallet worth $125, all fives and tens, how many of each bill does he have?

12. During a game of Monopoly, Kari tells her opponent that she has 15 bills that total $3900, all $500 and $100 bills. How many of each bill does she have?

13. The Rolling Stones and Madonna were the top two moneymakers from June 2006 to June 2007 in the music industry. The Rolling Stones earned $16 million more than Madonna, and together they earned $160 million. What were the earnings of each of these performers? (*Source: Forbes*)

14. The all-time top-grossing American movie is *Titanic*, which grossed $139.8 million more than the second place *Star Wars Episode IV: A New Hope*. Together the two movies grossed $1061.8 million. How much revenue was brought in by each movie? (*Source: The World Almanac and Book of Facts*, 2008)

15. The cashier's office at a college has collected $7020 from part-time students for a total of 42 courses. If each credit hour costs $45, how many 3-hour courses and how many 5-hour courses were paid for?

16. One day a college bookstore collected $6260 for the sale of 44 beginning algebra and calculus books. If beginning algebra books sold for $130 each and calculus books sold for $160, how many of each type of book did the bookstore sell?

17. In 2007, the top-grossing movie was *Spider-Man 3* and *Shrek the Third* was second. If *Shrek the Third* took in $15.8 million less than *Spider-Man 3* and their combined revenues were $657.2 million, find the revenue earned by each movie. (*Source:* The Internet Movie Database)

18. In 2007, *Rush Hour 3* had the year's top DVD rentals and *The Bourne Ultimatum* was second. If *The Bourne Ultimatum* took in $1.5 million less than *Rush Hour 3* and their combined revenues were $140.7 million, find the total U.S. home-video rental fees for each movie. (*Source:* End of boredom.com)

19. Troy has a checking account and a savings account with a combined balance of $4200. The balance in his savings account is four times the balance in his checking account. What are the balances of both accounts?

20. Susan has two credit cards with a combined balance of $2700. If she owes three times more on one of the cards than the other, how much does she owe on each card?

For Exercises 21–28, solve the geometric problems. See Example 6, Section 5.2, and Example 6, Section 5.3.

21. The width of a tennis court (along the doubles lines) is 42 feet less than the length. If the perimeter is 228 feet, what are the dimensions?

22. The perimeter of a college basketball court is 288 feet. If the length of the court is 44 feet longer than it is wide, what are the dimensions?

23. The perimeter of the base of the Castillo at Chichen Itza (a pyramid in Mexico) is 300 feet. If the length is equal to the width, what are the dimensions?

24. The perimeter of the colonnade in the Lincoln Memorial is 612 feet. It is 70 feet longer than it is wide. What are the dimensions of the colonnade?

25. A support beam is attached to a wall and to the bottom of a ceiling truss. The angle made with the truss on one side of the support beam is three times the angle on the other side of the support beam. Find the two angles.

26. A laser beam is aimed at a flat photocell at an angle. The angle on one side of the beam is 25 degrees more than the angle on the other side of the beam. Find the two angles.

27. A security camera is placed in the corner of a room near the ceiling. The camera is aimed so that the angle made with the wall on one side of the camera is 15 degrees less than the angle made with the wall on the other side of the camera. Find the two angles.

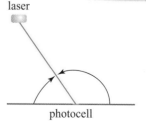

28. Used as a security device, a laser is placed a few inches above the floor on one wall of an office and is aimed at a photoreceptor in the corner of the office. At the receptor, the angle made by the beam with the wall on one side of the receptor is 12 degrees more than the angle made on the other side of the receptor. Find the two angles.

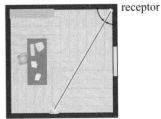

receptor

laser

For Exercises 29–36, solve the rate problems. See Examples 3 and 4.

29. John and Karen are traveling south in separate cars on the same interstate. Karen is traveling at 65 miles per hour; John, at 70 miles per hour. Karen passes Exit 38 at 1:15 P.M. John passes the same exit at 1:30 P.M. At what time will John catch up with Karen?

30. Diane and Curtis are running in the Cooper River Bridge Run. Diane passes the first checkpoint at 10:30 A.M; Curtis, at 11:00 A.M. While Curtis is pushing himself to run at 5 miles per hour, Diane is pacing herself at 4 miles per hour. When will Curtis catch up with Diane?

31. Sharon and Rae are riding bicycles on the same trail in the same direction. Sharon crosses a bridge at 12:18 P.M. Rae crosses the same bridge at 12:33 P.M. If Sharon is going 8 miles per hour and Rae is going 10 miles per hour, when will Rae catch up with Sharon?

32. Holly and Rebecca are jogging on the same road in the same direction. Holly passes a McDonald's at 8:10 A.M. Rebecca passes the same McDonald's at 8:20 A.M. If Holly is going 4 miles per hour and Rebecca is going 6 miles per hour, when will Rebecca catch up with Holly?

33. A boat travels 24 miles upstream in 3 hours. Going downstream, it can travel 36 miles in the same amount of time. Find the speed of the current and the speed of the boat in still water.

34. A boat travels 12 miles downstream in $1\frac{1}{2}$ hours. Going upstream, the boat takes 6 hours to return. Find the speed of the current and the speed of the boat in still water.

★ **35.** If a plane can travel 500 miles per hour with the wind and only 440 miles per hour against the wind, find the speed of the wind and the speed of the plane in still air.

★ **36.** If a plane can travel 200 miles per hour with the wind and only 120 miles per hour against the wind, find the speed of the wind and the speed of the plane in still air.

For Exercises 37–44, solve the investment problems. See Example 5.

37. Adam inherited $15,500. He invested part of the money at an APR of 4% and put the rest into a savings account at an APR of 3%. If his annual income from these investments is $585, how much did he invest at each rate?

38. Petunia invests some money in a plan that pays a 5% APR and invests twice as much money in a plan that pays an 8% APR. If the total interest from the two plans is $1029, how much was invested in each?

39. Ellyn owes $13,250 on two student loans. One loan has an APR of 5%; the other, an APR of 8%. If the total interest after one year is $727, what is the amount of each loan?

40. Enrique has two credit cards. One is a Visa at an APR of 14%, and the other is a department store card with an APR of 18%. If the total amount he owes is $1536 and the interest for one year is $245.36, what is the amount owed on each card?

41. Rhonda and Mike have two car loans totaling $16,250. One loan has an APR of 2.9%, and the other has an APR of 6%. If the total interest they paid after one year is $657.16, what is the amount owed on each car?

42. Corey borrowed a total of $250,000 in two loans. One loan is at an APR of 8%, and the other is at an APR of 18%. If he paid $23,000 in interest during the first year, how much was loaned at each rate?

43. Jason invested $20,000 in two accounts, one paying 7% and the other paying 9% in annual interest. If the total interest he earned for the year was $1550, how much did he invest in each?

44. George deposits some money into a bank account with an 8% APR, and Martha invests twice as much in a movie project that pays 21% annually. If the total interest they earn is $6000, how much did they invest in each?

For Exercises 45–52, solve the mixture problems. See Example 6.

45. How much of a 20% acid solution is to be mixed with 80 milliliters of a 15% acid solution to get an 18% solution?

46. How many gallons of milk containing 1% butterfat is to be mixed with 50 gallons of milk that is 6% butterfat to get milk that is 2% butterfat?

47. A pharmacist is preparing a 300-milliliter solution of 25% alcohol. If she has 15% and 45% alcohol solutions in stock, how much of each must she use?

48. During chemistry class, Dr. Ellis needs a solution of 20% HCl, but all he has is 10% and 25% HCl solutions. How many milliliters of each does he need to mix to get 150 milliliters of the 20% HCl solution?

49. A coffee shop is considering a new mixture of coffee beans. It will be created from Italian Roast beans costing $9.80 per pound and the Gold Coast Blend (GCB) beans costing $11.20 per pound. The two types of beans will be mixed to create 20 pounds of coffee to be sold for $10.43 per pound. How much of each type of bean should be used?

50. A store is selling two types of jelly beans. One kind is $1.20 per pound, and the other is $0.90 per pound. If the merchant wants to sell a mixture at $1.11 per pound, how many pounds of each must be used to make 30 pounds?

51. Cromers P-Nuts sells several types of peanuts and cashews. The Virginia Fancy Peanuts cost $2.59 per pound, and the California Salted Cashews cost $12.56 per pound. If a caterer is planning to serve mixed nuts, how much of each type would he need to make 7 pounds of a mixture that costs $5.44 per pound?

52. A candy store owner mixes candy worth $2.50 per pound with candy worth $3.50 per pound. How many pounds of each does he need if he wants 10 pounds of the mixture worth $2.90 per pound?

Review Exercises

Exercises 1–6 EQUATIONS AND INEQUALITIES

[2.6] *For Exercises 1–3, solve.*

1. $-6x \geq 30$

2. $x > 2x - 3$

3. $2x + 3 \leq 10 - 5x$

[4.6] *For Exercises 4 and 5, graph.*

4. $y > 3x + 2$

5. $x + y \leq -3$

[5.2] **6.** Use substitution to solve: $\begin{cases} x = 3 - y \\ 5x + 3y = 5 \end{cases}$

5.5 Solving Systems of Linear Inequalities

Objectives

① Graph the solution set of a system of linear inequalities.

② Solve applications involving a system of linear inequalities.

Objective ① **Graph the solution set of a system of linear inequalities.** In Section 2.6, we learned to solve linear inequalities in one variable. In Section 4.6, we learned how to graph linear inequalities in two variables. In this section, we develop a graphical approach to solving *systems of linear inequalities* in two variables.

Consider the system of linear inequalities $\begin{cases} x + 2y < 6 \\ 2x - y \geq 2 \end{cases}$. First, let's graph each inequality separately.

Note: Recall from Section 4.6 that we use a dashed line with $<$ or $>$ and that points on a dashed line are not in the solution set.

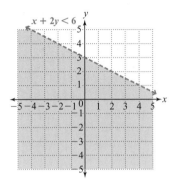

In the graph of $x + 2y < 6$, the shaded region contains all ordered pairs that make $x + 2y < 6$ true.

Note: Recall from Section 4.6 that we determine which side of the line to shade by choosing an ordered pair not on the line and checking to see if it makes the inequality true. If it does, we shade on the side of the line containing that ordered pair. If it does not, we shade the other side of the line.

Note: Recall from Section 4.6 that we use a solid line with $\leq$ or $\geq$ and that points on a solid line are part of the solution set.

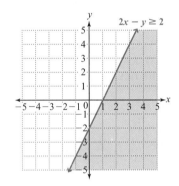

In the graph of $2x - y \geq 2$, ordered pairs in the shaded region and on the line itself make $2x - y \geq 2$ true.

Now we put the two graphs together on the same grid to determine the solution set for the system. A solution for a system of inequalities is an ordered pair that makes every inequality in the system true.

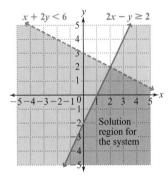

On our graph containing both $x + 2y < 6$ and $2x - y \geq 2$, the region where the shading overlaps contains ordered pairs that make both inequalities true. So all ordered pairs in this region are in the solution set for the system. Also, ordered pairs on the solid line for $2x - y \geq 2$ where it touches the region of overlap are in the solution set for the system, whereas ordered pairs on the dashed line for $x + 2y < 6$ are not. This implies the following procedure.

Procedure **Solving a System of Linear Inequalities in Two Variables**

To solve a system of linear inequalities in two variables, graph all of the inequalities on the same grid. The solution set for the system contains all ordered pairs in the region where the inequalities' solution sets overlap along with ordered pairs on the portion of any solid line that touches the region of overlap.

To check, we can select a point in the solution region such as $(3, 0)$ and verify that it makes both inequalities true.

First inequality: **Second inequality:**

$$x + 2y < 6 \qquad\qquad 2x - y \geq 2$$

$$3 + 2(0) \overset{?}{<} 6 \qquad\qquad 2(3) - 0 \overset{?}{\geq} 2 \qquad \text{Replace } x \text{ with 3 and } y \text{ with 0 in both inequalities.}$$

$$3 < 6 \quad \text{True.} \qquad\qquad 6 \geq 2 \quad \text{True.}$$

Because $(3, 0)$ makes both inequalities true, it is a solution to the system. Although we selected only one ordered pair in the solution region, remember that *every* ordered pair in that region is a solution.

Example 1 Graph the solution set for the system of inequalities.

a. $\begin{cases} x + 3y > 6 \\ y < 2x - 1 \end{cases}$

SOLUTION Graph the inequalities on the same grid. Because both lines are dashed, the solution set for the system contains only those ordered pairs in the region of overlap (the purple shaded region).

Note: Ordered pairs on dashed lines are not part of the solution region.

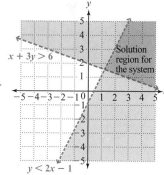

b. $\begin{cases} y < 2 \\ 5x - y \leq 4 \end{cases}$

SOLUTION Graph the inequalities on the same grid. The solution set for this system contains all ordered pairs in the region of overlap (purple shaded region) together with all ordered pairs on the portion of the solid blue line that touches the purple shaded region.

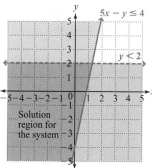

Connection Remember that for a system of two linear equations, a solution is a point of intersection of the graphs of the two equations. Similarly, for a system of linear inequalities, the solution set is a region of intersection of the graphs of the two inequalities.

Answers Your Turn 1

a.

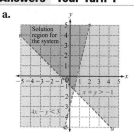

b. Solution region for the system (including the portion of the red line touching this region)

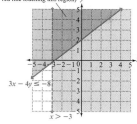

Your Turn 1 Graph the solution set for the system of inequalities.

a. $\begin{cases} x + y > -1 \\ 4x - y < 3 \end{cases}$

b. $\begin{cases} x > -3 \\ 3x - 4y \leq -8 \end{cases}$

Inconsistent Systems

Some systems of linear inequalities have no solution. We say that these systems are inconsistent.

Example 2   Graph the solution set for the system of inequalities.

$$\begin{cases} x - 4y \geq 8 \\ y \geq \dfrac{1}{4}x + 3 \end{cases}$$

SOLUTION   Graph the inequalities on the same grid.

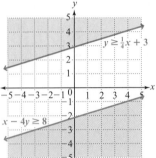

> **Connection**   Notice that the lines in Example 2 are parallel. Recall that in a system of linear equations, if the lines are parallel, the system is inconsistent because there is no point of intersection. With linear inequalities, the lines must be parallel *and* the shaded regions must not overlap for the system to be inconsistent.

The lines appear to be parallel. We can verify by writing $x - 4y \geq 8$ in slope–intercept form and comparing the slopes and y-intercepts.

$$x - 4y \geq 8$$
$$-4y \geq -x + 8 \quad \text{Subtract } x \text{ from both sides.}$$
$$y \leq \frac{1}{4}x - 2 \quad \begin{array}{l}\text{Divide both sides by } -4, \\ \text{which changes the direc-} \\ \text{tion of the inequality symbol.}\end{array}$$

The slopes are equal, and the y-intercepts differ; so the lines are in fact parallel. Because the lines are parallel and the shaded regions do not overlap, there is no solution region for this system. The system is inconsistent.

Objective ❷ Solve applications involving a system of linear inequalities.

Example 3   A home interiors store stocks two different-size prints from an artist. The manager wants to purchase at least 15 prints for the store. The artist sells the smaller print for $20 and the larger print for $30. The manager cannot spend more than $800 for the prints. Write a system of inequalities that describes the manager's order; then solve the system by graphing.

Understand   We must translate to a system of inequalities, then solve the system.

Plan and Execute   Let x represent the number of small prints and y represent the number of large prints ordered.

Relationship 1:   The manager wants to purchase at least 15 prints.

> The words *at least* indicate that the combined number of prints is to be greater than or equal to 15; so $x + y \geq 15$.

Relationship 2:   The manager cannot spend more than $800.

> Because the small prints cost $20 each, $20x$ describes the amount spent on small prints. Similarly, $30y$ describes the amount spent on large prints. The total cannot exceed $800; so $20x + 30y \leq 800$.

$$\text{Our system: } \begin{cases} x + y \geq 15 \\ 20x + 30y \leq 800 \end{cases}$$

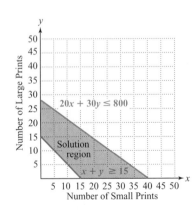

Answer   (See the graph.) Because the manager cannot order a negative number of prints, the solution set is confined to quadrant I. Any ordered pair in the solution region or on a portion of either line touching the solution region is a solution for the system. However, assuming that only whole prints can be purchased, only ordered pairs of whole numbers in the solution set, such as $(10, 15)$ and $(25, 10)$, are realistic.

Answer   Your Turn 3

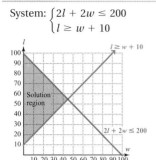

System: $\begin{cases} 2l + 2w \le 200 \\ l \ge w + 10 \end{cases}$

Check We will check one ordered pair in the solution region. The ordered pair (10, 15) indicates an order of 10 small prints and 15 large prints so that 25 prints are ordered (which is more than 15) that cost a total of $20(10) + 30(15) = \$650$ (which is less than \$800).

Your Turn 3 In designing a building with a rectangular base, an architect decides that the perimeter cannot exceed 200 feet. Also, the length will be at least 10 feet more than the width. Write a system of inequalities that describes the design requirements; then solve the system graphically.

Note: Because length and width must be positive numbers, the solution region is confined to quadrant 1.

5.5 Exercises   For Extra Help *MyMathLab*

Note: Exercises marked with a ★ represent challenging exercises.

1. How would you check a possible solution to a system of inequalities?

2. How do you determine the region that is the solution set for a system of linear inequalities?

3. What circumstances would cause a system of inequalities to have no solution?

4. Write a system of linear inequalities whose solution set is the entire first quadrant.

5. Write a system of linear inequalities whose solution set is the entire third quadrant.

6. Write a system of linear inequalities whose solution set is the entire fourth quadrant.

For Exercises 7–30, graph the solution set for the system of inequalities. See Examples 1 and 2.

7. $\begin{cases} x + y > 4 \\ x - y < 6 \end{cases}$

8. $\begin{cases} x - y > 2 \\ x + y > 4 \end{cases}$

9. $\begin{cases} x + y < -2 \\ x - y > 6 \end{cases}$

10. $\begin{cases} x - y < -5 \\ x + y < 3 \end{cases}$

11. $\begin{cases} 2x + y \ge -1 \\ x - y \le 5 \end{cases}$

12. $\begin{cases} x - y < -5 \\ 2x - y < -7 \end{cases}$

13. $\begin{cases} x + y < 3 \\ x - 2y \ge 2 \end{cases}$

14. $\begin{cases} x + 3y \le 6 \\ x - y > 5 \end{cases}$

15. $\begin{cases} 2x + y > 9 \\ y > \dfrac{1}{4}x \end{cases}$

16. $\begin{cases} 3x + 4y \le -9 \\ y < 3x \end{cases}$

17. $\begin{cases} y < x \\ y > -x + 1 \end{cases}$

18. $\begin{cases} y < x \\ y < 2x - 3 \end{cases}$

19. $\begin{cases} 3x > -4y \\ 3x + 4y \le -8 \end{cases}$

20. $\begin{cases} 2y - x \ge 6 \\ 2x - 4y \ge 5 \end{cases}$

21. $\begin{cases} y < 3x + 1 \\ 3x - y \le 4 \end{cases}$

22. $\begin{cases} x - 2y > 3 \\ 2x - 4y \le 20 \end{cases}$

23. $\begin{cases} x > 2 \\ 3x - 2y > 6 \end{cases}$

24. $\begin{cases} y \le -1 \\ x + 2y > 3 \end{cases}$

25. $\begin{cases} x \ge -3y \\ y \ge 2x \end{cases}$

26. $\begin{cases} x > 2y \\ x + y > 6 \end{cases}$

27. $\begin{cases} y > 2 \\ x < -1 \end{cases}$

28. $\begin{cases} y \ge -1 \\ x < 2 \end{cases}$

29. $\begin{cases} 2x + y \ge 1 \\ x - y > -1 \\ x > 2 \end{cases}$

30. $\begin{cases} x + y \ge 1 \\ x - y \ge -5 \\ y > -2 \end{cases}$

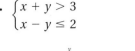
31. $\begin{cases} x + y > 3 \\ x - y \le 2 \end{cases}$

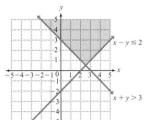

32. $\begin{cases} y < \dfrac{1}{2}x \\ y \le -3x + 1 \end{cases}$

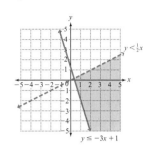

33. $\begin{cases} x < 3 \\ y > 4 \end{cases}$

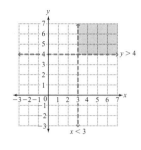

34. $\begin{cases} x > 3y \\ x + y \ge 2 \end{cases}$

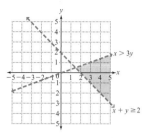

For Exercises 35–38, solve. see Example 3.

35. To be admitted to a certain college, prospective students must have a combined verbal and math score of at least 1100 on the SAT. To be admitted to the engineering program at this college, students must also have a math score of at least 500.

  a. Write two inequalities describing the requirements to be admitted to the college of engineering.

  b. The maximum score on each part of the SAT is 800. Write two inequalities describing these maximum scores.

  c. Solve the system of inequalities described by the four inequalities in parts a and b by graphing. Let the horizontal axis be verbal and the vertical axis be math.

  d. Give two different combinations of verbal and math scores that satisfy the admission requirements to the college of engineering.

36. An architect is designing a rectangular platform for an auditorium. The client wants the length of the platform to be at least twice the width, and the perimeter must not exceed 200 feet.

  a. Write a system of inequalities that describes the specifications for the platform.

  b. Solve the system by graphing. Let the horizontal axis be the width and the vertical axis be the length.

  c. Give two different combinations of length and width that satisfy the requirements of the client.

★**37.** A company sells two versions of software: the regular version and the deluxe version. The regular version sells for $15.95, and the deluxe version sells for $20.95. The company gives a bonus to any salesperson who sells at least 100 units and has at least $1700 in total sales.

    a. Write a system of inequalities that describes the requirements a salesperson must meet to receive the bonus.

    b. Solve the system by graphing. Let the horizontal axis be the regular version and the vertical axis be the deluxe version.

    c. Give two different combinations of number of units sold that meet the requirements for receiving the bonus.

★**38.** An investor is trying to decide how much to invest in two different stocks. She has decided that she will invest at most $5000. The "safe" stock is very stable and is projected to return about 5% over the next year. The other stock is more risky, but it could return about 9% if the company grows as expected. She wants her total dividend at the end of the year to be at least $300.

    a. Write a system of inequalities that describes the amount invested in the two stocks and the return on that investment.

    b. Solve the system by graphing. Let the horizontal axis be the "safe" stock and the vertical be the "risky" stock.

    c. Give two different combinations of investment amounts that return at least $300.

Collaborative Exercises   Maximizing the Profit

Linear programming is an area of mathematics that solves problems like the one described below. One of the fundamental elements of a linear programming model is the constraint. A constraint is simply an inequality that describes some limited resource required in the problem. For example, suppose you make two types of bicycles, style A and style B. Also suppose that style A requires 40 minutes of welding for each bicycle and style B requires 25 minutes of welding. If you have only 600 minutes of welding time available, the welding-time constraint would be $40A + 25B \le 600$, *where A represents the number of style A produced and B represents the number of style B produced.*

An aspiring artists' group is preparing for a fund-raising sale. The artists design and construct two types of antique-reproduction tables: a semicircular foyer table and a side table. The materials cost $100 for each foyer table and $200 for each side table. Both types of table require 2 hours of cutting and carving. The foyer table requires 3.5 hours to assemble, sand, and finish, and the side table requires 1.75 hours to assemble, sand, and finish. The group has $3300 to purchase all materials; 40 hours available for cutting and carving; and 63 hours available for assembling, sanding, and finishing. For each foyer table they sell, their profit will be $350. For each side table, their profit will be $215. How many of each table should they make to maximize their profit?

    1. For the artists' fund-raising problem, there are three basic constraints: (1) cost of materials, (2) cutting and carving, and (3) assembling/sanding/finishing. For each of these constraints, write an inequality similar to the one in the bicycle example. Let x represent the number of foyer tables and y the number of side tables.

    2. Because x and y represent numbers of tables, they cannot represent negative values. Write inequalities for these two additional constraints.

    3. Now graph the system of inequalities described by these five constraints.

4. The region that is the solution of the system of inequalities is called the *feasible region*, and it includes all points that satisfy the system. The goal is to find the optimal solution, which, according to linear programming, is one of the corner points of the feasible region. There should be five corner points: one at the origin, one on each axis, and two more in quadrant I. Determine the coordinates of each corner point. Note that a corner point lying on an axis is the x- or y-intercept for the line passing through that point. Because two lines intersect to form a corner point that is not on an axis, make a system out of their two equations; then solve the system to find their point of intersection.

5. Next, develop an algebraic expression describing the profit gained for selling x of the foyer tables and y of the side tables. This is called the *objective function*.

6. Using the objective function, test each of the points found in step 4 to see which one yields the maximum profit.

7. Using your solution, answer the following questions.
   a. How many of each type of table should the group produce?

   b. How many hours will be spent in each phase of production?

   c. How much money will they need to purchase the materials?

   d. What amount of profit will the group receive?

Review Exercises

Exercises 1–6 EXPRESSIONS

[1.5] *For Exercises 1 and 2, evaluate.*

1. 3^2

2. -10^2

[1.4] *Multiply.*

3. $(-1.46)(1000)$

4. $(-3.21)(-4.6)$

[1.7] *Simplify.*

5. $3.2x^2 + 6.3x^2$

6. $3x - 2(2x + 7)$

Chapter 5 Summary

Defined Terms

Section 5.1

System of equations
(p. 336)

Solution for a system of
equations (p. 336)

Consistent system of
equations (p. 340)

Inconsistent system of
equations (p. 340)

Procedures, Rules, and Key Examples

Procedures/Rules	Key Examples

SECTION 5.1 Solving Systems of Linear Equations Graphically

To verify or check a solution to a system of equations,
1. Replace each variable in each equation with its corresponding value.
2. Verify that each equation is true.

Example 1: Determine whether the ordered pair is a solution to the system.

$$\begin{cases} 2x + 3y = 8 \\ y = 5x - 3 \end{cases}$$

a. $(-2, 4)$

$$
\begin{array}{ll}
2x + 3y = 8 & y = 5x - 3 \\
2(-2) + 3(4) \stackrel{?}{=} 8 & 4 \stackrel{?}{=} 5(-2) - 3 \\
-4 + 12 \stackrel{?}{=} 8 & 4 \stackrel{?}{=} -10 - 3 \\
8 = 8 & 4 \neq -13
\end{array}
$$

$(-2, 4)$ is not a solution to the system.

b. $(1, 2)$

$$
\begin{array}{ll}
2x + 3y = 8 & y = 5x - 3 \\
2(1) + 3(2) \stackrel{?}{=} 8 & 2 \stackrel{?}{=} 5(1) - 3 \\
2 + 6 \stackrel{?}{=} 8 & 2 \stackrel{?}{=} 5 - 3 \\
8 = 8 & 2 = 2
\end{array}
$$

$(1, 2)$ is a solution to the system.

To solve a system of linear equations graphically,
1. Graph each equation.
   a. If the lines intersect at a single point, the coordinates of that point form the solution.
   b. If the lines are parallel, there is no solution.
   c. If the lines are identical, there are an infinite number of solutions, which are the coordinates of all points on that line.
2. Check your solution.

Example 2: Solve the system of equations graphically.

$$\begin{cases} 2x + 3y = 8 \\ y = 5x - 3 \end{cases}$$

Graph the two equations; then find the point of intersection.

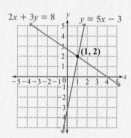

The two lines intersect at the point $(1, 2)$, so $(1, 2)$ is the solution to the system.

(continued)

Procedures/Rules	Key Examples

SECTION 5.1 Solving Systems of Linear Equations Graphically (*continued*)

To classify a system of two linear equations in two unknowns, write the equations in slope–intercept form and compare the slopes and y-intercepts.

1. **Consistent system with independent equations:** The system has a single solution at the point of intersection. The graphs are different. They have different slopes.
2. **Consistent system with dependent equations:** The system has an infinite number of solutions. The graphs are identical. They have the same slope and same y-intercept.
3. **Inconsistent system:** The system has no solution. The graphs are parallel lines. They have the same slope but different y-intercepts.

Example 3: Determine whether the system is consistent with independent equations, consistent with dependent equations, or inconsistent and discuss the number of solutions for the system.

$$\begin{cases} x + 3y = 6 \\ 2x + 6y = -5 \end{cases}$$

Solution: Write the equations in slope–intercept form.

$$\begin{array}{ll} x + 3y = 6 & 2x + 6y = -5 \\ 3y = -x + 6 & 6y = -2x - 5 \\ y = -\dfrac{1}{3}x + 2 & y = -\dfrac{1}{3}x - \dfrac{5}{6} \end{array}$$

Because the slopes are the same $\left(\text{both } -\dfrac{1}{3}\right)$ but the y-intercepts are different, this system is inconsistent. This means there is no solution to the system.

SECTION 5.2 Solving Systems of Linear Equations by Substitution

To find the solution of a system of two linear equations using the substitution method,

1. Isolate one of the variables in one of the equations.
2. In the other equation, substitute the expression you found in step 1 for that variable.
3. Solve this new equation. (It will now have only one variable.)
4. Using one of the equations containing both variables, substitute the value you found in step 3 for that variable and solve for the value of the other variable.
5. Check the solution in the original equations.

Example 1: Solve the system using the substitution method.

$$\begin{cases} y = 4x - 1 \\ 5x - y = 2 \end{cases}$$

Because y is already isolated in the first equation, we substitute $4x - 1$ in place of y in the second equation.

$$\begin{array}{ll} 5x - y = 2 & \\ 5x - (4x - 1) = 2 & \text{Substitute } 4x - 1 \text{ for } y. \\ 5x - 4x + 1 = 2 & \text{Distribute.} \\ x + 1 = 2 & \text{Combine like terms.} \\ x = 1 & \text{Subtract 1 from both sides.} \end{array}$$

Now solve for the value of y by substituting 1 in place of x in one of the equations. We will use $y = 4x - 1$.

$$\begin{array}{ll} y = 4x - 1 & \\ y = 4(1) - 1 & \text{Substitute 1 for } x. \\ y = 4 - 1 & \\ y = 3 & \end{array}$$

Solution: $(1, 3)$

SECTION 5.3 Solving Systems of Linear Equations by Elimination

To solve a system of two linear equations using the elimination method,

1. Write the equations in standard form ($Ax + By = C$).
2. Use the multiplication principle of equality to clear fractions or decimals (optional).

Example 1: Solve the system using the elimination method.

$$\begin{cases} \dfrac{1}{4}x + \dfrac{1}{2}y = -1 \\ 4x + y = 5 \end{cases}$$

(*continued*)

Procedures/Rules	Key Examples

SECTION 5.3 Solving Systems of Linear Equations by Elimination (*continued*)

3. If necessary, multiply one or both equations by a number (or numbers) so that they have a pair of terms that are additive inverses.
4. Add the equations. The result should be an equation in terms of one variable.
5. Solve the equation from step 4 for the value of that variable.
6. Using an equation containing both variables, substitute the value you found in step 5 for the corresponding variable and solve for the value of the other variable.
7. Check your solution in the original equations.

Solution: First, we clear the fractions in the first equation.

$$4\left(\frac{1}{4}x + \frac{1}{2}y\right) = 4(-1)$$
$$x + 2y = -4$$

Now we will eliminate y by multiplying the second equation by -2, then add the two equations.

$$
\begin{array}{l}
x + 2y = -4 \\
4x + y = 5
\end{array}
\xrightarrow{\text{Multiply by } -2.}
\begin{array}{l}
x + 2y = -4 \\
-8x - 2y = -10
\end{array}
$$

Add equations. $-7x + 0 = -14$

Divide both sides by -7. $\begin{array}{r} -7x = -14 \\ x = 2 \end{array}$

Now solve for the value of y by substituting 2 for x in one of the equations. We will use $4x + y = 5$.

$$
\begin{array}{ll}
4x + y = 5 & \\
4(2) + y = 5 & \text{Substitute 2 for } x. \\
8 + y = 5 & \text{Multiply.} \\
y = -3 & \text{Subtract 8 from both sides.}
\end{array}
$$

Solution: $(2, -3)$

SECTION 5.4 More Applications with Systems of Equations

To solve problems involving two unknowns using a system of equations,
1. Select two variables to represent the two unknowns.
2. Using the two relationships in the problem, write a system of two equations in terms of the two variables.
3. Solve the system.

Note: If there is an isolated variable in one of the equations, the substitution method is the better method. Otherwise the elimination method is best.

Example 1: A vendor at a sporting event sells small and large drinks. Small drinks sell for \$2, and large drinks sell for \$3. After the event, the vendor finds that 2400 drinks were sold for a total of \$5400. How many of each size were sold?

Relationship 1: The total number of drinks sold is 2400.

Translation: $x + y = 2400$

Relationship 2: The total revenue is \$5400.

Translation: $2x + 3y = 5400$

System: $\begin{cases} x + y = 2400 \\ 2x + 3y = 5400 \end{cases}$

Because there are no isolated variables, we will use the elimination method. We will eliminate y by multiplying the first equation by -3, then adding the two equations.

$$
\begin{array}{l}
x + y = 2400 \\
2x + 3y = 5400
\end{array}
\xrightarrow{\text{Multiply by } -3.}
\begin{array}{l}
-3x - 3y = -7200 \\
2x + 3y = 5400
\end{array}
$$

Add equations. $-x + 0 = -1800$

Divide both sides by -1. $x = 1800$

(*continued*)

Procedures/Rules	Key Examples

SECTION 5.4 More Applications with Systems of Equations (*continued*)

	We can now find the value of y by substituting 1800 into one of the equations.

$$x + y = 2400$$
$$1800 + y = 2400 \qquad \text{Substitute 1800 for } x.$$
$$y = 600 \qquad \text{Subtract 1800 from both sides.}$$

Answer: The vendor sold 1800 small drinks and 600 large drinks.

SECTION 5.5 Solving Systems of Linear Inequalities

To solve a system of linear inequalities in two variables, graph all of the inequalities on the same grid. The solution set for the system contains all ordered pairs in the region where the inequalities' solution sets overlap along with ordered pairs on the portion of any solid line that touches the region of overlap.

Example 1: Solve the system of inequalities.

$$\begin{cases} x + y > -2 \\ y \le 3x - 1 \end{cases}$$

Solution: Graph both inequalities. The region where the shading overlaps is the solution set for the system.

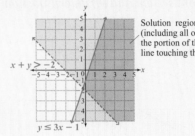

Solution region for the system (including all ordered pairs on the portion of the solid blue line touching this region).

Chapter 5 Review Exercises

For Exercises 1–5, answer true or false.

[5.1] **1.** If an ordered set of numbers satisfies one of the equations in a system of equations, it is a solution of the system.

[5.4] **2.** Every system of two equations in two unknowns has one solution or no solution.

[5.1] **3.** $(3, 2)$ is the solution for the system $\begin{cases} x - y = 1 \\ 3x + y = 11 \end{cases}$.

[5.1] **4.** When solving a system of two equations in two unknowns graphically, if the system has only one solution, it will be the coordinates of the point of intersection of the graphs of the equations.

[5.2] **5.** When using the substitution method to solve a system of two equations in two unknowns, we isolate the same variable in both equations.

For Exercises 6–10, complete the rule.

[5.1] **6.** To verify or check a solution to a system of equations,
  a. _____ each variable in each equation with its corresponding value.
  b. Verify that each equation is true.

[5.1] **7.** Systems of equations that have at least one solution are _____.

8. In an inconsistent system of two linear equations in two unknowns, the graphs of the equations are _____ lines; so the system has no solution.

9. Independent linear equations in two unknowns have _____ graphs. Dependent linear equations in two unknowns have _____ graphs.

10. To solve a system of two linear inequalities,
   a. Graph each inequality.
   b. The region of overlap for the two graphs is the _____ for the system of inequalities.

Exercises 11–54   EQUATIONS AND INEQUALITIES

[5.1] *For Exercises 11 and 12, determine whether the ordered pair is a solution to the given system of equations.*

11. $(4, 3)$; $\begin{cases} x - y = 1 \\ -x + y = -1 \end{cases}$

12. $(1, 1)$; $\begin{cases} 2x - y = 7 \\ x + y = 8 \end{cases}$

[5.1] *For Exercises 13 and 14, (a) determine whether the graph shows a consistent system with independent equations, a consistent system with dependent equations, or an inconsistent system and (b) determine how many solutions the system has.*

13.

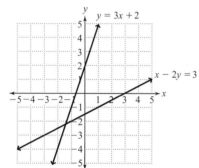

14.

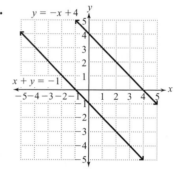

[5.1] *For Exercises 15 and 16, (a) determine whether the system of equations is consistent with independent equations, consistent with dependent equations, or inconsistent and (b) determine how many solutions the system has.*

15. $\begin{cases} x - 3y = 10 \\ 2x + 3y = 5 \end{cases}$

16. $\begin{cases} x - 5y = 10 \\ 2x - 10y = 10 \end{cases}$

[5.1] *For Exercises 17–22, solve the system graphically.*

17. $\begin{cases} y - x = 6 \\ 3x + 4y = -18 \end{cases}$

18. $\begin{cases} 3x - 5y = 1 \\ x + 2y = 4 \end{cases}$

19. $\begin{cases} 2x + y = 0 \\ 3x - y = 5 \end{cases}$

20. $\begin{cases} x - y = 3 \\ 2x + y = 12 \end{cases}$

21. $\begin{cases} 4x = y \\ 3x + y = -7 \end{cases}$

22. $\begin{cases} x - y = 4 \\ 2x + 3y = 3 \end{cases}$

[5.2] *For Exercises 23–30, solve the system of equations using substitution. Note that some systems may be inconsistent or consistent with dependent equations.*

23. $\begin{cases} x + y = 5 \\ x - y = 1 \end{cases}$

24. $\begin{cases} 3 + y = 12 \\ 3x = 12 \end{cases}$

25. $\begin{cases} 2x - y = 8 \\ y = -2 \end{cases}$

26. $\begin{cases} 3x + 10y = 2 \\ x - 2y = 6 \end{cases}$

27. $\begin{cases} y = -2x + 1 \\ 4x + 2y = 7 \end{cases}$

28. $\begin{cases} 2x + 5y = 8 \\ x - 10y = 9 \end{cases}$

29. $\begin{cases} y = 6 \\ x + y = 10 \end{cases}$

30. $\begin{cases} x - y = 5 \\ x + y = 7 \end{cases}$

[5.3] *For Exercises 31–38, solve the system of equations using elimination. Note that some systems may be inconsistent or consistent with dependent equations.*

31. $\begin{cases} x + y = 4 \\ x - y = -2 \end{cases}$

32. $\begin{cases} x + 3y = 6 \\ -x + 2y = -1 \end{cases}$

33. $\begin{cases} 3x - y = 1 \\ x - y = 5 \end{cases}$

34. $\begin{cases} 3x - 4y = 1 \\ x + y = -2 \end{cases}$

35. $\begin{cases} 3x + 2y = 4 \\ 2x - 3y = 7 \end{cases}$

36. $\begin{cases} -2x - 5y = 1 \\ 2x + 3y = 9 \end{cases}$

37. $\begin{cases} 0.25x + 0.75y = 4 \\ x + 3y = 16 \end{cases}$

38. $\begin{cases} \dfrac{1}{5}x - \dfrac{1}{3}y = 2 \\ x + y = 2 \end{cases}$

[5.5] *For Exercises 39–44, graph the solution set for the system of inequalities.*

39. $\begin{cases} x + y > -5 \\ x - y \geq -1 \end{cases}$

40. $\begin{cases} 3x + y < 2 \\ 2x - y \geq 3 \end{cases}$

41. $\begin{cases} 2x - y \leq 6 \\ 3x + y > 4 \end{cases}$

42. $\begin{cases} 2x + y \geq -4 \\ -3x + y \geq -1 \end{cases}$

43. $\begin{cases} 2x + y > -1 \\ 3x + y \leq 4 \end{cases}$

44. $\begin{cases} -3x + 4y < 12 \\ 2x - y \leq -3 \end{cases}$

[5.4] *For Exercises 45–54, solve.*

45. The sum of two numbers is 16. Their difference is 4. What are the two numbers?

46. If two angles are complementary, find the measures of each if the smaller angle is 10 degrees less than one-third of the larger angle.

47. The length of a rectangular plot of land is 25 feet more than twice the width. If the perimeter is 530 feet, what are the dimensions of the plot?

48. A movie theater charges $7.00 for adults and $4.25 for children. During a recent showing of *Ratatouille*, 139 tickets were sold for a total of $720. How many of each ticket were sold?

49. FedEx Office charges 8¢ per copy for black-and-white documents and 49¢ per copy for color documents. A document containing 50 pages that are a mixture of black-and-white and color pages is copied, costing $6.05. How many black-and-white pages and how many color pages does the document contain? (*Source:* FedEx Office)

50. Duke is beginning to invest for his son's college fund. He put part of a $5000 account in a fund that yields 7% and the rest in a fund that yields 9%. If he earned $380 in interest during the year, how much was invested in each fund?

51. Yolanda and Dee are traveling west in separate cars on the same highway. Yolanda is traveling at 60 miles per hour; Dee, at 70 miles per hour. Yolanda passes the I-95 exit at 4:00 P.M. Dee passes the same exit at 4:15 P.M. At what time will Dee catch up with Yolanda?

52. Suppose a small plane is traveling with the jet stream and can travel 450 miles in 3 hours. When the plane returns, it takes 5 hours to travel the same distance against the jet stream. What is the speed of the plane in still air?

53. How much of a 10% saline solution is needed to combine with a 60% saline solution to obtain 50 milliliters of a 30% saline solution?

[5.5] 54. A real estate course is offered in a continuing education program at a two-year college. The course is limited to 16 students. Also, to run the course, the college must receive at least $2750 in tuition. The college charges $250 per student if he or she lives in the same county as the college and $400 if he or she lives in another county (or out-of-state).

   a. Write a system of inequalities that describes the number of in-county students enrolled in the course versus the number of out-of-county students and the amount they paid.

   b. Solve the system.

Chapter 5 Practice Test

For Extra Help

CHAPTER
Test Prep
VIDEOS

Step-by-step test solutions are found on the Chapter Test Prep Videos available via the Video Resources on DVD, in *MyMathLab* , and on You Tube (search "CarsonElemAlg" and click on "Channels").

For Exercises 1 and 2, determine whether the given ordered pair is a solution to the given system of equations.

1. $(-1, 3)$; $\begin{cases} 3x + 2y = 3 \\ 4x - y = -7 \end{cases}$

2. $(-4, 3)$; $\begin{cases} 2x + 3y = -1 \\ x + 2y = -2 \end{cases}$

For Exercises 3–12, solve the system of equations using substitution or elimination. Note that some systems may be inconsistent or consistent with dependent equations.

3. $\begin{cases} 2x - 3y = 5 \\ x + y = 0 \end{cases}$

4. $\begin{cases} -2x + 9y = 4 \\ 4x - 3y = -3 \end{cases}$

5. $\begin{cases} 3x + 2y = 5 \\ 2x + 3y = 20 \end{cases}$

6. $\begin{cases} x + y = 3 \\ x + y = -1 \end{cases}$

7. $\begin{cases} 5x + 3y = 4 \\ -15x - 9y = -12 \end{cases}$

8. $\begin{cases} x + 3y = 1 \\ -2x - 7y = -3 \end{cases}$

9. $\begin{cases} 3x - 2y = 1 \\ -2x + 3y = -4 \end{cases}$

10. $\begin{cases} 6x + 3y = -45 \\ x + y = -7 \end{cases}$

11. $\begin{cases} x + y = 12 \\ x - y = 4 \end{cases}$

12. $\begin{cases} 2x + 2y = 44 \\ y + 2 = x - 4 \end{cases}$

For Exercises 13 and 14, graph the solution set for the system of inequalities.

13. $\begin{cases} 3y > 2x - 1 \\ x + 2y < -2 \end{cases}$

14. $\begin{cases} 2x + y < 2 \\ 6x + 3y > 2 \end{cases}$

For Exercises 15–20, solve.

15. The sum of two numbers is 5, and the difference is 3. Find the numbers.

16. Excedrin surveyed workers in various professions who get headaches at least once a year on the job. 9 more accountants than waiters/waitresses in the survey got a headache. If the combined number of accountants and waiters/waitresses that got a headache was 163, how many people in each of those professions got a headache? (*Source: The State* newspaper)

17. When asked what the Internet most resembled, three times as many people said a library as opposed to a highway. If 240 people were polled, how many people considered the Internet to be a library? (*Source: bLINK* magazine)

18. A boat traveling with the current took 3 hours to go 30 miles. The same boat went 12 miles in 3 hours against the current. What is the rate of the boat in still water?

19. A small business sold $1482 in charms for charm bracelets during the Christmas rush. Some were priced at $6 and some at $10. If 171 charms were sold, how many of each type were sold?

20. Janice invested $12,000 in two funds. One of the funds returned 6% interest and the other returned 8% interest after one year. If the total interest for the year was $880, how much did she invest in each fund?

Chapters 1–5 Cumulative Review Exercises

For Exercises 1–4, answer true or false.

[1.1] **1.** $2x + 3 = 7$ is an expression.

[1.6] **2.** Six less than 5 translates to $6 - 5$.

[1.3] **3.** $6(3 + 4) = 6 \cdot 3 + 6 \cdot 4$

[1.2] **4.** $5 \div 0$ is undefined.

For Exercises 5 and 6, fill in the blank.

[3.2] **5.** To determine the percent of increase, we need the amount of the increase. This is found by subtracting the initial amount from _____.

[4.2] **6.** To find a solution to an equation in two variables.
   a. Choose a value for one of the variables.
   b. Replace the corresponding variable with your chosen value.
   c. Solve the equation for the value of _____ _____.

Exercises 7–9 EXPRESSIONS

[1.5] **For Exercises 7–9, simplify.**

7. $-|6 - (4 + 5)|$

8. $18 - (2 \cdot 3^2)$

9. $\dfrac{6^2 + 3(4)}{2^4 - 16}$

Exercises 10–30 EQUATIONS AND INEQUALITIES

[4.3] **10.** Find the x- and y-intercepts for $4y = 8 - 2x$.

[4.2] **11.** Graph $y = -3x$.

[4.2] **12.** Find three solutions for $y = -x + 1$.

[4.6] **13.** Determine whether $(2, 3)$ is a solution for $4x - y < 7$.

[4.4] **14.** Determine the slope of the line $x - 2y = 6$.

[3.1] **15.** Solve: $\dfrac{4}{25} = \dfrac{100}{x}$

[4.7] 16. The table shows the top ten metro areas ranked by the number of traffic deaths. Determine whether the correspondence is a function and identify the domain and range. The metro areas are ranked by the number of traffic deaths per 100,000 residents.

City	Traffic Deaths
Orlando	18.8
Tampa–St. Petersburg	17.6
West Palm Beach	16.6
Austin	15.8
Las Vegas	15.4
Phoenix	15.2
Memphis	14.7
Jacksonville	14.6
Fort Lauderdale	14.5
Kansas City	14.2

(*Sources:* The Road Information Program; NHTSA)

[2.3] *For Exercises 17 and 18, solve and check.*

17. $7(r - 1) + 10 = 5(2 + r)$

18. $4x - 10 = 0.5(x - 6)$

[2.6] *For Exercises 19 and 20, solve the inequalities.*

19. $16x + 8 - 4x > -16 + 10x$

20. $10s - 3(4 - 5s) > s + 6(4 + s)$

[2.4] **21.** Solve $A = \pi r^2$ for r^2

For Exercises 22–24, solve the system using the given method.

[5.1] **22.** $\begin{cases} x + y = 1 \\ 2x - 3y = 12 \end{cases}$ by graphing

[5.2] **23.** $\begin{cases} 2x + 5y = 11 \\ x + 3y = 7 \end{cases}$ by substitution

[5.3] **24.** $\begin{cases} 7x + 3y = 11 \\ 2x - 5y = 9 \end{cases}$ by elimination

[5.5] **25.** Graph: $\begin{cases} x + y < 3 \\ x - y \geq 4 \end{cases}$

For Exercises 26–30, solve.

[3.2] **26.** Consider the following poll. If 1000 people were polled, how many would microwave leftovers almost daily?

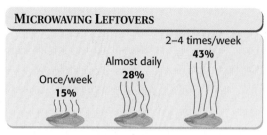

Source: Opinion Research for Tupperware

[3.2] **27.** According to the U.S. Department of Labor, out of the 6,782,000 people whose occupation is sales, 4,306,000 are women. What percent of people involved in sales are women? Round to the nearest tenth of a percent. (*Source:* U.S. Department of Labor, Bureau of Labor Statistics)

[3.2] **28.** For each book sold, an author receives 2% of the price. If a book costs $57, how much will the author receive?

[3.3]
[5.4] **29.** Charles has been saving his change. If he has 20 coins, all nickels and dimes totaling $1.75, how many of each coin does he have?

[3.3]
[5.4] **30.** According to a *Newsweek* article, public schools sponsored 114,000 more Boy Scout troops than did parent–teacher organizations. If 608,000 troops were sponsored in 2008 by the two groups combined, how many troops did each group sponsor?

Polynomials

6

In Chapter 1, we reviewed real numbers and introduced algebraic expressions. We learned that there are two actions we can perform on an expression: (1) We can evaluate an expression, or (2) we can rewrite an expression. In this chapter, we focus on polynomial expressions, exploring how to evaluate and rewrite these expressions. We also see that we can add, subtract, multiply, and divide polynomial expressions in much the same way that we add, subtract, multiply, and divide numbers.

6.1 Exponents and Scientific Notation

Objectives
1. Evaluate exponential forms with integer exponents.
2. Write scientific notation in standard form.
3. Write standard form numbers in scientific notation.

Objective ① Evaluate exponential forms with integer exponents.

Positive Integer Exponents

We learned in Section 1.5 that exponents indicate repeated multiplication of a base number.

$$2^4 = \underbrace{2 \cdot 2 \cdot 2 \cdot 2}_{\text{four factors of two}} = 16 \qquad (-5)^3 = \underbrace{(-5)(-5)(-5)}_{\text{three factors of negative five}} = -125$$

We also discovered some rules for evaluating exponential forms with negative bases.

Rule **Evaluating Exponential Forms with Negative Bases**

If the base of an exponential form is a negative number and the exponent is even, then the product is positive.

If the base is a negative number and the exponent is odd, then the product is negative.

What if the base is a fraction, as in $\left(\dfrac{2}{3}\right)^4$?

$$\left(\frac{2}{3}\right)^4 = \frac{2}{3} \cdot \frac{2}{3} \cdot \frac{2}{3} \cdot \frac{2}{3} = \frac{2^4}{3^4} = \frac{16}{81}$$

This example suggests the following conclusion and rule.

Conclusion If a fraction is raised to a power, we can write its numerator and denominator to that power.

Rule **Raising a Quotient to a Power**

If a and b are real numbers, where $b \neq 0$ and n is a natural number, then $\left(\dfrac{a}{b}\right)^n = \dfrac{a^n}{b^n}$.

Example 1 Evaluate each exponential form.

a. $(-3)^5$

SOLUTION $(-3)^5 = (-3)(-3)(-3)(-3)(-3) = -243$

b. -2^4

SOLUTION $-2^4 = -[2 \cdot 2 \cdot 2 \cdot 2] = -16$

Warning: Notice that -2^4 does not mean that four -2's are to be multiplied. Rather, it means find the additive inverse after multiplying four 2's.

c. $\left(\dfrac{4}{5}\right)^3$

SOLUTION $\left(\dfrac{4}{5}\right)^3 = \dfrac{4^3}{5^3} = \dfrac{64}{125}$

Your Turn 1 Evaluate each exponential form.

a. $(-4)^4$ **b.** -4^4 **c.** $\left(\dfrac{5}{8}\right)^2$

Nonpositive Integer Exponents

So far, we have considered evaluating exponential forms only with positive exponents. However, exponents can be positive, zero, or negative. What does an exponent of 0 indicate? Consider the following pattern.

$$2^4 = 2 \cdot 2 \cdot 2 \cdot 2 = 16$$
$$2^3 = 2 \cdot 2 \cdot 2 \quad\;\; = 8 \;\longleftarrow$$
$$2^2 = 2 \cdot 2 \qquad\quad = 4$$
$$2^1 = 2 \qquad\qquad\; = 2$$
$$2^0 = \qquad\qquad\quad\; = 1 \;\longleftarrow$$

Notice that as the exponent decreases, we can divide by 2 to determine the next result; that is, because 2^4 is 16, 2^3 is $16 \div 2$, which is 8.

To continue the pattern, when we go from 2^1 to 2^0, we divide the result for 2^1 by 2 to determine the result for 2^0, and $2 \div 2 = 1$.

The pattern suggests the following conclusion and rule.

Conclusion A nonzero base number raised to the 0 power is equal to 1. It can be shown that any nonzero number raised to the 0 power equals 1.

Rule **Zero as an Exponent**

If a is a real number and $a \neq 0$, then $a^0 = 1$.

If we continue the pattern, we can see what a negative exponent means.

$$2^0 = 1$$
$$2^{-1} = \dfrac{1}{2} \;\longleftarrow$$
$$2^{-2} = \dfrac{1}{2^2} = \dfrac{1}{4} \;\longleftarrow$$
$$2^{-3} = \dfrac{1}{2^3} = \dfrac{1}{8}$$

To continue the pattern, when we go from 2^0 to 2^{-1}, we divide the result for 2^0 by 2 to determine the result for 2^{-1}, and $1 \div 2 = \dfrac{1}{2}$.

To continue the pattern, when we go from 2^{-1} to 2^{-2}, we divide the result for 2^{-1} by 2 to determine the result for 2^{-2}, and

$$\dfrac{1}{2} \div 2 = \dfrac{1}{2} \cdot \dfrac{1}{2} = \dfrac{1}{2^2} = \dfrac{1}{4}.$$

The pattern suggests the following conclusion.

Conclusion An exponential form with a negative exponent is equal to its reciprocal with the exponent made positive.

It can be shown that our conclusion is true for all nonzero real-number bases.

Rule If a is a real number, where $a \neq 0$ and n is a natural number, then $a^{-n} = \dfrac{1}{a^n}$.

Answers **Your Turn 1**

a. 256   **b.** -256 **c.** $\dfrac{25}{64}$

Example 2 Rewrite with a positive exponent; then if the expression is numeric, evaluate it.

a. 10^0

SOLUTION $10^0 = 1$ $\qquad a^0 = 1$ for $a \neq 0$.

b. $(-2)^{-4}$

SOLUTION $(-2)^{-4} = \dfrac{1}{(-2)^4} = \dfrac{1}{(-2)(-2)(-2)(-2)} = \dfrac{1}{16}$

c. $(-5)^{-3}$

SOLUTION $(-5)^{-3} = \dfrac{1}{(-5)^3} = \dfrac{1}{(-5)(-5)(-5)} = \dfrac{1}{-125} = -\dfrac{1}{125}$

d. x^{-6}

SOLUTION $x^{-6} = \dfrac{1}{x^6}$ $\qquad$ Rewrite using $a^{-n} = \dfrac{1}{a^n}$.

e. $2a^{-3}$

SOLUTION $2a^{-3} = 2 \cdot \dfrac{1}{a^3} = \dfrac{2}{a^3}$ $\qquad$ **Note:** Only a is raised to the -3 power.

f. -6^{-2}

SOLUTION $-6^{-2} = -\dfrac{1}{6^2} = -\dfrac{1}{36}$ $\qquad$ **Note:** Only 6 is raised to the -2 power, not -6.

Your Turn 2 If necessary, rewrite with a positive exponent; then if the expression is numeric, evaluate it.

$\quad$ **a.** $(-8)^0$ $\qquad$ **b.** $(-3)^{-4}$ $\qquad$ **c.** $(-2)^{-3}$ $\qquad$ **d.** y^{-5} $\qquad$ **e.** $4b^{-5}$ $\qquad$ **f.** -2^{-4}

Warning: A common error is to confuse negative exponents with negative numbers. The value of $4^{-2} \neq -16$. Also, $4^{-2} \neq (-2)(4)$. The value of $4^{-2} = \dfrac{1}{4^2} = \dfrac{1}{16}$.

Answers **Your Turn 2**

a. 1 $\quad$ **b.** $\dfrac{1}{81}$ $\quad$ **c.** $-\dfrac{1}{8}$ $\quad$ **d.** $\dfrac{1}{y^5}$

e. $\dfrac{4}{b^5}$ $\quad$ **f.** $-\dfrac{1}{16}$

Calculator Tips

To evaluate exponential forms with negative exponents, such as $(-2)^{-4}$ in part b of Example 2, using a calculator that has $\boxed{\wedge}$ *for exponents and* $\boxed{(-)}$ *for negatives, type*

$$\boxed{(}\ \boxed{(-)}\ \boxed{2}\ \boxed{)}\ \boxed{\wedge}\ \boxed{(-)}\ \boxed{4}\ \boxed{\text{ENTER}}.$$

The result will be a decimal number, 0.0625. To see the fraction form, press $\boxed{\text{MATH}}$ *and select 1: Frac from the menu; then press* $\boxed{\text{ENTER}}$. *On some scientific calculators, use the* F $\blacktriangleleft\blacktriangleright$ D *function.*

If the calculator has $\boxed{y^x}$ *and* $\boxed{+/-}$ *keys, type*

$$\boxed{(}\ \boxed{2}\ \boxed{+/-}\ \boxed{)}\ \boxed{y^x}\ \boxed{4}\ \boxed{+/-}\ \boxed{=}.$$

Connection Remember that $\dfrac{1}{\frac{1}{8}}$ means $1 \div \dfrac{1}{8}$ and that

$1 \div \dfrac{1}{8} = 1 \cdot \dfrac{8}{1} = \dfrac{8}{1} = 8.$

What if a negative exponent is in the denominator of a fraction, as in $\dfrac{1}{2^{-3}}$? If we rewrite 2^{-3} in the denominator using $a^{-n} = \dfrac{1}{a^n}$, we have

$$\dfrac{1}{2^{-3}} = \dfrac{1}{\dfrac{1}{2^3}} = \dfrac{1}{\dfrac{1}{8}} = \dfrac{8}{1} = 8 \qquad \text{Rewrite the denominator using } a^{-n} = \dfrac{1}{a^n}.$$

Our example suggests the following rule.

Rule   If a is a real number, where $a \neq 0$ and n is a natural number, then $\dfrac{1}{a^{-n}} = a^n$.

Example 3   Rewrite with a positive exponent; then if the expression is numeric, evaluate it.

a. $\dfrac{1}{3^{-4}}$

SOLUTION $\dfrac{1}{3^{-4}} = 3^4 = 3 \cdot 3 \cdot 3 \cdot 3 = 81$ Rewrite using $\dfrac{1}{a^{-n}} = a^n$; then simplify.

b. $\dfrac{1}{x^{-5}}$

SOLUTION $\dfrac{1}{x^{-5}} = x^5$ Rewrite using $\dfrac{1}{a^{-n}} = a^n$.

Your Turn 3   Rewrite with a positive exponent; then if the expression is numeric, evaluate it.

a. $\dfrac{1}{4^{-3}}$

b. $\dfrac{1}{x^{-8}}$

What if a fraction is raised to a negative exponent, as in $\left(\dfrac{2}{5}\right)^{-2}$? If we rewrite the expression using $a^{-n} = \dfrac{1}{a^n}$, we have

$$\left(\frac{2}{5}\right)^{-2} = \frac{1}{\left(\dfrac{2}{5}\right)^2} = \frac{1}{\dfrac{4}{25}} = \frac{25}{4} = \left(\frac{5}{2}\right)^2.$$

Our example suggests that when we evaluate a fraction raised to a negative exponent, we can write the reciprocal of the fraction and change the sign of the exponent.

Rule   If a and b are real numbers, where $a \neq 0$ and $b \neq 0$ and n is a natural number, then
$$\left(\frac{a}{b}\right)^{-n} = \left(\frac{b}{a}\right)^{n}.$$

Example 4   Rewrite with a positive exponent; then if the expression is numeric, evaluate it.

a. $\left(\dfrac{3}{4}\right)^{-3}$

SOLUTION $\left(\dfrac{3}{4}\right)^{-3} = \left(\dfrac{4}{3}\right)^{3} = \dfrac{4^3}{3^3} = \dfrac{64}{27}$ Rewrite using $\left(\dfrac{a}{b}\right)^{-n} = \left(\dfrac{b}{a}\right)^{n}$; then simplify.

b. $\left(\dfrac{x}{y}\right)^{-4}$

SOLUTION $\left(\dfrac{x}{y}\right)^{-4} = \left(\dfrac{y}{x}\right)^{4} = \dfrac{y^4}{x^4}$ Rewrite using $\left(\dfrac{a}{b}\right)^{-n} = \left(\dfrac{b}{a}\right)^{n}$; then simplify.

Your Turn 4 Rewrite with a positive exponent; then if the expression is numeric, evaluate it.

a. $\left(-\dfrac{4}{5}\right)^{-3}$

b. $\left(\dfrac{a}{b}\right)^{-5}$

Objective ② Write scientific notation in standard form. Sometimes, we use very large or very small numbers, such as when we describe the vast distances between stars and galaxies or the tiny size of bacteria and atomic structures. For example, the distance from the Sun to the next nearest star, Proxima Centauri, is about 24,700,000,000,000 miles; a single streptococcus bacterium (photo at left) is about 0.00000075 meter in diameter. The large number of zero digits in these numbers makes them tedious to write. **Scientific notation** gives us a shorthand way to write such numbers.

Definition **Scientific notation:** A number expressed in the form $a \times 10^n$, where a is a decimal number with $1 \le |a| < 10$ and n is an integer.

The number 3.58×10^4 is in scientific notation because 3.58 is greater than or equal to 1 but less than 10 and the exponent 4 is an integer. The number 35.8×10^3 is not in scientific notation because 35.8 is greater than 10. The number 0.358×10^5 also is not in scientific notation because 0.358 is less than 1.

What does 3.58×10^4 mean?

$$
\begin{aligned}
3.58 \times 10^4 &= 3.58 \times 10 \times 10 \times 10 \times 10 \\
&= 35.8 \times 10 \times 10 \times 10 \\
&= 358 \times 10 \times 10 \\
&= 3580 \times 10 \\
&= 35{,}800
\end{aligned}
$$

Note: 3.58×10^4, 35.8×10^3, and 0.358×10^5 all represent the same number, 35,800. Only 3.58×10^4 is in scientific notation.

Notice that each multiplication by 10 moves the decimal point one place to the right. Because we multiplied by four factors of 10, the decimal point moved a total of four places to the right from where it started in 3.58.

$$3.58 \times 10^4 = 3\,5\,8\,0\,0\, = 35{,}800 \qquad \text{The decimal point moves four places to the right.}$$

Our example suggests that the exponent of the power of 10 determines the number of places that the decimal point moves.

Procedure **Changing Scientific Notation (Positive Exponent) to Standard Form**

To change from scientific notation with a positive integer exponent to standard form, move the decimal point to the right the number of places indicated by the exponent.

Example 5 Write each number in standard form.

a. 4.5×10^6

SOLUTION Multiplying 4.5 by 10^6 means that the decimal point will move six places to the right.

$$4.5 \times 10^6 = 4{,}500{,}000$$

Connection Because 10^6 is another way to represent 1,000,000, it is popular to read 4.5×10^6 as 4.5 million.

Answers Your Turn 4

a. $-\dfrac{125}{64}$ **b.** $\dfrac{b^5}{a^5}$

b. 9×10^5

SOLUTION Multiplying 9 by 10^5 means that the decimal point will move five places to the right.

$$9 \times 10^5 = 900{,}000$$

Your Turn 5 Write each number in standard form.

   a. 7.305×10^6 **b.** 8×10^7

What if the scientific notation contains a negative exponent, as in 8.45×10^{-2}? To evaluate 10^{-2}, we write it with a positive exponent, $\dfrac{1}{10^2}$.

$$8.45 \times 10^{-2} = 8.45 \times \frac{1}{10^2}$$

$$= \frac{8.45}{10^2}$$

$$= \frac{8.45}{100}$$

> **Note:** Multiplying 8.45 by 10^{-2} is equivalent to dividing 8.45 by 10^{-2}. Dividing by 10^2, which is 100, causes the decimal point to move two places to the left.

$$= 0.0845$$

0.0 8 4 5

Our example suggests the following procedure.

Procedure   **Changing Scientific Notation (Negative Exponent) to Standard Form**

To change from scientific notation with a negative exponent to standard form, move the decimal point to the left the same number of places as the absolute value of the exponent.

Example 6 Write each number in standard form.

a. 4.2×10^{-5}

SOLUTION Multiplying by 10^{-5} is equivalent to dividing by 10^5, which causes the decimal point to move five places to the *left*.

$$4.2 \times 10^{-5} = \frac{4.2}{10^5} = 0.000042$$

b. 7×10^{-9}

SOLUTION Multiplying by 10^{-9} is equivalent to dividing by 10^9, which causes the decimal point to move nine places to the left.

$$7 \times 10^{-9} = \frac{7}{10^9} = 0.000000007$$

Answers Your Turn 5
 a. 7,305,000   **b.** 80,000,000

Your Turn 6 Write each number in standard form.

   a. 7.24×10^{-6} **b.** 5×10^{-7}

Answers Your Turn 6
 a. 0.00000724   **b.** 0.0000005

Objective ❸ Write standard form numbers in scientific notation. We mentioned that the distance from the Sun to the next nearest star, Proxima Centauri, is about 24,700,000,000,000 miles. Suppose we want to express 24,700,000,000,000 in scientific notation. Because scientific notation begins with a decimal number greater than or equal to 1 but less than 10, our first step is to determine the position of the decimal point. The only place the decimal point can go to satisfy the criteria for scientific notation is between the 2 and 4 digits.

$$24{,}700{,}000{,}000{,}000$$

↑

The decimal goes here to express a decimal number greater than or equal to 1 but less than 10.

Notice that there are 13 place values to the right of the decimal point in its new position between the 2 and the 4. To make the scientific notation equal to the original number, we must account for the 13 place values by expressing them as a power of 10.

$$24{,}700{,}000{,}000{,}000 = 2.4700000000000 \times 10^{13}$$

The 13 places to the right of the new decimal position are expressed as a power of 10.

$$24{,}700{,}000{,}000{,}000 = 2.4700000000000 \times 10^{13} = 2.47 \times 10^{13}$$

These zeros to the right of the 7 are no longer needed.

Procedure | **Changing Standard Form to Scientific Notation**

To write a number greater than 1 in scientific notation,

1. Move the decimal point so that the number is greater than or equal to 1 but less than 10. (*Tip: Place the decimal point to the right of the first nonzero digit*).
2. Write the decimal number multiplied by 10^n, where n is the number of places between the new decimal position and the original decimal position.
3. Delete zeros to the right of the last nonzero digit.

Example 7   Write 986,000 in scientific notation.

Solution   Place the decimal to the right of the first nonzero digit, 9. Next, count the places to the right of this decimal position, which is five places. Write the 5 as the exponent of 10. Finally, delete all of the 0's to the right of the last nonzero digit, which is the 6 in this case.

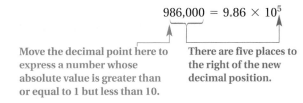

$$986,000 = 9.86 \times 10^5$$

Move the decimal point here to express a number whose absolute value is greater than or equal to 1 but less than 10.

There are five places to the right of the new decimal position.

Your Turn 7   Write the number in scientific notation.

a. 603,000,000

b. 92,500,000,000,000

We learned that a number written in scientific notation that has a negative exponent, such as 5×10^{-2}, has an absolute value less than 1. Therefore, a positive number less than 1, such as 0.00000075, has a negative exponent when written in scientific notation. To write 0.00000075 in scientific notation, we place the decimal between the 7 and 5 digits so that the number is greater than or equal to 1 but less than 10. Moving the decimal point to this position means that we have seven places to account for, which we indicate with a −7 exponent.

$$0.00000075 = 7.5 \times 10^{-7}$$

This suggests the following procedure.

Procedure   **Changing Standard Form to Scientific Notation**

To write a positive decimal number that is less than 1 in scientific notation,

1. Move the decimal point so that the number is greater than or equal to 1 but less than 10. (*Tip: Place the decimal point to the right of the first nonzero digit.*)

2. Write the decimal number multiplied by 10^n, where n is a negative integer whose absolute value is the number of places between the new decimal position and the original decimal position.

3. Delete zeros to the left of the first nonzero digit.

Example 8   Write 0.0000608 in scientific notation.

Solution   Place the decimal point between the 6 and 0 digits so that you have 6.08, which is a decimal number greater than 1 but less than 10. Because there are five decimal places between the original decimal position and the new position, the exponent is −5.

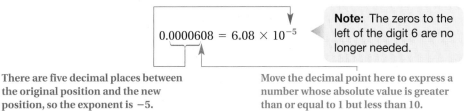

Note: The zeros to the left of the digit 6 are no longer needed.

$$0.0000608 = 6.08 \times 10^{-5}$$

There are five decimal places between the original position and the new position, so the exponent is −5.

Move the decimal point here to express a number whose absolute value is greater than or equal to 1 but less than 10.

Your Turn 8 Write in scientific notation.

a. 0.000172

b. 0.0000009034

6.1 Exercises   For Extra Help   MyMathLab

Note: Exercises marked with a ★ represent challenging exercises.

1. Explain how to rewrite an expression with a negative exponent so that its exponent is positive.

2. After simplifying -5^2, is the result positive or negative? Explain.

3. After simplifying 2^{-3}, is the result positive or negative? Explain.

4. For what kinds of numbers is scientific notation a useful notation?

5. How do we write a number that is greater than 1 in scientific notation?

6. How do we write a positive decimal number that is less than 1 in scientific notation?

For Exercises 7–20, evaluate the exponential form. See Example 1.

7. 4^0 **8.** 3^0 **9.** -2^3 **10.** -3^2 **11.** $(-5)^3$ **12.** $(-2)^5$ **13.** -4^2

14. -3^4 **15.** $\left(\dfrac{3}{4}\right)^4$ **16.** $\left(\dfrac{5}{6}\right)^2$ **17.** $(-0.2)^3$ **18.** $(-0.04)^3$ **19.** $(-2.3)^2$ **20.** $(-1.2)^2$

For Exercises 21–42, rewrite each expression with positive exponents; then if the expression is numeric, evaluate it. See Examples 2–4.

21. a^{-5} **22.** x^{-4} **23.** 2^{-3} **24.** 5^{-1} **25.** $(-4)^{-3}$ **26.** $(-1)^{-5}$

27. -2^{-4} **28.** -3^{-2} ★**29.** $(-0.2)^{-3}$ ★**30.** $(-0.1)^{-4}$ **31.** $5m^{-2}$ **32.** $6r^{-4}$

33. $-6x^{-7}$ **34.** $-8y^{-6}$ **35.** $\left(\dfrac{m}{n}\right)^{-5}$ **36.** $\left(\dfrac{p}{q}\right)^{-6}$ **37.** $\left(\dfrac{5}{7}\right)^{-2}$ **38.** $\left(\dfrac{1}{6}\right)^{-3}$

39. $\dfrac{1}{b^{-6}}$ **40.** $\dfrac{1}{c^{-2}}$ **41.** $\dfrac{1}{4^{-3}}$ **42.** $\dfrac{1}{3^{-4}}$

For Exercises 43–50, use a scientific or graphing calculator to evaluate each exponential form.

43. $(-24)^{-5}$ **44.** $(-45)^{-4}$ **45.** $\left(-\dfrac{5}{8}\right)^{-3}$ **46.** $\left(-\dfrac{4}{9}\right)^{-4}$

47. -0.05^{-4} **48.** -0.02^{-2} **49.** $(3.7)^{-6}$ **50.** $(2.3)^{-4}$

Find ⊗ *the mistake*   *For Exercises 51–54, find and explain the mistake; then work the problem correctly. See Examples 1 and 2.*

51. $-3^4 = (-3)(-3)(-3)(-3) = 81$ **52.** $(-5)^3 = -75$

53. $5^{-2} = -25$ **54.** $\left(\dfrac{3}{4}\right)^{-2} = \dfrac{9}{16}$

For Exercises 55–70, write the number in standard form. See Examples 5 and 6.

55. Scientists speculate that the universe is about 1.65×10^{10} years old.

56. The solar system formed approximately 4.6×10^9 years ago.

57. In empty space, light travels at a constant speed of approximately 2.998×10^8 meters per second.

58. The equatorial radius of the Sun is approximately 6.96×10^8 meters.

59. The equatorial radius of Earth is approximately 6.378×10^6 meters.

60. A light-year is the distance that light travels in a year's time and is equal to about 5.76×10^{12} miles.

61. Earth is about 9.292×10^7 miles from the Sun.

Of Interest

The light from the Sun takes about 8.2 minutes to reach Earth and about 4.26 years to reach Proxima Centauri. We would say that Earth is at a distance of about 8.2 light-minutes from the Sun and Proxima Centauri is at a distance of about 4.26 light-years.

62. The next nearest star to ours (the Sun) is Proxima Centauri, which is about 2.47×10^{13} miles from the Sun.

63. Mitochondria are organelles that are about 2.5×10^{-6} meter across.

64. The diameter of human DNA is about 2×10^{-9} meter.

65. The rest mass of a proton is about 1.67×10^{-27} kilogram.

66. Light with a wavelength of 7.6×10^{-11} meter is in the X-ray portion of the spectrum.

67. Light with a wavelength of 2.95×10^{-9} meter is in the ultraviolet portion of the spectrum.

68. Light with a wavelength of 4.5×10^{-7} meter is in the visible spectrum and is blue in color.

69. The mass of an alpha particle, which is emitted in the radioactive decay of plutonium 239, is 6.645×10^{-27} kilogram.

70. Protons have a positive electrical charge of 1.602×10^{-19} coulomb.

For Exercises 71–86, write the number in scientific notation. See Examples 7 and 8.

71. There are approximately 25,000,000,000,000 red blood cells in the average person's bloodstream.

72. A single red blood cell contains about 250,000,000 molecules of hemoglobin.

73. Human DNA consists of about 5,300,000,000 nucleotide pairs.

74. The Andromeda galaxy is about 2,140,000 light-years from the Sun.

75. Uranium 236 is a radioactive isotope of uranium and has a half-life of 23,420,000 years.

76. The population of the United States is about 304,000,000.

77. The most populated country in the world is China, with a population of about 1,300,000,000.

78. In 2008, the population of Earth was about 6,678,000,000.

79. Light with a wavelength of 0.000000586 meter is yellow in color.

80. Light with a wavelength of 0.000000712 meter is red in color.

81. Electrons have a mass of approximately 0.00000000000000000000000000000091094 kilogram.

82. The size of the HIV virus that causes AIDS is about 0.0000001 meter.

83. The most common diameter for the core of a fiber-optic cable is 0.0000625 meter.

84. A human hair is about 0.00005 meter wide.

85. Intel introduced an improved process for manufacturing computer chips, which allows the company to make transistors that are 0.00000005 meter in length.

86. On April 2, 2008, 0.00975 U.S. dollar equaled 1 japanese yen.

For Exercises 87 and 88, write the numbers in order from smallest to largest.

87. 7.5×10^6, 8.95×10^5, 9×10^7, 1.3×10^8, 7.2×10^6

88. 2.8×10^{-5}, 3.7×10^{-7}, 6.2×10^{-2}, 9.8×10^{-5}, 6.1×10^{-2}

★ **89.** In Exercises 66 and 67, we saw that light with a wavelength of 7.6×10^{-11} meter is in the X-ray portion of the spectrum and light with a wavelength of 2.95×10^{-9} meter is in the ultraviolet portion of the spectrum. About how many times greater is ultraviolet light's wavelength compared to X-rays?

★ **90.** In Exercises 65 and 69, we saw that a proton has a mass of 0.00000000000000000000000000167 kilogram and an alpha particle has a rest mass of about 6.645×10^{-27} kilogram, respectively. Which has the greater mass? About how many times greater is the mass of the larger of the two particles?

Review Exercises

Exercises 1 and 2 EXPRESSIONS

[1.7]   **1.** Use the distributive property to rewrite the expression $-4(y - 3)$.

[1.3]   **2.** Simplify $-13 + 2$.

Exercises 3–6 EQUATIONS AND INEQUALITIES

[3.2]   **3.** 40 is 15% of what?

[4.4]   **4.** Find the slope of the line that passes through $(2, -1)$ and $(-3, -2)$.

[4.4]   **5.** Find the slope and y-intercept for $2x - 3y = 6$.

[4.7]   **6.** Given the function $f(x) = x^3 - 5x^2 + 7$, find $f(-3)$.

6.2 Introduction to Polynomials

Objectives

1 Identify monomials.

2 Identify the coefficient and degree of a monomial.

3 Classify polynomials.

4 Identify the degree of a polynomial.

5 Evaluate polynomials.

6 Write polynomials in descending order of degree.

7 Combine like terms.

Objective 1 **Identify monomials.** We have explored some simple algebraic expressions (Chapter 1) and linear equations and inequalities (Chapters 2–5). We now move back to the expression level of our Algebra Pyramid and explore a particular class of expressions known as *polynomials*.

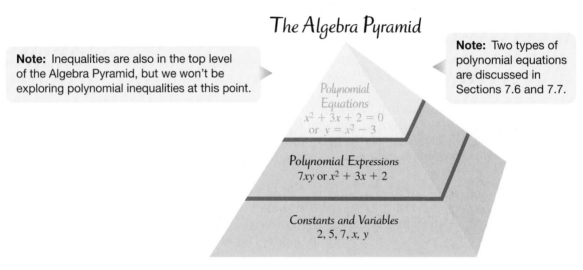

The Algebra Pyramid

Note: Inequalities are also in the top level of the Algebra Pyramid, but we won't be exploring polynomial inequalities at this point.

Note: Two types of polynomial equations are discussed in Sections 7.6 and 7.7.

Polynomial
Equations
$x^2 + 3x + 2 = 0$
or $y = x^2 - 3$

Polynomial Expressions
$7xy$ or $x^2 + 3x + 2$

Constants and Variables
$2, 5, 7, x, y$

First, we consider a special type of polynomial expression called a **monomial**.

Definition | ***Monomial:*** An expression that is a constant, a variable, or a product of a constant and variable(s) that are raised to whole number powers.

Implied in the definition of a monomial is that there is no addition or subtraction; so there is only one term.

Connection Although no variable factor appears with the constant 7, it can be expressed with a variable factor. Recall from Section 6.1 that any nonzero number raised to the 0 power simplifies to 1. Therefore, 7 can be written as $7x^0$ because $x^0 = 1$ (assuming that $x \neq 0$). We use this fact to help us understand the next objective concerning degree.

Example 1 Is the given expression a monomial? Explain.

a. 7

Answer 7 is a monomial because it is a constant.

b. $3x^2$

Answer $3x^2$ is a monomial because it is a product of a constant, 3, and a variable, x, which has a whole-number exponent of 2.

c. $9y$

Answer Remember that a number or variable with no apparent exponent has an understood 1 exponent; so $9y = 9y^1$. Therefore, $9y$ is a monomial because it is a product of a constant, 9, and a variable, y, which has a whole-number exponent of 1.

d. $-0.6xy^2$

Answer $-0.6xy^2$ is a monomial because it is a product of a constant, -0.6, and variables, x and y^2, which have whole-number exponents. As we saw in Example 1(c), we could write $-0.6xy^2$ as $-0.6x^1y^2$. Both of the variables' exponents, 1 and 2, are whole numbers.

e. y^4

Answer   It might seem that y^4 is not a monomial because no numerical factor is visible. However, y^4 is the same as $1y^4$. We don't need to write the 1 because 1 multiplied by an amount does not affect the amount.

f. $\dfrac{5}{x^2}$

Answer   The expression $\dfrac{5}{x^2}$ can be written as a product by writing the divisor with a negative exponent, $5x^{-2}$. However, when written as a product, the variable's exponent is not a whole number, which means it is not a monomial.

g. $4y^2 + 9y - 3$

Answer   The expression is not a monomial because it is not a product of a constant and variables. Instead, addition and subtraction are involved.

> **Note:** This means that a monomial can't have a variable as a divisior.

Your Turn 1   Is the given expression a monomial? Explain.

a. x^3 **b.** $-y^2$ **c.** $-\dfrac{2}{3}tu$ **d.** $\dfrac{4m^2}{n}$ **e.** 0.56 **f.** $3y + 7$

Objective ②   **Identify the coefficient and degree of a monomial.**   In our introduction to expressions in Section 1.7, we defined a **coefficient** as the numerical or constant factor in a term. The same is true for monomials. Another piece of information about a monomial is its **degree**, which relates to the exponents of the variables.

Definitions   *Coefficient of a monomial:* The numerical factor in a monomial.

Degree of a monomial: The sum of the exponents of all variables in a monomial.

Answers   **Your Turn 1**

a. x^3 is a monomial because it is a product of a constant, 1, and a variable, x, which has a whole-number exponent of 3.

b. $-y^2$ is a monomial because it is a product of a constant, -1, and a variable y, with a whole-number exponent of 2.

c. $-\dfrac{2}{3}tu$ is a monomial because it is a product of a constant, $-\dfrac{2}{3}$, and variables, t and u, both of which have the whole number 1 as an exponent.

d. $\dfrac{4m^2}{n}$ is not a monomial because it has a variable, n, as a divisor.

e. 0.56 is a monomial because 0.56 is a constant.

f. $3y + 7$ is not a monomial because it contains addition.

Example 2   Identify the coefficient and degree of each monomial.

a. $-5t^2$

Answer   The coefficient is -5 because it is the numerical factor. The degree is 2 because it is the exponent for the single variable t.

b. $-mn^4$

Answer   We can express $-mn^4$ as $-1m^1n^4$. In this form, we can see that the coefficient is -1 and that the exponents for the variables are 1 and 4. Because the degree is the sum of the variables' exponents, the degree is 5.

c. 7

Answer   Remember that $7 = 7x^0$, where x is any real number except 0. In this alternative form, we can see that 7 is the coefficient and 0 is the degree because it is the variable's exponent.

Your Turn 2   Identify the coefficient and degree of each monomial.

a. $-4x^4$ **b.** $-0.9n^2p^5$ **c.** $\dfrac{5}{8}ab$ **d.** 13

Answers   **Your Turn 2**

a. c: -4; d: 4   **b.** c: -0.9; d: 7

c. c: $\dfrac{5}{8}$; d: 2   **d.** c: 13; d: 0

Objective ❸ Classify polynomials.   Now we are ready to formally define a **polynomial**.

Definition   *Polynomial:* A monomial or an expression that can be written as a sum of monomials.

Examples of polynomials: $4x$ $4x + 8$ $2x^2 + 5xy + 8y$ $2x^3 - 5x^2 + 3x - 9$

> **Note:** Although it seems that the subtractions violate the word *sum* in the definition, the expression $2x^3 - 5x^2 + 3x - 9$ is a polynomial because it can be written as a sum of monomials by rewriting the subtractions as equivalent additions.
>
> $$2x^3 - 5x^2 + 3x - 9 = 2x^3 + (-5x^2) + 3x + (-9)$$
>
> Although those two forms are equivalent, $2x^3 - 5x^2 + 3x - 9$ is preferable because it is the simplest form. That is, it has the fewest symbols.

Notice that the polynomial $2x^3 - 5x^2 + 3x - 9$ has the same variable x in each variable term, whereas $2x^2 + 5xy + 8y$ has a mixture of two variables, x and y. We call a polynomial such as $2x^3 - 5x^2 + 3x - 9$ a **polynomial in one variable**.

Definition   *Polynomial in one variable:* A polynomial in which every variable term has the same variable.

$x^2 - 5x + 2$ is a polynomial in one variable, x.

$x^2 + 8y$ is a polynomial in two variables, x and y.

Special names have been given to some polynomials. We have already discussed monomials such as $4x$, which is a single-term polynomial. Notice that the prefix *mono* in the word *monomial* indicates "one" term and the prefix *poly* in *polynomial* indicates "many" terms. Continuing with the prefixes, a two-term polynomial such as $4x + 8$ is called a **binomial**, and a three-term polynomial such as $2y^2 + 5y + 8$ is called a **trinomial**. No special names are given to polynomials with more than three terms.

Definitions   *Binomial:* A polynomial containing two terms.

Trinomial: A polynomial containing three terms.

Example 3   Indicate whether the expression is a monomial, binomial, or trinomial, has no special polynomial name or is not a polynomial.

a. $2.9xy^3z$

Answer $2.9xy^3z$ is a monomial because it has a single term.

b. $-7a^2 + b$

Answer $-7a^2 + b$ is a biomial because it contains two terms.

c. $9t^2 + 3.4t - 8.2$

Answer $9t^2 + 3.4t - 8.2$ is a trinomial because it contains three terms.

d. $y^3 + 6y^2 - y + 15$

Answer   Although $y^3 + 6y^2 - y + 15$ is a polynomial, it has no special name because it has more then three terms.

e. $6x + \dfrac{3}{y}$

Answer $6x + \dfrac{3}{y}$ is not a polynomial because $\dfrac{3}{y}$ is not a monomial.

Your Turn 3   Indicate whether the expression is a monomial, binomial, or trinomial, has no special polynomial name or is not a polynomial.

   a. $b + 2$ **b.** $7r^3 - 9r^2 + r - 6$ **c.** $5.7x^4y$ **d.** $x^2 - 7xy - y^2$ **e.** $4m - \dfrac{3}{n^2}$

Objective ④ Identify the degree of a polynomial.   Previously, we defined the degree of a monomial to be the sum of the exponents of the variables. We can also talk about the **degree of a polynomial**.

Definition   *Degree of a polynomial:* The greatest degree of any of the terms in the polynomial.

Example 4   Identify the degree of each polynomial.

a. $x^3 + 12x^7 - 10x^2 + 9x - 14$

Answer   The degree is 7 because it is the greatest degree of all of the terms.

b. $12x^6 - 3x^5y^3 - xy^3 + 4y^2 - 16$

Answer   To determine the degree of the monomial $-3x^5y^3$, we add the exponents of its variables; so its degree is 8. Likewise, we add the exponents of the variables in $-xy^3$ and find that its degree is 4. Comparing these degrees with the degrees of the other terms, we see that 8 is the greatest degree. Therefore, 8 is the degree of the polynomial.

Your Turn 4   Identify the degree of each polynomial.

   a. $y^7 - 3y^2 - 8y^9 + 4y + 5$ **b.** $2x^5 - 9x^6y + x^3y^3 - 15y - 7$

Objective ⑤ Evaluate polynomials.   In Chapter 1, we learned that we can evaluate an expression. Let's now evaluate a polynomial expression. To evaluate an expression, we replace the variable(s) with given values and calculate a numerical result.

Example 5   Evaluate $x^3 - 5x^2 + 7$ when $x = -3$.

Solution

$$x^3 \quad - \quad 5x^2 + 7$$

$$(-3)^3 - 5(-3)^2 + 7 \qquad \text{Replace each } x \text{ with } -3.$$
$$= -27 - 5(9) + 7 \qquad \text{Simplify.}$$
$$= -27 - 45 + 7$$
$$= -65$$

Example 6   If we neglect air resistance, the polynomial $-16t^2 + h_0$ describes the height of a falling object after falling from an initial height h_0 for t seconds. It is said that Galileo Galilei dropped cannonballs from the Leaning Tower of Pisa to study gravitational effects. If the tower is 180 feet tall, what is the height of a cannonball after it falls 3 seconds?

Answers   **Your Turn 3**

a. binomial

b. no special polynomial name

c. monomial

d. trinomial

e. not a polynomial

Answers   **Your Turn 4**

a. 9   **b.** 7

SOLUTION Evaluate $-16t^2 + h_0$ when $t = 3$ and $h_0 = 180$.

$$-16(3)^2 + 180 \qquad \text{Replace } t \text{ with 3 and } h_0 \text{ with 180.}$$
$$= -16(9) + 180$$
$$= -144 + 180$$
$$= 36$$

After 3 seconds, the height of the cannonball is 36 feet.

Your Turn 6 Use the polynomial in Example 6 to find the height of an object after it falls 4 seconds from an initial height of 300 feet.

Of Interest

Galileo revolutionized scientific thought with his contributions to the understanding of gravity and the motion of objects. By constructing his own telescope, he made observations that confirmed Copernicus's theory that Earth and the planets revolve around the Sun. Although these beliefs came under great scrutiny, their accuracy holds true to this day.

Objective ⑥ **Write polynomials in descending order of degree.** Now let's consider some ways to rewrite a polynomial. Recall from Chapter 1 that rewriting an expression means that we write an equivalent expression by applying properties of arithmetic. For example, we can apply the commutative property of addition and rearrange the terms in $-8 + 4x^3 - 7x^2 + 2x$ to get an equivalent expression, $4x^3 - 7x^2 + 2x - 8$. Notice that the signs move with the terms. Also notice that the terms in $4x^3 - 7x^2 + 2x - 8$ are written in order from the greatest degree to the least degree.

Connection Because $4x^3 - 7x^2 + 2x - 8$ is a rearrangement of $-8 + 4x^3 - 7x^2 + 2x$, equating them forms the identity $4x^3 - 7x^2 + 2x - 8 = -8 + 4x^3 - 7x^2 + 2x$. This means that we can evaluate both expressions using any real number and both will give the same result.

$$\begin{array}{cccc} 4x^3 & -7x^2 & +2x & -8 \\ \uparrow & \uparrow & \uparrow & \uparrow \\ \text{Degree 3} & \text{Degree 2} & \text{Degree 1} & \text{Degree 0} \end{array}$$

A polynomial written this way is said to be in *descending order of degree*.

Procedure **Writing a Polynomial in Descending Order of Degree**

To write a polynomial in descending order of degree, place the highest degree term first, then the next highest degree term, and so on.

Writing a polynomial in more than one variable in descending order of degree can be complicated, so we will write only polynomials in one variable in descending order of degree.

Example 7 Write each polynomial in descending order of degree.

a. $9x^2 + 7x^4 - 2x^3 - 5 + 3x$

SOLUTION Rearrange the terms so that the highest degree term is first, then the next highest degree, and so on.

$$\begin{array}{ccccc} 7x^4 & -2x^3 & +9x^2 & +3x & -5 \\ \uparrow & \uparrow & \uparrow & \uparrow & \uparrow \\ \text{Degree 4} & \text{Degree 3} & \text{Degree 2} & \text{Degree 1} & \text{Degree 0} \end{array}$$

Answer $7x^4 - 2x^3 + 9x^2 + 3x - 5$

b. $-9y^3 + 2y^6 + y - 3y^4 + 12 - 6y^2$

Answer $2y^6 - 3y^4 - 9y^3 - 6y^2 + y + 12$

Answer **Your Turn 6**

44 feet

Your Turn 7 Write each polynomial in descending order of degree.

a. $2n^3 + n^5 - 9n^2 + 8 - 3n^4 + 4n$ **b.** $-4t^2 + 2t + 16 + 5t^3 - 13t^6$

Objective ⑦ Combine like terms. Another way we can rewrite a polynomial expression is by combining like terms. We learned in Section 1.7 that combining like terms is one way to rewrite an expression in simplest form. We also learned that like terms have the same variables raised to the same powers.

Example 8 Combine like terms and write the resulting polynomial in descending order of degree.

$$12x^3 + 5x^2 - 2x^3 - 13 + 6x + x^2 + 2 - 6x$$

Warning: When combining like terms, we *add only the coefficients*, keeping the exponents the same.

SOLUTION $12x^3 + 5x^2 - 2x^3 - 13 + 6x + x^2 + 2 - 6x$

$= 12x^3 - 2x^3 + 5x^2 + x^2 + 6x - 6x - 13 + 2$ Collect like terms as needed.

$= \quad\quad 10x^3 \quad + \quad 6x^2 \quad + \quad 0 \quad\quad -11$ Combine like terms.

$= 10x^3 + 6x^2 - 11$ Drop 0 and bring the terms together.

ALTERNATIVE SOLUTION Some people prefer to skip the collecting step and strike through like terms as they are combined to keep track of the terms that have been combined. It is helpful to combine the terms in descending order of degree so that the answer will be in descending order of degree.

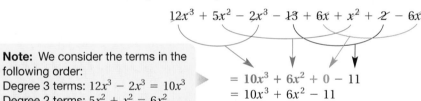

Note: We consider the terms in the following order:
Degree 3 terms: $12x^3 - 2x^3 = 10x^3$
Degree 2 terms: $5x^2 + x^2 = 6x^2$
Degree 1 terms: $6x - 6x = 0$
Degree 0 terms: $-13 + 2 = -11$

$= 10x^3 + 6x^2 + 0 - 11$
$= 10x^3 + 6x^2 - 11$

Your Turn 8 Combine like terms and write the resulting polynomial in descending order of degree.

a. $10x^3 - 4x^2 - 3x^3 + 8 - 5x^2 - 9 + 6x$
b. $3y^4 - 12y - y^2 + y - 6 + y^2 + 5y^4$

Remember that polynomials may have multiple variables. Multivariable monomials are like terms if they have the same variables raised to the same exponents.

Example 9 Combine like terms.

$$x^4 + 8x^2y - 7 - 13x^2y + 4y - 2xy^2 + 6y - x^4 + 2$$

SOLUTION Collecting the like terms and then combining them gives us the following:

$$x^4 + 8x^2y - 7 - 13x^2y + 4y - 2xy^2 + 6y - x^4 + 2$$
$$= x^4 - x^4 + 8x^2y - 13x^2y - 2xy^2 + 4y + 6y + 2 - 7$$
$$= \quad 0 \quad\quad - 5x^2y \quad - 2xy^2 \quad + 10y \quad\quad - 5$$
$$= -5x^2y - 2xy^2 + 10y - 5$$

ALTERNATIVE SOLUTION Instead of collecting the like terms, we strike through like terms in the given polynomial as they are combined:

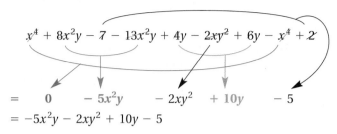

$$= \quad 0 \quad - 5x^2y \quad - 2xy^2 \quad + 10y \quad - 5$$
$$= -5x^2y - 2xy^2 + 10y - 5$$

Your Turn 9   Combine like terms.

Answers Your Turn 9
a. $2x^5 - 3x^3y^2 - 4$
b. $3t^3u - t^2u - 2u$

a. $-x^5 + 6x^3y^2 + 2y - 9x^3y^2 + 3x^5 - 4 - 2y$

b. $-6t^3u + 12tu - 4t^2u + 9t^3u - u + 3t^2u - 12tu - u$

6.2 Exercises   For Extra Help   *MyMathLab*   Math XL PRACTICE   WATCH   DOWNLOAD

Note: Exercises marked with a ★ represent challenging exercises.

1. How do we determine the degree of a monomial?

2. Explain the differences between a monomial, a binomial, and a trinomial.

3. How do we determine the degree of a polynomial with two or more terms?

4. Given a polynomial in one variable, what does *descending order of degree* mean?

5. Explain how to combine like terms.

6. Why do we say that combining the like terms in a polynomial simplifies it?

For Exercises 7–18, determine whether the expression is a monomial. See Example 1.

7. $-3x^2$

8. $5y^3$

9. $\dfrac{1}{5}$

10. $\dfrac{4}{7}$

11. $\dfrac{6m}{4y^3}$

12. $\dfrac{-8t}{5x}$

13. $4x^2 - 3x + 7$

14. $p^2 - q^2$

15. y

16. 0.76

17. $3m^4n^2$

18. $6u^3v^4w$

For Exercises 19–30, identify the coefficient and degree of each monomial. See Example 2.

19. $-5m^3$

20. $8p^2$

21. $-xy^4$

22. $-m^2n^6$

23. -9 **24.** 18 **25.** $4.2n^3p$ **26.** $-6.7uv^7$

27. $16abc$ **28.** $-8.1lkm$ **29.** y **30.** w

For Exercises 31–42, indicate whether the expression is a monomial, binomial, or trinomial or has no special polynomial name. If the expression is a polynomial, give the degree. See Examples 3 and 4.

31. $6x + 8y + 3z$ **32.** $17.3x^2 - 3x + 2.1$ **33.** $-7m^2n$ **34.** $18xy^3$

35. $5.2x^3 - 3x^2 + 4.1x - 11$ **36.** $x^3 - 4x^2 + 4x + 5$ **37.** $5u^3 - 16u$ **38.** $7k + 5k^3$

39. -21 **40.** 36 **41.** $\dfrac{25}{x} - x^2$ **42.** $y^2 - \dfrac{16}{y}$

For Exercises 43–52, identify the degree of each polynomial. See Example 4.

43. $19 - 7y^4 + 3y - 2y^3 - 7$ **44.** $6a^5 - 19a^3 + a^9 + 5a - 14$ **45.** $16z^4 + 7z^2 - z^5 - 4z^3 + z$

46. $11 + 4t^5 - 5t + 7t^3 - 18t^2 + t^8$ **47.** $2u^2 + 5u^3 - u^7 + 13u^4 - 3u$ **48.** $22j^3 + 5j^2 - 16j^4 + 21 - 14j$

49. $3ab^2 - 6a^2b^2 + 4a - 7b^3$ **50.** $6mn - 7m^3n^2 + 4m^2n^5 - 3m^4$

51. $-8x^4y + 5x^3y^3 - 5x^2y^6 + 3xy^7 + 2x^6$ **52.** $-2p^3q^3 + 5p^6q - 3p^2q^4 + 7p^4q^4 - 8q^5$

For Exercises 53–60, evaluate the polynomial using the given values. See Example 5.

53. $-3xy^2; x = -5, y = 2$ **54.** $-2x^2y; x = -1, y = 4$ **55.** $x^2 - 6x + 1; x = 3$

56. $n^2 - 8n - 3; n = -4$ **57.** $a^3 + 0.5ab + 2.4b;$ **58.** $m^3 + 3.6mn - 0.2n^2;$
                                        $a = -1, b = -2$ $m = -3, n = 4$

59. $\dfrac{1}{2}x^2 + \dfrac{2}{3}x + 3; x = 6$ **60.** $\dfrac{1}{4}a^2 + \dfrac{3}{2}a + 2; a = 4$

For Exercises 61–68, evaluate. See Example 6.

61. If we neglect air resistance, the polynomial $-16t^2 + h_0$ describes the height of a falling object after falling from an initial height h_0 for t seconds. A marble is dropped from a tower at a height of 50 feet.
   a. What is its height after 0.5 seconds?

   b. What is its height after 1.2 seconds?

62. The polynomial $2lw + 2lh + 2wh$ describes the surface area of a box.
   a. Find the surface area of a box with a length of 8 inches, a width of 6.5 inches, and a height of 4 inches.

   b. Find the surface area of a box with a length of 4 inches, a width of 3 inches, and a height of 2.5 inches.

63. The polynomial $5p^2 - 6p + 3$ describes the number of units sold for every 1000 people in a particular region based on the price of the product, which is represented by p.
   a. Find the number of units sold for every 1000 people if the price of each unit is \$3.

   b. Find the number of units sold for every 1000 people if the price of each unit is \$3.50.

64. The polynomial $7r^2 - 2r + 6$ describes the voltage in a circuit, where r represents the resistance in the circuit.
   a. Find the voltage if the resistance is 6 ohms.

   b. Find the voltage if the resistance is 8 ohms.

65. An engineer is designing a chemical storage tank that is capsule-shaped. The polynomial $\frac{4}{3}\pi r^3 + \pi r^2 h$ describes the volume of the tank.

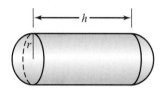

   a. Find the volume of the tank if the radius is 4 feet and the height is 20 feet. Round the result to the nearest tenth.

   b. Find the volume of the tank if the radius is 3 feet and the height is 15 feet. Round the result to the nearest tenth.

66. The polynomial $lwh - \pi r^2 h$ describes the volume of metal remaining in a block after a cylinder has been bored into the block of metal.

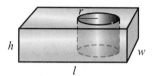

   a. Find the volume of metal remaining if the length is 15 inches, the width is 7 inches, the height is 4 inches, and the radius of the cylinder is 3 inches. Round to the nearest tenth.

   b. Find the volume of metal remaining if the length is 15 inches, the width is 7 inches, the height is 4 inches, and the radius of the cylinder is 2 inches. Round to the nearest tenth.

67. The polynomial $-0.2x^2 + 4.25x + 26$ describes the number of international visitors (in millions) to the United States each year from 1986 on, where x represents the number of years after 1986. ($x = 0$ means 1986, $x = 1$ means 1987, and so on.) (*Source:* Tourism Industries, International Trade Administration, Department of Commerce)
   a. How many international visitors came to the United States in 1986?

   b. How many international visitors came to the United States in 1993?

   c. How many international visitors came to the United States in 2003?

   d. If the model holds, predict the number of international visitors in 2012. Do you think the prediction is reasonable? Explain.

68. The polynomial $28x^3 + 25x^2 + 100x + 340$ describes the number of cell phone subscribers in the United States (in thousands of subscribers) each year since 1985, where x represents the number of years since 1985. ($x = 0$ means 1985, $x = 1$ means 1986, and so on.) (*Source:* The CTIA Semiannual Wireless Industry Survey)
   a. How many subscribers were in the United States in 1985?

   b. How many subscribers were in the United States in 1995?

   c. How many subscribers were in the United States in 2001?

   d. If the model holds, predict the number of subscribers in 2015. Do you think the prediction is reasonable? Explain.

For Exercises 69–74, write each polynomial in descending order of degree. See Example 7.

69. $7x^4 + 5x^7 - 8x + 14 - 3x^5$

70. $-4y^3 + 7y - 8y^5 - 6y^2 + 9 + 2y^4$

71. $4r - 3r^2 + 18r^5 + 7r^3 - 8r^6$

72. $u^4 + 7u^2 - u^9 + 15u + 27 - 3u^5 - 8u^3$

73. $20 - w^3 + 5w + 11w^2 - 12w^4$

74. $a - 6 + 7a^5 + 3a^2 - 4a^3$

For Exercises 75–94, combine like terms and write the resulting polynomial in descending order of degree. See Example 8.

75. $3x + 4 + 2x - 7$

76. $2y - 9 + 5y + 3$

77. $7 - 6a + 2a - 2$

78. $9 + 4b - 8b - 3$

79. $7x^2 + 2x + 4x - 9x^2$

80. $4a^3 + 3a^2 - 2a^2 - 7a^3$

81. $\frac{1}{2}x^2 + \frac{1}{3}x + \frac{4}{3}x + \frac{3}{2}x^2$

82. $\frac{2}{3}y^2 + \frac{1}{4}y + \frac{7}{4}y + \frac{5}{3}y^2$

83. $\frac{5}{3}a^2 + \frac{2}{5}a - \frac{1}{2}a^2 - \frac{5}{3}a$

84. $\frac{9}{4}n^2 - \frac{7}{6}n - \frac{2}{3}n^2 + \frac{3}{4}n$

85. $2x^2 + 5x - 7x^2 + 8 - 5x + 6$

86. $3m^5 + 7m^2 - 8m + 9m^5 - 7m - 7m^2$

87. $15 - 4y + 8y^2 - 4y - 3y^2 + 7 - y$

88. $6l - 5l^3 - 2l^4 + 7l^4 + 5l - l^3 + 4l^3$

89. $9k - 5k^2 + 6k^3 + 2k^2 + k^4 - 3k - 3k^4$

90. $11p - p^2 + 12p^3 + 5p^2 - 7p^3 + 20 - 4p$

91. $7a^2 - 5a^4 + 6a^3 - 5a^4 + 12 - 6a^3 + a + 4$

92. $-6c^9 + 7c^3 - 4c^5 + 8 - 14c^2 + 4c^5 - 7 - c^9$

93. $12v^3 + 19v^2 - 20 - v^5 - 5v^3 + 16v^2 - 20v^5 + v^3$

94. $b^3 - 13b^2 + b^4 + 6 - 2b^3 - 15b^4 - 17b^3 + 8$

For Exercises 95–106, combine like terms. See Example 9.

95. $6a - 3b - 2a + b$

96. $5x - y - 3x - 2y$

97. $3y^2 + 5y - 2y - 7y^2$

98. $2z^3 - 4z - 6z - 10z^3$

99. $2x^2 + 3y - 4x^2 + 6y - 5x^2 - 2x^2$

100. $3a^2 - 5b - 4a^2 + 7b - 5b + 6b$

★**101.** $\frac{1}{3}a^2 + \frac{3}{5}a - \frac{3}{2}a + \frac{5}{4}a^2 + \frac{1}{3}a$

★**102.** $\frac{7}{6}x^2 - \frac{3}{4}x - \frac{2}{5}x^2 - \frac{4}{5}x + \frac{4}{3}x$

103. $y^6 + 2yz^4 - 10yz + 3yz^4 - 3z^5 - 7z^3 + 4yz - 10$ **104.** $x^4 + xy^3 - 6xy + xy^3 + 3y^4 - 4y^2 + 3xy - 8$

105. $-3w^2z - 9 + 6wz^2 + 3w^2 - w^2z + 8 + 9w^2 - 7wz^2$ **106.** $-m^3n - 8mn + 4 - 5m^2n^2 + 3m^3n - 7 + mn - 16m^2n^2$

Puzzle Problem

Rearrange the numbers in the figure shown so that no two consecutive numbers are next to each other horizontally, vertically, or diagonally.

	1	
2	3	4
5	6	7
	8	

Review Exercises

Exercises 1–4   **EXPRESSIONS**

[1.3] *For Exercises 1 and 2, simplify.*

    1. $-\dfrac{3}{4} + \dfrac{1}{6}$ **2.** $-14.6 - (-10.3)$

[6.1]   **3.** Evaluate: -7^0

[6.1]   **4.** The Andromeda galaxy (the closest one to our Milky Way galaxy) contains at least 200,000,000,000 stars. Write this number in scientific notation.

Exercises 5 and 6   **EQUATIONS AND INEQUALITIES**

[2.3]   **5.** Solve and check: $\dfrac{1}{4}x = 3$

[3.5]   **6.** How many liters of a 40% solution must be added to 200 liters of a 20% solution to obtain a 35% solution?

6.3 Adding and Subtracting Polynomials

Objectives ① Add polynomials.

② Subtract polynomials.

Objective ① Add polynomials.

Understanding Polynomial Addition

We can add and subtract polynomials in the same way that we add and subtract numbers. In fact, polynomials are like whole numbers that are in an expanded form. Consider the polynomial $4x^2 + 3x + 6$. If we replace the x's with the number 10, we have the expanded form for the number 436.

$$4x^2 + 3x + 6$$

$$4 \cdot 10^2 + 3 \cdot 10 + 6 \qquad \text{Replacing } x \text{ with 10}$$
$$= 400 + 30 + 6$$
$$= 436$$

In our base-ten number system, each place value is a power of 10. We can think of polynomials as a variable-base number system. In other words, the place values are variables, where x^2 is like the hundreds place (10^2) and x is like the tens place (10^1). To add whole numbers, we add the digits in like place values. Polynomials are added in a similar way. However, instead of adding digits in like place values, we add like terms. Consider the following comparison.

Numeric addition:
$436 + 251 = 687$

$$
\begin{array}{r}
436 \\
+\,251 \\
\hline
687
\end{array}
$$

Note: In numeric addition, like place values are added.

Polynomial addition:
$(4x^2 + 3x + 6) + (2x^2 + 5x + 1) = 6x^2 + 8x + 7$

$$
\begin{array}{r}
4x^2 + 3x + 6 \\
+\,2x^2 + 5x + 1 \\
\hline
6x^2 + 8x + 7
\end{array}
$$

Note: In polynomial addition, like terms are added.

From this simple case, we see that polynomials are added by combining the like terms.

Procedure **Adding Polynomials**

To add polynomials, combine like terms.

Adding Polynomials

Although we stacked the preceding polynomials, we do not actually have to stack polynomials in order to add them.

Example 1 Add and write the resulting polynomial in descending order of degree.

a. $(2x^3 + 5x^2 + 4x + 1) + (6x^3 + 3x + 8)$

Note: Our thinking flows in this order:

Degree 3 terms:
$2x^3 + 6x^3 = 8x^3$

Degree 2 terms:
$5x^2$ has no like term, so it is copied into the answer.

Degree 1 terms:
$4x + 3x = 7x$

Degree 0 terms:
$1 + 8 = 9$

Solution Combine like terms. Notice that combining in order of degree places the resulting polynomial in descending order of degree.

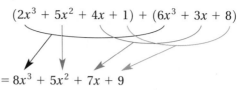

$$= 8x^3 + 5x^2 + 7x + 9$$

b. $(y^5 + 9y^3 - 3y - 7) + (2y^5 - 4y^3 + 3y + 5)$

Solution $(y^5 + 9y^3 - 3y - 7) + (2y^5 - 4y^3 + 3y + 5)$ Combine like terms.

$$= 3y^5 + 5y^3 \qquad + 0 \qquad -2$$
$$= 3y^5 + 5y^3 - 2$$

If we had stacked the polynomials in Example 1(a), we would have had to leave a blank space under the $5x^2$ because there was no x^2 term in the second polynomial. It is the same as having a 0 in that place.

$$
\begin{array}{r}
2x^3 + 5x^2 + 4x + 1 \\
+\,6x^3 + 0 + 3x + 8 \\
\hline
8x^3 + 5x^2 + 7x + 9
\end{array}
$$

As the polynomials get more complex, these blanks become more prevalent. This is why the stacking method is not a preferred method for adding polynomials.

Your Turn 1   Add and write the resulting polynomial in descending order of degree.

a. $(x^4 + 3x^3 + 4x + 2) + (5x^3 + 2x + 7)$

b. $(14n^5 - 6n^3 - 12n + 15) + (2n^5 - 4n^3 + 8n - 5)$

Note: Our thinking flows like this:

The t^4 terms:
$7t^4 + (-8t^4) = -1t^4 = -t^4$

The tu^2 terms:
$2.9tu^2 + (-7tu^2) = -4.1tu^2$

The tu terms:
$-\dfrac{3}{4}tu + \dfrac{1}{6}tu =$
$-\dfrac{9}{12}tu + \dfrac{2}{12}tu = -\dfrac{7}{12}tu$

The constant terms:
$11 + (-5) = 6$

Sometimes the polynomials contain terms that have several variables or coefficients that are fractions or decimals.

Example 2   Add and write the resulting polynomial in descending order of degree:

$$\left(7t^4 + 2.9tu^2 - \frac{3}{4}tu + 11\right) + \left(-8t^4 - 7tu^2 + \frac{1}{6}tu - 5\right)$$

Solution $\left(7t^4 + 2.9tu^2 - \dfrac{3}{4}tu + 11\right) + \left(-8t^4 - 7tu^2 + \dfrac{1}{6}tu - 5\right)$ Combine like terms.

$$= -t^4 - 4.1tu^2 - \frac{7}{12}tu + 6$$

Your Turn 2   Add and write the resulting polynomial in descending order of degree.

a. $(12x^3 - 6.2xy^2 - 0.3xy - y^2) + (x^3 + 7xy^2 - 0.9xy + 0.4y^2)$

b. $\left(a^4 - \dfrac{4}{5}ab^3 - \dfrac{2}{3}ab + 9\right) + \left(a^4 + \dfrac{1}{4}ab^3 + ab - 5\right)$

Answers Your Turn 1

a. $x^4 + 8x^3 + 6x + 9$

b. $16n^5 - 10n^3 - 4n + 10$

Answers Your Turn 2

a. $13x^3 + 0.8xy^2 - 1.2xy - 0.6y^2$

b. $2a^4 - \dfrac{11}{20}ab^3 + \dfrac{1}{3}ab + 4$

Example 3 Write an expression in simplest form for the perimeter of the rectangle shown.

$3x + 7$ [rectangle] $5x - 1$

Understand *Perimeter* means the total distance around the shape. Therefore, we need to add the lengths of all of the sides of the shape.

Plan In this case, the lengths of the sides are represented by polynomials. Therefore, we add the polynomials to represent the perimeter.

Execute Perimeter = Length + Width + Length + Width

$$= (5x - 1) + (3x + 7) + (5x - 1) + (3x + 7)$$
$$= 5x + 3x + 5x + 3x - 1 + 7 - 1 + 7$$
$$= 16x + 12$$

Answer The expression for the perimeter is $16x + 12$.

Check To check, we (1) choose a value for x and evaluate the original expressions for length and width, (2) determine the corresponding numeric perimeter, and (3) evaluate the perimeter expression using the same value for x and verify that we get the same numeric perimeter. Let's choose $x = 2$.

Length: $5x - 1$ Width: $3x + 7$ Perimeter:

$\quad 5(2) - 1$ $3(2) + 7$ The perimeter of the rectangle with a

$\quad = 10 - 1$ $= 6 + 7$ length of 9 and width of 13 is

$\quad = 9$ $= 13$ $9 + 13 + 9 + 13 = 44$.

Now evaluate the perimeter expression where $x = 2$; you should find that the result is 44.

$$\text{Perimeter expression:} \quad 16x + 12$$
$$16(2) + 12$$
$$= 32 + 12$$
$$= 44 \qquad \text{This agrees with our calculation above.}$$

Your Turn 3 Write an expression in simplest form for the perimeter of the following shape.

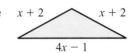
$x + 2$ $x + 2$ $4x - 1$

Objective ② Subtract polynomials.

Understanding Polynomial Subtraction

Subtracting polynomials is similar to subtracting signed numbers. When we subtract signed numbers, it is often simpler to write the subtraction statement as an equivalent addition statement. Consider the following comparison of numeric subtraction and polynomial subtraction.

Numeric subtraction: **Polynomial subtraction:**

$975 - 621$ $(9x^2 + 7x + 5) - (6x^2 + 2x + 1)$

We can stack by place: We can stack like terms:

$$\begin{array}{r} 975 \\ -621 \\ \hline 354 \end{array}$$

$$\begin{array}{r} 9x^2 + 7x + 5 \\ -(6x^2 + 2x + 1) \\ \hline 3x^2 + 5x + 4 \end{array}$$

Answer   Your Turn 3

$6x + 3$

In numeric subtraction, we subtract digits in the same place value, whereas in polynomial subtraction, we subtract like terms. Note that subtracting 1 from 5 is equivalent to combining 5 and -1, subtracting $2x$ from $7x$ is equivalent to combining $7x$ and $-2x$, and subtracting $6x^2$ from $9x^2$ is equivalent to combining $9x^2$ and $-6x^2$. This suggests that we can write polynomial subtraction as equivalent polynomial addition by changing the sign of each term in the second polynomial (the subtrahend), like this:

$$(9x^2 + 7x + 5) - (6x^2 + 2x + 1)$$
$$= (9x^2 + 7x + 5) + (-6x^2 - 2x - 1)$$
$$= 3x^2 + 5x + 4$$

Note: We are allowed to change the signs because of the distributive property. If we disregard the initial polynomial, we have
$$-(6x^2 + 2x + 1) = -1(6x^2 + 2x + 1).$$
The distributive property tells us that we can distribute the -1 (or minus sign) to each term inside the parentheses.
$$= -1 \cdot 6x^2 - 1 \cdot 2x - 1 \cdot 1$$
$$= -6x^2 - 2x - 1$$

Subtracting Polynomials

Our exploration suggests the following procedure for subtracting polynomials.

Procedure | **Subtracting Polynomials**

To subtract polynomials,

1. Write the subtraction statement as an equivalent addition statement.
   a. Change the operation symbol from a minus sign to a plus sign.
   b. Change the subtrahend (second polynomial) to its additive inverse. To get the additive inverse, change the sign of each term in the polynomial.
2. Combine like terms.

Connection The procedure for subtracting polynomials is the same as the procedure for subtracting signed numbers (Section 1.3). In both cases, we write subtraction as equivalent addition by changing the subtrahend to its additive inverse.

Example 4 Subtract and write the resulting polynomial in descending order of degree: $(7x^3 + 8x^2 + 6x + 9) - (3x^3 + 6x^2 + 5x + 1)$

SOLUTION Write an equivalent addition statement; then combine like terms.

$$(7x^3 + 8x^2 + 6x + 9) - (\quad 3x^3 + 6x^2 + 5x + 1)$$

Change the minus sign to a plus sign.   Change all signs in the subtrahend.

$$= (7x^3 + 8x^2 + 6x + 9) + (-3x^3 - 6x^2 - 5x - 1)$$
$$= 4x^3 + 2x^2 + x + 8 \qquad \text{Combine like terms.}$$

Connection The polynomial subtraction in Example 4 is equivalent to the following numeric subtraction:

$$\begin{array}{r} 7869 \\ -3651 \\ \hline 4218 \end{array}$$

Notice that the numeric result 4218 corresponds to the polynomial result $4x^3 + 2x^2 + x + 8$. We could use the stacking method for subtracting polynomials; however, as we saw with adding polynomials, it is not the best method for all cases because we often have missing terms or terms that are not like terms.

You may have noticed that the polynomials in Example 4 contained only plus signs. It is important to recognize that the polynomials may contain a mixture of signs and multiple variables. This is shown in Example 5.

Example 5 Subtract and write the resulting polynomial in descending order of degree.

a. $(15y^3 + 3y^2 + y - 2) - (7y^3 - 6y^2 + 5y - 8)$

SOLUTION Write an equivalent addition statement; then combine like terms.

$$(15y^3 + 3y^2 + y - 2) - (\quad 7y^3 - 6y^2 + 5y - 8)$$

Change the minus sign to a plus sign.   Change all signs in the subtrahend.

$$= (15y^3 + 3y^2 + y - 2) + (-7y^3 + 6y^2 - 5y + 8)$$
$$= 8y^3 + 9y^2 - 4y + 6 \qquad \text{Combine like terms.}$$

b. $(9.7x^5 - 2x^2y + xy^2 - 14.6xy - 7y^2) - (x^5 - 5.8xy^2 + 10.3xy - 15y^2)$

SOLUTION We write an equivalent addition statement, then combine like terms.

$$(9.7x^5 - 2x^2y + xy^2 - 14.6xy - 7y^2) - (\quad x^5 - 5.8xy^2 + 10.3xy - 15y^2)$$

Change the minus sign to a plus sign.   Change all signs in the subtrahend.

$$= (9.7x^5 - 2x^2y + xy^2 - 14.6xy - 7y^2) + (-x^5 + 5.8xy^2 - 10.3xy + 15y^2)$$
$$= 8.7x^5 - 2x^2y + 6.8xy^2 - 24.9xy + 8y^2$$

Note: The $-2x^2y$ term had no like term, so we rewrote it in the final expression.

Your Turn 5 Subtract and write the resulting polynomial in descending order of degree.

a. $(8t^4 + 5t^2 + 9t + 7) - (2t^4 + 5t^2 + t + 4)$

b. $(12.5x^4 - 15x^3 + 2x^2 - 9) - (13.1x^4 + 9x^3 - 6x^2 - 14)$

c. $(7a^5 - a^3b^2 + 9a^2b^2 - 2ab^2 + b - 18) - (2a^5 + a^3b^2 - ab^2 + 6b - 15)$

Answers Your Turn 5

a. $6t^4 + 8t + 3$

b. $-0.6x^4 - 24x^3 + 8x^2 + 5$

c. $5a^5 - 2a^3b^2 + 9a^2b^2 - ab^2 - 5b - 3$

6.3 Exercises   For Extra Help   *MyMathLab*

Math XP   PRACTICE   WATCH   DOWNLOAD

Note: Exercises marked with a ★ represent challenging exercises.

1. How do we add two polynomials?

2. What is the function of parentheses in the addition problem $(3x + 9) + (4x - 2)$?

3. How do we find the additive inverse of a polynomial?

4. How do we subtract two polynomials?

For Exercises 5–28, add and write the resulting polynomial in descending order of degree. See Examples 1 and 2.

5. $(3x + 2) + (5x - 1)$

6. $(5y + 4) + (3y + 1)$

7. $(8y + 7) + (2y + 5)$

8. $(3m - 4) + (5m + 1)$

9. $(2x + 3y) + (5x - 3y)$

10. $(4a - 3b) + (6a + 3b)$

11. $(2x + 5) + (3x^2 - 4x + 7)$

12. $(5p^2 + 3p - 1) + (4p + 8)$

13. $(z^2 - 3z + 7) + (5z^2 - 8z - 9)$

14. $(2w^2 - 5w - 1) + (4w^2 - 5w + 1)$

15. $(3r^2 - 2r + 10) + (2r^2 - 5r - 11)$

16. $(7m^2 + 8m - 1) + (-4m^2 - 3m - 2)$

17. $(4y^2 - 8y + 1) + (5y^2 + 8y + 2)$

18. $(5k^2 - k - 1) + (4k^2 + k + 7)$

19. $\left(x^2 - \dfrac{2}{3}x + 3 \right) + \left(2x^2 + \dfrac{3}{4}x - 2 \right)$

20. $\left(3a^2 + 3a - \dfrac{5}{6} \right) + \left(2a^2 - 5a + \dfrac{3}{4} \right)$

21. $(4r^2 + 2.7r - 3.6) + (-2r^2 - 1.3r - 4.2)$

22. $(5b^2 - 4.6b - 1.7) + (-4b^2 + 3.3b + 5.6)$

23. $(9a^3 + 5a^2 - 3a - 1) + (-4a^3 - 2a^2 + 6a - 2)$

24. $(4u^3 - 6u^2 + u + 11) + (-5u^3 - 3u^2 + u - 5)$

25. $(7p^3 - 9p^2 + 5p - 1) + (-4p^3 + 8p^2 + 2p + 10)$

26. $(12r^4 - 5r^2 + 8r - 15) + (-7r^4 + 3r^3 + 2r - 9)$

27. $(-5w^4 - 3w^3 - 8w^2 + w - 14) + (-3w^4 + 6w^3 + w^2 + 12w + 5)$

28. $(-r^4 + 3r^2 - 12r - 14) + (5r^4 - 2r^3 - 3r^2 + 10r - 5)$

★*For Exercises 29–38, add. See Example 2.*

29. $(a^3b^2 - 6ab^2 - 6a^2b + 5ab + 5b^2 - 6) + (-4a^3b^2 + 4ab^2 - 2a^2b - ab - 2b^2 - 2)$

30. $(x^2y^2 + 8xy^2 - 12x^2y - 4xy + 7y^2 - 9) + (-3x^2y^2 + 2xy^2 + 4x^2y - xy + 6y^2 + 4)$

31. $(-2u^4 + 6uv^3 - 8u^2v^3 + 7u^3v - u + v^2 + 9) + (-5u^4 - 6uv^3 + 9u^2v^3 + 8u - 14v^2 - 12)$

32. $(-13a^6 + a^3b^2 + 3ab^3 - 8b^3) + (11a^6 - 3a^2b^3 - 4ab^3 + 10ab^2 + 4b^3)$

33. $(-8mnp - 6m^2n^2 - 13mn^2p + 14m^2n - 12n + 6) + (-4nmp - 10m^2n^2 + mn^2p - 19n + 3)$

34. $(14abc - 2a^2b^2 - ab^2c + 4b - 9c - 3) + (-15abc + 5a^2b^2 - 13ab^2c + 7a - 12b + 6)$

35. $\left(4a^4 - \dfrac{2}{3}a^3b + \dfrac{3}{5}ab^3 - 3b^4\right) + \left(-2a^4 + \dfrac{1}{4}a^3b + \dfrac{1}{3}ab^3 + 2b^4\right)$

36. $\left(-5y^4 + \dfrac{3}{4}y^2z^2 - \dfrac{5}{6}yz^3 + 2z^4\right) + \left(3y^4 - \dfrac{1}{6}y^2z^2 - \dfrac{3}{8}yz^3 - 5z^4\right)$

37. $(3.6a^2bc + 2.3ab^2c^2 - 5.7a^2b^2c^2) + (-1.8a^2bc - 4.1ab^2c^2 + 3.3a^2b^2c^2)$

38. $(-6.8x^3yz^2 + 4.1x^2y^2z - 8.2xy^2z^2) + (5.2x^3yz^2 - 3.6x^2y^2z - 3.2xy^2z^2)$

For Exercises 39–42, write an expression for the perimeter in simplest form.
See Example 3.

39.

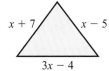

40.

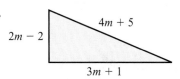

41.

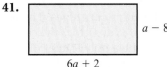

42.

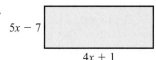

For Exercises 43–60, subtract and write the resulting polynomial in descending
order of degree. See Examples 4 and 5.

43. $(6x + 3) - (2x + 1)$

44. $(5m + 8) - (2m + 3)$

45. $(18a^2 + 3) - (5a^2 + 4)$

46. $(12b^2 + 4b) - (5b^2 + b)$

47. $(2x^2 - 3x + 4) - (6x + 2)$

48. $(7y^2 - 3y + 2) - (8y + 5)$

49. $(6z^2 - 3z + 1) - (4z^2 + 4z - 8)$

50. $(5y^2 - 3y - 10) - (5y^2 + 2y - 10)$

51. $(8a^2 + 11a - 12) - (-4a^2 + a + 3)$

52. $(y^2 + 3y + 6) - (-5y^2 + 3y - 8)$

53. $(6u^3 + 3u^2 - 8u + 5) - (7u^3 + 5u^2 - 2u - 3)$

54. $(-8t^3 + 3t^2 - 7t - 9) - (-10t^3 - 5t^2 + 3t - 11)$

55. $(-w^5 + 3w^4 + 8w^2 - w - 1) - (10w^5 + 3w^2 - w - 3)$

56. $(-s^4 + 7s^3 + 6s^2 - 13s - 11) - (3s^4 - 2s^2 - s + 5)$

57. $(-5p^4 + 8p^3 - 9p^2 + 3p + 1) - (6p^4 + 5p^3 + 4p^2 - 2p + 7)$

58. $(-6r^5 - 9r^4 + 5r^3 - 2r^2 + 8) - (5r^5 + 2r^4 + 7r^3 - 8r^2 - 9)$

59. $(16v^4 + 21v^2 + 4v - 8) - (5v^4 - 3v^3 - 2v^2 + 9v + 1)$

60. $(12q^4 + 5q^3 - 2q + 15) - (8q^4 - 14q^3 - 3q^2 + 15q + 8)$

For Exercises 61–68, subtract. See Example 5.

61. $(x^2 + xy + y^2) - (x^2 - xy + y^2)$

62. $(a^2 - 4ab + b^2) - (a^2 + 4ab + b^2)$

63. $(4xy + 5xz - 6yz) - (10xy - xz - 8yz)$

64. $(4ab - 5bc + 6ac) - (10ab - bc - ac)$

65. $(18p^3q^2 - p^3q^3 + 6pq^2 - 4pq + 16q^2 - 4) - (-p^3q^2 - p^3q^3 + 4pq^2 - 3pq - 6)$

66. $(15a^3y^4 + a^2y^2 + 5ay^3 - 7ay + 12y^2 - 11) - (-a^3y^4 + a^2y^2 + 2ay^3 - ay + 9)$

67. $(5.6a^2b - 2.3ab^2 - 4.3a^2b^2) - (3.7a^2b - 3.4ab^2 + 2.2a^2b^2)$

68. $(7.4m^3n - 6.7m^2n^2 + 2.6mn^3) - (3.6m^3n - 8.3m^2n^2 - 4.7mn^3)$

For Exercises 69–72, solve. (Hint: Profit = Revenue − Cost)

69. The polynomial $54.95x + 92.20y + 25.35z$ describes the revenue that a company generates from the sale of three different portable CD players. The expression $22.85x + 56.75y + 19.34z$ describes the cost of producing each CD player. Write a polynomial in simplest form that describes the company's net profit.

70. The polynomial $27.50x + 14.70y + 42.38z$ describes the revenue that a company generates from the sale of three different types of printer cartridges. The expression $12.75x + 5.25y + 27.42z$ describes the cost of producing each cartridge. Write a polynomial in simplest form that describes the company's net profit.

71. The polynomial $6.50a + 3.38b + 25.00c$ describes the revenue that a department store generates from the sale of three different candles. The expression $2.28a + 1.75b + 12.87c$ describes the cost the store pays to sell each candle. Write a polynomial in simplest form that describes the store's net profit.

72. The polynomial $10.55m + 14.75n + 27.50p$ describes the revenue that a pet store generates from the sale of three different litter boxes. The expression $5.73m + 8.26n + 15.22p$ describes the cost the store pays to sell each product. Write a polynomial in simplest form that describes the store's net profit.

Find
ⓧ *the mistake*

For Exercises 73 and 74, find and explain the mistake; then work the problem correctly.

73. $(3x^2 - 5x + 10) - (7x^2 - 3x + 5) = (3x^2 - 5x + 10) + (-7x^2 - 3x + 5)$
$$= 3x^2 - 5x + 10 - 7x^2 - 3x + 5$$
$$= -4x^2 - 8x + 15$$

74. $(7y^2 - 3y + 1) - (y^2 + 4y - 1) = (-7y^2 + 3y - 1) + (-y^2 - 4y + 1)$
$$= -7y^2 + 3y - 1 - y^2 - 4y + 1$$
$$= -8y^2 - y$$

Collaborative Exercises   Building Furniture and Profits

Suppose your group operates a furniture-manufacturing company that produces two different sizes of coffee table. Each larger coffee table sells for $350, and each smaller table sells for $280.

1. Write a monomial expression for the revenue if x is the number of large tables sold per month.

2. Write a monomial expression for the revenue if y is the number of small tables sold per month.

3. Write a polynomial expression that describes the total revenue generated from the sale of the two tables per month.

Suppose it costs $225 to produce each large table and $135 to produce each small table.

4. Write a monomial expression for the cost of producing x number of large tables per month.

5. Write a monomial expression for the cost of producing y number of small tables per month.

6. Your company also spends $18,500 per month to pay for the lease, utilities, and salaries. Write a polynomial expression that describes the total cost involved in production.

7. Using your polynomials for revenue and cost, write a polynomial in simplest form for the profit if x number of large tables are produced and sold and y number of small tables are produced and sold.

8. What does each coefficient in the polynomial for profit indicate?

9. Suppose in one month, 200 large tables are produced and sold and 400 small tables are produced and sold. Find the revenue, cost, and profit for the month.

10. Suppose in one month, 56 large tables are produced and sold and 65 small tables are produced and sold. Find the revenue, cost, and profit for the month. Explain the meaning of your answer.

11. Suppose in one month, only large tables are produced and sold. How many would have to be produced and sold to break even (profit is 0)?

12. Suppose only small tables are produced and sold in one month. How many would have to be produced and sold to break even?

Review Exercises

Exercises 1–5  EXPRESSIONS

For Exercises 1–4, evaluate.

[1.4] 1. $(2.3)(-7.5)$ **[1.4]** 2. $(-5)(-2)(1.5)$ **[6.1]** 3. 10^{-4} **[1.5]** 4. $10^3 \cdot 10^4$

[6.1] 5. Write 5.89×10^7 in standard form.

Exercise 6  EQUATIONS AND INEQUALITIES

[2.1] 6. The surface of the water in a rectangular swimming pool measures 20 meters by 15 meters. What is the surface area of the water?

6.4 Exponent Rules and Multiplying Monomials

Objectives

1. Multiply monomials.
2. Multiply numbers in scientific notation.
3. Simplify a monomial raised to a power.

In this section, we learn how to multiply monomials, multiply numbers in scientific notation, and simplify a monomial raised to a power. These simplifications require some new rules for exponents.

Objective 1 Multiply monomials. Because monomials contain variables in exponential form, to multiply monomials, we must develop a rule for multiplying exponential forms.

Multiplying Exponential Forms

Consider $2^3 \cdot 2^4$, which is a product of exponential forms. To simplify $2^3 \cdot 2^4$, we could follow the order of operations and evaluate the exponential forms first, then multiply.

$$2^3 \cdot 2^4 = 8 \cdot 16 = 128$$

However, there is an alternative. We can write the result in exponential form by first writing 2^3 and 2^4 in their factored forms.

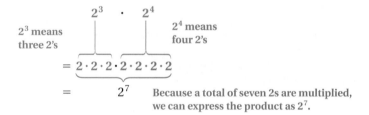

Notice that $2^7 = 128$, which agrees with our earlier calculation. Also notice that in the alternative method, the resulting exponent is the sum of the original exponents.

$$2^3 \cdot 2^4 = 2^{3+4} = 2^7$$

Conclusion To multiply exponential forms that have the same base, we can add the exponents and keep the same base.

This suggests the following rule.

Rule Product Rule for Exponents

If a is a real number and m and n are integers, then $a^m \cdot a^n = a^{m+n}$.

We apply this rule when multiplying monomials.

Example 1 Multiply $(x^5)(x^3)$.

SOLUTION Because the bases are the same, we can add the exponents and keep the same base.

$$(x^5)(x^3) = x^{5+3} = x^8$$

Your Turn 1 Multiply.

a. $3^2 \cdot 3^5$

b. $(c^7)(c^3)$

c. $t^5(t^3)$

Multiplying Monomials

Now let's consider multiplying monomials that have coefficients, such as $(4x^3)(2x^6)$. Because monomials are products by definition, we can use the commutative property of multiplication and write the following:

$$
\begin{aligned}
(4x^3)(2x^6) &= 4 \cdot x^3 \cdot 2 \cdot x^6 \\
&= 4 \cdot 2 \cdot x^3 \cdot x^6 \\
&= 8 \cdot x^{3+6} \quad \text{We multiply the coefficients and use the product} \\
&\qquad\qquad\quad \text{rule for exponents to simplify } x^3 \cdot x^6. \\
&= 8x^9
\end{aligned}
$$

What if the monomial factors have some variables that are different, as in $(5x^2)(7xy)$? If we expand the monomials in factored form and use the commutative property of multiplication, we have the following:

$$
\begin{aligned}
(5x^2)(7xy) &= 5 \cdot x^2 \cdot 7 \cdot x \cdot y \\
&= 5 \cdot 7 \cdot x^2 \cdot x \cdot y \\
&= 35 \cdot x^{2+1} \cdot y \\
&= 35x^3y
\end{aligned}
$$

Notice that the unlike variable base, y, is simply written unchanged in the result. Our examples suggest the following procedure.

Procedure **Multiplying Monomials**

To multiply monomials,

1. Multiply coefficients.
2. Add the exponents of the like bases.
3. Write any unlike variable bases unchanged in the product.

Example 2 Multiply.

a. $(9y^5)(3y^8)$

SOLUTION $(9y^5)(3y^8) = 9 \cdot 3y^{5+8}$ Multiply the coefficients and add the exponents of the like bases.

$$= 27y^{13}$$ Simplify the exponents.

b. $\left(-\dfrac{5}{6}a^2bc^3\right)\left(\dfrac{2}{3}a^5b^4\right)$

Note: Remember that a variable or number with no apparent exponent has an understood exponent of 1.

$$-\frac{5}{6}a^2bc^3 = -\frac{5}{6}a^2b^1c^3$$

SOLUTION $\left(-\dfrac{5}{6}a^2bc^3\right)\left(\dfrac{2}{3}a^5b^4\right) = -\dfrac{5}{\underset{3}{6}} \cdot \dfrac{\overset{1}{2}}{3}a^{2+5}b^{1+4}c^3$ Multiply the coefficients, add the exponents of the like bases, and write the unlike variable base c unchanged in the product.

$$= -\frac{5}{9}a^7b^5c^3$$ Simplify.

Your Turn 2 Multiply.

a. $(9n^4)(5n^8)$

b. $-2.5x^2y^3(-6xy^2z^3)$

Answers **Your Turn 1**

a. 3^7 b. c^{10} c. t^8

Answers **Your Turn 2**

a. $45n^{12}$ b. $15x^3y^5z^3$

Connection Multiplying distance measurements in area and volume is like multiplying monomials.

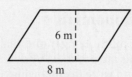 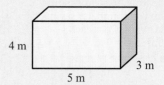

The area of the parallelogram:

$$A = bh$$
$$A = (8\,\text{m})\,(6\,\text{m})$$
$$A = 48\,\text{m}^2$$

The area calculation is similar to multiplying two monomials with m as the variable.

Multiply: $(8m)(6m) = 48m^2$

The volume of the box:

$$V = lwh$$
$$V = (5\,\text{m})(3\,\text{m})(4\,\text{m})$$
$$V = 60\,\text{m}^3$$

The volume calculation is similar to multiplying three monomials with m as the variable.

Multiply: $(5m)(3m)(4m) = 60m^3$

Example 3 Write an expression in simplest form for the volume of the box shown.

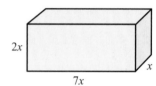

Understand We are given a box with side lengths that are monomial expressions.

Plan The volume of a box is found by multiplying the length, width, and height.

Execute $V = lwh$

$$V = (7x)\,(x)\,(2x)$$
$$V = 14x^3$$

Answer The expression for volume is $14x^3$.

Check Because $(7x)(x)(2x) = 14x^3$ is an identity, we can evaluate $(7x)$, (x), $(2x)$, and $14x^3$ using any chosen value for x and get the same result. Let's choose a value for x, such as 3, and try it.

$$(7x)(x)(2x) = 14x^3$$
$$[7(3)](3)[2(3)] \stackrel{?}{=} 14(3)^3$$
$$(21)(3)(6) \stackrel{?}{=} 14(27)$$
$$378 = 378 \qquad \text{This is true.}$$

Note: Choosing only one value does not guarantee that our simplification is correct, but it is a good indicator. Testing several values provides more conclusive assurance.

Objective ② Multiply numbers in scientific notation. We multiply numbers in scientific notation using the same procedure we used to multiply monomials. If we replace a with 10 in $(4a^3)\,(2a^6)$, we have a product of two numbers in scientific notation: $(4 \times 10^3)(2 \times 10^6)$. Following the same procedure for multiplying monomials, we multiply $4 \cdot 2$ to get 8 and then add the exponents of the like bases.

Monomials:

$$(4a^3)\,(2a^6) = 4 \cdot 2a^{3+6}$$
$$= 8a^9$$

Scientific notation:

$$(4 \times 10^3)\,(2 \times 10^6) = 4 \times 2 \times 10^{3+6}$$
$$= 8 \times 10^9$$

Note: The product 17.25×10^{11} is *not* in scientific notation because the absolute value of 17.25 is *not less than* 10. To fix this problem, we must move the decimal point one place to the left and account for this move by increasing the exponent by 1. If you are unconvinced that this is correct, consider the fact that 17.25×10^{11} is equal to 1,725,000,000,000. If we now write that standard form number in scientific notation, the proper position for the decimal point is between the 1 and the 7 digits, which means we must account for 12 places.

Example 4 Multiply and write the answers in scientific notation.

a. $(2.3 \times 10^5)(7.5 \times 10^6)$

SOLUTION Multiply 2.3 and 7.5; then add the exponents for base 10s.

$$(2.3 \times 10^5)(7.5 \times 10^6) = 2.3 \times 7.5 \times 10^{5+6}$$
$$= 17.25 \times 10^{11}$$
$$= 1.725 \times 10^{12}$$

b. $(4.2 \times 10^4)(6.5 \times 10^{-9})$

SOLUTION Multiply 4.2 and 6.5 and add the exponents for the base 10s.

$$(4.2 \times 10^4)(6.5 \times 10^{-9}) = 4.2 \times 6.5 \times 10^{4+(-9)}$$
$$= 27.3 \times 10^{-5}$$
$$= 2.73 \times 10^{-4} \quad \text{Rewrite in scientific notation.}$$

Your Turn 4 Multiply and write the answers in scientific notation.

a. $(4.2 \times 10^5)(2.8 \times 10^8)$ **b.** $(2.7 \times 10^{-3})(7.3 \times 10^{10})$

Objective 3 **Simplify a monomial raised to a power.** We now discuss how to simplify an expression such as $(3x^2)^4$, which is a monomial raised to a power. To simplify such an expression, we need to discuss two new rules of exponents: (1) raising a power to a power and (2) raising a product to a power.

Raising a Power to a Power

Consider a power raised to a power, such as $(2^3)^2$. To calculate $(2^3)^2$, we could follow the order of operations and evaluate the exponential form 2^3 within the parentheses first, then square the result.

$$(2^3)^2 = (2 \cdot 2 \cdot 2)^2 = 8^2 = 64$$

However, there is an alternative. The outside exponent, 2, indicates to multiply the inside exponential form 2^3 by itself.

$$(2^3)^2 = 2^3 \cdot 2^3$$
$$= 2^{3+3} \quad \text{Because this is a multiplication of exponential forms that have the same base, we can add the exponents.}$$
$$= 2^6$$

Notice that $2^6 = 64$. Also, in comparing $(2^3)^2$ with 2^6, notice that the exponent in 2^6 is the product of the original exponents.

$$(2^3)^2 = 2^{3 \cdot 2} = 2^6$$

Conclusion If an exponential form is raised to a power, we can multiply the exponents and keep the same base.

Rule **A Power Raised to a Power**

If a is a real number and m and n are integers, then $(a^m)^n = a^{mn}$.

Answers Your Turn 4
a. 1.176×10^{14} **b.** 1.971×10^8

Now let's consider a product raised to a power, such as $(3 \cdot 4)^2$. We could write the factored form twice and use the commutative property to rearrange the like bases.

$$(3 \cdot 4)^2 = (3 \cdot 4)(3 \cdot 4)$$

Because there are two factors of 3 and two factors of 4, we can write each of these in exponential form.

$$= 3 \cdot 3 \cdot 4 \cdot 4$$
$$= 3^2 \cdot 4^2$$

Conclusion If a product is raised to a power, we can evaluate the factors raised to that power.

Rule **Raising a Product to a Power**

If a and b are real numbers and n is an integer, then $(ab)^n = a^n b^n$.

Raising a Monomial to a Power

We apply these exponent rules to simplify an expression such as $(3x^2)^4$, which is a monomial raised to a power. Because the monomial $3x^2$ expresses a product, we can use the rule for raising a product to a power and raise the factors 3 and x^2 to the 4th power.

$$(3x^2)^4 = (3)^4(x^2)^4$$

To simplify $(x^2)^4$, we use the rule for raising a power to a power.

$$= 81x^{2 \cdot 4}$$
$$= 81x^8$$

Notice that the outside exponent is distributed to the coefficient and the variable(s) within the monomial. This suggests the following procedure.

Procedure **Simplifying a Monomial Raised to a Power**

To simplify a monomial raised to a power,

1. Evaluate the coefficient raised to that power.
2. Multiply each variable's exponent by the power.

Example 5 Simplify.

a. $(5a^6)^3$

SOLUTION $(5a^6)^3 = (5)^3 a^{6 \cdot 3}$

Write the coefficient, 5, raised to the 3rd power and multiply the variable's exponent by 3.

$$= 125a^{18}$$

Simplify.

b. $\left(-\dfrac{2}{3}xy^3z^5\right)^4$

Note: Recall that $x = x^1$.

SOLUTION $\left(-\dfrac{2}{3}xy^3z^5\right)^4 = \left(-\dfrac{2}{3}\right)^4 x^{1 \cdot 4} y^{3 \cdot 4} z^{5 \cdot 4}$

Write the coefficient, $-\dfrac{2}{3}$, raised to the 4th power and multiply each variable's exponent by 4.

$$= \dfrac{16}{81}x^4 y^{12} z^{20}$$

Simplify.

c. $(-0.4mn^6)^3$

SOLUTION $(-0.4mn^6)^3 = (-0.4)^3 m^{1 \cdot 3} n^{6 \cdot 3}$

$$= -0.064m^3 n^{18}$$

Your Turn 5 Simplify.

a. $(2y^4)^5$ 　　　　　**b.** $\left(-\dfrac{5}{3}t^2u^6\right)^3$ 　　　　　**c.** $(-0.2a^4b^5c)^6$

Now let's consider expressions that require us to use all of these exponent rules.

Example 6 Simplify.

a. $(3x^2)(4x^5)^3$

SOLUTION Because the order of operations is to simplify exponents before multiplying, we simplify $(4x^5)^3$ first, then multiply the result by $3x^2$.

$$
\begin{aligned}
(3x^2)(4x^5)^3 &= (3x^2)(4^3x^{5\cdot3}) \\
&= (3x^2)(64x^{15}) \quad \text{Simplify.} \\
&= 3 \cdot 64x^{2+15} \quad \text{Multiply coefficients and add} \\
&\qquad\qquad\qquad\quad \text{exponents of like variables.} \\
&= 192x^{17}
\end{aligned}
$$

b. $(-0.2a)^3(4ab)(1.5a^3c)^2$

SOLUTION We follow the order of operations and simplify the monomials raised to a power first, then multiply the monomials.

$$
\begin{aligned}
(-0.2a)^3(4ab)(1.5a^3c)^2 &= ((-0.2)^3a^{1\cdot3})(4ab)((1.5)^2a^{3\cdot2}c^{1\cdot2}) \\
&= (-0.008a^3)(4ab)(2.25a^6c^2) \quad \text{Simplify.} \\
&= -0.008 \cdot 4 \cdot 2.25a^{3+1+6}bc^2 \quad \text{Multiply coefficients} \\
&\qquad\qquad\qquad\qquad\qquad\qquad \text{and add exponents of} \\
&= -0.072a^{10}bc^2 \qquad\qquad\qquad\qquad \text{like variables.}
\end{aligned}
$$

Answers Your Turn 5

a. $32y^{20}$ 　**b.** $-\dfrac{125}{27}t^6u^{18}$

c. $0.000064a^{24}b^{30}c^6$

Answers Your Turn 6

a. $\dfrac{1}{18}y^{11}$ 　**b.** $60m^6n^5p^6$

Your Turn 6 Simplify.

a. $\left(\dfrac{1}{4}y^3\right)^2\left(\dfrac{8}{9}y^5\right)$ 　　　　　**b.** $(-5mn^4)(-2mp^2)^3(1.5m^2n)$

6.4 Exercises For Extra Help *MyMathLab*

Note: Exercises marked with a ★ represent challenging exercises.

1. What rule can be applied when two exponential forms that have the same base are multiplied?

2. Explain why $2^3 \cdot 5^2 \neq 10^5$.

3. Explain how to multiply monomials.

4. In multiplying the powers of 10 in $(8 \times 10^4)(6 \times 10^5)$, we add the exponents to get 10^9. However, the correct answer in scientific notation is 4.8×10^{10}. Why is the exponent 10 instead of 9?

5. How do we simplify $(x^3)^8$?

6. Explain how to simplify $(3x^4)^2$.

For Exercises 7–32, multiply. See Examples 1 and 2.

7. $a^3 \cdot a^4$

8. $b^2 \cdot b^3$

9. $2^3 \cdot 2^{10}$

10. $5^4 \cdot 5^3$

11. $a^3 \cdot a^2 b$

12. $y^2 \cdot y^3 z$

13. $2x \cdot 3x^2$

14. $5b \cdot 2b^3$

15. $(-2mn)(4m^2n)$

16. $(-3uv^2)(2u^3v)$

17. $\left(\dfrac{5}{8}st\right)\left(-\dfrac{2}{7}s^2t^7\right)$

18. $\left(\dfrac{3}{8}x^2y\right)\left(\dfrac{4}{9}xy^2\right)$

19. $(2.3ab^2c)(1.2a^2b^3c^5)$

20. $(-3.1mn)(2.4m^2n)$

21. $(r^2)(2r^2)(-3r)$

22. $(-2j)(-5j^3)(7j)$

23. $3xz^2(-4x^3z^2)(z^5)$

24. $-5mn^4(-m^3n^9)(4m^3n)$

25. $(5xyz)(9x^2y^4)(2x^3z^2)$

26. $(6hj^3k)(7h^5k)(hkj)$

27. $(5q^2r^4s^9)(-q^3rs^2)(-r^5s^{10})$

28. $(a^3b^3c^3)(-6a^3b^3c^3)(-a^4c^2)$

★**29.** $-5a^2\left(-\dfrac{1}{15}abc\right)\left(-\dfrac{3}{4}a^3b^2c\right)$

★**30.** $\left(-\dfrac{1}{4}l^2m\right)\left(\dfrac{5}{6}lm^3n\right)\left(-\dfrac{2}{3}l^5\right)$

★**31.** $(0.4wxy)(w^2y^2)(2.5x^2y^2)$

★**32.** $(0.4u^2v)(-3.2u^2v^3w^3)(0.2vw^9)$

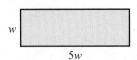

Find
the mistake

*For Exercises 33–36, find and explain the mistake; then work the
problem correctly. See Examples 1 and 2.*

33. $7x^3 \cdot 5x^4y = 35x^{12}y$

34. $-3m^2n \cdot 10m^5n = -30m^7n$

35. $(9x^3y^2z)(6xy^4) = 54x^3y^6z$

36. $(a^3b^2c)(ab^5) = a^4b^7$

*For Exercises 37 and 38, write an expression in simplest form for the area of the
figure. See Example 3.*

37.

w

$5w$

38.

h

$7h$

For Exercises 39 and 40, write an expression in simplest form for the volume of the figure. See Example 3.

39.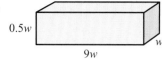

0.5w

9w

w

40.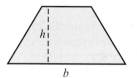

h

0.1h

0.2h

★**41.** The height of a trapezoid is half the base. The top side of the trapezoid is one-fourth the base.

   a. If *b* represents the length of the base, write expressions for the height and length of the top side of the trapezoid in terms of *b*.

   b. Using your expressions from part a, write an expression in simplest form for the area of the trapezoid. $\left[\text{Use } A = \frac{1}{2}h(a + b). \right]$

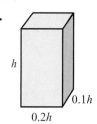

h

b

★**42.** The height of a box is three times the width. The length is five times the width.

   a. If *w* represents the width, write expressions for the height and length of the box in terms of *w*.

   b. Using your expressions from part a, write an expression in simplest form for the volume of the box.

   c. Using your expressions from part a, write an expression in simplest form for the surface area of the box. (Use $SA = 2lw + 2lh + 2wh$.)

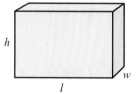

h

l

w

For Exercises 43–52, multiply and write your answer in scientific notation. See Example 4.

43. $(2 \times 10^5)(6 \times 10^3)$

44. $(9 \times 10^6)(5 \times 10^5)$

45. $(3.2 \times 10^5)(7.5 \times 10^8)$

46. $(4.5 \times 10^{10})(8.4 \times 10^7)$

47. $(2.9 \times 10^8)(6.3 \times 10^{-4})$

48. $(7.2 \times 10^{-2})(4.6 \times 10^7)$

49. $(4.23 \times 10^{-8})(2.7 \times 10^3)$

50. $(3.75 \times 10^6)(4.3 \times 10^{-12})$

51. $(9.2 \times 10^{-3})(6.1 \times 10^{-8})$

52. $(8.4 \times 10^{-6})(7.3 \times 10^{-8})$

53. Tennessee is roughly shaped like a parallelogram with a base of approximately 350 miles and a height of approximately 120 miles. Calculate the area of Tennessee and write your answer in scientific notation. (Use $A = bh$.)

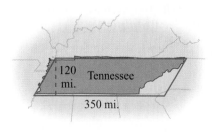

120 mi.

Tennessee

350 mi.

54. Idaho is roughly shaped like a triangle with a base of approximately 320 miles and a height of approximately 520 miles. Calculate the area of Idaho and write your answer in scientific notation. $\left(\text{Use } A = \dfrac{1}{2}bh. \right)$

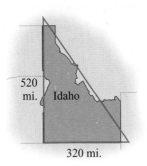

For Exercises 55 and 56, use the formula for the circumference of a circle: $C = 2\pi r$. (Use 3.14 to approximate π.) Write your answers in scientific notation. See Example 4.)

★ **55.** The path of Earth's orbit around the Sun is approximately a circle with a radius of about 9.3×10^7 miles. What is the circumference of Earth's orbit?

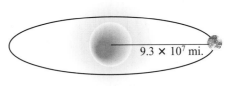

Not to scale.

★ **56.** The radius of Earth is approximately 6.37×10^6 meters. Calculate the circumference of the equator.

For Exercises 57 and 58, use the formula for the volume of a sphere: $V = \dfrac{4}{3}\pi r^3$.

(Use 3.14 to approximate π.) Write your answers in scientific notation. See Example 4.

★ **57.** The radius of the Sun is approximately 6.96×10^8 meters. Calculate the volume of the Sun.

★ **58.** The radius of Earth is approximately 6.37×10^6 meters. Calculate the volume of Earth.

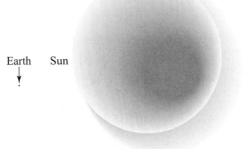

Earth   Sun

For Exercises 59 and 60, use the following information. The energy, in joules, of a single photon of light can be determined by $E = hv$, where h is the constant 6.626×10^{-34} joule-seconds and v is the frequency of the light in hertz. Write your answers in scientific notation. See Example 4.

59. Find the energy of a photon of red light with a frequency of 4.5×10^{14} Hz.

60. Find the energy of a photon of violet light with a frequency of 8.5×10^{14} Hz.

For Exercises 61 and 62, use the following information. Albert Einstein discovered that if a mass of m kilograms is converted to pure energy (as in a nuclear reaction), the amount of energy, in joules, that is released is described by $E = mc^2$, where c represents a constant speed of light, which is approximately 3×10^8 meters per second. Write your answers in scientific notation. See Example 4.

61. Suppose 2.4×10^{-10} kilogram of uranium is converted to energy in a nuclear reaction. How much energy is released?

62. Suppose 4.5×10^{-8} kilogram of uranium is converted to energy in a nuclear reaction. How much energy is released?

For Exercises 63–98, simplify. See Examples 5 and 6.

63. $(x^2)^3$

64. $(y^3)^5$

65. $(3^2)^4$

66. $(2^3)^2$

67. $(-h^2)^4$

68. $(-j^3)^2$

69. $(-y^2)^3$

70. $(-x^4)^5$

71. $(xy)^3$

72. $(ab)^3$

73. $(3x^2)^3$

74. $(4x^4)^2$

75. $(-2x^3y^2)^3$

76. $(-5a^4b^2)^2$

77. $\left(\frac{1}{4}m^2n\right)^3$

78. $\left(\frac{5}{6}u^3v\right)^2$

79. $(-2p^4q^4r^2)^6$

80. $(-6u^2v^4w^2)^3$

81. $(-0.2x^2y^5z)^3$

82. $(-0.3r^4st^3)^4$

83. $(2x^3)(3x^2)^2$

84. $(-3y^2)(2y^2)^3$

85. $(rs^2)(r^2s)^2$

86. $(m^2n)^3(mn)$

87. $(2ab)^2(5a^2b)^2$

88. $(4xy^3)^2(2x^2y)^2$

89. $\left(\frac{1}{4}a^2b^3c\right)^2\left(\frac{1}{4}a^2b^3c\right)^3$

90. $\left(-\frac{1}{2}s^2tu^4\right)^2\left(-\frac{1}{2}s^2tu^4\right)^5$

91. $5(3a)^2(2b)^3$

92. $2(2x)^4(4y)^3$

93. $-3(-2a^4b^2)^3(3a^3b)^2$

94. $-6(-2m^2n^4)^3(-4mn^3)^2$

95. $(6u)^2(u^2v)(-3uv)^2$

96. $(3x)^2(xy^3)(2x^3yz)^3$

97. $(0.6u)^2(u^2v)(-3uv)^2$

98. $(-0.2w)^4(-w^2z^3)^3(-4w^4z)^2$

Find
the mistake

For Exercises 99–102, find and explain the mistake; then work the problem correctly. See Example 5.

99. $(5x^3)^2 = 25x^5$

100. $(-4mn^5)^3 = -12m^3n^{15}$

101. $(2x^4y)^4 = 8x^8y$

102. $(-2ab^5)^3 = -8ab^{15}$

a. If there are about 10^{11} stars in a galaxy and about 10^{11} galaxies in the visible universe, about how many stars are in the visible universe?

b. Suppose one star out of every billion stars has a solar system with one planet that could support life. How many stars in the visible universe would have a solar system with one planet that could support life? (Hint: One out of every billion is one billionth, or 10^{-9}.)

Review Exercises

Exercises 1–6  **EXPRESSIONS**

[1.7] **1.** Simplify $2(3x + 4)$ using the distributive property.

[6.4] **2.** Simplify $-0.3t^2u \cdot 6t^3u$.

[1.7] **3.** Simplify $-10x - 12x$.

[6.3] **4.** Add $(3x + 4y - z) + (-7x - 4y - 3z)$.

[1.3] **5.** Add $(-4) + (-57)$.

[6.2] **6.** What is the degree of the monomial $-2r^5s$?

6.5 Multiplying Polynomials; Special Products

Objectives

1 Multiply a polynomial by a monomial.

2 Multiply binomials.

3 Multiply polynomials.

4 Determine the product when given special polynomial factors.

In this section, we use the rules for multiplying monomials to help us multiply multiple-term polynomials.

Objective **1** **Multiply a polynomial by a monomial.** To multiply polynomials, we use the distributive property, which we introduced in Section 1.7 and used in solving linear equations and inequalities in Chapter 2. For example, we have applied the distributive property in situations such as $2(3x + 4)$.

$$2(3x + 4) = 2 \cdot 3x + 2 \cdot 4$$
$$= 6x + 8$$

Notice that $2(3x + 4)$ is a product of a monomial 2 and a binomial $3x + 4$. This suggests the following procedure.

Procedure **Multiplying a Polynomial by a Monomial**

To multiply a polynomial by a monomial, use the distributive property to multiply each term in the polynomial by the monomial.

Connection Multiplying a polynomial by a monomial is like multiplying a multidigit number by a single-digit number. Consider $2 \cdot 34$.

$$\begin{array}{r} 34 \\ \times\ 2 \\ \hline 68 \end{array}$$

Notice that in the numerical multiplication, we distribute the 2 to each digit in the 34. In fact, if we expand 34 by expressing it as $30 + 4$, the problem takes the same form as $2(3x + 4)$, where $x = 10$.

$$\begin{aligned} 2(30 + 4) &= 2 \cdot 30 + 2 \cdot 4 & 2(3x + 4) &= 2 \cdot 3x + 2 \cdot 4 \\ &= 60 + 8 & &= 6x + 8 \\ &= 68 \end{aligned}$$

Now let's multiply a polynomial by a monomial that contains variables.

Example 1   Multiply.

a. $2y(3y^2 + 4y + 1)$

SOLUTION $2y \quad (3y^2 + 4y + 1)$ Multiply each term in the polynomial by $2y$.

$2y \cdot 3y^2$
$2y \cdot 4y$
$2y \cdot 1$

$$= 2y \cdot 3y^2 + 2y \cdot 4y + 2y \cdot 1$$
$$= 6y^3 + 8y^2 + 2y$$

Note: When multiplying the individual terms by the monomial, we multiply the coefficients and add the exponents of the like bases.

b. $-7x^2(5x^2 + 3x - 4)$

SOLUTION $-7x^2 \quad (5x^2 + 3x - 4)$ Multiply each term in the polynomial by $-7x^2$.

$-7x^2 \cdot 5x^2$
$-7x^2 \cdot 3x$
$-7x^2 \cdot (-4)$

$$= -7x^2 \cdot 5x^2 - 7x^2 \cdot 3x - 7x^2 \cdot (-4)$$
$$= -35x^4 - 21x^3 + 28x^2$$

Note: When we multiply a polynomial by a negative monomial, the signs of the resulting polynomial are opposite the signs in the original polynomial.

$$-7x^2(\ 5x^2\ +\ 3x\ -\ 4)$$
$$\downarrow \qquad \downarrow \qquad \downarrow$$
$$= -35x^4 - 21x^3 + 28x^2$$

LEARNING Strategy

VISUAL If you are a visual learner, try drawing lines connecting the terms that are to be multiplied.

---

c. $-0.3t^2u(6t^3u + tu^2 - 3t^3 + 0.2uv)$

SOLUTION $-0.3t^2u \quad (\ 6t^3u\ +\ tu^2\ -\ 3t^3\ +\ 0.2uv)$ Multiply each term in the polynomial by $-0.3t^2u$.

$-0.3t^2u \cdot 6t^3u$
$-0.3t^2u \cdot tu^2$
$-0.3t^2u \cdot (-3t^3)$
$-0.3t^2u \cdot 0.2uv$

$$= -0.3t^2u \cdot 6t^3u - 0.3t^2u \cdot tu^2 - 0.3t^2u \cdot (-3t^3) - 0.3t^2u \cdot 0.2uv$$
$$= -1.8t^5u^2 - 0.3t^3u^3 + 0.9t^5u - 0.06t^2u^2v$$

Note: When multiplying multivariable terms, it is helpful to multiply the coefficients first, then the variables in alphabetical order.

For $-0.3t^2u \cdot 6t^3u$, think $-0.3 \cdot 6 = -1.8$; $t^2 \cdot t^3 = t^5$; $u \cdot u = u^2$.

Result: $-1.8t^5u^2$

a. $4x^3(9x^2 - 7x + 3)$

b. $-3x^3yz(5x^2z - y^3z^2 + 8xz)$

Objective ② Multiply binomials.

Understanding Binomial Multiplication

Now that we have seen how to multiply a polynomial by a monomial, we are ready to multiply two binomials, which is like multiplying a pair of two-digit numbers. For example, consider $(12)(13)$ and compare this with $(x + 2)(x + 3)$.

$$
\begin{array}{r}
12 \\
\times\, 13 \\
\hline
36 \\
+\, 12 \\
\hline
156
\end{array}
$$

Note: We think to ourselves "3 times 2 is 6l, then 3 times 1 is 3," which creates the 36. We then move to the 1 in the tens place of the 13 and do the same thing. Because this 1 digit is in the tens place, it means 10; so when we multiply this 10 times 12, it makes 120. We usually omit writing the 0 in the ones place and write 12 in the next two places.

Again, the distributive property is the governing principle. We multiply each digit in one number by each digit in the other number and shift underneath as we move to each new place. The same process applies to the binomials. We can stack them as we just did. Notice, however, that we stack this only once to make the connection to numeric multiplication. As the polynomials get more complex, the stacking method becomes too tedious.

$$
\begin{array}{r}
x + 2 \\
x + 3 \\
\hline
3x + 6 \\
x^2 + 2x \\
\hline
x^2 + 5x + 6
\end{array}
$$

Note: We think "3 times 2 is 6, then 3 times x is $3x$." Now move to the x and think "x times 2 is $2x$, then x times x is x^2." Notice how we shifted so that the $2x$ and $3x$ line up. This is because they are like terms. It is the same as lining up the tens column when multiplying numbers. Note that the numeric result, 156, is basically the same as the algebraic result, $x^2 + 5x + 6$.

Notice that each term in $x + 2$ is multiplied by each term in $x + 3$. In general, this is how we multiply two polynomials.

Procedure   **Multiplying Polynomials**

To multiply two polynomials,

1. Multiply each term in the second polynomial by each term in the first polynomial.
2. Combine like terms.

Multiply Binomials

Example 2   Multiply $(x + 4)(x + 3)$.

SOLUTION   Multiply each term in $x + 3$ by each term in $x + 4$.

$$(x + 4)(x + 3)$$

$x \cdot 3$
$x \cdot x$
$4 \cdot x$
$4 \cdot 3$

Connection   If every term in both binomials is positive, the like terms are added and both signs in the resulting trinomial are plus signs. Noting these sign patterns will help when we factor in Chapter 7.

$$= x \cdot x + x \cdot 3 + 4 \cdot x + 4 \cdot 3$$
$$= x^2 + 3x + 4x + 12$$
$$= x^2 + 7x + 12 \qquad \text{Combine like terms: } 3x + 4x = 7x.$$
$$= x^2 + 7x + 12$$

The word *FOIL* is a popular way to remember the process of multiplying two binomials. *FOIL* stands for **First Outer Inner Last**. We will use Example 2 to demonstrate.

$(x + 4)(x + 3)$ First terms: $x \cdot x = x^2$

$(x + 4)(x + 3)$ Outer terms: $x \cdot 3 = 3x$

$(x + 4)(x + 3)$

   Inner terms: $4 \cdot x = 4x$

$(x + 4)(x + 3)$

   Last terms: $4 \cdot 3 = 12$

Example 3   Multiply $(3x + 5)(x - 4)$.

SOLUTION   Multiply each term in $x - 4$ by each term in $3x + 5$. (Think FOIL.)

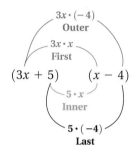

Connection   Notice that $3x + 5$ has a plus sign, whereas $x - 4$ has a minus sign. When multiplying binomials such as these, the last term of the resulting trinomial is always negative. The middle term can be positive or negative depending on the coefficients of the like terms.

      First   **Outer**   Inner   **Last**
$$= 3x \cdot x + 3x \cdot (-4) + 5 \cdot x + 5 \cdot (-4)$$
$$= 3x^2 \qquad\quad - 12x \quad + 5x \quad\; - 20$$
$$= 3x^2 \qquad\qquad\qquad - 7x \qquad\quad - 20$$
$$= 3x^2 - 7x - 20$$

Combine like terms: $-12x + 5x = -7x$.

In Example 4, we switch the signs in the binomials from Example 3. Look at how this changes the sign of the middle term in the resulting trinomial.

Example 4   Multiply $(3x - 5)(x + 4)$.

SOLUTION   Multiply each term in $x + 4$ by each term in $3x - 5$.

$$
\begin{array}{c}
3x \cdot 4 \\
3x \cdot x \\
(3x - 5) \quad (x + 4) \\
(-5) \cdot x \\
(-5) \cdot 4
\end{array}
$$

     First   Outer   Inner   Last
$$= 3x \cdot x + 3x \cdot 4 + (-5) \cdot x + (-5) \cdot 4$$
$$= 3x^2 \quad\; + 12x \quad - 5x \quad\; - 20$$
$$= 3x^2 \qquad\qquad + 7x \qquad\; - 20$$
$$= 3x^2 + 7x - 20$$

Combine like terms: $12x - 5x = 7x$.

Now consider a case when the signs between terms in the binomials are both negative.

Example 5   Multiply $(2x - 3)(4x - 5)$.

SOLUTION   Multiply each term in $4x - 5$ by each term in $2x - 3$. (Think *FOIL*.)

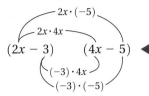

Connection   When multiplying binomials such as $2x - 3$ and $4x - 5$, notice that the last term in the resulting trinomial is positive because the product of two negatives is positive and the middle term is negative because the sum of two negatives is negative.

$$\begin{array}{cccc} \text{First} & \text{Outer} & \text{Inner} & \text{Last} \end{array}$$
$$= 2x \cdot 4x + 2x \cdot (-5) + (-3) \cdot 4x + (-3) \cdot (-5)$$
$$= 8x^2 \quad\quad - 10x \quad\quad - 12x \quad\quad + 15$$
$$= 8x^2 \quad\quad\quad - 22x \quad\quad\quad + 15 \quad \text{Combine like terms: } -10x - 12x = -22x.$$
$$= 8x^2 - 22x + 15$$

Keep in mind that Examples 2 through 5 do not illustrate all sign combinations that you could be given. The point of our discussion of signs is to note some patterns with the signs in the given binomial factors to see how they affect the resulting polynomial. Noticing these sign patterns now will help when we factor in Chapter 7.

Your Turn 5   Multiply.

a. $(n + 6)(n - 2)$ **b.** $(2y - 5)(y - 3)$

Connection   The product of two binomials can be shown in terms of geometry. Consider the rectangle shown with a length of $x + 7$ and a width of $x + 5$. We can describe its area in two ways: (1) the product of the length and width, $(x + 7)(x + 5)$, and (2) the sum of the areas of the four internal rectangles, $x^2 + 5x + 7x + 35$. Because these two approaches describe the same area, the two expressions must be equal.

	x	7
x	x^2	$7x$
5	$5x$	35

$$\begin{aligned} \text{Length} \cdot \text{width} &= \text{Sum of the areas of the} \\ &\quad\ \text{four internal rectangles} \\ (x + 7)(x + 5) &= x^2 + 5x + 7x + 35 \\ &= x^2 + 12x + 35 \quad \text{Combine like terms.} \end{aligned}$$

Objective 3   Multiply polynomials.   Now consider an example of multiplication involving a larger polynomial, such as a trinomial. Remember that no matter how many terms are in the polynomials, the process is to multiply every term in the second polynomial by every term in the first polynomial.

Example 6 Multiply.

a. $(x + 2)(3x^2 + 5x - 4)$

SOLUTION Multiply each term in $3x^2 + 5x - 4$ by each term in $x + 2$.

$$
\begin{array}{c}
x \cdot (-4) \\
x \cdot 5x \\
x \cdot 3x^2 \\
(x + 2) \quad (3x^2 + 5x - 4) \\
2 \cdot 3x^2 \\
2 \cdot 5x \\
2 \cdot (-4)
\end{array}
$$

Warning: FOIL does not make sense here because there are too many terms. FOIL handles only the four terms from two binomials.

$$
\begin{aligned}
&= x \cdot 3x^2 + x \cdot 5x + x \cdot (-4) + 2 \cdot 3x^2 + 2 \cdot 5x + 2 \cdot (-4) \\
&= 3x^3 \qquad\; + 5x^2 \; - 4x \qquad + 6x^2 \qquad + 10x \qquad\; - 8 \\
&= 3x^3 + 11x^2 + 6x - 8 \qquad \text{Combine like terms.}
\end{aligned}
$$

b. $(2x - 3)(4x^2 + 6x + 9)$

SOLUTION Multiply each term in $4x^2 + 6x + 9$ by each term in $2x - 3$.

$$
\begin{aligned}
&(2x - 3)(4x^2 + 6x + 9) \\
&= 2x(4x^2 + 6x + 9) - 3(4x^2 + 6x + 9) \\
&= 8x^3 + 12x^2 + 18x - 12x^2 - 18x - 27 \qquad \text{Distribute.} \\
&= 8x^3 - 27 \qquad\qquad\qquad\qquad\qquad\qquad \text{Combine like terms.}
\end{aligned}
$$

c. $(x^2 - 3x + 4)(x^2 + 2x - 5)$

SOLUTION Multiply each term in $x^2 + 2x - 5$ by each term in $x^2 - 3x + 4$.

$$
\begin{aligned}
&(x^2 - 3x + 4)(x^2 + 2x - 5) \\
&= x^2(x^2 + 2x - 5) - 3x(x^2 + 2x - 5) + 4(x^2 + 2x - 5) \\
&= x^4 + 2x^3 - 5x^2 - 3x^3 - 6x^2 + 15x + 4x^2 + 8x - 20 \qquad \text{Distribute.} \\
&= x^4 - x^3 - 7x^2 + 23x - 20 \qquad\qquad\qquad\qquad \text{Combine like terms.}
\end{aligned}
$$

Your Turn 6 Multiply.

a. $(x - 3)(x^2 - 6x + 2)$

b. $(x + 5)(x^2 - 5x + 25)$

c. $(y^2 + 4y - 2)(y^2 - 3y - 5)$

Objective 4 Determine the product when given special polynomial factors.

Multiplying Conjugates

In Objective 2, we discussed how the signs in the binomial factors affect the signs in the product in preparing for factoring in Chapter 7. To further prepare for factoring, it is helpful to note some patterns that occur when certain special polynomial factors are multiplied. First, we will multiply **conjugates**.

Definition   *Conjugates:* Binomials that differ only in the sign separating the terms.

The following binomial pairs are conjugates.

$$x + 9 \text{ and } x - 9 \qquad 2x + 3 \text{ and } 2x - 3 \qquad -6x + 5 \text{ and } -6x - 5$$

Let's multiply the conjugates $2x + 3$ and $2x - 3$ to see what pattern emerges.

$$
\begin{array}{rcll}
 & & \quad\text{First} \quad \text{Outer} \quad \text{Inner} \quad \text{Last} & \\
(2x + 3)(2x - 3) & = & 2x \cdot 2x + 2x \cdot (-3) + 3 \cdot 2x + 3 \cdot (-3) & \\
 & = & 4x^2 \quad -6x \quad + 6x \quad -9 & \\
 & = & 4x^2 + 0 - 9 & \text{Combine like terms: } -6x + 6x = 0. \\
 & = & 4x^2 - 9 &
\end{array}
$$

Note: The sum of the like terms is 0, so the resulting binomial is the square of the first term minus the square of the last term.

When conjugates are multiplied, the like terms are always additive inverses; so their sum is always 0. This suggests the following rule.

Rule **Multiplying Conjugates**

If a and b are real numbers, variables, or expressions, then $(a + b)(a - b) = a^2 - b^2$.

Using this rule allows us to multiply conjugates quickly.

Example 7 Multiply.

a. $(x + 3)(x - 3)$

SOLUTION $\quad (x + 3)(x - 3) = x^2 - 3^2 \qquad \text{Use } (a + b)(a - b) = a^2 - b^2.$
$$= x^2 - 9 \qquad \text{Simplify.}$$

b. $(3t - 8)(3t + 8)$

SOLUTION $\quad (3t - 8)(3t + 8) = (3t)^2 - 8^2 \qquad \text{Use } (a + b)(a - b) = a^2 - b^2.$
$$= 9t^2 - 64 \qquad \text{Simplify.}$$

Your Turn 7 Multiply.

a. $(m + 7)(m - 7)$

b. $(5y - 3)(5y + 3)$

Squaring a Binomial

Now let's look for a pattern when we square a binomial, as in $(3x + 4)^2$. Recall that squaring a number or an expression means to multiply that number or expression by itself; so $(3x + 4)^2 = (3x + 4)(3x + 4)$.

$$
\begin{array}{rcll}
 & & \quad\text{First} \quad \text{Outer} \quad \text{Inner} \quad \text{Last} & \\
(3x + 4)^2 = (3x + 4)(3x + 4) & = & 3x \cdot 3x + 3x \cdot 4 + 4 \cdot 3x + 4 \cdot 4 & \\
 & = & 9x^2 + 12x + 12x + 16 & \\
 & = & 9x^2 + 24x + 16 & \text{Combine like terms. } 12x + 12x = 24x
\end{array}
$$

Warning: Notice that $(3x + 4)^2$ does not equal $9x^2 + 16$.

Notice the pattern.

$$(3x + 4)^2 = 9x^2 + 24x + 16$$

This first term is the square of the first term in the binomial: $(3x)^2 = 9x^2$.

This last term is the square of the second term in the binomial: $(4)^2 = 16$.

This middle term is twice the product of the two terms in the binomial: $2(3x)(4) = 24x$.

Connection The square of a binomial can also be shown in terms of geometry. Consider the square shown, each side being $x + 3$. We can describe the area in two ways: (1) the length of a side squared, $(x + 3)^2$, or (2) the sum of the areas of the four internal rectangles, $x^2 + 3x + 3x + 3^2 = x^2 + 2 \cdot 3x + 9 = x^2 + 6x + 9$.

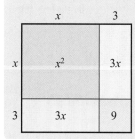

Side squared = Sum of the areas of the four internal rectangles.

$$(x + 3)^2 = x^2 + 2 \cdot 3x + 9$$
$$= x^2 + 6x + 9$$

Squaring a binomial with a minus sign, as in $(3x - 4)^2$, yields a similar pattern.

First   Outer   Inner   Last

$$(3x - 4)^2 = (3x - 4)(3x - 4) = 3x \cdot 3x + 3x \cdot (-4) + (-4) \cdot 3x + (-4) \cdot (-4)$$
$$= 9x^2 - 12x - 12x + 16$$
$$= 9x^2 - 24x + 16 \qquad \text{Combine like terms:}$$
$$-12x - 12x = -24x.$$

Warning: Notice that $(3x - 4)^2$ does *not* equal $9x^2 - 16$ or $9x^2 + 16$.

Notice the pattern.

$$(3x - 4)^2 = 9x^2 - 24x + 16$$

This first term is the square of the first term in the binomial: $(3x)^2 = 9x^2$.

This last term is the square of the second term in the binomial: $(-4)^2 = 16$.

This middle term is twice the product of the two terms in the binomial: $2(3x)(-4) = -24x$.

The patterns from our two examples suggest the following rules.

Rules | **Squaring a Binomial**

If a and b are real numbers, variables, or expressions, then

$$(a + b)^2 = a^2 + 2ab + b^2$$
$$(a - b)^2 = a^2 - 2ab + b^2$$

Example 8 Multiply.

a. $(5x + 3)^2$

SOLUTION $(5x + 3)^2 = (5x)^2 + 2(5x)(3) + (3)^2$ Use $(a + b)^2 = a^2 + 2ab + b^2$.
$$= 25x^2 + 30x + 9$$ Simplify.

b. $(6y - 7)^2$

SOLUTION $(6y - 7)^2 = (6y)^2 - 2(6y)(7) + (7)^2$ Use $(a - b)^2 = a^2 - 2ab + b^2$.
$$= 36y^2 - 84y + 49$$ Simplify.

Answers Your Turn 8
a. $9m^2 + 42m + 49$
b. $16t^2 - 40t + 25$

Your Turn 8 Multiply.

a. $(3m + 7)^2$ **b.** $(4t - 5)^2$

6.5 Exercises   For Extra Help   *MyMathLab*   Math XL
PRACTICE   WATCH   DOWNLOAD

Note: Exercises marked with a ★ represent challenging exercises.

1. What is the mathematical property applied in the first step when multiplying polynomials?

2. Explain, in general, how to multiply two polynomials.

3. Explain how to recognize that two binomials are conjugates.

4. What happens when conjugates are multiplied?

5. Explain why $(a + b)^2 \neq a^2 + b^2$.

6. Explain why $(a - b)^2 \neq a^2 - b^2$.

For Exercises 7–34, multiply the polynomial by the monomial. See Example 1.

7. $5(x + 3)$ **8.** $2(y + 7)$ **9.** $-6(2x - 4)$ **10.** $-4(3y - 5)$

11. $4n(7n + 2)$ **12.** $6n(2n + 5)$ **13.** $9a(a - b)$ **14.** $3y(y - z)$

15. $\dfrac{1}{8}m(2m - 5n)$ **16.** $\dfrac{1}{4}k(2k - 3l)$ **17.** $-0.3p^2(p^3 - 2q^2)$ **18.** $-0.4b^2(b^2 - 3c^2)$

19. $5x(4x^2 + 2x - 5)$ **20.** $5a(2a^2 - a + 6)$ **21.** $-3x^2(7x^3 - 5x + 1)$

22. $-4w^3(3w^3 + 7w^2 - 2)$ **23.** $-r^2s(r^2s^3 + 3rs^2 - s)$ **24.** $-x^2y^3(-2xy^2 + 3x^2y^5 - y)$

25. $-2a^2b^2(2a^3 - 6b + 3ab - a^2b^2)$ **26.** $-5hk^3(3k^3 - 2h^2k - 3hk^4 + 4h)$

27. $3abc(4a^2b^2c^2 - 2abc + 5)$ **28.** $5xy^2z(2xy^3 - 4x^2z^2 + 3)$

29. $4b\left(b^5 - \dfrac{1}{4}b^3 + \dfrac{1}{20}b^2 - \dfrac{1}{8}b + 4\right)$

30. $5r^2\left(r^4 - \dfrac{1}{20}r^3 - \dfrac{1}{5}r^2 + \dfrac{1}{10}r - 1\right)$

31. $-0.4rt^2(2.5r^3t - 8r^2t^2 + 4.1rt^3 - 2)$

32. $-0.25suv^3(10s^2u - 2u^2v - 18sv^4 + 4)$

33. $\dfrac{5}{6}x^2(30x^4y - 18x^3y^2 + 60xy^9)$

34. $\dfrac{2}{3}b^4(-30a^2 + 18ab^3 - 12a^2b^2)$

For Exercises 35–38, a larger rectangle is formed out of smaller rectangles.

    a. Write an expression in simplest form for the length (along the top).
    b. Write an expression in simplest form for the width (along the side).
    c. Write an expression that is the product of the length and width that you found in parts a and b.
    d. Write an expression in simplest form that is the sum of the areas of each of the smaller rectangles.
    e. Explain why the expressions in parts c and d are equivalent.

35.

36.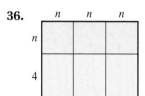

37.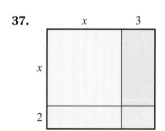

38.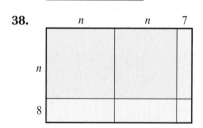

For Exercises 39–56, multiply the binomials. (Use FOIL.) See Examples 2–5.

39. $(x + 3)(x + 4)$ **40.** $(a + 4)(a + 9)$ **41.** $(x - 7)(x + 2)$ **42.** $(n + 7)(n - 5)$

43. $(y - 3)(y - 6)$ **44.** $(r - 4)(r - 5)$ **45.** $(2y + 5)(3y + 2)$ **46.** $(2x + 4)(4x + 3)$

47. $(5m - 3)(3m + 4)$ **48.** $(2x + 3)(3x - 2)$ **49.** $(3t - 5)(4t - 2)$ **50.** $(2t - 7)(3t - 1)$

51. $(5q - 3t)(3q + 4t)$ **52.** $(7k - 2j)(3k + 4j)$ **53.** $(7y + 3x)(2y + 4x)$ **54.** $(2a + 5b)(3a + 4b)$

55. $(a^2 + b)(a^2 - 2b)$ **56.** $(x^2 + 4y)(x^2 - 5y)$

For Exercises 57–72, multiply the polynomials. See Example 6.

57. $(x + 2)(x^2 - 2x + 1)$ **58.** $(x + 3)(x^2 - 3x + 2)$ **59.** $(a - 1)(3a^2 - a - 2)$

60. $(x - 1)(3x^2 - 2x + 1)$ **61.** $(3c + 2)(2c^2 - c - 3)$ **62.** $(2x + 3)(4x^2 + 3x - 5)$

63. $(2f + 3g)(2f^2 - 3fg - 9g^2)$

64. $(2m - 3n)(4m^2 - 6mn + 8n^2)$

65. $(3m - 2n)(9m^2 + 6mn + 4n^2)$

66. $(4a - b)(16a^2 + 4ab + b^2)$

67. $(2a + 5b)(4a^2 - 10ab + 25b^2)$

68. $(5m + 3n)(25m^2 - 15mn + 9n^2)$

69. $(x^2 + xy + y^2)(x^2 + 2xy - y^2)$

70. $(u^2 - u - 1)(u^2 + 2u + 1)$

71. $(x^2 + 10x + 25)(x^2 + 4x + 4)$

72. $(a^2 + 6a + 9)(a^2 - 8a + 16)$

For Exercises 73–80, state the conjugate of the given binomial. See Objective 4.

73. $x + 3$

74. $y + 8$

75. $4x - 2y$

76. $2g - 3h$

77. $4d + 3c$

78. $5m + 2n$

79. $-3j - k$

80. $-3x - 5y$

For Exercises 81–102, multiply using the rules for special products. See Examples 7 and 8.

81. $(x - 5)(x + 5)$

82. $(y + 4)(y - 4)$

83. $(2m - 5)(2m + 5)$

84. $(3p + 2)(3p - 2)$

85. $(x + y)(x - y)$

86. $(a + b)(a - b)$

87. $(8r - 10s)(8r + 10s)$

88. $(4b - 5c)(4b + 5c)$

89. $(-2x - 3)(-2x + 3)$

90. $(-h + 2k)(-h - 2k)$

91. $(x + 3)^2$

92. $(y + 5)^2$

93. $(4t - 1)^2$

94. $(9k - 2)^2$

95. $(m + n)^2$

96. $(a + b)^2$

97. $(2u + 3v)^2$

98. $(3r + 7s)^2$

99. $(9w - 4z)^2$

100. $(2q - 11c)^2$

101. $(9 - 5y)^2$

102. $(6 - 7t)^2$

For Exercises 103–106, write an expression in simplest form for the area.

103.

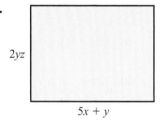

$2yz$

$5x + y$

104.

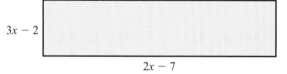

$3x - 2$

$2x - 7$

105.

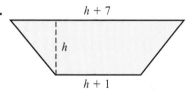

$h + 7$

h

$h + 1$

106.

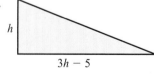

h

$3h - 5$

107. A rectangular room has a length that is 3 feet more than the width. Using w to represent the width, write an expression in simplest form for the area of the room in terms of w.

★ **108.** A circular metal plate for a machine has radius r. A smaller plate with radius $r - 2$ is to be used in another part of the machine. Write an expression in simplest form for the sum of the areas of the two circles in terms of r.

For Exercises 109 and 110, write an expression in simplest form for the volume.

109.

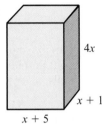

$4x$
$x + 1$
$x + 5$

110.

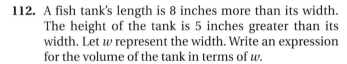

$a - 3$
$4a + 1$
a

111. A crate is designed so that the length is twice the width and the height is 3 feet more than the length. Let w represent the width. Write an expression for the volume in terms of w.

112. A fish tank's length is 8 inches more than its width. The height of the tank is 5 inches greater than its width. Let w represent the width. Write an expression for the volume of the tank in terms of w.

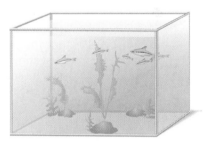

★ **113.** A cylinder has a height that is 2 inches more than the radius. Write an expression for the volume of the cylinder in terms of the radius.

114. A cone has a radius that is 5 centimeters less than the height. Write an expression for the volume of the cone in terms of the height.
★

Review Exercises

EXERCISES 1–4 EXPRESSIONS

For Exercises 1 and 2, simplify.

[1.4] **1.** $1.08 \div 2.4$

[1.3] **2.** $-6 - (-1)$

[6.1] **3.** Write 6.304×10^6 in standard form.

[6.4] **4.** Multiply $x^2 \cdot x^5$.

EXERCISES 5 AND 6 EQUATIONS AND INEQUALITIES

[2.4] **5.** Isolate r in the formula $\dfrac{5 - d}{r} = m$.

[3.3] **6.** The greater of two positive numbers is four times the smaller number. The difference between the numbers is 27. What are the numbers?

6.6 Exponent Rules and Dividing Polynomials

Objectives

1. Divide exponential forms with the same base.
2. Divide numbers in scientific notation.
3. Divide monomials.
4. Divide a polynomial by a monomial.
5. Use long division to divide polynomials.
6. Simplify expressions using rules of exponents.

In Section 6.4, we began by learning the product rule of exponents, which we used in multiplying polynomials. In this section, we follow a similar development and begin by learning the quotient rule of exponents so that we can divide polynomials.

Objective ① Divide exponential forms with the same base. In Section 6.4, we found that when we multiply exponential forms that have the same base, we can add the exponents and keep the same base. For example,

$$2^3 \cdot 2^4 = 2^{3+4} = 2^7.$$

Now consider how to reverse this process and divide exponential forms with the same base. Remember that multiplication and division are inverse operations, which means that they "undo" each other. A related division statement for the multiplication just shown is $2^7 \div 2^4 = 2^3$, which implies that in the division problem, the exponents can be subtracted. To verify this result, it is helpful to write the division in fraction form and use the definition of exponents.

$$\frac{2^7}{2^4} = \frac{2 \cdot 2 \cdot 2 \cdot 2 \cdot 2 \cdot 2 \cdot 2}{2 \cdot 2 \cdot 2 \cdot 2}$$

$$= \frac{2 \cdot 2 \cdot 2}{1}$$

$$= 2^3$$

> **Note:** We can divide out four of the common factors. Three 2's are left in the top, which can be expressed in exponential form.

This result supports our thinking that we can subtract the exponents.

$$2^7 \div 2^4 = 2^{7-4} = 2^3$$

Notice that the divisor's exponent is subtracted from the dividend's exponent. This order is important because subtraction is not a commutative operation.

Conclusion To divide exponential forms that have the same base, we can subtract the divisor's exponent from the dividend's exponent and keep the same base.

Rule Quotient Rule for Exponents

If m and n are integers and a is a real number, where $a \neq 0$, then $\dfrac{a^m}{a^n} = a^{m-n}$.

> **Note:** We cannot let a equal 0 because if a were replaced with 0, we would have $0^m \div 0^n$, and in Section 1.4, we concluded that $0 \div 0$ is indeterminate. Recall also that $\dfrac{a^m}{a^n}$ can be written as $a^m \div a^n$.

Example 1 Divide. $\dfrac{x^6}{x^2}$

> **Note:** Throughout the remainder of this section, assume that variables are not replaced by values that cause denominators to be 0.

SOLUTION Because the exponential forms have the same base, we can subtract the exponents and keep the same base.

$$\frac{x^6}{x^2} = x^6 \div x^2 = x^{6-2} = x^4$$

Your Turn 1 Divide.

a. $\dfrac{t^8}{t^2}$

b. $\dfrac{n^5}{n}$

c. $5^9 \div 5^2$

We can apply the quotient rule for exponential forms when the exponents are negative. After simplifying, if the exponent in the result is negative, we will rewrite the exponential form so that the exponent is positive. Recall the rule from Section 6.1:

If a is any real number not equal to 0 and n is a natural number, then $a^{-n} = \dfrac{1}{a^n}$.

Example 2 Divide and write the result with a positive exponent.

a. $\dfrac{x^6}{x^{-2}}$

SOLUTION $\dfrac{x^6}{x^{-2}} = x^{6-(-2)}$ Subtract exponents and keep the same base.

$= x^{6+2}$ **Note:** To evaluate the subtraction, we write an equivalent addition.

$= x^8$

b. $n^{-5} \div n^{-3}$

SOLUTION $n^{-5} \div n^{-3} = n^{-5-(-3)}$ Subtract the exponents and keep the same base.

$= n^{-5+3}$ Rewrite the subtraction as addition.

$= n^{-2}$ Simplify.

$= \dfrac{1}{n^2}$ Write with a positive exponent.

Your Turn 2 Divide and write the result with a positive exponent.

a. $\dfrac{y^3}{y^7}$

b. $m^{-2} \div m^{-6}$

c. $\dfrac{7^{-4}}{7^5}$

Objective 2 Divide numbers in scientific notation. We can use the quotient rule of exponents to divide numbers in scientific notation.

Example 3 Divide and write the result in scientific notation: $\dfrac{1.08 \times 10^9}{2.4 \times 10^4}$

Answers Your Turn 1
a. t^6 b. n^4 c. 5^7

Answers Your Turn 2
a. $\dfrac{1}{y^4}$ b. m^4 c. $\dfrac{1}{7^9}$

SOLUTION The decimal factors and powers of 10 can be separated into a product of two fractions. This allows us to calculate the decimal division and divide the powers of 10 separately.

$$\frac{1.08 \times 10^9}{2.4 \times 10^4} = \frac{1.08}{2.4} \times \frac{10^9}{10^4} \qquad \text{Separate the decimal numbers and powers of 10.}$$

$$= 0.45 \times 10^{9-4} \qquad \text{Divide decimal numbers and subtract exponents.}$$

$$= 0.45 \times 10^5$$

$$= 4.5 \times 10^4$$

Your Turn 3 Divide $\dfrac{1.82 \times 10^{-9}}{6.5 \times 10^4}$ and write the result in scientific notation.

Objective ③ Divide monomials. Dividing monomials is much like dividing numbers in scientific notation. Let's consider $\dfrac{24x^5}{6x^2}$. We can separate the coefficients and variables in the same way that we separated the decimal factors and powers of 10 when dividing numbers in scientific notation.

$$\frac{24x^5}{6x^2} = \frac{24}{6} \cdot \frac{x^5}{x^2}$$

We can now divide the coefficients and subtract the exponents of the like bases.

$$\frac{24x^5}{6x^2} = \frac{24}{6} \cdot \frac{x^5}{x^2} = 4x^{5-2} = 4x^3$$

This suggests the following procedure.

Procedure **Dividing Monomials**

To divide monomials,
1. Divide the coefficients or simplify them to fractions in lowest terms.
2. Use the quotient rule for the exponents with like bases.
3. Do not change unlike variable bases in the quotient.
4. Write the final expression so that all exponents are positive.

Example 4 Divide.

a. $\dfrac{16a^4b^7c}{20a^5b^2}$

SOLUTION $\dfrac{16a^4b^7c}{20a^5b^2} = \dfrac{16}{20} \cdot \dfrac{a^4}{a^5} \cdot \dfrac{b^7}{b^2} \cdot \dfrac{c}{1}$

$$= \frac{16}{20} \cdot a^{4-5} \cdot b^{7-2} \cdot c \qquad \begin{array}{l}\text{Divide coefficients and subtract expo-}\\ \text{nents of the like bases. The unlike}\\ \text{base, } c, \text{ will remain unchanged.}\end{array}$$

$$= \frac{4}{5} a^{-1} b^5 c$$

$$= \frac{4}{5} \cdot \frac{1}{a} \cdot \frac{b^5}{1} \cdot \frac{c}{1}$$

$$= \frac{4b^5c}{5a}$$

b. $\dfrac{10x^3y^2}{24x^3y}$

SOLUTION $\dfrac{10x^3y^2}{24x^3y} = \dfrac{10}{24} \cdot \dfrac{x^3}{x^3} \cdot \dfrac{y^2}{y}$

$\qquad\qquad = \dfrac{10}{24} \cdot x^{3-3} \cdot y^{2-1}$ Divide coefficients and subtract exponents of like bases.

$\qquad\qquad = \dfrac{5}{12}x^0y^1$

$\qquad\qquad = \dfrac{5}{12}y$

Your Turn 4   Divide.

a. $\dfrac{-32m^2n}{8mn}$ **b.** $\dfrac{56t^2u^6v}{16t^2u^4}$

Objective ④ **Divide a polynomial by a monomial.**   Let's now consider how to divide a polynomial by a monomial, as in $\dfrac{24x^5 + 18x^3}{6x^2}$. Recall that when fractions with a common denominator are added (or subtracted), the numerators are added and the denominator stays the same.

$$\dfrac{1}{7} + \dfrac{2}{7} = \dfrac{1+2}{7}$$

This process can be reversed so that a sum in the numerator of a fraction can be broken into fractions, with each addend over the same denominator.

$$\dfrac{1+2}{7} = \dfrac{1}{7} + \dfrac{2}{7}$$

Rule   **Dividing a Polynomial by a Monomial**

If a, b, and c are real numbers, variables, or expressions with $c \neq 0$, then

$$\dfrac{a+b}{c} = \dfrac{a}{c} + \dfrac{b}{c}.$$

We can apply this rule to the division problem $\dfrac{24x^5 + 18x^3}{6x^2}$.

$$\dfrac{24x^5 + 18x^3}{6x^2} = \dfrac{24x^5}{6x^2} + \dfrac{18x^3}{6x^2}$$ We now have a sum of monomial divisions, which we can simplify separately.

$$= 4x^{5-2} + 3x^{3-2}$$

$$= 4x^3 + 3x$$

Our illustration suggests the following procedure.

Procedure   **Dividing a Polynomial by a Monomial**

To divide a polynomial by a monomial, divide each term of the polynomial by the monomial.

Example 5 Divide.

a. $\dfrac{32y^6 + 24y^4 - 48y^2}{8y^2}$

Solution $\dfrac{32y^6 + 24y^4 - 48y^2}{8y^2} = \dfrac{32y^6}{8y^2} + \dfrac{24y^4}{8y^2} - \dfrac{48y^2}{8y^2}$ Divide each term in the polynomial by the monomial.

$\qquad\qquad\qquad\qquad\qquad\quad = 4y^4 + 3y^2 - 6y^0$

$\qquad\qquad\qquad\qquad\qquad\quad = 4y^4 + 3y^2 - 6$

b. $\dfrac{42t^7u^2 - 18t^4u - 10t}{2t^3u}$

Solution $\dfrac{42t^7u^2 - 18t^4u - 10t}{2t^3u} = \dfrac{42t^7u^2}{2t^3u} - \dfrac{18t^4u}{2t^3u} - \dfrac{10t}{2t^3u}$

Warning: A common error is to write the answer to Example 5(b) as $\dfrac{21t^4u - 9t - 5}{ut^2}$.

Note: When t is divided by t^3, because the denominator's exponent is greater than the numerator's exponent, the result is a negative exponent. Because u has no like base in the third term, it is written unchanged.

$\qquad\qquad\qquad\qquad\qquad\quad = 21t^4u - 9t^1u^0 - \dfrac{5t^{-2}}{u}$

$\qquad\qquad\qquad\qquad\qquad\quad = 21t^4u - 9t - \dfrac{5}{ut^2}$ Rewrite with positive exponents.

Your Turn 5 Divide.

a. $\dfrac{45x^6 - 27x^5 + 9x^3}{9x^3}$

b. $\dfrac{28m^2n^6 + 36m^4n - 40m^2}{4m^2n}$

Objective ⑤ Use long division to divide polynomials. To divide a polynomial by a polynomial, we can use long division. Consider the following numeric long division and think about the process.

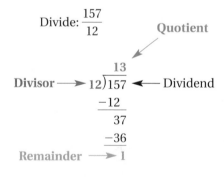

Divide: $\dfrac{157}{12}$

Because the answer has a remainder, we write the result as a mixed number, $13\dfrac{1}{12}$. Notice that we could check the answer by multiplying the divisor by the quotient, then adding the remainder. The result should be the dividend.

$$\text{Quotient} \cdot \text{Divisor} + \text{Remainder} = \text{Dividend}$$
$$13 \quad \cdot \quad 12 \quad + \quad 1 \quad = 157$$

Now let's consider polynomial division, which follows the same long division process.

Answers **Your Turn 5**

a. $5x^3 - 3x^2 + 1$

b. $7n^5 + 9m^2 - \dfrac{10}{n}$

Example 6 Divide $\dfrac{x^2 + 5x + 7}{x + 2}$.

SOLUTION Use long division. First, we determine what term will multiply by the first term in the divisor to equal the first term in the dividend. A clever way to determine this first term in the quotient is to divide the first term in the dividend by the first term in the divisor.

$$
\begin{array}{r}
x \\
x + 2 \overline{)x^2 + 5x + 7}
\end{array}
$$

Divide these first terms to determine the first term in the quotient: $x^2 \div x = x$.

Next, we multiply the divisor $x + 2$ by the x in the quotient.

multiply

$$
\begin{array}{r}
x \\
x + 2 \overline{)x^2 + 5x + 7} \\
x^2 + 2x
\end{array}
$$

Next, we subtract. Note that to subtract the binomial, we change the signs of the terms and then combine like terms. After combining terms, we bring down the next term in the dividend, which is 7.

$$
\begin{array}{r}
x \\
x + 2 \overline{)x^2 + 5x + 7} \\
-(x^2 + 2x)
\end{array}
$$

Change signs. ⟶

$$
\begin{array}{r}
x \\
x + 2 \overline{)x^2 + 5x + 7} \\
-x^2 - 2x \\
\hline
3x + 7
\end{array}
$$

Combine like terms and bring down the next term.

Now we repeat the process with $3x + 7$ as the dividend. The next term in the quotient is found by dividing the first term of $3x + 7$ by the first term in the divisor, $x + 2$. This gives $3x \div x = 3$. We then multiply the divisor by this 3.

multiply

$$
\begin{array}{r}
x + 3 \\
x + 2 \overline{)x^2 + 5x + 7} \\
-x^2 - 2x \\
\hline
3x + 7 \\
3x + 6
\end{array}
$$

Then we subtract. As before, we change the signs of the binomial to be subtracted.

$$
\begin{array}{r}
x + 3 \\
x + 2 \overline{)x^2 + 5x + 7} \\
-x^2 - 2x \\
\hline
3x + 7 \\
-(3x + 6)
\end{array}
$$

Change signs. ⟶

$$
\begin{array}{r}
x + 3 \\
x + 2 \overline{)x^2 + 5x + 7} \\
-x^2 - 2x \\
\hline
3x + 7 \\
-3x - 6 \\
\hline
1
\end{array}
$$

Combine like terms.

Notice that we have a remainder, 1. Recall that in the numeric version, we wrote the answer as a mixed number, $13\frac{1}{12}$, which means $13 + \frac{1}{12}$ and is in the form quotient $+ \dfrac{\text{remainder}}{\text{divisor}}$. With polynomials, we write a similar expression:

$$\underbrace{x + 3}_{\text{quotient}} + \underbrace{\frac{1 \longleftarrow \text{remainder}}{x + 2 \longleftarrow \text{divisor}}}$$

To check, we can multiply the quotient and the divisor, then add the remainder. The result should be the dividend.

$$\begin{aligned}(x + 3)(x + 2) + 1 &= x^2 + 2x + 3x + 6 + 1\\&= x^2 + 5x + 7 \qquad \text{It checks.}\end{aligned}$$

Procedure   **Dividing a Polynomial by a Polynomial**

To divide a polynomial by a polynomial, use long division. If there is a remainder, write the result in the following form:

$$\text{quotient} + \frac{\text{remainder}}{\text{divisor}}$$

In Example 6, the coefficients of the first term in the divisor and dividend were both 1. What if those coefficients are other than 1?

Example 7   Divide $\dfrac{15x^2 - 22x + 14}{3x - 2}$.

SOLUTION Begin by dividing the first term in the dividend by the first term in the divisor: $15x^2 \div 3x = 5x$.

$$\begin{array}{r} 5x \\ 3x - 2 \overline{)15x^2 - 22x + 14} \\ \underline{-(15x^2 - 10x)} \end{array}$$

Change signs. $\longrightarrow$

$$\begin{array}{r} 5x \\ 3x - 2 \overline{)15x^2 - 22x + 14} \\ \underline{-15x^2 + 10x} \\ -12x + 14 \end{array}$$

Combine like terms; then bring down the next term.

Determine the next part of the quotient by dividing $-12x$ by $3x$, which is -4, and repeat the multiplication and subtraction steps.

$$\begin{array}{r} 5x - 4 \\ 3x - 2 \overline{)15x^2 - 22x + 14} \\ \underline{-15x^2 + 10x} \\ -12x + 14 \\ \underline{-(-12x + 8)} \end{array}$$

Change signs. $\longrightarrow$

$$\begin{array}{r} 5x - 4 \\ 3x - 2 \overline{)15x^2 - 22x + 14} \\ \underline{-15x^2 + 10x} \\ -12x + 14 \\ \underline{+12x - 8} \\ 6 \end{array}$$

Combine like terms.

Answer $5x - 4 + \dfrac{6}{3x - 2}$

Answers   **Your Turn 7**

a. $x + 4$ **b.** $4x + 3 + \dfrac{7}{6x - 1}$

Your Turn 7   Divide.

a. $\dfrac{x^2 + 6x + 8}{x + 2}$ **b.** $\dfrac{24x^2 + 14x + 4}{6x - 1}$

Using a Place Marker in Long Division

In polynomial division, it is important that we write the terms in descending order of degree. If a term is missing, we write that term with a 0 coefficient as a place marker.

Example 8 Divide $\dfrac{-14x^2 - 9 + 8x^4}{2x - 1}$.

SOLUTION First, write the dividend in descending order of degree. Also, because degree 3 and degree 1 terms are missing in the dividend, we write those terms with 0 coefficients as placeholders. Divide $8x^4$ by $2x$ to get the first term in the quotient.

$$
\begin{array}{r}
4x^3 \\
2x - 1 \overline{)8x^4 + 0x^3 - 14x^2 + 0x - 9} \\
\underline{-(8x^4 - 4x^3)}
\end{array}
$$

Change signs. $\longrightarrow$

$$
\begin{array}{r}
4x^3 \\
2x - 1 \overline{)8x^4 + 0x^3 - 14x^2 + 0x - 9} \\
\underline{-8x^4 + 4x^3} \\
4x^3 - 14x^2
\end{array}
$$

Combine like terms; then bring down the next term.

Determine the next part of the quotient by dividing $4x^3$ by $2x$ and repeat the multiplication and subtraction steps.

$$
\begin{array}{r}
4x^3 + 2x^2 \\
2x - 1 \overline{)8x^4 + 0x^3 - 14x^2 + 0x - 9} \\
\underline{-8x^4 + 4x^3} \\
4x^3 - 14x^2 \\
\underline{-(4x^3 - 2x^2)}
\end{array}
$$

Change signs. $\longrightarrow$

$$
\begin{array}{r}
4x^3 + 2x^2 \\
2x - 1 \overline{)8x^4 + 0x^3 - 14x^2 + 0x - 9} \\
\underline{-8x^4 + 4x^3} \\
4x^3 - 14x^2 \\
\underline{-4x^3 + 2x^2} \\
-12x^2 + 0x
\end{array}
$$

Combine like terms; then bring down the next term.

Determine the next part of the quotient by dividing $-12x^2$ by $2x$ and repeat the multiplication and subtraction steps.

$$
\begin{array}{r}
4x^3 + 2x^2 - 6x \\
2x - 1 \overline{)8x^4 + 0x^3 - 14x^2 + 0x - 9} \\
\underline{-8x^4 + 4x^3} \\
4x^3 - 14x^2 \\
\underline{-4x^3 + 2x^2} \\
-12x^2 + 0x \\
\underline{-(-12x^2 + 6x)}
\end{array}
$$

Change signs. $\longrightarrow$

$$
\begin{array}{r}
4x^3 + 2x^2 - 6x \\
2x - 1 \overline{)8x^4 + 0x^3 - 14x^2 + 0x - 9} \\
\underline{-8x^4 + 4x^3} \\
4x^3 - 14x^2 \\
\underline{-4x^3 + 2x^2} \\
-12x^2 + 0x \\
\underline{+12x^2 - 6x} \\
-6x - 9
\end{array}
$$

Combine like terms; then bring down the next term.

Determine the final term of the quotient by dividing $-6x$ by $2x$, which is -3.

$$
\begin{array}{r}
4x^3 + 2x^2 - 6x - 3 \\
2x - 1 \overline{)8x^4 + 0x^3 - 14x^2 + 0x - 9} \\
\underline{-8x^4 + 4x^3} \\
4x^3 - 14x^2 \\
\underline{-4x^3 + 2x^2} \\
-12x^2 + 0x \\
\underline{12x^2 - 6x} \\
-6x - 9 \\
\underline{-(-6x + 3)}
\end{array}
$$

Change signs. $\longrightarrow$

$$
\begin{array}{r}
4x^3 + 2x^2 - 6x - 3 \\
2x - 1 \overline{)8x^4 + 0x^3 - 14x^2 + 0x - 9} \\
\underline{-8x^4 + 4x^3} \\
4x^3 - 14x^2 \\
\underline{-4x^3 + 2x^2} \\
-12x^2 + 0x \\
\underline{12x^2 - 6x} \\
-6x - 9 \\
\underline{+6x - 3} \\
-12
\end{array}
$$

Combine like terms.

Answer $4x^3 + 2x^2 - 6x - 3 + \dfrac{-12}{2x - 1}$, or $4x^3 + 2x^2 - 6x - 3 - \dfrac{12}{2x - 1}$

Answer **Your Turn 8**

$2x^2 + x - 4 + \dfrac{-1}{4x - 2}$ or

$2x^2 + x - 4 - \dfrac{1}{4x - 2}$

Your Turn 8 Divide $\dfrac{7 + 8x^3 - 18x}{4x - 2}$.

Objective ⑥ Simplify expressions using rules of exponents. The quotient rule for exponents completes our rules of exponents in this chapter. So far, the problems involving exponent rules have required only that we use one or two rules. Let's now consider problems that require us to use several of the exponent rules. Following is a summary of all of the rules of exponents.

Rule **Exponents Summary**

Assume that no denominators are 0, that a and b are real numbers, and that m and n are integers.

Zero as an exponent: $a^0 = 1$, where $a \neq 0$

0^0 is indeterminate.

Negative exponents: $a^{-n} = \dfrac{1}{a^n}$

$\dfrac{1}{a^{-n}} = a^n$

$\left(\dfrac{a}{b}\right)^{-n} = \left(\dfrac{b}{a}\right)^n$

Product rule for exponents: $a^m \cdot a^n = a^{m+n}$

Quotient rule for exponents: $\dfrac{a^m}{a^n} = a^{m-n}$

Raising a power to a power: $(a^m)^n = a^{mn}$

Raising a product to a power: $(ab)^n = a^n b^n$

Raising a quotient to a power: $\left(\dfrac{a}{b}\right)^n = \dfrac{a^n}{b^n}$

Example 9 Simplify. Write all answers with positive exponents.

a. $\left(\dfrac{n^{-5}}{n^2}\right)^3$

SOLUTION Following the order of operations, we simplify within the parentheses first, then consider the exponent outside the parentheses.

$$\left(\dfrac{n^{-5}}{n^2}\right)^3 = (n^{-5-2})^3 \qquad \text{Use the quotient rule for exponents.}$$

$$= (n^{-7})^3$$

$$= n^{-7 \cdot 3} \qquad \text{Use the rule for raising a power to a power.}$$

$$= n^{-21}$$

$$= \dfrac{1}{n^{21}} \qquad \text{Write with a positive exponent.}$$

b. $\dfrac{(m^2)^{-3}}{m^4 \cdot m^{-5}}$

SOLUTION $\dfrac{(m^2)^{-3}}{m^4 \cdot m^{-5}} = \dfrac{m^{2 \cdot (-3)}}{m^{4+(-5)}}$ In the numerator, use the rule for raising a power to a power. In the denominator, use the product rule for exponents.

$$= \dfrac{m^{-6}}{m^{-1}}$$

$$= m^{-6-(-1)} \qquad \text{Use the quotient rule for exponents.}$$

$$= m^{-6+1} \qquad \text{Write the subtraction as an equivalent addition.}$$

$$= m^{-5}$$

$$= \dfrac{1}{m^5} \qquad \text{Write with a positive exponent.}$$

c. $\dfrac{(4y^3)^2}{(2y^4)^5}$

SOLUTION $\dfrac{(4y^3)^2}{(2y^4)^5} = \dfrac{4^2 y^{3\cdot 2}}{2^5 y^{4\cdot 5}}$ Use the rule for raising a product to a power.

$= \dfrac{16y^6}{32y^{20}}$

$= \dfrac{16}{32} \cdot \dfrac{y^6}{y^{20}}$ Separate the coefficients and variables.

$= \dfrac{16}{32} \cdot y^{6-20}$ Use the quotient rule for exponents.

$= \dfrac{1}{2} \cdot y^{-14}$ Simplify the coefficient.

$= \dfrac{1}{2} \cdot \dfrac{1}{y^{14}}$ Write with a positive exponent.

$= \dfrac{1}{2y^{14}}$ Simplify.

Your Turn 9 Simplify. Write all answers with positive exponents.

a. $\left(\dfrac{x^2}{x^{-5}}\right)^{-4}$ **b.** $(y^{-2} \cdot y^{-3})^{-4} \div y^{-15}$ **c.** $\dfrac{(2x^4)^{-3}}{(x^2 y^3)^{-2}}$

Answers Your Turn 9

a. $\dfrac{1}{x^{28}}$ **b.** y^{35} **c.** $\dfrac{y^6}{8x^8}$

6.6 Exercises   For Extra Help   *MyMathLab*

PRACTICE   WATCH   DOWNLOAD

Note: Exercises marked with a ★ represent challenging exercises.

1. Explain how to divide exponential forms that have the same base.

2. What is the rule for rewriting a^{-n}, where a is a real number, $a \neq 0$, and n is a natural number, so that it has a positive exponent?

3. How do we divide two numbers in scientific notation?

4. How do we divide two monomials?

5. Explain how to divide a polynomial by a monomial.

6. When do we use a 0 placeholder in the process of dividing two polynomials by long division?

For Exercises 7–32, simplify using the rules of exponents. Write all answers with positive exponents. See Examples 1 and 2.

7. a^{-3}

8. n^{-7}

9. 2^{-5}

10. 3^{-4}

11. $\dfrac{y^8}{y^3}$

12. $\dfrac{n^4}{n}$

13. $\dfrac{3^4}{3^3}$

14. $\dfrac{6^{11}}{6^5}$

15. $\dfrac{4^3}{4^9}$

16. $\dfrac{5}{5^9}$

17. $\dfrac{x^3}{x^5}$

18. $\dfrac{m^4}{m^{10}}$

19. $a^8 \div a^6$

20. $u^5 \div u^3$

21. $w^7 \div w^{10}$

22. $j^6 \div j^{14}$

23. $\dfrac{a^{-2}}{a^5}$

24. $\dfrac{m^{-6}}{m^3}$

25. $\dfrac{r^6}{r^{-4}}$

26. $\dfrac{u^5}{u^{-12}}$

27. $\dfrac{y^{-7}}{y^{-15}}$

28. $\dfrac{x^{-3}}{x^{-8}}$

29. $\dfrac{p^{-5}}{p^{-2}}$

30. $\dfrac{n^{-11}}{n^{-5}}$

31. $\dfrac{t^{-4}}{t^{-4}}$

32. $\dfrac{x^{-6}}{x^{-6}}$

For Exercises 33–40, divide and write your answers in scientific notation. See Example 3.

33. $\dfrac{8.32 \times 10^5}{3.2 \times 10^6}$

34. $\dfrac{3.6 \times 10^6}{2.4 \times 10^9}$

35. $\dfrac{1.26 \times 10^{-4}}{2.1 \times 10^3}$

36. $\dfrac{1.32 \times 10^{-5}}{6.6 \times 10^{15}}$

37. $\dfrac{9.088 \times 10^1}{1.28 \times 10^{-9}}$

38. $\dfrac{9.964 \times 10^4}{1.88 \times 10^{-2}}$

39. $\dfrac{8.64 \times 10^{-3}}{3.2 \times 10^{-7}}$

40. $\dfrac{7.92 \times 10^{-3}}{2.2 \times 10^{-7}}$

For Exercises 41–44, solve. See Example 3.

41. If light travels at 3×10^8 meters per second, how long does it take light to travel the 1.5×10^{11} meters from the Sun to Earth?

42. In 2008, the population of the United States was approximately 3.04×10^8. Calculate the approximate number of people per square mile in the United States in 2008 if the estimated amount of land in the United States is 3.8×10^6 square miles. (*Source:* U.S. Bureau of the Census)

43. As of March 11, 2009, the national debt was about \$$1.128 \times 10^{13}$. If each of the 3.06×10^8 people in the United States contributed an equal share toward the debt, what would each person have to contribute to pay off the debt? (*Source:* U.S. National Debt Clock)

44. There are about 2.3×10^6 stone blocks in the Great Pyramid. If the total weight of the pyramid is about 1.3×10^{10} pounds and we assume that each block is of equal size, how much does each block weigh? (*Source:* Nova Online)

★ **For Exercises 45 and 46, use Einstein's formula E = mc², which describes the rest energy contained within a mass m, where c represents the speed of light, which is a constant with a value of 3 × 10⁸ meters per second. The units for energy are joules, which are equivalent to $\frac{kg \cdot m^2}{s^2}$. See Example 3.**

45. The first atom bomb released an energy equivalent of about 8.4×10^{13} joules. What amount of mass was converted to energy? Write your answer in scientific notation.

46. The largest atom bomb detonated was a hydrogen bomb tested by the Soviet Union in 1961. The bomb released 2.4×10^{17} joules of energy. What amount of mass was converted to energy in that explosion? (*Source:* Cesare Emiliani, *The Scientific Companion*; Wiley Popular Science, 1995)

For Exercises 47–58, divide the monomials. See Example 4.

47. $\frac{15x^5}{3x^2}$

48. $\frac{14a^6}{2a^4}$

49. $\frac{24x^2}{-6x^5}$

50. $\frac{-18a^6}{9a^{12}}$

51. $\frac{9m^3n^5}{-3mn}$

52. $\frac{-42x^5y^3}{6xy^2}$

53. $\frac{-24p^3q^5}{15p^3q^2}$

54. $\frac{48t^4u^7}{-18t^4u^2}$

55. $\frac{56x^2y^6}{42x^7y^2}$

56. $\frac{60a^8b^4}{66a^2b^9}$

57. $\frac{12a^4bc^3}{9a^6bc^2}$

58. $\frac{28p^5q^4r}{42p^2q^7r}$

For Exercises 59–76, divide the polynomial by the monomial. See Example 5.

59. $\frac{7a + 14b}{7}$

60. $\frac{4m + 16y}{4}$

61. $\frac{12x^3 - 6x^2}{3x}$

62. $\frac{18m^4 - 27m^2}{9m}$

63. $\frac{12x^3 + 8x}{4x^2}$

64. $\frac{24y^3 + 16y}{8y^2}$

65. $\frac{5x^2y^2 - 15xy^3}{5xy}$

66. $\frac{12k^4l^2 + 15k^2l^3}{3k^2l}$

67. $\frac{6abc^2 - 24a^2b^2c}{-3abc}$

68. $\frac{30x^3yz^5 - 15xyz^2}{-5xyz^2}$

69. $\frac{24x^3 + 16x^2 - 8x}{8x}$

70. $\frac{3x^3 + 6x^2 - 3x}{3x}$

71. $\frac{6x^3 - 12x^2 + 9x}{3x^2}$

72. $\frac{12y^4 - 16y^3 + 8y}{4y^2}$

73. $\frac{36u^3v^4 + 12uv^5 - 15u^2v^2}{3u^2v}$

74. $\frac{16hk^4 - 28hk - 4h^2k^2}{4hk^2}$

75. $\frac{y^5 + y^7 - 3y^8 + y}{y^3}$

76. $\frac{x^6 + x^8 - 4x^{10} + x^2}{x^4}$

Of Interest

The first atomic bomb, tested in 1945, used the process of nuclear *fission*, which involves the splitting of atoms of heavy elements, such as uranium and plutonium. The split atoms, in turn, split other atoms, resulting in a chain reaction in which some of the atomic mass is converted into energy.

In contrast, the hydrogen bomb uses nuclear *fusion*, which is how the stars release energy. In a fusion reaction, two nuclei are fused together to form a new atom. For example, two hydrogen nuclei can be forced together to form a helium nucleus. During this reaction, some of the mass of the two hydrogen nuclei is converted to pure energy so that the resulting helium nucleus has slightly less mass than the two original hydrogen nuclei.

77. The area of the parallelogram shown is described by the monomial $35mn^2$. Find the height.

height

$14n$

78. The area of the triangle shown is $18x^2y$. Find the base.

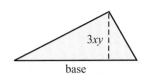

$3xy$

base

79. An engineer is designing a steel cover plate in the shape of a rectangle. The area of the plate is described by the polynomial $9x^2 - 15x + 12$, and the length must be $3x$.
   a. Find an expression for the width.

   b. Find the area, length, and width if $x = 2$.

80. The voltage in a circuit is described by the polynomial $36s^3 - 24s^2 + 42s$. The resistance is described by $6s$.
   a. Find an expression for the current.
   (*Hint:* Voltage = Current × Resistance)

   b. Find the voltage, current, and resistance if $s = 3$.

For Exercises 81–104, use long division to divide the polynomials. See Examples 6–8.

81. $\dfrac{x^2 + 7x + 12}{x + 3}$

82. $\dfrac{y^2 + 4y + 4}{y + 2}$

83. $\dfrac{m^2 - 8m + 15}{m - 3}$

84. $\dfrac{u^2 - 6u + 9}{u - 3}$

85. $\dfrac{x^2 - 17x + 64}{x - 5}$

86. $\dfrac{x^2 + 6x + 10}{x + 3}$

87. $\dfrac{3x^3 - 73x - 10}{x - 5}$

88. $\dfrac{2x^3 - 35x - 12}{x + 4}$

89. $\dfrac{2x^3 - x^2 + 5}{x - 2}$

90. $\dfrac{2x^3 - 2x^2 + 5}{x + 3}$

91. $\dfrac{p^3 + 125}{p + 5}$

92. $\dfrac{x^3 + 8}{x + 2}$

93. $\dfrac{12x^2 - 7x - 10}{4x - 5}$

94. $\dfrac{6a^2 + 2a - 28}{3a + 7}$

95. $\dfrac{14y^2 - 8y + 17}{7y + 3}$

96. $\dfrac{6x^2 + 7x + 5}{3x - 1}$

97. $\dfrac{4k^3 + 8k^2 + 6 - k}{k + 2}$

98. $\dfrac{y^3 + 2y^2 + 3 - 11y}{y + 2}$

99. $\dfrac{21u^3 - 19u^2 + 14u - 6}{3u + 2}$

100. $\dfrac{2v^3 + 5v^2 - 10v + 2}{2v - 3}$

101. $\dfrac{14b + b^3 - 6b^2 - 12}{b - 3}$

102. $\dfrac{-14x + x^2 + x^3 - 5}{x + 4}$

103. $\dfrac{y^3 - y + 6}{y + 2}$

104. $\dfrac{2b^4 + 3b^3 - 4b - 6}{2b + 3}$

105. The volume of a restaurant's walk-in refrigerator is described by $(6x^2 + 42x + 72)$ cubic feet. The width has to be 6 feet, and the length is described by $(x + 3)$ feet.
   a. Find an expression for the height of the refrigerator.

   b. Find the length, height, and volume of the refrigerator if $x = 4$.

106. In an architectural project, the specifications for a rectangular room call for the area to be $(10x^2 + 3x - 18)$ square feet. It is decided that the length should be described by the binomial $(2x + 3)$ feet.
   a. Find an expression for the width.

   b. Find the length, width, and area of the room if $x = 5$.

For Exercises 107–130, simplify. Write all answers with positive exponents.
See Example 9.

107. $(6m^2n^{-2})^2$

108. $(-4a^3b^{-2}c)^2$

109. $(4x^{-3}y^2)^{-3}$

110. $(3m^4n^{-3})^{-4}$

111. $(4rs^3)^3(2r^3s^{-1})^2$

112. $(-3abc^{-2})^2(2a^2b^{-3})^3$

113. $\left(\dfrac{m^6}{m^2}\right)^4$

114. $\left(\dfrac{p^7}{p^4}\right)^5$

115. $\left(\dfrac{x^{-3}}{x^4}\right)^2$

116. $\left(\dfrac{r^{-4}}{r^2}\right)^5$

117. $\left(\dfrac{x^2y^7}{x^5y^{-2}}\right)^3$

118. $\left(\dfrac{m^6n^5}{m^{-3}n^8}\right)^2$

119. $\left(\dfrac{x^{-3}}{x^4}\right)^{-2}$

120. $\left(\dfrac{p^{-5}}{p^3}\right)^{-3}$

121. $\left(\dfrac{3x^{-1}y^2}{z^2}\right)^3$

122. $\left(\dfrac{2x^{-2}y}{z^{-1}}\right)^4$

123. $\dfrac{(x^4)^{-3}}{x^{-5}\cdot x^3}$

124. $\dfrac{(y^3)^{-5}}{y^6\cdot y^{-8}}$

125. $\dfrac{4xy^{-2}z^2}{x^{-3}y^3z^{-1}}$

126. $\dfrac{3m^6n^{-1}p^{-2}}{m^{-1}n^{-3}p}$

127. $\dfrac{(2ab)^3}{12a^4b^5}$

128. $\dfrac{(2xy^2)^2}{16x^2y^3}$

129. $\dfrac{(8x^3)^3}{(4x^4)^5}$

130. $\dfrac{(3y^4)^2}{(6y^7)^3}$

Puzzle Problem

Fill the following blanks with the numbers 1, 2, 3, 4, 5, 6, 7, 8, and 9, using each number once so that the equation is true.

$$\square\square \div \square = \square\square \div \square = \square\square \div \square$$

Review Exercises

Exercises 1–4 **EXPRESSIONS**

[1.2] **1.** Find the prime factorization of 3024.

[6.4] **2.** Simplify: $-3x^2y\cdot 6x^2y^2$

[6.5] **3.** Simplify: $y(3x-4)$

[1.6] **4.** Translate to an algebraic expression: one number plus half of a second
number.

Exercises 5 and 6 **EQUATIONS AND INEQUALITIES**

[2.1] **5.** Find the area of a playground that is 17 feet by 20 feet.

[2.3] **6.** Solve and check: $3-y=27$

Chapter 6 Summary

Defined Terms

Section 6.1
Scientific notation (p. 400)

Section 6.2
Monomial (p. 408)
Coefficient of a monomial (p. 409)
Degree of a monomial (p. 409)
Polynomial (p. 410)

Polynomial in one variable (p. 410)
Binomial (p. 410)
Trinomial (p. 410)
Degree of a polynomial (p. 411)

Section 6.5
Conjugates (p. 443)

Procedures, Rules, and Key Examples

Procedures/Rules	Key Examples

SECTION 6.1   Exponents and Scientific Notation

If the base of an exponential form is a negative number and the exponent is even, then the product is positive. If the base is a negative number and the exponent is odd, then the product is negative.	**Example 1:** Evaluate. **a.** $(-2)^4 = (-2)(-2)(-2)(-2) = 16$ **b.** $(-2)^5 = (-2)(-2)(-2)(-2)(-2) = -32$
If a and b are real numbers, where $b \neq 0$ and n is a natural number, then $\left(\dfrac{a}{b}\right)^n = \dfrac{a^n}{b^n}$. If a is a real number and $a \neq 0$, then $a^0 = 1$. If a is a real number, where $a \neq 0$ and n is a natural number, then $a^{-n} = \dfrac{1}{a^n}$ and $\dfrac{1}{a^{-n}} = a^n$. If a and b are real numbers, where $a \neq 0$ and $b \neq 0$ and n is a natural number, then $\left(\dfrac{a}{b}\right)^{-n} = \left(\dfrac{b}{a}\right)^n$.	**Example 2:** Simplify. **a.** $\left(\dfrac{3}{4}\right)^2 = \dfrac{3^2}{4^2} = \dfrac{9}{16}$ **b.** $7^0 = 1$ **c.** $2^{-4} = \dfrac{1}{2^4} = \dfrac{1}{16}$ **d.** $\dfrac{1}{3^{-2}} = 3^2 = 9$ **e.** $\left(\dfrac{2}{3}\right)^{-4} = \left(\dfrac{3}{2}\right)^4 = \dfrac{81}{16}$
To change from scientific notation with a positive integer exponent to standard form, move the decimal point to the right the number of places indicated by the exponent.	**Example 3:** Write in standard form. **a.** $3.4 \times 10^5 = 340{,}000$
To change from scientific notation with a negative exponent to standard form, move the decimal point to the left the same number of places as the absolute value of the exponent.	**b.** $4.2 \times 10^{-5} = 0.000042$
To write a number greater than 1 in scientific notation, 1. Move the decimal point so that the number is greater than or equal to 1 but less than 10. (Place the decimal point to the right of the first nonzero digit.) 2. Write the decimal number multiplied by 10^n, where n is the number of places between the new decimal position and the original decimal position. 3. Delete zeroes to the right of the last nonzero digit.	**Example 4:** Write in scientific notation. $$7{,}230{,}000 = 7.23 \times 10^6$$

(continued)

Procedures/Rules	**Key Examples**

SECTION 6.1 Exponents and Scientific Notation (*continued*)

To write a positive decimal number that is less than 1 in scientific notation, 1. Move the decimal point so that the number is greater than or equal to 1 but less than 10. (Place the decimal point to the right of the first nonzero digit.) 2. Write the decimal number multiplied by 10^n, where n is a negative integer whose absolute value is the number of places between the new decimal position and the original decimal position. 3. Delete zeroes to the left of the first nonzero digit.	**Example 5:** Write in scientific notation. $$0.00000056 = 5.6 \times 10^{-7}$$

SECTION 6.2 Introduction to Polynomials

To write a polynomial in descending order of degree, place the highest degree term first, then the next highest degree term, and so on.	**Example 1:** Write in descending order. $$2x^2 + 4x^5 + 9 - 6x^3 = 4x^5 - 6x^3 + 2x^2 + 9$$

SECTION 6.3 Adding and Subtracting Polynomials

To add polynomials, combine like terms.	**Example 1:** Add. $$\begin{aligned}&(5x^3 + 12x^2 - 9x + 1) + (7x^2 - x - 13)\\ &= 5x^3 + 12x^2 + 7x^2 - 9x - x + 1 - 13\\ &= 5x^3 + 19x^2 - 10x - 12\end{aligned}$$

To subtract polynomials,
1. Write the subtraction statement as an equivalent addition statement.
  a. Change the operation symbol from a minus sign to a plus sign.
  b. Change the subtrahend (second polynomial) to its additive inverse. To get the additive inverse, change the sign of each term in the polynomial.
2. Combine like terms.
	Example 2: Subtract.
	
	$$\begin{aligned}&(6y^3 - 5y + 18) - (y^3 - 5y + 7)\\ &= (6y^3 - 5y + 18) + (-y^3 + 5y - 7)\\ &= 6y^3 - y^3 - 5y + 5y + 18 - 7\\ &= 5y^3 + 11\end{aligned}$$

SECTION 6.4 Exponent Rules and Multiplying Monomials

Product rule for exponents: If a is a real number and m and n are integers, then $a^m \cdot a^n = a^{m+n}$. To multiply monomials, 1. Multiply coefficients. 2. Add the exponents of the like bases. 3. Write any unlike variable bases unchanged in the product. If a is a real number and m and n are integers, then $(a^m)^n = a^{mn}$. If a and b are real numbers and n is an integer, then $$(ab)^n = a^n b^n.$$ To simplify a monomial raised to a power, 1. Evaluate the coefficient raised to that power. 2. Multiply each variable's exponent by the power.	**Example 1:** Multiply. **a.** $x^3 \cdot x^4 = x^{3+4} = x^7$ **b.** $-6x^4yz^2 \cdot 5x^5y^3 = -30x^{4+5}y^{1+3}z^2$ $\qquad\qquad\qquad\quad = -30x^9y^4z^2$ **Example 2:** Simplify. **a.** $(x^3)^4 = x^{3\cdot4} = x^{12}$ **b.** $(m^2n^3)^5 = m^{2\cdot5}n^{3\cdot5} = m^{10}n^{15}$ **c.** $(3x^2y^5)^4 = 3^4x^{2\cdot4}y^{5\cdot4} = 81x^8y^{20}$

(continued)

Procedures/Rules	Key Examples

SECTION 6.5 Multiplying Polynomials; Special Products

To multiply a polynomial by a monomial, use the distributive property to multiply each term in the polynomial by the monomial.

To multiply two polynomials,
1. Multiply each term in the second polynomial by each term in the first polynomial.
2. Combine like terms.

Special Products (a and b are real numbers, variables, or expressions):

Conjugates: $(a + b)(a - b) = a^2 - b^2$

Squaring a sum: $(a + b)^2 = a^2 + 2ab + b^2$

Squaring a difference: $(a - b)^2 = a^2 - 2ab + b^2$

Example 1: Multiply.
a. $3x(x^4 - 7x^2y + 9y^2 - 2)$
$= 3x \cdot x^4 - 3x \cdot 7x^2y + 3x \cdot 9y^2 - 3x \cdot 2$
$= 3x^5 - 21x^3y + 27xy^2 - 6x$

b. $(3x + 5)(2x - 1)$
$= 3x \cdot 2x + 3x \cdot (-1) + 5 \cdot 2x + 5(-1)$
$= 6x^2 - 3x + 10x - 5$
$= 6x^2 + 7x - 5$

c. $(4x + 3)(4x - 3) = 16x^2 - 9$

d. $(3y + 2)^2 = 9y^2 + 12y + 4$

e. $(3y - 2)^2 = 9y^2 - 12y + 4$

SECTION 6.6 Exponent Rules and Dividing Polynomials

Quotient rule for exponents: If m and n are integers and a is a real number, where $a \neq 0$, then $\dfrac{a^m}{a^n} = a^{m-n}$.

To divide monomials,
1. Divide the coefficients or simplify them to fractions in lowest terms.
2. Use the quotient rule for the exponents with like bases.
3. Do not change unlike variable bases in the quotient.
4. Write the final expression so that all exponents are positive.

If a, b, and c are real numbers, variables, or expressions with $c \neq 0$, then
$$\frac{a + b}{c} = \frac{a}{c} + \frac{b}{c}.$$

To divide a polynomial by a monomial, divide each term in the polynomial by the monomial.

To divide a polynomial by a polynomial, use long division. If there is a remainder, write the result in the following form:
$$\text{quotient} + \frac{\text{remainder}}{\text{divisor}}$$

Example 1: Divide.
a. $\dfrac{x^7}{x^2} = x^{7-2} = x^5$

b. $\dfrac{-39a^6b^5c^2}{13a^4b^5c}$
$= -3a^{6-4}b^{5-5}c^{2-1}$
$= -3a^2b^0c^1$
$= -3a^2(1)c$
$= -3a^2c$

c. $\dfrac{30y^6 + 45y^3}{5y^2} = \dfrac{30y^6}{5y^2} + \dfrac{45y^3}{5y^2}$
$= 6y^{6-2} + 9y^{3-2}$
$= 6y^4 + 9y$

d. $\dfrac{6x^2 + 7x - 17}{3x - 4}$

$$
\begin{array}{r}
2x + 5 \\
3x - 4 \overline{) 6x^2 + 7x - 17} \\
-6x^2 + 8x \\
\hline
15x - 17 \\
-15x + 20 \\
\hline
3
\end{array}
$$

Answer: $2x + 5 + \dfrac{3}{3x - 4}$

(continued)

Procedures/Rules	Key Examples

Exponents Summary

Assume that no denominators are 0, that a and b are real numbers, and that m and n are integers.

Example 1: Evaluate.

Zero as an exponent:
$$a^0 = 1, \text{ where } a \neq 0$$
$$0^0 \text{ is indeterminate.}$$

a. $3^0 = 1$ **b.** $x^0 = 1, x \neq 0$

Negative exponents:
$$a^{-n} = \frac{1}{a^n}$$

c. $3^{-4} = \frac{1}{3^4} = \frac{1}{81}$

$$\frac{1}{a^{-n}} = a^n$$

d. $\frac{1}{x^{-5}} = x^5$

$$\left(\frac{a}{b}\right)^{-n} = \left(\frac{b}{a}\right)^n$$

e. $\left(\frac{3}{x}\right)^{-2} = \left(\frac{x}{3}\right)^2$

Product rule for exponents:
$$a^m \cdot a^n = a^{m+n}$$

f. $x^3 \cdot x^5 = x^{3+5} = x^8$

Quotient rule for exponents:
$$\frac{a^m}{a^n} = a^{m-n}$$

g. $\frac{r^4}{r^{-2}} = r^{4-(-2)} = r^{4+2} = r^6$

Raising a power to a power:
$$(a^m)^n = a^{mn}$$

h. $(y^3)^5 = y^{3 \cdot 5} = y^{15}$

Raising a product to a power:
$$(ab)^n = a^n b^n$$

i. $(3x)^4 = 3^4 x^4 = 81x^4$

Raising a quotient to a power:
$$\left(\frac{a}{b}\right)^n = \frac{a^n}{b^n}$$

j. $\left(\frac{2}{a}\right)^5 = \frac{2^5}{a^5} = \frac{32}{a^5}$

Chapter 6 Review Exercises

For Exercises 1–6, answer true or false.

[6.1] **1.** A negative exponent will always make the coefficient negative.

[6.2] **2.** The degree of a binomial is two.

[6.4] **3.** To raise an exponential form to a power, multiply the exponents.

[6.5] **4.** Conjugates occur when both binomials contain subtraction.

[6.5] **5.** FOIL can be used for all types of polynomial multiplication.

[6.5] **6.** $(x - 4)^2 = x^2 + 16$

For Exercises 7–10, complete the rule.

[6.3] **7.** To add polynomials, _____ like terms.

[6.4] **8.** To multiply monomials:
   a. Multiply the _____.
   b. Add the _____ of the like bases.
   c. Rewrite any unlike variable bases in the product.

[6.4] **9.** $(a^m)^n = a^?$, where a is a real number and m and n are integers.

[6.5] **10.** $(a + b)^2 = a^2 + $ _____ $ + b^2$

[6.1] *For Exercises 11–14, evaluate the exponential form.*

11. $\left(\dfrac{2}{5}\right)^3$

12. -4^2

13. 5^{-2}

14. 13^{-1}

[6.1] *For Exercises 15–18, write the number in standard form.*

15. The radioactive half-life of uranium 238 is 4.5×10^9 years.

16. The temperature at the Sun's core is 1.38×10^7 degrees Celsius.

17. The mass of a hydrogen atom is 1.663×10^{-24} gram.

18. The mass of an electron is 2.006×10^{-30} pound.

[6.1] *For Exercises 19–22, write the number in scientific notation.*

19. Because atoms are so small, their weight is measured in atomic mass units (AMU). One atomic mass unit is 0.0000000000000000000001661 gram.

20. One molecule of NaCl, sodium chloride (table salt) is 0.00000000000000000000963 gram.

21. The speed of light is about 300,000,000 meters per second.

22. The approximate radius of Earth is 6,370,000 meters.

[6.2] *For Exercises 23–26, determine whether the expression is a monomial.*

23. $3xy^5$

24. $-\dfrac{1}{2}x$

25. $\dfrac{4}{ab^3}$

26. $2x^2 - 9$

[6.2] *For Exercises 27–30, identify the coefficient and degree of each monomial.*

27. $6x^4$

28. 27

29. $-2.6xy^3$

30. $-m$

[6.2] *For Exercises 31–34, indicate whether the expression is a monomial, binomial, or trinomial, has no special polynomial name or is not a polynomial.*

31. $4x^2 - 25$

32. $-st^5$

33. $3x - \dfrac{4}{y^2}$

34. $5x^3 - 6x^2 - 3x + 11$

[6.2] *For Exercises 35–38, identify the degree of each polynomial and write each polynomial in descending order of degree.*

35. $15 - 2x^9 - 3x + 21x^5 - 19x^3$

36. $22j^2 - 19 - j^4 + 5j$

37. $4u^5 - 18u^3 + 13u^4 - u - 21$

38. $16v^3 + 21v^4 - 2v + 6v^8 - 19 - v^2$

[6.2] *For Exercises 39–49, combine like terms. If possible, write the resulting polynomial in descending order of degree.*

39. $8y^5 - 3y^4 + 2y - 4y^5 - 2y + 8 + 7y^4$

40. $8l + 5l^4 - 2l^3 - l + 18l^5 + 6l^2 + 20$

41. $3m - 4 - m^2 - 2m^3 + 7 - 2m + 8m^2$

42. $-9 + 7y - 2y - 6 + 2y - 3y^2$

43. $a^2bc - 3ab^2c + 2ab^2c - 6$

44. $5xyz^2 - 4x^2yz - 7xyz^2 - 5x^2yz - 2x^2zy$

45. $6cd - 4cd^2 + 3cd^2 + 8c^3 + 5cd - 12c^3 + 2d + 8$

46. $4a^2bc + 5abc^3 - 8abc - 9a^5 - 2abc^3 - 4a^2bc + 8 + a^5$

47. $18jk - 2j^4 + 12jk^3 - j^4 - 8jk - 9jk^3 + 12 + k^2$

48. $4m^2 - 3mn + 2mn^2 - 8n^2 - 6mn - 4m^2 + 7 + mn^2$

49. $2a + 3b - 4ab - c - 18c - 20a + b - 4b + 8$

[6.3] *For Exercises 50–59, add or subtract and write the resulting polynomial in descending order of degree.*

50. $(5y - 2) + (6y - 8)$

51. $(8x^2 - 3) + (2x^2 + 3x - 1)$

52. $(5m - 1) - (2m + 7)$

53. $(5n + 8) - (2n^3 - 3n^2 + n + 5)$

54. $(2y - 4) + (4y + 8) - (6y - 5)$

55. $(3x^3 - 2x + 8) + (4x^2 - 3x - 1) - (4x^3 - 5x - 10)$

56. $(4p^3 + 2p^2 - 8p - 1) + (-5p^3 + 8p^2 - 2p + 10)$

57. $(2x^3 - 3x^2 - 2x - 1) - (18x^3 - 3x^2 + 2x + 1)$

58. $(5x^2 - 2xy + y^2) + (3x^2 + xy + 5y^2)$

59. $(m^2 + 5mn + 6) - (4m^2 - 3mn + 8)$

[6.4, 6.6] *For Exercises 60–74, simplify. Write all answers with positive exponents.*

60. $m \cdot m^4$

61. $2a \cdot 5a^8$

62. $(-3x^2y)(2x^4y^5)$

63. $(5x^2y)^2$

64. $(-6u^3)^2(2u^4)$

65. $(2x^2)(2x^2)^4$

66. $\dfrac{x^3}{x^{-5}}$

67. $\dfrac{u^{-1}}{u^{-8}}$

68. $\dfrac{s^{-5}}{s}$

69. $\left(\dfrac{1}{x^3}\right)^{-2}$

70. $(2a^3b^{-2})^{-3}$

71. $\dfrac{4hj^4k^{-2}}{(3hj^{-2})^2}$

72. $-5x^{-2}$

73. $8x^0 - (2x)^0$

74. $(-18x^4y^{-5})^0$

75. $2(x - 21)$

76. $-8(2x - 1)$

77. $4a(a - 3)$

78. $-6b(2b^2 - 4b - 1)$

79. $4m^3(-2m^4 - 3m^2 + m + 5)$

80. $-4abc^2(3a - 4ab^3 + 2abc^2 + 8)$

81. $(x + 5)(x - 1)$

82. $(y - 3)(y - 8)$

83. $(2m + 5)(6m - 1)$

84. $(5a + 2b)(3a - 2b)$

85. $(2x + 1)(2x - 1)$

86. $(x - 3)(x^2 + 2x + 5)$

87. $(3y - 1)(y^2 - 2y + 4)$

88. $(a^2 + ab + b^2)(a^2 - 2ab - b^2)$

89. $(4 - x)(4 + x)$

90. $(x - 6)^2$

91. $(3r - 5)^2$

92. $(6s + 2r)(6s - 2r)$

[6.5] **93. a.** Write an expression in simplest form for the area of the figure shown.

  b. Calculate the area if $x = 3$ feet.

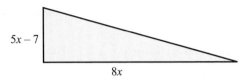

$5x - 7$

$8x$

[6.4] **94.** A park ranger is studying the number of trees per square meter in a national forest. The area in the study is rectangular and measures 9×10^3 meters by 2.5×10^4 meters. Find the area of this region. Write your answer in scientific notation.

[6.4] **95.** In a circuit, the current is measured to be 3.5×10^{-3} ampere. If the circuit has a resistance of 8.4×10^{-2} ohm, find the voltage. (*Hint:* Voltage = Current × Resistance) Write your answer in scientific notation.

[6.6] *For Exercises 96–101, divide.*

96. $\dfrac{4x - 20}{4}$

97. $\dfrac{5y^2 - 10y + 15}{-5}$

98. $\dfrac{2st + 20s^2t^4 - 4st^5}{2st}$

99. $\dfrac{a^3bc^3 - abc^2 + 2a^2bc^4}{abc^3}$

100. $\dfrac{6a^2b^2 + 3a^2b^3}{3a^2b^2}$

101. $\dfrac{2xyz^2 - 5x^2y^2z^2 + 10xy^2z}{5xy^3z}$

[6.6] **102. a.** The area of the parallelogram shown is described by the polynomial $54x^2 - 60x$. Find the height.

  b. Find the base, height, and area of this parallelogram if $x = 4$ inches.

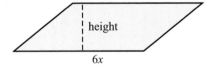

height

$6x$

[6.6] **103.** Proxima Centauri is approximately 2.47×10^{13} miles from the Sun. If light travels 5.881×10^{12} miles per year, how long does the light from Proxima Centauri take to reach Earth? (*Hint:* Distance = Rate × Time)

[6.6] *For Exercises 104–111, use long division to divide the polynomials.*

104. $\dfrac{z^2 + 8z + 15}{z + 3}$ **105.** $\dfrac{3m^2 - 17m + 11}{m - 5}$ **106.** $\dfrac{2y^2 - 7y - 8}{2y + 1}$ **107.** $\dfrac{6m^2 - 10m + 5}{3m - 2}$

108. $\dfrac{x^3 - 1}{x + 1}$ **109.** $\dfrac{8x^3 - 27}{2x - 3}$ **110.** $\dfrac{2s^3 - 2s - 3}{s - 1}$ **111.** $\dfrac{10y^4 - 5y^3 - 6y^2 + 7y - 2}{2y - 1}$

[6.6] 112. A storage room is to have an area described by $4x^2 + 23x - 35$. If the length is described by $x + 7$, find an expression for the width.

Chapter 6 Practice Test

For Extra Help

CHAPTER Test Prep VIDEOS

Step-by-step test solutions are found on the Chapter Test Prep Videos available via the Video Resources on DVD, in *MyMathLab*, and on YouTube (search "CarsonElemAlg" and click on "Channels").

For Exercises 1 and 2, evaluate the exponential form.

1. 2^{-3}

2. $\left(\dfrac{2}{3}\right)^{-2}$

3. Write 6.201×10^{-3} in standard form.

4. Write $275{,}000{,}000$ in scientific notation.

For Exercises 5 and 6, identify the degree.

5. $-7x^2y$

6. $4x^2 - 9x^4 + 8x - 7$

7. Evaluate $-6mn - n^3$, where $m = 4$ and $n = -2$.

8. Combine like terms and write the resulting polynomial in descending order of degree.
$-5x^4 + 7x^2 + 6x^2 - 5x^4 + 12 - 6x^3 + x^2$

For Exercises 9 and 10, add or subtract and write the resulting polynomial in descending order of degree.

9. $(3x^2 + 4x - 2) + (5x^2 - 3x - 2)$

10. $(7x^4 - 3x^2 + 4x + 1) - (2x^4 + 5x - 7)$

For Exercises 11–18, multiply.

11. $(4x^2)(3x^5)$

12. $(2ab^3c^7)(-a^5b)$

13. $(4xy^3)^2$

14. $3x(x^2 - 4x + 5)$

15. $-6t^2u(4t^3 - 8tu^2)$

16. $(n - 1)(n + 4)$

17. $(2x - 3)^2$

18. $(x + 2)(x^2 - 4x + 3)$

19. Write an expression in simplest form for the area of the shape shown.

$2n - 5$

$3n + 4$

For Exercises 20–25, simplify.

20. $\dfrac{x^9}{x^4}$

21. $\dfrac{(x^3)^{-2}}{x^4 \cdot x^{-5}}$

22. $\dfrac{(3y)^{-2}}{(x^3 y^2)^{-3}}$

23. $\dfrac{24x^5 + 18x^3}{6x^2}$

24. $\dfrac{x^2 - x - 12}{x + 3}$

25. $\dfrac{15x^2 - 22x + 14}{3x - 2}$

Chapters 1–6 Cumulative Review Exercises

For Exercises 1–4, answer true or false.

[5.1] **1.** (1, 1) is a solution for the system of equations
$$\begin{cases} x = y \\ 9y - 13 = -4x \end{cases}$$

[1.6] **2.** 6 less than a number can be translated to $6 - x$.

[4.5] **3.** $y - 2 = -3(x - 5)$ is an example of the point–slope form of a line.

[4.2] **4.** There is only one solution to any linear equation in two variables.

For Exercises 5 and 6, fill in the blank.

[2.3] **5.** To use the multiplication principle to clear a coefficient,
  a. Determine the coefficient you want to clear.
  b. Multiply both sides by the multiplicative inverse of that _____ or divide both sides by the _____.

[4.2] **6.** If an equation can be written in the form $y = mx$, where m is a real number other than 0, the x- and y-intercept will be at _____.

Exercises 7–18 **EXPRESSIONS**

[1.5] *For Exercises 7 and 8, simplify.*

7. $5 + 2(6 - 3^2)^2$

8. $|5 - 10| + 3(2) \div (-1)$

[3.2] **9.** 30.2% of what number is 18.12?

For Exercises 10–15, simplify.

[6.3] **10.** $(3y^2 - 2y + 10) + (2y^2 - 2y - 11)$

[6.3] **11.** $(2x + 1) - (-4x + 7)$

[6.5] **12.** $(2x + 3)(x - 7)$

[6.5] **13.** $(2x - 3)^2$

[6.6] **14.** $(3m^2 - 6m + 1) \div (3m)$

[6.6] **15.** $(x^2 + 5x + 7) \div (x + 2)$

16. $(3x^{-2}y^3)^{-3}$

17. $\left(\dfrac{2x^3}{y}\right)^2$

18. $(2x^2y^{-3})^2(-3x^4y^2)^3$

Exercises 19–30 EQUATIONS AND INEQUALITIES

[4.3] **19.** Graph: $3x + 2y = -6$

[4.6] **20.** Graph: $y > \dfrac{2}{3}x - 2$

For Exercises 21 and 22, solve.

[2.3] **21.** $10x - 3 + 5 = 8x + 26 + 4x$

[2.6] **22.** $3(x - 2) > 4(x + 1)$

[2.4] **23.** Solve for l in the equation $P = 2l + 2w$.

[5.2] **24.** Solve $\begin{cases} x - 3y = 3 \\ 3x - 2y = 5 \end{cases}$ by substitution.

[5.3] **25.** Solve $\begin{cases} 2x - 3y = 9 \\ 4x + 3y = -9 \end{cases}$ by elimination.

For Exercises 26–30, solve.

[3.5]
[5.4] **26.** The butcher at Kroger is combining 60 pounds of hamburger costing $2.50 per pound with some hamburger costing $3.10 per pound. How many pounds of the hamburger costing $3.10 does he need to add to make a mixture costing $2.65 per pound?

[3.5]
[5.4] **27.** A broker invests $\dfrac{1}{4}$ of a client's money in a government security paying 9% interest annually and the rest in a money market fund paying 13% interest annually. Find the amount invested in each if the total interest earned is $5400.

[3.2] **28.** Use the accompanying bar graph. If 1600 parents at a local high school were polled, how many could be expected to think that teens should not work?

[3.2] **29.** In Westchester County, New York, the average temperature is 68° in July and only 31.5° in January. What is the percent of decrease in temperature from July to January?

[3.3]
[5.4] **30.** The cashier's office at the college has collected $7020 from students for a total of 42 part-time classes. If each credit hour costs $45, how many three-hour classes and how many five-hour classes were paid for?

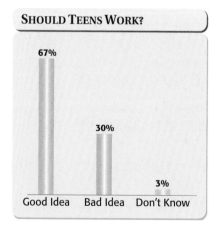

Source: Yankelovich Partners for Lutheran Brotherhood, U.S. Department of Labor

Factoring

In Chapter 6, we learned about a specific class of expressions called *polynomials*. We learned that polynomials can be added, subtracted, multiplied, and divided in much the same way as numbers. In Chapter 7, we explore polynomials further, and in Sections 7.1–7.5, we learn how to factor polynomial expressions. Factoring is another way to rewrite an expression and is one of the most important skills you will acquire in algebra. It is used extensively in simplifying the more complex expressions that we will study in future chapters. In Section 7.6, we use factoring to solve certain equations that contain polynomial expressions. Finally, in Section 7.7, we graph equations and functions containing polynomial expressions.

7.1 Greatest Common Factor and Factoring by Grouping

Objectives

1. List all possible factors for a given number.
2. Find the greatest common factor of a set of numbers or monomials.
3. Write a polynomial as a product of a monomial GCF and a polynomial.
4. Factor by grouping.

Often in mathematics, we need to consider the factors of a number or an expression. A number or an expression written as a product of factors is said to be in **factored form**.

Definition *Factored form:* A number or an expression written as a product of factors.

Following are some examples of factored form.

An integer written in factored form with integer factors: $28 = 2 \cdot 14$
A monomial written in factored form with monomial factors: $8x^5 = 4x^2 \cdot 2x^3$
A polynomial written in factored form with a monomial factor and a polynomial factor: $2x + 8 = 2(x + 4)$
A polynomial written in factored form with two polynomial factors:
$x^2 + 5x + 6 = (x + 2)(x + 3)$

Notice that we can check the factored form by multiplying the factors to equal the product. The process of writing an expression in factored form is called *factoring*.

Objective 1 **List all possible factors for a given number.** We begin our exploration of factoring by listing natural number factors when given an integer.

Example 1 List all natural number factors of 24.

SOLUTION To list all natural number factors, we can divide 24 by 1, 2, 3, and so on, writing each divisor and quotient pair as a product until we have all possible combinations.

> **Note:** The number 5 does not divide 24 evenly, and 6 is already in the $4 \cdot 6$ pair. When you reach a natural number that you already have listed, you can stop.

$1 \cdot 24$
$2 \cdot 12$
$3 \cdot 8$
$4 \cdot 6$

The natural number factors of 24 are 1, 2, 3, 4, 6, 8, 12, 24.

Your Turn 1 List all natural number factors of each given number.

a. 30

b. 42

Objective 2 **Find the greatest common factor of a set of numbers or monomials.** When factoring polynomials, the first step is to determine whether there is a monomial factor that is common to all of the terms in the polynomial. Further, we want that monomial factor to be the **greatest common factor** of the terms. Let's first consider how to find the greatest common factor of numbers.

Answers Your Turn 1
a. 1, 2, 3, 5, 6, 10, 15, 30
b. 1, 2, 3, 6, 7, 14, 21, 42

Greatest common factor (GCF): The largest natural number that divides all given numbers with no remainder.

For example, the greatest common factor of 12 and 18 is 6 because 6 is the largest number that divides into both 12 and 18 evenly.

Using Listing to Find the GCF of Numbers

One way to find the GCF of a given set of numbers is by listing factors.

Procedure **Listing Method for Finding GCF**

To find the GCF of a set of numbers by listing,

1. List all possible factors for each given number.
2. Search the lists for the largest factor common to all lists.

Example 2 Find the GCF of 24 and 60.

SOLUTION Factors of 24: 1, 2, 3, 4, 6, 8, 12, 24
Factors of 60: 1, 2, 3, 4, 5, 6, 10, 12, 15, 20, 30, 60
The GCF of 24 and 60 is 12.

Using Prime Factorization to Find the GCF

The listing method is a good method for smaller numbers, but for larger numbers, it is not the most efficient method to use. It turns out that we can use prime factorization, which we explained in Section 1.2, to find the GCF of a given set of numbers. Consider the prime factorizations for the numbers from Example 2.

$$24 = 2 \cdot 2 \cdot 2 \cdot 3 = 2^3 \cdot 3$$
$$60 = 2 \cdot 2 \cdot 3 \cdot 5 = 2^2 \cdot 3 \cdot 5$$

In Example 2, we found the GCF to be 12. Let's look at the prime factorization of 12 to see if we can discover a rule for finding the GCF.

$$12 = 2 \cdot 2 \cdot 3 = 2^2 \cdot 3$$

Notice that the 12 contains two factors of 2 and one factor of 3, which are the factors that are common to 24's and 60's prime factorizations. This suggests that we use only primes that are common to all factorizations involved. The factor 5 is not common to both 24's and 60's prime factorizations, so it is not included in the GCF.

$$24 = \boxed{2} \cdot \boxed{2} \cdot 2 \cdot \cancel{3} = 2^3 \cdot 3$$
$$60 = \boxed{2} \cdot \boxed{2} \cdot \cancel{3} \cdot 5 = 2^2 \cdot 3 \cdot 5$$

$$\text{GCF} = 2 \cdot 2 \cdot 3 = 2^2 \cdot 3 = 12$$

LEARNING Strategy

VISUAL If you are a visual learner, you may prefer to continue circling the common prime factors.

Notice that we could also compare exponents for a given common prime factor to determine how many of that factor should be included in the GCF. For example, 2^2 is included instead of 2^3 because 2^2 has the smaller exponent. This suggests the following procedure on the following page.

Prime Factorization Method for Finding GCF

To find the GCF of a given set of numbers by prime factorization,

1. Write the prime factorization of each given number in exponential form.
2. Create a factorization for the GCF that includes only those prime factors common to all factorizations, each raised to its smallest exponent in the factorizations.
3. Multiply the factors in the factorization created in step 2.

Note: If there are no common prime factors, the GCF is 1.

Example 3 Find the GCF.

a. 54 and 90

SOLUTION Write the prime factorization of 54 and 90 in factored form.

$$54 = 2 \cdot 3^3$$
$$90 = 2 \cdot 3^2 \cdot 5$$

The common prime factors are 2 and 3. Compare the exponents on each of these common factors. The smallest exponent of 2 is 1, and the smallest exponent of 3 is 2. The GCF is the product of 2^1 and 3^2.

$$GCF = 2^1 \cdot 3^2 = 2 \cdot 9 = 18$$

b. 3024 and 2520

SOLUTION Write the prime factorization of 3024 and 2520 in factored form.

$$3024 = 2^4 \cdot 3^3 \cdot 7$$
$$2520 = 2^3 \cdot 3^2 \cdot 5 \cdot 7$$

The common prime factors are 2, 3, and 7. Compare the exponents of each of these common factors. The smallest exponent of 2 is 3. The smallest exponent of 3 is 2. The smallest exponent of 7 is 1. The GCF is the product of 2^3, 3^2, and 7.

$$GCF = 2^3 \cdot 3^2 \cdot 7 = 8 \cdot 9 \cdot 7 = 504$$

Your Turn 3 Find the GCF.

  a. 180 and 600   **b.** 63, 84, and 105

GCF of Monomials

We use a similar approach to find the GCF of a set of monomials that have variables. The variables in the monomials are treated like prime factors.

Example 4 Find the GCF.

a. $x^3 y^3 z^4$ and $x^2 y^6$

SOLUTION By treating the variables like prime factors, each monomial is already written as a prime factorization. The common prime factors are x and y. The smallest exponent for the factor x is 2, and the smallest exponent for the factor y is 3. The GCF is the product of x^2 and y^3.

$$GCF = x^2 y^3$$

Answers Your Turn 3
  a. 60   **b.** 21

b. $24x^2y$ and $60x^3$

SOLUTION Write the prime factorization of each monomial, treating the variables like prime factors.

$$24x^2y = 2^3 \cdot 3 \cdot x^2 \cdot y$$
$$60x^3 = 2^2 \cdot 3 \cdot 5 \cdot x^3$$

The common prime factors are 2, 3, and x. The smallest exponent of 2 is 2. The smallest exponent of 3 is 1. The smallest exponent of x is 2. The GCF is the product of 2^2, 3, and x^2.

$$\text{GCF} = 2^2 \cdot 3 \cdot x^2 = 12x^2$$

Your Turn 4 Find the GCF.

    a. $a^3b^4c^3$ and $a^2c^5d^3$ **b.** $32a^2b$ and $40abc$ **c.** $35x^2$ and $18y$

Objective ③ Write a polynomial as a product of a monomial GCF and a polynomial. Recall that when we multiply a monomial by a polynomial, we apply the distributive property.

$$2(x + 4) = 2 \cdot x + 2 \cdot 4$$
$$= 2x + 8$$

Now let's reverse this process. Suppose we are given the product, $2x + 8$, and we want to write the factored form, $2(x + 4)$. Notice that terms in $2x + 8$ have 2 as a common factor. When writing factored form, we first determine the GCF of the terms. We then create a missing factor statement like this:

$$2x + 8 = 2(?)$$

Notice that the missing factor must be $x + 4$. To determine this missing factor, we divide the product $2x + 8$ by the known factor, 2.

$$2x + 8 = 2\,(?)$$
$$\frac{2x + 8}{2} = (?)$$
$$\frac{2x}{2} + \frac{8}{2} = (?)$$
$$x + 4 = (?)$$

Connection In Section 6.6, we learned that we divide a polynomial by a monomial by dividing each term in the polynomial by the monomial.

The factored form of $2x + 8$ is $2(x + 4)$, which suggests the following procedure.

Procedure   **Factoring a Monomial GCF Out of a Polynomial**

To factor a monomial GCF out of a given polynomial,

1. Find the GCF of the terms in the polynomial.
2. Rewrite the given polynomial as a product of the GCF and the quotient of the polynomial and the GCF.

$$\text{Polynomial} = \text{GCF}\left(\frac{\text{Polynomial}}{\text{GCF}}\right)$$

Example 5 Factor $12x^2 + 18x$.

SOLUTION

1. Find the GCF of $12x^2$ and $18x$.

   The GCF of $12x^2$ and $18x$ is $6x$.

   Note: 6 is the largest number that divides both 12 and 18 evenly, and x has the smaller exponent of the x^2 and x.

2. Write the given polynomial as the product of the GCF and the quotient of the polynomial and the GCF.

$$12x^2 + 18x = 6x\left(\frac{12x^2 + 18x}{6x}\right)$$

$$= 6x\left(\frac{12x^2}{6x} + \frac{18x}{6x}\right) \quad \text{Separate the terms.}$$

$$= 6x(2x + 3) \quad \text{Divide the terms by the GCF.}$$

Note: An alternative way of doing Example 5 is by inspection using the distributive property. Think of $12x^2 + 18x = 6x(\_\_\_ + \_\_\_)$, where $6x$ times the first blank gives $12x^2$ and $6x$ times the second blank gives $18x$. Because $6x \cdot 2x = 12x^2$ and $6x \cdot 3 = 18x$, the blanks are filled with $2x$ and 3. So $12x^2 + 18x = 6x(2x + 3)$.

Check We can check by multiplying the factored form using the distributive property.

$$6x(2x + 3) = 6x \cdot 2x + 6x \cdot 3 \quad \text{Distribute } 6x.$$

$$= 12x^2 + 18x \quad \text{The product is the original polynomial.}$$

Connection Keep in mind that when we factor, we are simply writing the original expression in a different form called *factored form*. When written equal to each other, the factored form and product make an identity, which means that every real number is a solution to the equation. Consider $12x^2 + 18x = 6x(2x + 3)$. Let's choose a value such as $x = 2$ and verify that it satisfies the equation. (Remember, we can select any real number.)

$$12x^2 + 18x = 6x(2x + 3)$$

$$12(2)^2 + 18(2) \overset{?}{=} 6(2)(2(2) + 3)$$

$$12(4) + 36 \overset{?}{=} 12(4 + 3)$$

$$48 + 36 \overset{?}{=} 12(7)$$

$$84 = 84 \qquad \text{It checks.}$$

Example 6 Factor $24x^2y - 60x^3$.

SOLUTION

1. Find the GCF of $24x^2y$ and $60x^3$.

   We found this GCF in Example 4(b) to be $12x^2$.

2. Write the given polynomial as the product of the GCF and the quotient of the polynomial and the GCF.

Note: Again, we could factor using the distributive property by thinking of $24x^2y - 60x^3 = 12x^3(\_\_\_ + \_\_\_)$. Because $12x^2 \cdot 2y = 24x^2y$ and $12x^2(-5x) = -60x^3$, the blanks represent $2y$ and $-5x$. So $24x^2y - 60x^3 = 12x^2(2y - 5x)$.

$$24x^2y - 60x^3 = 12x^2\left(\frac{24x^2y - 60x^3}{12x^2}\right)$$

$$= 12x^2\left(\frac{24x^2y}{12x^2} - \frac{60x^3}{12x^2}\right)$$

$$= 12x^2(2y - 5x)$$

Connection Remember that when we divide exponential forms that have the same base, we subtract exponents.

Check   Multiply the factored form using the distributive property.

$$12x^2(2y - 5x) = 12x^2 \cdot 2y - 12x^2 \cdot 5x$$
$$= 24x^2y - 60x^3$$

Your Turn 6   Factor.

a. $30xy - 45x$

b. $16r^2s^4 - 8r^3s^4 + 12rs^3$

Factoring When the First Term Is Negative

Consider factoring the expression $-6x + 10y$. Because the first term of the polynomial is negative, when we factor out the GCF, 2, the first term inside the parentheses is also negative, so that we have $2(-3x + 5y)$. However, it is considered undesirable to have a negative first term inside parentheses. We can avoid a negative first term in the parentheses by factoring the negative of the GCF; so in our example, we factor out -2. This changes the sign of each term inside the parentheses so that we have $-2(3x - 5y)$.

Example 7   Factor by factoring out the negative of the GCF:

$$-18x^4y^3 + 9x^2y^2z - 12x^3y$$

SOLUTION

1. Find the GCF of $-18x^4y^3$, $9x^2y^2z$, and $-12x^3y$.

   Because the first term in the polynomial is negative, we factor out the negative of the GCF to avoid a negative first term inside the parentheses. We factor out $-3x^2y$.

2. Write the given polynomial as the product of the GCF and the quotient of the polynomial and the GCF.

$$-18x^4y^3 + 9x^2y^2z - 12x^3y = -3x^2y\left(\frac{-18x^4y^3 + 9x^2y^2z - 12x^3y}{-3x^2y}\right)$$

$$= -3x^2y\left(-\frac{18x^4y^3}{-3x^2y} + \frac{9x^2y^2z}{-3x^2y} - \frac{12x^3y}{-3x^2y}\right)$$

$$= -3x^2y(6x^2y^2 - 3yz + 4x)$$

Check   Multiply the factored form using the distributive property.

$$-3x^2y(6x^2y^2 - 3yz + 4x) = -3x^2y \cdot 6x^2y^2 - 3x^2y \cdot (-3yz) - 3x^2y \cdot 4x$$
$$= -18x^4y^3 + 9x^2y^2z - 12x^3y$$

Your Turn 7   Factor by factoring out the negative of the GCF.

a. $-10x^3 - 15x^2 + 25x$

b. $-48a^4b^5 - 24a^3b^4 + 16ab^2c$

Before we discuss additional techniques of factoring, it is important to state that no matter what type of polynomial we are asked to factor, we always consider whether a monomial GCF (other than 1) can be factored out of the polynomial.

Factoring When the GCF Is a Polynomial

Sometimes in factoring, the GCF is a polynomial with more than one term.

Answers   Your Turn 6

a. $15x(2y - 3)$

b. $4rs^3(4rs - 2r^2s + 3)$

Answers   Your Turn 7

a. $-5x(2x^2 + 3x - 5)$

b. $-8ab^2(6a^3b^3 + 3a^2b^2 - 2c)$

Example 8 Factor $y(x + 2) + 7(x + 2)$.

SOLUTION Notice that this expression is a sum of two products, y and $(x + 2)$, and 7 and $(x + 2)$. Further, note that $(x + 2)$ is the GCF of the two products.

$$y(x + 2) + 7(x + 2) = (x + 2)\left(\frac{y(x + 2) + 7(x + 2)}{x + 2}\right)$$

$$= (x + 2)\left(\frac{y(x + 2)}{x + 2} + \frac{7(x + 2)}{x + 2}\right)$$

$$= (x + 2)(y + 7)$$

Note: The procedure is the same. We rewrite the given expression as the product of the GCF and the quotient of the expression and the GCF.

Your Turn 8 Factor $4a(a - 3) - b(a - 3)$.

Objective ④ **Factor by grouping.** The process of factoring out a polynomial GCF, as we did in Example 8, is an intermediate step in a process called *factoring by grouping*, which is a technique that we try when factoring a four-term polynomial such as $xy + 2y + 7x + 14$. The method is called *grouping* because we group pairs of terms and look for a common factor within each group or pair. We begin by pairing the first two terms as one group and the last two terms as a second group.

$$xy + 2y + 7x + 14 = (xy + 2y) + (7x + 14)$$

Note that the first two terms have a common factor of y and the last two terms have 7 as a common factor. If we factor y out of the first two terms and 7 out of the last two terms, we have the same expression that we factored in Example 8.

$$xy + 2y + 7x + 14 = (xy + 2y) + (7x + 14)$$
$$= y(x + 2) + 7(x + 2)$$
$$= (x + 2)(y + 7) \qquad \text{Factor out } (x + 2).$$

Procedure **Factoring by Grouping**

To factor a four-term polynomial by grouping,

1. Factor out any monomial GCF (other than 1) that is common to all four terms.
2. Group pairs of terms and factor the GCF out of each pair or group.
3. If there is a common binomial factor, factor it out.
4. If there is no common binomial factor, interchange the middle two terms and repeat the process. If there is still no common binomial factor, the polynomial cannot be factored by grouping.

Note: Sometimes the GCF is 1 or −1.

Example 9 Factor.

a. $6x^3 - 8x^2 + 3xy - 4y$

SOLUTION First, we look for a monomial GCF (other than 1). This polynomial does not have one. Because the polynomial has four terms, we now try to factor by grouping.

$$6x^3 - 8x^2 + 3xy - 4y = (6x^3 - 8x^2) + (3xy - 4y)$$
$$= 2x^2(3x - 4) + y(3x - 4)$$
$$= (3x - 4)(2x^2 + y)$$

Factor $2x^2$ out of $6x^3$ and $8x^2$; then factor y out of $3xy$ and $4y$.

Factor out $3x - 4$.

Answer **Your Turn 8**

$(a - 3)(4a - b)$

Note: When factoring by grouping and the third term is negative, the GCF of the third and fourth terms is usually negative.

Note: Now that the expression is in factored form, we no longer need the brackets.

b. $12mn^2 - 20mn - 24n^2 + 40n$

SOLUTION   Note that in this case, there is a monomial GCF, $4n$; so we first factor this GCF out of all four terms.

$$12mn^2 - 20mn - 24n^2 + 40n = 4n(3mn - 5m - 6n + 10)$$

Because the polynomial in the parentheses has four terms, we try to factor by grouping.

$$= 4n[(3mn - 5m) + (-6n + 10)]$$ Factor m out of 3 mn and $5m$; then
$$= 4n[m(3n - 5) - 2(3n - 5)]$$ factor -2 out of $-6n$ and 10.
$$= 4n(3n - 5)(m - 2)$$ Factor out $3n - 5$.

Your Turn 9   Factor.

  a. $6x^2 + 15xz + 2xy + 5yz$ **b.** $42a^2b - 56a^2 - 12ab + 16a$

Answers   Your Turn 9

a. $(2x + 5z)(3x + y)$
b. $2a(3b - 4)(7a - 2)$

7.1 Exercises   For Extra Help   *MyMathLab*
PRACTICE   WATCH   DOWNLOAD

Note: Exercises marked with a ★ represent challenging exercises.

1. Explain in your own words how to list all possible natural number factors of a number.

2. Define the GCF of a given set of numbers in your own words.

3. Given a set of terms, after finding the prime factorization of each term, how do you use those prime factors to create the GCF's factorization?

4. When factoring a monomial GCF out of the terms of a polynomial, after finding the GCF, how do you rewrite the polynomial?

5. What kinds of polynomials are factored by grouping?

6. When factoring by grouping, after factoring out a monomial GCF, what is the next step?

For Exercises 7–22, list all natural number factors of the given number. See Example 1.

7. 9   **8.** 49   **9.** 33   **10.** 21

11. 18   **12.** 12   **13.** 16   **14.** 81

15. 44   **16.** 45   **17.** 60   **18.** 36

19. 56   **20.** 64   **21.** 90   **22.** 84

For Exercises 23–34, find the GCF. See Examples 2–4.

23. 21, 30   **24.** 21, 35   **25.** 72, 80   **26.** 35, 60   **27.** 12, 42, 60   **28.** 10, 18, 36

29. $4xy, 6xy$ **30.** $6ab, 9ab$ **31.** $25h^2, 60h^4$ **32.** $25x^4, 10x^3$ **33.** $6a^5b, 15a^4b^2$ **34.** $7m^6n, 21m^5n^4$

For Exercises 35–62, factor by factoring out the GCF. See Examples 5 and 6.

35. $5c - 20$ **36.** $8y - 24$ **37.** $8x - 12$ **38.** $15x - 10$

39. $x^2 - x$ **40.** $y^2 + y$ **41.** $18z^6 - 12z^4$ **42.** $15k^3 - 24k^2$

43. $18p^3 - 15p^5$ **44.** $18r^4 - 27r^6$ **45.** $6a^2b - 3ab^2$ **46.** $9m^2n - 3mn^2$

47. $14uv^2 - 7uv$ **48.** $12x^2y^3 - 6xy^2$ **49.** $25xy - 50xz + 100x$ **50.** $8ab - 32ac + 40a$

51. $x^2y + xy^2 + x^3y^3$ **52.** $w^3v^2 + w^2v + wv^2$ **53.** $28ab^3c - 36a^2b^2c$ **54.** $30x^3yz^2 - 24x^2y^3z$

55. $20p^2q + 24pq - 16pq^2$ **56.** $6a^2c + 18abc - 12ac^2$ **57.** $3mn^5p^2 + 18mn^3p - 6mnp$

58. $4ax^3y^3 - 6ax^2y^2 + 4axy^2$ **59.** $105a^3b^2 - 63a^2b^3 + 84a^6b^4$ **60.** $21t^4u^3 - 105t^5u^3v + 63t^6u^2v^4$

61. $18x^4 - 9x^3 + 30x^2 - 12x$ **62.** $10g + 40g^3 - 100g^4 + 120g^5$

For Exercises 63–72, factor by factoring out the negative of the GCF.
See Example 7.

63. $-3x + 6y$ **64.** $-8x + 4y$ **65.** $-20a^2 - 15a$ **66.** $-30b^2 - 24b$

67. $-12a^4b + 20a^3b^3$ **68.** $-15x^4y + 21x^2y^2$ **69.** $-4x^2 - 8x + 16$ **70.** $-6x^2 - 18x + 24$

71. $-24x^3y^2z - 30x^2y^3z^4 + 12x^4y^5z^2$ **72.** $-21a^4bc^3 - 15a^3b^3c^3 + 27a^2b^2c^4$

For Exercises 73–80, factor out the polynomial GCF. See Example 8.

73. $y(a - 3) + 2(a - 3)$ **74.** $a(x - 4) + 5(x - 4)$ **75.** $4a(2m + 3n) + b(2m + 3n)$

76. $3m(3a + 5b) + n(3a + 5b)$ **77.** $5x(4m - 5) - 2(4m - 5)$ **78.** $3m(3a - 2) - 4(3a - 2)$

★**79.** $2r(6p + 5q) - 4s(6p + 5q)$ ★**80.** $6x(3y + 5z) - 3w(3y + 5z)$

For Exercises 81–104, factor by grouping. See Example 9(a).

81. $bx + 2b + cx + 2c$

82. $cx + cy + bx + by$

83. $am - an - bm + bn$

84. $xy - xw - yz + wz$

85. $x^3 + 2x^2 - 3x - 6$

86. $y^4 + 4y^3 - by - 4b$

87. $1 - m + m^2 - m^3$

88. $1 + x + x^2 + x^3$

89. $3xy + 5y + 6x + 10$

90. $2ab + 8a + 3b + 12$

91. $4b^2 - b + 4b - 1$

92. $x^3 - 4x^2 + x - 4$

93. $6 + 2b - 3a - ab$

94. $12 + 3y - 4x - xy$

95. $3ax + 6ay + 8by + 4bx$

96. $ac + 2ad + 2bc + 4bd$

97. $x^2y - x^2s - ry + rs$

98. $a^3m - a^3n - bm + bn$

99. $w^2 + 3wz + 5w + 15z$

100. $a^2 + 4ab + 3a + 12b$

101. $3st + 3ty - 2s - 2y$

102. $5ab + 5ac - 3b - 3c$

103. $ax^2 - 5y^2 + ay^2 - 5x^2$

104. $m^2p - 3n^2 + n^2p - 3m^2$

For Exercises 105–114, factor completely. See Example 9(b).

105. $2a^2 + 2ab + 6a + 6b$

106. $3m^2 + 3mn + 12m + 12n$

107. $12ab + 18a + 16b + 24$

108. $12pq + 8q + 30p + 20$

109. $24xz - 12xw - 6yz + 3yw$

110. $12ac + 8ad - 18bc - 12bd$

111. $3a^2y - 12a^2 + 9ay - 36a$

112. $4x^2y - 8x^2 + 20xy - 40x$

113. $2x^3 + 6x^2y + 10x^2 + 30xy$

114. $10uv^2 - 5v^2 + 30uv - 15v$

For Exercises 115 and 116, write an expression for the area of the shaded region; then factor completely.

115.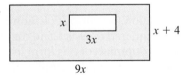

116.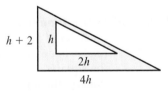

★117. The diagram shows the floor plan of a room with the hearth of a fireplace. Write an expression in factored form of the area of the room excluding the hearth, which is 4 feet by 1 foot.

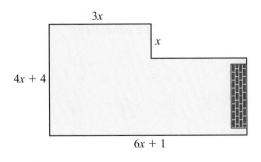

118. Write an expression in factored form for the area of the side of the house shown excluding the window.
★

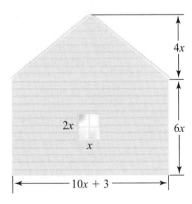

★ *For Exercises 119 and 120, write an expression for the volume of the object shown; then factor completely.*

119.

120.

Review Exercises

Exercises 1–6 EXPRESSIONS

[6.1]   **1.** Write 3.74×10^9 in standard form.

[6.1]   **2.** Write 45,600,000 in scientific notation.

[6.5] *For Exercises 3–6, multiply.*

  3. $(x + 2)(x + 5)$ **4.** $(x - 4)(x + 3)$ **5.** $(x - 5)(x - 3)$ **6.** $x(x - 7)$

7.2 Factoring Trinomials of the Form $x^2 + bx + c$

Objectives

1. Factor trinomials of the form $x^2 + bx + c$.
2. Factor out a monomial GCF; then factor the trinomial of the form $x^2 + bx + c$.

Objective 1   Factor trinomials of the form $x^2 + bx + c$.   In this section, we consider trinomials of the form $x^2 + bx + c$. Notice that the coefficient of the squared term is 1 and the trinomial is in descending order. In Section 7.3, we consider trinomials of this form in which the squared term has a coefficient other than 1. Following are some examples of trinomials of the form $x^2 + bx + c$.

$$x^2 + 5x + 6 \quad \text{or} \quad x^2 - 7x + 12 \quad \text{or} \quad x^2 + 2x - 15 \quad \text{or} \quad x^2 - 5x - 24$$

Trinomials of the form $x^2 + bx + c$ are often the result of the product of two binomials. Consider $(x + 2)(x + 3)$. Recall from Section 6.5 that we use FOIL to multiply two binomials. In the use of FOIL to multiply $(x + 2)(x + 3)$, pay attention to how the last term and the middle term of the resulting trinomial are generated.

$$\begin{array}{cccc} \text{F} & \text{O} & \text{I} & \text{L} \end{array}$$
$$(x + 2)(x + 3) = x^2 + 3x + 2x + 6$$
$$= x^2 \quad + 5x \quad + 6$$

Note: These two numbers multiply to equal the last term, 6, and add to equal the coefficient of the middle term, 5.

sum of   product
2 and 3   of 2 and 3

When we factor a trinomial of the form $x^2 + bx + c$, we reverse the FOIL process, using the fact that b is the sum of the last terms in the binomials and c is the product of the last terms in the binomials. For example, to factor $x^2 + 5x + 6$, we must find two binomials with last terms whose product is 6 and whose sum is 5.

$$x^2 + 5x + 6 = (x + \quad)(x + \quad)$$

The product of these numbers must be 6, and their sum must be 5.

If unsure of what those two numbers are, we can list all factor pairs of 6 and then find the pair whose sum is 5.

Factors Pairs of 6	Corresponding Sum	
$1 \cdot 6$	$1 + 6 = 7$	
$2 \cdot 3$	$2 + 3 = 5$	← This is the correct pair.

Notice that the 2 and 3 become the last terms in the binomials so that the factored form for $x^2 + 5x + 6$ looks like this:

$$x^2 + 5x + 6 = (x + 2)(x + 3)$$

This suggests the following procedure.

Procedure   **Factoring $x^2 + bx + c$**

To factor a trinomial of the form $x^2 + bx + c$,

1. Find two numbers with a product equal to c and a sum equal to b.
2. The factored trinomial will have the form $(x + \text{first number})(x + \text{second number})$.

Note: The signs in the binomial factors can be minus signs depending on the signs of b and c.

In the following examples, we explore cases where minus signs occur. Pay attention to how we determine the signs in the binomial factors.

Example 1   Factor.

a. $x^2 - 7x + 12$

SOLUTION   We must find a pair of numbers whose product is 12 and whose sum is -7. If two numbers have a positive product and negative sum, both must be negative. Following is a table listing the products and sums.

Product	Sum	
$(-1)(-12) = 12$	$-1 + (-12) = -13$	This is the correct combination, so -3 and -4 are the second
$(-2)(-6) = 12$	$-2 + (-6) = -8$	terms in each binomial of the
$(-3)(-4) = 12$	$-3 + (-4) = -7$	← factored form.

Answer $x^2 - 7x + 12 = (x - 3)(x - 4)$ ◄   **Note:** We could have first written $[x + (-3)][x + (-4)]$ to follow the procedure and then simplify.

Check We can check by multiplying the binomial factors to see if their product is the original polynomial.

$$(x - 3)(x - 4) = x^2 - 4x - 3x + 12$$ **Multiply the factors using FOIL.**
$$= x^2 - 7x + 12$$ **The product is the original polynomial.**

b. $x^2 + 2x - 15$

SOLUTION We must find a pair of numbers whose product is -15 and whose sum is 2. Because the product is negative, the two numbers must have different signs. Because the sum is positive, the number with the greater absolute value will be positive.

Product	Sum	
$(-1)(15) = -15$	$-1 + 15 = 14$	This is the correct combination, so
$(-3)(5) = -15$	$-3 + 5 = 2$ ←	-3 and 5 are the second terms in each binomial of the factored form.

Answer $x^2 + 2x - 15 = (x - 3)(x + 5)$

Check $(x - 3)(x + 5) = x^2 + 5x - 3x - 15$ **Multiply the factors using FOIL.**
$$= x^2 + 2x - 15$$ **The product is the original polynomial.**

c. $x^2 - 5x - 24$

SOLUTION We must find a pair of numbers whose product is -24 and whose sum is -5. Because the product is negative, the two numbers must have different signs. Because the sum also is negative, the number with the greater absolute value will be negative.

Product	Sum	
$(1)(-24) = -24$	$1 + (-24) = -23$	This is the correct combination, so 3 and -8 are the second terms
$(2)(-12) = -24$	$2 + (-12) = -10$	in each binomial of the factored
$(3)(-8) = -24$	$3 + (-8) = -5$ ←	form.

Note: In Examples 1(c)–3, we leave the checks to the reader.

Answer $x^2 - 5x - 24 = (x + 3)(x - 8)$

d. $x^2 + 3x + 4$

SOLUTION We must find a pair of numbers whose product is 4 and whose sum is 3. If two numbers have a positive product and a positive sum, both must be positive.

Product	Sum
$(1)(4) = 4$	$1 + 4 = 5$
$(2)(2) = 4$	$2 + 2 = 4$

Note: We listed all positive factors of 4 and found no combination whose sum is 3. This means that $x^2 + 3x + 4$ has no binomial factors with integer coefficients.

A polynomial such as $x^2 + 3x + 4$ that cannot be factored is like a prime number in that its factors are only 1 and the polynomial itself. We say that such a polynomial is *prime*. Notice that the factorization $(x + 4)(x - 1)$ gives the correct middle term, $3x$, but the last term is -4 instead of $+4$.

Answers Your Turn 1

a. $(y - 2)(y - 9)$
b. $(t - 4)(t + 7)$
c. $(n + 5)(n - 6)$
d. prime

Your Turn 1 Factor.

a. $y^2 - 11y + 18$ **b.** $t^2 + 3t - 28$ **c.** $n^2 - n - 30$ **d.** $x^2 - 2x - 6$

Now let's see how to factor trinomials containing two variables.

Example 2 Factor.

a. $x^2 - 10xy + 25y^2$

SOLUTION Because y is in the last term, it is helpful to rearrange the order of the variables in the middle term, writing the polynomial as $x^2 - 10yx + 25y^2$, so that we view the "coefficient" of $-10yx$ as $-10y$. We must find a pair of terms whose product is $25y^2$ and whose sum is $-10y$. These terms would have to be $-5y$ and $-5y$.

Answer $x^2 - 10xy + 25y^2 = (x - 5y)(x - 5y)$ or $(x - 5y)^2$

b. $a^2 + 2ab - 15b^2$

SOLUTION As in Example 2(a), we can view the coefficient of a in the middle term as $2b$. We must find a pair of terms whose product is $-15b^2$ and whose sum is $2b$. These terms would have to be $-3b$ and $5b$.

Answer $a^2 + 2ab - 15b^2 = (a - 3b)(a + 5b)$

Your Turn 2 Factor.

a. $m^2 + 7mn - 18n^2$ **b.** $x^2 - xy - 12y^2$

Objective ② **Factor out a monomial GCF; then factor the trinomial of the form $x^2 + bx + c$.**
Sometimes there is a monomial GCF (other than 1) in the three terms of a trinomial. Consider $2x^3 + 10x^2 + 12x$. Notice that the monomial $2x$ is the GCF of the terms. Factoring out this monomial, we have

$$2x^3 + 10x^2 + 12x = 2x(x^2 + 5x + 6).$$

Now we try to factor the trinomial within the parentheses. We look for two numbers whose product is 6 and whose sum is 5. Note that 2 and 3 work, so we can write the following:

$$2x^3 + 10x^2 + 12x = 2x(x^2 + 5x + 6)$$
$$= 2x(x + 2)(x + 3) \longleftarrow \boxed{\text{Factored form}}$$

Whenever factoring polynomials, the first step always should be to look for a monomial GCF among the terms.

Example 3 Factor.

a. $3mn^3 + 12mn^2 - 96mn$

SOLUTION First, we look for a monomial GCF (other than 1). Notice that the GCF of the terms is $3mn$. Factoring out this monomial, we have

$$3mn^3 + 12mn^2 - 96mn = 3mn(n^2 + 4n - 32).$$

Now try to factor the trinomial to two binomials. We must find a pair of numbers whose product is -32 and whose sum is 4.

Product	Sum	
$(-1)(32) = -32$	$-1 + 32 = 31$	This is the correct combination, so
$(-2)(16) = -32$	$-2 + 16 = 14$	-4 and 8 are the second terms in
$(-4)(8) = -32$	$-4 + 8 = 4$ $\longleftarrow$	each binomial.

Answer $3mn^3 + 12mn^2 - 96mn = 3mn(n - 4)(n + 8)$

b. $x^4 + 6x^3 + 14x^2$

Note: When $x^2 + 6x + 14$ is factored, the product must be positive 14 and the sum must be positive 6. This means that the second terms of both binomial factors must be positive. So neither -1 and -14 nor -2 and -7 are possible.

SOLUTION First, we factor out the monomial GCF, x^2.

$$x^4 + 6x^3 + 14x^2 = x^2(x^2 + 6x + 14)$$

Now we try to factor the trinomial to two binomials. We must find a pair of numbers whose product is 14 and whose sum is 6.

Product	Sum
$(1)(14) = 14$	$1 + 14 = 15$
$(2)(7) = 14$	$2 + 7 = 9$
$(-1)(-14) = 14$	$-1 + (-14) = -15$
$(-2)(-7) = 14$	$-2 + (-7) = -9$

Note: We listed all factor pairs of 14 and found no combination whose sum is 6. This means that $x^2 + 6x + 14$ has no binomial factors with integer terms; so $x^2(x^2 + 6x + 14)$ is the final factored form.

Answers Your Turn 3

a. $y^2(y - 9)(y - 1)$

b. $4ab(a - 7)(a + 2)$

c. $4x(y^2 - 2y + 3)$

Your Turn 3 Factor.

a. $y^4 - 10y^3 + 9y^2$ **b.** $4a^3b - 20a^2b - 56ab$ **c.** $4xy^2 - 8xy + 12x$

7.2 Exercises   **For Extra Help**   *MyMathLab* 

Note: Exercises marked with a ★ represent challenging exercises.

1. If given a trinomial in the form $x^2 + bx + c$, where $c > 0$ and $b > 0$, what can you conclude about the pair of numbers in the binomial factors?

2. If given a trinomial in the form $x^2 - bx + c$, where $c > 0$ and $b > 0$, what can you conclude about the pair of numbers in the binomial factors?

3. If given a trinomial in the form $x^2 + bx - c$, where $c > 0$ and $b > 0$, what can you conclude about the pair of numbers in the binomial factors?

4. If given a trinomial in the form $x^2 - bx - c$, where $c > 0$ and $b > 0$, what can you conclude about the pair of numbers in the binomial factors?

For Exercises 5–12, fill in the missing values in the factors. See Objective 1.

5. $x^2 + 7x + 10 = (x + 2)(x + \quad)$

6. $y^2 + 8y + 15 = (y + 5)(y + \quad)$

7. $n^2 - 12n + 20 = (n - 10)(n - \quad)$

8. $t^2 - 9t + 18 = (t - \quad)(t - 3)$

9. $m^2 - 2m - 24 = (m - \quad)(m + 4)$

10. $a^2 + a - 12 = (a + 4)(a - \quad)$

11. $u^2 + 13u - 30 = (u - \quad)(u + 15)$

12. $n^2 - 9n - 36 = (n - \quad)(n + 3)$

For Exercises 13–44, factor. If the polynomial is prime, so state. See Example 1.

13. $r^2 + 4r + 3$ **14.** $t^2 + 8t + 7$ **15.** $x^2 - 8x + 7$ **16.** $x^2 - 4x + 3$

17. $z^2 - 2z - 3$ **18.** $y^2 - 4y - 5$ **19.** $y^2 + 5y + 6$ **20.** $n^2 + 8n + 15$

21. $u^2 - 6u + 8$ **22.** $b^2 - 10b + 21$ **23.** $u^2 + u - 6$ **24.** $k^2 - 3k - 10$

25. $a^2 + 7a + 12$ **26.** $x^2 + 9x + 18$ **27.** $y^2 - 6y + 9$ **28.** $x^2 - 10x + 25$

29. $w^2 - w - 12$ **30.** $b^2 - 3b - 40$ **31.** $x^2 - x - 30$ **32.** $b^2 - 6b - 16$

33. $n^2 - 11n + 30$ **34.** $y^2 - 7y + 12$ **35.** $r^2 - 9r + 18$ **36.** $x^2 - 10x + 24$

37. $x^2 - 5x - 24$ **38.** $a^2 - 10a - 24$ **39.** $p^2 - 5p - 36$ **40.** $u^2 - 8u - 20$

41. $m^2 - 4m + 6$ **42.** $z^2 + 7z + 8$ **43.** $x^2 - 6x - 8$ **44.** $a^2 + 3a - 20$

For Exercises 45–52, factor the trinomials containing two variables. If the polynomial is prime, so state. See Example 2.

45. $p^2 - 10pq + 21q^2$ **46.** $x^2 - 9xy + 14y^2$ **47.** $a^2 - 6ab - 27b^2$ **48.** $m^2 + 2mn - 8n^2$

49. $x^2 - 14xy + 24y^2$ **50.** $t^2 - 11tu + 24u^2$ **51.** $r^2 - rs - 30s^2$ **52.** $h^2 - 8hk - 20k^2$

For Exercises 53–72, factor completely. See Example 3.

53. $4x^2 - 40x + 84$ **54.** $3m^2 - 33m + 54$ **55.** $2m^3 - 14m^2 + 12m$ **56.** $2k^2y - 18ky + 28y$

57. $3a^2b - 15ab - 72b$ **58.** $3a^2y + 6ay - 72y$ **59.** $4x^2 - 24x + 36$ **60.** $6a^2r + 12ar + 6r$

61. $n^4 + 5n^3 + 6n^2$ **62.** $r^4 + 6r^3 + 8r^2$ **63.** $7u^4 + 42u^3 + 35u^2$ **64.** $5x^5 + 20x^4 + 15x^3$

65. $6a^2b^2c - 36ab^2c + 48b^2c$ **66.** $3a^2b^2c - 24ab^2c + 45b^2c$ **67.** $3x^2 - 21xy + 30y^2$

68. $5m^2 - 30mn + 40n^2$ **69.** $2a^2b - 6ab^2 - 36b^3$ **70.** $7h^2k + 14hk^2 - 168k^3$

71. $4x^4y - 12x^3y^2 - 60x^2y^3$ **72.** $3a^3b^2 + 15a^2b^3 - 12ab^4$

Find
(X) the mistake   **For Exercises 73–76, find and correct the mistake. See Example 1.**

73. $x^2 - 5x - 6 = (x - 3)(x - 2)$ **74.** $x^2 - 5x + 6 = (x - 6)(x + 1)$

75. $x^2 - 3x - 4 = (x - 1)(x + 4)$ **76.** $x^2 + x + 2 = (x + 2)(x + 1)$

★ **For Exercises 77–80, find all natural number values of b that make the trinomial factorable.**

77. $x^2 + bx - 21$ **78.** $x^2 + bx - 10$ **79.** $x^2 + bx + 12$ **80.** $x^2 + bx + 18$

★*For Exercises 81–84, find all natural numbers c that make the trinomial factorable.*

81. $x^2 - 7x + c$ **82.** $x^2 - 5x + c$ **83.** $x^2 + 10x + c$ **84.** $x^2 + 11x + c$

85. The expression $h^2 + 6h + 8$ describes the area of the top (or bottom) of the crate shown, where h represents its height. The unknown expression for the length is the sum of h and an integer, and the expression for the width is the sum of h and a different integer. Find expressions for the length and width.

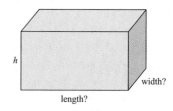

86. The expression $w^2 + w - 2$ describes the area of the top of the bar of gold shown, where w represents the width of the base of the bar. The unknown expression for the length of the top of the bar is the sum of w and an integer. The expression for the width of the top of the bar is the difference of w and an integer. Find the expressions for the length and width of the top of the bar.

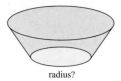

87. The expression $w^2 - 16w + 60$ describes the viewing area in the picture frame shown, where w represents the width of the frame. The unknown expression for the length of the viewing area is the difference of w and an integer. The expression for the width of the viewing area is the difference of w and a different integer. Find expressions for the length and width of the viewing area in the frame.

★**88.** The expression $\pi r^2 - 2\pi r + \pi$ describes the area occupied by the circular base of the papasan chair shown, where r represents the radius of the circle on which the chair rests. The expression that describes the radius of the circle that touches the floor is the difference of r and an integer. Find the expression that describes the radius of the circle that touches the floor.

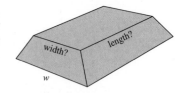

Review Exercises

Exercises 1–10 **EXPRESSIONS**

[7.1]   **1.** What is the GCF of $6x$ and $35y$?

[6.5] *For Exercises 2–4, multiply.*

   2. $(2x + 3)(5x + 1)$ **3.** $(3y - 4)(5y + 6)$ **4.** $4n(6n - 1)(3n - 2)$

[7.1] *For Exercises 5–10, factor completely.*

   5. $10x^2y^3 - 5x^3y - 20x^2y^2$ **6.** $8a^2b^4 - 16ab^2 - 12a^2b^3$

   7. $x(x + 2) - 3(x + 2)$ **8.** $z(z + 5) - 6(z + 5)$

   9. $15ac - 5ad - 3bc + bd$ **10.** $6ac + 4ad - 9bc - 6bd$

7.3 Factoring Trinomials of the Form $ax^2 + bx + c$, where $a \neq 1$

Objectives

  1 Factor trinomials of the form $ax^2 + bx + c$, where $a \neq 1$, by trial.

  2 Factor trinomials of the form $ax^2 + bx + c$, where $a \neq 1$, by grouping.

Objective 1 **Factor trinomials of the form $ax^2 + bx + c$, where $a \neq 1$, by trial.** In Section 7.2, we factored trinomials in which the coefficient of the squared term was 1. Now we focus on factoring trinomials in which the coefficient of the squared term is other than 1, such as the following:

$$3x^2 + 17x + 10 \qquad 8x^2 + 29x - 12$$

In general, like trinomials of the form $x^2 + bx + c$, trinomials of the form $ax^2 + bx + c$, where $a \neq 1$, may also have two binomial factors. When a is not 1, we must consider the factors of the ax^2 term and the factors of the c term. First, we develop a trial-and-error method for finding the factored form. Consider $3x^2 + 17x + 10$. Because all of the terms are positive, we know that all of the terms in the binomial factors will be positive.

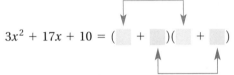

The *first* terms must multiply to equal $3x^2$. The logical factors are $3x$ and x.

$$3x^2 + 17x + 10 = (+)(+)$$

The *last* terms must multiply to equal 10. These factors could be 1 and 10, or 2 and 5.

We now try various combinations of these *first* and *last* terms in binomial factors and consider the resulting trinomial products. If the sum of the inner and outer terms matches the original trinomial, we know we have the correct combination in the binomial factors.

$$
\begin{aligned}
(3x + 10)(x + 1) &= 3x^2 + 3x + 10x + 10 = 3x^2 + 13x + 10 \\
(3x + 1)(x + 10) &= 3x^2 + 30x + x + 10 = 3x^2 + 31x + 10 \\
(3x + 5)(x + 2) &= 3x^2 + 6x + 5x + 10 = 3x^2 + 11x + 10
\end{aligned}
\left.\phantom{\begin{aligned}&\\&\\&\end{aligned}}\right\}
\begin{aligned}&\text{Incorrect}\\&\text{combinations}\end{aligned}
$$

$$(3x + 2)(x + 5) = 3x^2 + 15x + 2x + 10 = 3x^2 + 17x + 10 \qquad \text{Correct combination}$$

You may find that drawing lines to connect each product is helpful.

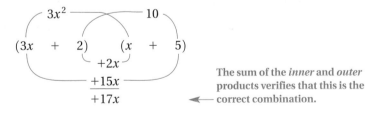

The sum of the *inner* and *outer* products verifies that this is the ← correct combination.

Our example suggests the following procedure.

Procedure **Factoring by Trial and Error**

To factor a trinomial of the form $ax^2 + bx + c$, where $a \neq 1$, by trial and error,

1. Look for a monomial GCF in all of the terms. If there is one, factor it out.

2. Write a pair of *first* terms whose product is ax^2.

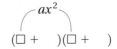
$$(\Box + \quad)(\Box + \quad)$$

3. Write a pair of *last* terms whose product is c.

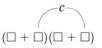
$$(\Box + \Box)(\Box + \Box)$$

4. Verify that the sum of the *inner* and *outer* products is bx (the middle term of the trinomial).

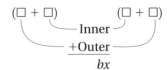

If the sum of the inner and outer products is *not* bx, try the following:

a. Exchange the last terms of the binomials from step 3; then repeat step 4.
b. For each additional pair of last terms, repeat steps 3 and 4.
c. For each additional pair of first terms, repeat steps 2–4.

Example 1   Factor.

a. $3x^2 + 5x + 2$

The *first* terms must multiply to equal $3x^2$.
These must be $3x$ and x.

SOLUTION $3x^2 + 5x + 2 = (\quad + \quad)(\quad + \quad)$

The *last* terms must multiply to equal 2, which can only be 1 and 2. Because the product of the last terms is positive and the sum of the outside and inside products must be $+$, both signs of the binomial must be $+$.

$(3x + 1)(x + 2) = 3x^2 + 6x + x + 2 = 3x^2 + 7x + 2$ Incorrect combination
$(3x + 2)(x + 1) = 3x^2 + 3x + 2x + 2 = 3x^2 + 5x + 2$ Correct combination

Answer $3x^2 + 5x + 2 = (3x + 2)(x + 1)$

b. $2x^2 - 7x + 6$

The *first* terms must multiply to $2x^2$.
These must be $2x$ and x.

SOLUTION $2x^2 - 7x + 6 = (\quad + \quad)(\quad - \quad)$

The *last* terms must multiply to equal $+6$, which could be 1 and 6, or 2 and 3. Because the product of the last terms is positive and the the sum of the outside and inside products is $-$, both signs of the binomial must be $-$.

Note: Because the original polynomial does not have a common factor, no factors can have a common factor. This means that $2x - 6$ and $2x - 2$ cannot be factors because both have a common factor of 2.

$(2x - 1)(x - 6) = 2x^2 - 12x - x + 6 = 2x^2 - 13x + 6$ Incorrect combination
$(2x - 6)(x - 1) = 2x^2 - 2x - 6x + 6 = 2x^2 - 8x + 6$ Incorrect combination
$(2x - 2)(x - 3) = 2x^2 - 6x - 2x + 6 = 2x^2 - 8x + 6$ Incorrect combination
$(2x - 3)(x - 2) = 2x^2 - 4x - 3x + 6 = 2x^2 - 7x + 6$ Correct combination

Answer $2x^2 - 7x + 6 = (2x - 3)(x - 2)$

c. $8x^2 + 29x - 12$

SOLUTION

The *first* terms must multiply to equal $8x^2$. These could be x and $8x$, or $2x$ and $4x$.

$$8x^2 + 29x - 12 = (+)(-)$$

The *last* terms must multiply to equal -12. Because -12 is negative, the last terms in the binomials must have different signs. We have already written the appropriate signs, so these factor pairs could be 1 and 12, 2 and 6, or 3 and 4.

Now we multiply binomials with various combinations of these first and last terms until we find a combination whose inner and outer products combine to equal $29x$.

Note: The products $(8x + 1)(x - 12)$ and $(8x - 1)(x + 12)$ differ only in the sign of the middle term. So unless a factorization is correct except for the sign of the middle term of the product, switching the signs of the binomial factors will not result in the correct factorization.

$$
\begin{aligned}
(8x + 1)(x - 12) &= 8x^2 - 96x + x - 12 = 8x^2 - 95x - 12 \\
(8x - 1)(x + 12) &= 8x^2 + 96x - x - 12 = 8x^2 + 95x - 12 \\
(8x + 12)(x - 1) &= 8x^2 - 8x + 12x - 12 = 8x^2 + 4x - 12 \\
(8x + 2)(x - 6) &= 8x^2 - 48x + 2x - 12 = 8x^2 - 46x - 12 \\
(8x + 6)(x - 2) &= 8x^2 - 16x + 6x - 12 = 8x^2 - 10x - 12 \\
(8x + 3)(x - 4) &= 8x^2 - 32x + 3x - 12 = 8x^2 - 29x - 12
\end{aligned}
$$
Incorrect combinations

Note: In the above factorizations, the factors $8x + 12$, $8x + 2$, and $8x + 6$ are not possible because each has a common factor.

Note: This middle term has the correct absolute value but the incorrect sign. (We are looking for $+29x$.) Switching the $+$ and $-$ signs in the binomials will give us $+29x$.

$$(8x - 3)(x + 4) = 8x^2 + 32x - 3x - 12 = 8x^2 + 29x - 12 \quad \text{Correct combination}$$

Answer $\quad 8x^2 + 29x - 12 = (8x - 3)(x + 4)$

Your Turn 1   Factor.

a. $5x^2 + 9x + 4$ **b.** $6c^2 - 19cd + 15d^2$ **c.** $6y^2 + 13y - 5$

Factoring Out a Monomial GCF First

Remember that we should always look to see if a monomial GCF can be factored out of the terms in the trinomial. This is step 1 for every factoring problem.

Example 2   Factor $40x^3 - 68x^2 + 12x$.

SOLUTION   First, we factor out the monomial GCF, $4x$.

$$40x^3 - 68x^2 + 12x = 4x(10x^2 - 17x + 3)$$

Answers   **Your Turn 1**

a. $(5x + 4)(x + 1)$
b. $(3c - 5d)(2c - 3d)$
c. $(2y + 5)(3y - 1)$

Now we factor the trinomial within the parentheses. The fact that 3 is positive and $-17x$ is negative indicates that both signs of the last terms in the binomial factors will be minus signs.

The *first* terms must multiply to equal $10x^2$. These could be x and $10x$, or $2x$ and $5x$.

$$4x(10x^2 - 17x + 3) = 4x(\boxed{} - \boxed{})(\boxed{} - \boxed{})$$

The *last* terms must multiply to equal 3. Because 3 is a prime number, its factors are 1 and 3.

Now we multiply binomials with various combinations of these first and last terms until we find a combination whose inner and outer products combine to equal $-17x$.

$$4x(10x - 3)(x - 1) = 4x(10x^2 - 10x - 3x + 3) = 4x(10x^2 - 13x + 3)$$
$$4x(10x - 1)(x - 3) = 4x(10x^2 - 30x - x + 3) = 4x(10x^2 - 31x + 3)$$ Incorrect combinations
$$4x(5x - 3)(2x - 1) = 4x(10x^2 - 5x - 6x + 3) = 4x(10x^2 - 11x + 3)$$

$$4x(5x - 1)(2x - 3) = 4x(10x^2 - 15x - 2x + 3) = 4x(10x^2 - 17x + 3)$$ Correct combination

Answer $40x^3 - 68x^2 + 12x = 4x(5x - 1)(2x - 3)$

Your Turn 2   Factor.

a. $16y^3 - 102y^2 + 36y$

b. $36m^2n - 120mn^2 - 21n^3$

Objective ② **Factor trinomials of the form $ax^2 + bx + c$, where $a \neq 1$, by grouping.**   The major drawback about using trial and error to factor is that it can be tedious to find the correct combination. An alternative method is to factor by grouping, which we introduced in Section 7.1. Recall that in the grouping method, we group pairs of terms in a four-term polynomial, then factor out the GCF from each pair of terms. Because a trinomial of the form $ax^2 + bx + c$ has only three terms, we split the bx term into two like terms to create a four-term polynomial that we can factor by grouping. To determine how to split the bx term, we use the fact that if $ax^2 + bx + c$ is factorable, then b will always equal the sum of a pair of factors of the product of a and c.

Consider again the trinomial $3x^2 + 17x + 10$, which we factored by trial and error at the beginning of this section. Notice that $a = 3$, $b = 17$, and $c = 10$; so the product of a and c is $(3)(10) = 30$. To split the bx term, which is $17x$ in this case, we look for a pair of factors of 30 whose sum is 17. It is helpful to list the factor pairs and their corresponding sums in a table.

Factors of ac	Sum of Factors of ac
$(1)(30) = 30$	$1 + 30 = 31$
$(2)(15) = 30$	$2 + 15 = 17$
$(3)(10) = 30$	$3 + 10 = 13$
$(5)(6) = 30$	$5 + 6 = 11$

Notice that 2 and 15 is the only factor pair of 30 whose sum is 17.

Now we can write $17x$ as $15x + 2x$ and then factor by grouping.

$$3x^2 + 17x + 10 = 3x^2 + 15x + 2x + 10$$
$$= 3x(x + 5) + 2(x + 5)$$
$$= (x + 5)(3x + 2)$$

Note: We also could have written $17x$ as $2x + 15x$ to get the same result.

Every trinomial factorable by grouping has only one factor pair of ac whose sum is b, which suggests the following procedure.

Answers   Your Turn 2

a. $2y(y - 6)(8y - 3)$

b. $3n(6m + n)(2m - 7n)$

Procedure

Factoring $ax^2 + bx + c$, where $a \neq 1$, by Grouping

To factor a trinomial of the form $ax^2 + bx + c$, where $a \neq 1$, by grouping,

1. Look for a monomial GCF in all of the terms. If there is one, factor it out.
2. Multiply a and c.
3. Find two factors of this product whose sum is b.
4. Write a four-term polynomial in which bx is written as the sum of two like terms whose coefficients are the two factors you found in step 3.
5. Factor by grouping.

Example 3 Factor by grouping.

a. $8x^2 - 22x + 5$

SOLUTION Notice that for this trinomial, $a = 8$, $b = -22$, and $c = 5$. We begin by multiplying a and c: $(8)(5) = 40$. Now we find two factors of 40 whose sum is -22. Notice that both factors must be negative. It is helpful to list the combinations in a table.

Factors of ac	Sum of Factors of ac	
$(-1)(-40) = 40$	$-1 + (-40) = -41$	
$(-2)(-20) = 40$	$-2 + (-20) = -22$	$\longleftarrow$ Correct
$(-4)(-10) = 40$	$-4 + (-10) = -14$	
$(-5)(-8) = 40$	$-5 + (-8) = -13$	

You do not need to list all possible combinations as we did here. We listed them to illustrate that only one combination is correct. We now write $-22x$ as $-2x - 20x$ and then factor by grouping.

$$8x^2 - 22x + 5 = 8x^2 - 2x - 20x + 5 \qquad \text{Write } -22x \text{ as } -2x - 20x.$$
$$= 2x(4x - 1) - 5(4x - 1) \qquad \begin{array}{l}\text{Factor } 2x \text{ out of } 8x^2 - 2x; \\ \text{factor } -5 \text{ out of } -20x + 5.\end{array}$$
$$= (4x - 1)(2x - 5) \qquad \text{Factor out } (4x - 1).$$

Check $(4x - 1)(2x - 5) = 8x^2 - 20x - 2x + 5 = 8x^2 - 22x + 5$

b. $6a^2 + 13a + 6$

SOLUTION For this trinomial, $a = 6$, $b = 13$, and $c = 6$. The product of $ac = 36$, so we need two factors of 36 whose sum is 13. Because the product is positive, the two factors have the same sign. Because the sum is positive, both factors are positive.

Factors of ac	Sum of Factors of ac	
$(1)(36) = 36$	$1 + 36 = 37$	
$(2)(18) = 36$	$2 + 18 = 20$	
$(3)(12) = 36$	$3 + 12 = 15$	
$(4)(9) = 36$	$4 + 9 = 13$	$\longleftarrow$ Correct

Now write $13a$ as $4a + 9a$ and then factor by grouping.

$$6a^2 + 13a + 6 = 6a^2 + 4a + 9a + 6 \qquad \text{Write } 13a \text{ as } 4a + 9a.$$
$$= 2a(3a + 2) + 3(3a + 2) \qquad \begin{array}{l}\text{Factor } 2a \text{ out of } 6a^2 + 4a; \\ \text{factor } 3 \text{ out of } 9a + 6.\end{array}$$
$$= (3a + 2)(2a + 3) \qquad \text{Factor out } (3a + 2).$$

Check $(3a + 2)(2a + 3) = 6a^2 + 9a + 4a + 6 = 6a^2 + 13a + 6$

c. $18y^4 + 15y^3 - 12y^2$

SOLUTION Notice that there is a monomial GCF, $3y^2$, that we can factor out.

$$18y^4 + 15y^3 - 12y^2 = 3y^2(6y^2 + 5y - 4)$$

Now we factor the trinomial within the parentheses. For this trinomial, $a = 6$, $b = 5$, and $c = -4$. The product of ac is $(6)(-4) = -24$. Now find two factors of -24 whose sum is 5. Because the product is negative, the two factors will have different signs. Because the sum is positive, the factor with the greater absolute value must be positive.

Factors of ac	Sum of Factors of ac	
$(-1)(24) = -24$	$-1 + 24 = 23$	
$(-2)(12) = -24$	$-2 + 12 = 10$	
$(-3)(8) = -24$	$-3 + 8 = 5$	⟵ Correct

Now write $5y$ as $-3y + 8y$ and then factor by grouping.

$$3y^2(6y^2 + 5y - 4) = 3y^2(6y^2 - 3y + 8y - 4) \qquad \text{Write } 5y \text{ as } -3y + 8y.$$
$$= 3y^2[3y(2y - 1) + 4(2y - 1)] \qquad \begin{array}{l}\text{Factor } 3y \text{ out of } 6y^2 - 3y;\\ \text{factor } 4 \text{ out of } 8y - 4.\end{array}$$
$$= 3y^2(2y - 1)(3y + 4) \qquad \text{Factor out } (2y - 1).$$

Check   The check is left to the reader.

Answers **Your Turn 3**

a. $(3x - 2)(2x - 5)$
b. $(3x - 4)(2x + 3)$
c. $4y(2x - 9)(x + 2)$

Your Turn 3   Factor by grouping.

a. $6x^2 - 19x + 10$ **b.** $6x^2 + x - 12$ **c.** $8x^2y - 20xy - 72y$

7.3 Exercises   For Extra Help   *MyMathLab*

Note: Exercises marked with a ★ represent challenging exercises.

1. In factoring a polynomial, your first step should be to look for a(n) _____.

2. Given a trinomial of the form $ax^2 + bx + c$, where $a \neq 1$, how do you determine the first terms in its binomial factors?

3. Given a trinomial of the form $ax^2 + bx + c$, where $a \neq 1$, how do you determine the last terms in its binomial factors?

4. When factoring a trinomial of the form $ax^2 + bx + c$, assuming there is no monomial GCF, what is the first step?

For Exercises 5–10, fill in the missing values in the factors. See Objective 1.

5. $3y^2 + 14y + 8 = (3y + \quad)(y + \quad)$

6. $6x^2 + 11x + 4 = (2x + \quad)(3x + \quad)$

7. $4t^2 + 19t - 30 = (\quad + 6)(\quad - 5)$

8. $8m^2 - 2m - 15 = (\quad - 3)(\quad + 5)$

9. $6x^2 - 25x + 14 = (2x - \quad)(\quad - 2)$

10. $9n^2 - 56n + 12 = (9n - \quad)(\quad - 6)$

For Exercises 11–42, factor completely. If prime, so indicate. See Examples 1 and 2.

11. $2j^2 + 5j + 2$

12. $2x^2 + 7x + 3$

13. $2y^2 - 3y - 5$

14. $3m^2 - 10m + 3$

15. $3m^2 - 10m + 8$

16. $3y^2 - 17y + 10$

17. $6a^2 + 13a + 7$

18. $4y^2 + 8y + 3$

19. $6p^2 + 2p + 1$

20. $6u^2 + 3u - 7$

21. $4a^2 - 19a + 12$

22. $4u^2 - 23u + 15$

23. $6x^2 + 19x + 15$

24. $8x^2 + 18x + 9$

25. $16d^2 - 14d - 15$

26. $18a^2 + 9a - 35$

27. $3p^2 + 13pq + 4q^2$

28. $2u^2 + 7uv + 6v^2$

29. $5k^2 - 7kh - 12h^2$

30. $3a^2 - 13ab - 30b^2$

31. $12a^2 - 40ab + 25b^2$

32. $6w^2 - 19wv + 10v^2$

33. $8m^2 - 27mn + 9n^2$

34. $9x^2 - 21xy + 10y^2$

35. $8x^2y - 4xy - y$

36. $6a^2b + 3ab + b$

37. $12y^2 + 24y + 9$

38. $6a^2 - 20a + 16$

39. $6ab^2 - 20ab + 14a$

40. $20xy^2 - 50xy + 30x$

41. $6w^3 + 16w^2v + 8wv^2$

42. $6x^3 + 51x^2y + 90xy^2$

For Exercises 43–74, factor by grouping. If prime, so state. See Example 3.

43. $3a^2 + 4a + 1$

44. $5m^2 + 11m + 2$

45. $2t^2 - 3t + 1$

46. $3h^2 - 5h + 2$

47. $3x^2 - 4x - 7$

48. $3l^2 + 2l - 5$

49. $2y^2 - y - 6$

50. $3a^2 + 2a - 8$

51. $10a^2 - 19a + 7$

52. $8m^2 - 10m + 3$

53. $8r^2 - 6r - 9$

54. $6x^2 + 5x - 6$

55. $20x^2 - 23x + 6$

56. $12c^2 - 28c + 15$

57. $6k^2 + 7jk + 2j^2$

58. $9x^2 + 9xy + 2y^2$

59. $5x^2 - 26xy + 5y^2$

60. $2t^2 - 3tu + u^2$

61. $10s^2 + st - 2t^2$

62. $10x^2 - 33xy - 7y^2$

63. $6u^2 + 5uv - 6v^2$

64. $14u^2 + uv - 4v^2$

65. $15y^2 - 13y - 8$

66. $8n^2 - 5n - 9$

67. $4k^2 - 14k + 12$

68. $12a^2 + 34a + 20$

69. $12m^3 + 10m^2 - 12m$

70. $36x^2y - 78xy + 36y$

71. $24x^3 - 64x^2y - 24xy^2$

72. $36x^3 - 93x^2y + 60xy^2$

73. $24x^3 - 20x^2y - 24xy^2$

74. $28m^2n + 2mn - 8n$

Find
the mistake

For Exercises 75–78, explain the mistake; then factor the trinomial correctly. See Examples 1 and 2.

75. $6x^2 + 13x + 5 = (6x + 1)(x + 5)$

76. $2x^2 + x - 3 = (2x - 3)(x + 1)$

77. $4n^2 + 12n + 8 = (4n + 8)(n + 1)$

78. $4h^2 - 3hk - 10k^2 = (h - 2)(4h + 5)$

For Exercises 79 and 80, given the area of the figure, factor to find possible expressions for the length and width. See Example 1.

79.

$\text{Area} = 15x^2 - 11x + 2$

80.

$\text{Area} = 2x^2 + 13x + 20$

81. Two adjacent plots of land are for sale. The area of the larger plot is described by $6w^2 - w - 2$ square feet, where w is the width of the smaller plot. Factor to find possible expressions for the dimensions of the larger plot.

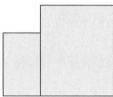

82. A cookie sheet has an area of $5h^2 + 8h - 85$ square inches, where h represents the height of the lip of the sheet. Factor to find possible expressions for the width and length of the cookie sheet.

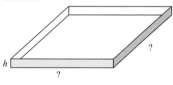

★ *For Exercises 83–88, find all natural numbers that can replace b and make the expression factorable. See Example 1.*

83. $2x^2 + bx + 6$

84. $3x^2 + bx + 10$

85. $5x^2 + bx - 9$

86. $2x^2 + bx - 15$

87. $6x^2 + bx - 7$

88. $4x^2 + bx - 5$

Review Exercises

Exercises 1–7 **EXPRESSIONS**

[6.5] *For Exercises 1–3, multiply.*

1. $(3x + 5)(3x - 5)$

2. $(x - 2)(x^2 + 2x + 4)$

3. $(2x - 5)^2$

For Exercises 4–7, factor completely.

[7.1] **4.** $16c^3d^2 - 24c^2d^4 + 36cd^2$

[7.1] **5.** $ab + 5a + 3b + 15$

[7.2] **6.** $m^2 + 2m - 24$

[7.2] **7.** $3x^3 + 9x^2 - 30x$

7.4 Factoring Special Products

Objectives
1. Factor perfect square trinomials.
2. Factor a difference of squares.
3. Factor a difference of cubes.
4. Factor a sum of cubes.

In Section 6.5, we explored some special products found by multiplying conjugates and squaring binomials. In this section, we see how to factor those special products.

Objective 1 **Factor perfect square trinomials.** First, we consider the product that is a result of squaring a binomial. The trinomial product is a perfect square. Recall the following rules for squaring a binomial that we developed in Section 6.5.

$$(a + b)^2 = a^2 + 2ab + b^2$$
$$(a - b)^2 = a^2 - 2ab + b^2$$

Consider $(2x + 3)^2$ and $(2x - 3)^2$. Using the rules, we have

$$(2x + 3)^2 = (2x)^2 + 2(2x)(3) + (3)^2 \qquad (2x - 3)^2 = (2x)^2 - 2(2x)(3) + (3)^2$$
$$= 4x^2 + 12x + 9 \qquad\qquad\qquad = 4x^2 - 12x + 9$$

To use these rules when factoring, we look at the terms in the given trinomial to determine whether the trinomial fits the form of a perfect square. If we were asked to factor $4x^2 + 12x + 9$, we would note that the first and last terms are perfect squares and that twice the product of their square roots equals the middle term.

The square root of $4x^2$ is $2x$.	The square root of 9 is 3.

$$4x^2 + 12x + 9$$

Twice the product of $2x$ and 3 is the middle term, $12x$.

Because $4x^2 + 12x + 9$ is a perfect square trinomial fitting the form $a^2 + 2ab + b^2$, where $a = 2x$ and $b = 3$, we can write the factored form as $(a + b)^2$, which is $(2x + 3)^2$.

Rule **Factoring Perfect Square Trinomials**

$$a^2 + 2ab + b^2 = (a + b)^2$$
$$a^2 - 2ab + b^2 = (a - b)^2$$

Example 1 Factor.

a. $9y^2 + 30y + 25$

SOLUTION This trinomial is a perfect square because the first and last terms are perfect squares and twice the product of their roots is the middle term.

The square root of $9y^2$ is $3y$.　　　The square root of 25 is 5.

$$9y^2 + 30y + 25$$

Twice the product of $3y$ and 5 is
$2(3y)(5) = 30y$, which is the middle term.

Answer $9y^2 + 30y + 25 = (3y + 5)^2$　Use $a^2 + 2ab + b^2 + (a + b)^2$, where $a = 3y$ and $b = 5$.

b. $36x^2 - 84x + 49$

SOLUTION This trinomial is a perfect square.

The square root of $36x^2$ is $6x$.　　　The square root of 49 is 7.

$$36x^2 - 84x + 49$$

Twice the product of $6x$ and 7 is
$2(6x)(7) = 84x$, which is the middle term.

Answer $36x^2 - 84x + 49 = (6x - 7)^2$　Use $a^2 - 2ab + b^2 = (a - b)^2$, where $a = 6x$ and $b = 7$.

Your Turn 1 Factor.

　a. $25n^2 + 20n + 4$　　　　　　　**b.** $49x^2 - 14x + 1$

Note that in the perfect square trinomial forms, both a and b can have variables. Also don't forget that as a first step, you should always look for a monomial GCF in the terms.

Example 2 Factor.

a. $16x^2 - 8xy + y^2$

SOLUTION $16x^2 - 8xy + y^2 = (4x - y)^2$　Use $a^2 - 2ab + b^2 = (a - b)^2$, where $a = 4x$ and $b = y$.

b. $4x^3y - 36x^2y + 81xy$

SOLUTION

$$4x^3y - 36x^2y + 81xy = xy(4x^2 - 36x + 81)$$　Factor out the monomial GCF, xy.
$$= xy(2x - 9)^2$$　Factor $4x^2 - 36x + 81$ using $a^2 - 2ab + b^2 = (a - b)^2$, where $a = 2x$ and $b = 9$.

Your Turn 2 Factor.

a. $9m^2 + 48mn + 64n^2$

b. $2xy^4 + 24xy^3 + 72xy^2$

Objective ② Factor a difference of squares. Another special product that we considered in Section 6.5 was found by multiplying conjugates. Recall that conjugates are binomials that differ only in the sign separating the terms. For example, $3x + 2$ and $3x - 2$ are conjugates. Note that multiplying conjugates produces a product that is a difference of squares.

$$(3x + 2)(3x - 2) = 9x^2 - 4$$

This term is the square of $3x$.   This term is the square of 2.

The rule for multiplying conjugates that we developed in Section 6.5 is:

$$(a + b)(a - b) = a^2 - b^2.$$

To turn that rule into a rule for factoring a difference of squares, we reverse it.

Rule **Factoring a Difference of Squares**

$$a^2 - b^2 = (a + b)(a - b)$$

Warning: A *sum* of squares $a^2 + b^2$ is prime and *cannot* be factored.

Example 3 Factor.

a. $25x^2 - 36y^2$

SOLUTION This binomial is a difference of squares because $25x^2 - 36y^2 = (5x)^2 - (6y)^2$. To factor it, we use the rule $a^2 - b^2 = (a + b)(a - b)$.

$$a^2 - b^2 = (a + b)(a - b)$$
$$25x^2 - 36y^2 = (5x)^2 - (6y)^2 = (5x + 6y)(5x - 6y) \qquad \text{Replace } a \text{ with } 5x \text{ and } b \text{ with } 6y.$$

b. $9x^4 - 16$

SOLUTION This binomial is a difference of squares because $9x^4 - 16 = (3x^2)^2 - (4)^2$.

$$9x^4 - 16 = (3x^2 + 4)(3x^2 - 4) \qquad \text{Use } a^2 - b^2 = (a + b)(a - b) \text{ with } a = 3x^2 \text{ and } b = 4.$$

c. $25y^7 - 64y^9$

SOLUTION The terms in this binomial have a monomial GCF, y^7.

$$25y^7 - 64y^9 = y^7(25 - 64y^2) \qquad \text{Factor out the monomial GCF, } y^7.$$
$$= y^7(5 + 8y)(5 - 8y) \qquad \text{Factor } 25 - 64y^2 \text{ using } a^2 - b^2 = (a + b)(a - b) \text{ with } a = 5 \text{ and } b = 8y.$$

Your Turn 3 Factor.

a. $n^2 - 49$

b. $9t^6 - 49u^4$

c. $50x^3 - 18x$

Sometimes the factors themselves can be factored.

Example 4   Factor $x^4 - 81$.

SOLUTION   This binomial is a difference of squares, where $a = x^2$ and $b = 9$.

$$x^4 - 81 = (x^2 + 9)(x^2 - 9) \qquad \text{Use } a^2 - b^2 = (a + b)(a - b).$$
$$= (x^2 + 9)(x + 3)(x - 3) \qquad \text{Factor } x^2 - 9 \text{ using } a^2 - b^2 = (a + b)(a - b)$$
$$\text{with } a = x \text{ and } b = 3.$$

Note: The factor $x^2 + 9$ is a sum of squares, which is prime.

Your Turn 4   Factor $y^4 - 16$.

Objective ③ Factor a difference of cubes.   Another common binomial form that we can factor is a difference of cubes. A difference of cubes is a result of a multiplication such as the following:

$$(x - 2)(x^2 + 2x + 4) = x^3 + 2x^2 + 4x - 2x^2 - 4x - 8$$
$$= x^3 - 8 \longleftarrow \text{Difference of cubes}$$

Writing factored form reverses this process so that we have

$$x^3 - 8 = (x - 2)(x^2 + 2x + 4).$$

This suggests the following rule for factoring a difference of cubes.

Rule   Factoring a Difference of Cubes

$$a^3 - b^3 = (a - b)(a^2 + ab + b^2)$$

Example 5   Factor.

a. $64y^3 - 27$

SOLUTION   This binomial is a difference of cubes because $64y^3 - 27 = (4y)^3 - (3)^3$. To factor, we use the rule $a^3 - b^3 = (a - b)(a^2 + ab + b^2)$ with $a = 4y$ and $b = 3$.

$$a^3 - b^3 = (a - b)(a^2 + a b + b^2)$$
$$64y^3 - 27 = (4y)^3 - (3)^3 = (4y - 3)((4y)^2 + (4y)(3) + (3)^2) \qquad \begin{array}{l}\text{Replace } a \text{ with} \\ 4y \text{ and } b \text{ with 3.}\end{array}$$
$$= (4y - 3)(16y^2 + 12y + 9) \qquad \text{Simplify.}$$

Note: This trinomial may seem like a perfect square. However, to be a perfect square, the middle term should be $2ab$. In this trinomial, we have only ab; so it cannot be factored.

b. $27m^4 - 8mn^3$

SOLUTION   The terms in this binomial have a monomial GCF, m.

$$27m^4 - 8mn^3 = m(27m^3 - 8n^3) \qquad \begin{array}{l}\text{Factor out the} \\ \text{monomial GCF, } m.\end{array}$$
$$= m(3m - 2n)((3m)^2 + (3m)(2n) + (2n)^2) \qquad \begin{array}{l}\text{Factor } 27m^3 - 8n^3, \text{ which} \\ \text{is a difference of cubes with}\end{array}$$
$$= m(3m - 2n)(9m^2 + 6mn + 4n^2) \qquad \begin{array}{l}a = 3m \text{ and } b = 2n. \\ \text{Simplify.}\end{array}$$

Answer   Your Turn 4

$(y^2 + 4)(y + 2)(y - 2)$

Your Turn 5 Factor.

a. $27 - u^3$

b. $xy^3 - 64x$

Objective ④ Factor a sum of cubes. A sum of cubes can be factored using a pattern similar to the difference of cubes. A sum of cubes is a result of a multiplication such as the following:

$$(x + 2)(x^2 - 2x + 4) = x^3 - 2x^2 + 4x + 2x^2 - 4x + 8$$
$$= x^3 + 8$$

Writing factored form reverses this process so that we have

$$x^3 + 8 = (x + 2)(x^2 - 2x + 4).$$

In general, we can write the following rule for factoring a sum of cubes.

Rule **Factoring a Sum of Cubes**

$$a^3 + b^3 = (a + b)(a^2 - ab + b^2)$$

The factored form of the sum or difference of cubes always has a binomial and a trinomial factor. The binomial factor consists of the cube roots of the terms of the sum or difference of cubes with the same sign. The first term of the trinomial factor is the square of the first term of the binomial factor, the second term of the trinomial factor is the opposite of the product of the terms of the binomial factor, and the third term of the trinomial factor is the square of the second term of the binomial factor.

LEARNING Strategy

AUDITORY TACTILE VISUAL A mnemonic for remembering the signs of the factorization of the sum or difference of cubes is SOAP.

$a^3 + b^3 =$
$(a + b)(a^2 - ab + b^2)$
$a^3 - b^3 =$
$(a - b)(a^2 + ab + b^2)$

S	O	A P
a	p	l o
m	p	w s
e	o	a i
	s	y t
	i	s i
	t	v
	e	e

Example 6 Factor.

a. $27x^3 + 8$

SOLUTION This binomial is a sum of cubes because $27x^3 + 8 = (3x)^3 + (2)^3$. To factor, we use the rule $a^3 + b^3 = (a + b)(a^2 - ab + b^2)$ with $a = 3x$ and $b = 2$.

$$a^3 \ + \ b^3 \ = \ (\,a + b\,)(\ a^2 \ - \ a\ b \ + \ b^2)$$
$$27x^3 + 8 = (3x)^3 + (2)^3 = (3x + 2)((3x)^2 - (3x)(2) + (2)^2) \quad \text{Replace } a \text{ with } 3x \text{ and } b \text{ with } 2.$$
$$= (3x + 2)(9x^2 - 6x + 4) \quad \text{Simplify.}$$

Note: This trinomial cannot be factored.

b. $5h + 40hk^3$

SOLUTION The terms in this binomial have a monomial GCF, $5h$.

$$5h + 40hk^3 = 5h(1 + 8k^3) \qquad \text{Factor out the monomial GCF, } 5h.$$
$$= 5h(1 + 2k)((1)^2 - (1)(2k) + (2k)^2) \qquad \text{Factor } 1 + 8k^3, \text{ which is a sum of cubes with } a = 1$$
$$= 5h(1 + 2k)(1 - 2k + 4k^2) \qquad \text{and } b = 2k.$$

Note: Because $1^2 = 1$, $1^3 = 1$, and so on, 1 is a perfect square, a perfect cube, and so on.

Answers Your Turn 5
a. $(3 - u)(9 + 3u + u^2)$
b. $x(y - 4)(y^2 + 4y + 16)$

Answers Your Turn 6
a. $(4 + m)(16 - 4m + m^2)$
b. $x^2(5x + 3)(25x^2 - 15x + 9)$

Your Turn 6 Factor.

a. $64 + m^3$

b. $125x^5 + 27x^2$

Note: Exercises marked with a ★ represent challenging exercises.

1. Explain in your own words how to recognize a perfect square trinomial.

2. How is the sign in the factored form of a perfect square trinomial determined?

3. The factors of a difference of squares are binomials known as ——————.

4. There is only one minus sign in the factors of a difference of cubes. Where is that minus sign placed in the factors?

5. There is only one minus sign in the factors of a sum of cubes. Where is that minus sign placed in the factors?

6. How are the a and b terms determined in a sum or difference of cubes?

For Exercises 7–24, factor the trinomials that are perfect squares. If the trinomial is not a perfect square, write **not a perfect square.** *See Examples 1 and 2.*

7. $x^2 + 14x + 49$

8. $y^2 + 6y + 9$

9. $b^2 - 8b + 16$

10. $m^2 - 6m + 9$

11. $n^2 + 12n + 144$

12. $y^2 + 6y + 36$

13. $25u^2 - 30u + 9$

14. $9m^2 - 24m + 16$

15. $100w^2 + 20w + 1$

16. $25a^2 + 10a + 1$

17. $y^2 + 2yz + z^2$

18. $w^2 + 2wv + v^2$

19. $4p^2 - 28pq + 49q^2$

20. $9r^2 - 12rs + 4s^2$

21. $16g^2 + 24gh + 9h^2$

22. $4y^2 - 12by + 9b^2$

23. $16t^2 + 80t + 100$

24. $54q^2 - 72q + 24$

For Exercises 25–44, factor the binomials that are the difference of squares. If prime, so state. See Examples 3 and 4.

25. $x^2 - 4$

26. $a^2 - 121$

27. $16 - y^2$

28. $49 - a^2$

29. $p^2 - q^2$

30. $m^2 - n^2$

31. $25u^2 - 16$

32. $16x^2 - 49$

33. $9x^2 + b^2$

34. $16x^2 + y^2$

35. $64m^2 - 25n^2$

36. $121p^2 - 9q^2$

37. $50x^2 - 32y^2$

38. $27a^2 - 48b^2$

39. $4x^2 + 100y^2$

40. $16x^2 + 36y^2$

41. $x^4 - y^4$

42. $a^4 - b^4$

43. $x^4 - 16$

44. $b^4 - 1$

For Exercises 45–64, factor the sum or difference of cubes. See Examples 5 and 6.

45. $n^3 - 27$

46. $y^3 - 64$

47. $x^3 + 27$

48. $y^3 + 8$

49. $x^3 - 1$

50. $r^3 - 1$

51. $m^3 + n^3$

52. $p^3 + q^3$

53. $27k^3 - 8$

54. $125a^3 - 64$

55. $27k^3 + 8$

56. $125b^3 + 27$

57. $c^3 - 64d^3$

58. $a^3 - 8b^3$

59. $125x^3 + 64y^3$

60. $27a^3 + 1000b^3$

61. $27x^3 - 64y^3$

62. $1000a^3 - 27b^3$

63. $8p^3 + q^3z^3$

64. $x^3y^3 + 27z^3$

For Exercises 65–80, factor. If prime, so state. See Examples 1–6.

65. $2x^2 - 50$

66. $5y^2 - 20$

67. $16x^2 - \dfrac{25}{49}$

68. $25y^2 - \dfrac{1}{36}$

69. $2u^3 - 2u$

70. $4x - 16x^3$

71. $y^5 - 16y^3b^2$

72. $a^4b^2 - a^6$

73. $50x^3 + 2x$

74. $27m^3 + 363mn^2$

75. $3y^3 - 24z^3$

76. $2x^3 - 54$

77. $c^3 - \dfrac{8}{27}$

78. $y^3 - \dfrac{1}{125}$

79. $16c^4 + 2cd^3$

80. $8d^4 - 64c^3d$

★*For Exercises 81–88, use the rules for a difference of squares, sum of cubes, or difference of cubes to factor completely. See Examples 3–6.*

81. $(2a - b)^2 - c^2$

82. $(3x - y)^2 - z^2$

83. $16 - 9(x - y)^2$

84. $25a^2 - 16(b + c)^2$

85. $x^3 + 27(y + z)^3$

86. $m^3 + 8(y + b)^3$

87. $(x + y)^3 - 64d^3$

88. $(t + u)^3 - 64$

For Exercises 89–92, find a natural number, b, that makes the expression a perfect square trinomial. See Objective 1.

89. $16x^2 + bx + 9$

90. $25x^2 + bx + 4$

91. $4x^2 - bx + 81$

92. $36x^2 - bx + 49$

For Exercises 93–96, find the natural number, c, that completes the perfect square trinomial. See Objective 1.

93. $x^2 + 10x + c$

94. $x^2 + 8x + c$

95. $9x^2 - 24x + c$

96. $4x^2 - 28x + c$

For Exercises 97–98, write a polynomial for the area of the shaded region; then factor completely.

★ **97.**

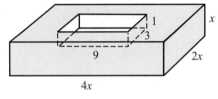

★ **98.**

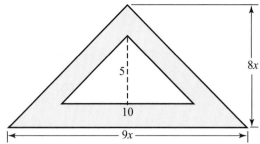

For Exercises 99–100, write a polynomial for the volume of the shaded region; then factor completely.

★ **99.**

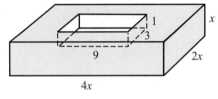

★ **100.**

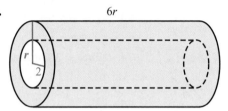

Review Exercises

Exercises 1 and 2 CONSTANTS AND VARIABLES

[7.1] **1.** List all natural number factors of 36.

[1.2] **2.** Find the prime factorization of 100.

Exercises 3–8 EXPRESSIONS

For Exercises 3–8, factor completely.

[7.1] **3.** $14x^3y^3 - 21x^2y^4 - 7xy^2$

[7.1] **4.** $x^2 + 4xy + 3x + 12y$

[7.2] **5.** $a^2 - 3ab - 18b^2$

[7.2] **6.** $3a^2b + 6ab - 45b$

[7.3] **7.** $6n^2 + 5n - 6$

[7.3] **8.** $12x^3y - 38x^2y^2 + 20xy^3$

7.5 Strategies for Factoring

Objective ① Factor polynomials. One of the difficulties in factoring expressions is determining which technique to use. In this section, we use the following general outline for factoring.

Procedure   **Factoring a Polynomial**

To factor a polynomial, first factor out any monomial GCF; then consider the number of terms in the polynomial. If the polynomial has

I. Four terms, try to factor by grouping.

II. Three terms, determine whether the trinomial is a perfect square.
     A. If the trinomial is a perfect square, consider its form.
         1. If in form $a^2 + 2ab + b^2$, the factored form is $(a + b)^2$.
         2. If in form $a^2 - 2ab + b^2$, the factored form is $(a - b)^2$.
     B. If the trinomial is not a perfect square, consider its form.
         1. If in form $x^2 + bx + c$, find two factors of c whose sum is b and write the factored form as $(x + \text{first number})(x + \text{second number})$.
         2. If in form $ax^2 + bx + c$, where $a \neq 1$, use trial and error. Or find two factors of ac whose sum is b; write these factors as coefficients of two like terms that, when combined, equal bx; and factor by grouping.

III. Two terms, determine whether the binomial is a difference of squares, sum of cubes, or difference of cubes.
     A. If given a binomial that is a difference of squares, $a^2 - b^2$, the factors are conjugates and the factored form is $(a + b)(a - b)$. Note that a sum of squares cannot be factored.
     B. If given a binomial that is a sum of cubes, $a^3 + b^3$, the factored form is $(a + b)(a^2 - ab + b^2)$.
     C. If given a binomial that is a difference of cubes, $a^3 - b^3$, the factored form is $(a - b)(a^2 + ab + b^2)$.

Note: Always check to see if any of the factors can be factored further.

Example 1   Factor.

a. $12x^2 + 28x - 5$

SOLUTION   There is no monomial GCF. Because this expression is a trinomial, we check to see if it is a perfect square. It is not a perfect square because the first and last terms are not perfect squares. It is a trinomial of the form $ax^2 + bx + c$, where $a \neq 1$. Therefore, we use trial and error or the alternative method that involves factoring by grouping. We will use trial and error here.

The *first* terms multiply to equal $12x^2$. The factors could be x and $12x$, $2x$ and $6x$, or $3x$ and $4x$.

$$12x^2 + 28x - 5 = (\quad + \quad)(\quad - \quad)$$

The *last* terms multiply to equal -5. We have already written in the appropriate signs, so these factors could be 1 and 5.

Note: Because $(x + 5)(12x - 1)$ did not give the middle term with the correct absolute value, but the incorrect sign, $(x - 5)(12x + 1)$ will also be incorrect. The same is true of $(12x + 5)(x - 1)$ and $(12x - 5)(x + 1)$.

Now list possible combinations of these first and last factor pairs and check by multiplying to see if the sum of the products of the inner and outer terms equals $28x$.

$$\left. \begin{array}{l} (x + 5)(12x - 1) = 12x^2 - x + 60x - 5 = 12x^2 + 59x - 5 \\ (12x + 5)(x - 1) = 12x^2 - 12x + 5x - 5 = 12x^2 - 7x - 5 \end{array} \right\}$$ Incorrect combinations

$$(2x + 5)(6x - 1) = 12x^2 - 2x + 30x - 5 = 12x^2 + 28x - 5$$ Correct combination

Answer $12x^2 + 28x - 5 = (2x + 5)(6x - 1)$

b. $7x^3 + 14x^2 - 105x$

SOLUTION

$$7x^3 + 14x^2 - 105x = 7x(x^2 + 2x - 15)$$ Factor out the monomial GCF, $7x$.

The trinomial in the parentheses is not a perfect square and is in the form $x^2 + bx + c$; so we look for two numbers whose product is -15 and whose sum is 2. Because the product is negative, the two factors must have different signs. Because the sum is positive, the factor with the greater absolute value will be the positive factor.

Product	Sum	
$(-1)(15) = -15$	$-1 + 15 = 14$	
$(-3)(5) = -15$	$-3 + 5 = 2$	Correct combination

Answer $7x^3 + 14x^2 - 105x = 7x(x - 3)(x + 5)$

c. $12m^4 - 48n^2$

SOLUTION

$$12m^4 - 48n^2 = 12(m^4 - 4n^2)$$ Factor out the monomial GCF, 12.

The binomial in the parentheses is a difference of squares in the form $a^2 - b^2$, where $a = m^2$ and $b = 2n$. Using $a^2 - b^2 = (a + b)(a - b)$ gives us

$$= 12(m^2 + 2n)(m^2 - 2n).$$ **Note:** Neither of these binomial factors can be factored further.

d. $10y^5 + 40y^3$

SOLUTION

$$10y^5 + 40y^3 = 10y^3(y^2 + 4)$$ Factor out the monomial GCF, $10y^3$.

The binomial in the parentheses is a sum of squares, which cannot be factored.

e. $45x^3y^3 - 120x^2y^4 + 80xy^5$

SOLUTION

$$45x^3y^3 - 120x^2y^4 + 80xy^5 = 5xy^3(9x^2 - 24xy + 16y^2)$$ Factor out the monomial GCF, $5xy^3$.

The trinomial in the parentheses is a perfect square in the form $a^2 - 2ab + b^2$, where $a = 3x$ and $b = 4y$. Using $a^2 - 2ab + b^2 = (a - b)^2$ gives us

$$= 5xy^3(3x - 4y)^2.$$

f. $x^5 - x^3 - 8x^2 + 8$

SOLUTION There is no monomial GCF common to all terms. Because this polynomial has four terms, we try to factor by grouping.

$$x^5 - x^3 - 8x^2 + 8 = x^3(x^2 - 1) - 8(x^2 - 1)$$
$$= (x^2 - 1)(x^3 - 8)$$

The two binomial factors can be factored further. The first binomial, $x^2 - 1$, is a difference of squares with $a = x$ and $b = 1$. The second binomial, $x^3 - 8$, is a difference of cubes with $a = x$ and $b = 2$.

$$= (x + 1)(x - 1)(x - 2)(x^2 + 2x + 4)$$

Answers   Your Turn 1
a. $(3y - 7)(8y + 1)$
b. $2t(t - 6)(t - 3)$
c. $5(3a + b)(3a - b)$
d. $3m^2n(n^2 + 5n + 2)$
e. $4a^2b(2a - 5b)^2$
f. $(2x + 3)(2x - 3)(3x - 2)$

Your Turn 1   Factor.

a. $24y^2 - 53y - 7$ **b.** $2t^3 - 18t^2 + 36t$
c. $45a^2 - 5b^2$ **d.** $3m^2n^3 + 15m^2n^2 + 6m^2n$
e. $16a^4b - 80a^3b^2 + 100a^2b^3$ **f.** $12x^3 - 8x^2 - 27x + 18$

7.5 Exercises   For Extra Help   *MyMathLab*

1. What should you look for first when factoring?

2. What are three different types of factorable binomials?

3. Describe one type of binomial that cannot be factored.

4. What are the two forms that a perfect square trinomial can have?

5. When do you factor by grouping?

6. How might you check a factored form to see if it is correct?

For Exercises 7–82, factor completely. If prime, so state. See Example 1.

7. $3xy^2 + 6x^2y$ **8.** $2a + ab$ **9.** $7a(x + y) - b(x + y)$ **10.** $(u + v)r + (u + v)s$

11. $2x^2 - 32$ **12.** $2x^2 - 2y^2$ **13.** $ax + ay + bx + by$ **14.** $ax + xy + ay + y^2$

15. $12a^3b^2c + 3a^2b^2c^2 + 5abc^3$ **16.** $15m^2n^3p + 2m^2n^2p^2 - 2mn^3p$ **17.** $x^2 + 8x + 15$

18. $b^2 + 9b + 14$ **19.** $x^4 - 16$ **20.** $t^4 - 81$ **21.** $x^2 + 25$

22. $a^2 + 49$ **23.** $15x^2 + 7x - 2$ **24.** $2a^2 - 5a - 12$ **25.** $ax^2 + 4ax + 4a$

26. $9a^2x + 12ax + 4x$ **27.** $6ab - 36ab^2$ **28.** $5ax - 25a^2x$ **29.** $x^2 + x + 2$

30. $x^2 + 3x + 4$ **31.** $x^2 - 49$ **32.** $x^2 - 25$ **33.** $p^2 + p - 30$

34. $y^2 - 6y - 40$

35. $u^3 - u$

36. $5m^2 - 45$

37. $2b^2 + 14b + 24$

38. $4y^2 + 36y + 80$

39. $6r^2 - 15r^3$

40. $18y^3 - 24y^4$

41. $14u^2 + 7u - 105$

42. $6r^2 - 26r - 20$

43. $4h^2 + 12h + 4$

44. $3m^2 + 9m + 3$

45. $5p^2 - 80$

46. $3r^2 - 108$

47. $3w^2 + 5w + 2$

48. $3m^2 - 8m + 5$

49. $8q^2 - 10q - 3$

50. $10r^2 + 9r - 7$

51. $12v^2 + 23v + 10$

52. $6x^2 + 13x + 6$

53. $2 - 50x^2$

54. $6 - 24x^2$

55. $80k^2 - 20k^2l^2$

56. $27x^3 - 3x^3y^2$

57. $2j^2 + j + 3$

58. $3k^2 - 2k + 1$

59. $50 - 20t + 2t^2$

60. $243 - 54x + 3x^2$

61. $3ax - 6ay - 8by + 4bx$

62. $2ax - 6bx - 3by + ay$

63. $x^3 - x^2 - x + 1$

64. $x^3 - 2x^2 - x + 2$

65. $4x^2 - 28x + 49$

66. $4m^2 + 20m + 25$

67. $2x^4 - 162$

68. $5a^4 - 3125$

69. $a^2 + 2ab + ab + 2b^2$

70. $x^2 - 3xy + xy - 3y^2$

71. $9 - 4m^2$

72. $64 - 9r^2$

73. $x^5 - 4xy^2$

74. $m^5 - 25mn^2$

75. $20x^2 + 3xy + 2y^2$

76. $20a^2 - 9a + 20$

77. $b^3 + 125$

78. $x^3 + 216$

79. $3y^3 - 24$

80. $4x^3 - 4$

81. $54x - 2xy^3$

82. $81n - 3m^3n$

For Exercises 83 and 84, given an expression for the area of each rectangle, find the length and width.

83.

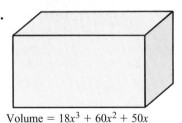

Area $= 6x^2 - 11x - 7$

84.

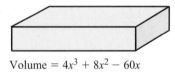

Area $= 25n^2 - 49$

For Exercises 85 and 86, given an expression for the volume of each box, find the length, width, and height.

85.

Volume $= 18x^3 + 60x^2 + 50x$

86.

Volume $= 4x^3 + 8x^2 - 60x$

87. A swimmer's goggles fall off a diving board that is 100 meters high. The height of the goggles after t seconds is given by $100 - 16t^2$. Factor $100 - 16t^2$.

88. An object is dropped from a cliff that is 400 feet high. The expression $400 - 16t^2$ gives the height of the falling object after t seconds. Factor $400 - 16t^2$.

89. The voltage in a circuit is the product of two factors, the resistance in the circuit and the current. If the voltage in a circuit is described by the expression $6ir + 15i + 8r + 20$, find the expressions for the current and resistance. (The expression for current will contain i, and the expression for resistance will contain r.)

90. If the voltage in the circuit described in Exercise 89 is changed so that it is described by the expression $35ir + 15i - 7r - 3$, what are the expressions for the current and resistance?

Review Exercises

 Exercises 1 and 2   **EXPRESSIONS**

[6.6] **1.** Divide: $\dfrac{2x^2 - 7x - 8}{2x + 1}$

[7.3] **2.** Factor $6y^2 + 23y - 4$.

 Exercises 3–5   **EQUATIONS AND INEQUALITIES**

[2.1] **3.** Find the area of a rectangular garden that is 14 feet by 16 feet.

[2.5] **4.** Translate to an equation; then solve. The product of four and the sum of a number and three is equal to twenty. Find the unknown number.

[3.3] **5.** Find three consecutive integers whose sum is 54.

7.6 Solving Quadratic Equations by Factoring

Objectives

1 Use the zero-factor theorem to solve equations containing expressions in factored form.

2 Solve quadratic equations by factoring.

3 Solve problems involving quadratic equations.

4 Use the Pythagorean theorem to solve problems.

Objective **1** **Use the zero-factor theorem to solve equations containing expressions in factored form.** So far in Chapters 6 and 7, we have rewritten polynomial expressions. We are now ready to work with polynomial equations, which we will solve using factoring. Notice that we are moving to the equation level of our Algebra Pyramid, which is built upon the foundation of expressions, variables, and constants.

The Algebra Pyramid

Note: Inequalities are also in the top level of the Algebra Pyramid, but we will not be exploring polynomial inequalities at this point.

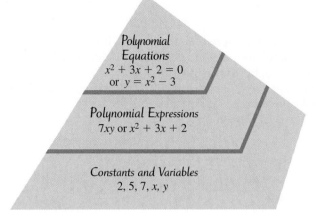

Connection
Throughout this book, after developing new types of expressions and ways of rewriting those expressions, we put those skills into the context of solving equations.

When solving polynomial equations by factoring, we make use of the **zero-factor theorem**, which states that if the product of two factors is 0, then one or the other of the two factors or both factors are equal to 0.

Rule **Zero-Factor Theorem**

If a and b are real numbers and $ab = 0$, then $a = 0$ or $b = 0$.

Example 1 Solve $(x + 2)(x - 3) = 0$.

SOLUTION In this equation, the product of two factors, $x + 2$ and $x - 3$, is equal to 0. According to the zero-factor theorem, one of the two factors or both factors must equal 0. In solving, we determine what values for x cause each factor to equal 0.

$$x + 2 = 0 \quad \text{or} \quad x - 3 = 0$$

Now solve each of the linear equations to get two solutions.

$$x = -2 \quad x = 3$$

Check Verify that -2 and 3 satisfy the original equation, $(x + 2)(x - 3) = 0$.

$$\text{For } x = -2: (-2 + 2)(-2 - 3) \stackrel{?}{=} 0 \qquad \text{For } x = 3: (3 + 2)(3 - 3) \stackrel{?}{=} 0$$
$$(0)(-5) = 0 \qquad\qquad\qquad\qquad (5)(0) = 0$$

Both -2 and 3 check; therefore, both are solutions.

Your Turn 1 Solve $(n - 6)(n + 1) = 0$.

Example 1 suggests the following procedure.

Procedure **Solving Equations with Two or More Factors Equal to 0**

To solve an equation in which two or more factors are equal to 0, use the zero-factor theorem.

1. Set each factor equal to zero.
2. Solve each of those equations.

Answer **Your Turn 1**
$n = 6$ or -1

Example 2   Solve.

a. $y(3y + 4) = 0$

SOLUTION   In this equation, the product of two factors, y and $3y + 4$, is equal to 0. We set each factor equal to 0 and then solve each of those equations.

$$y = 0 \quad \text{or} \quad 3y + 4 = 0$$

> **Note:** This equation is already solved.

$$3y = -4 \qquad \text{Subtract 4 from both sides to isolate } 3y.$$
$$y = -\frac{4}{3} \qquad \text{Divide both sides by 3 to isolate } y.$$

Check   Verify that 0 and $-\dfrac{4}{3}$ satisfy the original equation, $y(3y + 4) = 0$.

For $y = 0$: $(0)(3(0) + 4) \overset{?}{=} 0$ For $y = -\dfrac{4}{3}$: $\left(-\dfrac{4}{3}\right)\left[\dfrac{\overset{1}{3}}{1}\left(-\dfrac{4}{\underset{1}{3}}\right) + 4\right] \overset{?}{=} 0$

$(0)(4) = 0$ $\left(-\dfrac{4}{3}\right)[-4 + 4] \overset{?}{=} 0$

$\left(-\dfrac{4}{3}\right)(0) = 0$

Both 0 and $-\dfrac{4}{3}$ check; therefore, both are solutions.

b. $(x - 4)^2 = 0$

SOLUTION   Note that $(x - 4)^2 = 0$ can be written as $(x - 4)(x - 4) = 0$. Because the two factors are identical, there is no need to write two separate equations. We set $x - 4$ equal to 0 and then solve.

$$x - 4 = 0$$
$$x = 4 \qquad \text{Add 4 to both sides to isolate } x.$$

Check   Verify that 4 satisfies the original equation, $(x - 4)^2 = 0$.

$$(4 - 4)^2 \overset{?}{=} 0$$
$$(0)^2 = 0 \qquad \text{4 checks; therefore, it is a solution.}$$

c. $n(n + 6)(5n - 3) = 0$

SOLUTION   In this equation, the product of three factors, n, $n + 6$, and $5n - 3$, is 0. We set each factor equal to 0 and then solve each of those equations.

$$n = 0 \quad \text{or} \quad n + 6 = 0 \quad \text{or} \quad 5n - 3 = 0$$
$$n = -6 \qquad\qquad 5n = 3$$
$$n = \frac{3}{5}$$

Check   To check, we verify that 0, -6, and $\dfrac{3}{5}$ satisfy the original equation. We leave this check to the reader.

Your Turn 2   Solve.

a. $5x(3x - 2) = 0$ **b.** $(y + 3)^2 = 0$ **c.** $t(4t - 3)(t + 1) = 0$

Objective ② **Solve quadratic equations by factoring.** Now that we have seen how to use the zero-factor theorem to solve equations containing expressions that are in factored form, we are ready to consider equations that have an expression that is not in factored form. More specifically, we will use factoring to solve **quadratic equations in one variable**.

Definition | *Quadratic equation in one variable:* An equation that can be written in the form $ax^2 + bx + c = 0$, where a, b, and c are real numbers and $a \neq 0$.

The fact that a cannot equal zero means that the ax^2 term must be present for the equation to be a quadratic equation. A quadratic equation written in the form $ax^2 + bx + c = 0$ is said to be in *standard form*. Following are some examples of quadratic equations.

$$x^2 + 5x + 6 = 0 \qquad 3x^2 - 48 = 0 \qquad x^2 = x + 6$$

Note: This quadratic equation is in standard form.

Note: This quadratic equation is in standard form. It has no bx term because $b = 0$.

Note: Although not in standard form, this equation is still quadratic. Written in standard form, the equation would be $x^2 - x - 6 = 0$.

In Objective 1, we learned that to use the zero-factor theorem, we need an expression in factored form set equal to 0. So to solve a quadratic equation using factoring, it needs to be in standard form. If given an equation that is not in standard form, such as $x^2 = x + 6$, we first write it in standard form.

$$x^2 - x - 6 = 0 \qquad \text{Subtract } x \text{ and 6 from both sides to get standard form.}$$

After factoring $x^2 - x - 6$, we have $(x + 2)(x - 3) = 0$, which is the same equation that we solved in Example 1. This suggests the following procedure.

Procedure | **Solving Quadratic Equations Using Factoring**

To solve a quadratic equation using factoring,
1. Write the equation in standard form $(ax^2 + bx + c = 0)$.
2. Write the variable expression in factored form.
3. Use the zero-factor theorem to solve.

Example 3   Solve $2x^2 + 9x - 5 = 0$.

SOLUTION   This equation is in standard form, so we can simply factor the variable expression.

$$(2x - 1)(x + 5) = 0 \qquad \text{Factor } 2x^2 + 9x - 5.$$

Now use the zero-factor theorem to solve.

$$2x - 1 = 0 \quad \text{or} \quad x + 5 = 0$$
$$2x = 1 \qquad\qquad x = -5$$
$$x = \frac{1}{2}$$

Check   To check, we verify that $\frac{1}{2}$ and -5 satisfy the original equation. We leave this check to the reader.

Your Turn 3   Solve $12y^2 + 11y + 2 = 0$.

Sometimes we may have to factor out a monomial GCF.

Example 4   Solve $x^3 - 3x^2 - 10x = 0$.

SOLUTION

$$x(x^2 - 3x - 10) = 0 \qquad \text{Factor out the monomial GCF, } x.$$
$$x(x - 5)(x + 2) = 0 \qquad \text{Factor } x^2 - 3x - 10.$$

Now use the zero-factor theorem to solve.

$$x = 0 \quad \text{or} \quad x - 5 = 0 \quad \text{or} \quad x + 2 = 0$$
$$x = 5 \qquad\qquad x = -2$$

Check   Verify that 0, 5, and −2 each satisfy the original equation. We leave the check to the reader.

Your Turn 4   Solve $12x^3 + 20x^2 - 8x = 0$.

Rewrite Quadratic Equations in Standard Form

If a quadratic equation is not in standard form ($ax^2 + bx + c = 0$), we put it in standard form, then factor and use the zero-factor theorem.

Example 5   Solve.

a. $6y^2 - 5y = 4 - 28y$

SOLUTION   We first need to manipulate the equation so that it is in standard form.

$$6y^2 + 23y = 4 \qquad\qquad \text{Add 28y to both sides.}$$
$$6y^2 + 23y - 4 = 0 \qquad\qquad \text{Subtract 4 from both sides to get standard form.}$$
$$(6y - 1)(y + 4) = 0 \qquad\qquad \text{Factor using trial and error.}$$
$$6y - 1 = 0 \quad \text{or} \quad y + 4 = 0 \qquad \text{Use the zero-factor theorem to solve.}$$
$$6y = 1 \qquad\qquad y = -4$$
$$y = \frac{1}{6}$$

Check   We leave the check to the reader.

b. $x(x - 5) = -6$

> **Warning:** We cannot solve $x(x - 5) = -6$ by setting $x = -6$ or $x - 5 = -6$. For the *zero-factor* theorem to be used, one side of the equation must be 0.

SOLUTION

$$x(x - 5) + 6 = 0 \qquad\qquad \text{Add 6 to both sides so that the right-hand side is 0.}$$
$$x^2 - 5x + 6 = 0 \qquad\qquad \text{Multiply } x(x - 5) \text{ to get standard form.}$$
$$(x - 2)(x - 3) = 0 \qquad\qquad \text{Factor by looking for two numbers whose product is 6 and sum is } -5.$$
$$x - 2 = 0 \quad \text{or} \quad x - 3 = 0 \qquad \text{Use the zero-factor theorem to solve.}$$
$$x = 2 \qquad\qquad x = 3$$

Note: Although $x^3 - 3x^2 - 10x = 0$ is not a quadratic equation, after we factor out the x, the parentheses contain a quadratic form: $x^2 - 3x - 10$.

Answer   Your Turn 3

$y = -\dfrac{2}{3}$ or $-\dfrac{1}{4}$

Answer   Your Turn 4

$x = 0, \dfrac{1}{3},$ or -2

Note: We could have multiplied $x(x - 5)$ first and then moved the 6.

Note: We can not solve $(x + 3)(x - 5) = 20$ by setting $x + 3 = 20$ or $x - 5 = 20$.

Check   We leave the check to the reader.

c. $(x + 3)(x - 5) = 20$

SOLUTION

$$(x + 3)(x - 5) = 20$$
$$x^2 - 2x - 15 = 20 \qquad \text{Multiply } x + 3 \text{ and } x - 5.$$
$$x^2 - 2x - 35 = 0 \qquad \text{Subtract 20 from both sides.}$$
$$(x + 5)(x - 7) = 0 \qquad \text{Factor.}$$
$$x + 5 = 0 \quad \text{or} \quad x - 7 = 0 \qquad \text{Use the zero-factor theorem to solve.}$$
$$x = -5 \qquad\qquad x = 7$$

Check   We leave the check to the reader.

Your Turn 5   Solve.

  a. $15n^2 + 20n = -8 - 6n$ **b.** $x(x + 2) = 48$ **c.** $(x + 4)(x - 7) = -18$

Objective ③ Solve problems involving quadratic equations.   Many problems and applications are solved using quadratic equations. See Section 2.5 for a review of key words and their translations.

Example 6   The product of two consecutive odd natural numbers is 195. Find the numbers.

Understand   Odd natural numbers are $1, 3, 5, \ldots$. Note that adding 2 to a given odd natural number gives a consecutive odd natural number, which suggests this pattern:

First odd natural number: x

Consecutive odd natural number: $x + 2$

The word *product* means that two numbers are multiplied to equal 195.

Plan   Translate to an equation and then solve.

Execute $x(x + 2) = 195$ The equation has the product of x and $x + 2$ set equal to 195.

$$x(x + 2) - 195 = 0 \qquad \text{Subtract 195 from both sides.}$$
$$x^2 + 2x - 195 = 0 \qquad \text{Multiply } x(x + 2) \text{ to get the form } ax^2 + bx + c = 0.$$
$$(x - 13)(x + 15) = 0 \qquad \text{Factor by finding a pair of numbers whose product is } -195 \text{ and sum is 2.}$$
$$x - 13 = 0 \quad \text{or} \quad x + 15 = 0 \qquad \text{Use the zero-factor theorem to solve.}$$
$$x = 13 \qquad\qquad x = -15$$

Answer   Because -15 is not a natural number and 13 is a natural number, the first number is 13. This means that the consecutive odd natural number is 15.

Check   13 and 15 are consecutive odd natural numbers, and their product is 195; so the answer is correct.

Connection   In Section 3.3, we studied problems that dealt with the sum or difference of consecutive integers. The pattern for consecutive odd natural numbers is the same as for consecutive odd integers because every natural number is an integer.

Answers   Your Turn 5

a. $n = -\dfrac{2}{5}$ or $-\dfrac{4}{3}$

b. $x = 6$ or -8

c. $x = -2$ or $x = 5$

Answer   Your Turn 6

21 and 22

Your Turn 6   The product of two consecutive natural numbers is 462. Find the numbers.

Example 7 An architect is designing an addition to a house. The addition will be in the shape of a rectangle, and the room's floor will have an area of 270 square feet. The width of the room is to be 3 feet less than the length. Find the dimensions of the floor.

Understand The area of a rectangle is found by multiplying the length and width. We are given a relationship about the width and length, which we can translate as follows:

"The width is to be 3 feet less than the length."
Width $= l - 3$, where l represents length.

Plan Translate to an equation and then solve.

Execute $l(l - 3) = 270$ The equation has the product of the length, l, and width, $l - 3$, set equal to 270.

$l(l - 3) - 270 = 0$ Subtract 270 from both sides.

$l^2 - 3l - 270 = 0$ Multiply $l(l - 3)$ to get standard form.

$(l - 18)(l + 15) = 0$ Factor.

$l - 18 = 0$ or $l + 15 = 0$ Use the zero-factor theorem to solve.

$l = 18$ $l = -15$

Answer Because the problem involves room dimensions, only the positive number makes sense as an answer. If the length is 18 feet and the width is 3 feet less than that, the width is 15 feet.

Area $= 113.04$ cm^2

Check The width, 15 feet, is 3 feet less than the length, 18 feet. Also, the area of a 15-foot by 18-foot rectangle is 270 square feet.

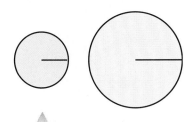

Your Turn 7 The two discs in the margin are designed so that the smaller of the two has an area of 113.04 square centimeters and a radius that is 8 centimeters less than the larger disc's radius. Find the radius of the larger disc. (Use 3.14 to approximate π.)

Note: This radius is 8 less than the other disc's radius.

Example 8 The equation $h = -16t^2 + h_0$ describes the height, h, in feet of an object after falling from an initial height h_0 for a time, t seconds. An object is dropped from a height of 600 feet. When the object reaches 344 feet, how much time has passed?

Understand We are given a formula, and we are to find the time it takes an object to drop from an initial height of 600 feet to a height of 344 feet.

Plan Replace the variables in the formula with the given values to get an equation in terms of t; then solve for t.

Execute $344 = -16t^2 + 600$ Replace h with 344 and h_0 with 600.

$-256 = -16t^2 + 0$ Subtract 600 from both sides.

$16t^2 - 256 = 0$ Add $16t^2$ to both sides.

$16(t^2 - 16) = 0$ Factor out the monomial GCF, 16.

$t^2 - 16 = 0$ Divide out the constant factor, 16.

$(t - 4)(t + 4) = 0$ Factor.

$t - 4 = 0$ or $t + 4 = 0$ Use the zero-factor theorem to solve.

$t = 4$ $t = -4$

Answer Because the problem involves the amount of time elapsed after an object was dropped, only the positive value, 4, makes sense. This means that it took the object 4 seconds to fall from 600 feet to 344 feet.

Check Verify that after 4 seconds, an object dropped from an initial height of 600 feet will descend to a height of 344 feet.

$$h = -16(4)^2 + 600$$
$$h = -256 + 600$$
$$h = 344$$

Answer **Your Turn 7**

14 cm

Your Turn 8   An object is dropped from a height of 225 feet above the ground. When the object is 81 feet from the ground, how much time has passed?

Objective ④ **Use the Pythagorean theorem to solve problems.**   One of the most popular theorems in mathematics is the Pythagorean theorem, named after the Greek mathematician Pythagoras. The theorem relates the side lengths of all right triangles. Recall that in a right triangle, the two sides that form the 90° angle are *legs* and the side directly across from the 90° angle is the *hypotenuse*.

Rule   **Pythagorean Theorem**

Given a right triangle, where a and b represent the lengths of the legs and c represents the length of the hypotenuse, $a^2 + b^2 = c^2$. Further, if a, b, and c represent the lengths of the sides of a triangle and $a^2 + b^2 = c^2$, then the triangle is a right triangle with hypotenuse c.

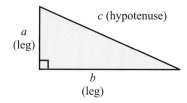

c (hypotenuse)

a (leg)

b (leg)

Of Interest

Pythagoras was a Greek mathematician who is believed to have been born around 569 B.C. and died in 500 B.C. The Pythagorean theorem is named after Pythagoras not because he discovered the relationship, as commonly thought, but because he and his followers are the first to have proved that the relationship is true for all right triangles. The relationship was known and used by other cultures prior to the Greeks; however, Pythagoras and others in the Greek culture believed that it was important to prove mathematical relationships.

We can use the Pythagorean theorem to find a missing length in a right triangle if we know the other two lengths.

Example 9   Find the length of the missing side.

3

?

4

Solution   Use the Pythagorean theorem, $a^2 + b^2 = c^2$. The hypotenuse, c, is the missing side length.

$$3^2 + 4^2 = c^2 \qquad \text{Substitute } a = 3 \text{ and } b = 4.$$

$$9 + 16 = c^2 \qquad \text{Simplify exponential forms.}$$

$$25 = c^2 \qquad \text{Add.}$$

$$0 = c^2 - 25 \qquad \text{Subtract 25 from both sides to get standard form.}$$

$$0 = (c - 5)(c + 5) \qquad \text{Factor.}$$

$$c - 5 = 0 \quad \text{or} \quad c + 5 = 0 \qquad \text{Use the zero-factor theorem.}$$

$$c = 5 \qquad\qquad c = -5$$

Because we are dealing with lengths, only the positive solution is sensible; so the missing length is 5.

Connection In Chapter 9, we will revisit the Pythagorean theorem. There we will use another technique involving square roots to solve equations such as $c^2 = 25$.

Answer   **Your Turn 8**

3 sec.

Your Turn 9 In the construction of a roof, three beams are used to form a right triangle frame. A 12-foot beam and a 5-foot beam are brought together to form a 90° angle. How long must the third beam be?

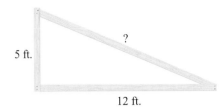
5 ft.
?
12 ft.

Answer Your Turn 9

13 ft.

7.6 Exercises

For Extra Help *MyMathLab*

PRACTICE   WATCH   DOWNLOAD

Note: Exercises marked with a ★ represent challenging exercises.

1. Explain the zero-factor theorem in your own words.

2. How do you use the zero-factor theorem to solve an equation that has an expression in factored form set equal to 0?

3. What is the first step in solving a quadratic equation?

4. Explain why we cannot write $(x + 2)(x - 2) = 32$ as $x + 2 = 32$ and $x - 2 = 32$.

For Exercises 5–16, solve using the zero-factor theorem. See Examples 1 and 2.

5. $(x + 5)(x + 2) = 0$

6. $(x - 7)(x + 3) = 0$

7. $(a + 3)(a - 4) = 0$

8. $(b - 7)(b + 2) = 0$

9. $(3x - 4)(2x + 3) = 0$

10. $(3a + 2)(2a - 5) = 0$

11. $x(x - 7) = 0$

12. $y(y + 3) = 0$

13. $m(m - 1)(m + 2) = 0$

14. $r(r - 2)(r + 3) = 0$

15. $b(b - 2)^2 = 0$

16. $m(m - 4)^2 = 0$

For 17–56, solve the quadratic equations. See Examples 3–5.

17. $y^2 - 4y = 0$

18. $p^2 - 3p = 0$

19. $6r^2 + 10r = 0$

20. $12c^2 + 9c = 0$

21. $x^2 - 4 = 0$

22. $y^2 - 25 = 0$

23. $x^2 + x - 6 = 0$

24. $b^2 - 2b - 8 = 0$

25. $n^2 - 6n + 9 = 0$

26. $p^2 - 16p + 64 = 0$

27. $3a^2 - 11a - 4 = 0$

28. $2d^2 + 7d - 4 = 0$

29. $6r^2 + 11r - 10 = 0$

30. $8c^2 - 6c - 9 = 0$

31. $x^2 - 4x = 21$

32. $u^2 + 6u = 27$

33. $p^2 = 3p - 2$

34. $x^2 = 6x - 8$

35. $v^2 = 9$

36. $m^2 = 16$

37. $2a^2 + 18a + 28 = 0$

38. $3x^2 - 6x - 45 = 0$

39. $x^3 - x^2 - 12x = 0$

40. $x^3 + 7x^2 + 10x = 0$

41. $12r^3 + 22r^2 - 20r = 0$

42. $16c^3 - 12c^2 - 18c = 0$

43. $3y^2 + 8 = 14y$

44. $6k^2 + 1 = 7k$

45. $4x^2 = 60 - x$ **46.** $4n^2 = 5 - 19n$ **47.** $7t = 12 + t^2$ **48.** $-9x = -20 - x^2$

49. $b(b - 5) = 14$ **50.** $m(m + 6) = -9$ **51.** $4x(x + 7) = -49$ **52.** $c(2c - 11) = -5$

53. $(x + 1)(x + 5) = -3$ **54.** $(a + 3)(a + 6) = -2$ **55.** $(a - 1)(a + 4) = 14$ **56.** $(b + 4)(b - 2) = 16$

For Exercises 57–72, translate to an equation and then solve. See Examples 6 and 7.

57. Find every number such that the square of the number added to 55 is the same as sixteen times the number.

58. Find every number such that triple the square of the number is equal to four times that number.

59. The product of two consecutive odd natural numbers is 143. Find the numbers.

60. The product of two consecutive natural numbers is 306. Find the numbers.

★**61.** The sum of the squares of two consecutive natural numbers is 365. Find the numbers.

★**62.** The difference of the squares of two consecutive even natural numbers is 60. Find the numbers.

63. The length of a rectangular garden is 9 meters more than the width. If the area is 252 square meters, find the dimensions of the garden.

64. The floor of the central chamber of the Lincoln Memorial has an area of 4440 square feet. If the length of the chamber is 14 feet more than the width, find the dimensions of the chamber floor. (*Source:* National Parks Service)

65. The largest billboard in Times Square's history has a length that is 10 feet more than three times its width. The area of the billboard is 9288 square feet. Find the dimensions of the billboard.

66. The length of a football field is 40 yards less than three times the width. The total area of the football field is 6400 square yards. Find the dimensions of the football field.

67. The design of the base of a small building calls for a rectangular shape with dimensions of 22 feet by 28 feet. The architect decides to change the shape of the building to a circle, but wants the base to have the same area. If we use $\frac{22}{7}$ to approximate π, what is the radius of the circular building?

68. A rectangle has a length of 14 centimeters and a width of 11 centimeters. Using $\frac{22}{7}$ as an approximation for π, find the radius of a circle with the same area as the rectangle.

69. A design on the front of a marketing brochure calls for a triangle with a base that is 6 centimeters less than the height. If the area of the triangle is to be 216 square centimeters, what are the lengths of the base and height?

70. The base of a triangular sign is 4 inches more than the height. If the area of the sign is 160 square inches, find the base and height.

71. A steel plate in the shape of a trapezoid has an area of 85.5 square inches. The dimensions are shown. Note that the length of the base is equal to the height. Calculate the height.

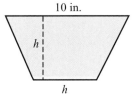

10 in.

h

h

72. The front elevation of one wing of a house is shown. Because of budget constraints, the total area of the front of this wing must be 352 square feet. The height of the triangular portion is 14 feet less than the base. Find the base length.

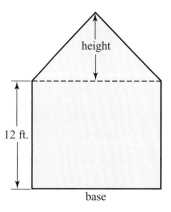

height

12 ft.

base

73. Find three consecutive even integers such that the product of the largest and second largest integers is 168.

74. Find three consecutive odd integers such that the product of the largest and second largest integers is 255.

For Exercises 75 and 76, use the formula $h = -16t^2 + v_0t + h_0$, where h is the final height in feet, t is the time of travel in seconds, v_0 is the initial velocity in feet per second, and h_0 is the initial height in feet of an object traveling upward. See Example 8.

★ **75.** A ball is thrown upward at 4 feet per second from a building 29 feet high. When will the ball be 9 feet above the ground?

★ **76.** A toy rocket is fired from the ground with an upward velocity of 200 feet per second. How many seconds will the rocket take to return to the ground?

For Exercises 77 and 78, use the formula $B = P\left(1 + \frac{r}{n}\right)^{nt}$, which is used to calculate the final balance of an investment or a loan after being compounded. Following is a list of what each variable represents: See Example 8.

    B represents the final balance.

    P represents the principal, which is the amount invested.

    r represents the annual interest rate as a decimal.

    t represents the number of years the principal is compounded.

    n represents the number of times per year the principal is compounded.

77. Carlita invests $4000 in an account that is compounded annually. If after two years her balance is $4840, what was the interest rate of the account?

78. Donovan invests $1000 in an account that is compounded semiannually (every six months). If after one year his balance is $1210, what is the interest rate of the account?

For Exercises 79–82, find the length of the hypotenuse. See Example 9.

79.
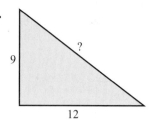

9
?
12

80.
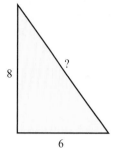

8
?
6

81.
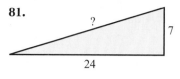

?
7
24

82.
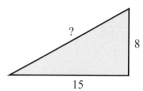

?
8
15

For Exercises 83–88, solve. See Example 9.

83. An artist welds a metal rod measuring 12 centimeters to a second rod measuring 35 centimeters to form a 90° angle. He wants a third rod to be welded to the ends of the first two rods to form a right triangle. What length must the third rod be?

84. A rectangular screen measures 27 inches by 36 inches. Find the diagonal distance between two opposing corners.

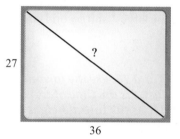

27
?
36

85. A person drives south 18 blocks then turns east and drives 24 blocks. Assuming that all city blocks are equal size and form a grid of streets that intersect at 90°, find the shortest distance in blocks that the person is from her original location.

86. To support a power pole, a wire is to be attached 24 feet from the ground, then staked 10 feet from the base of the pole. How long must the wire be?

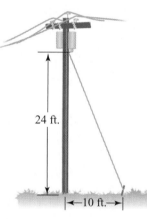

24 ft.
←10 ft.→

★**87.** When hurricanes threaten, a common practice is to tape windows to keep the glass from shattering. If a rectangular window is 24 inches wide and 51 inches diagonally, how long is the window?

★**88.** A sign is in the shape of a rectangle with a stripe painted diagonally across the sign. If the length of the stripe is 26 inches and the sign is 24 inches long, how wide is the sign?

Review Exercises

Exercises 1–6 EQUATIONS AND INEQUALITIES

[4.2] **1.** Is the equation $y = x^2 - 1$ a linear equation? Explain.

[4.2] **2.** Is $(1, -4)$ a solution for $4x + 3y = -8$?

[4.3] 3. Find the *x*- and *y*-intercepts for $2x - 5y = -20$.

[4.5] 4. Write the equation of the line passing through $(-1, 4)$ and $(-3, -2)$ in the form $Ax + By = C$, where *A*, *B*, and *C* are integers and $A > 0$.

[4.7] 5. For the function $f(x) = x^2 + 2$, find each of the following.
   a. $f(0)$ **b.** $f(-2)$

[4.7] 6. Graph: $f(x) = 2x - 3$

7.7 Graphs of Quadratic Equations and Functions

Objectives
1. Graph quadratic equations in the form $y = ax^2 + bx + c$.
2. Graph quadratic functions.

In Chapter 4, we graphed linear equations in two variables. Now we will graph **quadratic equations in two variables**.

Definition **Quadratic equation in two variables:** An equation that can be written in the form $y = ax^2 + bx + c$, where *a*, *b*, and *c* are real numbers and $a \neq 0$.

Following are some examples of quadratic equations in two variables.

$$y = x^2 \qquad y = 3x^2 + 2 \qquad y = x^2 - 4x + 1$$

We learned in Section 2.2 that every term in a linear equation has a degree of 1 or 0. Because quadratic equations always have a term with a degree of 2, they are non-linear, which means that their graphs will not be straight lines. What might their graphs look like?

Objective 1 **Graph quadratic equations in the form $y = ax^2 + bx + c$.** Recall that one way to graph an equation is to plot solutions to the equation in the coordinate plane. In Section 4.1, we said that a solution to an equation in two variables is an ordered pair (x, y) that satisfies the equation. For example, $(3, 9)$ is a solution to $y = x^2$ because $9 = (3)^2$ is true. Let's make a table of solutions for $y = x^2$, plot those solutions, and connect the points to see what the graph looks like.

x	y
−3	9
−2	4
−1	1
0	0
1	1
2	4
3	9

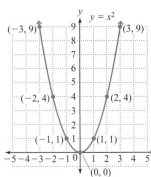

Note: Because we know that quadratic equations are nonlinear, we find many solutions to get a good sense of the shape of the graph.

The graph of $y = x^2$ is a *parabola*. In fact, the graph of every quadratic equation in two variables is a parabola.

Note: Parabolas are said to open up or down, as follows.

U   ∩
up   down

In the table of solutions for $y = x^2$, notice that when $x = 1$ or $-1, y = 1$; when $x = 2$ or $-2, y = 4$; and when $x = 3$ or $-3, y = 9$. In general, each x-coordinate and its additive inverse have the same y-coordinate. Looking at the graph, we see that the left side of the y-axis is the mirror image of the right side of the y-axis. Consequently, we say that this graph is symmetrical about the line $x = 0$ (the y-axis). Every parabola will have symmetry about a line called its **axis of symmetry**. Further, a parabola's axis of symmetry always passes through a point called the **vertex**.

Definitions

Axis of symmetry: A line that divides a graph into two symmetrical halves.

Vertex of a parabola: The lowest point on a parabola that opens up or the highest point on a parabola that opens down.

For $y = x^2$, the axis of symmetry is the line $x = 0$ (y-axis) and the vertex is $(0, 0)$. In Section 10.5, we will learn more about the vertex and axis of symmetry. For now, we graph by finding enough solutions to see where the parabola turns.

The axis of symmetry divides the graph into two halves that are mirror images of each other.

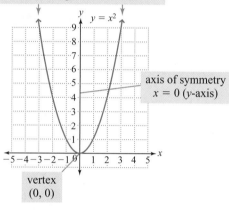

axis of symmetry
$x = 0$ (y-axis)

vertex
$(0, 0)$

Procedure   **Graphing Quadratic Equations**

To graph a quadratic equation,

1. Find ordered pair solutions and plot them in the coordinate plane. Continue finding and plotting solutions until the shape of the parabola can be clearly seen.
2. Connect the points to form a parabola.

Example 1   Graph.

a. $y = 3x^2 + 2$

SOLUTION   We complete a table of solutions, plot the solutions in the coordinate plane, and connect the points to form the graph.

Connection In a quadratic equation in the form $y = ax^2 + bx + c$, the constant c indicates the y-intercept just as the constant b does in a linear equation in the form $y = mx + b$.

Note: The constant 2 in $y = 3x^2 + 2$ indicates the y-intercept because replacing x with 0 leaves us with $y = 3(0)^2 + 2 = 2$.

x	y
-2	14
-1	5
0	2
1	5
2	14

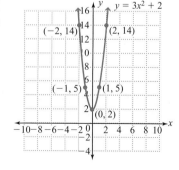

b. $y = x^2 + 2x - 8$

SOLUTION

Connection To find the x-intercepts of a graph, let $y = 0$. Thus, the solutions of $0 = x^2 + 2x - 8$ are the x-intercepts of the graph of $y = x^2 + 2x - 8$, which are $(-4, 0)$ and $(2, 0)$.

x	y
-4	0
-3	-5
-2	-8
-1	-9
0	-8
1	-5
2	0

Notice that in each equation we have graphed so far ($y = x^2$, $y = 3x^2 + 2$, and $y = x^2 + 2x - 8$), the coefficient of x^2 is a positive number. Also notice that the graph of each of those equations was a parabola that opens upward. Now let's consider quadratic equations in which the coefficient of x^2 is a negative number, as in $y = -2x^2 + 6x + 3$. We will see that the graphs of these equations open downward.

Example 2 Graph $y = -2x^2 + 6x + 3$.

SOLUTION Complete a table of solutions, plot the solutions, and connect the points to form a parabola.

x	y
-1	-5
0	3
1	7
2	7
3	3
4	-5

Note: The coefficient of x^2 is a negative number, -2, and the graph opens downward. In an equation of the form $y = ax^2 + bx + c$, if a is negative, the parabola opens downward.

Note: Because $x = 1$ and $x = 2$ have the same y-values, the x-coordinate of the vertex will be halfway between 1 and 2 at 1.5.

Rule **Opening of a Parabola**

Given an equation in the form $y = ax^2 + bx + c$, if $a > 0$, then the parabola opens upward; if $a < 0$, then the parabola opens downward.

Your Turn 2 Graph.

a. $y = x^2 - 5$ **b.** $y = 2x^2 - x + 1$ **c.** $y = -3x^2 + 6x + 2$

Objective 2 Graph quadratic functions. In Section 4.7, we learned how to graph linear functions. Now let's graph quadratic functions. A quadratic *equation* in two variables, $y = ax^2 + bx + c$, is a quadratic *function* that can be written in the form $f(x) = ax^2 + bx + c$. Notice that to change the equation in two variables to function notation, we replace y with $f(x)$.

Quadratic equation in two variables: $y = -2x^2 + 6x + 3$

Using function notation: $f(x) = -2x^2 + 6x + 3$

The graph of $f(x) = -2x^2 + 6x + 3$ is the same as $y = -2x^2 + 6x + 3$, which we graphed in Example 2. Recall from Section 4.7 that with function notation, we find ordered pairs by evaluating the function using various values of x. For example, the notation $f(1)$ means to find the value of the function when $x = 1$.

$$f(1) = -2(1)^2 + 6(1) + 3$$
$$= -2 + 6 + 3$$
$$= 7 \qquad \text{The ordered pair is } (1, 7).$$

Connection Finding $f(1)$ for $f(x) = -2x^2 + 6x + 3$ is the same as finding the y-value for $y = -2x^2 + 6x + 3$ when $x = 1$.

We would continue finding ordered pairs in this manner to produce the same table and graph that we produced in Example 2, which suggests the following procedure.

Procedure Graphing Quadratic Functions

To graph a quadratic function,

1. Find enough ordered pairs by evaluating the function for various values of x so that when those ordered pairs are plotted, the shape of the parabola can be clearly seen.

2. Connect the points to form the parabola.

Answers Your Turn 2

a.

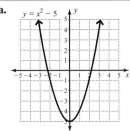

b.

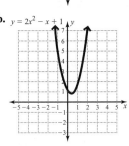

c.

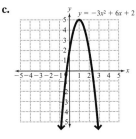

Example 3 Graph.

a. $f(x) = 3x^2 - 5$

SOLUTION Find enough ordered pairs to clearly see the graph; then connect the points to form the parabola.

x	y
-2	7
-1	-2
0	-5
1	-2
2	7

Connection This parabola opens upward. Given a quadratic function in the form $f(x) = ax^2 + bx + c$, if $a > 0$, the parabola opens upward.

b. $f(x) = -x^2 + 2x + 4$

SOLUTION

x	y
-2	-4
-1	1
0	4
1	5
2	4
3	1
4	-4

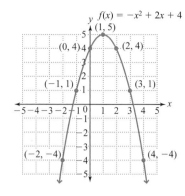

Connection This parabola opens downward. Given a quadratic function in the form $f(x) = ax^2 + bx + c$, if $a < 0$, the parabola opens downward.

Your Turn 3 Graph.

a. $f(x) = x^2 - 4$

b. $f(x) = -2x^2 - 8x - 3$

In Section 4.7, we learned that a graph represents a function if it passes the vertical line test. (A vertical line through any point in the domain touches the graph at only one point.) Because graphs of quadratic functions are parabolas that open up or down, they pass the vertical line test. There are also parabolas that open right or left, which do not pass the vertical line test, and therefore are not functions.

Example 4 Determine whether the graph is the graph of a function. Give the domain and range.

a.

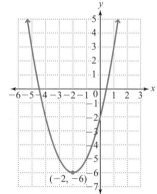

b.

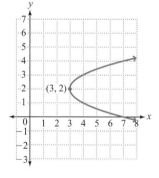

Answers Your Turn 3

a.

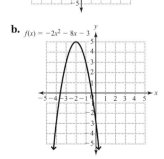

b.

SOLUTION a. This is a function because any vertical line intersects the graph in at most one point as illustrated with $x = -4$.

Domain: all real numbers or $(-\infty, \infty)$

Range: $\{y | y \geq -6\}$ or $[-6, \infty)$

SOLUTION **b.** This is not a function because a vertical line can be drawn that intersects the graph in more than one point as illustrated by $x = 5$.

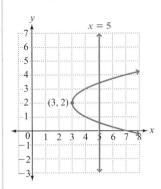

Domain: $\{x | x \geq 3\}$ or $[3, \infty)$

Range: all real numbers or $(-\infty, \infty)$

Your Turn 4 Determine whether the graph is the graph of a function. Give the domain and range.

a.

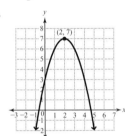

b.

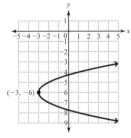

7.7 Exercises   For Extra Help   *MyMathLab*

PRACTICE   WATCH   DOWNLOAD

Note: Exercises marked with a ★ represent challenging exercises.

1. In your own words, describe what is meant by *symmetry* in a parabola.

2. Given an equation in the form $y = ax^2 + bx + c$, what indicates whether the graph opens up or down?

3. Given an equation in the form $y = ax^2 + bx + c$, what indicates the y-coordinate of the y-intercept?

4. Given that the graph of $y = ax^2 + bx + c$, where $a \neq 0$, is a parabola that opens upward or downward, explain why $f(x) = ax^2 + bx + c$ is a function.

For Exercises 5–8, use the graph of the parabola to determine the coordinates of the vertex. See Objective 1.

5.

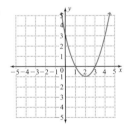

6.

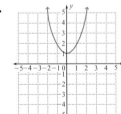

7.

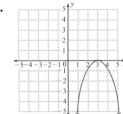

8.

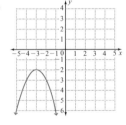

For Exercises 9–20, complete the table of solutions and then graph.
See Examples 1 and 2.

9. $y = 2x^2$

x	y
-2	
-1	
0	
1	
2	

10. $y = 3x^2$

x	y
-2	
-1	
0	
1	
2	

11. $y = -x^2$

x	y
-2	
-1	
0	
1	
2	

12. $y = -x^2 + 2$

x	y
-2	
-1	
0	
1	
2	

13. $y = x^2 - 3$

x	y
-2	
-1	
0	
1	
2	

14. $y = x^2 + 3$

x	y
-2	
-1	
0	
1	
2	

15. $y = -x^2 + 2x$

x	y
-1	
0	
1	
2	
3	

16. $y = -x^2 + 4x$

x	y
0	
1	
2	
3	
4	

17. $y = 2x^2 - 4x + 1$

x	y
-1	
0	
1	
2	
3	

18. $y = 2x^2 + 4x - 2$

x	y
-3	
-2	
-1	
0	
1	

19. $y = -3x^2 + 6x + 4$

x	y
-1	
0	
1	
2	
3	

20. $y = -2x^2 + 4x + 2$

x	y
-1	
0	
1	
2	
3	

For Exercises 21–36, graph. See Examples 1 and 2.

21. $y = x^2 + 1$

22. $y = x^2 - 4$

23. $y = -x^2 + 3$

24. $y = -x^2 + 2$

25. $y = 2x^2 - 5$ **26.** $y = 3x^2 - 4$ **27.** $y = -3x^2 + 2$ **28.** $y = -4x^2 + 5$

29. $y = x^2 + 4x - 1$ **30.** $y = x^2 - 6x + 3$ **31.** $y = -x^2 + 6x - 5$ **32.** $y = -x^2 + 4x - 5$

33. $y = 2x^2 - 4x - 3$ **34.** $y = 3x^2 + 6x - 1$ **35.** $y = -2x^2 + 8x - 5$ **36.** $y = -4x^2 - 12x - 3$

For Exercises 37–48, graph the function. See Example 3.

37. $f(x) = 3x^2$ **38.** $f(x) = 2x^2$ **39.** $f(x) = -x^2 + 1$ **40.** $f(x) = -x^2 - 1$

41. $f(x) = x^2 + 4x - 1$ **42.** $f(x) = x^2 - 6x + 5$ **43.** $f(x) = -x^2 - 4x - 3$ **44.** $f(x) = -x^2 + 6x - 2$

45. $f(x) = 3x^2 - 12x + 5$ **46.** $f(x) = 2x^2 + 6x - 1$ **47.** $f(x) = -2x^2 - 4x - 1$ **48.** $f(x) = -4x^2 + 8x - 3$

For Exercises 49–56, state whether the graph is the graph of a function. Give the domain and range. See Example 4.

49.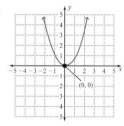

50.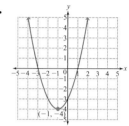

51.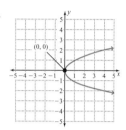

52.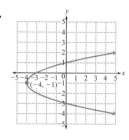

53.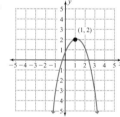

54.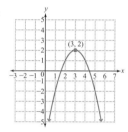

55.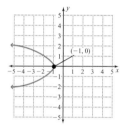

56.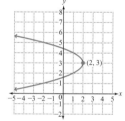

For Exercises 57–60, use a graphing calculator.

57. a. Enter the equation $y = x^2$; then graph. What are the coordinates of the y-intercept?

   b. Enter a second equation, $y = x^2 + 2$; then graph on the same grid. What are the coordinates of the y-intercept?

   c. On the same grid, graph $y = x^2 + 4$ and compare all three graphs. What do your explorations suggest about how c affects the graph of an equation in the form $y = x^2 + c$?

   d. On the same grid, graph $y = x^2 - 3$. Comparing this graph to the first three graphs, how did changing c to a negative value affect the graph?

58. a. Enter the equation $y = x^2$; then graph. What are the coordinates of the vertex?

b. Enter a second equation, $y = 2x^2$. Compare the graph of $y = 2x^2$ with the graph of $y = x^2$; then discuss their similarities and differences.

c. On the same grid, graph $y = 3x^2$ and compare all three graphs. What does this exploration suggest about how the size of a in $y = ax^2$ affects the graph?

d. In $y = ax^2$, if $a = \dfrac{1}{2}$, how will the graph be different from the graphs in parts a–c? Graph $y = \dfrac{1}{2}x^2$.

e. What do these explorations suggest about how the size of a affects the graph of an equation in the form $y = ax^2$?

59. a. Graph $y = x^2$ and $y = -x^2$ on the same grid. What is different about the two graphs? What does this suggest about the sign of a in an equation of the form $y = ax^2$?

b. Clearing the equations from part a, graph $y = 2x^2$ and $y = -2x^2$ together. Do these graphs confirm your conclusion from part a?

c. On a new grid, graph $y = \dfrac{1}{2}x^2$ and $y = -\dfrac{1}{2}x^2$. What is different about these graphs compared to those in parts a and b?

d. What do these explorations suggest about how the absolute value of a affects the graph of an equation in the form $y = ax^2$?

60. a. Enter the equation $y = x^2 + x + 1$; then graph. What are the coordinates of the y-intercept?

b. On the same grid, graph $y = x^2 + 2x + 1$. Compare the graph of $y = x^2 + x + 1$ with the graph of $y = x^2 + 2x + 1$; then discuss their similarities and differences.

c. Using the same grid, graph $y = x^2 + 3x + 1$. Compare this graph with the graphs of $y = x^2 + x + 1$ and $y = x^2 + 2x + 1$; then discuss similarities and differences.

d. On the same grid, graph $y = x^2 - x + 1$, $y = x^2 - 2x + 1$, and $y = x^2 - 3x + 1$. Compare these graphs with your graphs from parts a–c. What do these explorations suggest about how b affects the graph of an equation in the form $y = ax^2 + bx + c$?

Collaborative Exercises What Goes Up Must Come Down

Many sets of variables have a relationship that is roughly modeled by a parabolic shape. The following data give the high temperature (in degrees Fahrenheit, °F) and peak power load (in megawatts, MW) for 31 days throughout the calendar year.

1. On a sheet of graph paper, draw and label axes with the temperature along the horizontal axis and the peak load on the vertical axis. Then plot the points.

2. Draw a smooth parabola through the points that might best fit the data.

3. At what temperature does the minimum peak load seem to occur?

4. In your own words, describe what happens to the peak load as the temperature increases and decreases and explain why.

5. Use the model that you drew in part 2 to estimate the peak load on a day when the maximum temperature is 85.

6. Give another example of a situation that might be best modeled by a parabola.

Temp (°F)	Peak Load (MW)	Temp (°F)	Peak Load (MW)
69	94	95	141
71	95	46	142
59	100	43	147
65	100	98	150
67	102	96	151
60	104	97	155
74	105	100	158
79	105	43	165
56	112	38	173
84	113	102	176
52	114	36	193
88	115	103	199
50	120	105	204
90	123	106	226
92	132	34	228
92	132		

Review Exercises

Exercises 1–4 EXPRESSIONS

For Exercises 1 and 2, multiply.

[6.4] **1.** $3mn(m^2 - 4mn + 5)$ **[6.5]** **2.** $(x - 3)(2x + 1)$

For Exercises 3 and 4, factor.

[7.3] **3.** $3n^2 - 17n + 20$ **[7.4]** **4.** $x^2 - 25$

Exercises 5 and 6 EQUATIONS AND INEQUALITIES

[7.6] *For Exercises 5 and 6, solve.*

5. $h(h - 3) = 0$ **6.** $y^2 + 3y - 18 = 0$

Chapter 7 Summary

Defined Terms

Section 7.1
Factored form (p. 476)
Greatest common factor
(GCF) (p. 477)

Section 7.6
Quadratic equation in one
variable (p. 516)

Section 7.7
Quadratic equation in two
variables (p. 525)

Axis of symmetry (p. 526)
Vertex of a parabola
(p. 526)

Procedures, Rules, and Key Examples

Procedures/Rules	Key Examples

SECTION 7.1   Greatest Common Factor and Factoring by Grouping

To find the GCF of a set of numbers by listing,
1. List all possible factors for each given number.
2. Search the lists for the largest factor common to all lists.

Example 1: Find the GCF of 54 and 180 by listing.

Factors of 54: 1, 2, 3, 6, 9, 18, 27, 54
Factors of 180: 1, 2, 3, 4, 5, 6, 9, 10, 12, 15, 18, 20, 30, 36, 45, 60, 90, 180

GCF = 18

To find the GCF of a given set of numbers by prime factorization,
1. Write the prime factorization of each given number in exponential form.
2. Create a factorization for the GCF that includes only those prime factors common to all factorizations, each raised to its smallest exponent in the factorizations.
3. Multiply the factors in the factorization created in step 2.
Note: If there are no common prime factors, the GCF is 1.

Example 2: Find the GCF of 54 and 180 by prime factorization.

$$54 = 2 \cdot 3^3$$
$$180 = 2^2 \cdot 3^2 \cdot 5$$
$$GCF = 2 \cdot 3^2 = 18$$

To factor a monomial GCF out of a given polynomial,
1. Find the GCF of the terms in the polynomial.
2. Rewrite the given polynomial as a product of the GCF and the quotient of the polynomial and the GCF.

$$\text{Polynomial} = GCF\left(\frac{\text{Polynomial}}{GCF}\right)$$

Example 3: Factor $54x^3y - 180x^2yz$.

$$54x^3y - 180x^2yz$$
$$= 18x^2y\left(\frac{54x^3y - 180x^2yz}{18x^2y}\right)$$
$$= 18x^2y\left(\frac{54x^3y}{18x^2y} - \frac{180x^2yz}{18x^2y}\right)$$
$$= 18x^2y(3x - 10z)$$

To factor a four-term polynomial by grouping,
1. Factor out any monomial GCF (other than 1) that is common to all four terms.
2. Group pairs of terms and factor the GCF out of each pair or group.
3. If there is a common binomial factor, factor it out.
4. If there is no common binomial factor, interchange the middle two terms and repeat the process. If there is still no common binomial factor, the polynomial cannot be factored by grouping.

Example 4: Factor by grouping:

$$8x^2 - 20x - 6xy + 15y$$

$$8x^2 - 20x - 6xy + 15y$$
$$= (8x^2 - 20x) + (-6xy + 15y)$$
$$= 4x(2x - 5) - 3y(2x - 5)$$
$$= (2x - 5)(4x - 3y)$$

(continued)

Procedures/Rules	Key Examples

SECTION 7.2 Factoring Trinomials of the Form $x^2 + bx + c$

To factor a trinomial of the form $x^2 + bx + c$, 1. Find two numbers with a product equal to c and a sum equal to b. 2. The factored trinomial will have the form $(x + \text{first number})(x + \text{second number})$. *Note:* The signs in the binomial factors can be minus signs depending on the signs of b and c.	**Example 1:** Factor $x^2 + 9x + 20$. Find two numbers whose product is 20 and whose sum is 9. Product: Sum: $(1)(20) = 20$ $1 + 20 = 21$ $(2)(10) = 20$ $2 + 10 = 12$ $(4)(5) = 20$ $4 + 5 = 9$ Correct combination $$x^2 + 9x + 20 = (x + 4)(x + 5)$$ **Example 2:** Factor $x^2 + 2x - 15$. Product: Sum: $(-1)(15) = -15$ $-1 + 15 = 14$ $(-3)(5) = -15$ $-3 + 5 = 2$ Correct combination $$x^2 + 2x - 15 = (x - 3)(x + 5)$$

SECTION 7.3 Factoring Trinomials of the Form $ax^2 + bx + c$, where $a \neq 1$

To factor a trinomial of the form $ax^2 + bx + c$, where $a \neq 1$, by trial and error,
1. Look for a monomial GCF in all of the terms. If there is one, factor it out.
2. Write a pair of *first* terms whose product is ax^2.
3. Write a pair of *last* terms whose product is c.
4. Verify that the sum of the *inner* and *outer* products is bx (the middle term of the trinomial). If the sum of the inner and outer products is not bx, try the following:
  a. Exchange the last terms of the binomials from Step 3; then repeat Step 4.
  b. For each additional pair of last terms, repeat Steps 3 and 4.
  c. For each additional pair of first terms, repeat Steps 2–4.
	Example 1: Factor $4x^2 - 9x + 5$.
	Factors of 4 are 4 and 1, or 2 and 2.
	Factors of 5 are 1 and 5.
	Because the middle term $-9x$ is negative and the last term 5 is positive, we know that the second terms in each binomial factor will be negative.
	
	$$(x - 1)(4x - 5) = 4x^2 - 9x + 5$$

To factor a trinomial of the form $ax^2 + bx + c$, where $a \neq 1$, by grouping, 1. Look for a monomial GCF in all of the terms. If there is one, factor it out. 2. Multiply a and c. 3. Find two factors of this product whose sum is b. 4. Write a four-term polynomial in which bx is written as the sum of two like terms whose coefficients are the two factors you found in Step 3. 5. Factor by grouping.	**Example 2:** Factor $18y^3 + 12y^2 - 48y$. Factor out the monomial GCF, $6y$. $$18y^3 + 12y^2 - 48y = 6y(3y^2 + 2y - 8)$$ To factor the trinomial, multiply $3(-8) = -24$; then find two factors of -24 whose sum is 2. Note that -4 and 6 work. Write the middle term, $2y$, as $-4y + 6y$; then factor by grouping. $$\begin{aligned} &= 6y(3y^2 - 4y + 6y - 8) \\ &= 6y[y(3y - 4) + 2(3y - 4)] \\ &= 6y(3y - 4)(y + 2) \end{aligned}$$

(continued)

Procedures/Rules	Key Examples

SECTION 7.4 Factoring Special Products

Rules for factoring special products:

Perfect square trinomials: $a^2 + 2ab + b^2 = (a + b)^2$
$a^2 - 2ab + b^2 = (a - b)^2$

Difference of squares: $a^2 - b^2 = (a + b)(a - b)$

Note: A sum of squares, $a^2 + b^2$, cannot be factored.

Difference of cubes: $a^3 - b^3 = (a - b)(a^2 + ab + b^2)$

Sum of cubes: $a^3 + b^3 = (a + b)(a^2 - ab + b^2)$

Example 1: Factor.

a. $9x^2 + 30x + 25 = (3x + 5)^2$

b. $36m^2 - 60mn + 25n^2 = (6m - 5n)^2$

c. $4y^2 - 81 = (2y + 9)(2y - 9)$

d. $16p^2 + 1$ is prime.

e. $8x^3 - 27 = (2x - 3)(4x^2 + 6x + 9)$

f. $n^3 + 64 = (n + 4)(n^2 - 4n + 16)$

SECTION 7.5 Strategies for Factoring

To factor a polynomial, first factor out any monomial GCF; then consider the number of terms in the polynomial. If the polynomial has

I. Four terms, try to factor by grouping.

II. Three terms, determine whether the trinomial is a perfect square.
  A. If the trinomial is a perfect square, consider its form.
  1. If in form $a^2 + 2ab + b^2$, the factored form is $(a + b)^2$.
  2. If in form $a^2 - 2ab + b^2$, the factored form is $(a - b)^2$.
  B. If the trinomial is not a perfect square, consider its form.
  1. If in form $x^2 + bx + c$, find two factors of c whose sum is b and write the factored form as $(x + \text{first number})(x + \text{second number})$.
  2. If in form $ax^2 + bx + c$, where $a \neq 1$, use trial and error. Or find two factors of ac whose sum is b; write these factors as coefficients of two like terms that, when combined, equal bx; and factor by grouping.

III. Two terms, determine whether the binomial is a difference of squares, sum of cubes, or difference of cubes.
  A. If given a binomial that is a difference of squares $a^2 - b^2$, the factors are conjugates and the factored form is $(a + b)(a - b)$. Note that a sum of squares cannot be factored.
  B. If given a binomial that is a sum of cubes, $a^3 + b^3$, the factored form is $(a + b)(a^2 - ab + b^2)$.
  C. If given a binomial that is a difference of cubes, $a^3 - b^3$, the factored form is $(a - b)(a^2 + ab + b^2)$.

Note: Always check to see if any of the factors can be factored further.

The examples shown do not represent every possible type of problem. Each example contains a monomial GCF other than 1.

Example 1: Factoring by grouping:

$$8x^3 + 24x^2 - 2x^2y - 6xy$$
$$= 2x(4x^2 + 12x - xy - 3y)$$
$$= 2x[4x(x + 3) - y(x + 3)]$$
$$= 2x(x + 3)(4x - y)$$

Example 2: Factoring a perfect square trinomial:

$$75x^4 + 60x^3y + 12x^2y^2$$
$$= 3x^2(25x^2 + 20xy + 4y^2)$$
$$= 3x^2(5x + 2y)^2$$

Example 3: Factoring a difference of squares:

$$81y^5 - 16y = y(81y^4 - 16)$$
$$= y(9y^2 + 4)(9y^2 - 4)$$
$$= y(9y^2 + 4)(3y + 2)(3y - 2)$$

Example 4: Factoring a sum of cubes:

$$54n^4 + 16nm^3 = 2n(27n^3 + 8m^3)$$
$$= 2n(3n + 2m)(9n^2 - 6mn + 4m^2)$$

(continued)

Procedures/Rules	Key Examples

SECTION 7.6 Solving Quadratic Equations by Factoring

Zero-Factor Theorem

If a and b are real numbers and $ab = 0$, then $a = 0$ or $b = 0$.

To solve an equation in which two or more factors are equal to 0, use the zero-factor theorem.

1. Set each factor equal to zero.
2. Solve each of those equations.

Example 1: Solve $(2x - 5)(x + 4) = 0$.

$$2x - 5 = 0 \quad \text{or} \quad x + 4 = 0$$
$$2x = 5 \qquad\qquad x = -4$$
$$x = \frac{5}{2}$$

To solve a quadratic equation using factoring,

1. Write the equation in standard form $(ax^2 + bx + c = 0)$.
2. Write the variable expression in factored form.
3. Use the zero-factor theorem to solve.

Example 2: Solve $2x^2 = 3 - x$.

$$2x^2 = 3 - x$$
$$2x^2 + x = 3 \qquad \text{Add } x \text{ to both sides.}$$
$$2x^2 + x - 3 = 0 \qquad \text{Subtract 3 from both sides.}$$
$$(2x + 3)(x - 1) = 0 \qquad \text{Factor.}$$
$$2x + 3 = 0 \quad \text{or} \quad x - 1 = 0 \qquad \text{Use the zero-factor}$$
$$2x = -3 \qquad\qquad x = 1 \qquad \text{theorem.}$$
$$x = -\frac{3}{2}$$

Example 3: The length of a small rectangular building is to be 15 feet more than the width. The area of the base is to be 700 square feet. What must the dimensions of the base be?

Let w represent width. The length will be $w + 15$. Writing an equation for area, we have

$$w(w + 15) = 700$$
$$w^2 + 15w = 700 \qquad \text{Distribute } w.$$
$$w^2 + 15w - 700 = 0 \qquad \text{Subtract 700 from both sides.}$$
$$(w - 20)(w + 35) = 0 \qquad \text{Factor.}$$
$$w - 20 = 0 \quad \text{or} \quad w + 35 = 0 \qquad \text{Use the zero-factor theorem.}$$
$$w = 20 \qquad\qquad w = -35$$

Answer: The width is 20 ft., and the length is 35 ft.

Pythagorean Theorem

Given a right triangle, where a and b represent the lengths of the legs and c represents the length of the hypotenuse, $a^2 + b^2 = c^2$. Further, if a, b, and c, represent the lengths of the sides of a triangle and $a^2 + b^2 = c^2$, then the triangle is a right triangle with hypotenuse c.

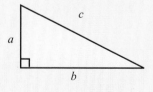

Example 4: Find the unknown length in the following right triangle.

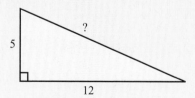

(continued)

Procedures/Rules	**Key Examples**

	Solution: Use the Pythagorean theorem, $a^2 + b^2 = c^2$.

$$5^2 + 12^2 = c^2 \qquad \text{$a = 5, b = 12$}$$
$$25 + 144 = c^2 \qquad \text{Simplify exponential forms.}$$
$$169 = c^2 \qquad \text{Add.}$$
$$0 = c^2 - 169 \qquad \text{Subtract 169 from both sides.}$$
$$0 = (c - 13)(c + 13) \qquad \text{Factor.}$$
$$c - 13 = 0, c + 13 = 0 \qquad \text{Use the zero-factor theorem.}$$
$$c = 13 \qquad c = -13$$

Answer: The unknown length is 13.

SECTION 7.7 Graphs of Quadratic Equations and Functions

To graph a quadratic equation,
1. Find ordered pair solutions and plot them in the coordinate plane. Continue finding and plotting solutions until the shape of the parabola can be clearly seen.
2. Connect the points to form a parabola.

Given an equation in the form $y = ax^2 + bx + c$, if $a > 0$, then the parabola opens upward; if $a < 0$, then the parabola opens downward.

Example 1: Graph $y = x^2 - 3$.

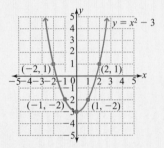

To graph a quadratic function,
1. Find enough ordered pairs by evaluating the function for various values of x so that when those ordered pairs are plotted, the shape of the parabola can be clearly seen.
2. Connect the points to form the parabola.

Example 2: Graph $f(x) = -x^2 + 4x - 1$.

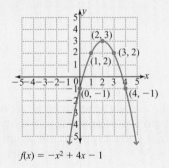

$$f(x) = -x^2 + 4x - 1$$

Chapter 7 Review Exercises

For Exercises 1–5, answer true or false.

[7.1]   **1.** The GCF of $24x^5$ and $36x^3$ is $6x^3$.

[7.1]   **2.** To factor a binomial, we use grouping.

[7.4]   **3.** $x^2 + 16 = (x + 4)(x + 4)$

[7.4]   **4.** $x^2 + 8x + 16 = (x + 4)^2$

[7.6]   **5.** If a and b are real numbers and $ab = 0$, then $a = 0$ or $b = 0$.

6. To factor a monomial out of a given polynomial,

    a. Find the _____ of the terms in the polynomial.

    b. Rewrite the given polynomial as a product of the GCF and the quotient of the _____ and the _____.

[7.2] **7.** Procedure: To factor a trinomial of the form $x^2 + bx + c$,

    a. Find two numbers with a product equal to _____ and a sum equal to _____.

    b. The factored trinomial will have the form $(x + \text{first number})(x + \text{second number})$.

[7.4]   **8.** $a^2 - b^2 = (\quad)(\quad)$ [7.4]   **9.** $a^3 - b^3 = (\quad)(\quad)$ [7.4]   **10.** $a^2 - 2ab + b^2 = (\quad)^2$

Exercises 11–18   CONSTANTS AND VARIABLES

[7.1] *For Exercises 11–18, list all of the natural number factors.*

11. 60   **12.** 9   **13.** 81   **14.** 44

15. 45   **16.** 49   **17.** 30   **18.** 33

Exercises 19–76   EXPRESSIONS

[7.1] *For Exercises 19–26, find the GCF.*

19. $21, 30$ **20.** $34, 51$ **21.** $12, 42, 60$ **22.** $10, 15, 45$

23. $10y^4, 25y^2$ **24.** $4x, 20xy$ **25.** $15m^2n, 3mn^5$ **26.** $8k, 22k^9$

[7.1] *For Exercises 27–34, factor out the GCF.*

27. $4x - 2$ **28.** $5m - 35m^3$ **29.** $y^3 - y$

30. $abc + a^2b^2c^2 - a^3b^3c^3$ **31.** $x^2y + xy^2 + x^3y^3$ **32.** $105a^3b^2 - 63a^2b^3 + 84a^6b^4$

33. $100k^4 + 120k^5 - 10k + 40k^3$ **34.** $18ab^3c - 36a^2b^2c$

[7.1] *For Exercises 35–42, factor by grouping.*

35. $ax + ay + bx + by$ **36.** $ax + 2a + bx + 2b$ **37.** $y^3 + 2y^2 + 3y + 6$ **38.** $y^4 + 4y^3 - by - 4b$

39. $xy + y + x + 1$ **40.** $ax^2 - 5y^2 + ay^2 - 5x^2$ **41.** $2b^3 - 2b^2 + b - 1$ **42.** $u^2 - 3u + 4uv - 12v$

[7.2] *For Exercises 43–50, factor.*

43. $x^2 - x - 12$ **44.** $x^2 + 14x + 45$ **45.** $n^2 - 6n + 8$ **46.** $a^2 + 3a - 20$

47. $h^2 + 51h + 144$ **48.** $y^2 - 10y - 24$ **49.** $4x^2 - 24x + 36$ **50.** $3m^2 - 33m + 54$

[7.3] *For Exercises 51–58, factor completely.*

51. $6x^2 + 3x - 7$

52. $2u^2 + 5u + 2$

53. $3m^2 - 10m + 3$

54. $5k^2 - 7kh - 12h^2$

55. $6a^2 - 20a + 16$

56. $8x^2y - 4xy - y$

57. $3p^2 - 13pq + 4q^2$

58. $24x^2 - 64xy - 24y^2$

[7.4] *For Exercises 59–66, factor.*

59. $v^2 - 8v + 16$

60. $u^2 + 6u + 9$

61. $4x^2 + 20x + 25$

62. $9y^2 - 12y + 4$

63. $x^2 - 4$

64. $25 - y^2$

65. $x^3 - 1$

66. $x^3 + 27$

[7.5] *For Exercises 67–76, factor completely.*

67. $6b^2 + b - 2$

68. $4ab - 24ab^2$

69. $y^2 + 25$

70. $3x^2 - 3y^2$

71. $x^4 - 81$

72. $7u^2 - 14u - 105$

73. $8x^3 - 27y^3$

74. $2 - 50y^2$

75. $3am - 6an - 8bn + 4bm$

76. $3m^2 + 9m + 27$

Exercises 77–94 EQUATIONS AND INEQUALITIES

[7.6] *For Exercises 77–84, solve.*

77. $(m + 3)(m - 4) = 0$

78. $y^2 - 4 = 0$

79. $x^2 - 5x + 6 = 0$

80. $x^2 - 4x = 21$

81. $m^2 = 9$

82. $x^2 + 3x = 18$

83. $y(y + 6) = -9$

84. $3n^2 - 11n = 4$

[7.6] *For Exercises 85–90, translate to an equation; then solve.*

85. Find every number such that five times the square of the number is equal to twice the number.

86. The sum of the squares of two consecutive natural numbers is 61. Find the numbers.

87. The product of a number and four times that same number is 100. Find all numbers.

88. Find the dimensions of a rectangle whose length is 6 more than its width and whose area is 91 square inches.

89. The product of two consecutive natural numbers is 110. Find the numbers.

90. Find the length of the missing side of the right triangle shown.

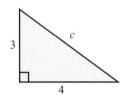

[7.7] *For Exercises 91–94, graph.*

91. $y = -5x^2$ **92.** $y = x^2 - 3$ **93.** $f(x) = x^2 + 2$ **94.** $f(x) = -3x^2 + 6x + 2$

Chapter 7 Practice Test

For Extra Help

CHAPTER
Test Prep
VIDEOS

Step-by-step test solutions are found on the Chapter Test Prep Videos available via the Video Resources on DVD, in *MyMathLab* , and on You Tube (search "CarsonElemAlg" and click on "Channels").

1. List all possible natural number factors of 80.

2. Find the GCF of $25m^2$ and $40m$.

Factor completely.

3. $5y - 30$ **4.** $6x^2y - 2y^2$ **5.** $ax - ay - bx + by$ **6.** $m^2 + 7m + 12$

7. $r^2 - 8r + 16$ **8.** $4y^2 + 20y + 25$ **9.** $3q^2 - 10q + 8$ **10.** $6x^2 - 23x + 20$

11. $ax^2 - 5ax - 24a$ **12.** $10n^3 + 38n^2 - 8n$ **13.** $c^2 - 25$ **14.** $x^2 + 4$

15. $2 - 50u^2$ **16.** $-4x^2 + 16$ **17.** $m^3 + 125$ **18.** $x^3 - 8$

Solve.

19. $a(a + 3) = 0$ **20.** $x^2 - 4x - 12 = 0$ **21.** $2n^2 + 7n = 15$

22. The product of two consecutive natural numbers is 72. Find the numbers.

23. The width of a rectangle is 5 feet less than the length. If the area is 36 square feet, find the length and width.

Graph.

24. $y = -2x^2 + 4$ **25.** $f(x) = x^2 - 6x + 5$

Chapters 1–7 Cumulative Review Exercises

For Exercises 1–4, answer true or false.

[6.4] **1.** $x^3 \cdot x^4 = x^{12}$

[7.1] **2.** The GCF of 12 and 5 is 1.

[1.7] **3.** The expression $5 + x$ can be simplified to equal $5x$.

[7.1] **4.** $4x + 12 = 2(2x + 6)$ is factored completely.

For Exercises 5 and 6, fill in the blank.

[7.4] **5.** To factor a difference of squares, we use the rule $a^2 - b^2 = $ _____ .

[7.6] **6.** To solve a quadratic equation using factoring,
   a. Write the equation in _____ form $(ax^2 + bx + c = 0)$.
   b. Write the expression in _____ form.
   c. Use the zero-factor theorem to solve.

Exercises 7–19 **EXPRESSIONS**

For Exercises 7–14, simplify.

[1.5] **7.** $14 - 2 \cdot 6 \cdot 3 + 8^2$

[1.5] **8.** $-6|5 + 3^2| - \sqrt{14 + 11}$

[6.3] **9.** $(5x^2 + 6x - 1) - (2x - 3)$

[6.4] **10.** $(6x^2)(3x^2y)$

[6.4] **11.** $(3x)(4x^3)^2$

[6.5] **12.** $(y - 8)(2y + 1)$

[6.6] **13.** $\dfrac{16h^3 + 4h^2 - 8h}{4h^2}$

[6.6] **14.** $(x^2 + 7x + 12) \div (x + 3)$

[7.1–7.5] **For Exercises 15–19, factor completely.**

15. $x^2 - 121$

16. $10x^2 + 40$

17. $x^2 - 8x + 16$

18. $2x^2 + 5x - 12$

19. $x^3 - 5x^2 + 5x - 25$

Exercises 20–30 **EQUATIONS AND INEQUALITIES**

For Exercises 20–22, solve.

[2.3] **20.** $\dfrac{3}{4} = x - \dfrac{2}{3}$

[2.3] **21.** $2.6 + 7a + 5 = 8a - 5.6$

[7.6] **22.** $x^2 - 3x = 28$

For Exercises 23–25, solve.

[2.4] **23.** Solve for x in the equation $2x - 3y = 7$.

[4.3] **24.** Find the x- and y-intercepts for $y = 2x - 6$.

[4.4] **25.** Find the slope of the line containing the points (6, 2) and (5, 2).

[4.2] **26.** Graph $y = 3x + 2$.

[2.6] *For Exercise 27 solve.* *a.* **Write the solution set using set-builder notation.**
  b. **Write the solution set using interval notation.**
  c. **Graph the solution set.**

27. $7x - 5 \le 3x + 11$

For Exercises 28–30, solve.

[3.3] **28.** Find three consecutive even integers whose sum is 24.
[5.4]

[3.5] **29.** Janice invests a total of $8000 in two different accounts. The first account re-
[5.4] turns 5%, while the second account returns 8%. If the total interest earned after one year is $565, what principal did she invest in each account?

[3.5] **30.** How much of a 35% HCl solution must be added to 50 milliliters of a 10%
[5.4] HCl solution to obtain a 20% solution?

Rational Expressions and Equations

8

Throughout this book, topics are developed in a specific pattern. We first define a new type of expression, then learn how to perform arithmetic operations on those expressions. Finally, we solve equations that contain those expressions. For example, in Chapter 6, we defined polynomials, then learned how to perform operations of arithmetic with the polynomials. In Chapter 7, we solved equations containing the polynomial expressions. We follow a similar pattern of development in Chapter 8. We first define rational expressions, then learn to perform arithmetic operations on those rational expressions, and finally solve equations that contain rational expressions.

8.1 Simplifying Rational Expressions

Objectives
1. Evaluate rational expressions.
2. Find numbers that cause a rational expression to be undefined.
3. Simplify rational expressions containing only monomials.
4. Simplify rational expressions containing multiterm polynomials.

Objective ① Evaluate rational expressions. Now that we have explored polynomials, we are ready to develop a new class of expressions called **rational expressions**.

Definition   *Rational expression:* An expression that can be written in the form

$$\frac{P}{Q}, \text{ where } P \text{ and } Q \text{ are polynomials and } Q \neq 0.$$

Connection Notice how the definition of a rational expression is like the definition of a rational number. Recall from Chapter 1 that a rational number is a number that can be expressed in the form $\frac{a}{b}$, where a and b are integers and $b \neq 0$. Some texts call rational expressions algebraic fractions.

Following are some rational expressions.

$$\frac{3x^5}{18x^2} \qquad \frac{7x}{x^2 - 9} \qquad \frac{x^2 + 2x - 15}{4x + 20}$$

As with any expression, we can evaluate or rewrite rational expressions. First, let's look at evaluating rational expressions. Remember that to evaluate an algebraic expression, we replace the variables with given values and then simplify the resulting numerical expression.

Example 1   Evaluate the expression $\dfrac{3x - 5}{x + 1}$, when

a. $x = 2$

SOLUTION $\dfrac{3x - 5}{x + 1} = \dfrac{3(2) - 5}{2 + 1}$ Replace x with 2.

$$= \frac{6 - 5}{3}$$

$$= \frac{1}{3}$$

b. $x = -0.4$

SOLUTION $\dfrac{3x - 5}{x + 1} = \dfrac{3(-0.4) - 5}{-0.4 + 1}$ Replace x with -0.4.

$$= \frac{-1.2 - 5}{0.6}$$

$$= \frac{-6.2}{0.6}$$

$$= -10.\overline{3} \quad \text{or} \quad -10\frac{1}{3}$$

c. $x = -1$

SOLUTION $\dfrac{3x - 5}{x + 1} = \dfrac{3(-1) - 5}{-1 + 1}$ Replace x with -1.

$$= \dfrac{-3 - 5}{0}$$

$$= \dfrac{-8}{0}, \quad \text{which is undefined.}$$

Your Turn 1   Evaluate the expression $\dfrac{x^2 - 9}{x - 2}$, when

a. $x = 4$.   **b.** $x = -3$.   **c.** $x = 2$.

Objective ② **Find numbers that cause a rational expression to be undefined.**   Note that in Example 1(c), the expression $\dfrac{3x - 5}{x + 1}$ is undefined when $x = -1$ because replacing x with -1 causes the denominator to be 0. We need to avoid values that cause a rational expression to be undefined, which means we need to identify and avoid values that make their denominators 0. This suggests the following procedure.

Procedure   **Finding Values That Make a Rational Expression Undefined**

To determine the value(s) that make a rational expression undefined,

1. Write an equation that has the denominator set equal to zero.
2. Solve the equation.

Example 2   Find every value for the variable that makes the expression undefined.

a. $\dfrac{5y}{3y - 2}$

SOLUTION

> **Note:** We do not consider the numerator because the rational expression is undefined only when the denominator is 0.

$3y - 2 = 0$ Set the denominator, $3y - 2$, equal to 0.

$3y = 2$ Add 2 to both sides.

$y = \dfrac{2}{3}$ Divide both sides by 3.

The expression $\dfrac{5y}{3y - 2}$ is undefined if y is replaced with $\dfrac{2}{3}$ because the denominator equals 0.

b. $\dfrac{7}{x^2 + x - 12}$

SOLUTION $x^2 + x - 12 = 0$ Set the denominator equal to 0.

$(x + 4)(x - 3) = 0$ Factor.

$x + 4 = 0$ or $x - 3 = 0$ Use the zero-factor theorem.

$x = -4$ $x = 3$

The original expression is undefined if x is replaced by -4 or 3.

c. $\dfrac{n-6}{n^2+9}$

SOLUTION $n^2 + 9 = 0$ Set the denominator equal to 0.

For all real-number values of n, $n^2 \geq 0$. Consequently $n^2 + 9 \neq 0$ for all real numbers. So $\dfrac{n-6}{n^2+9}$ is defined for all real numbers.

Your Turn 2   Find every value that can replace the variable in the expression and cause the expression to be undefined.

a. $\dfrac{y+4}{3y-1}$ **b.** $\dfrac{2x}{x^2+5x}$ **c.** $\dfrac{m}{m^2+1}$

Objective ③ **Simplify rational expressions containing only monomials.**   Now that we have defined rational expressions, we are ready to begin simplifying them. Recall from Section 1.2 that a fraction is in lowest terms when the only common factor of its numerator and denominator is 1. Also in Section 1.2, we used the rule $\dfrac{an}{bn} = \dfrac{a \cdot 1}{b \cdot 1} = \dfrac{a}{b}$, where $b \neq 0$ and $n \neq 0$, to reduce fractions. We can rewrite this rule so that it applies to rational expressions.

Rule $\dfrac{PR}{QR} = \dfrac{P \cdot 1}{Q \cdot 1} = \dfrac{P}{Q}$, where P, Q, and R are polynomials and Q and R are not 0.

The rule indicates that a factor common to both the numerator and denominator can be divided out of a rational expression. Consider the following comparison between simplifying a fraction and simplifying a similar rational expression.

$$\frac{10}{15} = \frac{2 \cdot 5}{3 \cdot 5} = \frac{2 \cdot 1}{3 \cdot 1} = \frac{2}{3} \qquad \frac{2x}{3x} = \frac{2 \cdot x}{3 \cdot x} = \frac{2 \cdot 1}{3 \cdot 1} = \frac{2}{3}$$

5 is the common factor here.

x is the common factor here.

Note: When working with rational expressions, we assume that the variables will not be replaced by any value that would cause the expression to be undefined or indeterminate.

This suggests the following procedure for simplifying rational expressions.

Procedure   **Simplifying Rational Expressions to Lowest Terms**

To simplify a rational expression to lowest terms,

1. Factor the numerator and denominator completely.

2. Divide out all common factors in the numerator and denominator.

3. Multiply the remaining factors in the numerator and the remaining factors in the denominator.

We first simplify rational expressions that are a monomial over a monomial.

Answers   Your Turn 2

a. $\dfrac{1}{3}$ **b.** 0 or -5

c. defined for all real numbers

Connection In Section 6.6 and Chapter 7, we divided monomials using $\dfrac{a^m}{a^n} = a^{m-n}$ and $a^{-n} = \dfrac{1}{a^n}$. In this chapter, we divide monomials (simplify to lowest terms) by factoring the numerator and/or denominator into prime factors, as we did in Section 1.2, for consistency of approach. As we will see later in this section, this approach is the only way rational expressions with multiterm polynomials in the numerator and/or denominator can be simplified to lowest terms.

Example 3 Simplify $\dfrac{3x^5}{15x^2}$.

SOLUTION Factor the numerator and denominator completely; then divide out all common factors.

$$\frac{3x^5}{15x^2} = \frac{3 \cdot x \cdot x \cdot x \cdot x \cdot x}{3 \cdot 5 \cdot x \cdot x}$$

Note: From here on, we omit this step with the 1's and simply highlight the common factors that are eliminated.

Note: The common factors are a single 3 and two x's. These form the GCF that we divide out, which is $3x^2$.

$$= \frac{1 \cdot x \cdot x \cdot x \cdot 1}{1 \cdot 5 \cdot 1}$$

$$= \frac{x^3}{5}$$

There are different styles for showing the process of dividing out common factors. For example, we could have used cancellation marks to show that the two x's and one 3 divide out in Example 3.

$$\frac{3x^5}{15x^2} = \frac{3 \cdot \cancel{x} \cdot \cancel{x} \cdot x \cdot x \cdot x}{3 \cdot 5 \cdot \cancel{x} \cdot \cancel{x}} = \frac{x^3}{5}$$

Develop a style that works best for you, keeping in mind that no matter the style, the bottom line is that you will be dividing out the GCF of the numerator and denominator to get a rational expression in lowest terms.

Example 4 Simplify.

a. $-\dfrac{4x^2}{28x^3}$

Connection
Using cancellation marks:

$$-\frac{4x^2}{28x^3} = -\frac{\cancel{2} \cdot \cancel{2} \cdot \cancel{x} \cdot \cancel{x}}{\cancel{2} \cdot \cancel{2} \cdot 7 \cdot \cancel{x} \cdot \cancel{x} \cdot x}$$

$$= -\frac{1}{7x}$$

SOLUTION $-\dfrac{4x^2}{28x^3} = -\dfrac{2 \cdot 2 \cdot x \cdot x}{2 \cdot 2 \cdot 7 \cdot x \cdot x \cdot x}$

Factor the numerator and denominator completely. Then eliminate the common factors, which are two 2s and two x's. Multiply the remaining factors.

$$= -\frac{1}{7x}$$

Note: Because all of the factors in the numerator divided out, the numerator is 1.

b. $\dfrac{36ab^3c}{54a^2b^3}$

SOLUTION $\dfrac{36ab^3c}{54a^2b^3} = \dfrac{2 \cdot 2 \cdot 3 \cdot 3 \cdot a \cdot b \cdot b \cdot b \cdot c}{2 \cdot 3 \cdot 3 \cdot 3 \cdot a \cdot a \cdot b \cdot b \cdot b}$

Factor the numerator and denominator completely. Then divide out the common factors, which are one 2, two 3s, one a, and three b's. Multiply the remaining factors.

$$= \frac{2c}{3a}$$

Your Turn 4 Simplify.

a. $-\dfrac{12t^6}{30t^2}$

b. $\dfrac{9n}{18n^3}$

c. $-\dfrac{72x^2y}{60x^2y^3z}$

Answers **Your Turn 4**

a. $-\dfrac{2t^4}{5}$ b. $\dfrac{1}{2n^2}$ c. $-\dfrac{6}{5y^2z}$

Connection Dividing out common factors when simplifying rational expressions with monomial numerators and/or denominators to lowest terms can be viewed as dividing monomials using the rules of exponents, as we did in Section 6.6.

Dividing out common factors:

$$\frac{12x^4y^2}{6x^2y} = \frac{2 \cdot 2 \cdot 3 \cdot x \cdot x \cdot x \cdot x \cdot y \cdot y}{2 \cdot 3 \cdot x \cdot x \cdot y}$$

$$= 2 \cdot x \cdot x \cdot y = 2x^2y$$

Using rules of exponents:

$$\frac{12x^4y^2}{6x^2y} = \frac{12}{6} \cdot \frac{x^4}{x^2} \cdot \frac{y^2}{y}$$

$$= 2x^{4-2}y^{2-1} = 2x^2y$$

The coefficients can be simplified as we do with ordinary fractions. Because we are dividing out two of the four x's and one of the two y's in the numerator, we are left with two x's and one y in the numerator. Subtracting the exponents accounts for eliminating the two common factors of x and the one common factor of y. This connection also applies when the greater exponent is in the denominator.

Dividing out common factors:

$$\frac{15a^2}{10a^5} = \frac{3 \cdot 5 \cdot a \cdot a}{2 \cdot 5 \cdot a \cdot a \cdot a \cdot a \cdot a}$$

$$= \frac{3}{2 \cdot a \cdot a \cdot a} = \frac{3}{2a^3}$$

Using rules of exponents:

$$\frac{15a^2}{10a^5} = \frac{15}{10} \cdot \frac{a^2}{a^5} = \frac{3}{2} \cdot a^{2-5} = \frac{3}{2}a^{-3}$$

$$= \frac{3}{2} \cdot \frac{1}{a^3} = \frac{3}{2a^3}$$

Notice that when dividing out like bases, the position of the base with the greater exponent determines whether we place the result in the numerator or denominator.

Objective ④ Simplify rational expressions containing multiterm polynomials. The rational expressions we have considered so far have had monomials in both the numerator and denominator. We now consider rational expressions that contain multiterm polynomials. Remember that we can divide out only common factors, so we must first factor polynomials completely.

Example 5 Simplify $\dfrac{9ab}{3a + 6}$.

SOLUTION $\dfrac{9ab}{3a + 6} = \dfrac{3 \cdot 3 \cdot a \cdot b}{3 \cdot (a + 2)}$

Factor the numerator and denominator completely. Then divide out the common factor, which is 3. Multiply the remaining factors.

$$= \frac{3ab}{a + 2}$$

Your Turn 5 Simplify.

a. $\dfrac{6xy}{3x + 12}$

b. $\dfrac{2a + 8}{10a^2}$

We now consider rational expressions with multiterm polynomials in both the numerator and denominator. Again, because we can divide out only common factors, we must first factor the polynomials.

Answers **Your Turn 5**

a. $\dfrac{2xy}{x + 4}$ b. $\dfrac{a + 4}{5a^2}$

Example 6 Simplify.

a. $\dfrac{2x^2 + 8x}{3x^2 + 12x}$

SOLUTION $\dfrac{2x^2 + 8x}{3x^2 + 12x} = \dfrac{2 \cdot x \cdot (x + 4)}{3 \cdot x \cdot (x + 4)}$ Factor the numerator and denominator completely. Then divide out the common factors, x and $x + 4$.

$$= \dfrac{2}{3}$$

b. $\dfrac{y^2 - 16}{y^2 + 4y - 32}$

SOLUTION $\dfrac{y^2 - 16}{y^2 + 4y - 32} = \dfrac{(y + 4)(y - 4)}{(y + 8)(y - 4)}$ Factor the numerator and denominator completely. Then divide out the common factor, $y - 4$.

$$= \dfrac{y + 4}{y + 8}$$

c. $\dfrac{2x^2 - 3x - 20}{2x^2 + x - 10} = \dfrac{(2x + 5)(x - 4)}{(2x + 5)(x - 2)}$ Factor the numerator and denominator completely. Then divide out the common factor, $2x + 5$.

$$= \dfrac{x - 4}{x - 2}$$

Warning: A common error is to divide out terms instead of factors. Recall that the terms of a polynomial are connected by plus and minus signs. You have factors only when multiplying. Below are examples of both the correct and incorrect way of simplifying a rational expression to lowest terms.

Correct: $\dfrac{x^2 + x - 6}{x^2 + 5x + 6} = \dfrac{\cancel{(x + 3)}(x - 2)}{\cancel{(x + 3)}(x + 2)} = \dfrac{x - 2}{x + 2}$ Divide out factors.

Incorrect: $\dfrac{x^2 + x - 6}{x^2 + 5x - 6} = \dfrac{\cancel{x^2} + \cancel{x} - \cancel{6}}{\cancel{x^2} + 5\cancel{x} - \cancel{6}} = \dfrac{1}{5}$ Do not divide out terms.

Your Turn 6 Simplify.

a. $\dfrac{5x^2 + 10x}{7x^2 + 14x}$

b. $\dfrac{y^2 - 36}{y^2 - 2y - 24}$

c. $\dfrac{3n^2 - 23n - 8}{3n^2 + 13n + 4}$

Recall from Chapter 6 that to factor multiterm polynomials completely, we must sometimes factor the first factorization we get as in Example 7.

Example 7 Simplify.

a. $\dfrac{6y^4 + 2y^3 - 4y^2}{36y^3 - 42y^2 + 12y}$

SOLUTION $\dfrac{6y^4 + 2y^3 - 4y^2}{36y^3 - 42y^2 + 12y} = \dfrac{2y^2(3y^2 + y - 2)}{6y(6y^2 - 7y + 2)}$ Factor out the monomial GCFs.

$$= \dfrac{2 \cdot y \cdot y \cdot (y + 1) \cdot (3y - 2)}{2 \cdot 3 \cdot y \cdot (2y - 1) \cdot (3y - 2)}$$ Factor the polynomial factors.

$$= \dfrac{y(y + 1)}{3(2y - 1)} \quad \text{or} \quad \dfrac{y^2 + y}{6y - 3}$$ Divide out the common factors 2, y, and $3y - 2$.

Answers   **Your Turn 6**

a. $\dfrac{5}{7}$ b. $\dfrac{y + 6}{y + 4}$ c. $\dfrac{n - 8}{n + 4}$

b. $\dfrac{3x^2 + 6x - 24}{x^4 - 16}$

SOLUTION Factor the numerator and denominator completely. Then divide out all common factors.

$$\frac{3x^2 + 6x - 24}{x^4 - 16} = \frac{3(x^2 + 2x - 8)}{(x^2 + 4)(x^2 - 4)}$$

Factor 3 out of $3x^2 + 6x - 24$ and factor $x^4 - 16$, which is a difference of squares.

$$= \frac{3 \cdot (x + 4) \cdot (x - 2)}{(x^2 + 4) \cdot (x + 2) \cdot (x - 2)}$$

Factor the polynomial factors. Notice that $x^2 - 4$ is a difference of squares.

$$= \frac{3(x + 4)}{(x^2 + 4)(x + 2)} \text{ or } \frac{3x + 12}{x^3 + 2x^2 + 4x + 8}$$

Divide out the common factor, $x - 2$.

Your Turn 7 Simplify.

a. $\dfrac{14x^3 + 70x^2 + 84x}{7x^4 - 14x^3 - 105x^2}$

b. $\dfrac{5x^2 - 5x - 60}{x^4 - 81}$

Binomial factors that are additive inverses can be tricky.

Example 8 Simplify $\dfrac{3x^2 - 7x - 20}{8 - 2x}$.

SOLUTION Factor the numerator and denominator completely. Then divide out all common factors.

$$\frac{3x^2 - 7x - 20}{8 - 2x} = \frac{(3x + 5)(x - 4)}{2(4 - x)}$$

It appears that there are no common factors. However, $x - 4$ and $4 - x$ are additive inverses, which is more apparent when $4 - x$ is written in descending order as $-x + 4$. We can get matching binomial factors by factoring -1 out of $x - 4$ or $4 - x$.

$$x - 4 = -1(-x + 4) = -1(4 - x) \quad \text{or} \quad 4 - x = -1(-4 + x) = -1(x - 4)$$

We can use either form. We will use $x - 4 = -1(4 - x)$.

Note: We could have factored the -1 out of the denominator:

$$\frac{(3x + 5)(x - 4)}{2(-1)(x - 4)} = \frac{3x + 5}{-2}$$

$$= \frac{(3x + 5)(-1)(4 - x)}{2(4 - x)}$$

Divide out the common factor, $4 - x$; then multiply the remaining factors.

$$= \frac{-3x - 5}{2} \quad \text{or} \quad -\frac{3x + 5}{2}$$

Example 8 also illustrates the rule of sign placement in a fraction or rational expression. In a negative fraction (or rational expression), the minus sign can be placed in the numerator or denominator or aligned with the fraction line. Note, however, that it is generally considered unsightly to write the minus sign in the denominator.

Answers Your Turn 7

a. $\dfrac{2(x + 2)}{x(x - 5)}$ or $\dfrac{2x + 4}{x^2 - 5x}$

b. $\dfrac{5(x - 4)}{(x^2 + 9)(x - 3)}$ or $\dfrac{5x - 20}{x^3 - 3x^2 + 9x - 27}$

Rule

$$-\frac{P}{Q} = \frac{-P}{Q} = \frac{P}{-Q},$$ where P and Q are polynomials and $Q \neq 0$.

Answer   Your Turn 8

$-\dfrac{x}{x + 6}$

Your Turn 8   Simplify $\dfrac{2x - x^2}{x^2 + 4x - 12}$.

8.1 Exercises   For Extra Help   *MyMathLab*
PRACTICE   WATCH   DOWNLOAD

1. Explain how to evaluate an expression.

2. In general, what causes a rational expression to be undefined?

3. How do you determine what values cause a rational expression to be undefined?

4. When simplifying a rational expression, we divide out all common _____ in the numerator and denominator.

5. Explain why you cannot divide out the y's in $\dfrac{x + y}{2y}$.

6. What is the first step when simplifying rational expressions?

For Exercises 7–14, evaluate the rational expression. See Example 1.

7. $\dfrac{4x^2}{9y}$ **a.** when $x = -2, y = 5$ **b.** when $x = 3, y = -4$ **c.** when $x = 0, y = 7$

8. $\dfrac{3x}{7y^2}$ **a.** when $x = 1, y = 2$ **b.** when $x = -2, y = 3$ **c.** when $x = -4, y = -2$

9. $\dfrac{2x}{x + 4}$ **a.** when $x = 4$ **b.** when $x = -4$ **c.** when $x = -2.4$

10. $\dfrac{4x}{x - 3}$ **a.** when $x = 3$ **b.** when $x = -3$ **c.** when $x = 2.5$

11. $\dfrac{2x + 5}{3x - 1}$ **a.** when $x = 3$ **b.** when $x = -2$ **c.** when $x = -1.3$

12. $\dfrac{3x + 2}{4x - 5}$ **a.** when $x = 2$ **b.** when $x = -1$ **c.** when $x = 1.6$

13. $\dfrac{x^2 - 4}{x + 3}$ **a.** when $x = 0$ **b.** when $x = -1$ **c.** when $x = -4.2$

14. $\dfrac{x^2 + 3}{2x + 1}$ **a.** when $x = 1$ **b.** when $x = -2$ **c.** when $x = -3.5$

For Exercises 15–26, find every value for the variable that makes the expression undefined. See Example 2.

15. $\dfrac{2x}{x - 3}$

16. $\dfrac{3y}{y + 5}$

17. $\dfrac{3}{x^2 - 9}$

18. $\dfrac{4r}{r^2 - 4}$

19. $\dfrac{6}{x^2 - 14x + 45}$

20. $\dfrac{2x + 3}{x^2 + 5x + 6}$

21. $\dfrac{2m}{m^2 - 3m}$

22. $\dfrac{2p + 3}{p^2 + 6p}$

23. $\dfrac{4a}{2a^2 - 3a - 5}$

24. $\dfrac{t}{2t^2 + 7t + 3}$

25. $\dfrac{x - 5}{x^2 + 16}$

26. $\dfrac{3x - 2}{y^2 + 36}$

For Exercises 27–78, simplify. See Examples 3–8.

27. $\dfrac{9x^2}{12xy}$

28. $\dfrac{14h^2k}{21h^3}$

29. $-\dfrac{4m^3n}{mn^2}$

30. $-\dfrac{5a^3b}{a^2b^4}$

31. $\dfrac{54x^4y}{36x^6yz^2}$

32. $-\dfrac{48t^5uv^4}{32\,tuv^2}$

33. $-\dfrac{15m^2n}{5m + 10n}$

34. $-\dfrac{28r^3t}{7r + 14t}$

35. $\dfrac{7}{21(x + 5)}$

36. $\dfrac{9}{18(a - 1)}$

37. $\dfrac{3(m - 2)}{5(m - 2)}$

38. $\dfrac{7(k + 2)}{15(k + 2)}$

39. $\dfrac{10x - 20}{15x - 30}$

40. $\dfrac{12y + 2}{18y + 3}$

41. $\dfrac{30a + 15b}{18a + 9b}$

42. $\dfrac{8m - 24n}{12m - 36n}$

43. $\dfrac{x^2 + 5x}{4x + 20}$

44. $\dfrac{ab - b^2}{2a - 2b}$

45. $\dfrac{m^2 - 6m}{m - 6}$

46. $\dfrac{x^2 + 5x}{x + 5}$

47. $\dfrac{5y^2 - 10y - 15}{xy^2 - 2xy - 3x}$

48. $\dfrac{3m^3 - 2m^2 + m}{6m^2 - 4m + 2}$

49. $\dfrac{x^2 - y^2}{x^2 + 2xy + y^2}$

50. $\dfrac{a^2 - b^2}{a^2 + 2ab + b^2}$

51. $\dfrac{x^2 - 5x}{x^2 - 7x + 10}$

52. $\dfrac{x^2 - 2x}{x^2 - 6x + 8}$

53. $\dfrac{t^2 - 9}{t^2 + 5t + 6}$

54. $\dfrac{k^2 - 4}{k^2 + k - 2}$

55. $\dfrac{4a^2 - 4a - 3}{6a^2 - a - 2}$

56. $\dfrac{3x^2 + 16x - 35}{5x^2 + 33x - 14}$

57. $\dfrac{6b^2 + b - 12}{10b^2 + 13b - 3}$

58. $\dfrac{8y^2 - 6y - 9}{12y^2 + y - 6}$

59. $\dfrac{x^3 - 16x}{x^3 + 6x^2 + 8x}$

60. $\dfrac{m^3 - 4m}{m^3 - 4m^2 + 4m}$

61. $\dfrac{px + py + qx + qy}{mx - nx + my - ny}$

62. $\dfrac{by - ay + bx - ax}{x^2 + ax + xy + ay}$

63. $\dfrac{6u^3 - 4u^2 - 3u + 2}{3u^2 + u - 2}$

64. $\dfrac{8n^3 + 20n^2 - 2n - 5}{4n^2 + 12n + 5}$

65. $\dfrac{x - 3}{3 - x}$

66. $\dfrac{5 - b}{b - 5}$

67. $\dfrac{4x - 4}{1 - x}$

68. $\dfrac{12 - 4m}{m - 3}$

69. $\dfrac{y^2 - 4}{2 - y}$

70. $\dfrac{a^2 - b^2}{b - a}$

71. $\dfrac{3 - w}{w^2 - 2w - 3}$

72. $\dfrac{1 - u}{u^2 - 3u + 2}$

73. $\dfrac{x^2 - 4}{x^3 + 8}$

74. $\dfrac{r^2 - s^2}{r^3 + s^3}$

75. $\dfrac{x^3 + y^3}{x^3 - x^2y + xy^2}$

76. $\dfrac{x^3 - 8}{x^3 + 2x^2 + 4x}$

77. $\dfrac{3 - m}{m^3 - 27}$

78. $\dfrac{4 - x}{x^3 - 64}$

Find the mistake

For Exercises 79–84, find and explain the mistake; then correct the mistake.

79. $-\dfrac{5x^2y}{10x^3y} = -\dfrac{x}{2}$

80. $\dfrac{8x^2y}{4x^3} = 2xy$

81. $\dfrac{x - \overset{3}{\cancel{6}}}{\underset{1}{\cancel{2}}y} = \dfrac{x - 3}{y}$

82. $\dfrac{y - \overset{3}{\cancel{12}}}{\underset{1}{\cancel{4}}y - 3} = \dfrac{y - 3}{y - 3} = 1$

83. $\dfrac{yx - x^2}{x - y} = \dfrac{x(y - x)}{x - y} = x$

84. $\dfrac{x^2 - x - 6}{x^3 + 2x^2 - 9x - 18} = \dfrac{(x + 2)(x - 3)}{(x + 2)(x^2 - 9)} = \dfrac{x - 3}{x^2 - 9}$

For Exercises 85–90, evaluate. See Example 1.

85. A person's body mass index (BMI) is a measure of the amount of body fat based on height, h, in inches, and weight, w, in pounds. The formula for BMI is

$$\text{BMI} = \frac{704.5w}{h^2}.$$

a. Use the formula to find the BMI of each person listed here.

Height	Weight	BMI
5' 4"	138	
5' 9"	155	
6' 2"	220	

b. Use the following table from the National Institutes of Health to classify each person in part a as underweight, normal weight, overweight, or obese.

Underweight:	BMI $\leq$ 18.5
Normal Weight:	BMI $=$ 18.5 $-$ 24.9
Overweight:	BMI $=$ 25 $-$ 29.9
Obese:	BMI $\geq$ 30

Connection In Section 2.5, Exercise 68, you translated the BMI formula from words to symbols.

86. Recall the following formula for slope from Chapter 4:

$$m = \frac{y_2 - y_1}{x_2 - x_1}$$

a. Find the slope of a line passing through the points $(3, 5)$ and $(-1, -4)$.

b. Find the slope of a line passing through the points $(-4, 7)$ and $(-2, -1)$.

Connection Although we hadn't formally defined rational expressions in Chapter 4, we were evaluating the rational expression $\frac{y_2 - y_1}{x_2 - x_1}$ to find slope.

87. Given the diameter in inches, d, and the number of revolutions per minute, N, of a revolving tool such as a circular saw or drill, the cutting speed, C, in feet per minute, of the tool can be found by the formula

$$C = \frac{\pi d N}{12}.$$

(*Source:* Robert D. Smith, *Mathematics for Machine Technology*, 5th ed., Cengage Learning, 2003)

Use the formula to complete the following table for a table saw. Leave answers in terms of π.

Blade Diameter	rpm	Cutting Speed
8 inches	800 rpm	
8 inches	1000 rpm	
10 inches	800 rpm	
10 inches	1000 rpm	

88. The cutting time, T, in minutes, that it takes a drill to cut through something depends on its speed of revolution, N, in rpm; the thickness of the material, L, in inches; and the tool feed, F, in inches per revolution. The formula for calculating cutting time is

$$T = \frac{L}{FN}.$$

(*Source:* Robert D. Smith, *Mathematics for Machine Technology*, 5th ed., Cengage Learning, 2003)

Use the formula to complete the following table for a drill operating at a speed of 480 rpm and a feed of 0.04 inches per revolution.

Material Thickness (inches)	Cut Time (minutes)
0.25	
0.5	
0.75	
1	

89. The formula for the volume of a cone is $V = \dfrac{1}{3}\pi r^2 h$.

a. Solve this formula for h so that we have a formula for finding the height of a cone given its volume and radius. (The formula for height of a cone will contain a rational expression.)

b. Suppose a cone has a volume of approximately 150.8 cubic centimeters with a radius of 4 centimeters. Find the height of the cone.

90. The formula for the volume of a pyramid is $V = \dfrac{1}{3} lwh$.

a. Solve this formula for h so that we have a formula for finding the height of a pyramid given its volume, length, and width.

b. Suppose a decorative onyx pyramid has a volume of 126 cubic inches, a width of 3.5 inches, and a length of 4.5 inches. Find the height of the pyramid.

For Exercises 91 and 92, simplify. See Examples 5–7.

91. In analyzing circuits, electrical engineers often have to simplify rational expressions. Suppose the following rational expression describes an electrical circuit.

$$\frac{s^2 - 9}{s^2 + 7s + 12}$$

Simplify the expression to lowest terms.

92. An engineer derives the following rational expression, which describes a certain circuit.

$$\frac{4s^2 - 6s}{2s^2 + 7s - 15}$$

Simplify the expression to lowest terms.

Collaborative Exercises Graphs of Rational Expressions

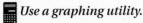 *Use a graphing utility.*

1. Graph $f(x) = \dfrac{x}{x - 3}$. The vertical line you see is called an *asymptote*.

   a. What is the equation of the asymptote?

   b. What do you think the asymptote indicates?

2. Graph $f(x) = \dfrac{5}{(x + 2)(x - 4)}$.

   a. Why are there two asymptotes?

   b. What are the equations of the two asymptotes?

   Note: The suggested window settings in Exercise 3 are for a TI-83 or TI-84. If a different graphing utility is used, the settings may need to be changed in order to see the hole.

3. Graph $f(x) = \dfrac{x^2 - x - 6}{x^2 + 3x + 2}$ with these window settings: $x_{\min} = -4.7$, $x_{\max} = 4.7$, $y_{\min} = -10$, and $y_{\max} = 10$.

   a. Are there any asymptotes? If so, write the equation for each.

   b. Rewrite $f(x) = \dfrac{x^2 - x - 6}{x^2 + 3x + 2}$ so that the numerator and denominator are in factored form.

   c. Find $f(-2)$.

   d. Look at the graph of $f(x) = \dfrac{x^2 - x - 6}{x^2 + 3x + 2}$ in the second quadrant. The blank space you see in the graph where $x = -2$ is a *removable discontinuity* and is often called a *hole*. How is a hole different from an asymptote?

4. Writing $\dfrac{x^2 - x - 6}{x^2 + 3x + 2}$ in simplest form, we have $\dfrac{x - 3}{x + 1}$. Graph $f(x) = \dfrac{x - 3}{x + 1}$. Compare the graph of $f(x) = \dfrac{x - 3}{x + 1}$ with the graph of $f(x) = \dfrac{x^2 - x - 6}{x^2 + 3x + 2}$. What do you notice about the graphs?

5. Given a rational function, how can you identify its asymptotes and holes?

Review Exercises

Exercises 1–6 EXPRESSIONS

[1.4] *For Exercises 1 and 2, multiply.*

1. $\dfrac{2}{3} \cdot \dfrac{6}{4}$

2. $-\dfrac{15}{16} \cdot \dfrac{4}{5}$

[1.4] *For Exercises 3 and 4, divide.*

3. $\dfrac{3}{8} \div \dfrac{5}{6}$

4. $-\dfrac{7}{9} \div \left(-\dfrac{5}{12}\right)$

[7.4] *For Exercises 5 and 6, factor completely.*

5. $4y^2 - 25$

[7.3] **6.** $3x^2 + 13x - 10$

8.2 Multiplying and Dividing Rational Expressions

Objectives
1. Multiply rational expressions.
2. Divide rational expressions.
3. Convert units of measurement using dimensional analysis.

Objective ❶ **Multiply rational expressions.** In Section 1.4, we used the rule $\dfrac{a}{b} \cdot \dfrac{c}{d} = \dfrac{ac}{bd}$, where $b \neq 0$ and $d \neq 0$, to multiply fractions. We can rewrite this rule so that it applies to multiplying rational expressions.

Rule **Multiplying Rational Expressions**

> $\dfrac{P}{Q} \cdot \dfrac{R}{S} = \dfrac{PR}{QS}$, where P, Q, R, and S are polynomials and $Q \neq 0$ and $S \neq 0$.

Also remember that when multiplying fractions, we can simplify after multiplying or we can simplify before multiplying by dividing out factors common to both the numerator and denominator. For example,

| Multiply, then eliminate common factors. | or | Eliminate common factors; then multiply. |

$$\dfrac{8}{15} \cdot \dfrac{3}{4} = \dfrac{24}{60} = \dfrac{2 \cdot 2 \cdot 2 \cdot 3}{2 \cdot 2 \cdot 3 \cdot 5} = \dfrac{2}{5} \qquad \dfrac{8}{15} \cdot \dfrac{3}{4} = \dfrac{2 \cdot 2 \cdot 2}{3 \cdot 5} \cdot \dfrac{3}{2 \cdot 2} = \dfrac{2}{5}$$

Multiplying rational expressions works the same way.

Procedure **Multiplying Rational Expressions:** To multiply rational expressions,
1. Factor each numerator and denominator completely.
2. Divide out any numerator factor with any matching denominator factor.
3. Multiply numerator by numerator and denominator by denominator.
4. Simplify as needed.

Example 1 Multiply $\dfrac{9x}{5y} \cdot \dfrac{15xy^2}{12}$.

SOLUTION $\dfrac{9x}{5y} \cdot \dfrac{15xy^2}{12} = \dfrac{3 \cdot 3 \cdot x}{5 \cdot y} \cdot \dfrac{3 \cdot 5 \cdot x \cdot y \cdot y}{2 \cdot 2 \cdot 3}$ Factor the numerators and denominators completely.

$= \dfrac{3 \cdot 3 \cdot x}{1} \cdot \dfrac{x \cdot y}{2 \cdot 2}$ Divide out the common factors, which are 3, 5, and y.

$= \dfrac{9x^2y}{4}$ Multiply the remaining numerator factors and denominator factors.

Your Turn 1 Multiply.

a. $-\dfrac{36m}{8n^4} \cdot \dfrac{14mn^2}{28n}$

b. $-\dfrac{9}{4xy^2} \cdot -\dfrac{20xz}{15y}$

Now consider rational expressions that contain multiterm polynomials.

Example 2 Multiply.

a. $-\dfrac{2x}{3x-9} \cdot \dfrac{5x-15}{20x^3}$

SOLUTION $-\dfrac{2x}{3x-9} \cdot \dfrac{5x-15}{20x^3}$

$= -\dfrac{2 \cdot x}{3 \cdot (x-3)} \cdot \dfrac{5 \cdot (x-3)}{2 \cdot 2 \cdot 5 \cdot x \cdot x \cdot x}$ Factor the numerators and denominators completely.

$= -\dfrac{1}{3} \cdot \dfrac{1}{2 \cdot x \cdot x}$ Divide out the common factors, which are 2, 5, x, and $(x-3)$.

$= -\dfrac{1}{6x^2}$ Multiply the remaining numerator factors and denominator factors.

b. $\dfrac{12-4x}{15x} \cdot \dfrac{5x^2}{2x-6}$

SOLUTION $\dfrac{12-4x}{15x} \cdot \dfrac{5x^2}{2x-6} = \dfrac{4 \cdot (3-x)}{3 \cdot 5 \cdot x} \cdot \dfrac{5 \cdot x \cdot x}{2 \cdot (x-3)}$

$= \dfrac{2 \cdot 2 \cdot (-1)(x-3)}{3 \cdot 5 \cdot x} \cdot \dfrac{5 \cdot x \cdot x}{2 \cdot (x-3)}$

$= \dfrac{2 \cdot (-1)}{3} \cdot \dfrac{x}{1}$ Divide out the common factors, which are 2, 5, x, and $(x-3)$.

$= \dfrac{-2x}{3}$ Multiply the remaining numerator factors and denominator factors.

> **Note:** We can factor 4 further. Also, because $3 - x$ and $x - 3$ are additive inverses, we can factor -1 out of one of them.

c. $\dfrac{x^2+x-6}{9x^3} \cdot \dfrac{12x^2-6x}{2x^2-5x+2}$

SOLUTION $\dfrac{x^2+x-6}{9x^3} \cdot \dfrac{12x^2-6x}{2x^2-5x+2}$

$= \dfrac{(x-2)(x+3)}{3 \cdot 3 \cdot x \cdot x \cdot x} \cdot \dfrac{2 \cdot 3 \cdot x \cdot (2x-1)}{(x-2)(2x-1)}$ Factor the numerators and denominators completely.

$= \dfrac{x+3}{3 \cdot x \cdot x} \cdot \dfrac{2}{1}$ Divide out the common factors, which are $(x-2)$, 3, x, and $(2x-1)$.

$= \dfrac{2(x+3)}{3x^2}$ or $\dfrac{2x+6}{3x^2}$ Multiply the remaining numerator factors and denominator factors.

Your Turn 2 Multiply.

a. $\dfrac{-8h^2}{2h+10} \cdot -\dfrac{6h+30}{18h^4}$

b. $-\dfrac{9y-3}{y^2+3y} \cdot \dfrac{2y+6}{2-6y}$

c. $\dfrac{m^2-25}{20m} \cdot \dfrac{30m+10}{3m^2+16m+5}$

Objective ② **Divide rational expressions.** In Section 1.4, we learned how to divide fractions using the rule $\dfrac{a}{b} \div \dfrac{c}{d} = \dfrac{a}{b} \cdot \dfrac{d}{c}$, where b, c, and d are not 0, which indicates to multiply by the reciprocal of the divisor. We can rewrite this rule so that it applies to dividing rational expressions.

Rule **Dividing Rational Expressions**

$\dfrac{P}{Q} \div \dfrac{R}{S} = \dfrac{P}{Q} \cdot \dfrac{S}{R}$, where P, Q, R, and S are polynomials and $Q \neq 0$, $R \neq 0$, and $S \neq 0$.

Following is a procedure for dividing rational expressions.

Procedure **Dividing Rational Expressions**

To divide rational expressions,

1. Write an equivalent multiplication statement with the reciprocal of the divisor.
2. Factor each numerator and denominator completely. (Steps 1 and 2 are interchangeable.)
3. Divide out any numerator factor with any matching denominator factor.
4. Multiply numerator by numerator and denominator by denominator.
5. Simplify as needed.

Example 3 Divide $-\dfrac{9x^2 y}{28z^3} \div \dfrac{12y}{7z}$.

SOLUTION

$-\dfrac{9x^2 y}{28z^3} \div \dfrac{12y}{7z} = -\dfrac{9x^2 y}{28z^3} \cdot \dfrac{7z}{12y}$
   Write an equivalent multiplication statement by changing the division sign to multiplication and changing the divisor to its reciprocal.

$= -\dfrac{3 \cdot 3 \cdot x \cdot x \cdot y}{2 \cdot 2 \cdot 7 \cdot z \cdot z \cdot z} \cdot \dfrac{7 \cdot z}{2 \cdot 2 \cdot 3 \cdot y}$
   Factor the numerators and denominators completely.

$= -\dfrac{3 \cdot x \cdot x}{2 \cdot 2 \cdot z \cdot z} \cdot \dfrac{1}{2 \cdot 2}$
   Divide out the common factors, which are 3, 7, y, and z.

$= -\dfrac{3x^2}{16z^2}$
   Multiply the remaining factors.

Your Turn 3 Divide.

a. $\dfrac{15a^3}{28b^2} \div \dfrac{10ab}{7}$

b. $-\dfrac{2t}{33u^3 v} \div \left(-\dfrac{8t^3 v}{21u^4} \right)$

Now consider dividing rational expressions that contain multiterm polynomials.

Answers **Your Turn 3**

a. $\dfrac{3a^2}{8b^3}$ b. $\dfrac{7u}{44t^2 v^2}$

Example 4 Divide.

a. $\dfrac{2x - 10}{15x^3} \div \dfrac{x^2 - 5x}{12x}$

SOLUTION

$\dfrac{2x - 10}{15x^3} \div \dfrac{x^2 - 5x}{12x} = \dfrac{2x - 10}{15x^3} \cdot \dfrac{12x}{x^2 - 5x}$

$\qquad\qquad Write an equivalent multiplication statement by changing the division sign to multiplication and changing the divisor to its reciprocal.$

$\qquad = \dfrac{2 \cdot (x - 5)}{3 \cdot 5 \cdot x \cdot x \cdot x} \cdot \dfrac{2 \cdot 2 \cdot 3 \cdot x}{x \cdot (x - 5)}$

Factor the numerators and denominators completely.

$\qquad = \dfrac{2}{5 \cdot x \cdot x} \cdot \dfrac{2 \cdot 2}{x}$

Divide out the common factors, which are $3, x,$ and $(x - 5)$.

$\qquad = \dfrac{8}{5x^3}$

Multiply the remaining factors.

b. $\dfrac{12x^2 + 3x}{2x^2 + 3x - 9} \div \dfrac{4x^2 + 9x + 2}{6x^3 + 18x^2}$

SOLUTION

$\dfrac{12x^2 + 3x}{2x^2 + 3x - 9} \div \dfrac{4x^2 + 9x + 2}{6x^3 + 18x^2}$

$= \dfrac{12x^2 + 3x}{2x^2 + 3x - 9} \cdot \dfrac{6x^3 + 18x^2}{4x^2 + 9x + 2}$

Write an equivalent multiplication statement by changing the division sign to multiplication and changing the divisor to its reciprocal.

$= \dfrac{3x(4x + 1)}{(2x - 3)(x + 3)} \cdot \dfrac{6x^2(x + 3)}{(4x + 1)(x + 2)}$

Factor the numerators and denominators completely.

$= \dfrac{3 \cdot x \cdot (4x + 1)}{(2x - 3)(x + 3)} \cdot \dfrac{2 \cdot 3 \cdot x \cdot x \cdot (x + 3)}{(4x + 1) \cdot (x + 2)}$

$= \dfrac{3 \cdot x}{2x - 3} \cdot \dfrac{2 \cdot 3 \cdot x \cdot x}{x + 2}$

Divide out the common factors, which are $(4x + 1)$ and $(x + 3)$.

$= \dfrac{18x^3}{(2x - 3)(x + 2)}$

Multiply the remaining factors.

Note: It is customary not to multiply out remaining binomial factors.

c. $\dfrac{8x^2 - 6x^3}{6xy} \div \dfrac{9x^2 - 16}{5y^2 - 10y}$

SOLUTION

$\dfrac{8x^2 - 6x^3}{6xy} \div \dfrac{9x^2 - 16}{5y^2 - 10y} = \dfrac{8x^2 - 6x^3}{6xy} \cdot \dfrac{5y^2 - 10y}{9x^2 - 16}$

Write an equivalent multiplication statement by changing the division sign to multiplication and changing the divisor to its reciprocal.

$= \dfrac{2x^2(4 - 3x)}{6xy} \cdot \dfrac{5y(y - 2)}{(3x + 4)(3x - 4)}$

Factor the numerators and denominators completely.

Note: $4 - 3x$ and $3x - 4$ are additive inverses, so we factored -1 out of $4 - 3x$.

$= \dfrac{2 \cdot x \cdot x \cdot (-1) \cdot (3x - 4)}{2 \cdot 3 \cdot x \cdot y} \cdot \dfrac{5 \cdot y \cdot (y - 2)}{(3x + 4) \cdot (3x - 4)}$

$= \dfrac{x \cdot (-1)}{3} \cdot \dfrac{5 \cdot (y - 2)}{3x + 4}$

Divide out the common factors, which are $2, x, y,$ and $(3x - 4)$.

$= \dfrac{-5x(y - 2)}{3(3x + 4)}$ or $\dfrac{-5xy + 10x}{9x + 12}$ or $\dfrac{10x - 5xy}{9x + 12}$

Multiply the remaining factors.

Your Turn 4 Divide.

a. $\dfrac{6yz^3}{2y - y^2} \div \dfrac{8yz}{6 - 3y}$

b. $\dfrac{mn - 3m}{14m^4} \div \dfrac{n}{7n + 7}$

c. $\dfrac{4y^2 - 25}{9y^2 + 18y} \div \dfrac{2y^2 + 3y - 5}{3y^4 - 3y^3}$

d. $\dfrac{3x^2 + 10x + 8}{18 - 6x} \div \dfrac{10 + 5x}{4x^3 - 12x^2}$

Objective ❸ Convert units of measurement using dimensional analysis. We can convert from one unit of a measurement to another using a method called *dimensional analysis*, which involves writing measurement facts as conversion factors. A conversion factor is a ratio of equivalent measures.

Measurement fact:	Conversion factor:
1 foot = 12 inches	$\dfrac{1 \text{ foot}}{12 \text{ inches}}$ or $\dfrac{12 \text{ inches}}{1 \text{ foot}}$
1 pound = 16 ounces	$\dfrac{1 \text{ pound}}{16 \text{ ounces}}$ or $\dfrac{16 \text{ ounces}}{1 \text{ pound}}$

Note: Because a conversion factor is a ratio of equivalent measures, its ratio always equals 1.

To use dimensional analysis, we multiply a measurement by appropriate conversion factors so that the undesired units divide out, leaving only the desired units of measurement. For example, if we want to convert 6 feet to inches, we multiply by the conversion factor $\dfrac{12 \text{ inches}}{1 \text{ foot}}$ so that the units of feet divide out, leaving inches.

$$6 \text{ ft.} = \dfrac{6 \text{ ft.}}{1} \cdot \dfrac{12 \text{ in.}}{1 \text{ ft.}} = 72 \text{ in.}$$

Connection The ratio of the conversion factor $\dfrac{12 \text{ in.}}{1 \text{ ft.}}$ is 1. So multiplying 6 ft. by this ratio means that the result, 72 in., is equal to 6 ft.

Procedure   **Using Dimensional Analysis to Convert between Units of Measurement**

To convert units using dimensional analysis, multiply the given measurement by conversion factors so that the undesired units divide out, leaving the desired units.

Example 5   Convert.

a. 500 ounces to pounds

Connection Dividing out units in dimensional analysis is like dividing out common factors when multiplying rational expressions.

SOLUTION   There are 16 ounces to 1 pound. We write this fact as a unit ratio and multiply so that ounces divides out, leaving pounds.

$$500 \text{ oz.} = \dfrac{500 \text{ oz.}}{1} \cdot \dfrac{1 \text{ lb.}}{16 \text{ oz.}} = \dfrac{500}{16} \text{ lb.} = 31.25 \text{ lb.}$$

b. 6 miles to yards

SOLUTION   There are 5280 feet to 1 mile and 3 feet to 1 yard. Thus, we write these facts as unit fractions so that miles and feet divide out, leaving yards.

$$6 \text{ mi.} = \dfrac{6 \text{ mi.}}{1} \cdot \dfrac{5280 \text{ ft.}}{1 \text{ mi.}} \cdot \dfrac{1 \text{ yd.}}{3 \text{ ft.}} = \dfrac{31{,}680}{3} \text{ yd.} = 10{,}560 \text{ yd.}$$

Answers   Your Turn 4

a. $\dfrac{9z^2}{4y}$ b. $\dfrac{(n - 3)(n + 1)}{2m^3n}$

c. $\dfrac{y^2(2y - 5)}{3(y + 2)}$ or $\dfrac{2y^3 - 5y^2}{3y + 6}$

d. $-\dfrac{2x^2(3x + 4)}{15}$ or $-\dfrac{6x^3 + 8x^2}{15}$

c. 40 miles per hour to feet per second

SOLUTION First, 40 miles per hour means $\dfrac{40 \text{ miles}}{1 \text{ hour}}$. Notice that we must convert miles to feet and hours to seconds. For the distance conversion, we use the fact that there are 5280 feet to 1 mile. For the time conversion, there are 60 minutes in 1 hour and 60 seconds in 1 minute.

$$\frac{40 \text{ miles}}{1 \text{ hour}} = \frac{40 \text{ mi.}}{1 \text{ hr.}} \cdot \frac{5280 \text{ ft.}}{1 \text{ mi.}} \cdot \frac{1 \text{ hr.}}{60 \text{ min.}} \cdot \frac{1 \text{ min.}}{60 \text{ sec.}} = \frac{211{,}200 \text{ ft.}}{3600 \text{ sec.}} = 58.\overline{6} \text{ ft./sec.}$$

Your Turn 5 Convert.

a. 192 ounces to pounds   **b.** 540 inches to yards   **c.** 44 feet per second into miles per hour

Answers   Your Turn 5
a. 12 lb.   **b.** 15 yd.   **c.** 30 mph

8.2 Exercises   For Extra Help   *MyMathLab*

Note: Exercises marked with a ★ represent challenging exercises.

1. When multiplying rational expressions, why is it important to factor the numerators and denominators completely?

2. If after factoring you determine that there are two binomial factors that are additive inverses, how can you get them to be common binomial factors?

3. Why are $\dfrac{x}{y}$ and $\dfrac{y}{x}$ multiplicative inverses (or reciprocals)?

4. To divide rational expressions, we write an equivalent multiplication with the _____ _____ of the divisor.

5. What conversion factor would be used to convert 3.8 pounds to ounces?

6. Explain how to convert one unit of measurement to another using dimensional analysis.

For Exercises 7–42, multiply. See Examples 1 and 2.

7. $\dfrac{m}{n} \cdot \dfrac{2m}{5n}$

8. $\dfrac{x}{y} \cdot \dfrac{3x}{2y}$

9. $\dfrac{r^3}{s^3} \cdot \dfrac{s^5}{r^6}$

10. $\dfrac{a^5}{b^3} \cdot \dfrac{b^2}{a^3}$

11. $\dfrac{3}{y} \cdot \dfrac{y^3}{3}$

12. $\dfrac{y^4}{7} \cdot \dfrac{7}{y^2}$

13. $-\dfrac{n}{rs^2} \cdot \dfrac{rs}{mn^3}$

14. $\dfrac{ab}{xy} \cdot -\dfrac{x}{a^3b^2}$

15. $-\dfrac{6r^2s}{5r^2s^2} \cdot -\dfrac{15r^4s^2}{2r^3s^2}$

16. $-\dfrac{14mn^2}{8m^2n} \cdot -\dfrac{16mz^2}{7n^2z}$

17. $\dfrac{5n^2}{15m} \cdot \dfrac{n}{2} \cdot \dfrac{3m^2n}{n^2}$

18. $\dfrac{a}{b^2} \cdot \dfrac{3b^2}{4a} \cdot \dfrac{ab^2}{9}$

19. $\dfrac{3b - 12}{14} \cdot \dfrac{21}{5b - 20}$

20. $\dfrac{9}{2a + 4} \cdot \dfrac{3a + 6}{15}$

21. $\dfrac{2x - 6}{3y + 12} \cdot \dfrac{2y + 8}{5x - 15}$

22. $\dfrac{3a + 12}{5b - 30} \cdot \dfrac{3b - 18}{4a + 16}$

23. $\dfrac{4x^2 - 25}{3x - 4} \cdot \dfrac{9x^2 - 16}{2x + 5}$

24. $\dfrac{16x^2 - 9}{4x + 3} \cdot \dfrac{25x^2 - 1}{5x + 1}$

25. $\dfrac{r^2 + 3r}{r^2 - 36} \cdot \dfrac{r^2 - 6r}{r^2 - 9}$

26. $\dfrac{m^2 + 5m}{m^2 - 16} \cdot \dfrac{m^2 - 4m}{m^2 - 25}$

27. $\dfrac{12q}{q^2 - 2q + 1} \cdot \dfrac{q^2 - 1}{24q^2}$

28. $\dfrac{36x^3}{x^2 - 25} \cdot \dfrac{x^2 - 10x + 25}{9x}$

29. $\dfrac{w^2 - 4w - 5}{w^2 - 1} \cdot \dfrac{w^2 - w - 20}{w^2 - 10w + 25}$

30. $\dfrac{y^2 - 2y - 24}{y^2 - 16} \cdot \dfrac{y^2 + 10y + 25}{y^2 - y - 30}$

31. $\dfrac{n^2 - 6n + 9}{n^2 - 9} \cdot \dfrac{n^2 + 4n + 3}{n^2 - 3n}$

32. $\dfrac{4u^2 + 4u + 1}{2u^2 + u} \cdot \dfrac{u^2 + 3u - 4}{2u^2 - u - 1}$

33. $\dfrac{6x^2 + x - 12}{2x^2 - 5x - 12} \cdot \dfrac{3x^2 - 14x + 8}{9x^2 - 18x + 8}$

34. $\dfrac{2x^2 + x - 15}{4x^2 + 11x - 3} \cdot \dfrac{4x^2 - 9x + 2}{2x^2 - 9x + 10}$

35. $\dfrac{ac + 3a + 2c + 6}{ad + a + 2d + 2} \cdot \dfrac{ad - 5a + 2d - 10}{ac + 2c + 3a + 6}$

36. $\dfrac{mn + 2m + 4n + 8}{np - 3n + 2p - 6} \cdot \dfrac{pq + 5p - 3q - 15}{mq + 4m + 4q + 16}$

37. $\dfrac{x^3 + 27}{x^3 - 3x^2 + 9x} \cdot \dfrac{x^2 - 7x + 10}{x^2 - 2x - 15}$

38. $\dfrac{x^2 + 6x + 8}{x^3 + 5x^2 + 25x} \cdot \dfrac{x^3 - 125}{x^2 - x - 20}$

39. $\dfrac{x^2 + 7xy + 10y^2}{x^2 + 6xy + 5y^2} \cdot \dfrac{x + 2y}{y} \cdot \dfrac{x + y}{x^2 + 4xy + 4y^2}$

40. $\dfrac{x^2 + 6xy + 9y^2}{x^2 + xy - 6y^2} \cdot \dfrac{x + 4y}{x} \cdot \dfrac{x - 2y}{x^2 + 7xy + 12y^2}$

41. $\dfrac{4x^2 y^3}{2x - 6} \cdot \dfrac{12 - 4x}{6x^3 y}$

42. $\dfrac{6b^3 c^3}{70 - 10b} \cdot \dfrac{5b - 35}{8b^2 c^5}$

43. $\dfrac{2x^2 - 11x + 12}{2x^2 + 11x - 21} \cdot \dfrac{3x^2 + 20x - 7}{4 - x}$

44. $\dfrac{3x^2 - 17x + 10}{3x^2 + 7x - 6} \cdot \dfrac{2x^2 + x - 15}{5 - x}$

For Exercises 45–74, divide. See Examples 3 and 4.

45. $\dfrac{x}{y^4} \div \dfrac{x^3}{y^2}$

46. $\dfrac{a^3}{b^3} \div \dfrac{a^5}{b^2}$

47. $\dfrac{2m}{3n^2} \div \dfrac{8m^3}{15n}$

48. $\dfrac{21c^2}{7d^3} \div \dfrac{8c^4}{12d^2}$

49. $-\dfrac{12w^2 y}{5z^2} \div \left(-\dfrac{12w^2}{y} \right)$

50. $-\dfrac{7a^2 b}{2c^2} \div \left(-\dfrac{7a^2}{b} \right)$

51. $\dfrac{8x^7 g^3}{15x^2 g^4} \div \left(-\dfrac{3xg^2}{4x^2 g^3} \right)$

52. $-\dfrac{10c^5 g^5}{12c^2 g^4} \div \dfrac{8cg^2}{4c^2 g}$

53. $\dfrac{4a - 8}{15a} \div \dfrac{3a - 6}{5a^2}$

54. $\dfrac{6b + 24}{7b^2} \div \dfrac{7b + 28}{14b^3}$

55. $\dfrac{4p + 12q}{3p - 6q} \div \dfrac{5p + 15q}{6p - 12q}$

56. $\dfrac{5r + 10s}{3r - 12s} \div \dfrac{6r + 12s}{4r - 16s}$

57. $\dfrac{a^2 - b^2}{x^2 - y^2} \div \dfrac{a + b}{x - y}$

58. $\dfrac{x + 4}{y - 3} \div \dfrac{x^2 - 16}{y^2 - 9}$

59. $\dfrac{x^2 + 16x + 64}{x^2 - 16x + 64} \div \dfrac{x + 8}{x - 8}$

60. $\dfrac{w + 3}{w - 3} \div \dfrac{w^2 + 6w + 9}{w^2 - 6w + 9}$

61. $\dfrac{x^2 + 7x + 10}{x^2 + 2x} \div \dfrac{x + 5}{x^2 - 4x}$

62. $\dfrac{t^2 - 2t}{t + 1} \div \dfrac{t^2 + 3t}{t^2 + 4t + 3}$

63. $\dfrac{a^2 - b^2}{a^2 + 2ab + b^2} \div \dfrac{a^2 - 3ab + 2b^2}{a^2 + 3ab + 2b^2}$

64. $\dfrac{3w^2 - 7w - 6}{w^2 - 9} \div \dfrac{3w^2 - 10w - 8}{3w^2 + 7w - 6}$

65. $\dfrac{3k + 2}{3k^4 + 2k^3} \div \dfrac{10k^2 + 3k - 1}{5k^3 - k^2}$

66. $\dfrac{5y - 3}{4y^2 - y - 3} \div \dfrac{5y^3 - 3y^2}{4y^2 + 3y}$

67. $\dfrac{8x^3 - 27}{3x^2 + 20x - 7} \div \dfrac{2x^2 - 7x + 6}{x^2 + 5x - 14}$

68. $\dfrac{x^2 + 6x - 16}{3x^2 - 10x + 8} \div \dfrac{2x^2 + 13x - 24}{27x^3 - 64}$

69. $\dfrac{x^2 - 2x - 8}{x^2 + 6x + 8} \div (x^2 - 3x - 4)$

70. $\dfrac{u^2 + 7u + 12}{u^2 + u - 6} \div (u^2 + u - 12)$

71. $\dfrac{ab + 3a + 2b + 6}{bc + 4b + 3c + 12} \div \dfrac{ac - 3a + 2c - 6}{bc + 4b - 4c - 16}$

72. $\dfrac{xy - 3x + 4y - 12}{xy + 6x - 3y - 18} \div \dfrac{xy + 5x + 4y + 20}{xy + 5x - 3y - 15}$

73. $\dfrac{b^2 - 16}{b^2 + b - 12} \div \dfrac{4 - b}{b^2 + 2b - 15}$

74. $\dfrac{y^2 - y - 20}{y^2 - 2y - 24} \div \dfrac{5 - y}{y^2 - 3y - 18}$

★ **For Exercises 75–80, perform the indicated operations. (Remember that the order of operations is to multiply or divide from left to right.)**

75. $\dfrac{y^2 + y - 6}{2y^2 - 3y - 2} \cdot \dfrac{2y^2 + 9y + 4}{y^2 + 7y + 12} \div \dfrac{4y + y^2}{7 + y}$

76. $\dfrac{2m^2 - 7m - 15}{4m^2 - 9} \cdot \dfrac{2m^2 - 5m - 3}{2m^2 - 7m + 3} \div \dfrac{m^2 - 5m}{2m - 1}$

77. $\dfrac{6x^2 - 5x - 6}{x^4 - 81} \div \dfrac{3x - 2x^2}{x^3 + 3x^2 + 9x + 27} \div \dfrac{3x + 2}{x^4}$

78. $\dfrac{12h^2 + 11h - 5}{h^4 - 16} \div \dfrac{h - 3h^2}{h^3 + 4h - 2h^2 - 8} \div \dfrac{4h + 5}{h^3}$

79. $\dfrac{6x^2}{x^2 - 9} \cdot \dfrac{3x^2 + x - 24}{20y - 6xy} \cdot \dfrac{x - 3}{3x - 8} \div \dfrac{3x^2 + 9x}{3x^2 - x - 30}$

80. $\dfrac{8y^3}{3y^2 - 4y - 4} \div \dfrac{8xy - 20x}{2y^2 - 9y + 10} \cdot \dfrac{y + 5}{7 - 4y} \cdot \dfrac{12y^2 - 13y - 14}{2y^2 + 10y}$

Find

ⓧ *the mistake*   **For Exercises 81–84, explain the mistake; then correct the mistake.**

81. $\dfrac{5y}{9} \div \dfrac{3y}{10} = \dfrac{y}{3} \cdot \dfrac{y}{2} = \dfrac{y^2}{6}$

82. $\dfrac{12x^3y}{5} \cdot \dfrac{3y}{6x^2} = \dfrac{2xy}{5} \cdot \dfrac{3y}{1} = \dfrac{6y^2}{5}$

83. $\dfrac{x^2 - 9}{15y} \cdot \dfrac{y - 2}{x + 3} = \dfrac{(x - 3)(x + 3)}{15y} \cdot \dfrac{y - 2}{x + 3}$

$\qquad = \dfrac{x - 3}{15} \cdot (-2)$

$\qquad = \dfrac{-2x + 6}{15}$

84. $\dfrac{x^2 + 3x + 2}{x^2 - 1} \cdot \dfrac{x - 1}{x + 1} = \dfrac{(x + 1)(x + 2)}{1} \cdot \dfrac{1}{x + 1}$

$\qquad = x + 2$

For Exercises 85–90, use dimensional analysis and the following facts to convert units of length. See Example 5.

$\qquad$ 1 foot = 12 inches $\qquad$ 1 yard = 3 feet $\qquad$ 1 mile = 5280 feet

85. 18.5 feet to inches

86. 24.2 miles to feet

87. 6 miles to yards

88. 2400 yards to miles

89. 8 yards to inches

90. 216 inches to yards

For Exercises 91–96, use dimensional analysis and the following values to convert units of weight. See Example 5.

$\qquad$ 1 pound = 16 ounces $\qquad$ 1 ton = 2000 pounds

91. 5.5 pounds to ounces

92. 36 ounces to pounds

93. 25,400 pounds to tons

94. 2.5 tons to pounds

95. 3 tons to ounces

96. 48,000 ounces to tons

For Exercises 97–104, use dimensional analysis. See Example 5.

97. The fastest winning speed at the Daytona 500 was 177.602 miles per hour by Buddy Baker in 1980. How many feet per second is this? (*Source:* www.daytonainternationalspeedway.com)

98. The speed of sound is 741.8 miles per hour at 32°F at sea level. How many feet per second is this?

99. A sprinter's best time in the 40-yard dash is 4.3 seconds. What is that speed in miles per hour?

100. A runner knows that she can run 400 yards in 87.2 seconds. If she can maintain that pace, how many minutes will it take her to run a mile?

101. The mean orbital velocity of Earth is 29.78 kilometers per second. How many kilometers does Earth orbit in one year? (Hint: One year is $365\frac{1}{4}$ days.)

(*Source:* Windows to the Universe)

102. The mean orbital velocity of Pluto is 4.74 kilometers per second. How many kilometers does Pluto orbit in one year? (*Source:* Windows to the Universe)

★**103.** One light-year is the distance light travels in a year. If light travels at a speed of approximately 186,000 miles per second, how many miles does light travel in one year?

★**104.** The light from the Sun takes about 8 minutes to reach Earth. Using the fact that light travels at a speed of about 186,000 miles per second, determine Earth's distance from the Sun in miles.

Puzzle Problem

If $\dfrac{1}{1000}$ of a light-year is about 9.46×10^9 kilometers, find the speed of light in meters per second.

Of Interest

The nearest star to the Sun is Proxima Centauri, which is about 4.2 light-years away. This means that when we look at Proxima Centauri, we are seeing an image of that star that is 4.2 years old. Some objects we see in the night sky are much farther from Earth. For example, the Andromeda galaxy is about 2,140,000 light-years from Earth, which means that when we look at it, we are seeing it as it was 2,140,000 years ago.

For Exercises 105–110, use dimensional analysis and the following table of exchange rates as of 1/11/08 to convert. See Example 5.

	USD($)	GBP(£)	CAN($)	EUR(€)
USD($)	1	1.9580	0.981258	1.47895
GBP(£)	0.510725	1	0.500951	0.755093
CAN($)	1.01910	1.99620	1	1.50761
EUR(€)	0.676155	1.32434	0.6633	1

Source: Federal Reserve Bank of New York

Note: Each number is the ratio of the row's unit to 1 of the column's unit. For example, 1.50761 means that there are 1.50761 Can$ to 1€.

105. Convert $500 into £ (Great Britain pound).

106. Convert $250 into Can$ (Canadian dollars).

107. Convert 450 € (euros) into U.S. dollars.

108. Convert 700 £ into U.S. dollars.

★**109.** Convert 650 £ to €.

110. Convert 400 Can$ to €.

Review Exercises

Exercises 1–6 EXPRESSIONS

For Exercises 1–6, simplify.

[1.3]   **1.** $\dfrac{1}{3} + \dfrac{2}{3}$

[1.3]   **2.** $\dfrac{1}{4} - \dfrac{3}{4}$

[1.7]   **3.** $\dfrac{1}{2}x + \dfrac{3}{2}x$

[1.7]   **4.** $\dfrac{3}{8}y - 5 - \dfrac{1}{8}y + 1$

[6.3]   **5.** $(6x^2 + x - 1) + (2x^2 - 7x - 3)$

[6.3]   **6.** $(3n^2 - 5n + 7) - (4n^2 - 5n + 3)$

8.3 Adding and Subtracting Rational Expressions with the Same Denominator

Objective ❶ Add or subtract rational expressions with the same denominator. Recall that when adding or subtracting fractions with the same denominator, we add or subtract the numerators and keep the same denominator.

$$\frac{3}{7} + \frac{1}{7} = \frac{3+1}{7} = \frac{4}{7} \text{ or } \frac{7}{9} - \frac{2}{9} = \frac{7-2}{9} = \frac{5}{9}$$

We add and subtract rational expressions the same way.

$$\frac{3x}{7} + \frac{x}{7} = \frac{3x+x}{7} = \frac{4x}{7} \text{ or } \frac{7x}{9} - \frac{2x}{9} = \frac{7x-2x}{9} = \frac{5x}{9}$$

Procedure   Adding or Subtracting Rational Expressions (Same Denominator)

To add or subtract rational expressions that have the same denominator,

1. Add or subtract the numerators and keep the same denominator.
2. Simplify to lowest terms. (Remember to factor the numerators and denominators completely in order to simplify.)

Example 1   Add $\dfrac{5x}{14} + \dfrac{x}{14}$.

SOLUTION $\dfrac{5x}{14} + \dfrac{x}{14} = \dfrac{5x+x}{14}$ Because the rational expressions have the same denominator, we add numerators and keep the same denominator.

$$= \frac{6x}{14}$$ Combine like terms.

$$= \frac{2 \cdot 3 \cdot x}{2 \cdot 7}$$ Factor.

$$= \frac{3x}{7}$$ Divide out the common factor, 2.

Your Turn 1   Subtract $\dfrac{2a^2}{15} - \dfrac{7a^2}{15}$.

Numerators with No Like Terms

In Example 1, the numerators were like terms. If the numerators have no like terms, we write a multiterm polynomial expression in the numerator of the sum or difference.

Answer   Your Turn 1

$-\dfrac{a^2}{3}$

Example 2   Subtract.

a. $\dfrac{y}{y + 5} - \dfrac{3}{y + 5}$

SOLUTION $\dfrac{y}{y + 5} - \dfrac{3}{y + 5} = \dfrac{y - 3}{y + 5}$ Subtract numerators and keep the same denominator. Because y and 3 are not like terms, we express their difference as a polynomial.

Warning: Although it may be tempting to do so, we *cannot* divide out the y's because they are terms, not factors.

b. $\dfrac{n^2}{n + 3} - \dfrac{9}{n + 3}$

SOLUTION $\dfrac{n^2}{n + 3} - \dfrac{9}{n + 3} = \dfrac{n^2 - 9}{n + 3}$ **Note:** The numerator can be factored, so we may be able to simplify.

$$= \dfrac{(n - 3)(n + 3)}{n + 3}$$ Factor the numerator.

$$= n - 3$$ Divide out the common factor, $n + 3$.

c. $\dfrac{4m}{15} - \left(-\dfrac{8}{15}\right)$

SOLUTION $\dfrac{4m}{15} - \left(-\dfrac{8}{15}\right) = \dfrac{4m - (-8)}{15}$

$$= \dfrac{4m + 8}{15}$$

Note: $4m + 8$ can be factored.

$$4m + 8 = 4(m + 2)$$

However, because there are no common factors in the denominator, we cannot simplify further.

Your Turn 2   Add or subtract.

a. $\dfrac{x}{x - 4} + \dfrac{6}{x - 4}$

b. $\dfrac{16n^2}{4n + 5} - \dfrac{25}{4n + 5}$

c. $\dfrac{5x^2}{7} - \left(-\dfrac{2x}{7}\right)$

Numerators That Are Multiterm Polynomials

So far, the numerators in the given rational expressions were monomials. Now look at some cases where the numerators in the given rational expressions are multiterm polynomials. Remember from Chapter 6 that to add polynomials, we combine like terms.

Example 3   Add.

a. $\dfrac{7y - 13}{5} + \dfrac{y + 8}{5}$

SOLUTION $\dfrac{7y - 13}{5} + \dfrac{y + 8}{5} = \dfrac{(7y - 13) + (y + 8)}{5}$ Combine $7y$ with y to get $8y$ and combine -13 with 8 to get -5.

$$= \dfrac{8y - 5}{5}$$ **Warning:** It may be tempting to divide out the 5, but the 5 in $8y - 5$ is not a factor. Because $8y - 5$ cannot be factored, we cannot simplify this rational expression.

b. $\dfrac{x^2 - 5x - 13}{x - 6} + \dfrac{3x - 11}{x - 6}$

SOLUTION $\dfrac{x^2 - 5x - 13}{x - 6} + \dfrac{3x - 11}{x - 6} = \dfrac{(x^2 - 5x - 13) + (3x - 11)}{x - 6}$

> **Note:** Because the numerator can be factored, we may be able to simplify.

$$= \dfrac{x^2 - 2x - 24}{x - 6} \qquad \text{Combine like terms in the numerator.}$$

$$= \dfrac{(x - 6)(x + 4)}{x - 6} \qquad \text{Factor the numerator.}$$

$$= x + 4 \qquad \text{Divide out the common factor, } x - 6.$$

Your Turn 3   Add.

a. $\dfrac{n + 7}{11n} + \dfrac{2 - 5n}{11n}$

b. $\dfrac{3t^2 - 5t}{t - 3} + \dfrac{t^2 - 8t + 3}{t - 3}$

Simplifying When the Denominator Is Factorable

Sometimes you also may be able to factor the denominator. If the denominator can be factored, you will need to write it in factored form in order to simplify.

Example 4   Add $\dfrac{4x^2 + 9x + 1}{2x^3 - 18x} + \dfrac{6x^2 + 23x + 5}{2x^3 - 18x}$.

SOLUTION $\dfrac{4x^2 + 9x + 1}{2x^3 - 18x} + \dfrac{6x^2 + 23x + 5}{2x^3 - 18x} = \dfrac{(4x^2 + 9x + 1) + (6x^2 + 23x + 5)}{2x^3 - 18x}$

$$= \dfrac{10x^2 + 32x + 6}{2x^3 - 18x} \qquad \text{Combine like terms in the numerator.}$$

$$= \dfrac{2(5x^2 + 16x + 3)}{2x(x^2 - 9)} \qquad \begin{array}{l}\text{Factor 2 out of the numerator and } 2x \text{ out} \\ \text{of the denominator.}\end{array}$$

$$= \dfrac{2(5x + 1)(x + 3)}{2x(x - 3)(x + 3)} \qquad \begin{array}{l}\text{Factor the trinomial in the numerator} \\ \text{and the difference of squares in the} \\ \text{denominator.}\end{array}$$

$$= \dfrac{5x + 1}{x(x - 3)} \text{ or } \dfrac{5x + 1}{x^2 - 3x} \qquad \begin{array}{l}\text{Divide out the common factors, 2 and} \\ x + 3.\end{array}$$

Your Turn 4   Add $\dfrac{x^3 + 3x^2 - 7x}{3x^4 - 12x^2} + \dfrac{5x - 2x^2}{3x^4 - 12x^2}$.

Subtracting with Numerators That Are Multiterm Polynomials

Now let's look at subtraction of rational expressions that have the same denominator and multiterm polynomials in the numerators. Recall from Chapter 6 that when subtracting polynomials, we write an equivalent addition with the additive inverse of the subtrahend.

Answers **Your Turn 3**

a. $\dfrac{9 - 4n}{11n}$ **b.** $4t - 1$

Answer **Your Turn 4**

$\dfrac{x - 1}{3x(x - 2)}$ or $\dfrac{x - 1}{3x^2 - 6x}$

Example 5   Subtract.

a. $\dfrac{7t^2 - t + 5}{t + 2} - \dfrac{3t^2 - 6t - 2}{t + 2}$

SOLUTION $\dfrac{7t^2 - t + 5}{t + 2} - \dfrac{3t^2 - 6t - 2}{t + 2} = \dfrac{(7t^2 - t + 5) - (3t^2 - 6t - 2)}{t + 2}$

Note: To write an equivalent addition, change the operation symbol from a minus sign to a plus sign and change all of the signs in the subtrahend (second) polynomial.

$$= \dfrac{(7t^2 - t + 5) + (-3t^2 + 6t + 2)}{t + 2}$$

$$= \dfrac{4t^2 + 5t + 7}{t + 2} \qquad \begin{array}{l}\text{Combine like terms.}\\ 4t^2 + 5t + 7 \text{ is prime.}\end{array}$$

b. $\dfrac{x^3 + 4x - 3}{5x^2 - 10x} - \dfrac{4x + 5}{5x^2 - 10x}$

SOLUTION $\dfrac{x^3 + 4x - 3}{5x^2 - 10x} - \dfrac{4x + 5}{5x^2 - 10x} = \dfrac{(x^3 + 4x - 3) - (4x + 5)}{5x^2 - 10x}$

$$= \dfrac{(x^3 + 4x - 3) + (-4x - 5)}{5x^2 - 10x} \qquad \text{Write the equivalent addition.}$$

Note: The numerator is a difference of cubes.

$$= \dfrac{x^3 - 8}{5x^2 - 10x} \qquad \text{Combine like terms.}$$

$$= \dfrac{(x - 2)(x^2 + 2x + 4)}{5x(x - 2)} \qquad \begin{array}{l}\text{Factor the numerator and}\\ \text{denominator. Divide out the}\\ \text{common factor, } x - 2.\end{array}$$

$$= \dfrac{x^2 + 2x + 4}{5x} \qquad x^2 + 2x + 4 \text{ is prime.}$$

Your Turn 5   Subtract.

Answers   Your Turn 5

a. $\dfrac{t - 1}{t + 7}$ b. $\dfrac{y^2 + 3y + 9}{3y + 2}$

a. $\dfrac{3t^2 - 4t - 7}{t^2 + 10t + 21} - \dfrac{2t^2 - 6t - 4}{t^2 + 10t + 21}$

b. $\dfrac{y^3 + y - 20}{3y^2 - 7y - 6} - \dfrac{y + 7}{3y^2 - 7y - 6}$

8.3 Exercises   For Extra Help   *MyMathLab*
PRACTICE   WATCH   DOWNLOAD

1. Two rational expressions with monomial numerators are added to form a rational expression whose numerator is a multiterm polynomial. Why would this happen?

2. After adding or subtracting two rational expressions, what is the first step in simplifying the resulting rational expression?

3. Two rational expressions that have numerators that are multiterm polynomials and the same denominator are to be subtracted. Explain the process.

4. After adding or subtracting two rational expressions, the resulting rational expression contains a multiterm polynomial in the numerator that can be factored. Does this mean that the rational expression can be simplified? Explain.

For Exercises 5–50, add or subtract. Simplify your answers to lowest terms. See Examples 1–5.

5. $\dfrac{3x}{10} + \dfrac{2x}{10}$

6. $\dfrac{2x}{9} + \dfrac{x}{9}$

7. $\dfrac{4a^2}{9} + \left(-\dfrac{a^2}{9}\right)$

8. $\dfrac{6x^3}{11} + \left(-\dfrac{x^3}{11}\right)$

9. $\dfrac{6}{a} - \left(-\dfrac{3}{a}\right)$

10. $\dfrac{3}{w} - \left(-\dfrac{2}{w}\right)$

11. $\dfrac{6x}{7y^2} - \dfrac{2x}{7y^2} - \dfrac{5x}{7y^2}$

12. $\dfrac{9a}{11x^2} - \dfrac{8a}{11x^2} - \dfrac{3a}{11x^2}$

13. $\dfrac{n}{n+5} + \dfrac{3}{n+5}$

14. $\dfrac{n}{n+2} + \dfrac{7}{n+2}$

15. $\dfrac{2r}{r+2} + \dfrac{r}{r+2}$

16. $\dfrac{5z}{b-6} + \dfrac{7z}{b-6}$

17. $\dfrac{1-2q}{3q} + \dfrac{5q+2}{3q}$

18. $\dfrac{6x-3}{4x} + \dfrac{2x+5}{4x}$

19. $\dfrac{19u+6}{5u} - \dfrac{4u+6}{5u}$

20. $\dfrac{8y+5}{3y} - \dfrac{2y+5}{3y}$

21. $\dfrac{2-2a}{2a-3} + \dfrac{8a-11}{2a-3}$

22. $\dfrac{4-2x}{3x-2} + \dfrac{14x-12}{3x-2}$

23. $\dfrac{w^2+5}{w+1} + \dfrac{6-w^2}{w+1}$

24. $\dfrac{x^2+2}{x+1} + \dfrac{6-x^2}{x+1}$

25. $\dfrac{4m+3}{m-1} - \dfrac{m+4}{m-1}$

26. $\dfrac{2x+5}{x+1} - \dfrac{x+8}{x+1}$

27. $\dfrac{1-3t}{2t-3} - \dfrac{8t-2}{2t-3}$

28. $\dfrac{3-4a}{5a-2} - \dfrac{2a-6}{5a-2}$

29. $\dfrac{2w+3}{w+5} + \dfrac{5w-2}{w+5} - \dfrac{w+6}{w+5}$

30. $\dfrac{6q+21}{q+1} - \dfrac{2q+8}{q+1} + \dfrac{6q-5}{q+1}$

31. $\dfrac{g}{g^2-9} - \dfrac{3}{g^2-9}$

32. $\dfrac{x}{x^2-49} + \dfrac{7}{x^2-49}$

33. $\dfrac{z^2+2z}{z-7} + \dfrac{21-12z}{z-7}$

34. $\dfrac{u^2+5u}{u+1} + \dfrac{2-2u}{u+1}$

35. $\dfrac{2s}{s^2+4s+4} - \dfrac{s-2}{s^2+4s+4}$

36. $\dfrac{v+8}{v^2+8v+16} - \dfrac{4}{v^2+8v+16}$

37. $\dfrac{y^2}{y^2+3y} - \dfrac{9}{y^2+3y}$

38. $\dfrac{x^2}{x^2-5x} - \dfrac{25}{x^2-5x}$

39. $\dfrac{x^2+3x}{x^2-x-12} + \dfrac{2x+6}{x^2-x-12}$

40. $\dfrac{y^2+2y}{y^2+5y+4} + \dfrac{5y+12}{y^2+5y+4}$

41. $\dfrac{t^2-4t}{t^2+10t+21} - \dfrac{-6t+3}{t^2+10t+21}$

42. $\dfrac{v^2+6}{v^2-3v-10} - \dfrac{2v+14}{v^2-3v-10}$

43. $\dfrac{3x^3}{x^2-2x+4} + \dfrac{24}{x^2-2x+4}$

44. $\dfrac{2x^3}{x^2-4x+16} + \dfrac{128}{x^2-4x+16}$

45. $\dfrac{x-5}{x^2-25} + \dfrac{2x+10}{x^2-25} + \dfrac{-2x}{x^2-25}$

46. $\dfrac{3x-3}{x^2-1} + \dfrac{2x+2}{x^2-1} + \dfrac{-4x}{x^2-1}$

47. $\dfrac{3x^2+2x}{x^2-2x-15} + \dfrac{3x-8}{x^2-2x-15} + \dfrac{2x+2}{x^2-2x-15}$

48. $\dfrac{m^2-6m}{m^2-5m+6} + \dfrac{2m-4}{m^2-5m+6} + \dfrac{6m-4}{m^2-5m+6}$

49. $\dfrac{2x^2+x-3}{x^2+6x+5} + \dfrac{x^2-2x+4}{x^2+6x+5} - \dfrac{2x^2+x+4}{x^2+6x+5}$

50. $\dfrac{3x^2-2x+3}{x^2+x-12} - \dfrac{x^2+x-2}{x^2+x-12} - \dfrac{x^2+4x-7}{x^2+x-12}$

Find

the mistake   *For Exercises 51–54, explain the mistake; then find the correct sum or difference.*

51. $\dfrac{4y}{5} + \dfrac{1}{5} = \dfrac{5y}{5} = y$

52. $\dfrac{7}{m} + \dfrac{n}{m} = \dfrac{7+n}{2m}$

53.
$$\dfrac{x+1}{3} - \dfrac{x-5}{3} = \dfrac{(x+1)-(x-5)}{3}$$
$$= \dfrac{(x+1)+(-x-5)}{3}$$
$$= -\dfrac{4}{3}$$

54.
$$\dfrac{x}{x-2} + \dfrac{x-4}{x-2} = \dfrac{x^2-4}{x-2}$$
$$= \dfrac{(x+2)(x-2)}{x-2}$$
$$= x+2$$

For Exercises 55–58, find the perimeter of the figure shown.

55.

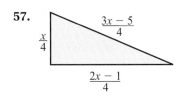

$\dfrac{3x}{4}$ $\dfrac{2x+9}{4}$

56.

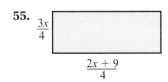

$\dfrac{5x}{2}$

57.

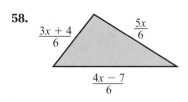

$\dfrac{x}{4}$ $\dfrac{3x-5}{4}$ $\dfrac{2x-1}{4}$

58.

$\dfrac{3x+4}{6}$ $\dfrac{5x}{6}$ $\dfrac{4x-7}{6}$

Review Exercises

Exercises 1–6 ◣◣◣ EXPRESSIONS

For Exercises 1–6, simplify.

[1.3]   **1.** $\dfrac{1}{2} + \dfrac{3}{5}$ [1.3]   **2.** $\dfrac{5}{6} - \dfrac{3}{8}$ [1.3]   **3.** $\dfrac{4}{7} - \left(-\dfrac{3}{5}\right)$ [1.7]   **4.** $\dfrac{1}{3}x - \dfrac{1}{6} + \dfrac{2}{5}x + \dfrac{3}{4}$

[6.3]   **5.** $\left(x^2 - 9x - \dfrac{5}{8}\right) + \left(x^2 - 3x + \dfrac{1}{6}\right)$ [6.3]   **6.** $\left(3y^2 + \dfrac{3}{4}y - 8\right) - \left(y^2 + \dfrac{4}{5}y + 2\right)$

8.4 Adding and Subtracting Rational Expressions with Different Denominators

Objectives

1. Find the LCD of two or more rational expressions.
2. Given two rational expressions, write equivalent rational expressions with their LCD.
3. Add or subtract rational expressions with different denominators.

Adding or subtracting rational expressions with different denominators follows three main stages: (1) finding the LCD, (2) rewriting the expressions with the LCD, and (3) adding or subtracting the numerators. We explore the first two stages in Objectives 1 and 2, then put them back in their proper context in Objective 3.

Objective ❶ Find the LCD of two or more rational expressions.   Remember that when adding or subtracting fractions with different denominators, we must first find a common denominator. It is helpful to use the least common denominator (LCD), which is the smallest number that is evenly divisible by all of the denominators. For example, given the fractions $\dfrac{1}{8}$ and $\dfrac{7}{12}$, the LCD is 24.

To help us find the LCD when denominators are more complex, as in rational expressions, let's consider a method involving prime factorizations. Consider the prime factorizations of the denominators, 8 and 12, compared to the prime factorization of the LCD, 24, to determine a procedure.

$$8 = 2^3 \qquad 12 = 2^2 \cdot 3$$

$$\text{LCD} = 2^3 \cdot 3 = 24$$

Both 8 and 12 have factors of 2, but 8 has the greatest number of 2s, which is why 2^3 is included in the factorization for the LCD instead of 2^2. Similarly, 12 has the greatest number of factors of 3 (a single factor, whereas 8 has no factors of 3), so a single factor of 3 is included in the factorization of the LCD. This suggests the following procedure.

Procedure Finding the LCD

To find the LCD of two or more rational expressions,

1. Find the prime factorization of each denominator.
2. Write a product that contains each unique prime factor the greatest number of times that factor occurs in any factorization. Or if you prefer to use exponents, write the product that contains each unique prime factor raised to the greatest exponent that occurs on that factor in any factorization.
3. Simplify the product found in step 2.

First, let's consider cases where the denominators are monomials.

Example 1 Find the LCD of $\dfrac{3}{4x^2}$ and $\dfrac{5}{6x^3}$.

SOLUTION We first factor the denominators $4x^2$ and $6x^3$ by writing their prime factorizations.

$$4x^2 = 2^2 \cdot x^2$$
$$6x^3 = 2 \cdot 3 \cdot x^3$$

The unique factors are 2, 3, and x. To generate the LCD, include 2, 3, and x the greatest number of times each appears in any of the factorizations.

The greatest number of times that 2 appears is twice (in $2^2 \cdot x^2$).
The greatest number of times that 3 appears is once (in $2 \cdot 3 \cdot x^3$).
The greatest number of times that x appears is three times (in $2 \cdot 3 \cdot x^3$).

$$\text{LCD} = 2^2 \cdot 3 \cdot x^3 = 12x^3$$

Your Turn 1 Find the LCD.

a. $\dfrac{y}{12x^5}$ and $\dfrac{5}{9x^2}$

b. $\dfrac{p}{10mn}$ and $\dfrac{7}{15n^2}$

Now consider denominators that are multiterm polynomials.

Example 2 Find the LCD.

a. $\dfrac{7}{5x - 15}$ and $\dfrac{5}{2x - 6}$

SOLUTION Factor the denominators $5x - 15$ and $2x - 6$.

$$5x - 15 = 5(x - 3)$$
$$2x - 6 = 2(x - 3)$$

The unique factors are 2, 5, and $x - 3$.
The greatest number of times that $x - 3$ appears in any of the factorizations is once.
The greatest number of times that 2 appears is once [in $2(x - 3)$].
The greatest number of times that 5 appears is once [in $5(x - 3)$].

$$\text{LCD} = 2 \cdot 5 \cdot (x - 3) = 10(x - 3) \text{ or } 10x - 30$$

Answers Your Turn 1

a. $36x^5$ b. $30mn^2$

b. $\dfrac{7}{x^2 + 10x + 25}$ and $\dfrac{17}{6x^4 + 30x^3}$

SOLUTION Factor $x^2 + 10x + 25$ and $6x^4 + 30x^3$.

$$x^2 + 10x + 25 = (x + 5)^2$$
$$6x^4 + 30x^3 = 6x^3(x + 5) = 2 \cdot 3 \cdot x^3(x + 5)$$

The factors are 2, 3, x, and $x + 5$.

The greatest number of times that 2 appears is once [in $2 \cdot 3 \cdot x^3(x + 5)$].

The greatest number of times that 3 appears is once [in $2 \cdot 3 \cdot x^3(x + 5)$].

The greatest number of times that $x + 5$ appears is twice [in $(x + 5)^2$].

The greatest number of times that x appears is three times [in $2 \cdot 3 \cdot x^3(x + 5)$].

$$\text{LCD} = 2 \cdot 3 \cdot x^3 \cdot (x + 5)^2 = 6x^3(x + 5)^2$$

Your Turn 2 Find the LCD.

a. $\dfrac{y}{3y + 6}$ and $\dfrac{5}{4y + 8}$

b. $\dfrac{9}{x^2 - 2x + 1}$ and $\dfrac{17}{2x^3 - 2x}$

Objective 2 **Given two rational expressions, write equivalent rational expressions with their LCD.** Writing equivalent rational expressions is much like writing equivalent fractions, which we reviewed in Section 1.2. Remember that to write an equivalent fraction, we multiply the numerator and denominator by the same nonzero number.

Example 3 Write $\dfrac{3}{4x^2}$ and $\dfrac{5}{6x^3}$ as equivalent rational expressions with their LCD.

SOLUTION In Example 1, we found the LCD to be $12x^3$. For each rational expression, we multiply both the numerator and denominator by an appropriate factor so that the denominator becomes $12x^3$.

$$\dfrac{3}{4x^2} = \dfrac{3 \cdot 3x}{4x^2 \cdot 3x} = \dfrac{9x}{12x^3}$$

$$\dfrac{5}{6x^3} = \dfrac{5 \cdot 2}{6x^3 \cdot 2} = \dfrac{10}{12x^3}$$

Note: We can determine the appropriate factor by dividing the LCD by the original denominator. In this case, $12x^3 \div 4x^2 = 3x$.

Now consider rational expressions with denominators that are multiterm polynomials. Rewriting these rational expressions with their LCD is usually easier if we leave their denominators in factored form.

Example 4 Write $\dfrac{7}{x^2 + 10x + 25}$ and $\dfrac{17}{6x^4 + 30x^3}$ as equivalent rational expressions with their LCD.

SOLUTION In Example 2(b), we found the LCD to be $6x^3(x + 5)^2$.

$$\frac{7}{x^2 + 10x + 25} = \frac{7}{(x + 5)^2}$$

Factor the denominator.

Note: Another way to determine the appropriate factor is to think of it as the factor in the LCD that is missing from the original denominator. In this case, $6x^3$ is in the LCD, but not in the original denominator.

$$= \frac{7 \cdot 6x^3}{(x + 5)^2 \cdot 6x^3}$$

Multiply the numerator and denominator by the same factor, $6x^3$, to get the LCD, $6x^3(x + 5)^2$.

$$= \frac{42x^3}{6x^3(x + 5)^2}$$

Note: Although we could multiply out $6x^3(x + 5)^2$, it is simpler to leave it in this form.

$$\frac{17}{6x^4 + 30x^3} = \frac{17}{6x^3(x + 5)}$$

Factor the denominator.

Note: There are two factors of $x + 5$ in the LCD, whereas only one factor of $x + 5$ appears in the original denominator.

$$= \frac{17 \cdot (x + 5)}{6x^3(x + 5) \cdot (x + 5)}$$

Multiply the numerator and denominator by the same factor, $x + 5$, to get the LCD, $6x^3(x + 5)^2$.

$$= \frac{17(x + 5)}{6x^3(x + 5)^2}$$

Answers   Your Turn 4

a. $\dfrac{3np}{30mn^2}$ and $\dfrac{14m}{30mn^2}$

b. $\dfrac{18x(x + 1)}{2x(x - 1)^2(x + 1)}$ and $\dfrac{17(x - 1)}{2x(x - 1)^2(x + 1)}$

Your Turn 4 Write equivalent rational expressions with their LCD. You found the LCD of these expressions in Your Turns 1b and 2b.

a. $\dfrac{p}{10mn}$ and $\dfrac{7}{15n^2}$

b. $\dfrac{9}{x^2 - 2x + 1}$ and $\dfrac{17}{2x^3 - 2x}$

Objective ③ Add or subtract rational expressions with different denominators. To add or subtract rational expressions with different denominators, we use the same process that we use to add or subtract fractions.

Procedure   Adding or Subtracting Rational Expressions with Different Denominators

To add or subtract rational expressions with different denominators,

1. Find the LCD.
2. Write each rational expression as an equivalent expression with the LCD.
3. Add or subtract the numerators and keep the LCD.
4. Simplify.

Example 5 Add or subtract as indicated.

a. $\dfrac{3}{4x^2} + \dfrac{5}{6x^3}$

Note: Because $9x$ and 10 are not like terms, we must indicate the addition as a polynomial expression. Also, because $9x + 10$ cannot be factored, we cannot simplify this rational expression.

SOLUTION We found the LCD for these expressions in Example 1. We then wrote equivalent expressions in Example 3. Now we add the numerators and keep the LCD.

$$\frac{3}{4x^2} + \frac{5}{6x^3} = \frac{9x}{12x^3} + \frac{10}{12x^3}$$

Rewrite with the LCD.

$$= \frac{9x + 10}{12x^3}$$

Add numerators.

b. $\dfrac{7}{x^2 + 10x + 25} - \dfrac{17}{6x^4 + 30x^3}$

SOLUTION   We found the LCD for these expressions in Example 2(b) and wrote equivalent expressions in Example 4. Now we combine the numerators and keep the LCD.

Note: Leaving the LCD in factored form helps us determine whether the final expression can be reduced. Because $42x^3 - 17x - 85$ cannot be factored, we cannot reduce this rational expression.

$$\dfrac{7}{x^2 + 10x + 25} - \dfrac{17}{6x^4 + 30x^3} = \dfrac{42x^3}{6x^3(x + 5)^2} - \dfrac{17(x + 5)}{6x^3(x + 5)^2}$$ Rewrite with the LCD.

Warning: Do not divide out the $(x + 5)$'s because $(x + 5)$ is not a factor of the numerator.
$$= \dfrac{42x^3 - 17(x + 5)}{6x^3(x + 5)^2}$$ Subtract numerators.

$$= \dfrac{42x^3 - 17x - 85}{6x^3(x + 5)^2}$$ Simplify in the numerator.

Now that we have seen all of the pieces of the process spread out over several examples, let's put them together using new rational expressions.

Example 6   Add or subtract as indicated.

a. $\dfrac{2x}{9y^3} + \dfrac{3}{4xy}$

SOLUTION   First, find the LCD.

$$4xy = 2^2 \cdot x \cdot y \qquad 9y^3 = 3^2 \cdot y^3 \qquad \text{LCD} = 2^2 \cdot 3^2 \cdot x \cdot y^3 = 36xy^3$$

$$\dfrac{2x}{9y^3} + \dfrac{3}{4xy} = \dfrac{2x(4x)}{9y^3(4x)} + \dfrac{3(9y^2)}{4xy(9y^2)}$$ Write equivalent rational expressions with the LCD, $36xy^3$.

$$= \dfrac{8x^2}{36xy^3} + \dfrac{27y^2}{36xy^3}$$

$$= \dfrac{8x^2 + 27y^2}{36xy^3}$$ Add numerators.

Note: The numerator cannot be factored, so we cannot reduce.

b. $\dfrac{4}{m - 5} - \dfrac{3}{m + 2}$

SOLUTION   First, find the LCD. Because both $m - 5$ and $m + 2$ are prime, the LCD is their product, $(m - 5)(m + 2)$.

$$\dfrac{4}{m - 5} - \dfrac{3}{m + 2} = \dfrac{4(m + 2)}{(m - 5)(m + 2)} - \dfrac{3(m - 5)}{(m + 2)(m - 5)}$$ Write equivalent rational expressions with the LCD, $(m - 5)(m + 2)$.

$$= \dfrac{4m + 8}{(m - 5)(m + 2)} - \dfrac{3m - 15}{(m - 5)(m + 2)}$$ Multiply numerators.

$$= \dfrac{(4m + 8) - (3m - 15)}{(m - 5)(m + 2)}$$ Subtract numerators.

$$= \dfrac{(4m + 8) + (-3m + 15)}{(m - 5)(m + 2)}$$ Rewrite the subtraction as addition.

$$= \dfrac{m + 23}{(m - 5)(m + 2)}$$ Combine like terms.

c. $\dfrac{x}{x+2} - \dfrac{10}{x^2-x-6}$

SOLUTION   First, find the LCD by factoring $x+2$ and x^2-x-6.

Factored forms: $x+2$ cannot be factored; $x^2-x-6 = (x+2)(x-3)$
LCD $= (x+2)(x-3)$

$$\dfrac{x}{x+2} - \dfrac{10}{x^2-x-6} = \dfrac{x}{x+2} - \dfrac{10}{(x+2)(x-3)}$$

Factor the denominators.

$$= \dfrac{x(x-3)}{(x+2)(x-3)} - \dfrac{10}{(x+2)(x-3)}$$

Write equivalent rational expressions with the LCD, $(x+2)(x-3)$.

$$= \dfrac{x^2-3x}{(x+2)(x-3)} - \dfrac{10}{(x+2)(x-3)}$$

Distribute.

$$= \dfrac{x^2-3x-10}{(x+2)(x-3)}$$

Subtract numerators.

$$= \dfrac{(x+2)(x-5)}{(x+2)(x-3)}$$

Factor the numerator.

$$= \dfrac{x-5}{x-3}$$

Divide out the common factor, $x+2$.

d. $\dfrac{x-1}{x^2+4x} - \dfrac{3x+1}{x^2-16}$

SOLUTION   To determine the LCD, we write the factored forms of x^2+4x and x^2-16.

Factored forms: $x^2+4x = x(x+4)$; $x^2-16 = (x+4)(x-4)$
LCD $= x(x+4)(x-4)$

$$\dfrac{x-1}{x^2+4x} - \dfrac{3x+1}{x^2-16} = \dfrac{x-1}{x(x+4)} - \dfrac{3x+1}{(x+4)(x-4)}$$

Factor the denominators. Write equivalent rational expressions with the LCD, $x(x+4)(x-4)$.

$$= \dfrac{(x-1)(x-4)}{x(x+4)(x-4)} - \dfrac{(3x+1)(x)}{(x+4)(x-4)(x)}$$

$$= \dfrac{x^2-5x+4}{x(x+4)(x-4)} - \dfrac{3x^2+x}{x(x+4)(x-4)}$$

Multiply in the numerator.

$$= \dfrac{(x^2-5x+4)-(3x^2+x)}{x(x+4)(x-4)}$$

Subtract numerators.

Note: Remember that to subtract polynomials, we write an equivalent addition.

$$= \dfrac{(x^2-5x+4)+(-3x^2-x)}{x(x+4)(x-4)}$$

$$= \dfrac{-2x^2-6x+4}{x(x+4)(x-4)}$$

Note: Although the numerator can be factored to $-2(x^2+3x-2)$, none of those factors match any factor in the denominator; so the result is in lowest terms.

Your Turn 6   Add or subtract as indicated.

a. $\dfrac{5}{6x^2y} + \dfrac{3}{8xy^3}$ **b.** $\dfrac{7}{x+3} - \dfrac{3}{x-4}$ **c.** $\dfrac{x}{x+3} - \dfrac{-6x+24}{x^2-x-12}$ **d.** $\dfrac{2}{x^2+x} - \dfrac{x-4}{x^2-1}$

If denominators are additive inverses, we can multiply the numerator and denominator of either rational expression by -1 to obtain the LCD.

Example 7 Add $\dfrac{5x}{x-3} + \dfrac{2}{3-x}$.

SOLUTION $\dfrac{5x}{x-3} + \dfrac{2}{3-x} = \dfrac{5x}{x-3} + \dfrac{2(-1)}{(3-x)(-1)}$

$= \dfrac{5x}{x-3} + \dfrac{-2}{x-3}$

$= \dfrac{5x-2}{x-3}$

Because $x-3$ and $3-x$ are additive inverses, we obtain the LCD by multiplying the numerator and denominator of one of the rational expressions by -1. We chose the second rational expression.

Your Turn 7 Add or subtract as indicated.

a. $\dfrac{7y}{2y-3} - \dfrac{4}{3-2y}$

b. $\dfrac{3x}{x-2} + \dfrac{6}{2-x}$

8.4 Exercises For Extra Help *MyMathLab*
PRACTICE   WATCH   DOWNLOAD

Note: Exercises marked with a ★ represent challenging exercises.

1. Explain how to find the LCD of two rational expressions.

2. Suppose the LCD of two fractions is $48x^3yz$. If one of the fraction's denominators is $8x^2y$, by what factor do you multiply when writing it as an equivalent fraction with the LCD? Explain how you determine this factor.

3. In general, if you are given rational expressions that contain multiterm polynomials in the denominator(s), what is your first step toward finding the LCD?

4. Instead of using the LCM as a common denominator, a student rewrites $\dfrac{3}{2x-y} + \dfrac{5}{y-2x}$ as $\dfrac{-3}{y-2x} + \dfrac{5}{y-2x}$. Explain what the student did.

5. Suppose you have found the LCD of two rational expressions and have written the equivalent rational expressions. If the problem is now at the stage $\dfrac{(7x+21)-(2x-10)}{(x-5)(x+3)}$, how do you proceed?

★ **6.** Suppose we have found the LCD of two rational expressions and have written the equivalent rational expressions. If the problem is now at the stage $\dfrac{(7x+21)-(2x-10)}{(x-5)(x+3)}$, what might the original problem have been? Explain.

For Exercises 7–32, find the least common denominator for the rational expressions and write equivalent rational expressions with the LCD. See Examples 1–4.

7. $\dfrac{2}{n}, \dfrac{3}{m}$

8. $\dfrac{6}{r}, \dfrac{5}{t}$

9. $\dfrac{2t}{a^2b^5}, \dfrac{3z}{a^3b^3}$

10. $\dfrac{6u}{h^4k^2}, \dfrac{4v}{h^3k^7}$

11. $\dfrac{2}{3a}, \dfrac{3}{5a^2}$

12. $\dfrac{5}{4b^2}, \dfrac{3}{5b}$

13. $\dfrac{3}{7x^2y}, \dfrac{5}{14xy^2}$

14. $\dfrac{2}{9mn^3}, \dfrac{7}{27m^3n}$

15. $\dfrac{8}{15m^3n^3}, \dfrac{7}{18m^2n}$

16. $\dfrac{5}{10x^2y^3}, \dfrac{3}{14x^2y^2}$

17. $\dfrac{3}{y+2}, \dfrac{4y}{y-2}$

18. $\dfrac{3x}{x-5}, \dfrac{2}{x+1}$

19. $\dfrac{3y}{4y-20}, \dfrac{7y}{6y-30}$

20. $\dfrac{2x}{6x-12}, \dfrac{3x}{8x-16}$

21. $\dfrac{5}{6x+18}, \dfrac{2}{x^2+3x}$

22. $\dfrac{3}{4w-8}, \dfrac{5}{w^2-2w}$

23. $\dfrac{2}{p^2-4}, \dfrac{3+p}{p+2}$

24. $\dfrac{5}{y^2-36}, \dfrac{y-2}{y-6}$

25. $\dfrac{4u}{(u+2)^2}, \dfrac{2}{3(u+2)}$

26. $\dfrac{7m}{(x-3)^2}, \dfrac{4}{7(x-3)}$

27. $\dfrac{2t-4}{t^2-36}, \dfrac{3t+2}{t^2-7t+6}$

28. $\dfrac{4a+2}{a^2-9}, \dfrac{2a-1}{a^2-a-6}$

29. $\dfrac{x+2}{x^2+3x-4}, \dfrac{x-3}{x^2+6x+8}$

30. $\dfrac{b-4}{b^2-2b-35}, \dfrac{b+3}{b^2+7b+10}$

31. $\dfrac{2}{x-1}, \dfrac{3}{x^2-3x+2}, \dfrac{5x}{x-2}$

32. $\dfrac{3}{y-3}, \dfrac{2}{y^2+2y-15}, \dfrac{4y}{y+5}$

For Exercises 33–70, add or subtract as indicated. See Examples 5–7.

33. $\dfrac{3a - 5}{9} - \dfrac{2a + 7}{6}$

34. $\dfrac{3x - 4}{6} - \dfrac{2x - 1}{4}$

35. $\dfrac{5}{2c} + \dfrac{1}{6c}$

36. $\dfrac{5}{8r} + \dfrac{3}{4r}$

37. $\dfrac{5}{x} - \dfrac{2}{x^3}$

38. $\dfrac{5a}{b^2} - \dfrac{3a}{b}$

39. $\dfrac{3}{y + 2} + \dfrac{5}{y - 4}$

40. $\dfrac{2}{c + 4} + \dfrac{3}{c + 3}$

41. $\dfrac{3}{x - 2} - \dfrac{2}{x + 2}$

42. $\dfrac{4}{x + 3} - \dfrac{3}{x - 3}$

43. $\dfrac{4}{x^2 - 25} + \dfrac{2}{x - 5}$

44. $\dfrac{5}{r^2 - 9} + \dfrac{4}{r + 3}$

45. $\dfrac{2p}{p^2 - 2p + 1} + \dfrac{8}{p - 1}$

46. $\dfrac{u}{u - 1} + \dfrac{2u}{u^2 - 2u + 1}$

47. $\dfrac{z + 5}{z^2 + z - 12} - \dfrac{z + 2}{z - 3}$

48. $\dfrac{a + 6}{a^2 + 8a + 15} - \dfrac{a - 3}{a + 3}$

49. $\dfrac{n + 2}{n - 4} + \dfrac{n - 3}{n + 3}$

50. $\dfrac{y - 8}{y + 5} + \dfrac{y - 9}{y - 6}$

51. $\dfrac{t + 2}{t + 4} - \dfrac{2t - 1}{t + 6}$

52. $\dfrac{3s + 1}{s - 2} - \dfrac{4s + 1}{s - 3}$

53. $\dfrac{c + 5}{c^2 + 10c + 25} + \dfrac{3}{2c + 10}$

54. $\dfrac{k + 3}{k^2 + 6k + 9} - \dfrac{7}{2k + 6}$

55. $\dfrac{5y + 6}{y^2 + 3y} + \dfrac{y}{y + 3}$

56. $\dfrac{2b - 8}{b^2 + 4b} + \dfrac{b}{b + 4}$

57. $\dfrac{3r}{r^2 - 1} - \dfrac{2r + 1}{r^2 - 2r + 1}$

58. $\dfrac{x + 1}{x^2 - 4x + 4} - \dfrac{4x}{x^2 + 3x - 10}$

59. $\dfrac{q + 5}{q^2 - 9} - \dfrac{1}{q^2 + 3q}$

60. $\dfrac{x - 2}{x^2 - 16} - \dfrac{1}{x^2 - 4x}$

61. $\dfrac{w - 7}{w^2 + 4w - 5} - \dfrac{w - 9}{w^2 + 3w - 10}$

62. $\dfrac{y - 1}{y^2 - 3y + 2} - \dfrac{y + 2}{y^2 + y - 2}$

63. $\dfrac{2v + 5}{v^2 - 16} - \dfrac{v - 9}{v^2 - v - 12}$

64. $\dfrac{4t}{t^2 - 9} - \dfrac{3t - 2}{t^2 - 8t + 15}$

65. $\dfrac{3}{t - 3} + \dfrac{6}{3 - t}$

66. $\dfrac{10}{2r - 1} - \dfrac{6}{1 - 2r}$

67. $\dfrac{4}{2x - 1} + \dfrac{x}{1 - 2x}$

68. $\dfrac{3}{t - 4} + \dfrac{t}{4 - t}$

69. $\dfrac{y}{3y - 6} - \dfrac{2}{2 - y}$

70. $\dfrac{x}{6x - 2} - \dfrac{3x}{1 - 3x}$

For Exercises 71–76, explain the mistake; then find the correct sum or difference.

71. $\dfrac{3}{x} - \dfrac{x}{2} = \dfrac{3-x}{x-2}$

72. $\dfrac{5}{2x} + \dfrac{3}{\cancel{10}x} = \dfrac{1}{2x} + \dfrac{3}{2x}$

$= \dfrac{4}{2x}$

$= \dfrac{2}{x}$

73. $\dfrac{3x}{x+2} - \dfrac{4x+7}{x+2} = \dfrac{3x - (4x+7)}{x+2}$

$= \dfrac{3x + (-4x+7)}{x+2}$

$= \dfrac{-x+7}{x+2}$

74. $\dfrac{3}{x} + \dfrac{2}{x+y} = \dfrac{3(+y)}{x(+y)} + \dfrac{2}{x+y}$

$= \dfrac{3+y+2}{x+y}$

$= \dfrac{5+y}{x+y}$

75. $\dfrac{2w}{x} + \dfrac{5w}{x} = \dfrac{7w}{2x}$

76. $\dfrac{2}{x} + \dfrac{3}{y} = \dfrac{6}{xy}$

For Exercises 77–82, solve.

77. Adam and Candace build and install cabinets in newly constructed homes. When they work together, the expression $\dfrac{t}{2}$ represents the portion of the job completed by Adam and $\dfrac{t}{3}$ represents the portion completed by Candace. Write a rational expression in simplest form that describes their combined work.

78. Two construction teams are combined in an effort to speed construction on a building. The expression $\dfrac{t}{5}$ describes the portion of the job that team 1 can complete and $\dfrac{t}{6}$ describes the portion of the job that team 2 can complete working independently. Write a rational expression in simplest form that describes their combined work.

79. An engineer uses the expression $\dfrac{1}{x+3}$ to describe the width of a rectangular steel plate she is designing for a machine. The expression $\dfrac{1}{x}$ describes the length. Write a rational expression in simplest form for the perimeter of the steel plate.

80. An electrical engineer is trying to determine the effects of using a thinner layer of a chemical on a circuit board. Currently, the thickness of the chemical layer is described by $\dfrac{1}{n-4}$. If the thinner layer is described by $\dfrac{1}{n-2}$, write a rational expression that describes the difference in the coating thickness.

81. Write a rational expression in simplest form for the area of the shaded region in the figure shown.

82. Write a rational expression in simplest form for the area of the figure shown.

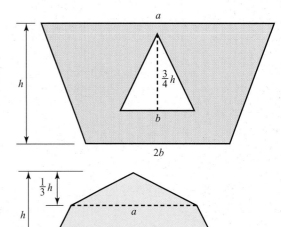

★ *For Exercises 83–92, use the order-of-operations agreement to simplify.*

83. $\dfrac{x}{4x^2 - 1} + \dfrac{2}{4x - 2} - \dfrac{3x + 1}{1 - 2x}$

84. $\dfrac{c}{4c^2 - 9} - \dfrac{2c + 1}{3 - 2c} + \dfrac{3}{6c + 9}$

85. $(x + y) \div \left(\dfrac{1}{x} + \dfrac{1}{y} \right)$

86. $\left(\dfrac{1}{a} + \dfrac{1}{b} \right) \div (a + b)$

87. $\dfrac{2}{a} + \dfrac{3}{4a + 8} \cdot \dfrac{4a}{5}$

88. $\dfrac{3}{x} + \dfrac{1}{x^2 + 3x} \cdot \dfrac{2x}{3}$

89. $\left(\dfrac{a + b}{a - b} + \dfrac{a - b}{a + b} \right) \div \dfrac{a + b}{a - b}$

90. $\left(\dfrac{2x - y}{2x + y} + \dfrac{2x + y}{2x - y} \right) \div \dfrac{2x - y}{2x + y}$

91. $\left(\dfrac{1}{x + 2} - \dfrac{3}{x^2 - 4} \right) \div \dfrac{3}{x - 2}$

92. $\dfrac{6}{x + 4} \div \left(\dfrac{1}{x - 3} - \dfrac{x + 1}{x^2 + x - 12} \right)$

Review Exercises

Exercises 1–6 **EXPRESSIONS**

For Exercises 1 and 2, divide.

[1.4] **1.** $\dfrac{1}{4} \div \dfrac{2}{3}$

[8.2] **2.** $\dfrac{m}{6} \div \dfrac{18}{m}$

For Exercises 3 and 4, factor completely.

[7.4] **3.** $4x^2 - 16$

[7.3] **4.** $6y^3 - 27y^2 - 15y$

[8.1] *For Exercises 5 and 6, simplify.*

5. $\dfrac{x^2 - 7x + 6}{x^2 - 4x - 12}$

6. $\dfrac{x + 3}{x^2 - 9}$

8.5 Complex Rational Expressions

Objective ➊ Simplify complex rational expressions. Sometimes a rational expression may contain rational expressions in its numerator or denominator. Such an expression is called a **complex rational expression** or a **complex fraction**.

Definition *Complex rational expression:* A rational expression that contains rational expressions in the numerator or denominator.

Here are some complex rational expressions:

$$\dfrac{\dfrac{3}{4}}{\dfrac{5}{6}} \qquad \dfrac{x + \dfrac{x}{3}}{\dfrac{y}{2}} \qquad \dfrac{7}{\dfrac{t+2}{t}} \qquad \dfrac{\dfrac{2}{n} + \dfrac{1}{4}}{\dfrac{1}{n} - \dfrac{n}{3}}$$

There are two common methods for simplifying complex rational expressions. One method is to rewrite the complex rational expression as a horizontal division problem. For example, the complex fraction $\dfrac{\dfrac{3}{4}}{\dfrac{5}{6}}$ can be expressed as the division problem $\dfrac{3}{4} \div \dfrac{5}{6}$, then simplified.

$$\dfrac{\dfrac{3}{4}}{\dfrac{5}{6}} = \dfrac{3}{4} \div \dfrac{5}{6} = \dfrac{3}{\overset{}{\underset{2}{4}}} \cdot \dfrac{\overset{3}{\cancel{6}}}{5} = \dfrac{9}{10}$$

Another method is to multiply the numerator and denominator of the complex fraction by the LCD of the numerator and denominator fractions. In this case, the denominators are 4 and 6, which means that the LCD is 12.

Connection Notice that the essence of this second method is writing an equivalent fraction by multiplying both the numerator and denominator by the same factor.

$$\dfrac{\dfrac{3}{4} \cdot 12}{\dfrac{5}{6} \cdot 12} = \dfrac{\dfrac{3}{\overset{}{4}} \cdot \dfrac{\overset{3}{\cancel{12}}}{1}}{\dfrac{5}{\overset{}{6}} \cdot \dfrac{\overset{2}{\cancel{12}}}{1}} = \dfrac{\dfrac{9}{1}}{\dfrac{10}{1}} = \dfrac{9}{10}$$

Procedure **Simplifying Complex Rational Expressions**

To simplify a complex rational expression, use one of the following methods.

Method 1
1. Simplify the numerator and denominator if needed.
2. Rewrite as a horizontal division problem.

Method 2
1. Multiply the numerator and denominator of the complex rational expression by their LCD.
2. Simplify.

Example 1 Simplify.

a. $\dfrac{\dfrac{2}{x}}{\dfrac{y}{3}}$

Method 1

Note: Method 1 is best used when the numerator and denominator are both a single rational expression, as in Example 1(a). Otherwise, method 2 is at least as easy as method 1.

$$\dfrac{\dfrac{2}{x}}{\dfrac{y}{3}} = \dfrac{2}{x} \div \dfrac{y}{3} \qquad \text{Write the complex fraction as a division problem.}$$

$$= \dfrac{2}{x} \cdot \dfrac{3}{y} \qquad \text{Write an equivalent multiplication problem.}$$

$$= \dfrac{6}{xy}$$

Method 2 Multiply the numerator and denominator of the rational expression by the LCD of all rational expressions involved. In the case, that LCD is $3x$.

$$\dfrac{\dfrac{2}{x}}{\dfrac{y}{3}} = \dfrac{\dfrac{2}{x} \cdot 3x}{\dfrac{y}{3} \cdot 3x} \qquad \text{Multiply the numerator and denominator by } 3x.$$

$$= \dfrac{\dfrac{2}{\cancel{x}} \cdot \dfrac{\overset{1}{\cancel{3x}}}{1}}{\dfrac{y}{\cancel{3}} \cdot \dfrac{\overset{1}{\cancel{3x}}}{1}} \qquad \text{Divide out the common factors.}$$

$$= \dfrac{6}{xy} \qquad \text{Simplify.}$$

b. $\dfrac{\dfrac{1}{4} + \dfrac{2}{3}}{\dfrac{3}{8} - \dfrac{1}{6}}$

Method 1 Simplify the numerator and denominator and then divide.

$$\dfrac{\dfrac{1}{4} + \dfrac{2}{3}}{\dfrac{3}{8} - \dfrac{1}{6}} = \dfrac{\dfrac{1(3)}{4(3)} + \dfrac{2(4)}{3(4)}}{\dfrac{3(3)}{8(3)} - \dfrac{1(4)}{6(4)}} \qquad \text{Write the numerator fractions as equivalent fractions with their LCD, 12, and write the denominator fractions with their LCD, 24.}$$

$$= \dfrac{\dfrac{3}{12} + \dfrac{8}{12}}{\dfrac{9}{24} - \dfrac{4}{24}}$$

$$= \dfrac{\dfrac{11}{12}}{\dfrac{5}{24}} \qquad \text{Add in the numerator and subtract in the denominator.}$$

$$= \frac{11}{12} \div \frac{5}{24}$$ Write the complex fraction as a division problem.

$$= \frac{11}{\underset{1}{\cancel{12}}} \cdot \frac{\overset{2}{\cancel{24}}}{5}$$ Write an equivalent multiplication problem, divide out the common factors, and then multiply.

$$= \frac{22}{5}$$

Method 2 Multiply the numerator and denominator of the rational expression by the LCD of all rational expressions involved. In this case, that LCD is 24.

$$\frac{\dfrac{1}{4} + \dfrac{2}{3}}{\dfrac{3}{8} - \dfrac{1}{6}} = \frac{\left(\dfrac{1}{4} + \dfrac{2}{3} \right) \cdot 24}{\left(\dfrac{3}{8} - \dfrac{1}{6} \right) \cdot 24}$$ Multiply the numerator and denominator by 24.

$$= \frac{\dfrac{1}{\cancel{4}} \cdot \dfrac{\overset{6}{\cancel{24}}}{1} + \dfrac{2}{\cancel{3}} \cdot \dfrac{\overset{8}{\cancel{24}}}{1}}{\dfrac{3}{\cancel{8}} \cdot \dfrac{\overset{3}{\cancel{24}}}{1} - \dfrac{1}{\cancel{6}} \cdot \dfrac{\overset{4}{\cancel{24}}}{1}}$$ Distribute the 24 to each fraction and divide out the common factors.

$$= \frac{6 + 16}{9 - 4}$$ Simplify the numerator and denominator.

$$= \frac{22}{5}$$

c. $$\dfrac{5 + \dfrac{x}{3}}{\dfrac{4}{x} - \dfrac{1}{2x}}$$

Method 1 Simplify the numerator and denominator and then divide.

$$\frac{5 + \dfrac{x}{3}}{\dfrac{4}{x} - \dfrac{1}{2x}} = \frac{\dfrac{5(3)}{1(3)} + \dfrac{x(1)}{3(1)}}{\dfrac{4(2)}{x(2)} - \dfrac{1(1)}{2x(1)}}$$ Write the numerator fractions as equivalent fractions with their LCD, 3, and write the denominator fractions with their LCD, $2x$.

$$= \frac{\dfrac{15}{3} + \dfrac{x}{3}}{\dfrac{8}{2x} - \dfrac{1}{2x}}$$

$$= \frac{\dfrac{15 + x}{3}}{\dfrac{7}{2x}}$$ Add in the numerator and subtract in the denominator.

$$= \frac{15 + x}{3} \div \frac{7}{2x}$$ Write the complex fraction as a division problem.

$$= \frac{15 + x}{3} \cdot \frac{2x}{7}$$ Write an equivalent multiplication problem and then multiply.

$$= \frac{30x + 2x^2}{21}$$

Method 2 Multiply the numerator and denominator of the rational expression by the LCD of all rational expressions involved. In this case, that LCD is $6x$.

$$\frac{5 + \dfrac{x}{3}}{\dfrac{4}{x} - \dfrac{1}{2x}} = \frac{\left(5 + \dfrac{x}{3}\right) \cdot 6x}{\left(\dfrac{4}{x} - \dfrac{1}{2x}\right) \cdot 6x}$$

Multiply the numerator and denominator by $6x$.

$$= \frac{\dfrac{5}{1} \cdot \dfrac{6x}{1} + \dfrac{x}{\overset{}{\underset{1}{3}}} \cdot \dfrac{\overset{2}{6x}}{1}}{\dfrac{4}{\underset{1}{x}} \cdot \dfrac{\overset{1}{6x}}{1} - \dfrac{1}{2x} \cdot \dfrac{\overset{3}{6x}}{1}}$$

Distribute the $6x$ to each fraction and divide out the common factors.

$$= \frac{30x + 2x^2}{24 - 3}$$

Simplify the numerator and denominator.

$$= \frac{30x + 2x^2}{21}$$

d. $\dfrac{1 - \dfrac{2}{x} - \dfrac{15}{x^2}}{1 + \dfrac{7}{x} + \dfrac{12}{x^2}}$

Method 1 Simplify the numerator and denominator and then divide.

$$\frac{1 - \dfrac{2}{x} - \dfrac{15}{x^2}}{1 + \dfrac{7}{x} + \dfrac{12}{x^2}} = \frac{1 \cdot \dfrac{x^2}{x^2} - \dfrac{2}{x} \cdot \dfrac{x}{x} - \dfrac{15}{x^2}}{1 \cdot \dfrac{x^2}{x^2} + \dfrac{7}{x} \cdot \dfrac{x}{x} + \dfrac{12}{x^2}}$$

Write the numerator fractions as equivalent fractions with their LCD, x^2, and write the denominator fractions with their LCD, also x^2.

$$= \frac{\dfrac{x^2}{x^2} - \dfrac{2x}{x^2} - \dfrac{15}{x^2}}{\dfrac{x^2}{x^2} + \dfrac{7x}{x^2} + \dfrac{12}{x^2}}$$

Simplify.

$$= \frac{\dfrac{x^2 - 2x - 15}{x^2}}{\dfrac{x^2 + 7x + 12}{x^2}}$$

Subtract in the numerator and add in the denominator.

$$= \frac{x^2 - 2x - 15}{x^2} \div \frac{x^2 + 7x + 12}{x^2}$$

Write the complex fraction as a division problem.

$$= \frac{x^2 - 2x - 15}{x^2} \cdot \frac{x^2}{x^2 + 7x + 12}$$

Write as an equivalent multiplication problem.

$$= \frac{(x - 5)(x + 3)}{\underset{1}{x^2}} \cdot \frac{\overset{1}{x^2}}{(x + 4)(x + 3)}$$

Factor, divide out the common factors, and then multiply the remaining factors.

$$= \frac{x - 5}{x + 4}$$

Method 2 Multiply the numerator and denominator of the rational expression by the LCD of all rational expressions involved. In this case, that LCD is x^2.

$$\frac{1 - \dfrac{2}{x} - \dfrac{15}{x^2}}{1 + \dfrac{7}{x} + \dfrac{12}{x^2}} = \frac{\left(1 - \dfrac{2}{x} - \dfrac{15}{x^2}\right) \cdot x^2}{\left(1 + \dfrac{7}{x} + \dfrac{12}{x^2}\right) \cdot x^2}$$

Multiply the numerator and denominator by x^2.

$$= \frac{1 \cdot x^2 - \dfrac{2}{\cancel{x}} \cdot \dfrac{\cancel{x^2}}{1} - \dfrac{15}{\cancel{x^2}} \cdot \dfrac{\cancel{x^2}}{1}}{1 \cdot x^2 + \dfrac{7}{\cancel{x}} \cdot \dfrac{\cancel{x^2}}{1} + \dfrac{12}{\cancel{x^2}} \cdot \dfrac{\cancel{x^2}}{1}}$$

Distribute the x^2 to each fraction and divide out the common factors.

$$= \frac{x^2 - 2x - 15}{x^2 + 7x + 12}$$

Simplify the numerator and denominator.

$$= \frac{(x - 5)(\cancel{x + 3})}{(x + 4)(\cancel{x + 3})}$$

Factor and divide out the common factor, $x + 3$.

$$= \frac{x - 5}{x + 4}$$

Your Turn 1 Simplify.

a. $\dfrac{\dfrac{a}{5}}{\dfrac{3}{b}}$

b. $\dfrac{2 - \dfrac{3}{4}}{\dfrac{4}{5} + \dfrac{1}{2}}$

c. $\dfrac{2 - \dfrac{1}{y}}{y - \dfrac{1}{3}}$

d. $\dfrac{1 - \dfrac{3}{x} - \dfrac{28}{x^2}}{1 - \dfrac{1}{x} - \dfrac{20}{x^2}}$

Answers Your Turn 1

a. $\dfrac{ab}{15}$ **b.** $\dfrac{25}{26}$

c. $\dfrac{6y - 3}{3y^2 - y}$ **d.** $\dfrac{x - 7}{x - 5}$

8.5 Exercises   For Extra Help *MyMathLab*

PRACTICE   WATCH   DOWNLOAD

1. Identify the numerator and denominator of the complex rational expression

$$\frac{\dfrac{4}{x + 3}}{\dfrac{3}{x}}.$$

2. Given the complex rational expression $\dfrac{\dfrac{2x}{x - 3}}{4}$, is the numerator $2x$ or $\dfrac{2x}{x - 3}$?

How could the expression be written so that the numerator and denominator are easily identified?

3. Write $\dfrac{2x}{x - 3} \div \dfrac{4 - x}{2x - 6}$ as a complex rational expression.

4. Write $\dfrac{5}{x-1} \div 7x$ as a complex rational expression.

5. Which of the two methods discussed in this section would you use to simplify
$$\dfrac{\dfrac{x}{6} - \dfrac{3}{4}}{\dfrac{2}{3} + x}?\, \text{Why?}$$

6. Which of the two methods discussed in this section would you use to simplify
$$\dfrac{\dfrac{x}{7}}{\dfrac{x}{x+1}}?\, \text{Why?}$$

For Exercises 7–48, simplify. See Example 1.

7. $\dfrac{\dfrac{3}{4}}{\dfrac{3}{2}}$

8. $\dfrac{\dfrac{2}{3}}{\dfrac{3}{2}}$

9. $\dfrac{\dfrac{m}{n}}{\dfrac{q}{p}}$

10. $\dfrac{\dfrac{a}{b}}{\dfrac{c}{d}}$

11. $\dfrac{\dfrac{1}{2} + \dfrac{3}{5}}{\dfrac{3}{2} - \dfrac{1}{5}}$

12. $\dfrac{\dfrac{2}{3} + \dfrac{1}{4}}{\dfrac{4}{3} - \dfrac{3}{4}}$

13. $\dfrac{5 + \dfrac{2}{a}}{3 + \dfrac{4}{a}}$

14. $\dfrac{4 + \dfrac{1}{x}}{3 + \dfrac{5}{x}}$

15. $\dfrac{\dfrac{1}{y} - 1}{\dfrac{1}{y} + 1}$

16. $\dfrac{x - \dfrac{1}{x}}{1 + \dfrac{1}{x}}$

17. $\dfrac{r - \dfrac{3}{4}}{\dfrac{1}{2} - r}$

18. $\dfrac{\dfrac{5}{6} - x}{x - \dfrac{2}{3}}$

19. $\dfrac{\dfrac{3}{x} - 1}{\dfrac{9}{x^2} - 1}$

20. $\dfrac{1 - \dfrac{2}{p}}{1 - \dfrac{4}{p^2}}$

21. $\dfrac{x + y}{\dfrac{1}{x} + \dfrac{1}{y}}$

22. $\dfrac{\dfrac{2}{a} + \dfrac{3}{b}}{a + b}$

23. $\dfrac{x + 2}{\dfrac{x^2 - 4}{x}}$

24. $\dfrac{\dfrac{a}{b} - 1}{a^2 - b^2}$

25. $\dfrac{\dfrac{6}{3x - 2}}{\dfrac{x}{2x + 1}}$

26. $\dfrac{\dfrac{5}{2x - 1}}{\dfrac{x}{x + 1}}$

27. $\dfrac{\dfrac{a}{a^2 - b^2}}{\dfrac{b}{a + b}}$

28. $\dfrac{\dfrac{x}{4x^2 - 1}}{\dfrac{5}{2x - 1}}$

29. $\dfrac{\dfrac{y^2 - 25}{5y}}{\dfrac{y^2 + 5y}{2y^3}}$

30. $\dfrac{\dfrac{x^2 + 4x}{y^2}}{\dfrac{x^2 - 16}{3y}}$

31. $\dfrac{2 + \dfrac{2}{u - 1}}{\dfrac{2}{u - 1}}$

32. $\dfrac{\dfrac{1}{w+1} - 1}{\dfrac{1}{w+1}}$

33. $\dfrac{\dfrac{a}{a-2} - 1}{1 + \dfrac{a}{a-2}}$

34. $\dfrac{1 - \dfrac{x}{x+2}}{\dfrac{x}{x+2} + 1}$

35. $\dfrac{\dfrac{x-3}{x} - 1}{\dfrac{x^2-9}{x} - x}$

36. $\dfrac{q - \dfrac{q^2-1}{q}}{1 - \dfrac{q-1}{q}}$

37. $\dfrac{\dfrac{1}{x-1} - \dfrac{1}{x+1}}{\dfrac{1}{x^2-1}}$

38. $\dfrac{\dfrac{16}{x^2-16}}{\dfrac{2}{x+4} - \dfrac{2}{x-4}}$

39. $\dfrac{1 + \dfrac{1}{y} - \dfrac{2}{y^2}}{\dfrac{1}{y} + \dfrac{2}{y^2}}$

40. $\dfrac{1 - \dfrac{3}{x}}{1 - \dfrac{2}{x} - \dfrac{3}{x^2}}$

41. $\dfrac{1 - \dfrac{1}{n}}{n - 2 + \dfrac{1}{n}}$

42. $\dfrac{y - 1 - \dfrac{6}{y}}{1 + \dfrac{2}{y}}$

43. $\dfrac{1 - \dfrac{1}{2y+1}}{\dfrac{1}{2y^2 - 5y - 3} - \dfrac{1}{y-3}}$

44. $\dfrac{1 - \dfrac{1}{f-3}}{\dfrac{1}{f^2-f-6} - \dfrac{1}{f+2}}$

45. $\dfrac{\dfrac{x^2 - 2x - 8}{x^2 - 16}}{\dfrac{x^2 + 2x}{x^2 + 3x - 4}}$

46. $\dfrac{\dfrac{v^2 + v - 2}{v^2 + 4v}}{\dfrac{v^2 - 4}{v^2 + 2v - 8}}$

47. $\dfrac{\dfrac{2}{x+h} - \dfrac{2}{x}}{h}$

48. $\dfrac{h}{\dfrac{3}{y+h} - \dfrac{3}{y}}$

Find the mistake

For Exercises 49–56, explain the mistake; then find the correct answer.

49. $\dfrac{x - \dfrac{1}{2}}{y - \dfrac{1}{2}} = \dfrac{x}{y}$

50. $\dfrac{\dfrac{2}{3}}{\dfrac{4}{6}} = \dfrac{\overset{1}{2}}{3} \cdot \dfrac{4}{\underset{3}{6}}$

$= \dfrac{4}{9}$

51. $\dfrac{\dfrac{m}{6}}{\dfrac{3}{m}} = \dfrac{\overset{1}{m}}{\underset{2}{6}} \div \dfrac{\overset{1}{3}}{\underset{1}{m}}$

$= \dfrac{1}{2}$

52. $\dfrac{\dfrac{1}{x} + \dfrac{1}{y}}{x^2 - y^2} = \dfrac{\dfrac{x}{1} \cdot \dfrac{1}{x} + \dfrac{y}{1} \cdot \dfrac{1}{y}}{x \cdot x^2 - y \cdot y^2}$

$= \dfrac{1 + 1}{x^3 - y^3}$

$= \dfrac{2}{x^3 - y^3}$

53. $\dfrac{c + \dfrac{1}{c}}{\dfrac{1}{c}} = \dfrac{c + \dfrac{c}{1} \cdot \dfrac{1}{c}}{\dfrac{c}{1} \cdot \dfrac{1}{c}}$

$= \dfrac{c + 1}{1}$

$= c + 1$

54. $\dfrac{1 + \dfrac{2}{r}}{1 - \dfrac{4}{r^2}} = \dfrac{r \cdot 1 + \dfrac{r}{1} \cdot \dfrac{2}{r}}{r \cdot 1 - \dfrac{r}{1} \cdot \dfrac{4}{r^2}}$

$= \dfrac{r + 2}{r - 4r}$

$= \dfrac{r + 2}{-3r}$

55. $\dfrac{\dfrac{3}{x}}{\dfrac{x}{6}} = \dfrac{3}{x} \div \dfrac{x}{6}$

$= \dfrac{3}{x} \cdot \dfrac{6}{x}$

$= \dfrac{18}{2x}$

$= \dfrac{9}{x}$

56. $\dfrac{\dfrac{1}{x} + \dfrac{1}{y}}{\dfrac{1}{x} - \dfrac{1}{y}} = \dfrac{\dfrac{xy}{1} \cdot \dfrac{1}{x} + \dfrac{xy}{1} \cdot \dfrac{1}{y}}{\dfrac{xy}{1} \cdot \dfrac{1}{x} - \dfrac{xy}{1} \cdot \dfrac{1}{y}}$

$= \dfrac{y + x}{y - x}$

$= -1$

For Exercises 57 and 58, average rate can be found using the following formula:

$$\text{Average rate} = \frac{\text{Total distance}}{\text{Total time}}$$

57. Suppose a person travels 10 miles in $\dfrac{1}{3}$ of an hour, then returns by traveling that same 10 miles in $\dfrac{1}{4}$ of an hour. What is that person's average rate?

58. Juan travels 50 miles in $\dfrac{3}{4}$ of an hour and then encounters a traffic jam so that the next 15 miles take $\dfrac{1}{3}$ of an hour. The last 40 miles of his trip take him $\dfrac{2}{3}$ of an hour. What was his average rate for the trip?

59. Given the area and length of a rectangle, the width can be found using the formula $w = \dfrac{A}{l}$. Suppose the area of a rectangle is described by $\dfrac{5n - 2}{8}$ square feet and the length is described by $\dfrac{n - 1}{6}$ feet. Find an expression for the width.

60. Given the area and base of a triangle, the height can be found using the formula $h = \dfrac{2A}{b}$. Suppose the area of a triangle is described by $\dfrac{n + 1}{6}$ and its base is described by $\dfrac{5n - 2}{9}$. Find an expression for the height.

61. In electrical circuits, if two resistors R_1 and R_2 are wired in parallel, the resistance of the circuit is found using the complex rational expression $\dfrac{1}{\dfrac{1}{R_1} + \dfrac{1}{R_2}}$.

  a. Simplify this complex rational expression.

  b. If a 100-ohm and 10-ohm resistor are wired in parallel, find the resistance of the circuit.
  c. Two 16-ohm speakers are wired in parallel; find the resistance of the circuit.

62. Suppose three resistors R_1, R_2, and R_3 are wired in parallel. The resistance of the circuit is found using the complex rational expression $\dfrac{1}{\dfrac{1}{R_1} + \dfrac{1}{R_2} + \dfrac{1}{R_3}}$.

a. Simplify this complex rational expression.

b. A 20-ohm, a 10-ohm, and a 40-ohm resistor are wired in parallel. Find the resistance of the circuit.

c. A 300-ohm, a 500-ohm, and a 200-ohm resistor are wired in parallel. Find the resistance of the circuit.

Review Exercises

Exercises 1–6 EQUATIONS AND INEQUALITIES

[2.3] *For Exercises 1–4, solve.*

1. $-6x = 42$

2. $9t - 8 = 19$

3. $\dfrac{x}{4} = \dfrac{5}{8}$

4. $\dfrac{3}{4}y - \dfrac{1}{3} = \dfrac{1}{6}y + 2$

[2.1] **5.** A runner travels 1.2 miles in 15 minutes. Find the runner's average rate in miles per hour.

[3.4] **6.** An eastbound car and westbound car pass each other on a highway. The eastbound car is traveling 40 miles per hour, and the westbound car is traveling 60 miles per hour. How much time elapses from the time they pass each other until they are 9 miles apart?

8.6 Solving Equations Containing Rational Expressions

Objective **①** **Solve equations containing rational expressions.** In Section 2.3, we learned that we can eliminate fractions from an equation by multiplying both sides of the equation by their LCD. For example, if given the equation $\dfrac{1}{2}x - \dfrac{3}{4} = \dfrac{2}{3}$, we can eliminate the fractions by multiplying both sides by 12, which is the LCD of $\dfrac{1}{2}$, $\dfrac{3}{4}$, and $\dfrac{2}{3}$.

$$\overset{6}{\cancel{12}} \cdot \frac{1}{\underset{1}{\cancel{2}}}x - \overset{3}{\cancel{12}} \cdot \frac{3}{\underset{1}{\cancel{4}}} = \overset{4}{\cancel{12}} \cdot \frac{2}{\underset{1}{\cancel{3}}}$$

$$6x - 9 = 8$$

Because the rewritten equation contains only integers, it is easier to solve. Similarly, we can use the LCD to eliminate rational expressions in an equation.

Example 1 Solve $\dfrac{x}{5} = \dfrac{x}{2} - \dfrac{1}{4}$.

SOLUTION Eliminate the rational expressions by multiplying both sides by the LCD, 20.

$$20\left(\frac{x}{5}\right) = 20\left(\frac{x}{2} - \frac{1}{4}\right) \qquad \text{Multiply both sides by 20.}$$

$$\overset{4}{20} \cdot \frac{x}{\underset{1}{5}} = \overset{10}{20} \cdot \frac{x}{\underset{1}{2}} - \overset{5}{20} \cdot \frac{1}{\underset{1}{4}} \qquad \begin{array}{l}\text{Use the distributive property to multiply every term}\\ \text{by 20 and then divide out the common factors.}\end{array}$$

$$4x = 10x - 5$$

$$-6x = -5 \qquad \text{Subtract } 10x \text{ from both sides.}$$

$$\frac{-6x}{-6} = \frac{-5}{-6} \qquad \text{Divide both sides by } -6.$$

$$x = \frac{5}{6} \qquad \text{Simplify both sides.}$$

Check Using the original equation, we replace x with $\dfrac{5}{6}$ and see if the equation is true. It is helpful to write $\dfrac{x}{5}$ as $\dfrac{1}{5}x$ and $\dfrac{x}{2}$ as $\dfrac{1}{2}x$.

$$\frac{1}{5}x = \frac{1}{2}x - \frac{1}{4}$$

$$\frac{1}{5}\left(\frac{\overset{1}{\cancel{5}}}{6}\right) \overset{?}{=} \frac{1}{2}\left(\frac{5}{6}\right) - \frac{1}{4} \qquad \text{Replace } x \text{ with } \frac{5}{6} \text{ and then simplify.}$$

$$\frac{1}{6} \overset{?}{=} \frac{5}{12} - \frac{3}{12}$$

$$\frac{1}{6} = \frac{2}{12} \qquad \text{True; therefore, } \frac{5}{6} \text{ is a solution.}$$

Now consider equations that contain rational expressions with variables in the denominator.

Example 2 Solve $\dfrac{1}{2} = \dfrac{3}{x} - \dfrac{1}{4}$.

SOLUTION Eliminate the rational expressions by multiplying both sides by the LCD, $4x$.

$$4x\left(\frac{1}{2}\right) = 4x\left(\frac{3}{x} - \frac{1}{4}\right) \qquad \text{Multiply both sides by } 4x.$$

$$\overset{2}{\cancel{4}}x \cdot \frac{1}{\underset{1}{2}} = 4\cancel{x} \cdot \frac{3}{\underset{1}{\cancel{x}}} - \cancel{4}x \cdot \frac{1}{\underset{1}{\cancel{4}}} \qquad \begin{array}{l}\text{Use the distributive property to multiply every term by } 4x\\ \text{and then divide out the common factors.}\end{array}$$

$$2x = 12 - x$$

$$3x = 12 \qquad \text{Add } x \text{ to both sides.}$$

$$x = 4 \qquad \text{Divide both sides by 3.}$$

Check We will leave the check to the reader.

Your Turn 2 Solve.

a. $\dfrac{x}{2} = 4 - \dfrac{x}{3}$

b. $\dfrac{1}{x} + \dfrac{3}{8} = \dfrac{5}{6}$

Extraneous Solutions

If an equation contains rational expressions with variables in the denominators, we must make sure that no solution causes any of the expressions in the equation to be undefined. In Example 2, if the solution had turned out to be 0, the expression $\dfrac{3}{x}$ would have been undefined. By inspecting the denominators of each rational expression, you can determine the values that will cause the expressions to be undefined before solving the equation. If by solving an equation, you obtain a number that causes an expression in the original equation to be undefined, we say that the number is an **extraneous solution** and we discard it.

Definition *Extraneous solution:* An apparent solution that does not solve its equation.

The following procedure summarizes the process of solving equations containing rational expressions.

Procedure **Solving Equations Containing Rational Expressions**

To solve an equation that contains rational expressions,

1. Eliminate the rational expressions by multiplying both sides of the equation by their LCD.
2. Solve the equation using the methods we learned in Chapters 2 (linear equations) and 7 (quadratic equations).
3. Check your solution(s) in the original equation. Discard any extraneous solutions.

Connection The equation in Example 3 is a proportion. We could have solved the equation by cross multiplying, which leads to the same equation as in the second step of the solution in Example 3.

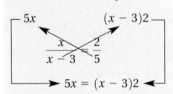

Warning: You can use this method only if the equation is a proportion.

Example 3 Solve. $\dfrac{x}{x-3} = \dfrac{2}{5}$.

SOLUTION Notice that if $x = 3$, then $\dfrac{x}{x-3}$ is undefined; so the solution cannot be 3. To solve, we first multiply both sides by the LCD, $5(x-3)$.

$$5(\overset{1}{\cancel{x-3}}) \cdot \dfrac{x}{\underset{1}{\cancel{x-3}}} = \overset{1}{\cancel{5}}(x-3) \cdot \dfrac{2}{\underset{1}{\cancel{5}}} \qquad \text{Divide out the common factors.}$$

$$5x = (x-3)2$$

$$5x = 2x - 6 \qquad \text{Distribute 2 to clear the parentheses.}$$

$$3x = -6 \qquad \text{Subtract } 2x \text{ from both sides.}$$

$$x = -2 \qquad \text{Divide both sides by 3.}$$

Check We will leave the check to the reader.

Your Turn 3 Solve $\dfrac{2}{x+5} = \dfrac{4}{x-1}$.

Answers Your Turn 2

a. $\dfrac{24}{5}$ b. $\dfrac{24}{11}$

Answer Your Turn 3

-11

In Examples 1–3, after eliminating the rational expressions, we were left with linear equations. Now let's look at some equations that transform to quadratic equations after we use the LCD to eliminate the rational expressions. Remember that when solving quadratic equations, we manipulate the equation so that one side is 0; then we factor and use the zero-factor theorem.

Example 4   Solve.

a. $\dfrac{3}{x} - 5 = x - 3$

SOLUTION   Notice that because the denominator is x, the solution cannot be 0.

$$x\left(\dfrac{3}{x} - 5\right) = x(x - 3) \qquad \text{Multiply both sides by the LCD, } x.$$

$$\overset{1}{\cancel{x}} \cdot \dfrac{3}{\underset{1}{\cancel{x}}} - x \cdot 5 = x \cdot x - x \cdot 3 \qquad \text{Distribute } x; \text{ then divide out common factors.}$$

$$3 - 5x = x^2 - 3x \qquad \begin{array}{l} \text{Because this equation is quadratic, we set it equal to} \\ \text{0 by subtracting 3 from and adding } 5x \text{ to both sides.} \end{array}$$

$$0 = x^2 - 3x + 5x - 3$$

$$0 = x^2 + 2x - 3 \qquad \text{Combine like terms.}$$

$$0 = (x + 3)(x - 1) \qquad \text{Factor.}$$

$$x + 3 = 0 \quad \text{or} \quad x - 1 = 0 \qquad \text{Use the zero-factor theorem.}$$

$$x = -3 \qquad\qquad x = 1$$

Check   Verify that both -3 and 1 make the original equation true.

$$\dfrac{3}{-3} - 5 \overset{?}{=} -3 - 3 \qquad\qquad\qquad \dfrac{3}{1} - 5 \overset{?}{=} 1 - 3$$

$$-1 - 5 \overset{?}{=} -6 \qquad\qquad\qquad\qquad 3 - 5 \overset{?}{=} -2$$

$$-6 = -6 \qquad \text{True} \qquad\qquad\qquad -2 = -2 \qquad \text{True}$$

Both -3 and 1 are solutions.

b. $\dfrac{x}{x + 1} = \dfrac{6}{3x - 1}$

SOLUTION   We could multiply both sides by the LCD, $(x + 1)(3x - 1)$, but because the equation is a proportion, we choose to cross multiply instead. Note that $\dfrac{1}{3}$ and -1 cannot be solutions.

$$\dfrac{x}{x + 1} = \dfrac{6}{3x - 1}$$

$$\overset{3x^2 - x}{\underset{\displaystyle \dfrac{x}{x + 1} \searrow\nearrow \dfrac{6}{3x - 1}}{}} {}^{6x + 6} \qquad \text{Cross multiply.}$$

$$3x^2 - x = 6x + 6 \qquad \begin{array}{l} \text{Because this equation is quadratic, we set it equal} \\ \text{to 0 by subtracting } 6x \text{ and 6 from both sides.} \end{array}$$

$$3x^2 - 7x - 6 = 0$$

$$(3x + 2)(x - 3) = 0 \qquad \text{Factor.}$$

$$3x + 2 = 0, x - 3 = 0 \qquad \text{Use the zero-factor theorem.}$$

$$3x = -2 \quad x = 3 \qquad \text{Solve.}$$

$$x = -\dfrac{2}{3}$$

Check   We will leave the check to the reader.

Your Turn 4 Solve.

a. $\dfrac{1}{3} - \dfrac{2}{x} = x + 4$

b. $\dfrac{x}{x + 1} = \dfrac{6}{4x + 1}$

Example 5 Solve.

a. $\dfrac{x^2}{x + 2} - 3 = \dfrac{x + 6}{x + 2} - 4$

SOLUTION By inspecting the denominators, notice that the solution cannot be -2.

$$(x + 2)\left(\dfrac{x^2}{x + 2} - 3\right) = (x + 2)\left(\dfrac{x + 6}{x + 2} - 4\right)$$

Multiply both sides by the LCD, $x + 2$.

$$\cancel{(x + 2)} \cdot \dfrac{x^2}{\cancel{x + 2}} - (x + 2) \cdot 3 = \cancel{(x + 2)} \cdot \dfrac{x + 6}{\cancel{x + 2}} - (x + 2) \cdot 4$$

Distribute $x + 2$ and then divide out the common factors.

$$x^2 - 3x - 6 = x + 6 - 4x - 8 \qquad \text{Combine like terms.}$$

$$x^2 - 3x - 6 = -3x - 2$$

$$x^2 - 4 = 0 \qquad \begin{array}{l}\text{Because the equation is quadratic, we set it equal to 0}\\ \text{by adding } 3x \text{ and 2 to both sides.}\end{array}$$

$$(x + 2)(x - 2) = 0 \qquad \text{Factor.}$$

$$x + 2 = 0 \quad \text{or} \quad x - 2 = 0 \qquad \text{Use the zero-factor theorem.}$$

$$x = -2 \qquad\qquad x = 2$$

Check In the original equation $\dfrac{x^2}{x + 2}$ and $\dfrac{x + 6}{x + 2}$ are both undefined when $x = -2$ so -2 is extraneous. Therefore, we need to check only 2.

$$\dfrac{2^2}{2 + 2} - 3 \overset{?}{=} \dfrac{2 + 6}{2 + 2} - 4 \qquad \begin{array}{l}\text{In the original equation, replace } x \text{ with 2}\\ \text{and then simplify.}\end{array}$$

$$\dfrac{4}{4} - 3 \overset{?}{=} \dfrac{8}{4} - 4$$

$$1 - 3 \overset{?}{=} 2 - 4$$

$$-2 = -2 \qquad \text{True, so 2 is a solution and the only solution.}$$

b. $\dfrac{5}{x - 3} = \dfrac{x}{x - 2} + \dfrac{x}{x^2 - 5x + 6}$

SOLUTION To determine the LCD of all of the denominators, we first factor quadratic form $x^2 - 5x + 6$.

$$\dfrac{5}{x - 3} = \dfrac{x}{x - 2} + \dfrac{x}{(x - 2)(x - 3)} \qquad \begin{array}{l}\text{Factor the denominator}\\ x^2 - 5x + 6.\end{array}$$

$$(x - 2)(x - 3)\left(\dfrac{5}{x - 3}\right) = (x - 2)(x - 3)\left(\dfrac{x}{x - 2} + \dfrac{x}{(x - 2)(x - 3)}\right)$$

Multiply both sides by the LCD, $(x - 2)(x - 3)$.

$$(x - 2)\cancel{(x - 3)} \cdot \dfrac{5}{\cancel{(x - 3)}} = \cancel{(x - 2)}(x - 3) \cdot \dfrac{x}{\cancel{(x - 2)}} + \cancel{(x - 2)}\cancel{(x - 3)} \cdot \dfrac{x}{\cancel{(x - 2)}\cancel{(x - 3)}}$$

Note: Inspecting the denominators after factoring, we see that neither 2 nor 3 can be a solution.

Distribute $(x - 2)(x - 3)$; then divide out the common factors.

$$(x - 2)5 = (x - 3)x + x$$

$$5x - 10 = x^2 - 3x + x \qquad \text{Distribute to clear the parentheses.}$$

$$5x - 10 = x^2 - 2x \qquad \text{Combine like terms.}$$

Answers Your Turn 4

a. $-\dfrac{2}{3}$ and -3 **b.** $-\dfrac{3}{4}$ and 2

$$0 = x^2 - 7x + 10$$

Because the equation is quadratic, we set it equal to 0 by subtracting $5x$ from and adding 10 to both sides.

$$0 = (x - 2)(x - 5)$$ Factor.

$$x - 2 = 0 \quad \text{or} \quad x - 5 = 0$$ Use the zero-factor theorem.

$$x = 2 \qquad\qquad x = 5$$

Check   In the original equation, note that $\dfrac{x}{x-2}$ is undefined when $x = 2$; so 2 is an extraneous solution. Consequently, 5 is the only solution. We will leave the check for 5 to the reader.

Your Turn 5   Solve.

a. $\dfrac{3x + 4}{x - 4} = \dfrac{x^2}{x - 4} + 3$

b. $\dfrac{2}{x + 5} - \dfrac{6}{x^3 + 5x^2} = \dfrac{1}{x}$

Warning: Make sure you understand the difference between performing operations with rational expressions and solving equations containing rational expressions. For example,

$\dfrac{5}{x} + \dfrac{1}{3}$ is an *expression* that means two rational expressions are to be added. Remember, we cannot solve expressions; we can only evaluate or rewrite them. To rewrite this expression, we begin by writing each rational expression with the LCD, $3x$.

$$\frac{5}{x} + \frac{1}{3}$$

$$= \frac{5(3)}{x(3)} + \frac{1(x)}{3(x)}$$

$$= \frac{15}{3x} + \frac{x}{3x}$$

$$= \frac{15 + x}{3x}$$

To solve an *equation* such as $\dfrac{3}{x} + \dfrac{1}{4} = \dfrac{1}{2}$, we first eliminate the rational expressions by multiplying both sides of the equation by the LCD, $4x$.

$$4x\left(\frac{3}{x} + \frac{1}{4}\right) = 4x\left(\frac{1}{2}\right).$$

$$12 + x = 2x$$

$$12 = x \qquad \text{Subtract } x \text{ from both sides.}$$

Answers   Your Turn 5

a. -4, (4 is extraneous)

b. 6 and -1

8.6 Exercises   **For Extra Help**   *MyMathLab*

Note: Exercises marked with a ★ represent challenging exercises.

1. How are rational expressions eliminated from an equation?

2. What is the LCD for the rational expressions in $\dfrac{x}{x^2 + 4x - 5} = \dfrac{5}{x} - 2x$?

3. What is an extraneous solution for an equation that contains rational expressions?

4. Why might an equation with rational expressions lead to one or more extraneous solutions?

5. What values might be extraneous solutions in the equation $\dfrac{3}{x-4} = \dfrac{5x}{x+2}$? Why?

6. What values might be extraneous solutions in the equation $\dfrac{x}{x^2+4x-5} = \dfrac{5}{x} - 2x$?

Why?

For Exercises 7–12, check the given values to see if they are solutions to the equation.

7. $\dfrac{3}{x} + \dfrac{2}{3} = \dfrac{5}{6}; x = 9$

8. $\dfrac{3}{4} - \dfrac{1}{x} = \dfrac{2}{3}; x = 12$

9. $\dfrac{4}{m} - \dfrac{2}{5} = \dfrac{3}{4m}; m = \dfrac{65}{8}$

10. $\dfrac{5}{6n} + \dfrac{1}{4} = \dfrac{2}{3n}; n = -\dfrac{2}{3}$

11. $\dfrac{1}{t+2} + \dfrac{1}{4} = \dfrac{t}{16}; t = -4$

12. $\dfrac{1}{y-3} - \dfrac{5}{6} = -\dfrac{y}{12}; y = -7$

For Exercises 13–62, solve and check. Identify any extraneous solutions.
See Examples 1–5.

13. $\dfrac{2n}{3} + \dfrac{3n}{2} = \dfrac{13}{3}$

14. $\dfrac{3x}{5} - \dfrac{x}{2} = \dfrac{19}{5}$

15. $3y - \dfrac{4y}{5} = 22$

16. $\dfrac{3t}{4} - 2 = \dfrac{t}{4}$

17. $\dfrac{r-7}{2} - 1 = \dfrac{r+9}{9} + \dfrac{1}{3}$

18. $\dfrac{d+5}{3} - 4 = \dfrac{d-8}{4} - \dfrac{1}{2}$

19. $\dfrac{4}{x} + \dfrac{3}{2x} = \dfrac{11}{6}$

20. $\dfrac{3}{2u} - \dfrac{1}{3} = \dfrac{5}{6u}$

21. $\dfrac{4-6t}{2t} + \dfrac{3}{5} = -\dfrac{2}{5t}$

22. $\dfrac{2x+3}{2x} = \dfrac{5}{12} + \dfrac{3x+2}{3x}$

23. $\dfrac{5}{20} = \dfrac{x-7}{x+2}$

24. $\dfrac{y-2}{y+3} = \dfrac{3}{8}$

25. $\dfrac{6}{y-2} = \dfrac{5}{y-3}$

26. $\dfrac{4}{3x-3} = \dfrac{7}{4x+1}$

27. $\dfrac{x+4}{x+2} - 5 = \dfrac{6}{x+2}$

28. $\dfrac{m+5}{m-3} - 5 = \dfrac{4}{m-3}$

29. $\dfrac{6f-5}{6} = \dfrac{2f-1}{2} - \dfrac{f+2}{2f+5}$

30. $\dfrac{2x-1}{8} = \dfrac{x}{4} - \dfrac{x-1}{5x-2}$

31. $\dfrac{2}{x^2-2x-3} - \dfrac{3}{x-3} = \dfrac{2}{x+1}$

32. $\dfrac{5}{t^2-9} = \dfrac{3}{t+3} - \dfrac{2}{t-3}$

33. $\dfrac{1}{u-4} + \dfrac{2}{u^2-16} = \dfrac{3}{u+4}$

34. $\dfrac{5}{h+5} - \dfrac{2}{h^2 + 2h - 15} = \dfrac{2}{h-3}$

35. $\dfrac{x}{2-3x} = \dfrac{1}{3x+2}$

36. $\dfrac{5p}{p+4} = \dfrac{p}{p-1}$

37. $\dfrac{5}{2x} + \dfrac{15}{2x+16} = \dfrac{6}{x+8}$

38. $\dfrac{4}{3x} - \dfrac{2}{x+4} = \dfrac{2}{3x+12}$

39. $\dfrac{2}{x^2-1} = \dfrac{-3}{7x+7}$

40. $\dfrac{3}{y^2-4} = \dfrac{-2}{5y+10}$

41. $\dfrac{3}{x+5} - \dfrac{x-1}{2x} = \dfrac{-3}{2x^2+10x}$

42. $\dfrac{-1}{3x^2+9x} = \dfrac{1}{x+3} - \dfrac{x-3}{3x}$

43. $\dfrac{3m+1}{m^2-9} - \dfrac{m+3}{m-3} = \dfrac{1-5m}{m+3}$

44. $\dfrac{2n+10}{n^2-1} - \dfrac{n-1}{n+1} + \dfrac{2n+1}{n-1} = 0$

45. $\dfrac{x+3}{x-1} + \dfrac{2}{x+3} = \dfrac{5}{3}$

46. $\dfrac{m+2}{m-2} - \dfrac{2}{m+2} = -\dfrac{7}{3}$

47. $1 - \dfrac{3}{x-2} = -\dfrac{12}{x^2-4}$

48. $\dfrac{1}{x-3} + \dfrac{1}{3} = \dfrac{6}{x^2-9}$

49. $\dfrac{2r}{r-2} - \dfrac{4r}{r-3} = -\dfrac{7r}{r^2-5r+6}$

50. $\dfrac{3a}{a^2-2a-15} - \dfrac{a}{a+3} = \dfrac{2a}{a-5}$

51. $\dfrac{4u^2+3u+4}{u^2+u-2} - \dfrac{3u}{u+2} = \dfrac{-2u-1}{u-1}$

52. $\dfrac{2z}{2z-3} - \dfrac{3z}{2z+3} = \dfrac{15-32z^2}{4z^2-9}$

53. $\dfrac{2k}{k+7} - 1 = \dfrac{1}{k^2+10k+21} + \dfrac{k}{k+3}$

54. $\dfrac{p-5}{p+5} - 2 + \dfrac{p+15}{p-5} = \dfrac{50}{p^2-25}$

55. $\dfrac{6}{t^2+t-12} = \dfrac{4}{t^2-t-6} + \dfrac{1}{t^2+6t+8}$

56. $\dfrac{7}{v^2-6v+5} - \dfrac{2}{v^2-4v-5} = \dfrac{3}{v^2-1}$

57. $\dfrac{x+1}{x^2-2x-15} - \dfrac{6}{x^2-3x-10} = \dfrac{2}{x^2+5x+6}$

58. $\dfrac{x+1}{x^2+2x-24} - \dfrac{3}{x^2-x-12} = \dfrac{3}{x^2+9x+18}$

★ **59.** $\dfrac{1}{d^2-5d+6} + \dfrac{d-2}{d^2-d-6} - \dfrac{d}{d^2-4} = 0$

★ **60.** $\dfrac{1}{y^2-5y+6} + \dfrac{y}{y^2-2y-3} = \dfrac{y+2}{y^2-y-2}$

★ **61.** $\dfrac{1}{a-3} - \dfrac{6}{a^2-1} = \dfrac{12}{a^3-3a^2-a+3}$

★ **62.** $\dfrac{1}{b+2} - \dfrac{2}{b^2-1} = \dfrac{-2}{b^3+2b^2-b-2}$

For Exercises 63 and 64, explain the mistake; then find the correct solution(s).

63.
$$\frac{y}{3} + \frac{y}{y-3} = \frac{3}{y-3}$$

$$3(y-3) \cdot \frac{y}{3} + 3(y-3) \cdot \frac{y}{y-3} = 3(y-3) \cdot \frac{3}{y-3}$$

$$y^2 - 3y + 3y = 9$$

$$y^2 - 9 = 0$$

$$(y+3)(y-3) = 0$$

$$y = -3, 3$$

64.
$$\frac{x}{x+3} + \frac{3}{x+3} = 0$$

$$(x+3) \cdot \frac{x}{x+3} + (x+3) \cdot \frac{3}{x+3} = (x+3) \cdot 0$$

$$x + 3 = 0$$

$$x = -3$$

65. Explain how the LCD is used differently in $\frac{3x}{4} - \frac{5}{6}$ and in $\frac{3x}{4} - \frac{5}{6} = \frac{x}{2}$.

66. Explain how the LCD is used differently in $\frac{5x}{8} + \frac{1}{3}$ and in $\frac{5x}{8} + \frac{1}{3} = \frac{x}{6}$.

For Exercises 67–70, use the formula for the total resistance R in an electrical circuit with resistors $R_1, R_2, R_3, \ldots, R_n$ that are wired in parallel.

$$\frac{1}{R} = \frac{1}{R_1} + \frac{1}{R_2} + \frac{1}{R_3} + \cdots + \frac{1}{R_n}$$

67. Two resistors are wired in parallel, one of which is 100 ohms. If the total resistance is to be 80 ohms, what must the value be of the other resistor?

★**68.** Three resistors are wired in parallel, two of which are equal in value. If the third resistor is 200 ohms and the total resistance is to be 20 ohms, what are the values of the other two resistors?

★**69.** Two resistors are wired in parallel. One resistor is to have a value that is 10 ohms more than the other resistor, and the total resistance is to be 12 ohms. Find the value of both resistors.

★**70.** Two resistors are wired in parallel. One resistor is to have a value that is 20 ohms less than the other resistor, and the total resistance is to be 24 ohms. Find the value of both resistors.

In optics, a lens can be used to bend light from an object through a focal point to reproduce an inverted image that is larger (or smaller) than the original object. The following formula describes how the image of an object is affected by a lens, where o represents the object's distance from the lens, i represents the image's distance from the lens, and f represents the focal length of the lens.

Of Interest
The lens in our eye focuses images on the retina in the back of the eyeball. The image is inverted on our retina. Our brains then correct the inverted images so that we do not see the world upside down.

Formula: $\dfrac{1}{o} + \dfrac{1}{i} = \dfrac{1}{f}$

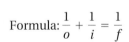

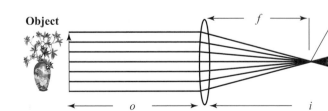

71. An object is 40 feet from a lens with a focal length of 2 feet. What is the image's distance from the lens?

72. An object is placed 50 centimeters from your eye. Your lens focuses the image of the object onto your retina, which is about 2.5 centimeters from your lens. What is the focal length of your lens?

★ **73.** A lens is being designed so that the focal length will be 5 millimeters less than the image length. Suppose an object is placed 60 millimeters from the lens. What are the focal length and image length?

★ **74.** A lens is being designed so that the image length is 9 millimeters more than the focal length. Suppose an object is placed 40 millimeters from the lens. What are the focal length and image length?

Review Exercises

Exercises 1–6 EQUATIONS AND INEQUALITIES

[2.1] **1.** Andrew notes that he traveled a distance of 120 miles in $1\frac{3}{4}$ hours. What was his average rate in miles per hour?

[3.1] **2.** A grocery store has a 10.5-ounce can of soup on sale for \$0.88. Write a unit ratio in simplest form of the price to capacity. Interpret the answer.

[3.1] **3.** Daniel knows that he can paint 800 square feet of wall space in 4 hours. If he were able to maintain that rate, how long would it take him to paint 1200 square feet?

[3.3] **4.** The sum of two positive consecutive even integers is 146. What are the numbers?

[3.3] **5.** A rectangular eraser has a perimeter of 36 millimeters. If the length is three less than twice the width, find the length and width.

[3.4] **6.** Two joggers pass each other going in opposite directions on a path. If one jogger is traveling at a rate of 6 miles per hour and the other is traveling at a rate of 4 miles per hour, how long will it take them to be 1.5 miles apart?

8.7 Applications with Rational Expressions, Including Variation

Objectives

① Use tables to solve problems with two unknowns involving rational expressions.

② Solve problems involving direct variation.

③ Solve problems involving inverse variation.

Objective ① **Use tables to solve problems with two unknowns involving rational expressions.** In Chapter 3, we used tables to organize information involving two unknown amounts. Now let's look at some similar problems that lead to equations containing rational expressions.

Problems Involving Work

Tables are helpful in problems involving two or more people (or machines) working together to complete a task. In these problems, we are given each person's rate of work and are asked to find the time for them to complete the task if they work together. For each person involved, the rate of work, time at work, and amount of the task completed are related as follows:

$$\boxed{\text{Person's rate of work}} \cdot \boxed{\text{Time at work}} = \boxed{\text{Amount of the task completed by that person}}$$

Because the people are working together, the sum of each individual's amount of the task completed equals the whole task.

$$\boxed{\text{Amount completed by one person}} + \boxed{\text{Amount completed by the other person}} = \boxed{\text{Whole task}}$$

Example 1   Karen and Jeff own a cleaning business. Karen can clean an average-size house in 2 hours. Jeff can clean a similar house in 3 hours. How long will it take them to clean an average-size house working together?

Understand   Karen cleans at a rate of 1 house in 2 hours, or $\frac{1}{2}$ of a house per hour.

Jeff cleans at a rate of 1 house in 3 hours, or $\frac{1}{3}$ of a house per hour.

Category	Rate of Work	Time at Work	Amount of Task Completed
Karen	$\frac{1}{2}$	t	$\frac{1}{2}t$ or $\frac{t}{2}$
Jeff	$\frac{1}{3}$	t	$\frac{1}{3}t$ or $\frac{t}{3}$

Note: Because they are working together on the same job for the same amount of time, we let t represent that amount of time.

Note: Multiplying the rate of work and the time at work gives an expression of the amount of work completed. For example, if Karen works 4 hours at a rate of $\frac{1}{2}$ of a house every hour, she can clean $4 \cdot \frac{1}{2} = 2$ houses.

The total job in this case is 1 house, so we can write an equation that combines the individual expressions for the task completed and set this sum equal to 1 house.

Plan and Execute   Karen's amount completed + Jeff's amount completed = 1 house

$$\frac{t}{2} + \frac{t}{3} = 1$$

$$6\left(\frac{t}{2} + \frac{t}{3}\right) = 6(1) \qquad \text{Multiply both sides by the LCD, 6.}$$

$$\overset{3}{6} \cdot \frac{t}{\underset{1}{2}} + \overset{2}{6} \cdot \frac{t}{\underset{1}{3}} = 6 \qquad \text{Distribute and then divide out common factors.}$$

$$3t + 2t = 6$$

$$5t = 6 \qquad \text{Combine like terms.}$$

$$t = \frac{6}{5} \qquad \text{Divide both sides by 5.}$$

Answer   Working together, it takes Karen and Jeff $\frac{6}{5}$, or $1\frac{1}{5}$, hours to clean an average-size house.

Check   Karen cleans $\frac{1}{2}$ of a house per hour; so if she works alone $1\frac{1}{5}$ hours, she cleans $\frac{1}{2} \cdot \frac{6}{5} = \frac{3}{5}$ of a house. Jeff cleans $\frac{1}{3}$ of a house per hour; so in $1\frac{1}{5}$ hours, he cleans $\frac{1}{3} \cdot \frac{6}{5} = \frac{2}{5}$ of a house. Combining their individual amounts, we see that in $1\frac{1}{5}$ hours, they clean $\frac{3}{5} + \frac{2}{5} = \frac{5}{5} = 1$ house.

Your Turn 1   Terrell owns a landscaping company. Every Monday, he services the same group of clients, and it takes him 5 hours to complete the service. On one occasion, he hired a student, Jarod, to service those same clients, and it took Jarod 6 hours. Working together, how much time would it take them to service the clients?

Motion Problems

Recall that the formula for calculating distance, given the rate of travel and time of travel, is $d = rt$. If we isolate r, we have $r = \frac{d}{t}$. If we isolate t, we have $t = \frac{d}{r}$. These equations suggest that we use rational expressions when describing rate or time.

Example 2   Warren runs 3 miles away from the gym, turns around, and runs an average of 2 miles per hour faster on the return trip. If the total time of his run is $1\frac{1}{4}$ hours, what is his speed in the outbound leg and the inbound leg of his run?

Understand   We are to find the rates for each leg of Warren's run. Because this situation involves two different rates, we use a table to organize the distance, rate, and time of each leg of the run.

Category	Distance	Rate	Time
Outbound	3 miles	r	$\dfrac{3}{r}$
Inbound	3 miles	$r + 2$	$\dfrac{3}{r + 2}$

Note: Warren runs 2 mph faster during the inbound leg of his run; so we let r represent the outbound rate and add 2 mph to r for the inbound rate.

Note: Because $d = rt$, to describe time, we divide the distance by the rate: $t = \dfrac{d}{r}$.

Because the total time of the trip is $1\frac{1}{4}$ $\left(\text{or } \frac{5}{4}\right)$ hours, we can write an equation that is the sum of the outbound and inbound times.

Answer   Your Turn 1

$2\frac{8}{11}$ hr.

Plan and Execute   Outbound time + Inbound time = $\dfrac{5}{4}$ hours

$$\frac{3}{r} + \frac{3}{r+2} = \frac{5}{4}$$

$$4r(r+2)\left(\frac{3}{r} + \frac{3}{r+2}\right) = 4r(r+2)\left(\frac{5}{4}\right)$$ Multiply both sides by the LCD, $4r(r+2)$.

$$4\cancel{r}(r+2)\cdot\frac{3}{\cancel{r}} + 4r\cancel{(r+2)}\cdot\frac{3}{\cancel{r+2}} = \cancel{4}r(r+2)\cdot\frac{5}{\cancel{4}}$$ Distribute and then divide out common factors.

$$12(r+2) + 12r = 5r(r+2)$$

$$12r + 24 + 12r = 5r^2 + 10r$$ Distribute.

$$24r + 24 = 5r^2 + 10r$$ Combine like terms.

$$0 = 5r^2 - 14r - 24$$ Subtract $24r$ and 24 from both sides.

$$0 = (5r + 6)(r - 4)$$ Factor.

$$5r + 6 = 0 \quad \text{or} \quad r - 4 = 0$$ Use the zero-factor theorem.

$$5r = -6 \qquad\qquad r = 4$$

$$r = -\frac{6}{5}$$

Answer   Although $-\dfrac{6}{5}$ is a solution to the equation, a negative rate does not make sense in this situation; so we do not consider it as an answer. Therefore, the rate of the outbound leg is 4 miles per hour. Because the rate of the inbound leg is 2 miles per hour faster, the rate of speed must be 6 miles per hour.

Check   Verify that traveling 3 miles at 4 miles per hour, then 3 miles at 6 miles per hour takes a total of $1\dfrac{1}{4}$ hours.

$$\text{Outbound time} = \frac{3}{4}\,\text{hour} \qquad \text{Inbound time} = \frac{3}{6} = \frac{1}{2}\,\text{hour}$$

$$\text{Total time} = \frac{3}{4} + \frac{1}{2} = \frac{3}{4} + \frac{2}{4} = \frac{5}{4} = 1\frac{1}{4}\,\text{hour}$$

Your Turn 2   A plane travels 1200 miles against the jet stream, causing its airspeed to be decreased by 20 miles per hour. On the return flight, the plane travels with the jet stream so that its airspeed is increased by 20 miles per hour. If the total flight time of the round trip is $6\dfrac{1}{3}$ hours, what is the plane's rate in still air?

Objective ❷ Solve problems involving direct variation.   Suppose a vehicle travels at a constant rate of 30 miles per hour. Using the formula $d = rt$, with r replaced by 30, we have $d = 30t$. In the following table we use $d = 30t$ to determine distances for various values of time.

Time t	Distance Traveled $d = 30t$
1 hour	30 miles
2 hours	60 miles
3 hours	90 miles

Answer   Your Turn 2

380 mph

From the table, we see that as the time of travel increases, so does the distance traveled. Or, more formally, as values of t increase, so do values of d. In $d = 30t$, the two variables, d and t, are said to be in **direct variation** or are *directly proportional*.

Definition   *Direct variation:* Two variables, y and x, are in direct variation if $y = kx$, where k is a constant.

In words, direct variation is written as y *varies directly as* x or y *is directly proportional to* x, and these phrases translate to $y = kx$. Often the constant, k, is not given in problems involving variation, so the first objective is to find its value.

Example 3   Suppose y varies directly as x. When $y = 9$, $x = 4$. Find y when $x = 7$.

SOLUTION   Translating y *varies directly as* x, we have $y = kx$. We need to find the value of the constant k. We replace y with 9 and x with 4 in $y = kx$, then solve for k.

$$9 = k \cdot 4 \qquad \text{Replace } y \text{ with 9 and } x \text{ with 4.}$$
$$2.25 = k \qquad \text{Divide both sides by 4.}$$

Now we can replace k with 2.25 in $y = kx$ so that we have $y = 2.25x$. We can now use this equation to find y when $x = 7$.

$$y = 2.25(7) = 15.75$$

Example 4   The distance a vehicle travels varies directly with the amount of fuel it uses. A family traveled 207 miles in its van, using 9 gallons of fuel. How many gallons are required to travel 368 miles?

Understand   Translating *the distance a vehicle travels varies directly with the amount of fuel*, we write $d = kf$, where d represents distance and f represents the amount of fuel.

Plan   Use $d = kf$, replacing d with 207 miles and f with 9 gallons, to solve for the value of k. Then use that value in $d = kf$ to solve for the number of gallons required to travel 368 miles.

Execute $207 = k \cdot 9$
$$23 = k \qquad \text{Divide both sides by 9.}$$

Connection Consider the units of measurement. Because miles $= k \cdot$ gallons, when we isolate k, we have $k = \dfrac{\text{miles}}{\text{gallon}}$. The constant k represents the miles per gallon of this van, which is 23 miles per gallon. The number of miles a vehicle can travel per gallon of fuel is also called its *fuel economy*.

Replacing k with 23 in $d = kf$, we have $d = 23f$. We use $d = 23f$ to solve for f when d is 368 miles.

$$368 = 23f$$
$$16 = f \qquad \text{Divide both sides by 23.}$$

Answer   To travel 368 miles, the van needs 16 gallons of fuel.

Check   If the van's fuel economy is 23 miles per gallon, then with 16 gallons, the van will travel $23(16) = 368$ miles.

Your Turn 4   At a meat market, the price per pound, or unit price, of hamburger is constant. The amount a customer pays is directly proportional to the quantity purchased. Juan notes that a 2.5-pound package is priced at $7.20. If he wants to buy 8 pounds of hamburger, how much will he pay?

Answer   Your Turn 4

$23.04

Objective ③ Solve problems involving inverse variation. Suppose we know that the distance to a particular place is 60 miles. Given a rate, we can calculate the travel time to that place using the formula $t = \dfrac{d}{r}$. For example, if we travel at a rate of 10 miles per hour, it will take $\dfrac{60 \text{ mi}}{10 \text{ mph}} = 6$ hours to drive the 60 miles. In the following table, we use $t = \dfrac{60}{r}$ to see the relationship between rate and travel time for a fixed distance of 60 miles.

Rate r	Travel time $t = \dfrac{60}{r}$
10 mph	6 hours
20 mph	3 hours
30 mph	2 hours

From the table, we see that as speed increases, the travel time decreases. More formally, as values of r increase, values of t decrease. In $t = \dfrac{60}{r}$, the two variables, t and r, are said to be in **inverse variation** or are *inversely proportional*.

Definition   ***Inverse variation:*** Two variables, y and x, are in inverse variation if $y = \dfrac{k}{x}$, where k is a constant.

In words, inverse variation is written as y *varies inversely as* x or y *is inversely proportional to* x, and these phrases translate to $y = \dfrac{k}{x}$.

Example 5   A technician applies a constant voltage to a circuit that has a variable resistor. Varying the resistance causes the current to vary inversely. At a resistance of 4 ohms, the current is measured to be 8 amperes. Find the current when the resistance is 12 ohms.

Understand   Because the current and resistance vary inversely, we can write $i = \dfrac{k}{R}$, where i represents the current, R represents the resistance, and k represents the constant voltage.

Plan   To determine the value of the constant k, use the fact that when the resistance is 4 ohms, the current is 8 amperes. Then use the constant to find the current when the resistance is 12 ohms.

Execute $8 = \dfrac{k}{4}$

$$4 \cdot 8 = \overset{1}{\cancel{4}} \cdot \dfrac{k}{\underset{1}{\cancel{4}}}$$ Multiply both sides by 4 to eliminate the 4 in the denominator.

$$32 = k$$ **Note:** This means that the voltage in the circuit is a constant 32 volts.

Replacing k with 32 in $i = \dfrac{k}{R}$, we have $i = \dfrac{32}{R}$. We use $i = \dfrac{32}{R}$ to solve for i when R is 12 ohms.

$$i = \frac{32}{12}$$

$$i = 2\frac{2}{3} \text{ or } 2.\overline{6}$$

Answer With a resistance of 12 Ω, the current is $2\dfrac{2}{3}$ or $2.\overline{6}$ A.

Check Recall from the exercises in Section 2.3 (page 129) that the formula for voltage is $V = iR$. Using this formula, we can verify that a current of $2\dfrac{2}{3}$ A with a resistance of 12 Ω yields a voltage of 32 V.

$$V = \left(2\frac{2}{3}\right)12 = 32$$

Your Turn 5 If the wavelength of a wave remains constant, the velocity, v, of a wave is inversely proportional to its period, T. In an experiment, waves are created in a fluid so that the period is 5 seconds and the velocity is 6 centimeters per second. If the period is increased to 8 seconds, what is the velocity?

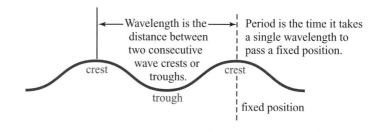

Wavelength is the distance between two consecutive wave crests or troughs.

Period is the time it takes a single wavelength to pass a fixed position.

crest

crest

trough

fixed position

Answer   Your Turn 5

3.75 cm/sec.

8.7 Exercises   For Extra Help   *MyMathLab*

PRACTICE   WATCH   DOWNLOAD

Note: Exercises marked with a ★ represent challenging exercises.

1. A person can complete a task in x hours. Write an expression that describes the portion of the task she completes in 1 hour.

2. If x represents the amount of time one person takes to complete a task and y represents the amount of time for a second person to complete the same task, write an expression that describes the portion of the task completed while both people work together for one unit of time.

3. If a vehicle travels a distance of 100 miles at a particular rate, r, write an expression that describes the amount of time it takes the vehicle to travel that 100 miles.

4. A cyclist travels 10 miles at an average rate, r, then travels 5 miles per hour faster during the return trip. Write an expression that describes the total time of the trip.

5. If two quantities vary directly, as one of the quantities increases, what happens to the other quantity?

6. If two quantities are inversely proportional, as one quantity increases, what happens to the other quantity?

For Exercises 7–14, solve the problems involving work. Use a table to organize the information and then solve. See Example 1.

7. Juan can paint a 3600-square-foot house in 12 days, and Carissa can do the same job in only 8 days. How long will it take them to paint a 3600-square-foot house if they work together?

8. After a snowfall, Franklin can shovel the snow from the walkway in 6 minutes and Shawnna can do it 8 minutes. How long will it take them to clear the walkway if they work together?

9. The Smith children are responsible for mowing the lawn. If it takes Anna 20 minutes and Christina 30 minutes, assuming that they have two lawn mowers, how long will it take them to mow the lawn working together?

10. A large field is being cleared to begin construction on a new library. To remove the dirt, two bulldozers are being used. The larger one can do the job in 6 days, while the smaller one takes 9 days. If both bulldozers are used, how long will the clearing take?

11. When filling a new swimming pool, the builder knows that it will take 6 hours using the smaller pipe. The larger pipe can do the job in only 4 hours. How long will it take the builder to fill the swimming pool if both pipes are used?

12. Karmin is planting vincas for her summer garden. Last summer, it took her 4 hours to complete the job. Her sister Kim planted the fall garden in only 3 hours. If the sisters work together, how long will the planting take?

13. To earn money for college, Peter and Susan address wedding invitations using calligraphy. Peter can complete a job of 150 invitations in 6 hours, and Susan can complete the same number in 5 hours. Working together, how long will it take them to address 150 invitations?

14. Normally, it takes Professor Gunther 3 hours to grade a class of his students' final exam essays. When his graduate assistant grades the essays, it takes her 5 hours. How long will it take them to grade the essays together?

For Exercises 15–22, solve the motion problems. Use a table to organize the information and then solve. See Example 2.

15. During an afternoon practice, two track team members took the same amount of time to run 18 miles and 12 miles, respectively. If one of the track team members ran 2 miles per hour faster than the other, find the rate of each team member.

16. If an airplane travels 1050 miles in the same amount of time that an automobile travels 150 miles and the speed of the airplane is 50 mph more than six times the speed of the car, how fast is each moving?

17. A jet flies 90 miles with a 20 mph tailwind in the same amount of time as it takes to fly 80 miles against the wind. What is the jet's speed in still air?

18. The river current is 8 miles per hour. A jet ski can travel 34 miles downstream in the same time it takes to go 18 miles upstream. What is the speed of the jet ski with no current?

19. A kayaker rows 12 miles upstream before turning around and returning to the launch area. The current in the stream is 2 miles per hour, and the total trip took 4.5 hours. What is the average speed of the kayaker in still water?

20. The Quad Cities offer tours on the Mississippi River. If the *Quad City Queen* riverboat travels upstream 120 miles and then returns and the river current is measured at 4 miles per hour, what is the speed of the *Quad City Queen* if the total travel time is 22.5 hours?

21. After training for a year, a cyclist discovers that she has increased her average rate by 4 miles per hour and can travel 15 miles in the same time that it used to take her to travel 12 miles. What was her old average rate? What is her new average rate?

22. During target practice, it is determined that a heavily armed bomber flying from base to a target 600 miles away required the same flight time as it did to fly 720 miles at a rate 30 miles per hour faster. What was the rate for each trip?

For Exercises 23–36, solve the problems involving direct variation.
See Examples 3 and 4.

23. If y varies directly as x and $y = 45$ when $x = 9$, what is the value of y when $x = 6$?

24. If a varies directly as b and $a = 15$ when $b = 27$, what is a when $b = 54$?

25. If x varies directly as y and $x = 4$ when $y = 1$, what is the value of y when $x = 18$?

26. If m varies directly as n and $m = 18$ when $n = 4$, what is the value of n when $m = 20$?

27. At the Walmart Supercenter, the cost of center cut chuck roast is constant. The purchase price increases with the quantity purchased. Darlene notices that a 1.5-pound package is priced at $6.44. If she wants to buy 6 pounds of center cut chuck roast, how much will it cost?

28. At Kroger, the cost of corn is constant. The purchase price increases with the quantity purchased. Uviv notices that 5 ears of corn cost $0.99. If she wants to buy 8 ears of corn, how much will it cost?

29. The distance a car can travel varies directly with the amount of gas it uses. On the highway, a Jeep Wrangler travels 112 miles using 7 gallons of fuel. How many gallons are required to travel 240 miles?

30. The distance a car can travel varies directly with the amount of gas it uses. On the highway, a BMW M3 convertible travels 92 miles using 4 gallons of fuel. How many gallons are required to travel 207 miles?

31. The weight of an object is directly proportional to its mass. A person who weighs 161 pounds has a mass of 5 slugs. Find the weight of a person with a mass of 4.5 slugs.

32. The weight of an object is directly proportional to its mass. An object with a mass of 55 kilograms has a weight of 539 newtons. Find the weight of a person with a mass of 60 kilograms.

Of Interest

Weight is the force due to gravity. The gravitational pull of Earth exerts a constant acceleration on all objects. In Exercise 31, k is the constant of acceleration in terms of American units (feet per second squared). In Exercise 32, k is in metric units (meters per second squared).

33. The distance a spring stretches varies directly as tension is applied to the spring. If a tension of 550 pounds stretches a spring 8 inches, how far will a tension of 1375 pounds stretch the spring?

34. The frequency of the vibrating string on a stringed instrument is directly proportional to the tension of the string. Suppose the tension of a string is 300 pounds with a frequency of 440 vibrations per second (vps). If we increase the string tension to 340 pounds by turning one of the tuning keys, what will be the frequency of the string?

35. The volume of a gas is directly proportional to temperature. Suppose a balloon containing air has a volume of 525 cubic inches at a temperature of 75°F. If the temperature drops to 60°F, what will be the volume of the balloon?

★ **36.** In physics, the energy, E, of a photon of light is directly proportional to the frequency of the light, represented by v (pronounced "nu"). A "hard" X-ray with a frequency of 2×10^{19} hertz has energy of 1.325×10^{-14} joules. What energy would a gamma ray with a frequency of 4×10^{22} hertz have?

For Exercises 37–48, solve the problems involving inverse variation. See Example 5.

37. If y varies inversely as x and $y = 4$ when $x = 10$, what is y when x is 8?

38. If m varies inversely as n and m is 6 when n is 1.5, what is m when n is 9?

39. If a varies inversely as b and b is 4 when a is 7, what is a when b is 2?

40. If x varies inversely as y and $y = 5$ when $x = 2$, what is y when x is 5?

41. The pressure that a gas exerts against the walls of a container is inversely proportional to the volume of the container. A gas is inside a cylinder with a piston at one end that can vary the volume of the cylinder. When the volume of the cylinder is 60 cubic inches, the pressure inside is 20 pounds per square inch (psi). Find the pressure of the gas if the piston compresses the gas to a volume of 40 cubic inches.

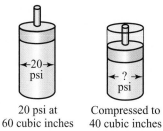

20 psi at   Compressed to
60 cubic inches 40 cubic inches

42. Find the pressure in the cylinder from Exercise 41 if the piston compresses the gas to a volume of 20 cubic inches.

43. A constant voltage is applied to a circuit that has a variable resistor. Varying the resistance causes the current to vary inversely. At a resistance of 6 ohms, the current is measured to be 10 amperes. Find the current when the resistance is 15 ohms.

44. Find the current of the circuit in Exercise 43 when the resistance is 20 ohms.

45. On a stringed instrument such as a guitar, the frequency of the vibrating string is inversely proportional to the string's length. Suppose the length of a string on a guitar is 25.5 inches with a frequency of 440 vibrations per second (vps). If we shorten the string to a length of 17 inches by placing a finger on the fret board, what will be the frequency of the string?

46. If the tension, density, and length of a string are held constant, the frequency varies inversely with the diameter. If the frequency of a string with a diameter of 0.2 millimeter is 500 vibrations per second (vps), find the frequency of a string with the same tension, density, and length if the diameter is 0.5 millimeter.

47. If the wavelength of a wave remains constant, the velocity, v, of a wave is inversely proportional to its period, T. In an experiment, waves are created in a fluid so that the period is 7 seconds and the velocity is 4 centimeters per second. If the period is increased to 12 seconds, what is the velocity?

48. The weight, W, that can be supported by a horizontal beam varies inversely with the length, L, of the beam. If a beam 12 feet long can support a weight of 600 pounds, how much weight can a beam 18 feet long support?

49. The weight of an object is inversely proportional to the square of the distance of the object from the center of Earth. At the surface of Earth, the space shuttle weighs 4.5 million pounds. It is then 4000 miles from the center of Earth. How much will the space shuttle weigh when it is 50 miles from the surface of Earth?

50. The intensity of light is inversely proportional to the square of the distance from the source. A light meter 6 feet from a lightbulb measures the intensity to be 16 foot-candles. What is the intensity at 20 feet from the bulb?

★**51.** The following table lists data from a business of the number of units it supplied to the market and the corresponding price per unit.

Supply (in thousands)	Price per Unit
150	$1.25
175	$1.50
200	$2.00
250	$2.75

  a. Plot the ordered pairs of data as points in the coordinate plane. Connect the points to form a graph.

  b. Are supply and price directly proportional or inversely proportional? Explain.

  c. Do the data represent a function? Explain.

★ **52.** The following table lists data for a business of the demand for its product and the price per unit.

Demand (in thousands)	Price per Unit
150	$2.25
175	$1.50
200	$1.00
250	$0.75

a. Plot the ordered pairs of data as points in the coordinate plane. Connect the points to form a graph.
b. Are demand and price directly proportional or inversely proportional? Explain.

c. Do the data represent a function? Explain.

Review Exercises

Exercises 1–6 EQUATIONS AND INEQUALITIES

[3.3] **1.** Find two consecutive integers so that twice the first plus four times the second is equal to 34.

[4.2] **2.** Determine whether $(3, -1)$ is a solution to the equation $-2x + 3y = -1$.

[4.2] **3.** Graph $y = -3x - 5$.

[4.3] **4.** Find the x- and y-intercepts of $y - 4x = 6$.

[4.7] *For Exercises 5 and 6, given $f(x) = -3x + 2$, find each of the following.*

5. $f\left(-\dfrac{3}{4}\right)$

6. $f(2 - a)$

Chapter 8 Summary

Defined Terms

Section 8.1
Rational expression
(p. 548)

Section 8.5
Complex rational
expression (p. 588)

Section 8.6
Extraneous solution
(p. 598)

Section 8.7
Direct variation (p. 609)
Inverse variation (p. 610)

Formulas

Direct variation: $y = kx$ (p. 609)

Inverse variation: $y = \dfrac{k}{x}$ (p. 610)

$d = rt$ (p. 607)

$t = \dfrac{d}{r}$ (p. 607)

$r = \dfrac{d}{t}$ (p. 607)

Procedures, Rules, and Key Examples

Procedures/Rules	Key Examples

SECTION 8.1 Simplifying Rational Expressions

To determine the value(s) that make a rational expression undefined,
1. Write an equation that has the denominator set equal to zero.
2. Solve the equation.

Example 1: Find every value for the variable that makes the expression undefined.

a. $\dfrac{5x}{x - 6}$

Solution: $x - 6 = 0$
$\qquad\qquad x = 6$

b. $\dfrac{n}{n^2 - 4n - 5}$

Solution: $n^2 - 4n - 5 = 0$
$(n - 5)(n + 1) = 0$
$n - 5 = 0 \quad$ or $\quad n + 1 = 0$
$n = 5 \qquad\qquad\quad n = -1$

To simplify a rational expression to lowest terms,
1. Factor the numerator and denominator completely.
2. Divide out all common factors in the numerator and denominator.
3. Multiply the remaining factors in the numerator and the remaining factors in the denominator.

Example 2: Simplify.

a. $\dfrac{18x^3y}{24x^2yz} = \dfrac{2 \cdot 3 \cdot 3 \cdot x \cdot x \cdot x \cdot y}{2 \cdot 2 \cdot 2 \cdot 3 \cdot x \cdot x \cdot y \cdot z}$

$\qquad\quad = \dfrac{3 \cdot x}{2 \cdot 2 \cdot z}$

$\qquad\quad = \dfrac{3x}{4z}$

b. $\dfrac{5m^2 - 15m}{m^2 - 9} = \dfrac{5 \cdot m \cdot (m - 3)}{(m + 3)(m - 3)}$

$\qquad\qquad\quad = \dfrac{5m}{m + 3}$

(continued)

Procedures/Rules	Key Examples

SECTION 8.2 Multiplying and Dividing Rational Expressions

To multiply rational expressions,
1. Factor each numerator and denominator completely.
2. Divide out any numerator factor with any matching denominator factor.
3. Multiply numerator by numerator and denominator by denominator.
4. Simplify as needed.

Example 1: Multiply.

a.
$$-\frac{8u^3}{3t} \cdot \frac{9t^2}{10u^3} =$$

$$-\frac{2 \cdot 2 \cdot 2 \cdot u \cdot u \cdot u}{3 \cdot t} \cdot \frac{3 \cdot 3 \cdot t \cdot t}{2 \cdot 5 \cdot u \cdot u \cdot u}$$

$$= -\frac{2 \cdot 2}{1} \cdot \frac{3 \cdot t}{5}$$

$$= -\frac{12t}{5}$$

b.
$$\frac{x^2 + 4x}{x^2 - 3x - 10} \cdot \frac{3x + 6}{12x}$$

$$= \frac{x \cdot (x + 4)}{(x + 2) \cdot (x - 5)} \cdot \frac{3 \cdot (x + 2)}{2 \cdot 2 \cdot 3 \cdot x}$$

$$= \frac{(x + 4)}{(x - 5)} \cdot \frac{1}{2 \cdot 2}$$

$$= \frac{x + 4}{4(x - 5)} \text{ or } \frac{x + 4}{4x - 20}$$

To divide rational expressions,
1. Write an equivalent multiplication statement with the reciprocal of the divisor.
2. Factor each numerator and denominator completely. (Steps 1 and 2 are interchangeable.)
3. Divide out any numerator factor with any matching denominator factor.
4. Multiply numerator by numerator and denominator by denominator.
5. Simplify as needed.

Example 2: Divide.

a.
$$\frac{3n^3}{14m} \div \frac{12mn^2}{7m}$$

$$= \frac{3n^3}{14m} \cdot \frac{7m}{12mn^2}$$

$$= \frac{3 \cdot n \cdot n \cdot n}{2 \cdot 7 \cdot m} \cdot \frac{7 \cdot m}{2 \cdot 2 \cdot 3 \cdot m \cdot n \cdot n}$$

$$= \frac{n}{2} \cdot \frac{1}{2 \cdot 2 \cdot m}$$

$$= \frac{n}{8m}$$

b.
$$\frac{r - 4}{3r^2 - 17r + 10} \div \frac{r^2 - 3r - 4}{3r^2 + r - 2}$$

$$= \frac{r - 4}{3r^2 - 17r + 10} \cdot \frac{3r^2 + r - 2}{r^2 - 3r - 4}$$

$$= \frac{r - 4}{(3r - 2) \cdot (r - 5)} \cdot \frac{(3r - 2) \cdot (r + 1)}{(r - 4) \cdot (r + 1)}$$

$$= \frac{1}{r - 5}$$

To convert units using dimensional analysis, multiply the given measurement by conversion factors so that the undesired units divide out, leaving the desired units.

Example 3: Convert.

a. 20 feet to inches

$$20 \text{ ft.} = \frac{20 \text{ ft.}}{1} \cdot \frac{12 \text{ in.}}{1 \text{ ft.}} = 240 \text{ in.}$$

b. 6 miles to yards

$$6 \text{ mi.} = \frac{6 \text{ mi.}}{1} \cdot \frac{5280 \text{ ft.}}{1 \text{ mi.}} \cdot \frac{1 \text{ yd.}}{3 \text{ ft.}} = 10,560 \text{ yd.}$$

(continued)

Procedures/Rules	Key Examples

SECTION 8.3 Adding and Subtracting Rational Expressions with the Same Denominator

To add or subtract rational expressions that have the same denominator,
1. Add or subtract the numerators and keep the same denominator.
2. Simplify to lowest terms. (Remember to factor the numerators and denominators completely in order to simplify.)

Example 1: Add or subtract.

a. $\dfrac{7y}{12} + \dfrac{y}{12} = \dfrac{7y + y}{12} = \dfrac{8y}{12} = \dfrac{2y}{3}$

b. $\dfrac{u}{u + 2} - \dfrac{3}{u + 2} = \dfrac{u - 3}{u + 2}$

c. $\dfrac{7x^2 + 5x}{x + 1} - \dfrac{x + 3}{x + 1}$

$= \dfrac{(7x^2 + 5x) - (x + 3)}{x + 1}$

$= \dfrac{(7x^2 + 5x) + (-x - 3)}{x + 1}$

$= \dfrac{7x^2 + 4x - 3}{x + 1}$

$= \dfrac{(7x - 3)(x + 1)}{x + 1}$

$= 7x - 3$

SECTION 8.4 Adding and Subtracting Rational Expressions with Different Denominators

To find the LCD of two or more rational expressions,
1. Find the prime factorization of each denominator.
2. Write a product that contains each unique prime factor the greatest number of times that factor occurs in any factorization. Or if you prefer to use exponents, write the product that contains each unique prime factor raised to the greatest exponent that occurs on that factor in any factorization.
3. Simplify the product found in step 2.

Example 1: Find the LCD.

a. $\dfrac{5}{6t}$ and $\dfrac{1}{8t^2}$

$$6t = 2 \cdot 3 \cdot t \quad 8t^2 = 2^3 \cdot t^2$$
$$\text{LCD} = 2^3 \cdot 3 \cdot t^2 = 24t^2$$

b. $\dfrac{2}{5t - 15}$ and $\dfrac{t}{t^2 - 9}$

$$5t - 15 = 5(t - 3)$$
$$t^2 - 9 = (t - 3)(t + 3)$$
$$\text{LCD} = 5(t - 3)(t + 3)$$

To add or subtract rational expressions with different denominators,
1. Find the LCD.
2. Write each rational expression as an equivalent expression with the LCD.
3. Add or subtract the numerators and keep the LCD.
4. Simplify.

Example 2: Add or subtract.

a. $\dfrac{5}{6t} + \dfrac{1}{8t^2} = \dfrac{5(4t)}{6t(4t)} + \dfrac{1(3)}{8t^2(3)}$

$= \dfrac{20t}{24t^2} + \dfrac{3}{24t^2}$

$= \dfrac{20t + 3}{24t^2}$

b. $\dfrac{2}{5t - 15} - \dfrac{t}{t^2 - 9}$

$= \dfrac{2}{5(t - 3)} - \dfrac{t}{(t - 3)(t + 3)}$

$= \dfrac{2(t + 3)}{5(t - 3)(t + 3)} - \dfrac{t(5)}{(t - 3)(t + 3)(5)}$

$= \dfrac{2t + 6}{5(t - 3)(t + 3)} - \dfrac{5t}{5(t - 3)(t + 3)}$

$= \dfrac{2t + 6 - 5t}{5(t - 3)(t + 3)}$

$= \dfrac{-3t + 6}{5(t - 3)(t + 3)}$

(continued)

Procedures/Rules	Key Examples

SECTION 8.5 Complex Rational Expressions

To simplify a complex rational expression, use one of the following methods.

Method 1

1. Simplify the numerator and denominator if needed.
2. Rewrite as a horizontal division problem.

Example 1: Simplify $\dfrac{\dfrac{5t}{t+1}}{\dfrac{10}{t^2-1}}$.

$$\dfrac{\dfrac{5t}{t+1}}{\dfrac{10}{t^2-1}} = \dfrac{5t}{t+1} \div \dfrac{10}{t^2-1}$$

$$= \dfrac{5t}{t+1} \cdot \dfrac{t^2-1}{10}$$

$$= \dfrac{5 \cdot t}{t+1} \cdot \dfrac{(t+1)(t-1)}{2 \cdot 5}$$

$$= \dfrac{t}{1} \cdot \dfrac{t-1}{2}$$

$$= \dfrac{t^2-t}{2}$$

Method 2

1. Multiply the numerator and denominator of the complex rational expression by their LCD.
2. Simplify.

Example 2: Simplify $\dfrac{\dfrac{5}{6}-\dfrac{t}{2}}{\dfrac{t}{4}+\dfrac{1}{3}}$.

$$\dfrac{\dfrac{5}{6}-\dfrac{t}{2}}{\dfrac{t}{4}+\dfrac{1}{3}} = \dfrac{12\left(\dfrac{5}{6}-\dfrac{t}{2}\right)}{12\left(\dfrac{t}{4}+\dfrac{1}{3}\right)}$$

$$= \dfrac{\dfrac{\overset{2}{\cancel{12}}}{1}\cdot\dfrac{5}{\cancel{6}} - \dfrac{\overset{6}{\cancel{12}}}{1}\cdot\dfrac{t}{\cancel{2}}}{\dfrac{\overset{3}{\cancel{12}}}{1}\cdot\dfrac{t}{\cancel{4}} + \dfrac{\overset{4}{\cancel{12}}}{1}\cdot\dfrac{1}{\cancel{3}}}$$

$$= \dfrac{10-6t}{3t+4}$$

SECTION 8.6 Solving Equations Containing Rational Expressions

To solve an equation that contains rational expressions,

1. Eliminate the rational expressions by multiplying both sides of the equation by their LCD.
2. Solve the equation using the methods we learned in Chapters 2 (linear equations) and 7 (quadratic equations).
3. Check your solution(s) in the original equation. Discard any extraneous solutions.

Example 1: Solve

$$\dfrac{x^2-6}{x-3} + 2 = \dfrac{x^2-2x}{x-3} - x.$$

Note that x cannot be 3 because it would cause the denominators to be 0, making those expressions undefined.

$$(x-3)\left(\dfrac{x^2-6}{x-3} + 2\right)$$

$$= (x-3)\left(\dfrac{x^2-2x}{x-3} - x\right)$$

(continued)

Procedures/Rules	Key Examples

SECTION 8.6 Solving Equations Containing Rational Expressions (*continued*)

$$(x-3) \cdot \frac{x^2 - 6}{x-3} + (x-3) \cdot 2$$

$$= (x-3) \cdot \frac{x^2 - 2x}{x-3} - (x-3) \cdot x$$

$$x^2 - 6 + 2x - 6 = x^2 - 2x - x^2 + 3x$$
$$x^2 + 2x - 12 = x$$
$$x^2 + x - 12 = 0$$
$$(x - 3)(x + 4) = 0$$
$$x - 3 = 0 \quad \text{or} \quad x + 4 = 0$$
$$x = 3 \qquad\qquad x = -4$$

3 is extraneous.
−4 is the solution. We leave the check to the reader.

SECTION 8.7 Applications with Rational Expressions, Including Variation

If y varies directly as x, we translate to an equation $y = kx$, where k is a constant.

Example 1: The price of photocopies is directly proportional to the number of copies produced. If 20 copies cost $1.40, how much will 150 copies cost?

Direct variation means that $y = kx$. In this case, the price of photocopies is y and the number of copies is x. We can use the fact that 20 copies cost $1.40 in $y = kx$ to find k.

$$1.40 = k(20)$$
$$0.07 = k \qquad \text{Divide by 20.}$$

Now use $y = kx$ again with $k = 0.07$ to solve for y when $x = 150$ copies.

$$y = (0.07)(150)$$
$$y = \$10.50$$

If y varies inversely as x, we translate to an equation $y = \dfrac{k}{x}$, where k is a constant.

Example 2: Given that $y = 20$ when $x = 8$, if y varies inversely as x, find x when $y = 9$.

Inverse variation means that $y = \dfrac{k}{x}$. We can use the fact that $y = 20$ and $x = 8$ to find k.

$$20 = \frac{k}{8}$$
$$160 = k \qquad \text{Multiply by 8.}$$

Now use $y = \dfrac{k}{x}$ again with $k = 160$ to solve for x when $y = 9$.

$$9 = \frac{160}{x}$$
$$9x = 160 \qquad \text{Multiply by } x.$$
$$x = 17\frac{7}{9} \qquad \text{Divide by 9.}$$

Chapter 8 Review Exercises

For Exercises 1–5, answer true or false.

[8.1]   **1.** The rational expression $\dfrac{2x-1}{x}$ is undefined when $x = \dfrac{1}{2}$.

[8.1]   **2.** A fraction is in lowest terms when the GCF of the numerator and denominator is 1.

[8.1]   **3.** The expression $\dfrac{2-x}{x-2}$ simplifies to -1.

[8.1]   **4.** The expression $-x + 5$ can be rewritten as $-1(x - 5)$.

[8.3]   **5.** $\dfrac{2x}{3} + \dfrac{x}{3} = \dfrac{3x}{3} = x$

For Exercises 6–10, complete the rule.

[8.2]   **6.** To divide rational expressions,
   a. Write an equivalent _____ statement with the reciprocal of the divisor.
   b. Factor each numerator and denominator completely. (Steps 1 and 2 are interchangeable.)
   c. _____ out any numerator factor with any matching denominator factor.
   d. Multiply numerator by numerator and denominator by denominator.
   e. Simplify as needed.

[8.3]   **7.** To add or subtract rational expressions that have the same denominator,
   a. Add or subtract the _____ and keep the same _____.
   b. Simplify to lowest terms. (Remember to factor the numerators and denominators completely in order to reduce.)

[8.5]   **8.** A _____ rational expression is a rational expression that contains rational expressions in the numerator or denominator.

[8.5]   **9.** One method of simplifying a complex rational expression is as follows:
   a. Simplify the _____ and _____ of the complex rational expression.
   b. Divide.

[8.6]   **10.** To solve an equation that contains rational expressions,
   a. Eliminate the rational expressions by multiplying through by their _____.
   b. Solve the equation using the methods we learned in Chapters 2 (linear equations) and 7 (quadratic equations).
   c. Check your solution(s) in the original equation.

Exercises 11–70   ◢ EXPRESSIONS

[8.1] For Exercises 11 and 12, evaluate the rational expression.

11. $\dfrac{4x+5}{2x}$ when   **a.** $x = 3$ **b.** $x = -1.3$ **c.** $x = -2$

12. $\dfrac{5x}{x+2}$ when   **a.** $x = 3$ **b.** $x = -3$ **c.** $x = 1.2$

[8.1] *For Exercises 13–16, find every value that can replace the variable in the expression and cause the expression to be undefined.*

13. $\dfrac{9y}{5-y}$

14. $\dfrac{2x}{x^2-4}$

15. $\dfrac{6}{x^2+5x-6}$

16. $\dfrac{2x+3}{x^2+6x}$

[8.1] *For Exercises 17–24, simplify to lowest terms.*

17. $\dfrac{3(x+5)}{9}$

18. $\dfrac{2y+12}{3y+18}$

19. $\dfrac{x^2+6x}{4x+24}$

20. $\dfrac{xy-y^2}{3x-3y}$

21. $\dfrac{a^2-b^2}{a^2-2ab+b^2}$

22. $\dfrac{3x+9}{x^4-81}$

23. $\dfrac{x^2-y^2}{y-x}$

24. $\dfrac{1-w}{w^2+2w-3}$

[8.2] *For Exercises 25–32, multiply.*

25. $\dfrac{x}{y}\cdot\dfrac{7x}{4y}$

26. $-\dfrac{2n^2}{8m^2n}\cdot\dfrac{24mp}{10n^2p^5}$

27. $\dfrac{6}{2m+4}\cdot\dfrac{3m+6}{15}$

28. $\dfrac{x^2+3x}{x^2-16}\cdot\dfrac{x^2-4x}{x^2-9}$

29. $\dfrac{n^2-6n+9}{n^2-9}\cdot\dfrac{n^2+4n+3}{n^2-3n}$

30. $\dfrac{4r^2+4r+1}{r+2r^2}\cdot\dfrac{2r}{2r^2-r-1}$

31. $\dfrac{4m}{m^2-2m+1}\cdot\dfrac{m^2-1}{16m^2}$

32. $\dfrac{w^2-2w-24}{w^2-16}\cdot\dfrac{(w+5)(w+4)}{w^2-w-30}$

[8.2] *For Exercises 33–40, divide.*

33. $\dfrac{y^2}{ab}\div\dfrac{x}{y}$

34. $-\dfrac{7y^2}{10b^2}\div-\dfrac{21y^2}{25b^2}$

35. $\dfrac{v+1}{3}\div\dfrac{3v+3}{18}$

36. $\dfrac{2a+4}{5}\div\dfrac{4a+8}{25a}$

37. $\dfrac{x^2-y^2}{c^2-d^2}\div\dfrac{x-y}{c+d}$

38. $\dfrac{b^3-6b^2+8b}{6b}\div\dfrac{2b-4}{10b+40}$

39. $\dfrac{3j+2}{5j^2-j}\div\dfrac{6j^2+j-2}{10j^2+3j-1}$

40. $\dfrac{u^2-2u-8}{u^2+3u+2}\div(u^2-3u-4)$

[8.3] *For Exercises 41–48, add or subtract as indicated.*

41. $\dfrac{5x}{18} + \dfrac{x}{18}$

42. $\dfrac{2r}{r+3} + \dfrac{r}{r+3}$

43. $\dfrac{8m+5}{6m^2} - \dfrac{10m+5}{6m^2}$

44. $\dfrac{4x+5}{x+1} - \dfrac{x+4}{x+1}$

45. $\dfrac{2x}{x+y} - \dfrac{x-y}{x+y}$

46. $\dfrac{2r-5s}{r^2-s^2} - \dfrac{2r-6s}{r^2-s^2}$

47. $\dfrac{9x^2-24x+16}{2x+1} + \dfrac{2x^2+3x-9}{2x+1}$

48. $\dfrac{q^2+2q}{q-7} - \dfrac{12q-21}{q-7}$

[8.4] *For Exercises 49–54, write equivalent rational expressions with their LCD.*

49. $\dfrac{2}{4c}, \dfrac{3}{3c^3}$

50. $\dfrac{5}{xy^3}, \dfrac{3}{x^2y^2}$

51. $\dfrac{4}{m-1}, \dfrac{4y}{m+1}$

52. $\dfrac{3}{4h-8}, \dfrac{5}{h^2-2h}$

53. $\dfrac{x}{x^2-1}, \dfrac{2+x}{x+1}$

54. $\dfrac{2}{p^2-4}, \dfrac{p}{p^2+4p+4}$

[8.4] *For Exercises 55–62, add or subtract as indicated.*

55. $\dfrac{3a-4}{18} - \dfrac{2a+5}{12}$

56. $\dfrac{4}{x+3} - \dfrac{5}{x-3}$

57. $\dfrac{3}{y-2} + \dfrac{5}{y-4}$

58. $\dfrac{5}{x^2-4} + \dfrac{4}{x+2}$

59. $\dfrac{6t}{(t-3)^2} - \dfrac{3t}{2t-6}$

60. $\dfrac{a+6}{a^2+7a+12} - \dfrac{a-3}{a+3}$

61. $\dfrac{10}{2y-1} - \dfrac{5}{1-2y}$

62. $\dfrac{3}{3x-12} + \dfrac{15}{x^2-16}$

[8.5] *For Exercises 63–70, simplify.*

63. $\dfrac{\dfrac{a}{b}}{\dfrac{x}{y}}$

64. $\dfrac{\dfrac{1}{3}-\dfrac{1}{2}}{\dfrac{1}{2}-\dfrac{1}{3}}$

65. $\dfrac{\dfrac{5}{3x+5}}{\dfrac{x}{x-2}}$

66. $\dfrac{\dfrac{x+3}{x^2-9}}{x}$

67. $\dfrac{\dfrac{x}{4x^2 - 1}}{\dfrac{5}{2x + 1}}$

68. $\dfrac{r - \dfrac{r^2 - 1}{r}}{1 - \dfrac{r - 1}{r}}$

69. $\dfrac{\dfrac{1}{x^2} + \dfrac{1}{y^2}}{\dfrac{7}{xy}}$

70. $\dfrac{\dfrac{1}{h + 1} - 1}{\dfrac{1}{h + 1}}$

Exercises 71–84 EQUATIONS AND INEQUALITIES

[8.6] *For Exercises 71–78, solve and check. Identify any extraneous solutions.*

71. $\dfrac{3x}{2} - \dfrac{x}{5} = \dfrac{19}{5}$

72. $\dfrac{3t}{6} - 2 = \dfrac{t}{6}$

73. $\dfrac{2x + 1}{4x} = \dfrac{5}{12} + \dfrac{3x - 2}{3x}$

74. $\dfrac{y + 5}{y - 3} - 5 = \dfrac{4}{y - 3}$

75. $\dfrac{x}{x - 2} = \dfrac{6}{x - 1}$

76. $\dfrac{6g - 5}{6} = \dfrac{2g - 1}{2} - \dfrac{g + 2}{2g + 5}$

77. $\dfrac{1}{p - 4} + \dfrac{1}{4} = \dfrac{8}{p^2 - 16}$

78. $\dfrac{3}{r - 2} - \dfrac{4}{r + 3} = -\dfrac{6}{r^2 + r - 6}$

[8.7] *For Exercises 79–84, solve.*

79. After giving a math test, Yolonda can grade 20 tests in 60 minutes and Shawn can do it 68 minutes. How long will it take them to grade the tests if they work together?

80. During an afternoon practice, two football team members took the same amount of time to run sprints of 40 yards and 50 yards, respectively. If one of the team members ran 2 yards per second faster than the other, find the rate of each team member.

81. If a varies directly as b and $a = 15$ when $b = 27$, what is a when $b = 54$?

82. If x varies inversely as y and $y = 5$ when $x = 2$, what is y when x is 5?

83. The distance a car can travel varies directly with the amount of gas it uses. On the highway, a car travels 80 miles using 4 gallons of fuel. How many gallons are required to travel 200 miles?

84. A constant voltage is applied to a circuit that has a variable resistor. Varying the resistance causes the current to vary inversely. At a resistance of 8 ohms, the current is measured to be 12 amperes. Find the current when the resistance is 20 ohms.

Chapter 8 Practice Test

For Extra Help

CHAPTER
Test Prep
VIDEOS

Step-by-step test solutions are found on the Chapter Test Prep Videos available via the Video Resources on DVD, in *MyMathLab*, and on You Tube (search "CarsonElemAlg" and click on "Channels").

For Exercises 1 and 2, find the value(s) that can replace the variable in the expression and cause the expression to be undefined.

1. $\dfrac{2x}{x-7}$

2. $\dfrac{8-m}{m^2-16}$

For Exercises 3 and 4, simplify to lowest terms.

3. $\dfrac{12-3x}{x^2-8x+16}$

4. $\dfrac{x-y}{x^2-y^2}$

For Exercises 5 and 6, write equivalent rational expressions with their LCD.

5. $\dfrac{6}{fg^3}, \dfrac{2}{f^2}$

6. $\dfrac{5x}{x^2-9}, \dfrac{2}{x^2+6x+9}$

For Exercises 7–18, perform the indicated operation.

7. $\dfrac{4a^2b^3}{15x^3y} \cdot \dfrac{25x^5y}{16ab}$

8. $-\dfrac{9x^3y^4}{16ab^2} \div \dfrac{45x^5y^2}{14a^7b^9}$

9. $\dfrac{4ab-8b}{x^2} \div \dfrac{2b-ab}{x^3}$

10. $\dfrac{12x^2-6x}{x^2+6x+5} \cdot \dfrac{2x^2+10x}{4x^2-1}$

11. $\dfrac{2x}{2x+3} + \dfrac{5x}{2x+3}$

12. $\dfrac{x}{x^2-9} - \dfrac{3}{x^2-9}$

13. $\dfrac{6}{x^2-4} + \dfrac{x-3}{x^2-2x}$

14. $\dfrac{4}{x-1} + \dfrac{5}{x+2}$

15. $\dfrac{5}{x-4} - \dfrac{2}{x+1}$

16. $\dfrac{x}{2x+4} - \dfrac{2}{x^2+2x}$

17. $\dfrac{2+\dfrac{1}{y}}{3-\dfrac{1}{y}}$

18. $\dfrac{\dfrac{x-y}{2}}{\dfrac{x^2-y^2}{4}}$

For Exercises 19–22, solve the equation. Identify any extraneous solutions.

19. $\dfrac{3x-1}{4} - \dfrac{7}{6} = \dfrac{2}{3}$

20. $\dfrac{4}{5m-1} = \dfrac{2}{2m-1}$

21. $\dfrac{y}{y+2} - \dfrac{y+6}{y^2-4} + \dfrac{2}{y-2} = 0$

22. $\dfrac{g}{g-1} = \dfrac{8}{g+2}$

For Exercises 23–25, solve.

23. If Erin can paint a room in 3 hours and her husband can paint the same room in 2 hours, how long will it take them to paint the room together?

24. Jake runs 3 miles away from the gym, turns around, and runs an average of 2 miles per hour slower on the return trip. If the total time of his run is $1\frac{1}{4}$ hours, what was his speed in the outbound leg and the inbound leg of his run?

25. The weight of an object is directly proportional to its mass. Suppose an object with a mass of 40 kilograms weighs 392 newtons. How much would an object with a mass of 54 kilograms weigh?

Chapter 1–8 Cumulative Review Exercises

For Exercises 1–3, answer true or false.

[2.2]   **1.** The equation $y = x^2 - 3$ is a linear equation.

[4.3]   **2.** The x-intercept for $y = 2x$ is $(0, 2)$.

[8.4]   **3.** The LCD of $\dfrac{2}{x^2 - 25}$ and $\dfrac{4x}{x + 5}$ is $(x^2 - 25)(x + 5)$.

For Exercises 4–7, fill in the blank.

[8.3]   **4.** To add or subtract rational expressions that have the same denominator,
        a. Add or subtract the numerators and keep the same _____.
        b. Simplify to lowest terms. (Remember to factor the numerators and denominators completely in order to simplify.)

[8.1]   **5.** Explain in your own words what it means to simplify a rational expression to lowest terms.

[6.4]   **6.** To multiply monomials,
        a. _____ coefficients.
        b. _____ the exponents of the like bases.
        c. Rewrite any unlike bases in the product.

[8.5]   **7.** A _____ is a rational expression that contains rational expressions in the numerator or denominator.

Exercises 8–19   **EXPRESSIONS**

[1.5] *For Exercises 8 and 9, simplify.*

  8. $5^2 - (-4^2) - 20$ **9.** $16 \div (-8) + 2^3$

For Exercises 10 and 11, simplify.

[6.6] **10.** $\dfrac{15a^4b}{-5a^{-9}b}$

[6.5] **11.** $(m + 3)(m^2 - 6m + 1)$

For Exercises 12–14, factor completely.

[7.1] **12.** $ax - 2x + a - 2$

[7.3] **13.** $3y^2 - 3y - 60$

[7.5] **14.** $-3h^2 + 75$

For Exercises 15–19, simplify.

[8.2] **15.** $\dfrac{3x - 6}{4} \cdot \dfrac{12}{x - 2}$

[8.2] **16.** $\dfrac{8}{m - 4} \div \dfrac{4}{m^2 - 16}$

[8.3] **17.** $\dfrac{x}{x^2 - 9} - \dfrac{3}{x^2 - 9}$

[8.4] **18.** $\dfrac{4}{x^2 - 49} + \dfrac{2}{x^2 - 7x}$

[8.5] **19.** $\dfrac{\dfrac{y^2 - y - 12}{4y}}{\dfrac{y - 4}{8}}$

Exercises 20–30   EQUATIONS AND INEQUALITIES

[2.3] **20.** Solve and check: $20 + 24x = 8 - 3(5 - 9x)$

[4.4] **21.** For the equation $y = \dfrac{2}{3}x + 1$,

      a. Find the slope.   **b.** Graph.

[7.6] **22.** Solve $2x^2 - 13x = 7$.

[5.3] **23.** Solve the system using elimination:
$$\begin{cases} 3x - 2y = 7 \\ 4x - 3y = 10 \end{cases}$$

[8.6] *For Exercises 24 and 25, solve.*

24. $\dfrac{u^2}{u - 3} - \dfrac{12}{u - 3} = 8$

25. $\dfrac{12}{r} = \dfrac{4}{r - 4}$

For Exercises 26–30, solve.

[3.4] **26.** Two planes leaving Seattle at the same time are going in opposite directions. If the first plane travels 600 miles per hour and the second plane travels 450 miles per hour, how long will it be until the two planes are 2625 miles apart?

[3.5] **27.** Jackson invested $7000. He put some of the $7000 into an account earning
[5.4]   9% interest annually and the rest in an account earning 7%. If he earned $600 in interest after one year, how much did he invest at each rate?

[3.5] **28.** How many ounces of a 90% balsamic dressing need to be added to 6 ounces
[5.4]   of a 70% balsamic dressing to get an 80% balsamic dressing?

[8.7] **29.** Jake can clean the garage in 8 hours, and his brother can do it in 6 hours. Working together, how long will it take them to clean the garage?

[8.7] **30.** If y varies directly as x and $y = 7$ when $x = 2$, find x when y is 21.

Roots and Radicals

9

In Section 1.5, we reviewed the basic rules of square roots. In this chapter, we explore square roots and radicals in more detail. We begin in Section 9.1 with evaluating and simplifying square root expressions, much like how we started with evaluating and simplifying polynomial expressions in the first part of Chapter 6. Then in Sections 9.2–9.4, we move to rewriting expressions containing square roots by performing arithmetic operations with them. Once we have explored arithmetic operations with radical expressions, we begin solving equations that contain radicals. We focus only on square roots in Sections 9.1–9.5. Once the foundational understanding of square roots is established, we explore some higher roots in Section 9.6.

9.1 Square Roots and Radical Expressions

Objectives

1. Find the principal square root of a given number.
2. Approximate an irrational root.
3. Simplify radical expressions.
4. Solve problems involving square roots.

In this section through Section 9.4, we focus on the expression portion of the Algebra Pyramid and explore square root and radical expressions.

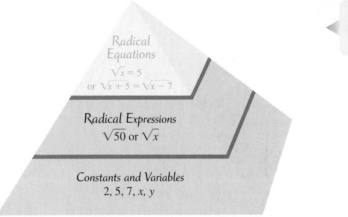

The Algebra Pyramid

Radical Equations
$\sqrt{x} = 5$
or $\sqrt{x} + 5 = \sqrt{x} - 7$

Radical Expressions
$\sqrt{50}$ or $\sqrt{x}$

Constants and Variables
$2, 5, 7, x, y$

Note: We explore radical equations in Section 9.5.

More specifically, in this section, we review some basic rules for evaluating and simplifying square root expressions.

Objective 1 Find the principal square root of a given number. In Section 1.5, we reviewed square roots and said that a square root of a given number is a number that, when squared, is the given number. We also noted that there are two square roots for every positive real number: a positive square root and a negative square root. For example, the square roots of 25 are 5 and -5 because $5^2 = 25$ and $(-5)^2 = 25$. We can write both roots more compactly as ± 5.

Also remember from Section 1.5 that the *radical* symbol, $\sqrt{\ }$, is used to indicate the positive, or *principal*, root of a number with more than one root. For example, the expression $\sqrt{25}$ means the positive, or *principal*, square root of the *radicand* 25 so that $\sqrt{25} = 5$. To indicate the negative square root of 25, we write $-\sqrt{25}$ so that $-\sqrt{25} = -5$.

Rule **Rules for Square Roots**

- Every positive number has two square roots: a positive root and a negative root.
- The radical symbol, $\sqrt{\ }$, denotes only the positive (principal) square root.
- The square root of a negative number is *not* a real number.
- $\sqrt{\dfrac{a}{b}} = \dfrac{\sqrt{a}}{\sqrt{b}}$, where $a \geq 0$ and $b > 0$.

Example 1 Evaluate the square root.

a. $\sqrt{225}$

SOLUTION $\sqrt{225} = 15$

b. $\sqrt{\dfrac{100}{121}}$

SOLUTION $\sqrt{\dfrac{100}{121}} = \dfrac{\sqrt{100}}{\sqrt{121}} = \dfrac{10}{11}$

c. $\sqrt{0.\overline{16}}$

> **Note:** The square root of a decimal has one-half as many decimal places as the radicand.

SOLUTION $\sqrt{0.\overline{16}} = 0.4$

d. $-\sqrt{81}$

> **Note:** Some people find it helpful to think of $-\sqrt{81}$ as $-1 \cdot \sqrt{81} = -1 \cdot 9 = -9$.

SOLUTION $-\sqrt{81} = -9$

e. $\sqrt{-36}$

> **Note:** In Chapter 10, we define the square root of a negative number using a set of numbers called the imaginary numbers.

SOLUTION not a real number

Your Turn 1 Evaluate the square root.

a. $\sqrt{169}$ **b.** $\sqrt{\dfrac{49}{81}}$ **c.** $\sqrt{-0.64}$ **d.** $-\sqrt{0.04}$

Objective ② **Approximate an irrational root.** Note that all of the square roots we have considered so far have been rational. Some roots, such as $\sqrt{2}$, are called *irrational* because we cannot express their exact value using rational numbers. In fact, writing $\sqrt{2}$ with the radical sign is the only way we can express its exact value. However, we can *approximate* $\sqrt{2}$ using rational numbers. We can use a table of roots (see inside back cover) or a calculator to find a decimal approximation for irrational square roots.

For example, using a table or calculator, we find that $\sqrt{2} \approx 1.414213562$, which we can round to various places as desired.

Approximating to two decimal places: $\sqrt{2} \approx 1.41$
Approximating to three decimal places: $\sqrt{2} \approx 1.414$

> **Note:** Remember that the symbol $\approx$ means "approximately equal to."

🖩 Calculator Tip

To approximate the square root of a number using some scientific calculators, enter the number and press $\boxed{\sqrt{}}$ *. Then you may need to press* $\boxed{=}$ *or* $\boxed{\text{ENTER}}$ *. To approximate a square root using other scientific calculators or graphing calculators, press* $\boxed{\text{2nd}}$ *and then* $\boxed{x^2}$ *(to access the* $\sqrt{}$ *function), enter the number, and press* $\boxed{\text{ENTER}}$ *. Note that some graphing calculators show an open parenthesis with the* $\sqrt{}$ *function, in which case you must type a closed parenthesis before pressing* $\boxed{\text{ENTER}}$ *.*

Example 2 Approximate the following irrational numbers using a calculator or the table on the page inside the back cover entitled Powers and Roots. Round to three decimal places.

a. $\sqrt{14}$

SOLUTION $\sqrt{14} \approx 3.742$

b. $-\sqrt{38}$

SOLUTION $-\sqrt{38} \approx -6.164$

Your Turn 2 Approximate the following irrational numbers. Round to three decimal places.

a. $\sqrt{19}$ **b.** $-\sqrt{54}$

Objective ③ Simplify radical expressions. Let's simplify radical expressions that contain variables. For now, we consider only radicands that are perfect squares.

Example 3 Simplify the radical expression. Assume that variables represent non-negative numbers.

Note: We assume that the variables represent nonnegative numbers because the principal square root of a nonnegative number is a nonnegative number and is not a real number when the radicand is negative.

a. $\sqrt{x^2}$

SOLUTION $\sqrt{x^2} = x$ Because $(x)^2 = x^2$

b. $\sqrt{x^8}$

SOLUTION $\sqrt{x^8} = x^4$ Because $(x^4)^2 = x^8$

c. $\sqrt{a^{10}}$

SOLUTION $\sqrt{a^{10}} = a^5$ Because $(a^5)^2 = a^{10}$

d. $\sqrt{y^{34}}$

SOLUTION $\sqrt{y^{34}} = y^{17}$ Because $(y^{17})^2 = y^{34}$

Your Turn 3 Simplify the radical expression. Assume that variables represents non-negative numbers.

a. $\sqrt{x^{12}}$ **b.** $\sqrt{y^{16}}$

Example 3 suggests the following conclusion.

Conclusion The square root of an exponential form with an even exponent has the same base raised to half of the original exponent.

Frequently, radicands contain both numbers and variables.

Example 4 Simplify the radical expression. Assume that variables represent positive numbers.

Note: Here, we assume that the variables represent positive numbers because the denominator of a fraction (part d) cannot be zero.

a. $\sqrt{16x^2}$

SOLUTION $\sqrt{16x^2} = 4x$ Because $(4x)^2 = 4^2x^2 = 16x^2$

b. $\sqrt{81x^6}$

SOLUTION $\sqrt{81x^6} = 9x^3$ Because $(9x^3)^2 = 81x^6$

c. $\sqrt{36x^{36}y^{64}}$

SOLUTION $\sqrt{36x^{36}y^{64}}$

$= 6x^{18}y^{32}$ Because $(6x^{18}y^{32})^2 = 36x^{36}y^{64}$

Connection In Section 6.4, we learned to simplify an exponential form raised to a power by multiplying the exponent by the power.

$$(9x^3)^2 = 9^{1\cdot2}x^{3\cdot2}$$
$$= 9^2x^6 = 81x^6$$

Answers Your Turn 2
a. 4.359 **b.** −7.348

Answers Your Turn 3
a. x^6 **b.** y^8

Connection Notice the similarity in the method for finding the square root of a fraction and squaring a fraction. To find the square root, we find the square root of the numerator and denominator. To square a fraction, we square the numerator and denominator.

d. $\sqrt{\dfrac{25m^4}{64n^2}}$

SOLUTION $\sqrt{\dfrac{25m^4}{64n^2}} = \dfrac{\sqrt{25m^4}}{\sqrt{64n^2}} = \dfrac{5m^2}{8n}$ Because $\left(\dfrac{5m^2}{8n}\right)^2 = \dfrac{(5m^2)^2}{(8n)^2} = \dfrac{25m^4}{64n^2}$

Your Turn 4 Simplify the radical expression. Assume that variables represent positive numbers.

a. $\sqrt{9x^{10}}$ **b.** $-\sqrt{36x^8}$ **c.** $\sqrt{49x^{18}y^{24}}$ **d.** $\sqrt{\dfrac{t^6}{9u^4}}$

Objective 4 **Solve problems involving square roots.** Many problems in math, science, and engineering can be solved using formulas that contain radicals.

Example 5

a. Ignoring air resistance, the velocity of an object, v, in meters per second can be found after the object has fallen h meters by using the formula $v = -\sqrt{19.6h}$. Find the velocity of a stone that has fallen 30 meters after being dropped from a cliff. Round the answer to the nearest tenth.

Understand   We are to find the velocity of an object after it has fallen 30 meters.

Plan   Use the formula $v = -\sqrt{19.6h}$, replacing h with 30.

Execute $v = -\sqrt{19.6(30)}$ Replace h with 30.

$\qquad\qquad v = -\sqrt{588}$ Multiply within the radical.

$\qquad\qquad v \approx -24.2$ Evaluate the square root.

Note: A negative velocity indicates that the object is traveling down.

Answer   After falling 30 meters, the stone is traveling at a velocity of -24.2 meters per second.

Check   We can verify the calculations, which we will leave to the reader.

b. The *period* of a pendulum is the amount of time it takes the pendulum to swing from the point of release to the opposite extreme, then back to the point of release. The period, T, measured in seconds, can be found using the formula

$$T = 2\pi\sqrt{\dfrac{L}{9.8}},$$

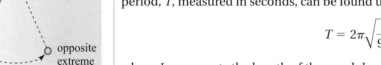

point of release   opposite extreme

where L represents the length of the pendulum in meters. Find the period of a pendulum that is 0.5 meter long. Round the answer to the nearest hundredth.

Understand   We are to find the period of a pendulum that is 0.5 meter long.

Plan   Use the formula $T = 2\pi\sqrt{\dfrac{L}{9.8}}$, approximating π with 3.14 and replacing L with 0.5.

Execute $T \approx 2(3.14)\sqrt{\dfrac{0.5}{9.8}}$ Simplify.

$\qquad\qquad T \approx 6.28\sqrt{0.051}$

$\qquad\qquad T \approx 6.28(0.226)$ Approximate the square root.

$\qquad\qquad T \approx 1.42$

Answer   The period of a pendulum that is 0.5 meter long is 1.42 seconds.

Check   We can verify the calculations, which we will leave to the reader.

Answers   Your Turn 4

a. $3x^5$ **b.** $-6x^4$ **c.** $7x^9y^{12}$ **d.** $\dfrac{t^3}{3u^2}$

Your Turn 5

a. A skydiver jumps from a plane and puts her body into a dive position so that air resistance is minimized. Find her velocity after she has fallen 100 meters. Round the answer to the nearest tenth.

b. Find the period of a pendulum that is 0.2 meter long. Round your answer to the nearest thousandth. Use $\pi \approx 3.14$.

In Section 7.6, we used the Pythagorean theorem, $c^2 = a^2 + b^2$, to find missing side lengths of a right triangle. Conversely, if the sides of a triangle satisfy the Pythagorean theorem, the triangle is a right triangle. Solving the formula for c, we get $c = \sqrt{a^2 + b^2}$, which gives the length of the hypotenuse if we know the length of the legs.

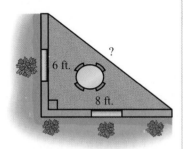

6 ft.

8 ft.

?

Example 6   A building contractor wants to make sure that two walls of a building meet at a right angle. If he marks a point 6 feet from the corner of one wall and a point 8 feet from the corner of the other wall, what should the distance be between those two points?

Understand   We know the two legs of a right triangle, and we want to find the hypotenuse.

Plan   Use the formula $c = \sqrt{a^2 + b^2}$ and replace a with 6 and b with 8.

Execute $c = \sqrt{6^2 + 8^2}$ Replace a with 6 and b with 8.

$c = \sqrt{36 + 64}$ Simplify within the radical.

$c = \sqrt{100}$

$c = 10$

Answer   If the angle is a right angle, the distance between the points should be 10 feet.

Check   We can verify the calculations, which we will leave to the reader.

Answers   Your Turn 5

a. -44.3 meters per second

b. ≈ 0.897 second

Answer   Your Turn 6

$\sqrt{80} \approx 8.94$ ft.

Your Turn 6   A newly transplanted tree is supported by three ropes that are attached to the tree at a point 8 feet above the ground. These ropes are tied to pegs that are driven into the ground 4 feet from the base of the tree. Find the length of one of the ropes. Round to the nearest hundredth if necessary.

9.1 Exercises   For Extra Help   *MyMathLab*   Math XP PRACTICE WATCH DOWNLOAD

Note: Exercises marked with a ★ represent challenging exercises.

1. Give an example of a number with a rational square root and an example of a number with an irrational square root. Discuss the difference between a rational square root and an irrational square root.

2. How can we express the exact value of an irrational square root?

3. Why are there two square roots for every positive real number?

4. What is the principal square root of a number?

5. Why is the square root of any negative number not a real number?

6. Given an expression of the form $\sqrt{x^n}$, if n is an even natural number, write an expression for the square root.

For Exercises 7–14, find all square roots of the given number. See Objective 1.

7. 36   **8.** 64   **9.** 121   **10.** 81   **11.** 196   **12.** 400

13. 225   **14.** 289

For Exercises 15–40, evaluate the square root if possible. See Example 1.

15. $\sqrt{49}$ **16.** $\sqrt{100}$ **17.** $\sqrt{361}$ **18.** $\sqrt{225}$ **19.** $-\sqrt{36}$ **20.** $-\sqrt{169}$

21. $\sqrt{-81}$ **22.** $\sqrt{-121}$ **23.** $\sqrt{\dfrac{25}{36}}$ **24.** $\sqrt{\dfrac{36}{169}}$ **25.** $\sqrt{\dfrac{16}{9}}$ **26.** $\sqrt{\dfrac{25}{144}}$

27. $\sqrt{-\dfrac{25}{49}}$ **28.** $\sqrt{-\dfrac{81}{25}}$ **29.** $-\sqrt{\dfrac{169}{25}}$ **30.** $-\sqrt{\dfrac{49}{16}}$ **31.** $\sqrt{0.36}$ **32.** $\sqrt{0.81}$

33. $\sqrt{0.04}$ **34.** $\sqrt{0.09}$ **35.** $\sqrt{-0.64}$ **36.** $\sqrt{-2.25}$ **37.** $\sqrt{0.0121}$ **38.** $\sqrt{0.0009}$

39. $-\sqrt{0.0049}$ **40.** $-\sqrt{0.1089}$

For Exercises 41–48, classify each square root as rational or irrational. See Objective 2.

41. $\sqrt{3}$ **42.** $\sqrt{24}$ **43.** $\sqrt{16}$ **44.** $\sqrt{4}$ **45.** $\sqrt{25}$ **46.** $\sqrt{36}$

47. $\sqrt{5}$ **48.** $\sqrt{6}$

For Exercises 49–60, use a calculator or the Powers and Roots table to approximate the square root. Round to three decimal places. See Example 2.

49. $\sqrt{3}$ **50.** $\sqrt{8}$ **51.** $\sqrt{12}$ **52.** $\sqrt{15}$ **53.** $-\sqrt{6}$ **54.** $-\sqrt{7}$

55. $\sqrt{11}$ **56.** $\sqrt{10}$ **57.** $\sqrt{51}$ **58.** $\sqrt{40}$ **59.** $\sqrt{29}$ **60.** $\sqrt{13}$

For Exercises 61–74, simplify. Assume that variables represent positive numbers. See Examples 3 and 4.

61. $\sqrt{a^2}$ **62.** $\sqrt{b^4}$ **63.** $\sqrt{x^6}$ **64.** $\sqrt{b^{10}}$ **65.** $\sqrt{64x^4}$ **66.** $\sqrt{4n^2}$

67. $\sqrt{64a^2b^6}$ **68.** $\sqrt{100x^2y^6}$ **69.** $\sqrt{0.81a^2b^6}$ **70.** $\sqrt{0.25a^6b^{10}}$ **71.** $\sqrt{\dfrac{16}{25}r^2s^2}$ **72.** $\sqrt{\dfrac{4}{25}x^8y^6}$

73. $\sqrt{\dfrac{36a^2b^4}{49m^2}}$ **74.** $\sqrt{\dfrac{25x^4y^8}{81z^6}}$

For Exercises 75 and 76, use the following information. Ignoring air resistance, the velocity of an object, v, in meters per second can be found after the object has fallen h meters by using the formula $v = -\sqrt{19.6h}$. See Example 5.

75. Find the velocity of a rock that has fallen 16 meters after being dropped from a cliff. Round to the nearest thousandth.

76. Find the velocity of a ball that has fallen 5 meters after being dropped from a roof. Round to the nearest thousandth.

For Exercises 77 and 78, use the following information. On a clear day, the distance a person can see to the horizon can be calculated using the formula $d = \sqrt{1.8a}$, where d represents the distance to the horizon in miles and a represents the person's altitude above the ground in feet. See Example 5.

77. Taipei 101 Tower in Taipei, Taiwan, is the tallest building in the world at 1667 feet. On December 25, 2004, Alain Robert, "the French Spider-Man", scaled to the top of the tower in approximately 4 hours. Calculate how far Robert could see to the horizon. Round to the nearest tenth. (*Source: France Today*)

78. "The French Spider-Man" referred to in Exercise 77 scaled the 1060-foot Eiffel Tower on December 31, 1996. Calculate how far Robert could see to the horizon. Round to the nearest tenth. (*Source: France Today*)

For Exercises 79–82, use the following information. The period, T (in seconds), of a pendulum can be found using the formula $T = 2\pi\sqrt{\dfrac{L}{9.8}}$, where L represents the length of the pendulum in meters. See Example 5.

79. Find the period of a pendulum that is 2.45 meters long. Give an exact answer and an approximation rounded to the nearest hundredth of a second.

80. Find the period of a pendulum that is 3.528 meters long. Give an exact answer and an approximation rounded to the nearest hundredth of a second.

81. In 1851, Leon Foucault built a large pendulum in the dome of the Pantheon in Paris to demonstrate the rotation of Earth. The length of that pendulum was 67 meters. What was the pendulum period rounded to the nearest tenth of a second? (*Source:* Wikipedia)

82. The length of the Foucault pendulum in the Houston Museum of Natural Science is 18.8 meters. Calculate its period rounded to the nearest tenth of a second. (*Source:* Houston Museum of Natural Science)

For Exercises 83 and 84, use the following information. The speed of a car can be determined by the length of the skid marks after the car brakes by using the formula $S = \sqrt{30Dfn}$ where D represents the length of the skid mark in feet, f is the drag factor of the surface, n is the braking efficiency as a percent (written as a decimal), and S represents the speed of the car in miles per hour. (Source: www.harristechnical.com). See Example 5.

83. Find the speed of a car if the skid length measures 40 feet long, the drag factor of the surface is 0.75 and the braking efficiency is 100%.

84. Find the speed of a car if the skid length measures 100 feet long, the drag factor of the surface is 0.30 and the braking efficiency is 30%.

For Exercises 85 and 86, use the Pythagorean theorem. If the result is irrational, round to the nearest tenth. See Example 6.

85. Three pieces of lumber are to be connected to form a right triangle that will be part of the frame for a roof. If the horizontal piece is to be 12 feet and the vertical piece is to be 5 feet, how long must the connecting piece be?

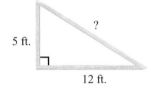

86. A fitness instructor decides to create a ropes course with a zip line. She wants the line to connect from a 40-foot-tall tower to the ground at a point 100 feet from the base of the tower. Assuming that the tower and ground form a right angle, find the length of the zip line.

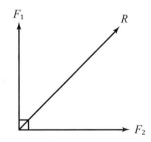

For Exercises 87 and 88, use the following information. Two forces, F_1 and F_2, acting on an object at a 90° angle will pull the object with a resultant force, R, at an angle between F_1 and F_2. (See the figure.) The value of the resultant force can be found by the formula $R = \sqrt{F_1^2 + F_2^2}$. If the result is irrational, round to three decimal places. See Example 5.

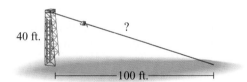

87. Find the resultant force if $F_1 = 9\,N$ and $F_2 = 12\,N$.

Note: The unit of force in the metric system is the newton (abbreviated N).

88. Find the resultant force if $F_1 = 6.2\,N$ and $F_2 = 4.5\,N$

89. The number of associate degrees, in thousands, conferred by institutions of higher education since 1970 can be approximated with the equation $y = 52\sqrt{x} + 252$, where x represents the number of years after 1970. (*Source:* National Center for Education Statistics)

   a. How many associate degrees were awarded in 1980?

   b. How many associate degrees were awarded in 2006?

90. The number of earthquakes worldwide since 2000 with a magnitude of 3.0 to 3.9 can be approximated using the equation $y = 1570\sqrt{x} + 4784$, where x represents the number of years after 2000. (*Source:* USGS Earthquake Hazards Program)

   a. Approximate how many earthquakes with a magnitude of 3.0 to 3.9 occurred in 2002.

   b. Approximate how many earthquakes with a magnitude of 3.0 to 3.9 occurred in 2008.

Collaborative Exercises   At the Root of It All

Recall that b is the square root of a if $a = b \cdot b = b^2$. How might we find a square root of a number (besides using the square root key on a calculator)?

One method for evaluating square roots is called the **bisection method.** *To find the square root of a given number with this method, we use a pair of factors of the given number to find other pairs of factors that are closer together. The process ends when the factors are so close that they are nearly identical.*

Here's how it works. Let N represent the number whose square root we want to find.

   a. Find two integer factors of N. The closer these factors are to each other, the quicker this process works.

   b. Find the average of the two factors from step a. (Find the average by adding the two factors and dividing the sum by 2.)

   c. Divide N by the average found in step b.

   d. The result of Step c and the average from step b are two new factors of N, which you can verify by multiplying them together. (Note that they may not be integers and they should be closer together than the factor pair with which you started this process.) Using the two new factors, begin again with step b. Repeat steps b and c until the two numbers are identical out to the desired number of decimal places.

1. Use this process to evaluate $\sqrt{300}$ so that the two factors are the same out to two decimal places.

2. Now use the $\sqrt{}$ function on your calculator to evaluate $\sqrt{300}$. Are the answers close?

3. Discuss how you would use a similar method to find the cube root of a number. Remember that b is a cube root of a if $a = b \cdot b \cdot b = b^3$. Also remember that the average of three numbers is the sum of the numbers divided by 3.

Review Exercises

Exercises 1–6   EXPRESSIONS

[1.5]　**1.** Follow the order of operations to simplify.

    a. $\sqrt{16 \cdot 9}$ **b.** $\sqrt{16} \cdot \sqrt{9}$

For Exercises 2–5, multiply.

[6.4]　**2.** $x^5 \cdot x^3$ **[6.4]**　**3.** $(-9m^3n)(5mn^2)$ **[6.5]**　**4.** $4x^2(3x^2 - 5x + 1)$ **[6.5]**　**5.** $(7y - 4)(3y + 5)$

[6.5]　**6.** Write an expression in simplest form for the area of the figure shown.

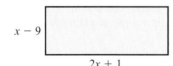

$x - 9$

$2x + 1$

9.2 Multiplying and Simplifying Square Roots

Objectives

1. Multiply square root expressions.
2. Use the product rule of square roots to simplify square roots with radicands that contain a perfect square factor.
3. Use the product rule of square roots to simplify square roots that contain variables.

In this section, we explore some ways to simplify expressions containing radicals. In particular, we explore expressions that involve multiplication of radicals.

Objective ① Multiply square root expressions. Consider the expression $\sqrt{4} \cdot \sqrt{25}$. The usual approach is to multiply the roots. However, we also can multiply the radicands, then find the root of the product.

$$\text{Find the roots first: } \sqrt{4} \cdot \sqrt{25} = 2 \cdot 5 = 10$$

or

$$\text{Multiply the radicands first: } \sqrt{4} \cdot \sqrt{25} = \sqrt{100} = 10$$

Both approaches give the same result, which suggests the following rule.

Rule **Product Rule for Square Roots**

$$\sqrt{a} \cdot \sqrt{b} = \sqrt{a \cdot b}, \text{ where } a \geq 0 \text{ and } b \geq 0$$

The product rule for square roots is particularly useful when we cannot easily determine the initial square roots. In such situations, applying the rule may lead to a square root that is easier to evaluate.

Example 1 Simplify. Assume that variables represent nonnegative numbers.

a. $\sqrt{3} \cdot \sqrt{27}$

> **Connection** Example 1(a) illustrates that the product of two irrational numbers can be a rational number.

SOLUTION Both $\sqrt{3}$ and $\sqrt{27}$ are irrational numbers. Using the product rule for square roots to multiply the radicands leads to a radicand that is a perfect square and avoids having to approximate roots.

$$\sqrt{3} \cdot \sqrt{27} = \sqrt{3 \cdot 27} \quad \text{Use the product rule for square roots to write the radicands under a common radical.}$$
$$= \sqrt{81} \quad \text{Multiply the radicands.}$$
$$= 9 \quad \text{Evaluate the square root.}$$

b. $2\sqrt{3} \cdot 5\sqrt{12}$

> **Connection** Multiplying two radical expressions is similar to multiplying two monomials. For example, $2\sqrt{3} \cdot 5\sqrt{12}$ is similar to $2x \cdot 5y$, which simplifies to $10xy$.

SOLUTION The numbers 2 and 5 are coefficients. Multiply coefficients by coefficients and radicands by radicands.

$$2\sqrt{3} \cdot 5\sqrt{12} = 2 \cdot 5 \cdot \sqrt{3} \cdot \sqrt{12} \quad \text{Use the commutative property to rearrange the factors.}$$
$$= 10 \cdot \sqrt{3 \cdot 12} \quad \text{Use the product rule for square roots.}$$
$$= 10\sqrt{36} \quad \text{Multiply in the radical.}$$
$$= 10 \cdot 6 \quad \text{Evaluate the square root.}$$
$$= 60 \quad \text{Multiply the root by the coefficient.}$$

c. $3x^3\sqrt{8x} \cdot 2x^2\sqrt{2x^3}$

SOLUTION The coefficients are $3x^3$ and $2x^2$. Multiply coefficients by coefficients and radicands by radicands.

$$3x^3\sqrt{8x} \cdot 2x^2\sqrt{2x^3} = 3x^3 \cdot 2x^2 \cdot \sqrt{8x} \cdot \sqrt{2x^3}$$ Use the commutative property to rearrange the factors.

$$= 6x^5\sqrt{8x \cdot 2x^3}$$ Multiply the coefficients and use the product rule for square roots.

$$= 6x^5\sqrt{16x^4}$$ Multiply in the radical.

$$= 6x^5 \cdot 4x^2$$ Evaluate the square root.

$$= 24x^7$$ Multiply the root by the coefficient.

Your Turn 1   Simplify. Assume that variables represent nonnegative numbers.

a. $\sqrt{2} \cdot \sqrt{32}$ **b.** $3\sqrt{20} \cdot 4\sqrt{5}$ **c.** $4y^4\sqrt{12y^5} \cdot 2y^2\sqrt{3y}$

Answers **Your Turn 1**

a. 8   **b.** 120   **c.** $48y^9$

Objective ❷ **Use the product rule of square roots to simplify square roots with radicands that contain a perfect square factor.**   If a radicand such as 12 is not a perfect square, yet contains a factor that is a perfect square (4, in this case), the square root ($\sqrt{12}$) is not considered to be in simplest form. We can use the product rule for square roots to simplify such square roots. We can write $\sqrt{12}$ as $\sqrt{4 \cdot 3}$ and then find the root of the perfect square factor, 4.

Note: Recall that the perfect squares are 1, 4, 9, 16, 25, 36, 49, 64, 81, 100, . . .

$$\sqrt{12} = \sqrt{4 \cdot 3} = \sqrt{4} \cdot \sqrt{3} = 2 \cdot \sqrt{3} \text{ or } 2\sqrt{3}$$

When we simplify a square root in this way, it is best to find the root of the greatest perfect square factor. Otherwise, we have to keep simplifying. Consider the expression $\sqrt{32}$. The radicand, 32, has two factors that are perfect squares, 4 and 16; so the expression could be rewritten two ways.

$$\sqrt{32} = \sqrt{4 \cdot 8} = \sqrt{4} \cdot \sqrt{8} = 2\sqrt{8}$$ Not yet simplified

or

$$\sqrt{32} = \sqrt{16 \cdot 2} = \sqrt{16} \cdot \sqrt{2} = 4\sqrt{2}$$ Simplified

Both results are correct in that they are equal to $\sqrt{32}$. However, $4\sqrt{2}$ is in simplest form, whereas $2\sqrt{8}$ is not. To continue simplifying $2\sqrt{8}$, we have

$$2\sqrt{8} = 2\sqrt{4 \cdot 2} = 2\sqrt{4} \cdot \sqrt{2} = 2 \cdot 2 \cdot \sqrt{2} = 4\sqrt{2}$$ Simplified

Procedure   **Simplifying Square Roots with a Perfect Square Factor in the Radicand**

To simplify a square root with a radicand that contains a perfect square factor,

1. Write the radicand in factored form so that one of the factors is the greatest perfect square factor.
2. Use the product rule of square roots to separate the two factors into two radical expressions.
3. Find the square root of the perfect square factor and leave the other factor in a radical.

Example 2   Simplify.

a. $\sqrt{45}$

SOLUTION $\sqrt{45} = \sqrt{9 \cdot 5}$ The greatest perfect square factor of 45 is 9, so we write 45 as $9 \cdot 5$.

$$= \sqrt{9} \cdot \sqrt{5}$$ Use the product rule of square roots to separate the factors into two radicals.

$$= 3\sqrt{5}$$ Find the square root of 9 and leave 5 in a radical.

b. $2\sqrt{150}$

SOLUTION $2\sqrt{150} = 2\sqrt{25 \cdot 6}$ The greatest perfect square factor of 150 is 25, so we write 150 as $25 \cdot 6$.

$= 2 \cdot \sqrt{25} \cdot \sqrt{6}$ Use the product rule of square roots to separate the factors into two radicals.

$= 2 \cdot 5 \cdot \sqrt{6}$ Find the square root of 25 and leave 6 in a radical.

$= 10\sqrt{6}$ Multiply 5 by the coefficient 2.

Your Turn 2   Simplify.

   a. $\sqrt{48}$ **b.** $7\sqrt{60}$

      If the greatest perfect square factor of a particular radicand is not obvious, try using the prime factorization of the radicand. Consider $\sqrt{45}$ from Example 2(a). Suppose we did not notice that 9 is the greatest perfect square factor of 45. In the prime factorization of 45, which is $3 \cdot 3 \cdot 5$, the pair of 3s indicate a perfect square, 9. In fact, in any prime factorization, an even number of the same factor indicates a perfect square.

Example 3   Simplify.

a. $\sqrt{112}$

SOLUTION   Because the greatest perfect square factor of 112 is not obvious, we use the prime factorization of 112, which is $2 \cdot 2 \cdot 2 \cdot 2 \cdot 7$. The four 2s indicate that a perfect square factor of 16 is in 112, so we can write 112 as $16 \cdot 7$.

$$\sqrt{112} = \sqrt{16 \cdot 7}$$ Write 112 as $16 \cdot 7$.

$$= \sqrt{16} \cdot \sqrt{7}$$ Use the product rule of square roots to separate the factors into two radicals.

$$= 4\sqrt{7}$$ Find the square root of 16 and leave 7 in a radical.

b. $5\sqrt{432}$

Connection We have discussed that a perfect square has an even number of prime factors. The prime factorization of its square root will have *half* the number of the same prime factors. Consider the perfect square 144. Its prime factorization has four 2s and two 3s and is written $2 \cdot 2 \cdot 2 \cdot 2 \cdot 3 \cdot 3$. The square root of 144, which is 12, has only two 2s and one 3 in its prime factorization, which is half the number of 2s and 3s in the prime factorization of 144.

SOLUTION   The prime factorization of 432 is $2 \cdot 2 \cdot 2 \cdot 2 \cdot 3 \cdot 3 \cdot 3$. The four 2s and pair of 3s multiply to equal the perfect square 144.

$$5\sqrt{432} = \sqrt{144 \cdot 3}$$ Write 432 as $144 \cdot 3$.

$$= 5 \cdot \sqrt{144} \cdot \sqrt{3}$$ Use the product rule of square roots to separate the factors into two radicals.

$$= 5 \cdot 12 \cdot \sqrt{3}$$ Find the square root of 144 and leave 3 in a radical.

$$= 60\sqrt{3}$$ Multiply 12 by the coefficient 5.

Your Turn 3   Simplify.

   a. $\sqrt{294}$ **b.** $3\sqrt{108}$

Objective ③ Use the product rule of square roots to simplify square roots that contain variables. We can use the product rule of square roots to simplify radical expressions that contain variables. We will rewrite the variable expression as a product of a perfect square and another factor.

Answers   **Your Turn 2**

a. $4\sqrt{3}$ **b.** $14\sqrt{15}$

Answers   **Your Turn 3**

a. $7\sqrt{6}$ **b.** $18\sqrt{3}$

Example 4   Simplify. Assume that variables represent nonnegative numbers.

a. $\sqrt{12x^3}$

SOLUTION $\sqrt{12x^3} = \sqrt{4x^2 \cdot 3x}$

> The greatest perfect square factor of $12x^3$ is $4x^2$, so begin by writing $12x^3$ as $4x^2 \cdot 3x$.

$$= \sqrt{4x^2} \cdot \sqrt{3x}$$

> Use the product rule of square roots to separate the factors into two radicals.

$$= 2x\sqrt{3x}$$

> Find the square root of $4x^2$ and leave $3x$ in a radical.

b. $\sqrt{72x^4y}$

SOLUTION $\sqrt{72x^4y} = \sqrt{36x^4 \cdot 2y}$

> The greatest perfect square factor of $72x^4y$ is $36x^4$, so begin by writing $72x^4y$ as $36x^4 \cdot 2y$.

$$= \sqrt{36x^4} \cdot \sqrt{2y}$$

> Use the product rule of square roots to separate the factors into two radicals.

$$= 6x^2\sqrt{2y}$$

> Find the square root of $36x^4$ and leave $2y$ in a radical.

c. $2x^4\sqrt{63x^5}$

SOLUTION $2x^4\sqrt{63x^5} = 2x^4\sqrt{9x^4 \cdot 7x}$

> The greatest perfect square factor of $63x^5$ is $9x^4$, so begin by writing $63x^5$ as $9x^4 \cdot 7x$.

$$= 2x^4 \cdot \sqrt{9x^4} \cdot \sqrt{7x}$$

> Use the product rule of square roots to separate the factors into two radicals.

$$= 2x^4 \cdot 3x^2 \cdot \sqrt{7x}$$

> Find the square root of $9x^4$ and leave $7x$ in a radical.

$$= 6x^6\sqrt{7x}$$

> Multiply $3x^2$ by the coefficient $2x^4$.

Your Turn 4   Simplify. Assume that variables represent nonnegative numbers.

a. $\sqrt{25x^3}$ **b.** $\sqrt{48x^2y^3}$ **c.** $4y^3\sqrt{18y^5}$

Answers   **Your Turn 4**

a. $5x\sqrt{x}$ **b.** $4xy\sqrt{3y}$
c. $12y^5\sqrt{2y}$

9.2 Exercises   For Extra Help   *MyMathLab*   Math XL
PRACTICE   WATCH   DOWNLOAD

1. Why is the product rule for square roots helpful in simplifying $\sqrt{3} \cdot \sqrt{12}$?

2. Using the product rule for square roots to simplify $\sqrt{2} \cdot \sqrt{3}$ gives $\sqrt{6}$. Although 6 is not a perfect square, using the product rule for square roots is preferable over multiplying the approximate square roots of 2 and 3. Why?

3. Explain why the expression $\sqrt{28}$ is not in simplest form.

4. Explain in your own words how to simplify a square root containing a radicand with a perfect square factor.

For Exercises 5–30, simplify. Assume that variables represent nonnegative numbers. See Example 1.

5. $\sqrt{2} \cdot \sqrt{32}$ **6.** $\sqrt{3} \cdot \sqrt{12}$ **7.** $2\sqrt{5} \cdot 3\sqrt{20}$ **8.** $3\sqrt{2} \cdot 4\sqrt{8}$

9. $\sqrt{8m} \cdot \sqrt{8m}$ **10.** $\sqrt{5x^2} \cdot \sqrt{5x^2}$ **11.** $\sqrt{3x} \cdot \sqrt{27x^5}$ **12.** $\sqrt{8y^3} \cdot \sqrt{2y}$

13. $\sqrt{6y^3} \cdot \sqrt{6y}$ **14.** $\sqrt{3m^4} \cdot \sqrt{27m^6}$ **15.** $\sqrt{5} \cdot \sqrt{20p^8}$ **16.** $\sqrt{6} \cdot \sqrt{24y^6}$

17. $\sqrt{72k^3} \cdot \sqrt{2k^7}$ **18.** $\sqrt{8m^3} \cdot \sqrt{2m}$ **19.** $\sqrt{25x^3} \cdot \sqrt{x^3}$ **20.** $\sqrt{9r^3} \cdot \sqrt{r^5}$

21. $4\sqrt{5yz^3} \cdot 2\sqrt{45yz}$ **22.** $3\sqrt{50u^3v^2} \cdot 2\sqrt{2uv^4}$ **23.** $-2\sqrt{28g^3} \cdot \sqrt{7gt^4}$

24. $\sqrt{12pv^3} \cdot -3\sqrt{27pv^5}$ **25.** $2\sqrt{13w^3x^2z} \cdot 3\sqrt{13wz}$ **26.** $4\sqrt{r^3s^2t} \cdot 5\sqrt{4rs^8t^7}$

27. $3cy\sqrt{16c^3y} \cdot 2c^2y\sqrt{cy^{11}}$ **28.** $4j^2k\sqrt{25j^4k^3} \cdot 2jk^2\sqrt{4j^6k^3}$ **29.** $-2xyz\sqrt{2x^3y^2z} \cdot 4x^2yz^2\sqrt{8xy^4z^7}$

30. $5q^2rs^2\sqrt{27qrs} \cdot -3qr^2s^2\sqrt{3q^3rs^5}$

For Exercises 31–54, simplify. Assume that variables represent nonnegative numbers. See Examples 2–4.

31. $\sqrt{12}$ **32.** $\sqrt{20}$ **33.** $\sqrt{24}$ **34.** $\sqrt{40}$ **35.** $\sqrt{63}$

36. $\sqrt{125}$ **37.** $4\sqrt{27}$ **38.** $6\sqrt{50}$ **39.** $\sqrt{8a^4}$ **40.** $\sqrt{27a^2}$

41. $\sqrt{20m^3}$ **42.** $\sqrt{50v^5}$ **43.** $\sqrt{48p^2q}$ **44.** $\sqrt{60x^2y^3}$ **45.** $\sqrt{120x^2y^5}$

46. $\sqrt{128x^2y^3}$ **47.** $5\sqrt{153k^4m^7}$ **48.** $3\sqrt{24u^3w^4}$ **49.** $4xy\sqrt{40x^2y^4}$ **50.** $2ab\sqrt{18a^6b^2}$

51. $-2a^2b^3\sqrt{54a^7b^2}$ **52.** $-3xy^2\sqrt{54x^4y^5}$ **53.** $-6x^3y\sqrt{48x^7y^5}$ **54.** $-4x^4y^2\sqrt{28x^3y^5}$

Find
ⓧ *the mistake*

For Exercises 55–58, explain the mistake; then simplify correctly. Assume that variables represent nonnegative numbers.

55. $\sqrt{32x} = \sqrt{16 \cdot 2x}$ **56.** $\sqrt{32x^3} = \sqrt{4x^2 \cdot 8x}$
$\qquad = 4\sqrt{2x}$ $\qquad\qquad = 2x\sqrt{8x}$

57. $\sqrt{12x^4} = \sqrt{4x^4 \cdot 3}$ **58.** $\sqrt{40x^6} = \sqrt{4x^6 \cdot 10}$
$\qquad = 4x^2\sqrt{3}$ $\qquad\qquad = 2\sqrt{10x^3}$

For Exercises 59–62, solve for the missing value. Assume that variables represent nonnegative numbers.

59. $2\sqrt{7} = \sqrt{?}$ **60.** $2\sqrt{5} = \sqrt{?}$ **61.** $2ab^2\sqrt{5b} = \sqrt{?}$ **62.** $4mn^2\sqrt{2mn} = \sqrt{?}$

For Exercises 63–66, find the area of each figure.

63.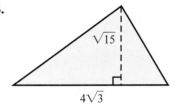

64.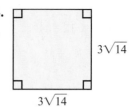

65.

66.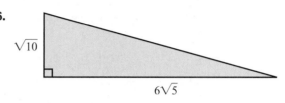

A telemarketer calls a house and speaks to the mother of three children. The telemarketer asks the mother, "How old are you?" The woman answers, "41." The marketer then asks, "How old are your children?" The woman replies, "The product of their ages is our house number, 1296, and all of their ages are perfect squares." What are the ages of the children?

Review Exercises

Exercises 1–6 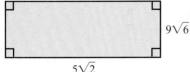 EXPRESSIONS

[1.7] **1.** Combine like terms: $5x^4 - 2x + 9x - 18x^4$

[6.6] *For Exercises 2 and 3, divide.*

2. $\dfrac{4x^9}{2x^6}$

3. $\dfrac{5r^5}{10r}$

For Exercises 4–6, factor.

[7.2] **4.** $x^2 - 7x + 12$

[7.3] **5.** $6y^2 - 5y - 21$

[7.4] **6.** $n^2 - 36$

9.3 Dividing and Simplifying Square Roots

Objectives

① Use the quotient rule for square roots to simplify square roots in the numerator and denominator of a fraction.

② Rationalize denominators containing square roots.

In Section 9.2, we learned that if a radicand has perfect square factors, the radical expression is not in simplest form and we use the product rule of square roots to simplify those expressions. In this section, we see that if a fraction has a radical in the denominator, as in $\dfrac{\sqrt{50}}{\sqrt{2}}$ or $\dfrac{1}{\sqrt{2}}$, the expression is not in simplest form and we use the quotient rule of square roots to simplify such fractions.

Objective ① **Use the quotient rule for square roots to simplify square roots in the numerator and denominator of a fraction.** First, let's consider fractions with square roots in both the numerator and denominator, such as $\dfrac{\sqrt{100}}{\sqrt{25}}$. We can follow the order of operations and divide the roots. However, we also can divide the radicands, then find the square root of the quotient.

$$\text{Find the roots first: } \dfrac{\sqrt{100}}{\sqrt{25}} = \dfrac{10}{5} = 2$$

or

$$\text{Divide the radicands first: } \dfrac{\sqrt{100}}{\sqrt{25}} = \sqrt{\dfrac{100}{25}} = \sqrt{4} = 2$$

Both approaches give the same result, which suggests the following rule.

Rule **Quotient Rule for Square Roots**

$$\dfrac{\sqrt{a}}{\sqrt{b}} = \sqrt{\dfrac{a}{b}}, \text{ where } a \geq 0 \text{ and } b > 0$$

The quotient rule for square roots is useful in a division of square roots when the radicands are not perfect squares but their quotient is a perfect square.

Example 1 Simplify. Assume that variables represent positive numbers.

a. $\dfrac{\sqrt{50}}{\sqrt{2}}$

SOLUTION $\dfrac{\sqrt{50}}{\sqrt{2}} = \sqrt{\dfrac{50}{2}}$ Use the quotient rule for square roots to write the radicands under a common radical.

$\phantom{\dfrac{\sqrt{50}}{\sqrt{2}}} = \sqrt{25}$ Divide the radicands.

$\phantom{\dfrac{\sqrt{50}}{\sqrt{2}}} = 5$

b. $\dfrac{\sqrt{48x^3}}{\sqrt{6x}}$

SOLUTION $\dfrac{\sqrt{48x^3}}{\sqrt{6x}} = \sqrt{\dfrac{48x^3}{6x}}$ Use the quotient rule of square roots to write the radicands under a common radical.

$\phantom{\dfrac{\sqrt{48x^3}}{\sqrt{6x}}} = \sqrt{8x^2}$ Divide the radicands.

$\phantom{\dfrac{\sqrt{48x^3}}{\sqrt{6x}}} = \sqrt{4x^2 \cdot 2}$ Use the product rule for square roots to simplify.

$\phantom{\dfrac{\sqrt{48x^3}}{\sqrt{6x}}} = 2x\sqrt{2}$

Your Turn 1 Simplify. Assume that variables represent positive numbers.

a. $\dfrac{\sqrt{48}}{\sqrt{3}}$

b. $\dfrac{\sqrt{84x^5y}}{\sqrt{7x}}$

Answers Your Turn 1
a. 4   b. $2x^2\sqrt{3y}$

The quotient rule can also be used in "reverse," where $\sqrt{\dfrac{a}{b}} = \dfrac{\sqrt{a}}{\sqrt{b}}$, when a quotient is not a perfect square but its numerator or denominator is.

Example 2   Simplify. Assume that variables represent nonnegative numbers.

a. $\sqrt{\dfrac{100}{9}}$

SOLUTION $\sqrt{\dfrac{100}{9}} = \dfrac{\sqrt{100}}{\sqrt{9}}$ Because both the numerator and denominator are perfect squares, use the quotient rule for square roots to write the numerator and denominator in separate radicals.

$= \dfrac{10}{3}$

b. $\sqrt{\dfrac{15}{16}}$

SOLUTION $\sqrt{\dfrac{15}{16}} = \dfrac{\sqrt{15}}{\sqrt{16}}$ Because the denominator is a perfect square, use the quotient rule for square roots to write the numerator and denominator in separate radicals.

$= \dfrac{\sqrt{15}}{4}$ **Note:** Because the radical is not in the denominator, this expression is in simplest form.

c. $\sqrt{\dfrac{x^2}{49}}$

SOLUTION $\sqrt{\dfrac{x^2}{49}} = \dfrac{\sqrt{x^2}}{\sqrt{49}}$ Because the denominator is a perfect square, use the quotient rule for square roots to write the numerator and denominator in separate radicals.

$= \dfrac{x}{7}$

Your Turn 2   Simplify. Assume that variables represent positive numbers.

a. $\sqrt{\dfrac{121}{64}}$ b. $\sqrt{\dfrac{19}{25}}$ c. $\sqrt{\dfrac{1}{4x^6}}$

Sometimes, we may be able to simplify the fraction in a radical. If so, it is usually best to simplify the fraction first, then use the quotient rule to determine the root.

Example 3   Simplify. Assume that variables represent positive numbers.

a. $\sqrt{\dfrac{128}{50}}$

SOLUTION $\sqrt{\dfrac{128}{50}} = \sqrt{\dfrac{64}{25}}$ Simplify the fraction by dividing out the common factor, 2.

$= \dfrac{\sqrt{64}}{\sqrt{25}}$ Use the quotient rule for square roots to write the numerator and denominator in separate radicals.

$= \dfrac{8}{5}$

Answers   Your Turn 2

a. $\dfrac{11}{8}$ b. $\dfrac{\sqrt{19}}{5}$ c. $\dfrac{1}{2x^3}$

b. $\sqrt{\dfrac{5x^7}{45x}}$

SOLUTION $\sqrt{\dfrac{5x^7}{45x}} = \sqrt{\dfrac{x^6}{9}}$ Simplify the rational expression by dividing out the common factor, $5x$.

$\qquad\qquad\quad = \dfrac{\sqrt{x^6}}{\sqrt{9}}$ Use the quotient rule for square roots to write the numerator and denominator in separate radicals.

$\qquad\qquad\quad = \dfrac{x^3}{3}$

Your Turn 3   Simplify. Assume that variables represent positive numbers.

a. $\sqrt{\dfrac{10}{490}}$ **b.** $\sqrt{\dfrac{243x}{3x^5}}$

Some simplifications may require the use of the product rule for square roots.

Example 4   Simplify. Assume that variables represent nonnegative numbers.

a. $\sqrt{\dfrac{80}{9}}$

SOLUTION $\sqrt{\dfrac{80}{9}} = \dfrac{\sqrt{80}}{\sqrt{9}}$ Because the denominator is a perfect square, use the quotient rule for square roots to write the numerator and denominator in separate radicals.

$\qquad\qquad\quad = \dfrac{\sqrt{16 \cdot 5}}{3}$ To simplify the numerator, write 80 as a product of its greatest perfect square factor, 16, and 5.

$\qquad\qquad\quad = \dfrac{\sqrt{16} \cdot \sqrt{5}}{3}$ Use the product rule of square roots to separate the factors.

$\qquad\qquad\quad = \dfrac{4\sqrt{5}}{3}$

b. $\sqrt{\dfrac{144x^3}{50}}$

SOLUTION $\sqrt{\dfrac{144x^3}{50}} = \sqrt{\dfrac{72x^3}{25}}$ Simplify the fraction by dividing out the common factor, 2.

$\qquad\qquad\quad = \dfrac{\sqrt{72x^3}}{\sqrt{25}}$ Use the quotient rule for square roots to write the numerator and denominator in separate radicals.

$\qquad\qquad\quad = \dfrac{\sqrt{36x^2 \cdot 2x}}{5}$ To simplify the numerator, write $72x^3$ as a product of its greatest perfect square factor, $36x^2$ and $2x$.

$\qquad\qquad\quad = \dfrac{\sqrt{36x^2} \cdot \sqrt{2x}}{5}$ Use the product rule of square roots to separate the factors.

$\qquad\qquad\quad = \dfrac{6x\sqrt{2x}}{5}$

Answers   **Your Turn 3**

a. $\dfrac{1}{7}$ **b.** $\dfrac{9}{x^2}$

Answers   **Your Turn 4**

a. $\dfrac{2\sqrt{11}}{5}$ **b.** $\dfrac{3x\sqrt{x}}{2}$

Your Turn 4   Simplify. Assume that variables represent positive numbers.

a. $\sqrt{\dfrac{44}{25}}$ **b.** $\sqrt{\dfrac{54x^6}{24x^3}}$

Objective ❷ Rationalize denominators containing square roots. As we stated at the beginning of this section, if the denominator of a fraction contains a radical, the fraction is not considered to be in simplest form. Now we simplify fractions such as $\dfrac{1}{\sqrt{2}}$ that have an irrational number in the denominator. Our goal is to *rationalize* the denominator, which means to rewrite the expression so that it has a rational number in the denominator.

In general, we multiply the fraction by a well-chosen 1 so that the radical is eliminated. In the case of a square root in the denominator, we multiply it by a factor that makes the radicand a perfect square, which allows us to eliminate the square root. For example, to rationalize $\dfrac{1}{\sqrt{2}}$, we could multiply by $\dfrac{\sqrt{2}}{\sqrt{2}}$ because the product's denominator is the square root of a perfect square.

$$\frac{1}{\sqrt{2}} = \frac{1}{\sqrt{2}} \cdot \frac{\sqrt{2}}{\sqrt{2}} = \frac{\sqrt{2}}{\sqrt{4}} = \frac{\sqrt{2}}{2}$$

Note: We are not changing the value of $\dfrac{1}{\sqrt{2}}$ because we are multiplying it by 1 in the form of $\dfrac{\sqrt{2}}{\sqrt{2}}$.

Any factor that produces a perfect square radicand works. For example, we could have multiplied $\dfrac{1}{\sqrt{2}}$ by $\dfrac{\sqrt{8}}{\sqrt{8}}$.

$$\frac{1}{\sqrt{2}} = \frac{1}{\sqrt{2}} \cdot \frac{\sqrt{8}}{\sqrt{8}} = \frac{\sqrt{8}}{\sqrt{16}} = \frac{\sqrt{4 \cdot 2}}{4} = \frac{2\sqrt{2}}{4} = \frac{\sqrt{2}}{2}$$

Notice, however, that multiplying by $\dfrac{\sqrt{2}}{\sqrt{2}}$ required fewer steps to simplify because it produced a smaller perfect square radicand than $\dfrac{\sqrt{8}}{\sqrt{8}}$. Our example suggests the following procedure.

Procedure | **Rationalizing Square Root Denominators**

To rationalize a denominator containing a single square root, multiply the fraction by a 1 so that the product's denominator has a radicand that is a perfect square, then simplify.

Example 5 Rationalize the denominator. Assume that variables represent positive numbers.

a. $\dfrac{2}{\sqrt{3}}$

SOLUTION $\dfrac{2}{\sqrt{3}} = \dfrac{2}{\sqrt{3}} \cdot \dfrac{\sqrt{3}}{\sqrt{3}}$ Multiply by $\dfrac{\sqrt{3}}{\sqrt{3}}$.

$= \dfrac{2\sqrt{3}}{\sqrt{9}}$ Simplify.

$= \dfrac{2\sqrt{3}}{3}$

b. $\sqrt{\dfrac{3}{8}}$

SOLUTION $\sqrt{\dfrac{3}{8}} = \dfrac{\sqrt{3}}{\sqrt{8}}$ Use the quotient rule of square roots to separate the numerator and denominator into two radicals.

$= \dfrac{\sqrt{3}}{\sqrt{8}} \cdot \dfrac{\sqrt{2}}{\sqrt{2}}$ Multiply by $\dfrac{\sqrt{2}}{\sqrt{2}}$.

> **Note:** We chose to multiply by $\dfrac{\sqrt{2}}{\sqrt{2}}$ because it leads to a smaller perfect square than other choices, such as $\dfrac{\sqrt{8}}{\sqrt{8}}$. Multiplying by $\dfrac{\sqrt{8}}{\sqrt{8}}$ produces the same final answer but requires more steps.

$= \dfrac{\sqrt{6}}{\sqrt{16}}$ Simplify.

$= \dfrac{\sqrt{6}}{4}$

Warning: Although 6 and 4 have a common factor of 2, we cannot divide out that factor because 6 is a radicand whereas 4 is not. We never divide out factors common to a radicand and a number that is not a radicand.

c. $\dfrac{x^5\sqrt{3}}{\sqrt{x}}$

SOLUTION $\dfrac{x^5\sqrt{3}}{\sqrt{x}} = \dfrac{x^5\sqrt{3}}{\sqrt{x}} \cdot \dfrac{\sqrt{x}}{\sqrt{x}}$ Multiply by $\dfrac{\sqrt{x}}{\sqrt{x}}$.

$= \dfrac{x^5\sqrt{3x}}{\sqrt{x^2}}$ Simplify.

$= \dfrac{x^5\sqrt{3x}}{x}$ $\sqrt{x^2} = x$

$= x^4\sqrt{3x}$ Simplify to lowest terms.

Your Turn 5   Rationalize the denominator. Assume that variables represent positive numbers.

a. $\dfrac{1}{\sqrt{7}}$ **b.** $\sqrt{\dfrac{7}{12}}$ **c.** $\dfrac{m^3\sqrt{5}}{\sqrt{m}}$

Just as a fraction is not in simplest form unless it is reduced to lowest terms, a radical expression is not in simplest form unless certain conditions are met.

Procedure   **Simplest Form of a Square Root**

To ensure that an expression containing a square root is in simplest form, check that

- The radicand has no perfect square factors.
- The radicand has no fractions.
- The denominator has no radicals.

Answers   Your Turn 5

a. $\dfrac{\sqrt{7}}{7}$ **b.** $\dfrac{\sqrt{21}}{6}$ **c.** $m^2\sqrt{5m}$

9.3 Exercises   For Extra Help   *MyMathLab*

Note: Exercises marked with a ★ represent challenging exercises.

1. Explain why each of the following expressions is not in simplest form.

   a. $\sqrt{24}$

   b. $\sqrt{\dfrac{3}{16}}$

   c. $\dfrac{5}{\sqrt{3}}$

2. The quotient rule for square roots states that $\dfrac{\sqrt{a}}{\sqrt{b}} = \sqrt{\dfrac{a}{b}}$, where $a \geq 0$ and $b > 0$. Why can a be equal to 0, whereas b must be strictly greater than 0?

3. Explain in your own words how to rationalize a denominator.

4. Although $\dfrac{1}{\sqrt{3}} = \dfrac{\sqrt{3}}{3}$, explain why $\dfrac{\sqrt{3}}{3}$ is considered simplest form.

For Exercises 5–46, simplify. Assume that variables represent positive numbers. See Examples 1–4.

5. $\dfrac{\sqrt{20}}{\sqrt{5}}$

6. $\dfrac{\sqrt{72}}{\sqrt{8}}$

7. $\dfrac{\sqrt{8}}{\sqrt{2}}$

8. $\dfrac{\sqrt{18}}{\sqrt{2}}$

9. $\dfrac{\sqrt{72}}{\sqrt{18}}$

10. $\dfrac{\sqrt{75}}{\sqrt{3}}$

11. $\dfrac{\sqrt{80}}{\sqrt{5}}$

12. $\dfrac{\sqrt{98}}{\sqrt{2}}$

13. $\dfrac{\sqrt{48}}{\sqrt{2}}$

14. $\dfrac{\sqrt{90}}{\sqrt{2}}$

15. $\dfrac{\sqrt{54}}{\sqrt{3}}$

16. $\dfrac{\sqrt{60}}{\sqrt{5}}$

17. $\dfrac{\sqrt{x^3}}{\sqrt{x}}$

18. $\dfrac{\sqrt{n^5}}{\sqrt{n}}$

19. $\dfrac{\sqrt{48a^7}}{\sqrt{3a}}$

20. $\dfrac{\sqrt{72x^3}}{\sqrt{2x}}$

21. $\dfrac{\sqrt{24m^9n}}{\sqrt{3m}}$

22. $\dfrac{\sqrt{54x^5y^2}}{\sqrt{3xy}}$

23. $\dfrac{\sqrt{5tu}}{\sqrt{45t^3u}}$

24. $\dfrac{\sqrt{5x^3y^2}}{\sqrt{20x^7y^2}}$

25. $\dfrac{\sqrt{72a^5b}}{\sqrt{6a^3b^7}}$

26. $\dfrac{\sqrt{60m^3n^9}}{\sqrt{3m^7n}}$

27. $\sqrt{\dfrac{9}{64}}$

28. $\sqrt{\dfrac{49}{16}}$

29. $-\sqrt{\dfrac{16}{81}}$

30. $-\sqrt{\dfrac{25}{36}}$

31. $\sqrt{\dfrac{a^6}{100}}$

32. $\sqrt{\dfrac{b^8}{49}}$

33. $\sqrt{\dfrac{y^2}{25x^2}}$

34. $\sqrt{\dfrac{81}{4y^2}}$

35. $-\sqrt{\dfrac{128}{2}}$

36. $-\sqrt{\dfrac{150}{6}}$

37. $\sqrt{\dfrac{44}{99}}$

38. $\sqrt{\dfrac{80}{45}}$

39. $-\sqrt{\dfrac{27}{100}}$

40. $-\sqrt{\dfrac{90}{16}}$

41. $\sqrt{\dfrac{80}{5x^6}}$

42. $\sqrt{\dfrac{2n^8}{162}}$

43. $\sqrt{\dfrac{16m^3n}{m^5n^3}}$

44. $\sqrt{\dfrac{81w^3z}{wz^7}}$

45. $\sqrt{\dfrac{60a^3b}{45ab}}$

46. $\sqrt{\dfrac{24xy}{27xy^3}}$

X *the mistake*   **For Exercises 47–50, explain the mistake; then simplify correctly.**

47. $\sqrt{\dfrac{16}{4}} = 4$

48. $\sqrt{\dfrac{81}{9}} = 9$

49. $\dfrac{\sqrt{50}}{\sqrt{2}} = 25$

50. $\dfrac{\sqrt{48}}{\sqrt{3}} = 16$

For Exercises 51–70, simplify by rationalizing the denominator.
Assume that variables represent positive numbers. See Example 5.

51. $\dfrac{1}{\sqrt{3}}$ **52.** $\dfrac{1}{\sqrt{5}}$ **53.** $\dfrac{3}{\sqrt{8}}$ **54.** $\dfrac{5}{\sqrt{12}}$ **55.** $\dfrac{3}{\sqrt{a}}$ **56.** $\dfrac{6}{\sqrt{b}}$

57. $\sqrt{\dfrac{36}{7}}$ **58.** $\sqrt{\dfrac{81}{5}}$ **59.** $\sqrt{\dfrac{m}{n}}$ **60.** $\sqrt{\dfrac{p}{q}}$ **61.** $\dfrac{\sqrt{7x^2}}{\sqrt{50}}$ **62.** $\dfrac{\sqrt{3x^2}}{\sqrt{32}}$

63. $\dfrac{\sqrt{8}}{\sqrt{56}}$ **64.** $\dfrac{\sqrt{10}}{\sqrt{20}}$ **65.** $\dfrac{x^2\sqrt{3}}{\sqrt{x}}$ **66.** $\dfrac{r^6\sqrt{2}}{\sqrt{r}}$ **67.** $\sqrt{\dfrac{2}{7}}$ **68.** $\sqrt{\dfrac{15}{2}}$

69. $\sqrt{\dfrac{a}{b^3}}$ **70.** $\sqrt{\dfrac{m}{n^3}}$

X *the mistake*   **For Exercises 71 and 72, explain the mistake; then simplify correctly.**

71. $\dfrac{\sqrt{3}}{\sqrt{2}} = \dfrac{\sqrt{3}}{\sqrt{2}} \cdot \dfrac{2}{2} = \dfrac{2\sqrt{3}}{2}$

72. $\sqrt{\dfrac{7}{3}} = \dfrac{\sqrt{7}}{\sqrt{3}} \cdot \dfrac{\sqrt{3}}{\sqrt{3}} = \dfrac{\sqrt{21}}{9}$

★ **For Exercises 73–76, find the expression that can replace the highlighted box and**
make the equation true. Assume that variables represent positive numbers.

73. $\sqrt{\dfrac{18x^5}{}} = 3x$ **74.** $\sqrt{\dfrac{25x}{}} = \dfrac{5}{4x^2}$ **75.** $\sqrt{\dfrac{}{45xy^3}} = \dfrac{x}{3y}$ **76.** $\sqrt{\dfrac{}{8x}} = 3$

77. In Section 9.1, we used the formula $T = 2\pi\sqrt{\dfrac{L}{9.8}}$ to determine the period of a pendulum, where T is the period in seconds and L is the length in meters. Rewrite the formula so that the denominator is rationalized.

78. The formula $t = \sqrt{\dfrac{h}{16}}$ can be used to find the distance an object has fallen, where t is the time in seconds and h is the distance the object has fallen in feet. Rewrite the formula so that the denominator is rationalized.

79. The formula $s = \sqrt{\dfrac{3V}{h}}$ can be used to find the length, s, of each side of the base of a pyramid having a square base, volume V in cubic feet, and height h in feet.
   a. Rationalize the denominator in the formula.

   b. The volume of the Great Pyramid at Giza in Egypt is approximately 83,068,742 cubic feet, and its height is 449 feet. Find the length of each side of its base.

80. The formula $r = \sqrt{\dfrac{3V}{\pi h}}$ can be used to find the radius, r, of the base of a right circular cone, where V is the volume and h is the height.
   a. Rationalize the denominator.

   b. If the volume of a right cicular cone is 108π cubic feet and the height is 9 feet, find the radius of the base.

81. In AC circuits, voltage is often expressed as a *root-mean-square*, or *rms*, value. The formula for calculating the rms voltage given the maximum voltage value, V_{m}, is $V_{\text{rms}} = \dfrac{V_{\text{m}}}{\sqrt{2}}$.
   a. Rationalize the denominator in the formula.

   b. Given a maximum voltage of 163 V, write an expression for the rms voltage.

   c. Use a calculator to approximate the rms voltage in part b rounded to the nearest tenth.

82. The velocity, in meters per second, of a particle can be determined by the formula $v = \sqrt{\dfrac{2E}{m}}$, where E represents the kinetic energy, in joules, of the particle and m represents the mass, in kilograms, of the particle.
   a. Rationalize the denominator in the formula.

   b. A particle with a mass of 1×10^{-6} kilogram has 2.4×10^{7} joules of kinetic energy. Write a radical expression in simplest form for its velocity.

   c. Use a calculator to approximate the velocity in part b rounded to the nearest tenth.

83. Given $f(x) = \dfrac{6}{x}$, find each of the following. Express your answer in simplest form.

    a. $f(\sqrt{3})$ **b.** $f(\sqrt{2})$ **c.** $f(\sqrt{5})$

84. Given $g(x) = \dfrac{5\sqrt{2}}{x}$, find each of the following. Express your answer in simplest form.

    a. $g(\sqrt{6})$ **b.** $g(\sqrt{10})$ **c.** $g(\sqrt{22})$

 For Exercises 85–88, use a graphing calculator.

85. Graph $f(x) = \dfrac{1}{\sqrt{x}}$; then graph $g(x) = \dfrac{\sqrt{x}}{x}$.

    a. What do you notice about the two graphs? What does this indicate about the two functions?

    b. Simplify $f(x)$ by rationalizing the denominator. What do you notice?

86. Graph $f(x) = -\dfrac{1}{\sqrt{x}}$; then graph $g(x) = -\dfrac{\sqrt{x}}{x}$.

    a. What do you notice about the two graphs? What does this indicate about the two functions?

    b. Simplify $f(x)$ by rationalizing the denominator. What do you notice?

87. Graph $f(x) = \dfrac{1}{\sqrt{x}}$; then graph $g(x) = -\dfrac{1}{\sqrt{x}}$.

    a. What do you notice about the two graphs?

    b. What do the graphs suggest about how the minus sign affects the graph?

88. Given $f(x) = \dfrac{1}{\sqrt{x}}$, $g(x) = \dfrac{2}{\sqrt{x}}$, and $h(x) = \dfrac{3}{\sqrt{x}}$.

    a. What do you notice about the graphs?

    b. What do the graphs suggest about how increasing the constant in the numerator affects the graph?

Review Exercises

Exercises 1–6   **EXPRESSIONS**

For Exercises 1–6, simplify.

[1.5]   **1.** $\sqrt{0.01}$ **[1.5]**   **2.** $7(4^2 - 3 \cdot 2) + 8^0$

[1.7]   **3.** $4x + 3y - 2x + 7y$ **[1.7]**   **4.** $9t^2 - 3t - 2t^3 + 3t - 5 + t^2$

[6.3]   **5.** $(8y^2 - 5y + 7) + (y^2 + 3y - 10)$ **[6.3]**   **6.** $(3n^3 + 10n^2 - 2n - 5) - (8n^3 - 2n + 9)$

9.4 Addition, Subtraction, and Mixed Operations with Square Roots

Objectives

1. Add or subtract square roots.
2. Use the distributive property in expressions containing square roots.
3. Rationalize denominators that have a sum or difference with a square root term.

Objective 1 **Add or subtract square roots.** Adding or subtracting square roots is essentially the same as combining like terms. Remember that the distributive property is at work when we combine like terms because we factor out the common variable, which leaves a sum or difference of the coefficients.

$$3x + 4x = (3 + 4)x \quad \text{Factor out the variable } x; \text{ then add the coefficients.}$$
$$= 7x$$

Now consider a problem in which a square root is like the variable x in the preceding example.

$$3\sqrt{5} + 4\sqrt{5} = (3 + 4)\sqrt{5} \quad \text{Factor out } \sqrt{5}; \text{ then add the coefficients.}$$
$$= 7\sqrt{5}$$

Variable terms are like terms if the variables and their exponents are identical. Similarly, square roots with identical radicands are said to be **like radicals**.

Definition **Like radicals:** Two radical expressions with identical radicands (and indexes).

> **Note:** Currently, we work only with square roots that have an identical index of 2. When we study higher roots in Section 9.6, indexes become important to consider.

Procedure **Adding Like Radicals**

To add or subtract like radicals, add or subtract the coefficients and keep the radicals the same.

Example 1 Simplify.

a. $5\sqrt{2} - 9\sqrt{2}$

SOLUTION $5\sqrt{2} - 9\sqrt{2} = (5 - 9)\sqrt{2}$
$$= -4\sqrt{2}$$

b. $12\sqrt{3} - \sqrt{3}$

SOLUTION $12\sqrt{3} - \sqrt{3} = (12 - 1)\sqrt{3}$
$$= 11\sqrt{3}$$

c. $5\sqrt{x} + \sqrt{x} - 12\sqrt{x}$

SOLUTION $5\sqrt{x} + \sqrt{x} - 12\sqrt{x} = (5 + 1 - 12)\sqrt{x}$
$$= -6\sqrt{x}$$

d. $4\sqrt{2} - 5\sqrt{3} + 7\sqrt{3} - 11\sqrt{2}$

SOLUTION $\quad 4\sqrt{2} - 5\sqrt{3} + 7\sqrt{3} - 11\sqrt{2} = 4\sqrt{2} - 11\sqrt{2} - 5\sqrt{3} + 7\sqrt{3}$ Group like radicals.

$$= (4 - 11)\sqrt{2} + (-5 + 7)\sqrt{3}$$

$$= -7\sqrt{2} + 2\sqrt{3}$$

Your Turn 1   Simplify.

  a. $\sqrt{5} + 13\sqrt{5}$ **b.** $\sqrt{x} - 7\sqrt{x} + 3\sqrt{x}$ **c.** $6\sqrt{3} + 2\sqrt{6} - 2\sqrt{3} - 8\sqrt{6}$

In a problem involving addition or subtraction of radicals, if the radicands are not identical, it may be possible to simplify one or more of the radicals so that they are identical.

Example 2   Simplify.

a. $7\sqrt{3} + \sqrt{12}$

SOLUTION $\quad 7\sqrt{3} + \sqrt{12} = 7\sqrt{3} + \sqrt{4 \cdot 3}$ Factor out the perfect square factor, 4.

$$= 7\sqrt{3} + \sqrt{4} \cdot \sqrt{3}$$ Use the product rule of square roots to separate the radicals.

$$= 7\sqrt{3} + 2\sqrt{3}$$ Simplify.

$$= 9\sqrt{3}$$ Because the radicals are like radicals, add the coefficients.

b. $3\sqrt{50} - 7\sqrt{32}$

SOLUTION

$$3\sqrt{50} - 7\sqrt{32} = 3\sqrt{25 \cdot 2} - 7\sqrt{16 \cdot 2}$$ Simplify the radicals by finding a perfect square factor in each radicand.

$$= 3 \cdot \sqrt{25} \cdot \sqrt{2} - 7 \cdot \sqrt{16} \cdot \sqrt{2}$$ Use the product rule of square roots.

$$= 3 \cdot 5\sqrt{2} - 7 \cdot 4\sqrt{2}$$ Evaluate the square roots.

$$= 15\sqrt{2} - 28\sqrt{2}$$ Multiply $3 \cdot 5$ and $7 \cdot 4$.

$$= -13\sqrt{2}$$ Combine like radicals.

Your Turn 2   Simplify.

  a. $2\sqrt{5} - \sqrt{45}$ **b.** $5\sqrt{12} + 2\sqrt{27}$

Objective ② **Use the distributive property in expressions containing square roots.**   Suppose we must multiply a sum (or difference) of radicals by a radical, such as $2\sqrt{5}(\sqrt{7} + 4\sqrt{6})$. We can use the distributive property the same way we did when multiplying a multi-term polynomial by a monomial.

Example 3   Multiply $2\sqrt{5}(\sqrt{7} + 4\sqrt{6})$.

SOLUTION $\quad 2\sqrt{5}(\sqrt{7} + 4\sqrt{6}) = 2\sqrt{5} \cdot \sqrt{7} + 2\sqrt{5} \cdot 4\sqrt{6}$ Use the distributive property.

$$= 2\sqrt{5 \cdot 7} + 2 \cdot 4 \cdot \sqrt{5 \cdot 6}$$ Use the product rule of square roots.

$$= 2\sqrt{35} + 8\sqrt{30}$$ Multiply the coefficients and multiply in each radical.

Answers   **Your Turn 1**
a. $14\sqrt{5}$ **b.** $-3\sqrt{x}$
c. $4\sqrt{3} - 6\sqrt{6}$

Answers   **Your Turn 2**
a. $-\sqrt{5}$ **b.** $16\sqrt{3}$

Your Turn 3 Multiply $2\sqrt{3}(4\sqrt{5} - \sqrt{6})$.

Now consider multiplying radical expressions that are similar to binomials, such as $(\sqrt{2} - 3)(\sqrt{2} + 5)$. Again, we use the distributive property just as we did when multiplying two binomials. Recall that we used FOIL to remember the steps.

Example 4 Multiply $(\sqrt{2} - 3)(\sqrt{2} + 5)$.

$$
\underbrace{(\sqrt{2} - 3)(\sqrt{2} + 5)}
$$
$$\sqrt{2}\cdot 5$$
$$\sqrt{2}\cdot\sqrt{2}$$
$$-3\cdot\sqrt{2}$$
$$-3\cdot 5$$

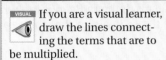
SOLUTION

$$
\begin{array}{cccc}
\textbf{First} & \textbf{Outer} & \textbf{Inner} & \textbf{Last}
\end{array}
$$

$$(\sqrt{2} - 3)(\sqrt{2} + 5) = \sqrt{2}\cdot\sqrt{2} + \sqrt{2}\cdot 5 - 3\cdot\sqrt{2} - 3\cdot 5 \qquad \text{Use FOIL.}$$
$$= \sqrt{2\cdot 2} + 5\sqrt{2} - 3\sqrt{2} - 15 \qquad \text{Multiply.}$$
$$= \sqrt{4} + 5\sqrt{2} - 3\sqrt{2} - 15 \qquad \text{Multiply } 2\cdot 2 \text{ the first radical.}$$
$$= 2 + 5\sqrt{2} - 3\sqrt{2} - 15 \qquad \text{Simplify } \sqrt{4}.$$
$$= -13 + 2\sqrt{2} \qquad \text{Combine like terms and like radicals.}$$

Your Turn 4 Multiply $(7 - \sqrt{5})(3 + \sqrt{5})$.

Radicals in Conjugates

Radical expressions can be conjugates. Like binomial conjugates, conjugates involving radicals differ only in the sign separating the terms. For example, $7 - \sqrt{3}$ and $7 + \sqrt{3}$ are conjugates. Let's explore what happens when we multiply conjugates containing radicals.

Example 5 Multiply $(7 - \sqrt{3})(7 + \sqrt{3})$.

SOLUTION

$$(7 - \sqrt{3})(7 + \sqrt{3}) = 7\cdot 7 + \sqrt{3}\cdot 7 - \sqrt{3}\cdot 7 - \sqrt{3}\cdot\sqrt{3} \qquad \text{Use FOIL.}$$
$$= 49 + 7\sqrt{3} - 7\sqrt{3} - \sqrt{3\cdot 3} \qquad \text{Multiply.}$$
$$= 49 + 7\sqrt{3} - 7\sqrt{3} - \sqrt{9} \qquad \text{Multiply } 3\cdot 3 \text{ in the last radical.}$$
$$= 49 + 7\sqrt{3} - 7\sqrt{3} - 3 \qquad \text{Simplify } \sqrt{9}.$$
$$= 46 \qquad \text{Combine like terms and like radicals.}$$

Note: After multiplying, the like radicals are additive inverses and are eliminated when combined.

Alternatively, recall the rule that we developed in Section 6.5 for multiplying conjugates: $(a + b)(a - b) = a^2 - b^2$. Using this rule streamlines the multiplication.

Answer   Your Turn 3

$8\sqrt{15} - 6\sqrt{2}$

Answer   Your Turn 4

$16 + 4\sqrt{5}$

Notice that when we multiply conjugates involving radicals, the product does not contain a radical. We use this fact to rationalize denominators in the next objective.

$$(7 - \sqrt{3})(7 + \sqrt{3}) = (7)^2 - (\sqrt{3})^2 \qquad \text{Use the rule } (a + b)(a - b) = a^2 - b^2.$$
$$= 49 - 3 \qquad \text{Square each term.}$$
$$= 46$$

Your Turn 5 Multiply $(1 + \sqrt{2})(1 - \sqrt{2})$.

Objective ③ Rationalize denominators that have a sum or difference with a square root term.
In Example 5, we saw that the product of two conjugates containing radicals did not contain any radicals. We can use this fact to rationalize a denominator that is a sum or difference containing a square root term.

For example, let's rationalize the denominator in $\dfrac{5}{7 - \sqrt{3}}$. Multiplying the numerator and denominator by $7 + \sqrt{3}$, which is the conjugate of the denominator, clears the radical from the denominator, thus rationalizing it.

$$\frac{5}{7 - \sqrt{3}} = \frac{5}{7 - \sqrt{3}} \cdot \frac{7 + \sqrt{3}}{7 + \sqrt{3}} \qquad \begin{array}{l} \text{Multiply the numerator and denominator by } 7 + \sqrt{3}, \\ \text{which is the conjugate of } 7 - \sqrt{3}. \end{array}$$

$$= \frac{5(7 + \sqrt{3})}{46} \qquad \text{Multiply the denominators. (See Example 5.)}$$

$$= \frac{35 + 5\sqrt{3}}{46} \qquad \text{Distribute 5 in the numerator.}$$

Procedure **Rationalizing a Denominator Containing a Sum or Difference**

To rationalize a denominator containing a sum or difference with a square root term, multiply the numerator and denominator by the conjugate of the denominator.

Example 6 Rationalize the denominator and simplify. Assume that variables represent positive numbers.

a. $\dfrac{9}{\sqrt{2} + 7}$

Note: Remember that when we multiply both numerator and denominator by the same amount, we are multiplying by 1, which does not change the fraction.

SOLUTION

$$\frac{9}{\sqrt{2} + 7} = \frac{9}{\sqrt{2} + 7} \cdot \frac{\sqrt{2} - 7}{\sqrt{2} - 7} \qquad \begin{array}{l} \text{Multiply the numerator and denominator} \\ \text{by the conjugate of the denominator.} \end{array}$$

$$= \frac{9(\sqrt{2} - 7)}{(\sqrt{2})^2 - (7)^2} \qquad \begin{array}{l} \text{Multiply in the numerator and use} \\ (a + b)(a - b) = a^2 - b^2 \text{ in the denominator.} \end{array}$$

$$= \frac{9\sqrt{2} - 63}{2 - 49} \qquad \begin{array}{l} \text{Distribute in the numerator and simplify} \\ \text{the squares in the denominator.} \end{array}$$

$$= \frac{9\sqrt{2} - 63}{-47} \qquad \begin{array}{l} \text{We can simplify the negative denominator} \\ \text{by factoring out } -1 \text{ in the numerator} \\ \text{and denominator.} \end{array}$$

$$= \frac{-1(63 - 9\sqrt{2})}{-1(47)} \qquad \begin{array}{l} \text{After factoring out the } -1, \text{ the signs of} \\ \text{the terms change. Because 63 is now} \\ \text{positive, we write it first.} \end{array}$$

$$= \frac{63 - 9\sqrt{2}}{47} \qquad \text{Divide out the common factor } -1.$$

Answer **Your Turn 5**

-1

b. $\dfrac{2\sqrt{3}}{\sqrt{6} - \sqrt{2}}$

SOLUTION $\dfrac{2\sqrt{3}}{\sqrt{6} - \sqrt{2}} = \dfrac{2\sqrt{3}}{\sqrt{6} - \sqrt{2}} \cdot \dfrac{\sqrt{6} + \sqrt{2}}{\sqrt{6} + \sqrt{2}}$ Multiply the numerator and denominator by the conjugate of the denominator.

$= \dfrac{2\sqrt{3}(\sqrt{6} + \sqrt{2})}{(\sqrt{6})^2 - (\sqrt{2})^2}$ Multiply in the numerator and use $(a + b)(a - b) = a^2 - b^2$ in the denominator.

$= \dfrac{2\sqrt{18} + 2\sqrt{6}}{6 - 2}$ Distribute in the numerator and simplify the squares in the denominator.

$= \dfrac{2\sqrt{9 \cdot 2} + 2\sqrt{6}}{4}$ Simplify $\sqrt{18}$ by factoring out a perfect square factor in 18.

$= \dfrac{2 \cdot 3\sqrt{2} + 2\sqrt{6}}{4}$ Simplify $\sqrt{9 \cdot 2}$ by finding the square root of 9.

Note: We cannot add $3\sqrt{2}$ and $\sqrt{6}$ because they are not like radicals.

$= \dfrac{2(3\sqrt{2} + \sqrt{6})}{2(2)}$ Factor 2 out of the terms in the numerator and factor the denominator.

$= \dfrac{3\sqrt{2} + \sqrt{6}}{2}$ Divide out the common factor 2.

c. $\dfrac{6}{\sqrt{x} - 5}$

SOLUTION $\dfrac{6}{\sqrt{x} - 5} = \dfrac{6}{\sqrt{x} - 5} \cdot \dfrac{\sqrt{x} + 5}{\sqrt{x} + 5}$ Multiply the numerator and denominator by the conjugate of the denominator.

$= \dfrac{6(\sqrt{x} + 5)}{(\sqrt{x})^2 - (5)^2}$

$= \dfrac{6\sqrt{x} + 30}{x - 25}$ Distribute in the numerator and simplify the squares in the denominator.

Answers Your Turn 6

a. $9\sqrt{5} - 18$

b. $\dfrac{\sqrt{10} + \sqrt{6}}{2}$

c. $\dfrac{3\sqrt{x} - 12}{x - 16}$

Your Turn 6 Rationalize the denominator and simplify. Assume that variables represent positive numbers.

a. $\dfrac{9}{\sqrt{5} + 2}$ **b.** $\dfrac{\sqrt{2}}{\sqrt{5} - \sqrt{3}}$ **c.** $\dfrac{3}{\sqrt{x} + 4}$

9.4 Exercises For Extra Help *MyMathLab*

Note: Exercises marked with a ★ represent challenging exercises.

1. What must be identical in order to add or subtract two square root expressions to get a single square root? What can be different?

2. Add $4x + 9x$; then add $4\sqrt{2} + 9\sqrt{2}$. Discuss the similarities in the process.

3. Multiply $(x + 3)(x + 2)$; then multiply $(\sqrt{5} + 3)(\sqrt{5} + 2)$. Discuss the similarities of the process. What differences do you note in the results?

4. Explain in your own words how to rationalize a denominator that has a sum or difference with a square root term.

For Exercises 5–24, simplify. Assume that variables represent nonnegative numbers. See Example 1.

5. $4\sqrt{3} + 2\sqrt{3}$

6. $3\sqrt{2} + 2\sqrt{2}$

7. $8\sqrt{7} - 5\sqrt{7}$

8. $5\sqrt{6} - 4\sqrt{6}$

9. $3\sqrt{10} + \sqrt{10} - 2\sqrt{10}$

10. $12\sqrt{5} - 9\sqrt{5} + \sqrt{5}$

11. $4\sqrt{2} - 15\sqrt{2} + \sqrt{2}$

12. $10\sqrt{3} - \sqrt{3} - 6\sqrt{3}$

13. $-8\sqrt{x} + 5\sqrt{x} - 9\sqrt{x}$

14. $2\sqrt{t} - 7\sqrt{t} - \sqrt{t}$

15. $\dfrac{\sqrt{5}}{6} - \dfrac{3\sqrt{5}}{4}$

16. $\dfrac{\sqrt{3}}{2} - \dfrac{4\sqrt{3}}{5}$

17. $\dfrac{4\sqrt{2}}{9} + \dfrac{5\sqrt{2}}{6}$

18. $\dfrac{7\sqrt{6}}{4} + \dfrac{2\sqrt{6}}{3}$

19. $0.7\sqrt{n} - 0.9\sqrt{n}$

20. $2.8\sqrt{3y} + 1.3\sqrt{3y}$

21. $6\sqrt{5} - 2\sqrt{3} - 3\sqrt{5} + 5\sqrt{3}$

22. $8\sqrt{7} - 4\sqrt{10} - 2\sqrt{7} + 9\sqrt{10}$

23. $3\sqrt{x} - 7\sqrt{y} + 2\sqrt{x} - 5\sqrt{y}$

24. $8\sqrt{a} + 4\sqrt{b} - 12\sqrt{b} - 3\sqrt{a}$

For Exercises 25–54, simplify. Assume that variables represent nonnegative numbers. See Example 2.

25. $\sqrt{12} - \sqrt{75}$

26. $\sqrt{12} + \sqrt{27}$

27. $\sqrt{72} + \sqrt{32}$

28. $\sqrt{20} + \sqrt{45}$

29. $\sqrt{3} - 8\sqrt{12}$

30. $\sqrt{5} - 4\sqrt{20}$

31. $3\sqrt{5} - \sqrt{20}$

32. $\sqrt{27} - 5\sqrt{3}$

33. $\sqrt{96} - 2\sqrt{24}$

34. $\sqrt{32} - 2\sqrt{8}$

35. $50\sqrt{3} - 2\sqrt{12}$

36. $6\sqrt{12} - 2\sqrt{27}$

37. $5\sqrt{18} - 7\sqrt{50}$

38. $2\sqrt{3} - 8\sqrt{27}$

39. $\sqrt{28} - \sqrt{63} + \sqrt{7}$

40. $\sqrt{44} + \sqrt{11} - \sqrt{99}$

41. $5\sqrt{27} - 4\sqrt{3} - \sqrt{12}$

42. $4\sqrt{45} + 2\sqrt{20} - \sqrt{125}$

43. $-5\sqrt{48} + 6\sqrt{27} - 2\sqrt{12} + 2\sqrt{3}$

44. $2\sqrt{12} + 5\sqrt{27} - \sqrt{48} + 2\sqrt{3}$

45. $\sqrt{12r} - 2\sqrt{27r} - 4\sqrt{75r}$

46. $-5\sqrt{32k} + 3\sqrt{50k} - \sqrt{18k}$

47. $\sqrt{\dfrac{7}{36}} - \sqrt{\dfrac{7}{25}}$

48. $\sqrt{\dfrac{3}{49}} - \sqrt{\dfrac{3}{25}}$

★49. $\sqrt{\dfrac{5}{16}} + \sqrt{\dfrac{20}{9}}$

★50. $\sqrt{\dfrac{28}{49}} - \sqrt{\dfrac{63}{4}}$

★51. $5\sqrt{24x^3} + 7\sqrt{6x^3}$

★52. $8\sqrt{27n^3} - \sqrt{12n^3}$

★53. $\sqrt{18t^5} - 4\sqrt{50t^5}$

★54. $3\sqrt{32y^5} - 7\sqrt{18y^5}$

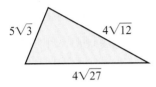
For Exercises 55–58, explain the mistake; then simplify correctly. Assume that variables represent nonnegative numbers.

55. $5\sqrt{2} + 3\sqrt{2} = 8\sqrt{4}$

56. $4\sqrt{y} + \sqrt{y} = 5\sqrt{2y}$

57. $9\sqrt{6x} - \sqrt{2x} = 8\sqrt{4x}$

58. $\sqrt{5} - \sqrt{3} = \sqrt{2}$

For Exercises 59 and 60, find the perimeter of the shape.

59.

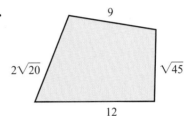

60.

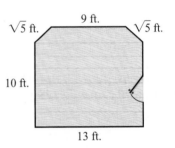

61. Crown molding, which is placed at the top of a wall, is to be installed around the perimeter of the room shown.

   a. Write an expression in simplest form for the perimeter of the room.

   b. Use a calculator to approximate the perimeter, rounded to the nearest tenth.

   c. If crown molding costs $1.89 per foot, how much will the crown molding cost for this room?

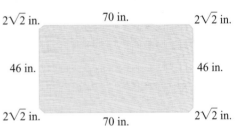

62. A tabletop is to be fitted with veneer strips along the sides.

   a. Write an expression in simplest form for the perimeter of the tabletop.

   b. Use a calculator to approximate the perimeter, rounded to the nearest tenth.

   c. If the veneer costs $0.54 per inch, how much will the veneer for the tabletop cost? Round your answer to the nearest cent.

For Exercises 63–104, multiply. Assume that variables represent nonnegative numbers. See Examples 3–5.

63. $\sqrt{2}(\sqrt{2} + \sqrt{3})$

64. $\sqrt{3}(\sqrt{3} - \sqrt{2})$

65. $\sqrt{5}(\sqrt{2} - \sqrt{10})$

66. $\sqrt{6}(\sqrt{5} + \sqrt{3})$

67. $\sqrt{10}(2\sqrt{7} + 2\sqrt{3})$

68. $\sqrt{13}(5\sqrt{2} + 3\sqrt{3})$

69. $3\sqrt{6}(4 - 2\sqrt{7})$

70. $2\sqrt{5}(\sqrt{6} - 4)$

71. $4\sqrt{2}(2\sqrt{7} - 5\sqrt{6})$

72. $3\sqrt{5}(7\sqrt{2} + 4\sqrt{15})$

73. $2\sqrt{3}(4\sqrt{6} + 2\sqrt{8})$

74. $3\sqrt{2}(4\sqrt{6} + 6\sqrt{10})$

75. $3\sqrt{x}(2\sqrt{x} + 3)$

76. $5\sqrt{y}(7 - 4\sqrt{y})$

77. $6\sqrt{5n}(\sqrt{5} - 2\sqrt{10n})$

78. $4\sqrt{6x}(9\sqrt{3x} + \sqrt{2})$

79. $(7 - 4\sqrt{3})(2 + \sqrt{3})$

80. $(3 - \sqrt{7})(2 - 3\sqrt{7})$

81. $(\sqrt{5} + 4)(\sqrt{2} - 1)$

82. $(\sqrt{2} + 5)(\sqrt{7} - 8)$

83. $(\sqrt{2} + \sqrt{3})(\sqrt{5} - \sqrt{7})$

84. $(\sqrt{3} - \sqrt{5})(\sqrt{2} + \sqrt{7})$

85. $(2\sqrt{3} + \sqrt{5})(\sqrt{3} - 2\sqrt{5})$

86. $(3\sqrt{2} + \sqrt{3})(\sqrt{2} - 5\sqrt{3})$

87. $(\sqrt{6x} - 4)(\sqrt{3x} + 5)$

88. $(\sqrt{5y} + 2)(\sqrt{10y} + 8)$

89. $(7\sqrt{a} - \sqrt{2b})(\sqrt{a} - 5\sqrt{2b})$

90. $(\sqrt{3p} - \sqrt{6q})(\sqrt{4p} - \sqrt{2q})$

91. $(\sqrt{6} + \sqrt{2})(\sqrt{6} - \sqrt{2})$

92. $(\sqrt{7} + \sqrt{3})(\sqrt{7} - \sqrt{3})$

93. $(2\sqrt{3} - 5)(2\sqrt{3} + 5)$

94. $(3\sqrt{2} + 1)(3\sqrt{2} - 1)$

95. $(5\sqrt{x} + 7)(5\sqrt{x} - 7)$

96. $(6\sqrt{t} + 3)(6\sqrt{t} - 3)$

97. $(\sqrt{2} - \sqrt{7})^2$

98. $(\sqrt{3} + \sqrt{2})^2$

99. $(\sqrt{6} + \sqrt{3})^2$

100. $(\sqrt{5} - \sqrt{10})^2$

101. $(3\sqrt{n} + 4)^2$

102. $(4\sqrt{m} - 9)^2$

103. $(6\sqrt{2x} - 5\sqrt{y})^2$

104. $(8\sqrt{3t} - 7\sqrt{2u})^2$

105. The sheet metal panel shown is to be bent and riveted onto the nose of a plane.
   a. What is the exact area of the panel?

   b. Use a calculator to approximate the area to the nearest tenth.

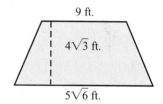
9 ft.
$4\sqrt{3}$ ft.
$5\sqrt{6}$ ft.

106. A circular disc has a radius of $2 + \sqrt{3}$ inches.
   a. Find the exact area of the disc.

   b. Use a calculator to approximate the area to the nearest tenth.

For Exercises 107–128, rationalize the denominator and simplify. Assume that variables represent positive numbers. See Example 6.

107. $\dfrac{3}{\sqrt{2} + 1}$

108. $\dfrac{3}{\sqrt{5} + 2}$

109. $\dfrac{4}{2 - \sqrt{3}}$

110. $\dfrac{2}{4 - \sqrt{15}}$

111. $\dfrac{5}{\sqrt{2} + \sqrt{3}}$

112. $\dfrac{7}{\sqrt{6} + \sqrt{5}}$

113. $\dfrac{4}{1 - \sqrt{5}}$

114. $\dfrac{6}{2 + \sqrt{3}}$

115. $\dfrac{\sqrt{3}}{\sqrt{3} - 1}$

116. $\dfrac{\sqrt{5}}{1 - \sqrt{5}}$

117. $\dfrac{2\sqrt{3}}{\sqrt{3} - 4}$

118. $\dfrac{2\sqrt{5}}{4 + \sqrt{5}}$

119. $\dfrac{4\sqrt{3}}{\sqrt{7} + \sqrt{2}}$

120. $\dfrac{\sqrt{8}}{\sqrt{4} + \sqrt{3}}$

121. $\dfrac{8\sqrt{2}}{4\sqrt{2} - \sqrt{3}}$

122. $\dfrac{5\sqrt{3}}{\sqrt{2} + 4\sqrt{6}}$

123. $\dfrac{6\sqrt{y}}{\sqrt{y}+1}$ **124.** $\dfrac{4\sqrt{x}}{\sqrt{x}+1}$ **125.** $\dfrac{3\sqrt{t}}{\sqrt{t}+2\sqrt{u}}$ **126.** $\dfrac{3\sqrt{m}}{\sqrt{n}-\sqrt{m}}$

127. $\dfrac{\sqrt{2y}}{\sqrt{x}-\sqrt{6y}}$ **128.** $\dfrac{\sqrt{14h}}{\sqrt{2h}+\sqrt{k}}$

129. The resistance in a circuit is found to be $\dfrac{5\sqrt{2}}{3+\sqrt{6}}$ Ω (ohms). Rationalize the denominator.

 130. Two charged particles q_1 and q_2 are separated by a distance of 8 centimeters. The values of the charges are $q_1 = 3 \times 10^{-6}$ coulomb and $q_2 = 1 \times 10^{-6}$ coulomb. Each charged particle exerts an electrical field. At a point between the two particles x centimeters away from q_1, the electric fields cancel each other so that the value of the fields at the point x is 0.

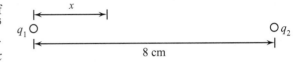

a. Use the formula $x = \dfrac{l}{1+\sqrt{\dfrac{q_2}{q_1}}}$ to find the distance from q_1 at which the electric field is canceled, where l is the distance separating the particles. Write the distance with a rationalized denominator.

b. Use a calculator to approximate the distance, rounded to the nearest tenth.

Review Exercises

Exercises 1–6   EQUATIONS AND INEQUALITIES

For Exercises 1–3, solve and check.

[2.3] **1.** $2(x-6)+7x=5x+8$ **[7.6]** **2.** $x^2-36=0$ **[7.6]** **3.** $x^2-5x+6=0$

[4.2] **4.** Graph: $y=-\dfrac{1}{3}x+4$

[4.4] **5.** For the equation $5y-2x=10$, find the slope and the y-intercept.

[4.7] **6.** Does the graph shown represent a function? Explain.

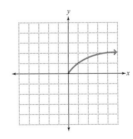

9.5 Solving Radical Equations

Objective ① Use the squaring principle to solve radical equations. We now explore how to solve radical equations.

Definition *Radical equation:* An equation containing at least one radical expression whose radicand has a variable.

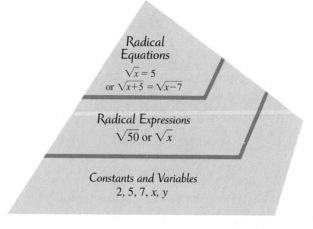

The Algebra Pyramid

Radical
Equations
$\sqrt{x} = 5$
or $\sqrt{x+5} = \sqrt{x-7}$

Radical Expressions
$\sqrt{50}$ or $\sqrt{x}$

Constants and Variables
2, 5, 7, x, y

Note: We move upward on the Algebra Pyramid, making equations out of the expressions we've studied.

To solve radical equations, we use a new principle of equality called the *squaring principle of equality.*

Rule Squaring Principle of Equality

$$\text{If } a = b, \text{ then } a^2 = b^2.$$

We apply the squaring principle in equations that have an isolated radical. When we use this principle, we square both sides of the equation, which eliminates the radical, leaving us with its radicand.

Isolated Radicals

First, we consider some equations in which the radical is isolated.

Example 1   Solve $\sqrt{x} = 9$.

SOLUTION   Note that we can "see" that the solution must be 81 because the square root of 81 is 9. However, we can use the squaring principle to reveal this solution.

$$(\sqrt{x})^2 = (9)^2 \quad \text{Square both sides.}$$
$$x = 81 \quad \text{Simplify each side.}$$

Check $\sqrt{81} \stackrel{?}{=} 9$

$$9 = 9 \quad \text{True}$$

The squaring principle of equality is peculiar in that it has the potential to turn a false equation into a true equation. Consider the false equation $5 = -5$. If we apply the principle and square both sides, we generate a true equation.

$$5 = -5 \qquad \longleftarrow \boxed{\text{False equation}}$$
$$(5)^2 = (-5)^2$$
$$25 = 25 \qquad \longleftarrow \boxed{\text{True equation}}$$

Because squaring both sides of an equation can turn a false equation into a true equation, we must be careful to check our solutions when we use the squaring principle to solve an equation, as in Example 2.

Example 2   Solve $\sqrt{x} = -9$.

SOLUTION   Notice that we can see that there is no real-number solution for this equation because the principal square root of a real number cannot be a negative number. Recall that the principal square root is the nonnegative root only. However, if we apply the squaring principle, we are incorrectly led to believe that there is a real-number solution.

Warning: As we will see in the check, this result is *not* a solution. ▶

$$(\sqrt{x})^2 = (-9)^2 \qquad \text{Square both sides.}$$
$$x = 81 \qquad \text{Simplify each side.}$$

Check   By checking 81 in the original equation, we see that it is not a solution.

$$\sqrt{81} \overset{?}{=} -9$$
$$9 = -9 \qquad \text{False. There is no real-number solution for this equation.}$$

In Example 2, we saw that 81 did not check in the original equation. In Section 8.6, we learned that such a number is called an *extraneous solution*. We now solve some radical equations that are a little more complicated using the following procedure. Some equations may have extraneous solutions.

Procedure   **Solving Radical Equations**

To solve a radical equation,

1. Isolate the radical if necessary. (If there is more than one radical term, isolate one of them.)
2. Square both sides of the equation.
3. If all radicals have been eliminated, solve. If a radical term remains, isolate it and square both sides again.
4. Check each solution in the origonal equation. Any apparent solution that does not check is an extraneous solution.

Example 3   Solve.

a. $\sqrt{x - 7} = 8$

SOLUTION $(\sqrt{x - 7})^2 = (8)^2 \qquad \text{The radical is isolated, so square both sides.}$
$$x - 7 = 64 \qquad \text{Simplify each side.}$$
$$x = 71 \qquad \text{Add 7 to both sides.}$$
Check $\sqrt{71 - 7} \overset{?}{=} 8$
$$\sqrt{64} \overset{?}{=} 8$$
$$8 = 8 \qquad \text{True}$$

b. $\sqrt{6x - 1} = \sqrt{x + 2}$

SOLUTION $(\sqrt{6x - 1})^2 = (\sqrt{x + 2})^2$ Each radical is isolated, so square both sides.

$$6x - 1 = x + 2 \qquad \text{Simplify each side.}$$

$$5x - 1 = 2 \qquad \text{Subtract } x \text{ from both sides.}$$

$$5x = 3 \qquad \text{Add 1 to both sides.}$$

$$x = \frac{3}{5} \qquad \text{Divide both sides by 5.}$$

Check $\sqrt{6\left(\dfrac{3}{5}\right) - 1} \stackrel{?}{=} \sqrt{\dfrac{3}{5} + 2}$

$$\sqrt{\frac{18}{5} - \frac{5}{5}} \stackrel{?}{=} \sqrt{\frac{3}{5} + \frac{10}{5}}$$

$$\sqrt{\frac{13}{5}} = \sqrt{\frac{13}{5}} \qquad \text{True}$$

Your Turn 3   Solve.

   a. $\sqrt{x - 5} = 6$ **b.** $\sqrt{x + 3} = -7$ **c.** $\sqrt{8x + 5} = \sqrt{2x + 7}$

Multiple Solutions

Radical equations may have multiple solutions if, after using the squaring principle, we are left with a quadratic form.

Example 4   Solve $x + 2 = \sqrt{5x + 16}$.

SOLUTION $(x + 2)^2 = (\sqrt{5x + 16})^2$ The radical is isolated, so square both sides.

$$x^2 + 4x + 4 = 5x + 16 \qquad \text{Use FOIL to square } x + 2. \text{ Simplify the right side.}$$

$$x^2 - x + 4 = 16 \qquad \text{Subtract } 5x \text{ from both sides.}$$

$$x^2 - x - 12 = 0 \qquad \begin{array}{l}\text{Because the equation is quadratic, subtract 16 from} \\ \text{both sides so that one side of the equation is 0.}\end{array}$$

$$(x + 3)(x - 4) = 0 \qquad \text{Factor.}$$

$$x + 3 = 0 \quad \text{or} \quad x - 4 = 0 \qquad \text{Use the zero-factor theorem.}$$

$$x = -3 \qquad\qquad x = 4$$

Checks $-3 + 2 \stackrel{?}{=} \sqrt{5(-3) + 16}$ $4 + 2 \stackrel{?}{=} \sqrt{5(4) + 16}$

$$-1 \stackrel{?}{=} \sqrt{-15 + 16} \qquad\qquad 6 \stackrel{?}{=} \sqrt{20 + 16}$$

$$-1 \stackrel{?}{=} \sqrt{1} \qquad\qquad\qquad 6 \stackrel{?}{=} \sqrt{36}$$

$$-1 = 1 \quad \text{False} \qquad\qquad 6 = 6 \quad \text{True}$$

Because -3 does not check, it is an extraneous solution. The only solution is 4.

Your Turn 4   Solve $\sqrt{3x + 13} = x + 3$.

Radicals Not Isolated

Now we consider radical equations in which the radical term is not isolated. In such equations, we must first isolate the radical term.

Answers   Your Turn 3

a. 41

b. no real-number solution

c. $\dfrac{1}{3}$

Answer   Your Turn 4

1, (-4 is extraneous)

Example 5   Solve.

a. $\sqrt{2x + 6} - 3 = 1$

SOLUTION   For the squaring principle to eliminate the radical effectively, we first isolate $\sqrt{2x + 6}$.

$$\sqrt{2x + 6} - 3 = 1$$

$$\sqrt{2x + 6} = 4 \qquad \text{Add 3 to both sides of the equation to isolate the radical.}$$

$$(\sqrt{2x + 6})^2 = 4^2 \qquad \text{Square both sides.}$$

$$2x + 6 = 16 \qquad \text{Simplify each side.}$$

$$2x = 10 \qquad \text{Subtract 6 from both sides.}$$

$$x = 5 \qquad \text{Divide both sides by 2.}$$

Check $\sqrt{2(5) + 6} - 3 = 1$

$$\sqrt{10 + 6} - 3 = 1$$

$$\sqrt{16} - 3 = 1$$

$$4 - 3 = 1$$

$$1 = 1 \qquad \text{True}$$

b. $\sqrt{x + 1} - 2x = x + 1$

SOLUTION

$$\sqrt{x + 1} - 2x = x + 1$$

$$\sqrt{x + 1} = 3x + 1 \qquad \text{Add } 2x \text{ to both sides to isolate the radical.}$$

$$(\sqrt{x + 1})^2 = (3x + 1)^2 \qquad \text{Square both sides.}$$

$$x + 1 = 9x^2 + 6x + 1 \qquad \text{Simplify the left side. Use FOIL to square } 3x + 1.$$

$$0 = 9x^2 + 5x \qquad \begin{array}{l}\text{Because the equation is quadratic, subtract } x \text{ and 1}\\ \text{from both sides so that one side of the equation is 0.}\end{array}$$

$$0 = x(9x + 5) \qquad \text{Factor.}$$

$$x = 0 \quad \text{or} \quad 9x + 5 = 0 \qquad \text{Use the zero-factor theorem.}$$

$$9x = -5$$

$$x = -\frac{5}{9}$$

Checks $\sqrt{0 + 1} - 2(0) \overset{?}{=} 0 + 1$ $\sqrt{-\dfrac{5}{9} + 1} - 2\left(-\dfrac{5}{9}\right) \overset{?}{=} -\dfrac{5}{9} + 1$

$$\sqrt{1} - 0 \overset{?}{=} 1 \qquad\qquad\qquad \sqrt{-\frac{5}{9} + \frac{9}{9}} - 2\left(-\frac{5}{9}\right) \overset{?}{=} -\frac{5}{9} + \frac{9}{9}$$

$$1 = 1 \qquad \text{True} \qquad\qquad\qquad \sqrt{\frac{4}{9}} + \frac{10}{9} \overset{?}{=} \frac{4}{9}$$

$$\frac{2}{3} + \frac{10}{9} \overset{?}{=} \frac{4}{9}$$

$$\frac{6}{9} + \frac{10}{9} \overset{?}{=} \frac{4}{9}$$

Note: This false equation indicates that $-\dfrac{5}{9}$ is an extraneous solution.

$$\frac{16}{9} = \frac{4}{9} \qquad \text{False}$$

The solution is 0.

Your Turn 5 Solve.

a. $\sqrt{3x + 1} + 3 = 7$

b. $\sqrt{5x^2 + 6x - 7} + 3x = 5x + 1$

Using the Squaring Principle Twice

Some equations require that we use the squaring principle twice to eliminate all radicals.

Example 6 Solve $\sqrt{x + 21} = \sqrt{x} + 3$.

SOLUTION Because one of the radicals is isolated, we square both sides.

Warning: Square both sides of the equation, not each term of the equation. Note that $(\sqrt{x} + 3)^2 \neq x + 9$.

$$(\sqrt{x + 21})^2 = (\sqrt{x} + 3)^2 \qquad \text{Square both sides.}$$

$$x + 21 = x + 3\sqrt{x} + 3\sqrt{x} + 9 \qquad \text{Simplify the left side. Use FOIL to square } \sqrt{x} + 3.$$

$$x + 21 = x + 6\sqrt{x} + 9 \qquad \text{Combine like radicals.}$$

$$12 = 6\sqrt{x} \qquad \text{Subtract } x \text{ and 9 from both sides to isolate the radical expression.}$$

$$2 = \sqrt{x} \qquad \text{Divide both sides by 6.}$$

$$(2)^2 = (\sqrt{x})^2 \qquad \text{Square both sides.}$$

$$4 = x$$

Check $\quad \sqrt{4 + 21} \overset{?}{=} \sqrt{4} + 3$

$$\sqrt{25} \overset{?}{=} 2 + 3$$

$$5 = 5 \qquad \text{True}$$

Answers Your Turn 5

a. 5 **b.** 2, (-4 is extraneous)

Answer Your Turn 6

0 and 4

Your Turn 6 Solve $\sqrt{2x + 1} = \sqrt{x} + 1$.

9.5 Exercises

For Extra Help **MyMathLab**

Note: Exercises marked with a ★ represent challenging exercises.

1. Explain why all potential solutions to radical equations must be checked.

2. What is an extraneous solution?

3. Explain why there is no real-number solution for the radical equation $\sqrt{x} = -6$.

4. Show why $(\sqrt{a})^2 = a$, assuming that $a \geq 0$.

5. Given the radical equation $\sqrt{x + 2} + 3x = 4x - 1$, what is the first step in solving the equation? Why?

6. Give an example of a radical equation that would require you to use the squaring principle of equality twice in solving the equation. Explain why the principle would have to be used twice.

For Exercises 7–52, solve. Identify any extraneous solutions. See Examples 1–6.

7. $\sqrt{x} = 2$

8. $\sqrt{y} = 5$

9. $\sqrt{x} = -4$

10. $\sqrt{x} = -1$

11. $\sqrt{n-1} = 4$ **12.** $\sqrt{x+3} = 7$ **13.** $\sqrt{t-7} = 2$ **14.** $\sqrt{m+8} = 1$

15. $\sqrt{3x-2} = 4$ **16.** $\sqrt{2x+5} = 3$ **17.** $\sqrt{2x+17} = 4$ **18.** $\sqrt{3y-2} = 3$

19. $\sqrt{2n-8} = -3$ **20.** $\sqrt{5x-1} = -6$ **21.** $\sqrt{u-3} - 10 = 1$ **22.** $\sqrt{y+1} - 4 = 2$

23. $\sqrt{y-6} + 2 = 9$ **24.** $\sqrt{r-5} + 6 = 10$ **25.** $\sqrt{x-5} - 2 = 3$ **26.** $\sqrt{x-3} - 2 = 0$

27. $\sqrt{2-3x} = \sqrt{5}$ **28.** $\sqrt{4-2x} = \sqrt{6}$ **29.** $\sqrt{4x+5} = \sqrt{6x-5}$ **30.** $\sqrt{x+1} = \sqrt{2x-3}$

31. $\sqrt{4a-20} = \sqrt{a-2}$ **32.** $\sqrt{10n-3} = \sqrt{4n+3}$ **33.** $\sqrt{y-4} = y-4$ **34.** $\sqrt{4x+1} = x+1$

35. $\sqrt{k+4} = k-8$ **36.** $\sqrt{5n-1} = 4-2n$ **37.** $y-1 = \sqrt{2y-2}$ **38.** $3+x = \sqrt{7+3x}$

39. $\sqrt{2x-1} + 2 = x$ **40.** $\sqrt{4y+1} + 5 = y$ **41.** $\sqrt{10n+4} - 3n = n+1$

42. $\sqrt{y+11} + 3y = 4y-1$ **43.** $\sqrt{3-13h} + 2h = 5h-1$ **44.** $\sqrt{4x-1} + x = x-1$

45. $2 + \sqrt{x} = \sqrt{2x+7}$ **46.** $2 + \sqrt{x} = \sqrt{5x+4}$ **47.** $\sqrt{3x} - 4 = \sqrt{x-2}$

48. $\sqrt{x+5} = \sqrt{4x} - 1$ **49.** $\sqrt{x+5} = 4 - \sqrt{x-3}$ **50.** $\sqrt{b+8} = 6 - \sqrt{b-4}$

51. $\sqrt{7x+1} - \sqrt{x-1} = 4$ **52.** $\sqrt{5x-1} - \sqrt{x+2} = 1$

Find

(X) *the mistake*   **For Exercises 53–56, explain the mistake; then solve correctly.**

53. $\sqrt{x} = -9$
       $(\sqrt{x})^2 = (-9)^2$
          $x = 81$

54. $\sqrt{x} = -2$
       $(\sqrt{x})^2 = (-2)^2$
          $x = 4$

55. $\sqrt{x} - 2 = 0$
    $\sqrt{x} - 2 + 2 = 0 + 2$
          $\sqrt{x} = 2$

56. $\sqrt{x+3} = 4$
     $(\sqrt{x+3})^2 = 4^2$
          $x + 9 = 16$
     $x + 9 - 9 = 16 - 9$
          $x = 7$

For Exercises 57–60, given the period of a pendulum, find the length of the pendulum. Use the formula $T = 2\pi\sqrt{\dfrac{L}{9.8}}$, where T is the period in seconds and L is the length in meters. Round to the nearest thousandth if necessary.

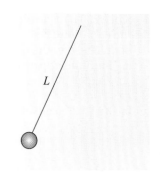

57. 2π seconds

58. 6π seconds

59. π seconds

60. $\dfrac{\pi}{3}$ seconds

For Exercises 61–64, find the distance an object has fallen. Use the formula $t = \sqrt{\dfrac{h}{16}}$, where t is the time in seconds and h is the distance the object has fallen in feet.

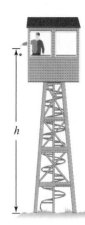

61. 0.3 second

62. 0.5 second

63. 3 seconds

64. $\dfrac{1}{4}$ second

For Exercises 65–68, find the length of the skid marks if a car slammed on the brakes going the given speed. Use the formula $S = 2\sqrt{2L}$, where L represents the length of the skid mark in feet and S represents the speed of the car in miles per hour.

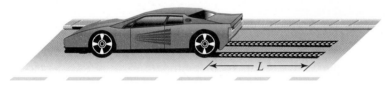

65. 30 mph

66. 60 mph

67. 45 mph

68. 75 mph

For Exercises 69–72, use the following information. Two forces, F_1 and F_2, acting on an object at a 90-degree angle will pull the object with a resultant force, R, at an angle between F_1 and F_2. (See the figure.) The value of the resultant force can be found by the formula $R = \sqrt{F_1^2 + F_2^2}$, and the unit is the newton, N.

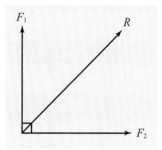

69. Find F_1 if $R = 5$ N and $F_2 = 3$ N.

70. Find F_1 if $R = 10$ N and $F_2 = 8$ N.

★ 71. Find F_2 if $R = 3\sqrt{5}$ N and $F_1 = 3$ N.

★ 72. Find F_2 if $R = 2\sqrt{13}$ N and $F_1 = 6$ N.

73. Since 2002, the sales of a company can be approximated by the equation $S = 3.7\sqrt{n + 44.63} + 512$. The sales, S, are in millions of dollars, and the number of years since 2002 is represented by n.
   a. Find the approximate sales in 2010. Round to the nearest million.

   b. In what year were the sales approximately $537.5 million?

74. Since 2004, the profits of a major corporation can be approximated by the equation $P = 4.2\sqrt{n + 3.63} + 213$. The profits, P, are in millions of dollars, and the number of years since 2004 is represented by n.

a. Find the approximate profits in 2007. Round to the nearest million.

b. In what year were the profits approximately $226 million?

In Exercises 75–80, we consider graphs of equations containing square roots to visualize their solutions.

75. Following is a table of incomplete ordered pairs for the equation $y = \sqrt{x}$.

a. Complete the table.

x	0	1	4	9	16	
y	0	1	2			5

b. Graph the ordered pairs in the coordinate plane; then connect the points to make a curve.

c. Does the graph extend below or to the left of the origin? Explain.

d. Does the curve represent a function? Explain.

76. Following is a table of incomplete ordered pairs for the equation $y = \sqrt{x} + 2$.

a. Complete the table.

x	0	1	4	9	16	
y	2	3	4			7

b. Graph the ordered pairs in the coordinate plane; then connect the points to make a curve.

c. Does the graph extend below or to the left of the origin? Explain.

d. Does the curve represent a function? Explain.

77. Use a graphing calculator to graph $y = \sqrt{x}$, $y = 2\sqrt{x}$, and $y = 3\sqrt{x}$. Based on your observations, as you increase the size of the coefficient, what happens to the graph?

78. Use a graphing calculator to compare the graph of $y = \sqrt{x}$ to the graph of $y = -\sqrt{x}$. What effect does the negative sign have on the graph?

79. Use a graphing calculator to graph $y = \sqrt{x}$, $y = \sqrt{x} + 2$, and $y = \sqrt{x} - 2$. Based on your observations, what effect does adding or subtracting a constant have on the graph of $y = \sqrt{x}$?

80. Based on your conclusions from Exercises 77–79, without graphing, describe the graph of $y = -2\sqrt{x} + 3$. Use a graphing calculator to graph the equation to confirm your description.

Review Exercises

Exercises 1–6 ◢◣ EXPRESSIONS

For Exercises 1–3, evaluate.

[1.5] **1.** 3^4

[1.5] **2.** $(-0.2)^3$

[6.1] **3.** $\left(\dfrac{2}{5}\right)^{-4}$

For Exercises 4–6, use the rules of exponents to simplify.

[6.4] **4.** $(x^3)(x^5)$

[6.4] **5.** $(n^4)^6$

[6.6] **6.** $\dfrac{y^7}{y^3}$

9.6 Higher Roots and Rational Exponents

Objectives

1. Evaluate roots higher than square roots.
2. Simplify radicals using the product rule of roots.
3. Simplify radicals using the quotient rule of roots.
4. Simplify expressions containing rational exponents.
5. Simplify expressions using rules of exponents.

In this section, we build on what we have learned about square roots and explore higher roots and rational exponents (exponents that are fractions).

Objective 1 Evaluate roots higher than square roots. Remember that a square root of a given number is a number that, when squared, equals the given number. Similarly, a *cube root* of a given number is a number that, when cubed, equals the given number. For example, the cube root of 8 is 2 because $2^3 = 8$. We can write a similar definition for any root.

Definition *nth root:* The number b is the nth root of a number a if $b^n = a$.

> **Note:** Recall that the symbol $\sqrt{}$ indicates the principal square root, so it is understood to have an index of 2.

For example, 2 is the 4th root of 16 because $2^4 = 16$. Also, 2 is the 5th root of 32 because $2^5 = 32$. To indicate a root other than a square root with a radical, we write an *index* in the radical symbol. We write the nth root of a number a as $\sqrt[n]{a}$, where n is the index. Examples of radicals indicating cube, 4th, and 5th roots follow.

> **Note:** The 3 is an index indicating that this radical is a cube root.

$\sqrt[3]{8} = 2$ because $2^3 = 8$

$\sqrt[4]{16} = 2$ because $2^4 = 16$

$\sqrt[5]{32} = 2$ because $2^5 = 32$

Example 1 Evaluate the radical. Assume that variables represent nonnegative numbers.

a. $\sqrt[3]{64}$

SOLUTION $\sqrt[3]{64} = 4$ Because $4^3 = 4 \cdot 4 \cdot 4 = 64$

b. $\sqrt[3]{-125}$

SOLUTION $\sqrt[3]{-125} = -5$ Because $(-5)^3 = (-5)(-5)(-5) = -125$

c. $\sqrt[4]{81}$

SOLUTION $\sqrt[4]{81} = 3$ Because $3^4 = 3 \cdot 3 \cdot 3 \cdot 3 = 81$

d. $\sqrt[4]{-16}$

Connection Remember that when finding the square root of an exponential form, we could divide the exponent by 2. Examples 2(e) and 2(f) suggest that when finding the nth root of an exponential form, we can divide the exponent by the index, n. We formalize this rule later in this section.

SOLUTION   There is no real number that can be raised to the 4th power to equal a negative number.

e. $\sqrt[3]{x^{12}}$

SOLUTION $\sqrt[3]{x^{12}} = x^4$ Because $(x^4)^3 = x^{12}$

f. $\sqrt[5]{x^{30}}$

SOLUTION $\sqrt[5]{x^{30}} = x^6$ Because $(x^6)^5 = x^{30}$

Example 1 suggests the following rules.

Rule   **Evaluating *n*th Roots**

When evaluating a radical expression $\sqrt[n]{a}$, the sign of a and the index n will determine possible outcomes.

If a is positive, then $\sqrt[n]{a} = b$, where b is positive and $b^n = a$.

If a is negative and n is even, then the root is not a real number.

If a is negative and n is odd, then $\sqrt[n]{a} = b$, where b is negative and $b^n = a$.

Your Turn 1   Evaluate. Assume that variables represent nonnegative numbers.

   a. $\sqrt[3]{-8}$ **b.** $\sqrt[4]{256}$ **c.** $\sqrt[4]{-625}$ **d.** $\sqrt[4]{x^{20}}$

📟 Calculator Tips

To evaluate roots other than square roots on a calculator, use the $\sqrt[x]{}$ function. On some calculators, the $\sqrt[x]{}$ function is above the $\boxed{\wedge}$ *key. To evaluate $\sqrt[4]{81}$, you type* $\boxed{4}$ $\boxed{\text{2nd}}$ $\boxed{\wedge}$ $\boxed{8}$ $\boxed{1}$ $\boxed{\text{ENTER}}$.

On some graphing calculators, you access $\sqrt[x]{}$ by pressing the $\boxed{\text{MATH}}$ *key and selecting $\sqrt[x]{}$ from the menu. To evaluate $\sqrt[4]{81}$ on a graphing calculator, you type* $\boxed{4}$ $\boxed{\text{MATH}}$ *select $\sqrt[x]{}$* $\boxed{8}$ $\boxed{1}$ $\boxed{\text{ENTER}}$.

On some scientific calculators, you use the $\sqrt[x]{y}$ function, which is usually above the $\boxed{y^x}$ *key. If $\sqrt[x]{y}$ is above* $\boxed{y^x}$ *, you must press* $\boxed{\text{2nd}}$ *, then* $\boxed{y^x}$ *. For example, to evaluate $\sqrt[4]{81}$, type* $\boxed{8}$ $\boxed{1}$ $\boxed{\text{2nd}}$ $\boxed{y^x}$ $\boxed{4}$ $\boxed{=}$.

Answers   **Your Turn 1**

a. -2 **b.** 4

c. not a real number   **d.** x^5

Objective 2   **Simplify radicals using the product rule of roots.**   Remember that the product rule of square roots allowed us to multiply the radicands ($\sqrt{2} \cdot \sqrt{8} = \sqrt{16} = 4$) or to split a radicand into factors and evaluate the square root of one of the factors ($\sqrt{50} = \sqrt{25 \cdot 2} = \sqrt{25} \cdot \sqrt{2} = 5\sqrt{2}$). We now extend this rule to encompass all roots.

Rule   **General Product Rule of Roots**

$$\sqrt[n]{a} \cdot \sqrt[n]{b} = \sqrt[n]{a \cdot b}, \text{ where } a \geq 0 \text{ and } b \geq 0$$

Warning: The general product rule applies only to roots with the same index, which means that we cannot multiply the radicands in an expression such as $\sqrt[3]{6} \cdot \sqrt[5]{4}$ because the indexes are different.

Example 2 Simplify. Assume that variables represent nonnegative numbers.

a. $\sqrt[3]{3} \cdot \sqrt[3]{9}$

SOLUTION $\sqrt[3]{3} \cdot \sqrt[3]{9} = \sqrt[3]{3 \cdot 9}$ Use the general product rule of roots to write the radicands under one radical.

$= \sqrt[3]{27}$ Multiply the radicands under a common cube root.

$= 3$ Find the cube root of 27.

b. $\sqrt[4]{x^3} \cdot \sqrt[4]{x^5}$

SOLUTION $\sqrt[4]{x^3} \cdot \sqrt[4]{x^5} = \sqrt[4]{x^3 \cdot x^5}$ Use the general product rule of roots to write the radicands under one radical.

$= \sqrt[4]{x^8}$ Multiply the radicands under a common fourth root.

$= x^2$ Because $x^8 = (x^2)^4$, the fourth root is x^2.

Your Turn 2 Simplify. Assume that variables represent nonnegative numbers.

a. $\sqrt[4]{4} \cdot \sqrt[4]{4}$ **b.** $\sqrt[6]{x^4} \cdot \sqrt[6]{x^8}$

We now use the general product rule of roots to simplify higher-order roots that contain a perfect nth root factor.

Example 3 Simplify. Assume that variables represent nonnegative numbers.

a. $\sqrt[3]{54}$

SOLUTION Because the radical is a cube root, we find the greatest perfect cube factor of 54, which is 27.

$\sqrt[3]{54} = \sqrt[3]{27 \cdot 2}$ Write 54 as $27 \cdot 2$.

$= \sqrt[3]{27} \cdot \sqrt[3]{2}$ Use the product rule to separate the factors into two radicals.

$= 3\sqrt[3]{2}$ Find the cube root of 27.

b. $\sqrt[3]{x^{11}}$

SOLUTION Because the radical is a cube root, we find the greatest perfect cube factor of x^{11}, which is x^9.

$\sqrt[3]{x^{11}} = \sqrt[3]{x^9 \cdot x^2}$ Write x^{11} as $x^9 \cdot x^2$.

$= \sqrt[3]{x^9} \cdot \sqrt[3]{x^2}$ Use the product rule to separate the factors into two radicals.

$= x^3 \sqrt[3]{x^2}$ Find the cube root of x^9.

c. $\sqrt[4]{64x^6}$

SOLUTION Because the radical is a fourth root, we find a factor of $64x^6$ that is an expression raised to the fourth power.

$\sqrt[4]{64x^6} = \sqrt[4]{16x^4 \cdot 4x^2}$ Write $64x^6$ as $16x^4 \cdot 4x^2$.

$= \sqrt[4]{16x^4} \cdot \sqrt[4]{4x^2}$ Use the product rule to separate the factors into two radicals.

$= 2x\sqrt[4]{4x^2}$ Find the fourth root of $16x^4$.

Answers Your Turn 2

a. 2 **b.** x^2

Answers Your Turn 3

a. $3\sqrt[4]{2}$ **b.** $x^2\sqrt[4]{x^3}$ **c.** $2x^2\sqrt[3]{3x}$

Your Turn 3 Simplify. Assume that variables represent nonnegative numbers.

a. $\sqrt[4]{162}$ **b.** $\sqrt[4]{x^{11}}$ **c.** $\sqrt[3]{24x^7}$

Objective ③ Simplify radicals using the quotient rule of roots. Remember that we used the quotient rule of square roots to simplify a square root containing a fraction. We can use the general quotient rule of roots similarly.

Rule **General Quotient Rule of Roots**

$$\sqrt[n]{\frac{a}{b}} = \frac{\sqrt[n]{a}}{\sqrt[n]{b}}, \text{ where } a \geq 0 \text{ and } b > 0$$

Example 4 Simplify.

a. $\sqrt[4]{\dfrac{81}{256}}$

SOLUTION $\sqrt[4]{\dfrac{81}{256}} = \dfrac{\sqrt[4]{81}}{\sqrt[4]{256}}$ Use the quotient rule of roots to write separate fourth roots in the numerator and denominator.

$= \dfrac{3}{4}$ Find the fourth root of the numerator and denominator.

b. $\sqrt[3]{\dfrac{54}{125}}$

SOLUTION $\sqrt[3]{\dfrac{54}{125}} = \dfrac{\sqrt[3]{54}}{\sqrt[3]{125}}$ Use the quotient rule of roots to write separate cube roots in the numerator and denominator.

$= \dfrac{\sqrt[3]{27 \cdot 2}}{5}$ Because 54 is not a cube, find a factor that is a cube, which is 27, and find the cube root of 125.

$= \dfrac{\sqrt[3]{27} \cdot \sqrt[3]{2}}{5}$ Use the product rule in the numerator.

$= \dfrac{3\sqrt[3]{2}}{5}$ Simplify the cube root of 27.

Your Turn 4 Simplify.

a. $\sqrt[3]{\dfrac{2}{27}}$ **b.** $\sqrt[4]{\dfrac{48}{625}}$

Objective ④ Simplify expressions containing rational exponents.

Rational Exponents with a Numerator of 1

In Section 6.1, we saw that a negative integer can be an exponent. We now consider exponents that are fractions, or rational exponents, which indicate roots. For example, an exponent of $^{1/2}$ indicates a square root. One way to see this connection is to recognize that the rules of exponents must apply to all types of numbers. If we multiply two exponential forms that have the same base, we can add their exponents and keep the same base. Consider the following:

$$3^{1/2} \cdot 3^{1/2} = 3^{1/2+1/2} = 3^1 = 3$$

Multiplying $\sqrt{3}$ by itself gives the same result:

$$\sqrt{3} \cdot \sqrt{3} = 3$$

Answers Your Turn 4
a. $\dfrac{\sqrt[3]{2}}{3}$ **b.** $\dfrac{2\sqrt[4]{3}}{5}$

These two results suggest that a base raised to the 1/2 power and the square root of that same base are equivalent expressions, or mathematically: $a^{1/2} = \sqrt{a}$. Note that the denominator, 2, of the rational exponent corresponds to the index of the root. (Remember that square roots have an invisible index of 2.) This relationship holds for other roots.

Rule | **Rational Exponents with a Numerator of 1**

$a^{1/n} = \sqrt[n]{a}$, where n is a natural number other than 1

If a is positive, then the root is positive.

If a is negative and n is odd, then the root is negative.

If a is negative and n is even, then the root is not a real number.

📟 Calculator Tips

We use the ⌃ key to evaluate a number raised to a rational exponent. For example, to evaluate $81^{1/4}$, we

type [8] [1] [⌃] [(] [1]

[÷] [4] [)] [ENTER].

Some scientific calculators have a [yˣ] *key. To evaluate*

$81^{1/4}$ *using* [yˣ]*, we type* [8]

[1] [yˣ] [1] [aᵇ/c] [4] [=].

Example 5 Rewrite using radicals; then simplify.

a. $36^{1/2}$

SOLUTION $36^{1/2} = \sqrt{36}$ Write the rational exponent as a radical.

$\qquad\qquad\quad = 6$ Evaluate the root.

b. $81^{1/4}$

SOLUTION $81^{1/4} = \sqrt[4]{81}$ Write the rational exponent as a radical.

$\qquad\qquad\quad = 3$ Evaluate the root.

c. $(-125)^{1/3}$

SOLUTION $(-125)^{1/3} = \sqrt[3]{-125}$ Write the rational exponent as a radical.

$\qquad\qquad\qquad\quad = -5$ Evaluate the root.

d. $(-64)^{1/4}$

SOLUTION $(-64)^{1/4} = \sqrt[4]{-64}$ Write the rational exponent as a radical and evaluate the root.

$\sqrt[4]{-64}$ is not a real number.

Your Turn 5 Rewrite using radicals; then simplify.

a. $49^{1/2}$ **b.** $625^{1/4}$ **c.** $(-64)^{1/3}$ **d.** $(-81)^{1/4}$

Rational Exponents with a Numerator Other Than 1

So far, we have considered rational exponents with 1 in the numerator. What if the numerator of a rational exponent is other than 1, as in $8^{2/3}$? Let's explore how we might rewrite $8^{2/3}$. Notice that we could write the fraction 2/3 as a product.

$$8^{2/3} = 8^{2 \cdot 1/3} \quad \text{or} \quad 8^{2/3} = 8^{1/3 \cdot 2}$$

Using the rule of exponents $(n^a)^b = n^{ab}$ in reverse, we can write

$$8^{2/3} = 8^{2 \cdot 1/3} = (8^2)^{1/3} \quad \text{or} \quad 8^{2/3} = 8^{1/3 \cdot 2} = (8^{1/3})^2.$$

Applying the rule $a^{1/n} = \sqrt[n]{a}$, we can write

$$8^{2/3} = 8^{2 \cdot 1/3} = (8^2)^{1/3} = \sqrt[3]{8^2} \quad \text{or} \quad 8^{2/3} = 8^{1/3 \cdot 2} = (8^{1/3})^2 = (\sqrt[3]{8})^2.$$

Answers **Your Turn 5**

a. $\sqrt{49}, 7$ **b.** $\sqrt[4]{625}, 5$

c. $\sqrt[3]{-64}, -4$

d. $\sqrt[4]{-81}$, not a real number

Notice that the denominator of the rational exponent becomes the index in the radical. The numerator of the rational exponent can be written as the exponent of the radicand or as an exponent for the entire radical.

Rule   **General Rule for Rational Exponents**

$a^{m/n} = \sqrt[n]{a^m} = (\sqrt[n]{a})^m$, where $a \geq 0$ and m and n are natural numbers other than 1

Example 6   Rewrite using radicals; then simplify.

a. $8^{2/3}$

SOLUTION $8^{2/3} = \sqrt[3]{8^2}$ Rewrite.   or $8^{2/3} = (\sqrt[3]{8})^2$ Rewrite.

$= \sqrt[3]{64}$ Square the radicand. $= (2)^2$ Evaluate the root.

$= 4$ Evaluate the root. $= 4$ Evaluate the exponential form.

b. $625^{3/4}$

Warning: In the expression $-9^{5/2}$, the negative sign is not part of the base. It tells us to find the opposite of the value of the exponential form. Mathematically,

$-9^{5/2} = -(9^{5/2}) = -(\sqrt{9})^5$.

You also can think of $-9^{5/2}$ as $-1 \cdot 9^{5/2}$. Using the order of operations, we calculate $9^{5/2}$ first, then multiply by -1.

$-9^{5/2} = -1 \cdot 9^{5/2}$

$= -1 \cdot (\sqrt{9})^5$

SOLUTION $625^{3/4} = (\sqrt[4]{625})^3$ Rewrite.

$= (5)^3$ Evaluate the root.

$= 125$ Evaluate the exponential form.

c. $-9^{5/2}$

SOLUTION $-9^{5/2} = -(\sqrt{9})^5$ Rewrite.

$= -(3)^5$ Evaluate the root.

$= -243$ Evaluate the exponential form.

> **Note:** Using $a^{m/n} = \sqrt[n]{a^m}$ leads to the expression $625^{3/4} = \sqrt[4]{625^3}$. Because calculating 625^3 is tedious without a calculator, we use the other form of the rule, $a^{m/n} = (\sqrt[n]{a})^m$, instead.

Your Turn 6   Rewrite using radicals; then simplify.

a. $32^{3/5}$ **b.** $-49^{3/2}$

Negative Rational Exponents

In Section 6.1, we developed negative integer exponents. Remember that $a^{-b} = \dfrac{1}{a^b}$. For example, $2^{-3} = \dfrac{1}{2^3}$. The rule tells us that we can rewrite an exponential form having a negative exponent by inverting the base and changing the sign of the exponent. The same rule applies to negative rational exponents.

Rule   **Negative Rational Exponents**

$a^{-m/n} = \dfrac{1}{a^{m/n}}$, where $a \neq 0$, and m and n are natural numbers with $n \neq 1$

Example 7   Rewrite using radicals; then simplify.

a. $81^{-1/2}$

SOLUTION $81^{-1/2} = \dfrac{1}{81^{1/2}}$ Rewrite the exponential form with a positive exponent by inverting the base and changing the sign of the exponent.

$= \dfrac{1}{\sqrt{81}}$ Write the rational exponent in radical form.

$= \dfrac{1}{9}$ Evaluate the square root.

Answers   Your Turn 6

a. $(\sqrt[5]{32})^3$, 8

b. $-(\sqrt{49})^3$, -343

b. $16^{-3/4}$

SOLUTION $16^{-3/4} = \dfrac{1}{16^{3/4}}$ Rewrite the exponential form with a positive exponent by inverting the base and changing the sign of the exponent.

$$= \dfrac{1}{\left(\sqrt[4]{16}\right)^3}$$ Write the rational exponent in radical form.

$$= \dfrac{1}{(2)^3}$$ Evaluate the radical.

$$= \dfrac{1}{8}$$ Simplify the exponential form.

c. $\left(\dfrac{16}{25}\right)^{-1/2}$

SOLUTION $\left(\dfrac{16}{25}\right)^{-1/2} = \dfrac{1}{\left(\dfrac{16}{25}\right)^{1/2}}$ Rewrite the exponential form with a positive exponent by inverting the base and changing the sign of the exponent.

$$= \dfrac{1}{\sqrt{\dfrac{16}{25}}}$$ Write the rational exponent in radical form.

$$= \dfrac{1}{\dfrac{4}{5}}$$ Evaluate the square root.

$$= \dfrac{5}{4}$$ Simplify the complex fraction.

Note: Our solution in part c suggests the following extension of the rule for negative rational exponents:

$$\left(\dfrac{a}{b}\right)^{-m/n} = \left(\dfrac{b}{a}\right)^{m/n}$$

where $a \neq 0$ and $b \neq 0$.

Your Turn 7   Rewrite using radicals; then simplify.

a. $49^{-1/2}$ **b.** $-27^{-2/3}$ **c.** $\left(\dfrac{16}{81}\right)^{-3/4}$

Objective ⑤ Simplify expressions using rules of exponents. We now apply the rules for exponents that we learned in Chapter 6 to expressions containing rational exponents.

Example 8   Simplify so that the result has a positive rational exponent. Assume that variables represent positive numbers.

a. $n^{5/6} \cdot n^{-1/6}$

SOLUTION $n^{5/6} \cdot n^{-1/6} = n^{5/6+(-1/6)}$ Use the product rule for exponents. (Add the exponents.)

$$= n^{4/6}$$ Add the exponents.

$$= n^{2/3}$$ Simplify the rational exponent.

b. $\dfrac{k^{4/9}}{k^{-1/9}}$

SOLUTION $\dfrac{k^{4/9}}{k^{-1/9}} = k^{4/9-(-1/9)}$ Use the quotient rule for exponents. (Subtract the exponents.)

$$= k^{4/9+1/9}$$ Rewrite the subtraction as addition.

$$= k^{5/9}$$ Add the exponents.

Answers Your Turn 7

a. $\dfrac{1}{\sqrt{49}}, \dfrac{1}{7}$ **b.** $-\dfrac{1}{\left(\sqrt[3]{27}\right)^2}, -\dfrac{1}{9}$

c. $\dfrac{1}{\left(\sqrt[4]{\dfrac{16}{81}}\right)^3}$ or $\left(\sqrt[4]{\dfrac{81}{16}}\right)^3, \dfrac{27}{8}$

c. $(y^{-3/4})^{2/3}$

SOLUTION $(y^{-3/4})^{2/3} = y^{-3/4 \cdot 2/3}$ Use the power rule for exponents. (Multiply the exponents.)

$= y^{-1/2}$ Multiply the exponents.

$= \dfrac{1}{y^{1/2}}$ Rewrite with a positive exponent.

d. $\left(x^6 y^{\frac{2}{5}}\right)^{\frac{3}{2}}$

SOLUTION $(x^6 y^{\frac{2}{5}})^{\frac{3}{2}} = (x^6)^{\frac{3}{2}}(y^{\frac{2}{5}})^{\frac{3}{2}}$ Use the product to a power rule.

$= x^{6 \cdot \frac{3}{2}} \cdot y^{\frac{2}{5} \cdot \frac{3}{2}}$ Use the power rule for exponents.

$= x^9 y^{\frac{3}{5}}$ Multiply the exponents.

Your Turn 8   Simplify so that the result has a positive rational exponent. Assume that variables represent positive numbers.

a. $t^{-1/10} \cdot t^{3/10}$ **b.** $x^{-2/7} \div x^{-4/7}$ **c.** $(m^{10})^{-4/5}$ **d.** $(a^{\frac{3}{4}} b^9)^{\frac{2}{3}}$

John Wallis (1616–1703) was one of the first mathematicians to explain rational and negative exponents. Despite Wallis's considerable genius on the subjects of mathematics and physics, his work was overshadowed by that of fellow Englishman Isaac Newton. Newton added to what Wallis began with exponents and roots and popularized the notation that is used today.

Answers   **Your Turn 8**

a. $t^{1/5}$ **b.** $x^{2/7}$ **c.** $\dfrac{1}{m^8}$ **d.** $a^{1/2} b^6$

9.6 Exercises   For Extra Help   **MyMathLab**

Note: Exercises marked with a ★ represent challenging exercises.

1. Explain why 5 is the cube root of 125.

2. Of what number is 4 the fourth root? Explain.

3. If $\sqrt[n]{a}$ is negative, what must be true about n? Explain.

4. To simplify $64^{3/4}$, which is easier to use: $a^{m/n} = \sqrt[n]{a^m}$ or $a^{m/n} = (\sqrt[n]{a})^m$? Explain.

5. Explain why $-81^{1/4}$ is a real number but $(-81)^{1/4}$ is not.

6. After $25^{-3/2}$ is simplified, will the result be positive or negative? Explain.

For Exercises 7–10, find the cube root. See Objective 1.

7. 64   **8.** 27   **9.** -125 **10.** -512

For Exercises 11–14, find the positive fourth root. See Objective 1.

11. 81   **12.** 16   **13.** 625   **14.** 10,000

For Exercises 15–58, simplify. Assume that variables represent nonnegative numbers. See Examples 1–4.

15. $\sqrt[3]{216}$

16. $\sqrt[3]{125}$

17. $\sqrt[4]{256}$

18. $\sqrt[4]{16}$

19. $\sqrt[5]{32}$

20. $\sqrt[5]{243}$

21. $\sqrt[4]{x^8}$

22. $\sqrt[5]{x^{15}}$

23. $\sqrt[3]{r^{18}}$

24. $\sqrt[4]{b^{16}}$

25. $\sqrt[3]{-64}$

26. $\sqrt[5]{-32}$

27. $\sqrt[6]{-64}$

28. $\sqrt[4]{-81}$

29. $\sqrt[3]{25} \cdot \sqrt[3]{5}$

30. $\sqrt[3]{2} \cdot \sqrt[3]{4}$

31. $\sqrt[5]{4} \cdot \sqrt[5]{8}$

32. $\sqrt[4]{2} \cdot \sqrt[4]{8}$

33. $\sqrt[4]{3} \cdot \sqrt[4]{27}$

34. $\sqrt[5]{16} \cdot \sqrt[5]{64}$

35. $\sqrt[6]{x^3} \cdot \sqrt[6]{x^3}$

36. $\sqrt[3]{y} \cdot \sqrt[3]{y^2}$

37. $\sqrt[5]{m^6} \cdot \sqrt[5]{m^4}$

38. $\sqrt[3]{x} \cdot \sqrt[3]{x^2}$

39. $\sqrt[3]{4y^5} \cdot \sqrt[3]{16y}$

40. $\sqrt[4]{27y^2} \cdot \sqrt[4]{3y^6}$

41. $\sqrt[3]{32}$

42. $\sqrt[4]{32}$

43. $\sqrt[5]{64}$

44. $\sqrt[3]{108}$

45. $\sqrt[3]{x^8}$

46. $\sqrt[4]{y^9}$

47. $\sqrt[3]{24x^8}$

48. $\sqrt[3]{54x^4}$

49. $\sqrt[5]{64x^4}$

50. $\sqrt[3]{81x^5}$

51. $\sqrt[3]{\dfrac{27}{64}}$

52. $\sqrt[4]{\dfrac{16}{81}}$

53. $\sqrt[3]{-\dfrac{27}{125}}$

54. $\sqrt[3]{-\dfrac{216}{64}}$

55. $\sqrt[3]{\dfrac{32}{125}}$

56. $\sqrt[3]{\dfrac{81}{8}}$

57. $\sqrt[5]{\dfrac{3}{32}}$

58. $\sqrt[4]{\dfrac{14}{625}}$

For Exercises 59–76, rewrite using radicals; then simplify. See Examples 5–7.

59. $16^{1/2}$

60. $36^{1/2}$

61. $(-81)^{1/4}$

62. $(-16)^{1/4}$

63. $(-32)^{1/5}$

64. $(-27)^{1/3}$

65. $4^{3/2}$

66. $27^{2/3}$

67. $-16^{3/4}$

68. $-81^{3/2}$

69. $4^{-3/2}$

70. $27^{-2/3}$

71. $-16^{-5/4}$

72. $-8^{-2/3}$

73. $\left(\dfrac{16}{81}\right)^{-3/4}$

74. $\left(\dfrac{9}{25}\right)^{-3/2}$

75. $\left(\dfrac{32}{243}\right)^{-2/5}$

76. $\left(\dfrac{64}{81}\right)^{-3/2}$

For Exercises 77–94, simplify so that the result has a positive rational exponent. Assume that variables represent positive numbers. See Example 8.

77. $x^{1/3} \cdot x^{2/3}$

78. $y^{2/5} \cdot y^{1/5}$

79. $(h^{1/7})^{2/3}$

80. $(b^{1/5})^{2/3}$

81. $\dfrac{m^{3/5}}{m^{1/5}}$

82. $\dfrac{a^{3/4}}{a^{2/4}}$

83. $\dfrac{1}{k^{-2/7}}$

84. $\dfrac{2}{p^{-1/2}}$

85. $n^{-3/4} \cdot n^{-2/4}$

86. $p^{-1/5} \cdot p^{-3/5}$

87. $(s^{2/3})^{-3/4}$ **88.** $(u^{-2/3})^{3/5}$ **89.** $t^{3/5} \div t^{2/5}$ **90.** $w^{4/5} \div w^{3/5}$ **91.** $(x^{1/2}y^{2/3})^{3/4}$

92. $(m^{2/5}n^{1/3})^{1/2}$ **93.** $\left(\dfrac{x^{\frac{3}{8}}}{y^{\frac{3}{4}}}\right)^{\frac{2}{3}}$ **94.** $\left(\dfrac{a^{\frac{5}{8}}}{b^{\frac{5}{3}}}\right)^{\frac{2}{5}}$

Find
⊗ the mistake   *For Exercises 95–98, explain the mistake; then simplify. Assume that variables represent nonnegative numbers.*

95. $\sqrt[3]{16x^4} = \sqrt[3]{8} \cdot \sqrt[3]{2x^4}$
$\qquad\qquad = 2\sqrt[3]{2x^4}$

96. $\sqrt[3]{64x^3} = 8x$

97. $64^{3/2} = (\sqrt[3]{64})^2$
$\qquad\quad\; = 4^2$
$\qquad\quad\; = 16$

98. $\sqrt[3]{x^5} \cdot \sqrt[3]{x} = \sqrt[3]{x^6}$
$\qquad\qquad\quad = \sqrt[3]{x^2}$

For Exercises 99–106, use a calculator to approximate each of the following to three decimal places.

99. $\sqrt[3]{20}$ **100.** $\sqrt[4]{15}$ **101.** $\sqrt[4]{12}$ **102.** $\sqrt[4]{125}$

103. $\sqrt[3]{81}$ **104.** $\sqrt{27}$ **105.** $\sqrt[3]{246}$ **106.** $\sqrt[5]{100}$

For Exercises 107–114, use a calculator to evaluate the expression.

107. $1296^{3/4}$ **108.** $512^{2/3}$ **109.** $7776^{2/5}$ **110.** $1024^{3/5}$

111. $512^{4/3}$ **112.** $3125^{6/5}$ **113.** $343^{-2/3}$ **114.** $729^{-5/6}$

For Exercises 115–118, use the formula $B = P(1 + r)^t$ to find the final balance in an account after a principal P was invested and it grew at a rate r for t years.

115. $P = \$1200, r = 8\%, t = \dfrac{2}{3}$ year

116. $P = \$4000, r = 6.5\%, t = \dfrac{3}{4}$ year

117. $P = \$800, r = 3.9\%, t = 2\dfrac{1}{4}$ years

118. $P = \$6000, r = 4\%, t = 3\dfrac{2}{3}$ years

★ *For Exercises 119–128, use the rules of roots and exponents to simplify.*

119. $\sqrt[3]{7} - 5\sqrt[3]{7} + 2\sqrt[3]{7}$

120. $8\sqrt[4]{6} + \sqrt[4]{6} - 12\sqrt[4]{6}$

121. $14\sqrt[4]{2} + \sqrt[3]{2} - 6\sqrt[3]{2} + \sqrt[4]{2}$

122. $11\sqrt[4]{8} + 7\sqrt[5]{8} - \sqrt[4]{8} + 9\sqrt[5]{8}$

123. $\sqrt[3]{9}(\sqrt[3]{6} - 5)$

124. $\sqrt[4]{4}(\sqrt[4]{12} + 3)$

125. $(\sqrt[3]{4} - 2)(\sqrt[3]{4} + 6)$

126. $(\sqrt[5]{8} - 2)(\sqrt[5]{8} + 6)$

127. $(\sqrt[4]{8} + 3)^2$

128. $(\sqrt[3]{9} - 5)^2$

Puzzle Problem

Rewrite $\sqrt[m]{\sqrt[n]{\sqrt[p]{x^y}}}$ in exponential form.

Review Exercises

 Exercises 1 and 2   **EXPRESSIONS**

[7.4]   **1.** Factor: $4x^2 - 4y^2$

[8.1]   **2.** Simplify: $\dfrac{6x - 12}{x^2 - 7x + 10}$

 Exercises 3–6   **EQUATIONS AND INEQUALITIES**

[2.5]   **3.** Twice the sum of a number and 8 is equal to 18. Translate to an equation; then find the number.

For Exercises 4 and 5, solve.

[7.6]   **4.** $2x^2 - 9x = 5$

[9.5]   **5.** $\sqrt{x + 1} = 5$

[5.1–5.3]   **6.** Solve the system. $\begin{cases} 2x + 3y = -6 \\ x - 3y = -12 \end{cases}$

Chapter 9 Summary

Defined Terms

Section 9.4
Like radicals (p. 656)

Section 9.5
Radical equation (p. 665)

Section 9.6
*n*th root (p. 673)

Procedures, Rules, and Key Examples

Procedures/Rules	Key Examples

SECTION 9.1 Square Roots and Radical Expressions

Every positive number has two square roots, a positive root and a negative root.

The radical symbol, $\sqrt{\ }$, denotes only the positive (principal) square root.
The square root of a negative number is not a real number.

$$\sqrt{\dfrac{a}{b}} = \dfrac{\sqrt{a}}{\sqrt{b}}, \text{ where } a \geq 0 \text{ and } b > 0$$

Note: Throughout this chapter, when simplifying square roots with radicands that have variables, we assume that those variables represent nonnegative numbers.

Example 1: Find all square roots of 36.
Answer: 6 and -6

Example 2: Simplify.
a. $\sqrt{36} = 6$
b. $\sqrt{-25}$ is not a real number.
c. $-\sqrt{121} = -11$
d. $\sqrt{\dfrac{49}{100}} = \dfrac{\sqrt{49}}{\sqrt{100}} = \dfrac{7}{10}$
e. $\sqrt{81x^2} = 9x$
f. $\sqrt{16x^6} = 4x^3$

SECTION 9.2 Multiplying and Simplifying Square Roots

Product rule for square roots:

$$\sqrt{a} \cdot \sqrt{b} = \sqrt{a \cdot b}, \text{ where } a \geq 0 \text{ and } b \geq 0$$

To simplify a square root with a radicand that contains a perfect square factor,
1. Write the radicand in factored form so that one of the factors is the greatest perfect square factor.
2. Use the product rule of square roots to separate the two factors into two radical expressions.
3. Find the square root of the perfect square factor and leave the other factor in a radical.

Example 1: Simplify.
a. $\sqrt{2} \cdot \sqrt{50} = \sqrt{2 \cdot 50} = \sqrt{100} = 10$
b. $\sqrt{3x} \cdot \sqrt{12x^3} = \sqrt{3x \cdot 12x^3}$
$\qquad = \sqrt{36x^4} = 6x^2$

Example 2: Simplify.
a. $\sqrt{50} = \sqrt{25 \cdot 2} = \sqrt{25} \cdot \sqrt{2} = 5\sqrt{2}$
b. $\sqrt{48x^3} = \sqrt{16x^2 \cdot 3x} = \sqrt{16x^2} \cdot \sqrt{3x}$
$\qquad = 4x\sqrt{3x}$

SECTION 9.3 Dividing and Simplifying Square Roots

Quotient rule for square roots:

$$\dfrac{\sqrt{a}}{\sqrt{b}} = \sqrt{\dfrac{a}{b}}, \text{ where } a \geq 0 \text{ and } b > 0$$

For Examples 1 and 2, assume that variables represent positive numbers.

Example 1: Simplify.
a. $\dfrac{\sqrt{45}}{\sqrt{5}} = \sqrt{\dfrac{45}{5}} = \sqrt{9} = 3$
b. $\dfrac{\sqrt{50x^5}}{\sqrt{2x}} = \sqrt{\dfrac{50x^5}{2x}} = \sqrt{25x^4} = 5x^2$

(continued)

Procedures/Rules	Key Examples

SECTION 9.3   Dividing and Simplifying Square Roots (*continued*)

Example 2: Simplify.

a. $\sqrt{\dfrac{20}{9}} = \dfrac{\sqrt{20}}{\sqrt{9}} = \dfrac{\sqrt{4\cdot 5}}{3} = \dfrac{2\sqrt{5}}{3}$

b. $\sqrt{\dfrac{18x^5}{8x^3}} = \sqrt{\dfrac{9x^2}{4}} = \dfrac{\sqrt{9x^2}}{\sqrt{4}} = \dfrac{3x}{2}$

To rationalize a denominator containing a single square root, multiply the fraction by a 1 so that the product's denominator has a radicand that is a perfect square; then simplify.

Example 3: Rationalize the denominator.

a. $\dfrac{7}{\sqrt{3}} = \dfrac{7}{\sqrt{3}} \cdot \dfrac{\sqrt{3}}{\sqrt{3}} = \dfrac{7\sqrt{3}}{3}$

b. $\sqrt{\dfrac{5}{12}} = \dfrac{\sqrt{5}}{\sqrt{12}} \cdot \dfrac{\sqrt{3}}{\sqrt{3}} = \dfrac{\sqrt{15}}{\sqrt{36}} = \dfrac{\sqrt{15}}{6}$

Simplest Form of a Square Root
To ensure that an expression containing a square root is in simplest form, check that:
- the radicand has no perfect square factors.
- the radicand had no fractions.
- the denominator has no radicals.

Example 4: The expressions $\sqrt{32}$, $\sqrt{\dfrac{1}{3}}$, and $\dfrac{x}{\sqrt{10}}$ are not in simplest form.

SECTION 9.4   Addition, Subtraction, and Mixed Operations with Square Roots

To add or subtract like radicals, add or subtract the coefficients and keep the radicals the same.

Example 1: Simplify.

a. $5\sqrt{2} + 4\sqrt{2} = (5+4)\sqrt{2} = 9\sqrt{2}$

b. $2\sqrt{x} - 7\sqrt{x} = (2-7)\sqrt{x} = -5\sqrt{x}$

Example 2: Simplify the radicals; then combine like radicals.

$$
\begin{aligned}
9\sqrt{3} - \sqrt{12} + 6\sqrt{75} \\
&= 9\sqrt{3} - \sqrt{4\cdot 3} + 6\sqrt{25\cdot 3} \\
&= 9\sqrt{3} - 2\sqrt{3} + 6\cdot 5\sqrt{3} \\
&= 9\sqrt{3} - 2\sqrt{3} + 30\sqrt{3} \\
&= 37\sqrt{3}
\end{aligned}
$$

Example 3: Multiply; then simplify.

$$
\begin{aligned}
(\sqrt{3} - 4)(\sqrt{3} + 6) & \\
= \sqrt{3}\cdot\sqrt{3} + \sqrt{3}\cdot 6 - 4\cdot\sqrt{3} - 4\cdot 6 \quad &\text{Use FOIL.} \\
= \sqrt{9} + 6\sqrt{3} - 4\sqrt{3} - 24 \quad &\text{Multiply.} \\
= 3 + 6\sqrt{3} - 4\sqrt{3} - 24 \quad &\text{Simplify } \sqrt{9}. \\
= -21 + 2\sqrt{3} \quad &\text{Combine like radicals and like terms.}
\end{aligned}
$$

To rationalize a denominator containing a sum or difference with a square root term, multiply the numerator and denominator by the conjugate of the denominator.

Example 4: Rationalize the denominator in $\dfrac{6}{4 - \sqrt{5}}$.

$$
\begin{aligned}
\dfrac{6}{4 - \sqrt{5}} &= \dfrac{6}{4 - \sqrt{5}} \cdot \dfrac{4 + \sqrt{5}}{4 + \sqrt{5}} \\
&= \dfrac{6\cdot 4 + 6\cdot\sqrt{5}}{16 - 5} \\
&= \dfrac{24 + 6\sqrt{5}}{11}
\end{aligned}
$$

(*continued*)

Procedures/Rules	Key Examples

SECTION 9.5 Solving Radical Equations

Squaring principle of equality: If $a = b$, then $a^2 = b^2$.

To solve a radical equation,
1. Isolate the radical, if necessary. (If there is more than one radical term, isolate one of them.)
2. Square both sides of the equation.
3. If all radicals have been eliminated, solve. If a radical term remains, isolate it and square both sides again.
4. Check each solution in the original equation. Any solution that does not check is an extraneous solution.

Example 1: Solve $\sqrt{x - 5} = 7$.

$$(\sqrt{x - 5})^2 = 7^2 \qquad \text{Square both sides.}$$
$$x - 5 = 49$$
$$x = 54 \qquad \text{Add 5 to both sides.}$$

Check: $\sqrt{54 - 5} = 7$
$$\sqrt{49} = 7 \qquad \text{True}$$

Example 2: Solve $\sqrt{n + 14} = n + 2$.

$$(\sqrt{n + 14})^2 = (n + 2)^2 \qquad \text{Square both sides.}$$
$$n + 14 = n^2 + 4n + 4$$
$$0 = n^2 + 3n - 10 \qquad \begin{array}{l}\text{Subtract } n \text{ and } 14 \\ \text{from both sides.}\end{array}$$
$$0 = (n + 5)(n - 2) \qquad \text{Factor.}$$
$$n + 5 = 0 \quad \text{or} \quad n - 2 = 0 \qquad \begin{array}{l}\text{Use the zero-factor} \\ \text{theorem.}\end{array}$$
$$n = -5 \qquad\qquad n = 2$$

Check -5: $\sqrt{-5 + 14} = -5 + 2$
$$\sqrt{9} = -3 \qquad \text{False}$$

Check 2: $\sqrt{2 + 14} = 2 + 2$
$$\sqrt{16} = 4 \qquad \text{True}$$

The only solution is 2. (-5 is an extraneous solution.)

Example 3: Solve $3 + \sqrt{t} = \sqrt{t + 21}$.

$$(3 + \sqrt{t})^2 = (\sqrt{t + 21})^2 \qquad \text{Square both sides.}$$
$$9 + 6\sqrt{t} + t = t + 21$$
$$6\sqrt{t} = 12 \qquad \begin{array}{l}\text{Subtract 9 and } t \text{ from} \\ \text{both sides.}\end{array}$$
$$\sqrt{t} = 2 \qquad \text{Divide both sides by 6.}$$
$$(\sqrt{t})^2 = 2^2 \qquad \text{Square both sides.}$$
$$t = 4$$

Check: $3 + \sqrt{4} = \sqrt{4 + 21}$
$$3 + 2 = \sqrt{25}$$
$$5 = 5 \qquad \text{True}$$

SECTION 9.6 Higher Roots and Rational Exponents

When evaluating a radical expression $\sqrt[n]{a}$, the sign of a and the index n will determine possible outcomes.
If a is positive, then $\sqrt[n]{a} = b$, where b is positive and $b^n = a$.
If a is negative and n is even, then the root is not a real number.
If a is negative and n is odd, then $\sqrt[n]{a} = b$, where b is negative and $b^n = a$.

General product rule of roots:

$$\sqrt[n]{a} \cdot \sqrt[n]{b} = \sqrt[n]{a \cdot b}, \text{ where } a \geq 0 \text{ and } b \geq 0$$

Example 1: Simplify.

a. $\sqrt[3]{64} = 4$

b. $\sqrt[4]{-16}$ is not a real number.

c. $\sqrt[5]{-32} = -2$

Example 2: Simplify using the product rule of roots.

a. $\sqrt[3]{2} \cdot \sqrt[3]{5} = \sqrt[3]{2 \cdot 5} = \sqrt[3]{10}$

b. $\sqrt[4]{27x^3} \cdot \sqrt[4]{3x} = \sqrt[4]{27x^3 \cdot 3x}$
$$= \sqrt[4]{81x^4} = 3x$$

c. $\sqrt[4]{48} = \sqrt[4]{16 \cdot 3} = \sqrt[4]{16} \cdot \sqrt[4]{3} = 2\sqrt[4]{3}$

d. $\sqrt[3]{54n^4} = \sqrt[3]{27n^3 \cdot 2n}$
$$= \sqrt[3]{27n^3} \cdot \sqrt[3]{2n} = 3n\sqrt[3]{2n}$$

(continued)

Procedures/Rules	Key Examples

SECTION 9.6 Higher Roots and Rational Exponents (*continued*)

General quotient rule of roots:

$$\sqrt[n]{\frac{a}{b}} = \frac{\sqrt[n]{a}}{\sqrt[n]{b}}, \text{ where } a \geq 0 \text{ and } b > 0$$

$a^{1/n} = \sqrt[n]{a}$, where n is a natural number other than 1
If a is positive, then the root is positive.
If a is negative and n is odd, then the root is negative.
If a is negative and n is even, then the root is not a real number.

$a^{m/n} = \sqrt[n]{a^m} = (\sqrt[n]{a})^m$, where $a \geq 0$ and m and n are natural numbers other than 1

$a^{-m/n} = \dfrac{1}{a^{m/n}}$, where $a \neq 0$, and m and n are natural numbers with $n \neq 1$

Example 3: Simplify using the quotient rule of roots.

a. $\sqrt[4]{\dfrac{16}{81}} = \dfrac{\sqrt[4]{16}}{\sqrt[4]{81}} = \dfrac{2}{3}$

b. $\sqrt[3]{\dfrac{24}{125}} = \dfrac{\sqrt[3]{24}}{\sqrt[3]{125}} = \dfrac{\sqrt[3]{8 \cdot 3}}{5} = \dfrac{2\sqrt[3]{3}}{5}$

Example 4: Evaluate.

a. $25^{1/2} = \sqrt{25} = 5$
b. $(-27)^{1/3} = \sqrt[3]{-27} = -3$
c. $(-16)^{1/4} = \sqrt[4]{-16}$; not a real number
d. $64^{2/3} = (\sqrt[3]{64})^2 = 4^2 = 16$

e. $16^{-3/4} = \dfrac{1}{16^{3/4}} = \dfrac{1}{(\sqrt[4]{16})^3} = \dfrac{1}{2^3} = \dfrac{1}{8}$

Chapter 9 Review Exercises

For Exercises 1–5, answer true or false.

[9.1] **1.** Every positive number has two square roots, a positive root and a negative root

[9.2] **2.** $\sqrt{80} = 4\sqrt{5}$

[9.3] **3.** To rationalize $\dfrac{3}{\sqrt{8}}$, we could multiply by $\dfrac{\sqrt{2}}{\sqrt{2}}$.

[9.5] **4.** Every radical equation has a single solution.

[9.5] **5.** It is necessary to check all potential solutions to radical equations.

For Exercises 6–10, complete the rule.

[9.3] **6.** To rationalize a denominator containing a single square root,
 a. _____ the fraction by a well chosen form of 1 so that the product's denominator has a radicand that is a perfect square.
 b. Simplify.

[9.4] **7.** Like radicals are two radical expression with identical _____ and indexes.

[9.4] **8.** To add or subtract like radicals, add or subtract the _____ and keep the radicals the same.

[9.4] **9.** To rationalize a denominator containing a sum or difference with a square root term, multiply the numerator and denominator by the _____ of the denominator.

[9.5] **10.** An apparent solution to a radical equation that does not solve the equation is considered _____ .

[9.1] *For Exercises 11 and 12, find all square roots of the given number.*

11. 121

12. 49

[9.1] *For Exercises 13–16, evaluate the square root if possible.*

13. $\sqrt{169}$

14. $-\sqrt{49}$

15. $\sqrt{-36}$

16. $\sqrt{\dfrac{1}{25}}$

[9.1] *For Exercises 17 and 18, classify each square root as rational or irrational.*

17. $\sqrt{7}$

18. $\sqrt{100}$

[9.1–9.3] *For Exercises 19–46, simplify. Assume that variables represent positive numbers.*

19. $\sqrt{81x^2}$

20. $\sqrt{25m^2n^4}$

21. $\sqrt{\dfrac{25}{16}a^4b^{10}}$

22. $\sqrt{\dfrac{81x^4y^2}{64a^8}}$

23. $\sqrt{3}\cdot\sqrt{12}$

24. $\sqrt{8xy^3}\cdot\sqrt{2xy}$

25. $\sqrt{10x^2}\cdot\sqrt{10x^2}$

26. $\sqrt{2m}\cdot\sqrt{18m^3}$

27. $\sqrt{8w^3}\cdot\sqrt{2wz^4}$

28. $\sqrt{x^2yz^3}\cdot\sqrt{100yz}$

29. $\sqrt{20}$

30. $\sqrt{48}$

31. $\sqrt{18}$

32. $\sqrt{120}$

33. $\sqrt{6x^4b^3}$

34. $\sqrt{10x^6}$

35. $\sqrt{36u^5v^4}$

36. $\sqrt{12x^4}$

37. $\dfrac{\sqrt{100}}{\sqrt{25}}$

38. $\dfrac{\sqrt{11}}{\sqrt{36}}$

39. $\dfrac{\sqrt{9}}{\sqrt{121}}$

40. $\sqrt{\dfrac{m^4}{81}}$

41. $\sqrt{\dfrac{1}{16m^4}}$

42. $\sqrt{\dfrac{5h^5}{20h}}$

43. $\sqrt{\dfrac{50}{16}}$

44. $\sqrt{\dfrac{18y}{8y^3}}$

45. $-\sqrt{\dfrac{18}{2}}$

46. $\dfrac{\sqrt{51a^3b^4}}{\sqrt{3ab^4c^2}}$

[9.3] *For Exercises 47–52, simplify by rationalizing the denominator. Assume that variables represent positive numbers.*

47. $\dfrac{1}{\sqrt{2}}$

48. $\dfrac{3}{\sqrt{3}}$

49. $\sqrt{\dfrac{4}{7}}$

50. $\dfrac{\sqrt{5x^2}}{\sqrt{2}}$

51. $\sqrt{\dfrac{17}{3y^2}}$

52. $\dfrac{\sqrt{7}}{\sqrt{21}}$

[9.4] For Exercises 53–66, simplify.

53. $4\sqrt{3} - 2\sqrt{3}$

54. $5\sqrt{2} + \sqrt{2}$

55. $2\sqrt{6} - 5\sqrt{6} + \sqrt{6}$

56. $\dfrac{\sqrt{2}}{6} - \dfrac{\sqrt{2}}{3}$

57. $\sqrt{3} + \sqrt{27}$

58. $\sqrt{20} + \sqrt{75}$

59. $\sqrt{20} + 2\sqrt{5}$

60. $3\sqrt{48} - 2\sqrt{12}$

61. $\sqrt{50} - \sqrt{98}$

62. $6\sqrt{2} + 5\sqrt{18}$

63. $8\sqrt{20} - \sqrt{80} + 6\sqrt{5}$

64. $\sqrt{44} - \sqrt{24} + \sqrt{11}$

65. $\sqrt{\dfrac{6}{49}} - \sqrt{\dfrac{6}{25}}$

66. $\sqrt{\dfrac{2}{25}} + \sqrt{\dfrac{2}{121}}$

[9.4] For Exercises 67–82, multiply. Assume that variables represent nonnegative numbers.

67. $\sqrt{5}(\sqrt{3} + \sqrt{2})$

68. $\sqrt{3}(\sqrt{2} - \sqrt{6})$

69. $\sqrt{7}(\sqrt{3} + 2\sqrt{7})$

70. $\sqrt{10}(2\sqrt{7} + 2\sqrt{3})$

71. $3\sqrt{6}(2 - 3\sqrt{6})$

72. $3\sqrt{2}(4\sqrt{2} + 3\sqrt{8})$

73. $(\sqrt{2} - \sqrt{3})(\sqrt{5} + \sqrt{7})$

74. $(\sqrt{2} - 4)(\sqrt{5} + 2)$

75. $(\sqrt{6x} + 2)(\sqrt{2x} - 1)$

76. $(\sqrt{5a} + \sqrt{3b})(\sqrt{5a} - \sqrt{3b})$

77. $(4\sqrt{r} - \sqrt{s})(4\sqrt{r} + \sqrt{s})$

78. $(2\sqrt{3} - \sqrt{5})(2\sqrt{3} + \sqrt{5})$

79. $(\sqrt{2} + 1)^2$

80. $(\sqrt{2} - \sqrt{5})^2$

81. $(3\sqrt{t} - 2)^2$

82. $(4\sqrt{x} + \sqrt{y})^2$

[9.4] For Exercises 83 and 84, write an expression in simplest form for the area of the figure.

83.

$\sqrt{6}$ $5\sqrt{10}$

84.

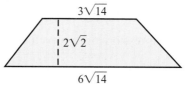

$3\sqrt{14}$ $2\sqrt{2}$ $6\sqrt{14}$

[9.4] For Exercises 85–88, rationalize the denominator and simplify.

85. $\dfrac{4}{\sqrt{2} - \sqrt{3}}$

86. $\dfrac{1}{4 + \sqrt{3}}$

87. $\dfrac{1}{2 - \sqrt{3}}$

88. $\dfrac{2\sqrt{3}}{3\sqrt{2} - 2\sqrt{3}}$

Exercises 89–107 EQUATIONS AND INEQUALITIES

[9.5] For Exercises 89–100, solve. Identify any extraneous solutions.

89. $\sqrt{y} = 7$

90. $\sqrt{m} = -2$

91. $\sqrt{w - 1} = 3$

92. $\sqrt{2x + 3} = 1$

93. $\sqrt{3y - 1} = 4$

94. $\sqrt{y + 1} - 4 = 2$

95. $\sqrt{2 - x} = \sqrt{5}$ **96.** $\sqrt{x + 2} = \sqrt{2x - 3}$ **97.** $\sqrt{h + 6} = h + 4$

98. $n + 1 = \sqrt{5n - 1}$ **99.** $1 + \sqrt{x} = \sqrt{2x + 1}$ **100.** $\sqrt{3x + 1} = 2 - \sqrt{3x}$

[9.5] *For Exercises 101 and 102, use the following information. The speed of a car can be determined by the length of the skid marks after the car brakes by using the formula $S = 2\sqrt{2L}$, where L represents the length of the skid mark in feet and S represents the speed of the car in miles per hour.*

101. **a.** Write an expression of the exact speed of a car if the length of the skid marks measures 40 feet.

     b. Approximate the speed to the nearest tenth.

102. Find the length of skid marks if a driver brakes hard at a speed of 50 miles per hour.

For Exercises 103 and 104, use the formula $T = 2\pi\sqrt{\dfrac{L}{9.8}}$, where T is the period of a pendulum in seconds and L is the length of the pendulum in meters.

[9.1] **103.** Find the exact period of a pendulum with a length of 2.45 meters.

[9.5] **104.** Suppose the period of the pendulum is $\dfrac{2\pi}{3}$ seconds. Find the length rounded to the nearest thousandth.

For Exercises 105 and 106, use the formula $t = \sqrt{\dfrac{h}{16}}$, where t is the time in seconds and h is the distance an object has fallen in feet.

[9.1] **105.** **a.** Write an expression in simplest form of the time an object takes to fall 40 feet.

     b. Approximate the time to the nearest hundredth.

[9.5] **106.** If an object falls for 0.5 second, what distance does it fall?

[9.1] **107.** Three pieces of lumber are to be connected to form a right triangle that will be part of the frame for the roof of a doghouse. If the horizontal piece is to be 4 feet and the vertical piece is to be 3 feet, how long must the connecting piece be?

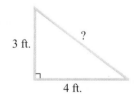

3 ft.

?

4 ft.

[9.6] *For Exercises 108–119, simplify. Assume that variables represent nonnegative numbers.*

108. $\sqrt[5]{32}$

109. $\sqrt[3]{64}$

110. $\sqrt[4]{-36}$

111. $\sqrt[3]{-8}$

112. $\sqrt[3]{8} \cdot \sqrt[3]{27}$

113. $\sqrt[4]{4} \cdot \sqrt[4]{4}$

114. $\sqrt[3]{4x} \cdot \sqrt[3]{2x^2}$

115. $\sqrt[5]{2y^4} \cdot \sqrt[5]{16y}$

116. $\sqrt[3]{\dfrac{64}{125}}$

117. $\sqrt[3]{-\dfrac{27}{216}}$

118. $\sqrt[4]{8}(\sqrt[4]{2} - 5)$

119. $(\sqrt[3]{12} + 7)^2$

[9.6] *For Exercises 120–127, rewrite using radicals; then simplify.*

120. $25^{1/2}$

121. $8^{1/3}$

122. $-32^{2/5}$

123. $-16^{3/4}$

124. $(-27)^{1/3}$

125. $\left(\dfrac{25}{9}\right)^{3/2}$

126. $64^{-4/3}$

127. $(-125)^{-2/3}$

[9.6] *For Exercises 128–131, simplify so that the result has a positive rational exponent. Assume that variables represent positive numbers.*

128. $n^{2/5} \cdot n^{1/5}$

129. $(h^{-3/4})^{1/5}$

130. $\dfrac{x^{2/5}}{x^{4/5}}$

131. $y^{3/5} \div y^{2/5}$

[9.6] *For Exercises 132–134, use the formula $B = P(1 + r)^t$ to find the final balance in an account after a principal P was invested and it grew at a rate r for t years.*

132. $P = \$4000, r = 8\%, t = \dfrac{1}{4}$ year

133. $P = \$1600, r = 4.5\%, t = 2\dfrac{3}{4}$ years

134. $P = \$3000, r = 6\%, t = 1\dfrac{1}{2}$ years

Chapter 9 Practice Test

For Extra Help

CHAPTER
Test Prep
VIDEOS

Step-by-step test solutions are found on the Chapter Test Prep Videos available via the Video Resources on DVD, in *MyMathLab*, and on You Tube (search "CarsonElemAlg" and click on "Channels").

For Exercises 1 and 2, evaluate the square root if possible.

1. $\sqrt{-25}$

2. $\sqrt{100}$

For Exercises 3–14, simplify. Assume that variables represent positive numbers.

3. $\sqrt{36x^4y^2}$

4. $\sqrt{20}$

5. $\sqrt{2} \cdot \sqrt{8}$

6. $\sqrt{27x^2y^5}$

7. $\dfrac{\sqrt{5}}{\sqrt{49}}$

8. $\dfrac{\sqrt{36}}{\sqrt{9}}$

9. $\sqrt{\dfrac{6x}{24x^3}}$

10. $5\sqrt{2} + 7\sqrt{2}$

11. $\sqrt{5} - \sqrt{125}$

12. $5\sqrt{3}(2\sqrt{2} + 3\sqrt{3})$

13. $(\sqrt{3} - \sqrt{2})(\sqrt{5} + \sqrt{2})$

14. $(\sqrt{3} - 1)^2$

15. Write an expression for the area of the figure.

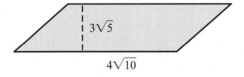

$3\sqrt{5}$

$4\sqrt{10}$

For Exercises 16 and 17, rationalize the denominator and simplify.

16. $\dfrac{5}{\sqrt{3}}$

17. $\dfrac{3}{\sqrt{2} + \sqrt{5}}$

For Exercises 18 and 19, solve the equation. Identify any extraneous solutions.

18. $\sqrt{x - 1} = 5$

19. $\sqrt{11 - 2x} = x + 2$

20. The speed of a car can be determined by the length of the skid marks after the car brakes by using the formula $S = \sqrt{30Dfn}$, where D represents the length of the skid mark in feet, f is the drag factor of the surface, n is the braking efficiency as a percent (written as a decimal), and S represents the speed of the car in miles per hour.

   a. Find the speed of a car if the skid length measures 60 feet long, the drag factor is 0.50, and the braking efficiency is 100%.

   b. Find the length of the skid marks of a car traveling 60 miles per hour on a surface with a drag factor of 0.30 and braking efficiency of 100%.

For Exercises 21–25, simplify.

21. $\sqrt[4]{16}$

22. $\sqrt[5]{-32}$

23. $\sqrt[5]{8x^2} \cdot \sqrt[5]{4x^3}$

24. $36^{1/2}$

25. $-4^{3/2}$

Chapter 1–9 Cumulative Review Exercises

For Exercises 1–3, answer true or false.

[2.2] **1.** Every linear equation has at least one solution.

[2.2] **2.** An identity is an expression that cannot be simplified.

[6.4] **3.** $(2^3)^2 = 2^5$

For Exercises 4–6, fill in the blank.

[1.7] **4.** If a, b, and c are any real numbers, the distributive property is $a(b + c) =$ _____.

[7.6] **5.** The zero-factor theorem states that if a and b are real numbers and $ab = 0$, then $a =$ _____ or $b =$ _____.

[7.6] **6.** If a and b represent the length of the legs of a right triangle and c represents the length of the hypotenuse, the Pythagorean theorem can be written as _____.

Exercises 7–19   **EXPRESSIONS**

For Exercises 7–9, simplify.

[1.5] **7.** $18 \div 3 + (2 - 4)^2 + 1$

[6.3] **8.** $(10y^2 - 8y + 5) - (6y^2 + 3y + 7)$

[6.5] **9.** $(5x - 6)^2$

For Exercises 10–12, factor completely.

[7.3] **10.** $6x^2 + 13x + 5$

[7.3] **11.** $12x^3 - 28x^2 + 8x$

[7.4] **12.** $m^3 - m$

For Exercises 13–18, simplify. For Exercises 15 and 16, assume that variables represent positive numbers.

[8.2] **13.** $\dfrac{3 - x}{x} \div \dfrac{9 - x^2}{x + 3}$

[8.4] **14.** $\dfrac{x}{x - 7} + \dfrac{x + 3}{x^2 - 4x - 21}$

[9.2] **15.** $\sqrt{8m^3} \cdot \sqrt{2m}$

[9.3] **16.** $\dfrac{\sqrt{48x^5}}{\sqrt{4x}}$

[9.4] **17.** $(5 - \sqrt{6})(4 + \sqrt{2})$

[9.4] **18.** $\sqrt{12} + \sqrt{27} + \sqrt{36}$

[9.3] **19.** Rationalize the denominator. $\dfrac{5}{\sqrt{3}}$

For Exercises 20–24, solve. Identify any extraneous solutions.

[2.3]   **20.** $3(x + 4) - 1 = 7x - (2x + 9)$

[7.6]   **21.** $x^2 = 5x - 6$

[8.6]   **22.** $3 + \dfrac{9 - 2x}{8x} = \dfrac{5}{2x}$

[5.3]   **23.** $\begin{cases} x - y = 14 \\ x + y = -2 \end{cases}$

[9.5]   **24.** $\sqrt{4x + 17} = x + 5$

[5.5]   **25.**   Graph the solution set. $\begin{cases} y < 2 \\ 5x < 4 + y \end{cases}$

[4.2]   **26.**   Graph: $y = 3x - 1$

For Exercises 27–30, solve.

[3.1]   **27.**   A telephone survey was conducted by the AFL-CIO to determine the percentage of women that work. It showed that 66% of working mothers tended to work more than 40 hours a week. The telephone survey was conducted with 1250 working women over the age of 18. How many of the women in the survey worked more than 40 hours a week? (*Source:* AFL-CIO, as reported by the Associated Press)

[7.6]   **28.**   Three boards are to be joined to form a right triangle as shown. Find the length of the missing board.

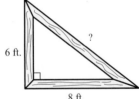

6 ft.
8 ft.

[3.4]
[5.4]   **29.**   A river's current is 8 miles per hour. If a boat can travel 34 miles downstream in the same time it takes to go 26 miles upstream, what is the speed of the boat in still water?

[3.3]
[5.4]   **30.**   A hardware store sells two different-sized cans of paint. One Saturday, the store sold twice as many large-size cans as small-size cans. If the small size sells for $8.95 and the large size sells for $15.95, how many of each size were sold if the combined sales were $694.45?

Quadratic Equations 10

In Section 7.6, we introduced quadratic equations and solved them by factoring. However, because not all quadratic equations can be solved by factoring, we need more powerful methods. In this chapter, we explore three other methods for solving quadratic equations, each of which uses square roots. In the first method, we solve quadratic equations using a new principle of equality called the *square root principle*. Like factoring, this first method has some limitations, so we develop a more powerful method called *completing the square*. Although we can complete the square to solve any quadratic equation, doing so is rather tedious. So we develop a third method that uses a general formula, called the *quadratic formula*, to solve any quadratic equation.

10.1 The Square Root Principle

Objectives

① Use the square root principle to solve quadratic equations.

② Determine the distance between two points in the coordinate plane.

Objective ① **Use the square root principle to solve quadratic equations.** In Section 7.6, we solved quadratic equations such as $x^2 = 25$ by subtracting 25 from both sides, factoring, and then using the zero-factor theorem. Let's recall that process.

$$x^2 - 25 = 0$$
$$(x - 5)(x + 5) = 0 \qquad \text{Factor}$$
$$x - 5 = 0 \quad \text{or} \quad x + 5 = 0 \qquad \text{Use the zero-factor theorem.}$$
$$x = 5 \qquad\qquad x = -5 \qquad \text{Isolate } x \text{ in each equation.}$$

Another approach to solving $x^2 = 25$ involves square roots. Notice that the solutions to this equation must be numbers that can be squared to equal 25. These numbers are the square roots of 25, which are 5 and -5. This suggests a new rule called the *square root principle*.

Rule **Square Root Principle**

Note: The expression $\pm\sqrt{a}$ is read "plus or minus the square root of a."

If $x^2 = a$, where a is a real number, then $x = \sqrt{a}$ or $x = -\sqrt{a}$. It is common to indicate the positive and negative solutions by writing $\pm\sqrt{a}$.

For example, if $x^2 = 49$, then $x = \sqrt{49} = 7$ or $x = -\sqrt{49} = -7$. Or we could simply write $x = \pm\sqrt{49} = \pm 7$.

Solve Equations in the Form $x^2 = a$

The square root principle is especially useful for solving equations in the form $x^2 = a$ when a is not a perfect square.

Example 1 Solve $x^2 = 32$.

SOLUTION $x^2 = 32$

Connection In Section 9.2, we learned how to simplify square roots of numbers that have perfect square factors.

$$x = \pm\sqrt{32} \qquad \text{Use the square root principle.}$$
$$x = \pm\sqrt{16 \cdot 2} \qquad \text{Simplify by factoring out a perfect square.}$$
$$x = \pm 4\sqrt{2}$$

Note: Remember, the $\pm$ symbol means that the two solutions are $4\sqrt{2}$ and $-4\sqrt{2}$.

We can check the two solutions using the original equation.

Check $\quad 4\sqrt{2}: (4\sqrt{2})^2 \overset{?}{=} 32$ *Check* $\quad -4\sqrt{2}: (-4\sqrt{2})^2 \overset{?}{=} 32$

                $16 \cdot 2 = 32$ True $16 \cdot 2 = 32$ True

For the remaining examples, the checks will be left to the reader.

Your Turn 1 Solve.

   a. $x^2 = 36$ **b.** $x^2 = 45$

Answers **Your Turn 1**

a. ± 6 **b.** $\pm 3\sqrt{5}$

Sometimes we have to use the addition or multiplication principles of equality to isolate x^2 before we can use the square root principle.

Solve Equations in the Form $x^2 + b = a$

If an equation is in the form $x^2 + b = a$, we first use the addition principle of equality to isolate x^2 and then use the square root principle.

Example 2 Solve.

a. $x^2 - 2 = 96$

SOLUTION $x^2 - 2 = 96$

$$x^2 = 98 \qquad \text{Add 2 to both sides to isolate } x^2.$$

$$x = \pm\sqrt{98} \qquad \text{Use the square root principle.}$$

$$x = \pm\sqrt{49 \cdot 2} \qquad \text{Simplify by factoring out a perfect square.}$$

$$x = \pm 7\sqrt{2}$$

b. $x^2 + 20 = 11$

SOLUTION $x^2 + 20 = 11$

$$x^2 = -9 \qquad \text{Subtract 20 from both sides to isolate } x^2.$$

$$x = \pm\sqrt{-9} \qquad \text{Use the square root principle.}$$

Note: We discuss square roots of negative numbers in Section 10.4.

Answer Because the square root of a negative number is not a real number, we say that this equation has no real-number solution.

Your Turn 2 Solve.

   a. $x^2 + 15 = 96$ **b.** $x^2 - 6 = 42$ **c.** $x^2 + 12 = 8$

Solve Equations in the Form $ax^2 = b$

To solve equations in the form $ax^2 = b$, we use the multiplication principle of equality to isolate x^2 by dividing both sides of the equation by a.

Example 3 Solve.

c. $9x^2 = 27$

SOLUTION $9x^2 = 27$

$$x^2 = 3 \qquad \text{Divide both sides by 9 to isolate } x^2.$$

$$x = \pm\sqrt{3} \qquad \text{Use the square root principle.}$$

b. $5x^2 = 12$

SOLUTION $5x^2 = 12$

$$x^2 = \frac{12}{5} \qquad \text{Divide both sides by 5 to isolate } x^2.$$

$$x = \pm\sqrt{\frac{12}{5}} \qquad \text{Use the square root principle.}$$

$$x = \pm\frac{\sqrt{12}}{\sqrt{5}} \cdot \frac{\sqrt{5}}{\sqrt{5}} \qquad \text{Rationalize the denominator.}$$

$$x = \pm\frac{\sqrt{60}}{5}$$

$$x = \pm\frac{\sqrt{4 \cdot 15}}{5} \qquad \text{Simplify by factoring out a perfect square.}$$

$$x = \pm\frac{2\sqrt{15}}{5}$$

Answers **Your Turn 2**

a. ± 9 **b.** $\pm 4\sqrt{3}$

c. no real-number solution

Your Turn 3 Solve.

a. $5x^2 = 60$

b. $2x^2 = 25$

Solve Equations in the Form $ax^2 + b = c$

We solve an equation in the form $ax^2 + b = c$ by using both the addition and multiplication principles of equality to isolate x^2 before using the square root principle.

Example 4 Solve $3x^2 + 5 = 59$.

SOLUTION $3x^2 + 5 = 59$

$$3x^2 = 54 \qquad \text{Subtract 5 from both sides.}$$
$$x^2 = 18 \qquad \text{Divide both sides by 3.}$$
$$x = \pm\sqrt{18} \qquad \text{Use the square root principle.}$$
$$x = \pm\sqrt{9 \cdot 2} \qquad \text{Simplify by factoring out a perfect square.}$$
$$x = \pm3\sqrt{2}$$

Your Turn 4 Solve.

a. $2x^2 + 11 = 65$

b. $9x^2 - 8 = 20$

Solve Equations in the Form $(ax + b)^2 = c$

In an equation in the form $(ax + b)^2 = c$, notice that the expression $ax + b$ is squared. We can use the square root principle to eliminate the square by thinking of the principle as follows: If $(\text{binomial})^2 = c$, then binomial $= \pm\sqrt{c}$.

Example 5 Solve.

a. $(x + 3)^2 = 16$

SOLUTION $(x + 3)^2 = 16$

$$x + 3 = \pm\sqrt{16} \qquad \text{Use the square root principle.}$$
$$x + 3 = \pm4$$
$$x = -3 \pm 4 \qquad \text{Subtract 3 from both sides.}$$
$$x = -3 + 4 \quad \text{or} \quad x = -3 - 4 \qquad \text{Simplify by separating the two solutions.}$$
$$x = 1 \qquad\qquad x = -7$$

b. $(2x - 5)^2 = 6$

SOLUTION $(2x - 5)^2 = 6$

$$2x - 5 = \pm\sqrt{6} \qquad \text{Use the square root principle.}$$
$$2x = 5 \pm \sqrt{6} \qquad \text{Add 5 to both sides to isolate } 2x.$$
$$x = \frac{5 \pm \sqrt{6}}{2} \qquad \text{Divide both sides by 2.}$$

c. $(6x - 7)^2 = -20$

SOLUTION $(6x - 7)^2 = -20$

$$6x - 7 = \pm\sqrt{-20} \qquad \text{Use the square root principle.}$$

Answers **Your Turn 3**

a. $\pm2\sqrt{3}$ b. $\pm\dfrac{5\sqrt{2}}{2}$

Answers **Your Turn 4**

a. $\pm3\sqrt{3}$ b. $\pm\dfrac{2\sqrt{7}}{3}$

Because the square root of a negative number is not a real number, there is no real-number solution to this equation.

Your Turn 5 Solve.

a. $(x - 5)^2 = 9$ b. $(3x + 1)^2 = 5$ c. $(5x + 2)^2 = -4$

Objective 2 **Determine the distance between two points in the coordinate plane.** We can use the Pythagorean theorem along with the square root principle to derive a formula for finding the distance between two points in the coordinate plane. First, consider a numerical case. Suppose we want to calculate the distance along a straight line connecting the points $(0, 0)$ and $(3, 4)$. Let's plot the points and draw the line. We'll label the unknown distance d. See the figure at left.

Notice that we can draw a right triangle with the length d as the hypotenuse. We can use the x- and y-axes to determine the lengths of the legs and then use the Pythagorean theorem to calculate the distance d. The vertical length is 4, and the horizontal length is 3. Recall the Pythagorean theorem

$$c^2 = a^2 + b^2,$$

where c is the length of the hypotenuse and a and b are the lengths of the legs.

To calculate the distance d, we replace a with 3, b with 4, and c with d.

$$d^2 = 3^2 + 4^2$$
$$d^2 = 9 + 16$$
$$d^2 = 25$$
$$d = \pm\sqrt{25}$$
$$d = \pm 5$$

Because d represents a distance, the reasonable answer is positive 5.

Now let's develop a formula that we can use to calculate the distance between *any* two points. We follow the same procedure as in our numerical example; however, instead of ordered pairs containing specific numbers, we use two points labeled (x_1, y_1) and (x_2, y_2). The subscripts are used to indicate which point is which. For convenience, we place the two points in the first quadrant.

We first write expressions describing the length of each leg of the triangle.

LEARNING Strategies

TACTILE If you are a tactile learner, to convince yourself that the result of 5 units is accurate, measure the hypotenuse of the triangle in the graph. To do this, place the edge of a sheet of paper along the x- or y-axis and mark off 5 units. Use this as your ruler to measure the hypotenuse of the triangle. You should see that it is exactly 5 units in length.

Note: If y_2 were 6 and y_1 were 2, the distance between them would be $6 - 2 = 4$. Therefore, to calculate the length of the vertical leg, we calculate $y_2 - y_1$.

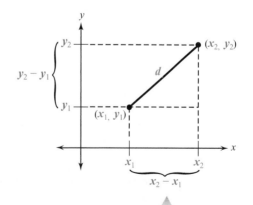

Note: If x_2 were 4 and x_1 were 1, the distance between them would be $4 - 1 = 3$. Therefore, to calculate the length of the horizontal leg, we calculate $x_2 - x_1$.

Answers **Your Turn 5**

a. 8 or 2 b. $\dfrac{-1 \pm \sqrt{5}}{3}$

c. no real-number solution

Now we can use the Pythagorean theorem, replacing a with $x_2 - x_1$, b with $y_2 - y_1$, and c with d. Then we isolate d using the square root principle.

$$c^2 = a^2 + b^2$$
$$d^2 = (x_2 - x_1)^2 + (y_2 - y_1)^2$$
$$d = \pm\sqrt{(x_2 - x_1)^2 + (y_2 - y_1)^2} \qquad \text{Use the square root principle to isolate } d.$$

Because d is a distance, it must be positive. So in the formula, we use only the positive value.

Procedure   **Using the Distance Formula**

To calculate the distance, d, between two points with coordinates (x_1, y_1) and (x_2, y_2), use the following formula:

$$d = \sqrt{(x_2 - x_1)^2 + (y_2 - y_1)^2}$$

Example 6   Find the distance between $(4, 2)$ and $(-3, -1)$. If the distance is an irrational number, give the exact expression and an approximation rounded to three decimal places.

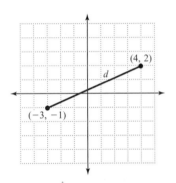

Note: The distance formula holds no matter what quadrants the points are in.

SOLUTION   Use the distance formula with $(4, 2)$ as (x_1, y_1) and $(-3, -1)$ as (x_2, y_2).

$$d = \sqrt{(x_2 - x_1)^2 + (y_2 - y_1)^2}$$
$$d = \sqrt{(-3 - 4)^2 + (-1 - 2)^2} \qquad \text{Replace the variables with the corresponding numbers.}$$
$$d = \sqrt{(-7)^2 + (-3)^2} \qquad \text{Subtract within parentheses.}$$
$$d = \sqrt{49 + 9} \qquad \text{Evaluate the exponential forms; then add.}$$
$$d = \sqrt{58} \qquad \longleftarrow \text{Exact answer}$$
$$d \approx 7.616 \qquad \longleftarrow \text{Approximate answer}$$

It doesn't matter which ordered pair is (x_1, y_1) and which is (x_2, y_2). To illustrate, consider Example 6 again with $(-3, -1)$ as (x_1, y_1) and $(4, 2)$ as (x_2, y_2).

$$d = \sqrt{(x_2 - x_1)^2 + (y_2 - y_1)^2}$$
$$d = \sqrt{(4 - (-3))^2 + (2 - (-1))^2}$$
$$d = \sqrt{(7)^2 + (3)^2} = \sqrt{49 + 9} = \sqrt{58} \approx 7.616$$

Note: In the Example 6 solution, the results in the parentheses were -7 and -3, and here we have 7 and 3. When we square -7 or 7, the result is the same, 49. Likewise, when we square -3 or 3, the result is 9. This is why it doesn't matter which coordinates are labeled (x_1, y_1) and which are labeled (x_2, y_2).

Your Turn 6   Determine the distance between the given points. If the distance is an irrational number, express the exact distance; then approximate the distance rounded to three decimal places.

   a. $(8, 2)$ and $(3, -4)$ **b.** $(6, -5)$ and $(0, -1)$

Answers   Your Turn 6

a. $\sqrt{61} \approx 7.810$

b. $2\sqrt{13} \approx 7.211$

Note: Exercises marked with a ★ represent challenging exercises.

1. Explain why there are two real-number solutions to an equation in the form $x^2 = a$, assuming that $a > 0$.

2. Explain why there are no real-number solutions to the equation $x^2 + 14 = 5$.

3. The expression $3 \pm 2\sqrt{5}$ indicates two solutions. What are they?

4. Are the solutions $3 \pm 2\sqrt{5}$ rational or irrational? Explain.

5. Write a formula for the solutions of $ax^2 - b = c$ by solving for x.

6. Write a formula for the solutions of $(ax - b)^2 = c$ by solving for x.

For Exercises 7–20, solve and check. See Example 1.

7. $x^2 = 16$

8. $x^2 = 36$

9. $r^2 = 196$

10. $t^2 = 400$

11. $y^2 = \dfrac{4}{9}$

12. $t^2 = \dfrac{1}{25}$

13. $c^2 = 0.36$

14. $y^2 = 1.44$

15. $k^2 = 12$

16. $m^2 = 20$

17. $t^2 = 50$

18. $r^2 = 27$

19. $w^2 = -25$

20. $c^2 = -49$

For Exercises 21–50, solve and check. First, use the addition or multiplication principles of equality to isolate the squared term. See Examples 2–4.

21. $x^2 - 100 = 0$

22. $x^2 - 49 = 0$

23. $n^2 - 7 = 42$

24. $y^2 - 5 = 59$

25. $h^2 + 2 = 3$

26. $u^2 - 6 = -5$

27. $y^2 - 7 = 29$

28. $k^2 + 5 = 30$

29. $4n^2 = 36$

30. $5y^2 = 125$

31. $-3t^2 = -75$

32. $-7y^2 = -28$

33. $2x^2 = 22$

34. $7x^2 = 42$

35. $25t^2 = 9$

36. $16d^2 = 49$

37. $-6h^2 = -16$

38. $3k^2 = 16$

39. $\dfrac{2}{3}x^2 = \dfrac{3}{4}$

40. $\dfrac{3}{4}m^2 = \dfrac{4}{5}$

41. $2x^2 + 1 = 3$

42. $4x^2 + 5 = 21$

43. $6y^2 + 1 = 13$

44. $5n^2 + 9 = 24$

45. $9k^2 - 17 = 10$ **46.** $2y^2 - 15 = 7$ **47.** $\frac{9}{4}y^2 + 5 = 8$ **48.** $\frac{6}{5}m^2 - 5 = 15$

49. $0.5t^2 - 0.4 = 0.32$ **50.** $0.4p^2 + 0.6 = 0.744$

For Exercises 51–70, solve and check. Use the square root principle to eliminate the square. See Example 5.

51. $(x + 3)^2 = 16$ **52.** $(y + 5)^2 = 4$ **53.** $(2n + 3)^2 = 36$ **54.** $(3h - 1)^2 = 49$

55. $(m - 7)^2 = 12$ **56.** $(t - 5)^2 = 28$ **57.** $(4k - 1)^2 = 40$ **58.** $(3x - 7)^2 = 50$

59. $(2y + 5)^2 = 18$ **60.** $(5m - 3)^2 = 48$ **61.** $(m - 8)^2 = -1$ **62.** $(t - 2)^2 = -5$

63. $\left(y - \frac{2}{3}\right)^2 = \frac{4}{9}$ **64.** $\left(x + \frac{3}{4}\right)^2 = \frac{25}{16}$ **65.** $\left(\frac{2}{3}d - \frac{1}{4}\right)^2 = \frac{1}{9}$ **66.** $\left(\frac{3}{4}h - \frac{5}{6}\right)^2 = \frac{1}{16}$

67. $(0.3x + 2.2)^2 = 0.49$ **68.** $(0.2n - 5.1)^2 = 1.69$ **69.** $(1.5x - 3.7)^2 = 1.96$ **70.** $(1.8t + 4.8)^2 = 0.09$

Find
(X) *the mistake*   *For Exercises 71–74, explain the mistake; then find the correct solution.*

71. $x^2 - 3 = 6$
$\qquad x^2 = 9$
$\qquad x = \sqrt{9}$
$\qquad x = 3$

72. $x^2 = 20$
$\qquad x = \sqrt{20}$
$\qquad x = 2\sqrt{5}$

73. $(x - 3)^2 = -2$
$\qquad x - 3 = \pm\sqrt{2}$
$\qquad x = 3\pm\sqrt{2}$

74. $(x - 1)^2 = -5$
$\qquad x - 1 = \pm\sqrt{5}$
$\qquad x = 1\pm\sqrt{5}$

For Exercises 75–80, solve; then use a calculator to approximate the irrational solutions rounded to three decimal places. See Examples 1–5.

75. $p^2 = 90$ **76.** $r^2 = 45$ **77.** $y^2 - 5 = 10$

78. $x^2 - 12 = 0$ **79.** $(n - 2)^2 = 3$ **80.** $(m - 1)^2 = 5$

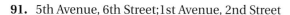

For Exercises 81–90, determine the distance between the two points. If the distance is an irrational number, write the exact answer; then use a calculator to approximate the distance rounded to three decimal places. See Example 6.

81. $(-7, -4)$ and $(-2, 8)$

82. $(-11, -3)$ and $(1, -3)$

83. $(0, 0)$ and $(3, 4)$

84. $(-5, -8)$ and $(3, 7)$

85. $(7, 3)$ and $(2, 5)$

86. $(3, 4)$ and $(5, 7)$

87. $(-4, 5)$ and $(5, -3)$

88. $(3, 8)$ and $(-2, -1)$

89. $(6, 9)$ and $(2, 3)$

90. $(3, -8)$ and $(9, -12)$

For Exercises 91–94, use the map of a city shown to find the distance between the indicated intersections. Give an exact answer and an approximation rounded to the nearest thousandth.

91. 5th Avenue, 6th Street; 1st Avenue, 2nd Street

92. 3rd Avenue, 1st Street; 8th Avenue, 4th Street

93. 4th Avenue, 3rd Street; 2nd Avenue, 2nd Street

94. 8th Avenue, 2nd Street; 6th Avenue, 1st Street

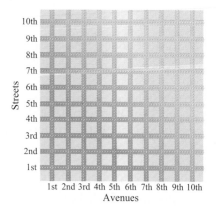

For Exercises 95–104, solve.

95. A square sheet of metal has an area of 196 square inches. What is the length of each side?

96. A severe thunderstorm warning is issued by the National Weather Service for a square area covering 14,400 square miles. What is the length of each side of the square area?

97. The length of a swimming pool is three times the width. If the depth is a constant 4 feet and the volume is 4800 cubic feet, find the length and width.

98. The longest advertising poster produced had a length that was 137 times its width. It was produced in Rome, Italy. The area of the poster was 548 square meters. Find the length and width of this poster. (*Source: Guinness World Records*)

99. The Arecibo radio telescope is like a giant satellite dish that covers a circular area of approximately $23,256.25\pi$ square meters. Find the diameter of the dish.

Of Interest

The Arecibo radio telescope is the world's largest radio telescope, which is a telescope that "listens" to radiation and signals emitted by objects in space. It was constructed in a natural depression near Arecibo, Puerto Rico.

100. A field is planted in a circular pattern. A watering device is to be constructed with pipe in a line extending from the center of the field to the edge of the field. The pipe is set on wheels so that it can rotate around the field and cover the entire field with water. If the area of the field is 7225π square feet, how long will the watering pipe be?

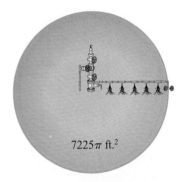

7225π ft.^{2}

101. A large storage tank is in the shape of a circular cylinder. It is 15 feet high, and its volume is 1696.5 cubic feet. What is the radius of the tank?

102. The largest trash can ever created was in the shape of a circular cylinder and was made by Bresco in Baltimore, Maryland. Its height was 5.38 meters, and its volume was 56.6 cubic meters. Find the diameter of this world-record-winning trash can. (*Source: Guinness World Records*)

103. A square piece of a sheet of wood has been cut as shown. If the area of the piece that was removed was 256 square inches, find x.

removed piece

256 in.^{2} 2 in.

2 in. x

x

104. The area of the hole in the washer shown is 25π square millimeters. Find r, the radius of the washer.

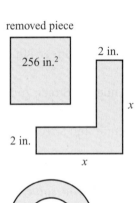

4 mm r

In physics, if an object is in motion, it has kinetic energy. The formula $E = \dfrac{1}{2}mv^2$ *is used to calculate the kinetic energy, E, of an object with a mass m and velocity v. If the mass is measured in kilograms and the velocity is in meters per second, the kinetic energy will be in units called joules (J).*

105. Suppose an object with a mass of 50 kilograms has 400 joules of kinetic energy. Find its velocity.

106. In a crash test, a vehicle with a mass of 1200 kilograms is found to have kinetic energy of 117,600 joules just before impact. Find the velocity of the vehicle just before impact.

For Exercises 107–110, use the formula $d = 16t^2$, *where d is the distance an object falls in feet and t is the time of the fall in seconds.*

107. Suppose a light cover falls from a 9-foot ceiling. How long does the cover take to hit the floor?

108. A construction worker tosses a scrap piece of lumber from the roof of a house. How long does the piece of lumber take to reach the ground 25 feet below?

★ **109.** A toy rocket is launched straight up and reaches a height of 180 feet in 1.5 seconds, then plummets back to the ground. Determine the total time of the rocket's round-trip.

 110. A lead weight is dropped from a tower at a height of 150 feet. How much time
★ passes for the weight to be halfway to the ground? Write an exact answer; then approximate to the nearest tenth of a second.

For Exercises 111–114, use the Pythagorean theorem, which is $a^2 + b^2 = c^2$,
where c represents the length of the hypotenuse of any right triangle and a and b
represent the lengths of the legs.

111. A baseball diamond is a square with a side length of 90 feet. What is the direct distance from first base to third base? Give the exact answer and an approximation rounded to the nearest tenth.

112. A ramp to a building entrance is constructed so that it rises 2 feet over a distance of 24 feet. How long is the ramp? Give the exact answer and an approximation rounded to the nearest tenth.

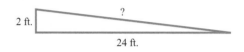

113. Andrea bought a 42-inch Hitachi plasma television. While at the store, she wrote down the height of the screen as 20 inches but forgot to make note of the width. Now she is trying to decide where to put it in her house, and she needs the dimensions. What is the width of the television? Give an exact answer and an approximation rounded to the nearest tenth. (*Source:* Hitachi)

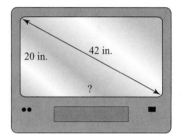

114. One end of a 40-foot wire is attached to a power pole, and the other end is staked into the ground 15 feet from the base of the pole. How high up the pole is the wire attached? Give an exact answer and an approximation rounded to the nearest tenth.

Review Exercises

Exercises 1–5 EXPRESSIONS

[1.7] **1.** Evaluate $\left(\dfrac{b}{2}\right)^2$ when $b = 12$.

[6.5] *For Exercises 2 and 3, multiply.*

    2. $(x + 4)^2$ **3.** $(y - 7)^2$

[7.2] **4.** Factor: $x^2 + 6x + 9$ [9.2] **5.** Simplify assuming that x represents nonnegative values: $\sqrt{9x^2}$

Exercise 6 EQUATION AND INEQUALITIES

[7.6] **6.** Solve: $x^2 + 8x = -16$

10.2 Solving Quadratic Equations by Completing the Square

Objectives

1 Complete the square.

2 Solve quadratic equations by completing the square.

In Section 10.1, we used the square root principle to solve equations like $(x + 2)^2 = 9$ and $(x - 4)^2 = 25$. Notice that the expressions $(x + 2)^2$ and $(x - 4)^2$ are squares. In this section, we use a technique called *completing the square* to rewrite expressions as squares in an equation so that we can use the square root principle to solve the equation.

Objective 1 **Complete the square.** Suppose we are given an expression such as $x^2 + 6x$, which is an "incomplete" square. If we add 9 to $x^2 + 6x$, we get $x^2 + 6x + 9$, which is a perfect square trinomial whose factored form is $(x + 3)^2$. When we add 9 to $x^2 + 6x$, we say that we are *completing the square*. To determine how to complete the square for an expression in the form $x^2 + bx$, look for a pattern in the following table.

Incomplete Square	Completed Square	Factored Form
$x^2 + 6x$	$x^2 + 6x + 9$	$(x + 3)^2$
$x^2 + 8x$	$x^2 + 8x + 16$	$(x + 4)^2$
$x^2 - 10x$	$x^2 - 10x + 25$	$(x - 5)^2$
$x^2 - 12x$	$x^2 - 12x + 36$	$(x - 6)^2$

Notice that squaring half of the coefficient of x gives the constant term in the completed square. For example:

$$x^2 + 6x + 9 \qquad\qquad x^2 - 10x + 25$$

Half of 6 is 3, which when squared, is 9.

Half of -10 is -5, which when squared, is 25.

Also notice that half of the coefficient of x is the constant in the factored form.

$$x^2 + 6x + 9 = (x + 3)^2 \qquad\qquad x^2 - 10x + 25 = (x - 5)^2$$

Half of 6 is 3, which is the constant in the factored form.

Half of -10 is -5, which is the constant in the factored form.

This suggests the following procedure for completing the square.

Procedure Completing the Square

Note: In order to complete the square, the coefficient of x^2 must be 1.

Given an expression in the form $x^2 + bx$, to complete the square, add the constant term $\left(\dfrac{b}{2}\right)^2$.

Note: The factored form for the completed square is $x^2 + bx + \left(\dfrac{b}{2}\right)^2 = \left(x + \dfrac{b}{2}\right)^2$.

Example 1 Add a term to the expression to complete the square; then write the completed square in factored form.

a. $x^2 + 16x$

SOLUTION Completed square: $x^2 + 16x + 64$
Factored form: $(x + 8)^2$

To complete the square, we add the square of half of 16. Half of 16 is 8; the square of 8 is 64.

b. $x^2 - 5x$

SOLUTION Completed square: $x^2 - 5x + \dfrac{25}{4}$
Factored form: $\left(x - \dfrac{5}{2}\right)^2$

We add the square of half of -5. Half of -5 is $-\dfrac{5}{2}$; the square of $-\dfrac{5}{2}$ is $\dfrac{25}{4}$.

c. $x^2 - \dfrac{3}{4}x$

SOLUTION Completed square: $x^2 - \dfrac{3}{4}x + \dfrac{9}{64}$
Factored form: $\left(x - \dfrac{3}{8}\right)^2$

We add the square of half of $-\dfrac{3}{4}$. Half of $-\dfrac{3}{4}$ is $-\dfrac{3}{8}$; the square of $-\dfrac{3}{8}$ is $\dfrac{9}{64}$.

Answers Your turn 1

a. $x^2 - 10x + 25; (x - 5)^2$

b. $x^2 + x + \dfrac{1}{4}; \left(x + \dfrac{1}{2}\right)^2$

c. $x^2 - \dfrac{5}{6}x + \dfrac{25}{144}; \left(x - \dfrac{5}{12}\right)^2$

Your Turn 1 Add a term to the expression to complete the square; then write the completed square in factored form.

a. $x^2 - 10x$ **b.** $x^2 + x$ **c.** $x^2 - \dfrac{5}{6}x$

Objective ② **Solve quadratic equations by completing the square.** To solve a quadratic equation by completing the square, we need the equation to be in the form $x^2 + bx = c$. For example, consider the quadratic equation $x^2 + 16x = 5$. Notice that the left–hand side of the equation is an incomplete square, $x^2 + 16x$. As we saw in Example 1, we can complete the square by adding 64. Because of the addition principle of equality, we must add 64 to both sides of the equation.

Note: The expression you get after completing the square is not equivalent to the original expression. Therefore, when solving quadratic equations by completing the square, you must add the same number, $\left(\dfrac{b}{2}\right)^2$, to both sides of the equation.

$$x^2 + 16x + 64 = 5 + 64$$

We can now factor the left-hand side of the equation, then use the square root principle as we did in Section 10.1 to finish solving the equation.

$(x + 8)^2 = 69$ Write the left-hand side in factored form.

$x + 8 = \pm\sqrt{69}$ Use the square root principle to eliminate the square.

$x = -8 \pm \sqrt{69}$ Subtract 8 from both sides to isolate x.

Example 2 Solve by completing the square.

a. $x^2 + 8x + 5 = 0$

SOLUTION We first write the equation in the form $x^2 + bx = c$.

Note: We found 16 by squaring half of 8.

$$\left(\dfrac{8}{2}\right)^2 = 4^2 = 16$$

$x^2 + 8x = -5$ Subtract 5 from both sides to get the form $x^2 + bx = c$.

$x^2 + 8x + 16 = -5 + 16$ Complete the square by adding 16 to both sides.

$(x + 4)^2 = 11$ Factor.

$x + 4 = \pm\sqrt{11}$ Use the square root principle.

$x = -4 \pm \sqrt{11}$ Subtract 4 from both sides to isolate x.

b. $x^2 + 12x + 7 = 11$

SOLUTION $x^2 + 12x = 4$

Subtract 7 from both sides to get the form $x^2 + bx = c$.

$$x^2 + 12x + 36 = 4 + 36$$

Complete the square by adding 36 to both sides.

$$(x + 6)^2 = 40$$

Factor.

$$x + 6 = \pm\sqrt{40}$$

Use the square root principle.

$$x = -6 \pm \sqrt{40}$$

Subtract 6 from both sides to isolate x.

$$x = -6 \pm \sqrt{4 \cdot 10}$$

Simplify the square root.

$$x = -6 \pm 2\sqrt{10}$$

c. $x^2 - 5x + 3 = 4$

SOLUTION $x^2 - 5x = 1$

Subtract 3 from both sides to get the form $x^2 + bx = c$.

$$x^2 - 5x + \frac{25}{4} = 1 + \frac{25}{4}$$

Complete the square by adding $\frac{25}{4}$ to both sides.

$$\left(x - \frac{5}{2}\right)^2 = \frac{29}{4}$$

Factor.

$$x - \frac{5}{2} = \pm\sqrt{\frac{29}{4}}$$

Use the square root principle.

$$x = \frac{5}{2} \pm \sqrt{\frac{29}{4}}$$

Add $\frac{5}{2}$ to both sides to isolate x.

$$x = \frac{5}{2} \pm \frac{\sqrt{29}}{2}$$

Simplify the square root.

$$x = \frac{5 \pm \sqrt{29}}{2}$$

Combine the fractions.

> **Note:** We found $\frac{25}{4}$ by squaring half of -5.
> $$\left(\frac{-5}{2}\right)^2 = \frac{25}{4}$$

Your Turn 2 Solve by completing the square.

  a. $x^2 + 10x + 1 = 0$ **b.** $x^2 - 14x + 3 = 10$ **c.** $x^2 - 3x - 1 = 2$

Leading Coefficient Other Than 1

Now let's look at quadratic equations in which the x^2 term has a coefficient other than 1, such as $3x^2 + 6x = 8$. Remember that to complete the square, we need the equation in the form $x^2 + bx = c$, where the coefficient of the x^2 term is 1. In the case of $3x^2 + 6x = 8$, dividing both sides of the equation by 3 $\left(\text{or multiplying both sides by } \frac{1}{3}\right)$ gives x^2 a coefficient of 1.

$$\frac{3x^2 + 6x}{3} = \frac{8}{3}$$

Divide both sides by 3.

$$x^2 + 2x = \frac{8}{3}$$

Answers Your Turn 2
a. $-5 \pm 2\sqrt{6}$ **b.** $7 \pm 2\sqrt{14}$
c. $\dfrac{3 \pm \sqrt{21}}{2}$

We can now solve by completing the square.

$$x^2 + 2x + 1 = \frac{8}{3} + 1 \qquad \text{Add 1 to both sides to complete the square.}$$

$$(x + 1)^2 = \frac{11}{3} \qquad \text{Factor.}$$

$$x + 1 = \pm\sqrt{\frac{11}{3}} \qquad \text{Use the square root principle.}$$

$$x = -1 \pm \sqrt{\frac{11}{3}} \qquad \text{Subtract 1 from both sides to isolate } x.$$

$$x = -1 \pm \frac{\sqrt{11}}{\sqrt{3}} \cdot \frac{\sqrt{3}}{\sqrt{3}} \qquad \text{Rationalize the denominator.}$$

$$x = -1 \pm \frac{\sqrt{33}}{3}$$

We can now write a procedure for solving any quadratic equation by completing the square.

Procedure

Solving Quadratic Equations by Completing the Square

To solve a quadratic equation by completing the square,

1. Write the equation in the form $x^2 + bx = c$.
2. Complete the square.
3. Write the completed square in factored form.
4. Use the square root principle to eliminate the square.
5. Isolate the variable.
6. Simplify as needed.

Example 3   Solve by completing the square.

a. $4x^2 - 20x = 6$

SOLUTION

$$\frac{4x^2 - 20x}{4} = \frac{6}{4} \qquad \text{Divide both sides by 4 so the coefficient of } x^2 \text{ is 1.}$$

$$x^2 - 5x = \frac{3}{2} \qquad \text{Simplify.}$$

$$x^2 - 5x + \frac{25}{4} = \frac{3}{2} + \frac{25}{4} \qquad \text{Add } \frac{25}{4} \text{ to both sides to complete the square.}$$

$$\left(x - \frac{5}{2}\right)^2 = \frac{31}{4} \qquad \text{Factor.}$$

$$x - \frac{5}{2} = \pm\sqrt{\frac{31}{4}} \qquad \text{Use the square root principle.}$$

$$x = \frac{5}{2} \pm \frac{\sqrt{31}}{2} \qquad \text{Add } \frac{5}{2} \text{ to both sides and simplify the square root.}$$

$$x = \frac{5 \pm \sqrt{31}}{2} \qquad \text{Combine the fractions.}$$

b. $3x^2 + 7x + 5 = 4$

SOLUTION $3x^2 + 7x + 5 = 4$

$$3x^2 + 7x = -1$$ Subtract 5 from both sides.

$$\frac{3x^2 + 7x}{3} = \frac{-1}{3}$$ Divide both sides by 3 so the coefficient of x^2 is 1.

$$x^2 + \frac{7}{3}x = -\frac{1}{3}$$ Simplify.

$$x^2 + \frac{7}{3}x + \frac{49}{36} = -\frac{1}{3} + \frac{49}{36}$$ Add $\frac{49}{36}$ to both sides to complete the square.

$$\left(x + \frac{7}{6}\right)^2 = \frac{37}{36}$$ Factor.

$$x + \frac{7}{6} = \pm\sqrt{\frac{37}{36}}$$ Use the square root principle.

$$x = -\frac{7}{6} \pm \frac{\sqrt{37}}{6}$$ Subtract $\frac{7}{6}$ from both sides and simplify the square root.

$$x = \frac{-7 \pm \sqrt{37}}{6}$$ Combine the fractions.

Note To complete the square, we square half of $\frac{7}{3}$:

$$\left(\frac{1}{2} \cdot \frac{7}{3}\right)^2 = \left(\frac{7}{6}\right)^2 = \frac{49}{36}$$

Answers   Your Turn 3

a. 2 or $\frac{1}{4}$

b. $-1 \pm \frac{3\sqrt{5}}{5}$ or $\frac{-5 \pm 3\sqrt{5}}{5}$

Your Turn 3   Solve by completing the square.

a. $4x^2 - 9x + 7 = 5$ **b.** $5x^2 + 10x = 4$

10.2 Exercises   For Extra Help   *MyMathLab*

PRACTICE   WATCH   DOWNLOAD

Note: Exercises marked with a ★ represent challenging exercises.

1. Given an expression in the form $x^2 + bx$, explain how to complete the square.

2. Discuss whether the solutions to $x^2 + 6x = 2$ are rational or irrational.

3. Consider the equation $x^2 - 6x + 8 = 0$. Which is a better method for solving the equation, factoring and then using the zero-factor theorem or completing the square? Explain.

4. Consider the equation $x^2 + 5x - 2 = 0$. Which is a better method for solving the equation, factoring and then using the zero-factor theorem or completing the square? Explain.

For Exercises 5–18: a. Add a term to the expression to complete the square. (Make it a perfect square.)
b. Write the completed square in factored form. See Example 1.

5. $x^2 + 10x$ **6.** $x^2 + 12x$ **7.** $y^2 - 6y$ **8.** $t^2 - 14t$

9. $d^2 - 12d$ **10.** $w^2 - 16w$ **11.** $x^2 - 5x$ **12.** $n^2 - 13n$

13. $y^2 + 9y$

14. $n^2 + 7n$

15. $v^2 + \dfrac{1}{6}v$

16. $u^2 + \dfrac{1}{2}u$

17. $s^2 - \dfrac{3}{4}s$

18. $b^2 + \dfrac{3}{2}b$

For Exercises 19–32, solve by completing the square. See Example 2.

19. $y^2 + 10y = -16$

20. $x^2 + 10x = -24$

21. $r^2 - 2r - 48 = 0$

22. $c^2 - 6c - 16 = 0$

23. $w^2 + w = 20$

24. $p^2 + 3p = 18$

25. $k^2 = 9k - 18$

26. $a^2 = 3a + 10$

27. $b^2 - 2b - 11 = 4$

28. $n^2 + 10n - 20 = 4$

29. $x^2 + 4x = 9$

30. $z^2 + 2z = 6$

31. $q^2 - 6q - 11 = 0$

32. $j^2 - 4j - 5 = -4$

For Exercises 33–50, solve by completing the square. Begin by writing the equation in the form $x^2 + bx = c$. See Example 3.

33. $2t^2 - 5 = -3t$

34. $3m^2 + 2m = 8$

35. $2g^2 + g - 11 = -5$

36. $4l^2 + l - 30 = 30$

37. $6w^2 - 6 = -5w$

38. $6x^2 - 10 = 11x$

39. $6d^2 - 13d - 20 = 8$

40. $8z^2 + 6z - 3 = 6$

41. $u^2 + \dfrac{1}{2}u = \dfrac{3}{2}$

42. $y^2 + \dfrac{1}{3}y = \dfrac{2}{3}$

43. $2h^2 - 6h = 5$

44. $3s^2 + 6s = -1$

45. $2x^2 + 6x = -3$

46. $3x^2 + 12x = 5$

47. $5k^2 + k - 2 = 0$

48. $3x^2 + 5x - 3 = 0$

49. $2x^2 = 4x + 3$

50. $2x^2 = -6x + 5$

51.

$$2x^2 - 2x = 1$$
$$x^2 - x = 1$$
$$x^2 - x + \frac{1}{4} = 1 + \frac{1}{4}$$
$$\left(x - \frac{1}{2}\right)^2 = \frac{5}{4}$$
$$x - \frac{1}{2} = \pm\frac{\sqrt{5}}{2}$$
$$x = \frac{1 \pm \sqrt{5}}{2}$$

52.

$$x^2 + 4x = 5$$
$$x^2 + 4x + 4 = 5 + 4$$
$$(x + 2)^2 = 9$$
$$x + 2 = \sqrt{9}$$
$$x + 2 = 3$$
$$x = -2 + 3$$
$$x = 1$$

For Exercises 53–62, solve.

53. A rectangular living room is 2 feet longer than it is wide. If the area of the room is 168 square feet, what are its dimensions?

54. A rectangular playground is designed so that the length is 20 feet more than the width. If the playground's area is to be 2400 square feet, what are the dimensions of the playground?

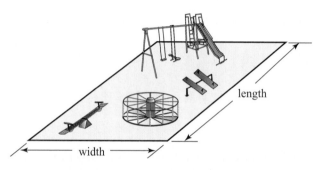

55. A ramp forms a right triangle. If the base of the ramp is 7 feet longer that the height and the hypotenuse is 13 feet, what are the dimensions of the base and height?

56. Two identical right triangles are placed together to form the frame of a roof. If the height of each triangle is 4 feet less than its base and its hypotenuse is 20 feet, what are the dimensions of the base and height?

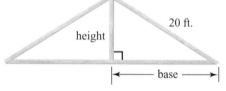

57. When building a rectangular gate, John wants to add a 10-foot diagonal brace for better support. If the base of the gate is 2 feet longer than the height, what are the dimensions of the base and height?

58. Ellen lives on a corner lot. She finds it annoying that the neighborhood children have been cutting across her lawn instead of walking around the yard. If the distance across the lawn is 40 feet and the longer part of the sidewalk is twice the shorter length, how many feet are the children saving by cutting across the lawn? Round your answer to the nearest tenth of a foot.

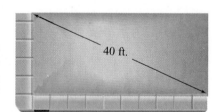

★ **59.** The metal panel shown is to have a total area of 825 square inches. Find the length x.

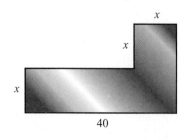

★ **60.** A side view of a concrete bridge support footing is shown. The side of the footing is to have a total area of 80 square feet. Find the length of x.

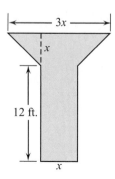

★ **61.** A plastic panel is to have a rectangular hole cut as shown.
   a. Find w so that the area remaining after the hole is cut is 520 square inches.
   b. Find the length and width of the plastic panel.

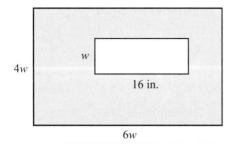

★ **62.** An 8-centimeter-wide groove with a height of h is to be cut into a steel block as shown.
   a. Find h so that the volume remaining in the block after the groove is cut is 400 cubic centimeters.
   b. Find the height and width of the block.

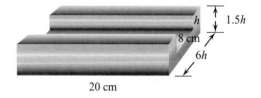

Puzzle Problem

Without using a calculator, which of the following numbers is a perfect square? (*Hint:* Write a list of smaller perfect squares and look for a pattern.)

9,456,804,219,745,618
2,512,339,789,576,516
7,602,985,471,286,543
4,682,715,204,643,182

Review Exercises

Exercises 1–5  EXPRESSIONS

[6.5]   **1.** Multiply: $(x + 4)^2$

For Exercises 2 and 3, factor.

[7.3]   **2.** $6x^2 - 7x - 5$ **[7.4]**   **3.** $x^2 - 10x + 25$

[9.1]   **4.** Evaluate $\dfrac{-b + \sqrt{b^2 - 4ac}}{2a}$ when $a = 2$, $b = -3$, and $c = 1$.

[9.1]   **5.** Evaluate $\dfrac{-b - \sqrt{b^2 - 4ac}}{2a}$ when $a = 2$, $b = -3$, and $c = 1$.

Exercise 6 EQUATIONS AND INEQUALITIES

[10.1]   **6.** Solve: $x^2 = 12$

10.3 Solving Quadratic Equations Using the Quadratic Formula

Objectives

1 Solve quadratic equations using the quadratic formula.

2 Use the discriminant to determine the number of real solutions that a quadratic equation will have.

In this section, we solve quadratic equations using a formula called the *quadratic formula*. This formula is much easier to use than completing the square.

Objective ① **Solve quadratic equations using the quadratic formula.** To derive the quadratic formula, we begin with the general form of the quadratic equation, $ax^2 + bx + c = 0$, and assume that $a \neq 0$. We follow the procedure for solving a quadratic equation by completing the square.

$$ax^2 + bx + c = 0$$

$$ax^2 + bx = -c \qquad \text{Subtract } c \text{ from both sides.}$$

$$\frac{ax^2 + bx}{a} = \frac{-c}{a} \qquad \text{Divide both sides by } a \text{ so that the coefficient of } x^2 \text{ is 1.}$$

$$x^2 + \frac{b}{a}x = -\frac{c}{a} \qquad \text{Simplify.}$$

Note: To complete the square, we square half of $\frac{b}{a}$.

$$\left(\frac{1}{2} \cdot \frac{b}{a}\right)^2 = \left(\frac{b}{2a}\right)^2 = \frac{b^2}{4a^2}$$

$$x^2 + \frac{b}{a}x + \frac{b^2}{4a^2} = -\frac{c}{a} + \frac{b^2}{4a^2} \qquad \text{Complete the square.}$$

$$\left(x + \frac{b}{2a}\right)^2 = -\frac{4ac}{4a^2} + \frac{b^2}{4a^2} \qquad$$
On the left side, write the factored form. On the right side, rewrite $-\frac{c}{a}$ with the common denominator $4a^2$ so that we can combine the rational expressions.

$$\left(x + \frac{b}{2a}\right)^2 = \frac{b^2 - 4ac}{4a^2} \qquad$$
Combine the rational expressions. For simplicity, we rearrange the order of the b^2 and $-4ac$ terms in the numerator.

$$x + \frac{b}{2a} = \pm\sqrt{\frac{b^2 - 4ac}{4a^2}} \qquad$$
Use the square root principle to eliminate the square.

$$x + \frac{b}{2a} = \pm\frac{\sqrt{b^2 - 4ac}}{2a} \qquad \text{Simplify the square root in the denominator.}$$

$$x = -\frac{b}{2a} \pm \frac{\sqrt{b^2 - 4ac}}{2a} \qquad \text{Subtract } \frac{b}{2a} \text{ from both sides to isolate } x.$$

$$x = \frac{-b \pm \sqrt{b^2 - 4ac}}{2a} \qquad \text{Combine rational expressions.}$$

This final equation is the *quadratic formula*. It can be used to solve any quadratic equation simply by replacing a, b, and c with the corresponding numbers from the given equation.

Procedure **Using the Quadratic Formula**

To solve a quadratic equation in the form $ax^2 + bx + c = 0$, where $a \neq 0$, use the quadratic formula:

$$x = \frac{-b \pm \sqrt{b^2 - 4ac}}{2a}$$

Note: The given quadratic equation must be in the form $ax^2 + bx + c = 0$ in order to identify a, b, and c for use in the quadratic formula.

Example 1   Solve using the quadratic formula.

a. $2x^2 + 9x - 5 = 0$

SOLUTION   Because this equation is in the form $ax^2 + bx + c = 0$, where $a = 2$, $b = 9$, and $c = -5$, we can use the quadratic formula, $x = \frac{-b \pm \sqrt{b^2 - 4ac}}{2a}$.

Connection In Example 3 of Section 7.6, we solved the equation $2x^2 + 9x - 5 = 0$ by factoring.

$$x = \frac{-9 \pm \sqrt{9^2 - 4(2)(-5)}}{2(2)}$$ Replace a with 2, b with 9, and c with -5.

$$x = \frac{-9 \pm \sqrt{81 + 40}}{4}$$

$$x = \frac{-9 \pm \sqrt{121}}{4}$$

$$x = \frac{-9 \pm 11}{4}$$ Split up the $\pm$ to calculate the two solutions.

$$x = \frac{-9 + 11}{4} \quad \text{or} \quad x = \frac{-9 - 11}{4}$$

$$x = \frac{2}{4} \qquad\qquad x = \frac{-20}{4}$$

$$x = \frac{1}{2} \qquad\qquad x = -5$$

Connection Notice that in this example, the radicand, 121, is a perfect square, which causes the solutions to be two rational numbers. As we consider more examples in this section, notice how the radicand determines the type of solutions.

b. $x^2 + 12x + 7 = 11$

SOLUTION First, we need the equation in the form $ax^2 + bx + c = 0$.

Connection In Example 2(b) of Section 10.2, we solved $x^2 + 12x + 7 = 11$ by completing the square.

$$x^2 + 12x - 4 = 0 \quad \text{Subtract 11 from both sides to get the form } ax^2 + bx + c = 0.$$

Now use the quadratic formula, $x = \dfrac{-b \pm \sqrt{b^2 - 4ac}}{2a}$.

$$x = \frac{-12 \pm \sqrt{12^2 - 4(1)(-4)}}{2(1)}$$ Replace a with 1, b with 12, and c with -4.

$$x = \frac{-12 \pm \sqrt{144 + 16}}{2}$$

$$x = \frac{-12 \pm \sqrt{160}}{2}$$

$$x = \frac{-12 \pm \sqrt{16 \cdot 10}}{2}$$ Simplify the square root.

$$x = \frac{-12 \pm 4\sqrt{10}}{2}$$

Connection Notice that the radicand, 160, is not a perfect square, which causes the solutions to be two irrational numbers.

$$x = \frac{-12}{2} \pm \frac{4\sqrt{10}}{2}$$ Separate into two rational expressions in order to simplify.

$$x = -6 \pm 2\sqrt{10}$$ Simplify by dividing out the 2.

c. $5x^2 = 12$

Connection In Example 3(b) of Section 10.1, we solved $5x^2 = 12$ using the principle of square roots.

SOLUTION First, we need the equation in the form $ax^2 + bx + c = 0$.

$$5x^2 - 12 = 0 \quad \text{Subtract 12 from both sides to get the form } ax^2 + bx + c = 0.$$

Note: There is no x term.

Now use the quadratic formula, $x = \dfrac{-b \pm \sqrt{b^2 - 4ac}}{2a}$.

$$x = \frac{-0 \pm \sqrt{0^2 - 4(5)(-12)}}{2(5)}$$ Replace a with 5, b with 0, and c with -12.

$$x = \frac{\pm\sqrt{240}}{10}$$

$$x = \frac{\pm\sqrt{16 \cdot 15}}{10}$$ Simplify the square root.

$$x = \frac{\pm 4\sqrt{15}}{10}$$

$$x = \frac{\pm 2\sqrt{15}}{5}$$ Simplify to lowest terms.

Your Turn 1 Solve using the quadratic formula.

a. $3x^2 + 7x - 20 = 0$ **b.** $5x^2 + 2 = -8x$ **c.** $9x^2 = 50$

Notice that each equation in Example 1 has two solutions. Example 1(a) has two rational solutions. Examples 1(b) and 1(c) have two irrational solutions. A quadratic equation can also have only one solution or no real-number solution, as we will see in Example 2.

Example 2 Solve using the quadratic formula.

a. $12x + 9 = -4x^2$

SOLUTION $4x^2 + 12x + 9 = 0$ Add $4x^2$ to both sides to get the form $ax^2 + bx + c = 0$.

$$x = \frac{-12 \pm \sqrt{12^2 - 4(4)(9)}}{2(4)}$$ In the quadratic formula, replace a with 4, b with 12, and c with 9.

$$x = \frac{-12 \pm \sqrt{144 - 144}}{8}$$

$$x = \frac{-12 \pm \sqrt{0}}{8}$$

$$x = -\frac{3}{2}$$ Simplify the square root and write the fraction in lowest terms.

> **Connection** Notice that the radicand is 0, which causes this equation to have only one solution.

b. $3x^2 - x + 5 = 0$

SOLUTION The equation is in the form $ax^2 + bx + c = 0$, so we use the quadratic formula.

$$x = \frac{-(-1) \pm \sqrt{(-1)^2 - 4(3)(5)}}{2(3)}$$ In the quadratic formula, replace a with 3, b with −1, and c with 5.

$$x = \frac{1 \pm \sqrt{1 - 60}}{6}$$

$$x = \frac{1 \pm \sqrt{-59}}{6}$$

> **Connection** Notice that the radicand is negative, which causes this equation to have no real-number solutions.

Because $\sqrt{-59}$ is not a real number, there are no real-number solutions.

Your Turn 2 Solve using the quadratic formula

a. $x^2 - 8x + 16 = 0$ **b.** $4x^2 + 1 = 3x$

Answers Your Turn 1

a. $\frac{5}{3}$ or -4 **b.** $\frac{-4 \pm \sqrt{6}}{5}$

c. $\pm\frac{5\sqrt{2}}{3}$

Answers Your Turn 2

a. 4 **b.** no real-number solution

Choosing a Method for Solving Quadratic Equations

We have learned several methods for solving quadratic equations. The following table summarizes the methods and conditions that make each method a wise choice.

Methods for Solving Quadratic Equations

Method	When the Method is Beneficial
1. Factoring (Section 7.6)	Use when the quadratic equation can be easily factored.
2. Square root principle (Section 10.1)	Use when the quadratic equation can be easily written in the form $ax^2 = c$.
3. Completing the square (Section 10.2)	This is rarely the best method.
4. Quadratic formula (Section 10.3)	Use when factoring is not easy.

Example 3   Solve using the most appropriate method.

a. $3x^2 + 5 = 149$

SOLUTION   This equation can easily be written in the form $ax^2 = c$, so use the square root principle.

$$3x^2 + 5 = 149$$
$$3x^2 = 144 \qquad \text{Subtract 5 from both sides.}$$
$$x^2 = 48 \qquad \text{Divide both sides by 3.}$$
$$x = \pm\sqrt{48} \qquad \text{Apply the square root principle.}$$
$$x = \pm 4\sqrt{3} \qquad \text{Simplify.}$$

b. $6x^2 - x - 15 = 0$

SOLUTION   This equation cannot be written in the form $ax^2 = c$, so try factoring.

$$6x^2 - x - 15 = 0$$
$$(3x - 5)(2x + 3) = 0 \qquad \text{Factor.}$$
$$3x - 5 = 0 \quad 2x + 3 = 0 \qquad \text{Use the zero-factor theorem.}$$
$$3x = 5 \qquad 2x = -3$$
$$x = \frac{5}{3} \qquad x = -\frac{3}{2} \qquad \text{Solve for } x.$$

c. $x^2 - 4x + 2 = 0$

SOLUTION   This equation cannot be written in the form $ax^2 = c$, nor can it be factored, so use the quadratic formula.

$$x^2 - 4x + 2 = 0$$
$$x = \frac{-(-4) \pm \sqrt{(-4)^2 - 4(1)(2)}}{2(1)} \qquad \begin{array}{l}\text{Substitute into the quadratic formula} \\ \text{using } a = 1, b = -4, \text{ and } c = 2.\end{array}$$
$$x = \frac{4 \pm \sqrt{16 - 8}}{2} \qquad \text{Simplify.}$$
$$x = \frac{4 \pm \sqrt{8}}{2}$$
$$x = \frac{4 \pm 2\sqrt{2}}{2}$$
$$x = 2 \pm \sqrt{2}$$

Your Turn 3   Solve using the most appropriate method.

a. $2x^2 + 4 = 200$ **b.** $2x^2 + x - 28 = 0$ **c.** $x^2 + 6x = 4$

Answers   Your Turn 3

a. $\pm 7\sqrt{2}$ **b.** $-4, \dfrac{7}{2}$

c. $-3 \pm \sqrt{13}$

Objective ② Use the discriminant to determine the number of real solutions that a quadratic equation will have. In the Connection boxes for Examples 1 and 2, we pointed out how the radicand in the quadratic formula affects the solutions to a given quadratic equation. The expression $b^2 - 4ac$, which is the radicand, is called the **discriminant**.

Definition *Discriminant:* The radicand $b^2 - 4ac$ in the quadratic formula.

We use the discriminant to determine the number of real solutions to a quadratic equation. It also can tell us whether those solutions are rational or irrational.

Procedure **Using the Discriminant**

To determine the number and type of solutions of a quadratic equation in the form $ax^2 + bx + c$, where a, b, and c are rational numbers and $a \neq 0$, evalute the discriminant $b^2 - 4ac$.

Note: When the descriminant is 0, the solution is $\dfrac{-b \pm \sqrt{0}}{2a} = \dfrac{-b}{2a}$.

If the **discriminant is positive**, the equation has two real-number solutions. The solutions will be rational if the discriminant is a perfect square and irrational otherwise.

If the **discriminant is 0**, then the equation has one rational solution.

If the **discriminant is negative**, the equation has no real-number solutions.

Example 4 Use the discriminant to determine the number of real solutions for the equation. If the solution(s) are real numbers, indicate whether they are rational or irrational.

a. $2x^2 - 9x = 6$

SOLUTION First, write the equation in the form $ax^2 + bx + c = 0$.

$$2x^2 - 9x - 6 = 0 \quad \text{Subtract 6 from both sides to get the form } ax^2 + bx + c = 0.$$

Now evaluate the discriminant, $b^2 - 4ac$.

$$(-9)^2 - 4(2)(-6) \quad \text{Replace } a \text{ with 2, } b \text{ with } -9, \text{ and } c \text{ with } -6.$$
$$= 81 + 48$$
$$= 129$$

Warning: 129 is the value of the discriminant, not a solution for the equation $2x^2 - 9x - 6 = 0$.

Because the discriminant is positive, this equation has two real-number solutions. Also, because 129 is not a perfect square, the solutions are irrational numbers.

b. $x^2 + \dfrac{9}{16} = -\dfrac{3}{2}x$

SOLUTION $x^2 + \dfrac{3}{2}x + \dfrac{9}{16} = 0 \quad \text{Add } \dfrac{3}{2}x \text{ to both sides to get the form } ax^2 + bx + c = 0.$

To work with integer values for a, b, and c, we multiply both sides of the equation by the LCD, which is 16.

$$16 \cdot x^2 + \dfrac{16}{1} \cdot \dfrac{3}{2}x + \dfrac{16}{1} \cdot \dfrac{9}{16} = 16 \cdot 0$$

$$16x^2 + 24x + 9 = 0$$

Now evaluate the discriminant, $b^2 - 4ac$.

$$(24)^2 - 4(16)(9) \qquad \text{Replace } a \text{ with 16, } b \text{ with 24, and } c \text{ with 9.}$$
$$= 576 - 576$$
$$= 0$$

Note: Because the discriminant is 0, the solution is

$$-\frac{b}{2a} = -\frac{24}{2(16)} = -\frac{24}{32} = -\frac{3}{4}.$$

Because the discriminant is zero, there is only one rational solution for this equation.

c. $0.3x^2 - 0.7x + 0.8 = 0$

Note: We could avoid calculations with decimal numbers by multiplying the original equation through by 10 so that it becomes

$$3x^2 - 7x + 8 = 0.$$

SOLUTION Evaluate the discriminant, $b^2 - 4ac$.

$$(-0.7)^2 - 4(0.3)(0.8) \qquad \text{Replace } a \text{ with 0.3, } b \text{ with } -0.7, \text{ and } c \text{ with 0.8.}$$
$$= 0.49 - 0.96$$
$$= -0.47$$

Because the discriminant is negative, there are no real-number solutions for this equation.

Answers Your Turn 4

a. discriminant is positive; two irrational solutions

b. discriminant is negative; no real-number solutions

c. discriminant is 0; one rational solution

Your Turn 4 Use the discriminant to determine the number of real solutions for the equation. If the solution(s) are real numbers, indicate whether they are rational or irrational.

a. $3x^2 - 7x - 2 = 0$

b. $x^2 + \frac{2}{3} = \frac{1}{2}x$

c. $0.3x^2 - 0.6x = -0.3$

10.3 Exercises   For Extra Help   *MyMathLab*

Note: Exercises marked with a ★ represent challenging exercises.

1. Explain the general plan for deriving the quadratic formula.

2. Discuss the advantages and disadvantages of using the quadratic formula to solve quadratic equations.

3. When might you choose factoring over the quadratic formula as a method for solving quadratic equations?

4. Are there quadratic equations that cannot be solved using factoring? Explain.

5. What part of the quadratic formula is the discriminant?

6. Why does a quadratic equation have no real-number solutions if the value of its discriminant is negative?

For Exercises 7–14, rewrite each quadratic equation in the form $ax^2 + bx + c = 0$; then identify a, b, and c. See Objective 1.

7. $x^2 - 4x + 9 = 0$

8. $x^2 + 5x - 12 = 0$

9. $2x^2 - 7x = 5$

10. $4x^2 + 9x = -16$

11. $0.5x^2 = x - 2.4$

12. $x - 10.5 = 0.6x^2$

13. $\frac{2}{3}x = -\frac{1}{2}x^2 + 8$

14. $\frac{5}{6}x^2 + 3 = \frac{1}{4}x$

For Exercises 15–38, solve using the quadratic formula. See Examples 1 and 2.

15. $x^2 + 9x + 18 = 0$

16. $x^2 + 7x - 18 = 0$

17. $x^2 + 6x = 16$

18. $x^2 - 2x = 8$

19. $x^2 + 3x = 4$

20. $x^2 + 5x = -6$

21. $x^2 - 3x = 0$

22. $x^2 + 5x = 0$

23. $x^2 - 10x = -25$

24. $x^2 + 6x = -9$

25. $x^2 - x - 1 = 0$

26. $x^2 + 3x - 5 = 0$

27. $3x^2 - 2x - 1 = 0$

28. $2x^2 - x - 3 = 0$

29. $6x^2 + 2 = -7x$

30. $8x^2 - 3 = 10x$

31. $3x^2 + 10x + 5 = 0$

32. $2x^2 + x - 5 = 0$

33. $4x^2 + 4 = 9x$

34. $5x^2 - 11x = -3$

35. $6x^2 - 3x = 4$

36. $2x^2 - 1 = 5x$

37. $3x^2 + 3 = -2x$

38. $2m^2 + 2 = -2m$

For Exercises 39–48, solve using the quadratic formula. (Hint: You might clear the fractions or decimals first by multiplying both sides by an appropriate chosen number.) See Examples 1 and 2.

39. $2x^2 + 0.2x = 0.04$

40. $3x^2 - 12.9x - 7.2 = 0$

41. $x^2 + \dfrac{5}{2}x - \dfrac{3}{2} = 0$

42. $x^2 + \dfrac{1}{3}x - \dfrac{2}{3} = 0$

43. $x^2 - \dfrac{9}{4} = 0$

44. $x^2 - \dfrac{16}{49} = 0$

45. $\dfrac{1}{3}x^2 - \dfrac{1}{2}x - \dfrac{1}{12} = 0$

46. $\dfrac{1}{5}x^2 - 2 = \dfrac{1}{2}x$

47. $0.06x^2 - 0.4x + 0.03 = 0$

48. $2.4x^2 + 6.3x - 4.5 = 0$

Find

 the mistake   **For Exercises 49–52, explain the mistake; then solve correctly.**

49. Solve $2x^2 - 3x - 5 = 0$ using the quadratic formula.

$$\frac{-3 \pm \sqrt{(-3)^2 - (4)(2)(-5)}}{2(2)} = \frac{-3 \pm \sqrt{49}}{4}$$

$$= \frac{-3 \pm 7}{4}$$

$$= 1, -\frac{5}{2}$$

50. Solve $x^2 - 2x - 5 = 0$ using the quadratic formula.

$$\frac{2 \pm \sqrt{(-2)^2 - (4)(1)(-5)}}{2(1)} = \frac{2 \pm \sqrt{4 + 20}}{2}$$

$$= \frac{2 \pm \sqrt{24}}{2}$$

51. Solve $x^2 - 2x + 3 = 0$ using the quadratic formula.

$$\frac{-(-2) \pm \sqrt{(-2)^2 - (4)(1)(3)}}{2(1)} = \frac{2 \pm \sqrt{4 - 12}}{2}$$

$$= \frac{2 \pm \sqrt{-8}}{2}$$

$$= \frac{2 \pm 2\sqrt{2}}{2}$$

$$= 1 \pm \sqrt{2}$$

52. Solve $x^2 - 4 = 0$ using the quadratic formula.

$$\frac{-(-4) \pm \sqrt{(-4)^2 - (4)(1)(0)}}{2(1)} = \frac{4 \pm \sqrt{16 - 0}}{2}$$

$$= \frac{4 \pm \sqrt{16}}{2}$$

$$= \frac{4 \pm 4}{2}$$

$$= 4, 0$$

For Exercises 53–62, indicate which of the following methods is the best choice for solving the given equation: factoring, using the square root principle, or using the quadratic formula. Then solve the equation. See Example 3.

53. $x^2 - 80 = 0$

54. $x^2 = 20$

55. $x^2 - 8x + 16 = 0$

56. $x^2 - 5x = 14$

57. $x^2 - 4x + 1 = 0$

58. $x^2 = 4x + 41$

59. $(x + 5)^2 = 16$

60. $(2x + 1)^2 = 49$

61. $x^2 + 4x = 0$

62. $x^2 - 3x = 0$

For Exercises 63–72, use the discriminant to determine the number of real solutions for the equation. If the solution(s) are real numbers, indicate whether they are rational or irrational. See Example 4.

63. $2y^2 - 15y + 18 = 0$

64. $2x^2 - 3x - 20 = 0$

65. $x^2 - 6x + 6 = 0$

66. $5z^2 + 2z - 10 = 0$

67. $2x^2 - 8x + 8 = 0$

68. $x^2 + 10x = -25$

69. $x^2 + 4x + 9 = 0$

70. $6a^2 - 4a + 2 = 0$

71. $\frac{1}{2}x^2 - 2x = -2$

72. $\frac{1}{3}x^2 = -2x - 3$

★**73.** For the equation $16x^2 - 8x + c = 0$,
  a. Find c so that the equation has only one rational number solution.

  b. Find the range of values of c for which the equation has two real-number solutions.

  c. Find the range of values of c for which the equation has no real-number solution.

★**74.** For the equation $9x^2 + 12x + c = 0$,
  a. Find c so that the equation has only one rational number solution.

  b. Find the range of values of c for which the equation has two real-number solutions.

  c. Find the range of values of c for which the equation has no real-number solution.

★ **75.** For the equation $ax^2 + 6x + 1 = 0$,

    a. Find a so that the equation has only one rational number solution.

    b. Find the range of values of a for which the equation has two real-number solutions.

    c. Find the range of values of a for which the equation has no real-number solution.

★ **76.** For the equation $ax^2 + 4x - 1 = 0$,

    a. Find a so that the equation has only one rational number solution.

    b. Find the range of values of a for which the equation has two real-number solutions.

    c. Find the range of values of a for which the equation has no real-number solution.

★ **77.** For the equation $8x^2 + bx + 2 = 0$,

    a. Find b so that the equation has only one rational number solution.

    b. Find all positive values of b for which the equation has two real-number solutions.

    c. Find all positive values of b for which the equation has no real-number solution.

★ **78.** For the equation $4x^2 + bx + 9 = 0$,

    a. Find b so that the equation has only one rational number solution.

    b. Find all positive values of b for which the equation has two real-number solutions.

    c. Find all positive values of b for which the equation has no real-number solution.

For Exercises 79–88, translate to a quadratic equation; then solve using the quadratic formula.

79. The sum of the square of a positive integer and six times its consecutive integer is equal to 33. Find the integers.

80. The difference of the square of a positive integer and the next largest consecutive integer is equal to 89. Find the integers.

81. A right triangle exists with side lengths that are three consecutive integers. Use the Pythagorean theorem to find the lengths of those sides. (Remember that the hypotenuse in a right triangle is always the longest side).

82. A right triangle exists with side lengths that are consecutive even integers. Use the Pythagorean theorem to find the lengths of those sides. (Remember that the hypotenuse in a right triangle is always the longest side).

83. The length of a rectangular air filter is 3 inches less than twice the width. Find the length and width of the filter if the area is 350 square inches.

84. A small access door for a crawl space is designed so that the width is 0.5 feet more than the length. Find the length and width of the access door if its area is 7.5 square feet.

85. An architect is experimenting with two different shapes of a room as shown.

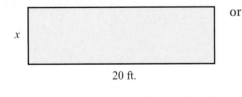 or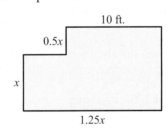

    a. Find x so that the rooms have the same area.

    b. Complete the dimensions for the L-shaped room.

86. A cylinder is to be made so that its volume is equal to that of a sphere with a radius of 3 inches. If a cylinder is to have a height of 4 inches, find its radius.

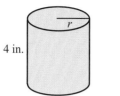 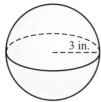

4 in.   3 in.

87. The expression $0.5n^2 + 2.5n$ describes the gross income from the sale of a particular software product, where n is the number of units sold. The expression $4.5n$ describes the cost of producing each unit. Find the number of units that must be produced and sold for the company to break even. (To break even means that the gross income and cost are the same.)

88. An economist and marketing manager discover that the expression $2n^2 + 5n$ models the price of a CD based on the demand for it, where n is the number of units (in millions) that the market demands. The expression $-0.5n^2 + 17.1$ describes the price of the CD based on the number of units (also in millions) supplied to the market.
   a. Find the number of units that need to be demanded and supplied so that the price based on demand is equal to the price based on supply. When the number of units demanded by the market is the same as the number of units supplied to the market, the product is said to be at equilibrium.

   b. What is the price of the CD at equilibrium?

89. Walmart reached number one in the Fortune 500 in 2002. It remained there through 2007. Walmart's revenue, in billions, can be approximated using the equation $R = 1.91y^2 + 15y + 227.4$ where R represents the revenue each year from 2002 to 2007 and y represents the number of years since 2002. (2002 corresponds to $y = 0$.) (*Source*: http://money.cnn.com/magazines/fortune)
   a. In what year between 2002 and 2007 did Walmart's revenue reach $244 billion?
   b. In what year between 2002 and 2007 did Walmart's revenue reach $318 billion?

90. The average price that farmers received for 100 pounds of veal for the years 2000–2007 can be approximated using the equation $R = 1.56y^2 - 3.29y + 103.39$ where R represents the revenue received and y is the number of years after 2000. (2000 corresponds to $y = 0$) (*Source: The World Almanac and Book of Facts*, 2008)
   a. In what year between 2000 and 2007 did farmers receive $107 for 100 pounds of veal?
   b. In what year between 2000 and 2007 did farmers receive $126 for 100 pounds of veal?

91. The equation $P = 0.013a^2 + 0.58a - 17.27$ approximately models the percentage of American women who have high blood pressure, where P represents the percent and a represents the age group in years. (*Source:* American Heart Association)
   a. Find the age, to the nearest year, at which 36% of the American women have high blood pressure.

   b. Find the age, to the nearest year, at which 5.8% of the American women have high blood pressure.

92. The equation $P = 0.006a^2 + 0.74a - 11.11$ approximately models the percentage of American men who have high blood pressure, where P represents the percent and a represents the age group in years. (*Source:* American Heart Association)
   a. Find the age, to the nearest year, at which 48% of the American men have high blood pressure.

   b. Find the age, to the nearest year, at which 22% of the American men have high blood pressure.

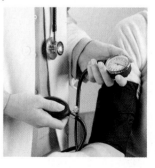

In the equation shown, A, B, C, D, and E are five consecutive positive integers where $A < B < C < D < E$. What are the integers?

$$A^2 + B^2 + C^2 = D^2 + E^2$$

Review Exercises

Exercises 1–6 EXPRESSIONS

For Exercises 1–6, simplify.

[6.3]   **1.** $(3x + 2) + (-4x + 5)$ [6.3]   **2.** $(-3x - 1) - (-2x + 5)$ [6.5]   **3.** $-3x(2x - 1)$

[6.5]   **4.** $(y - 7)(y + 7)$ [9.2]   **5.** $\sqrt{20}$ [9.4]   **6.** $\dfrac{2\sqrt{3}}{\sqrt{3} + 1}$

10.4 Complex Numbers

Objectives

1. Write imaginary numbers using i.
2. Perform arithmetic operations with complex numbers.
3. Solve quadratic equations that have solutions that are complex numbers.

In our study of square roots, we have said that the square root of a negative number is not a real number. In this section, we learn about the *imaginary number system* in which square roots of negative numbers are expressed using a notation involving the letter i.

Objective 1   Write imaginary numbers using i.   Using the product rule of square roots, any square root of a negative number can be rewritten as a product of a real number and an **imaginary unit**, which we express as i.

Definition   ***Imaginary unit:*** The number represented by i, where $i = \sqrt{-1}$ and $i^2 = -1$.

A number that can be expressed as a product of a real number and the imaginary unit is called an **imaginary number**.

Definition   ***Imaginary number:*** A number that can be expressed in the form bi, where b is a real number and i is the imaginary unit.

Example 1   Write each imaginary number as a product of a real number and i.

a. $\sqrt{-9}$

Warning: $\sqrt{a} \cdot \sqrt{b} \neq \sqrt{ab}$ for $a < 0$ and $b < 0$.

SOLUTION $\sqrt{-9} = \sqrt{-1 \cdot 9}$ Factor out -1 in the radicand.

$\qquad\qquad = \sqrt{-1} \cdot \sqrt{9}$ Use the product rule of square roots.

$\qquad\qquad = i \cdot 3$

$\qquad\qquad = 3i$

b. $\sqrt{-15}$

Note: We could write $\sqrt{15}i$, but we write $i\sqrt{15}$ so that it is clear that i is not part of the radicand.

SOLUTION $\sqrt{-15} = \sqrt{-1 \cdot 15}$ Factor out -1 in the radicand.

$= \sqrt{-1} \cdot \sqrt{15}$ Use the product rule of square roots.

$= i\sqrt{15}$

c. $\sqrt{-24}$

Note: For clarity in a product with the imaginary unit, we write integer factors first, then i, then square root factors.

SOLUTION $\sqrt{-24} = \sqrt{-1 \cdot 24}$ Factor out -1 in the radicand.

$= \sqrt{-1} \cdot \sqrt{24}$ Use the product rule of square roots.

$= i\sqrt{4 \cdot 6}$ Use the product rule again to simplify further.

$= 2i\sqrt{6}$

Example 1 suggests the following procedure.

Procedure   **Rewriting Imaginary Numbers**

To write an imaginary number $\sqrt{-n}$ in terms of the imaginary unit i,

1. Separate the radical into two factors, $\sqrt{-1} \cdot \sqrt{n}$.
2. Replace $\sqrt{-1}$ with i.
3. Simplify $\sqrt{n}$.

Your Turn 1   Write each imaginary number as a product of a real number and i.

a. $\sqrt{-36}$ **b.** $\sqrt{-22}$ **c.** $\sqrt{-50}$

Objective 2   Perform arithmetic operations with complex numbers.   We now have two distinct sets of numbers, the set of real numbers and the set of imaginary numbers. There is yet another set of numbers, called the set of **complex numbers**, that contains both the real and imaginary numbers.

Definition   **Complex number:** A number that can be expressed in the form $a + bi$, where a and b are real numbers and i is the imaginary unit.

When written in the form $a + bi$, a complex number is said to be in *standard form.* Following are some examples of complex numbers written in standard form.

$$2 + 3i \qquad 5 - 9i \qquad -7.9 - 3i\sqrt{5}$$

Note that if $a = 0$, then the complex number is purely an imaginary number, such as these:

$$-7i \qquad 5i\sqrt{2}$$

If $b = 0$, then the complex number is a real number.

The following Venn diagram shows how the set of complex numbers contains both the real numbers and the imaginary numbers.

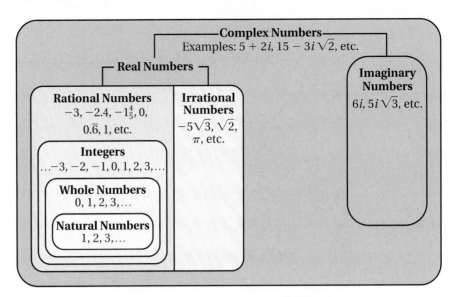

We can perform arithmetic operations with complex numbers. In general, we treat the complex numbers just like polynomials, where i is like a variable.

Adding and Subtracting Complex Numbers

Example 2 Add or subtract.

a. $(-9 + 2i) + (4 - 15i)$

SOLUTION We add complex numbers just like we add polynomials—by combining like terms.

$$(-9 + 2i) + (4 - 15i) = -5 - 13i$$

b. $(-1 + 8i) - (6 - 7i)$

SOLUTION We subtract complex numbers just like we subtract polynomials—by writing an equivalent addition and changing the signs in the second complex number.

$$(-1 + 8i) - (6 - 7i) = (-1 + 8i) + (-6 + 7i)$$
$$= -7 + 15i$$

Your Turn 2 Add or subtract.

  a. $(4 - 2i) + (-7 - 5i)$ **b.** $(-4 + 2i) - (-6 - 5i)$

Multiplying Complex Numbers

We multiply complex numbers the same way that we multiply monomials and binomials. However, we must be careful when simplifying because these products may contain i^2, which is equal to -1.

Example 3 Multiply.

a. $(6i)(-7i)$

SOLUTION Multiply the same way that we multiply monomials.

$$(6i)(-7i) = -42i^2 \qquad \text{Multiply coefficients. Add exponents of } i.$$
$$= -42(-1) \qquad \text{Replace } i^2 \text{ with } -1.$$
$$= 42$$

b. $(4i)(9 - 5i)$

SOLUTION Multiply the same way that we multiply a binomial by a monomial.

$$\begin{aligned}
(4i)(9 - 5i) &= 36i - 20i^2 & \text{Distribute.} \\
&= 36i - 20(-1) & \text{Replace } i^2 \text{ with } -1. \\
&= 36i + 20 & \text{Write in standard form.} \\
&= 20 + 36i
\end{aligned}$$

c. $(8 - 3i)(2 + i)$

SOLUTION Multiply the same way that we multiply binomials.

$$\begin{aligned}
(8 - 3i)(2 + i) &= 16 + 8i - 6i - 3i^2 & \text{Use FOIL.} \\
&= 16 + 2i - 3(-1) & \text{Combine like terms and replace } i^2 \text{ with } -1. \\
&= 16 + 2i + 3 & \text{Simplify.} \\
&= 19 + 2i & \text{Write in standard form.}
\end{aligned}$$

d. $(3 - i)(3 + i)$

Connection Recall that expressions of the form $a + b$ and $a - b$ are conjugates.

SOLUTION Note that these complex numbers are conjugates.

$$\begin{aligned}
(3 - i)(3 + i) &= 9 + 3i - 3i - i^2 & \text{Use FOIL.} \\
&= 9 - (-1) & \text{Combine like terms and replace } i^2 \text{ with } -1. \\
&= 9 + 1 \\
&= 10
\end{aligned}$$

Note: The product of the two complex numbers is a real number.

Your Turn 3 Multiply.

a. $(-4i)(-6i)$ **b.** $(-3i)(2 - i)$ **c.** $(5 - 9i)(4 + 7i)$ **d.** $(8 + 3i)(8 - 3i)$

The complex numbers that we multiplied in Example 3(d) and Your Turn 3d are called **complex conjugates**.

Definition   *Complex conjugates:* The complex conjugate of a complex number $a + bi$ is $a - bi$.

Other examples of complex conjugates follow.

$$5 + 6i \text{ and } 5 - 6i \qquad 9 - 7i \text{ and } 9 + 7i$$

Example (3d) and Your Turn 3d also illustrate the fact that the product of complex conjugates is always a real number. In Sections 6.5 and 9.4, we found the products of conjugates using $(a + b)(a - b) = a^2 - b^2$. Consequently, the product of complex conjugates is $(a + bi)(a - bi) = a^2 - (bi)^2 = a^2 - b^2i^2 = a^2 - b^2(-1) = a^2 + b^2$.

Dividing Complex Numbers

We have established that the product of complex conjugates is a real number. We can use that fact when the divisor in a division problem is a complex number. The process is similar to rationalizing denominators.

Answers   **Your Turn 3**

a. -24 **b.** $-3 - 6i$
c. $83 - i$ **d.** 73

Example 4   Write $\dfrac{2 + 3i}{6 - i}$ in standard form.

SOLUTION   Remember that when we rationalize a denominator that is a binomial with a term that is a square root, we multiply both the numerator and denominator by the conjugate of the denominator. In this case, we multiply by $6 + i$.

$$\dfrac{2 + 3i}{6 - i} = \dfrac{2 + 3i}{6 - i} \cdot \dfrac{6 + i}{6 + i} \qquad \text{Multiply numerator and denominator by the complex conjugate of the denominator.}$$

$$= \dfrac{12 + 2i + 18i + 3i^2}{36 - i^2}$$

$$= \dfrac{12 + 20i + 3(-1)}{36 - (-1)} \qquad \text{Simplify.}$$

$$= \dfrac{12 + 20i - 3}{36 + 1}$$

$$= \dfrac{9 + 20i}{37}$$

$$= \dfrac{9}{37} + \dfrac{20}{37}i \qquad \text{Write in standard form.}$$

> **Connection**   Remember that we rationalize denominators to clear undesired square root expressions from a denominator. The imaginary unit i represents a square root expression, $\sqrt{-1}$, which is why we rationalize denominators that contain i.

Your Turn 4   Write $\dfrac{4 + i}{5 - 2i}$ in standard form.

Objective ③ Solve quadratic equations that have solutions that are complex numbers.   When solving a quadratic equation that has a negative discriminant, now we can write its solutions using complex numbers.

Example 5   Solve. Write the solutions in standard form.

a. $x^2 + 9 = 0$

SOLUTION   We use the principle of square roots.

$$x^2 + 9 = 0$$
$$x^2 = -9 \qquad \text{Subtract 9 from both sides.}$$
$$x = \pm\sqrt{-9} \qquad \text{Use the square root principle.}$$
$$x = \pm 3i \qquad \text{Rewrite the imaginary number using } i.$$

> **Note:** Technically $\pm 3i$ is written as $0 \pm 3i$ in standard form, but that is rarely done.

b. $(x - 5)^2 = -12$

SOLUTION $(x - 5)^2 = -12$
$$x - 5 = \pm\sqrt{-12} \qquad \text{Use the square root principle.}$$
$$x - 5 = \pm 2i\sqrt{3} \qquad \text{Simplify the square root.}$$
$$x = 5 \pm 2i\sqrt{3} \qquad \text{Add 5 to both sides.}$$

c. $2x^2 + 4x + 11 = 6$

SOLUTION   We use the quadratic formula.

$$2x^2 + 4x + 5 = 0 \qquad \text{Write in standard form by subtracting 6 from both sides.}$$

Answer   Your Turn 4

$\dfrac{18}{29} + \dfrac{13}{29}i$

Now use the quadratic formula with $a = 2$, $b = 4$, and $c = 5$.

$$x = \frac{-4 \pm \sqrt{(4)^2 - 4(2)(5)}}{2(2)}$$

$$x = \frac{-4 \pm \sqrt{16 - 40}}{4}$$

$$x = \frac{-4 \pm \sqrt{-24}}{4}$$

$$x = \frac{-4 \pm 2i\sqrt{6}}{4} \qquad \text{Simplify the square root.}$$

$$x = \frac{-2 \pm i\sqrt{6}}{2} \qquad \text{Divide out the common factor, 2.}$$

$$x = -1 \pm \frac{\sqrt{6}}{2}i \qquad \text{Write in standard form.}$$

Answers Your Turn 5

a. $\pm 3i\sqrt{2}$ **b.** $6 \pm 9i$

c. $\dfrac{1}{3} \pm \dfrac{\sqrt{17}}{3}i$

Your Turn 5 Solve.

a. $x^2 + 18 = 0$

b. $(x - 6)^2 + 81 = 0$

c. $3x^2 - 2x + 15 = 9$

10.4 Exercises

For Extra Help *MyMathLab*

1. As an imaginary number, what does i represent?

2. Is every real number a complex number? Explain.

3. Is every complex number an imaginary number? Explain.

4. Explain how to add complex numbers.

5. Explain how to subtract complex numbers.

6. Is the expression $\dfrac{5 - 4i}{3}$ in standard form for a complex number? Explain.

For Exercises 7–22 write the imaginary number using i. See Example 1.

7. $\sqrt{-16}$ **8.** $\sqrt{-49}$ **9.** $\sqrt{-2}$ **10.** $\sqrt{-3}$ **11.** $\sqrt{-8}$ **12.** $\sqrt{-12}$

13. $\sqrt{-20}$ **14.** $\sqrt{-24}$ **15.** $\sqrt{-27}$ **16.** $\sqrt{-75}$ **17.** $\sqrt{-125}$ **18.** $\sqrt{-63}$

19. $\sqrt{-45}$ **20.** $\sqrt{-80}$ **21.** $\sqrt{-147}$ **22.** $\sqrt{-360}$

For Exercises 23–38, add or subtract. See Example 2.

23. $(4 + 3i) + (-3 + 2i)$

24. $(2 - 3i) + (4 - 5i)$

25. $(3 + 2i) + (5 - 6i)$

26. $(7 + i) + (5 - 2i)$

27. $(-4 + 6i) - (2 + i)$

28. $(5 + i) - (-2 - 2i)$

29. $(10 - 3i) - (-5i)$

30. $(4 + 3i) - (-2i)$

31. $(2 + 3i) + (-15 - 5i)$

32. $(9 - 7i) + (-8 - i)$

33. $(-5 - 9i) - (-5 - 9i)$

34. $(6 - i) - (-4 + 2i)$

35. $(4 + i) - (2 - 3i) + (6 - 8i)$ **36.** $(-4 + 2i) - (6 + i) + (9 + 3i)$ **37.** $(5 - 2i) - (3 - 4i) - 6i$

38. $-2i - (4 - 3i) - (8 + 4i)$

For Exercises 39–54, multiply. See Example 3.

39. $(2i)(3i)$

40. $(6i)(i)$

41. $(-8i)(2i)$

42. $(4i)(-3i)$

43. $2i(5 - 6i)$

44. $3i(7 - i)$

45. $-3i(2 + 4i)$

46. $-2i(5 - 3i)$

47. $(2 + i)(3 - i)$

48. $(5 - 2i)(1 + i)$

49. $(6 + 5i)(5 - 3i)$

50. $(4 - 3i)(2 + 8i)$

51. $(3 + 2i)(3 - 2i)$

52. $(2 + 7i)(2 - 7i)$

53. $(3 + i)^2$

54. $(4 - 2i)^2$

For Exercises 55–64, write in standard form. See Example 4.

55. $\dfrac{3}{2 + i}$

56. $\dfrac{4}{5 + i}$

57. $\dfrac{2i}{3 - 4i}$

58. $\dfrac{4i}{2 - 3i}$

59. $\dfrac{3 - 7i}{1 - i}$

60. $\dfrac{3 + i}{2 - i}$

61. $\dfrac{1 + 2i}{4 + 3i}$

62. $\dfrac{1 + 2i}{3 + 4i}$

63. $\dfrac{3 + 2i}{4 - i}$

64. $\dfrac{5 - 6i}{3 + 2i}$

For Exercises 65–84, solve. Write the solutions in standard form. See Example 5.

65. $x^2 + 16 = 0$

66. $x^2 + 4 = 0$

67. $x^2 + 20 = 0$

68. $x^2 + 24 = 0$

69. $(x + 4)^2 = -16$

70. $(x - 5)^2 = -25$

71. $(x - 3)^2 = -8$

72. $(x + 2)^2 = -20$

73. $x^2 - 2x + 5 = 0$

74. $x^2 + 6x + 25 = 0$

75. $5 = -4x^2 - 8x$

76. $4x^2 + 5 = 8x$

77. $x^2 = -2x - 8$

78. $x^2 - 4x = -11$

79. $x^2 - 4x + 5 = -2$

80. $x^2 + 4x + 10 = -2$

81. $25x^2 + 49 = 0$

82. $16x^2 + 25 = 0$

83. $2x^2 - 3x + 5 = 0$

84. $3x^2 = 2x - 1$

For Exercises 85–88, solve using the quadratic formula.

85. A financial analyst reports that the equation $P = -t^2 + 3t + 12$ models the projected impact of a new product on the company's profit over the next 5 years, where P represents the company's profit in millions of dollars and t represents the time in years the product is on the market. He indicates that the model predicts that the profit will peak at 20 million. Show why he's wrong.

86. A scientist reports that the equation $P = -t^2 + 5t + 10$ models the number of bacteria in a culture after t hours in a controlled environment, where P represents the number of bacteria in millions and t represents the time in hours the culture is in the environment. In her report, she indicates that the model predicts that the population will peak at 18 million. Show why she's wrong.

87. A football is kicked straight up from a height of 3 feet with an initial speed of 30 feet per second. The equation $h = -16t^2 + 30t + 3$ describes the height, in feet, of the football, where h represents the height in feet and t represents the time, in seconds, from when the ball is kicked until it hits the ground.
   a. After how many seconds is the ball at a height of 10 feet? If necessary, round to the nearest hundredth of a second.

   b. After how many seconds is the ball at a height of 25 feet?

88. A baseball 4 feet above ground level is hit with an initial speed of 78 feet per second. The equation $h = -16t^2 + 78t + 4$ describes the height, in feet, of the ball where h represents the height and t represents the time, in seconds, from when the ball is hit until it hits the ground.
   a. After how many seconds is the ball at a height of 96 feet?

   b. After how many seconds is the ball at a height of 100 feet?

Review Exercises

Exercise 1 EXPRESSIONS

[1.7]   **1.** Evaluate $-\dfrac{b}{2a}$ when a is 2 and b is 8.

Exercises 2–6 EQUATIONS AND INEQUALITIES

[4.3]   **2.** Find the x- and y-intercepts for $2x + 3y = 6$.   **[4.7]**   **3.** If $f(x) = 2x^2 - 3x + 1$, find $f(-1)$.

For Exercises 4–6, graph.

[4.2]   **4.** $x = -3$ **[7.7]**   **5.** $y = x^2 - 3$ **[7.7]**   **6.** $f(x) = 3x^2$

10.5 Graphing Quadratic Equations

Objectives

1. Given an equation in the form $y = ax^2 + bx + c$, determine whether the corresponding parabola opens upward or downward.

2. Find the x- and y-intercepts.

3. Find the coordinates of the vertex and write the equation of the axis of symmetry.

4. Solve applications involving parabolas.

In Section 7.7, we learned that the graph of every quadratic equation is a parabola. We also learned that a parabola has a *vertex* and an *axis of symmetry*. The vertex is the lowest point on a parabola that opens upward or the highest point on a parabola that opens downward. The axis of symmetry is a line that divides a graph into two symmetrical halves. We graphed quadratic equations by finding enough ordered pair solutions to determine the shape of the parabola. Let's use this method to graph the quadratic equation $y = x^2$.

x	y
-3	9
-2	4
-1	1
0	0
1	1
2	4
3	9

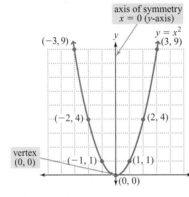

Note: The axis of symmetry passes through the vertex.

Finding many solutions is time-consuming. In this section, we learn more about parabolas in an effort to streamline our method for graphing quadratic equations.

Objective 1 **Given an equation in the form $y = ax^2 + bx + c$, determine whether the corresponding parabola opens upward or downward.** In Section 7.7, we learned that parabolas can open upward or downward. Given an equation in the form $y = ax^2 + bx + c$, the sign of a determines whether the graph opens upward or downward according to the following rule.

Rule **Opening of a Parabola**

Given an equation in the form $y = ax^2 + bx + c$, if $a > 0$, then the parabola opens upward; if $a < 0$, then the parabola opens downward.

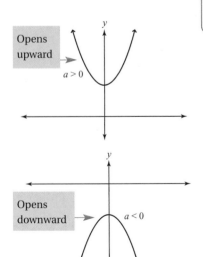

Referring to our graph of $y = x^2$ at the beginning of the section, we see that the parabola opens upward. This is because in $y = x^2$, $a = 1$, which is positive. If we change the sign of a to a negative number, as in $y = -x^2$, the graph opens downward.

x	y
-3	-9
-2	-4
-1	-1
0	0
1	-1
2	-4
3	-9

Note: Because $y = -x^2$ opens downward, the vertex is the highest point, which is $(0, 0)$. The axis of symmetry is the y-axis ($x = 0$).

Objective ② **Find the x- and y-intercepts.** Now let's consider the x- and y-intercepts of parabolas. In Chapter 4, we learned that x- and y-intercepts are points where a graph crosses an axis. Because an x-intercept is on the x-axis, the y-coordinate of the point is zero. Similarly, because a y-intercept is on the y-axis, the x-coordinate of the point is zero.

Procedure **Finding x- and y-Intercepts**

To find the x-intercept(s), replace y with 0 and solve for x.
To find the y-intercept, replace x with 0 and solve for y.

Connection To find the x-intercepts of $y = ax^2 + bx + c$, let $y = 0$. Therefore, the solutions of $0 = ax^2 + bx + c$ are the x-intercepts of the graph of $y = ax^2 + bx + c$.

Recall that quadratic equations in the form $ax^2 + bx + c = 0$ can have two, one, or no real-number solutions, which means that their graphs can have two, one, or no x-intercepts. The following figure illustrates every possibility.

LEARNING Strategy

AUDITORY If you are an auditory learner, think of how the language of the procedures for finding x- and y-intercepts sounds like opposites: to find $x \ldots$, we replace $y \ldots$, and to find $y \ldots$, we replace x.

Two x-intercepts: Here, $0 = ax^2 + bx + c$ has two real-number solutions.

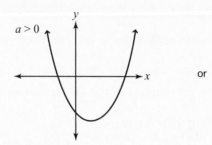

or

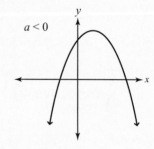

One x-intercept: Here, $0 = ax^2 + bx + c$ has one real-number solution.

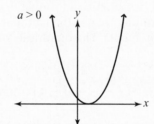

or

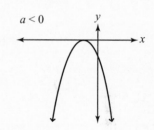

Connection In Section 10.3, we saw that if a quadratic equation has a positive discriminant, it has two real solutions. A discriminant of 0 indicates one real solution. A negative discriminant indicates no real solution. Therefore, parabolas with two x-intercepts have a positive discriminant; with one x-intercept, a discriminant of 0; with no x-intercept, a negative discriminant.

No x-intercepts: Here, $0 = ax^2 + bx + c$ has no real-number solution.

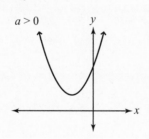

or

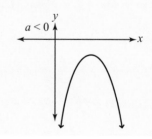

Although the number of x-intercepts may vary, every equation in the form $y = ax^2 + bx + c$ has one y-intercept, which is always $(0, c)$.

Example 1 Find the x- and y-intercepts of $y = x^2 + 2x - 8$.

SOLUTION First, we find the x-intercept by letting $y = 0$.

$$0 = x^2 + 2x - 8$$
$$0 = (x - 2)(x + 4) \qquad \text{Factor.}$$
$$x - 2 = 0 \quad \text{or} \quad x + 4 = 0 \qquad \text{Use the zero-factor theorem.}$$
$$x = 2 \qquad\qquad x = -4$$

x-intercepts: $(2, 0)$ and $(-4, 0)$

Now for the y-intercepts, we let $x = 0$.

$$y = (0)^2 + 2(0) - 8$$
$$y = -8$$

y-intercept: $(0, -8)$

Your Turn 1 Find the x- and y-intercepts.

a. $y = 2x^2 - 3x - 5$

b. $y = 3x^2 - 5x + 1$

Objective ③ Find the coordinates of the vertex and write the equation of the axis of symmetry.

Finding the Vertex

Recall that the vertex of a parabola is the lowest point on a parabola that opens up or the highest point on a parabola that opens down. Given an equation in the form $y = ax^2 + bx + c$, we can use a formula to determine the x-coordinate of the vertex. To derive this formula, we use the fact that the x-coordinate of the vertex is halfway between the x-intercepts. The quadratic formula tells us that the coordinates of the x-intercepts are

$$\left(\frac{-b + \sqrt{b^2 - 4ac}}{2a}, 0\right) \quad \text{and} \quad \left(\frac{-b - \sqrt{b^2 - 4ac}}{2a}, 0\right).$$

Because the vertex is halfway between the x-intercepts, we can average the x-coordinates of the x-intercepts to derive the formula for the x-coordinate of the vertex.

$$\left[\frac{-b + \sqrt{b^2 - 4ac}}{2a} + \frac{-b - \sqrt{b^2 - 4ac}}{2a}\right] \div 2 \qquad \begin{array}{l}\text{The average of the two } x\text{-coordinates}\\ \text{is their sum divided by 2.}\end{array}$$

$$= \left[\frac{-b - b}{2a}\right] \div 2 \qquad \begin{array}{l}\text{The square roots are additive}\\ \text{inverses, so their sum is 0.}\end{array}$$

$$= \left[\frac{-2b}{2a}\right] \div 2 \qquad \begin{array}{l}\text{Combine like terms in the}\\ \text{numerators.}\end{array}$$

$$= \frac{-b}{a} \div 2 \qquad \text{Sinplify the rational expression.}$$

$$= -\frac{b}{2a} \qquad \text{The vertex's } x\text{-coordinate}$$

Finding the Vertex of a Parabola

Given an equation in the form $y = ax^2 + bx + c$, to determine the vertex of the corresponding parabola,

1. Find the x-coordinate using the formula $x = -\dfrac{b}{2a}$.

2. Find the y-coordinate by substituting the x-coordinate into the given equation.

Example 2   For the equation $y = 2x^2 - 12x + 5$, find the vertex.

SOLUTION   First, find the x-coordinate of the vertex using $-\dfrac{b}{2a}$.

$$-\frac{(-12)}{2(2)} = \frac{12}{4} = 3 \qquad \text{Replace } a \text{ with 2 and } b \text{ with } -12.$$

Now find the y-coordinate of the vertex by replacing x with 3 in $y = 2x^2 - 12x + 5$.

$$y = 2(3)^2 - 12(3) + 5$$
$$y = 18 - 36 + 5$$
$$y = -13$$

The vertex is at the point with coordinates $(3, -13)$.

Your Turn 2   For the equation $y = 3x^2 + 18x + 25$, find the vertex.

Finding the Axis of Symmetry

Now let's learn how to write the equation of a parabola's axis of symmetry, which is a vertical line that divides a parabola into two symmetrical halves. Because the axis of symmetry passes through the vertex of a parabola, the equation of the axis of symmetry has x equal to the x-coordinate of the vertex. For example, in Example 2, we found the vertex of $y = 2x^2 - 12x + 5$ to be $(3, -13)$; so the axis of symmetry is a vertical line passing through $(3, -13)$, and its equation is $x = 3$.

Finding the Axis of Symmetry of a Parabola

Given an equation in the form $y = ax^2 + bx + c$, the equation of the axis of symmetry is $x = -\dfrac{b}{2a}$.

Now let's use all of the features that we have discussed in this section to graph quadratic equations.

Example 3   For the equation $y = -x^2 - 2x + 3$,

a. Determine whether the graph opens upward or downward.

b. Find the x- and y-intercepts.

c. Find the vertex and axis of symmetry.

d. Graph.

Answer   **Your Turn 2**

$(-3, -2)$

SOLUTION

a. This parabola opens downward because $a = -1$, which is negative.

b. We find the x-intercepts by letting $y = 0$ and then solving the quadratic equation.

$$0 = -x^2 - 2x + 3$$

Replace y with 0. We use the quadratic formula to solve.

$$x = \frac{-(-2) \pm \sqrt{(-2)^2 - 4(-1)(3)}}{2(-1)}$$

Substitute $a = -1, b = -2$, and $c = 3$ into the quadratic formula.

$$x = \frac{2 \pm \sqrt{16}}{-2}$$

Simplify.

$$x = \frac{2 \pm 4}{-2}$$

Simplify the square root.

$$x = \frac{2 + 4}{-2} = \frac{6}{-2} = -3 \quad \text{or} \quad x = \frac{2 - 4}{-2} = \frac{-2}{-2} = 1$$

Separate the plus and minus cases.

The x-intercepts are $(-3, 0)$ and $(1, 0)$.

Because $y = -x^2 - 2x + 3$ is in the form $y = ax^2 + bx + c$, the y-intercept is $(0, c)$, which in this case is $(0, 3)$.

c. We find the x-coordinate of the vertex using $-\dfrac{b}{2a}$.

$$-\frac{(-2)}{2(-1)} = \frac{2}{-2} = -1$$

Now find the y-coordinate by replacing x with -1 in $y = -x^2 - 2x + 3$.

$$y = -(-1)^2 - 2(-1) + 3$$
$$y = -1 + 2 + 3$$
$$y = 4$$

The vertex is at the point with coordinates $(-1, 4)$.

The axis of symmetry is $x = -1$.

d. Now graph.

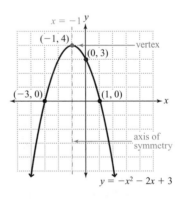

Note: The fact that the solutions are integers means that we could have easily solved this equation by factoring.

$$0 = -x^2 - 2x + 3$$
$$0 = -1(x^2 + 2x - 3)$$
$$0 = -1(x + 3)(x - 1)$$
$$x + 3 = 0 \quad \text{or} \quad x - 1 = 0$$
$$x = -3 \qquad\qquad x = 1$$

Answers   Your Turn 3

a. upward

b. x-intercepts:
$(2 + 0.5\sqrt{2}, 0)$ and
$(2 - 0.5\sqrt{2}, 0)$ or $\approx (2.7, 0)$
and $(1.3, 0)$
y-intercept: $(0, 7)$

c. vertex: $(2, -1)$
axis of symmetry: $x = 2$

d.

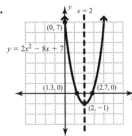

Your Turn 3   For the equation $y = 2x^2 - 8x + 7$,

a. Determine whether the graph opens upward or downward.

b. Find the x- and y-intercepts.

c. Find the vertex and axis of symmetry.

d. Graph.

Objective ④ Solve applications involving parabolas. Parabolas often model real-world events, such as the relationship between the height of an object propelled straight up and the time it is in the air.

Example 4 A toy rocket is launched straight up with an initial velocity of 40 feet per second. The equation $h = -16t^2 + 40t$ describes the height of the rocket, where h represents the height in feet and t represents the time in seconds from when the rocket is being launched until it hits the ground.

a. Graph the equation.

b. What is the maximum height the rocket reaches?

c. Approximate the amount of time the rocket is in the air.

SOLUTION

a. To graph, we find the vertex and intercepts. In the equation, t is like x and h is like y.

$$t\text{-coordinate of the vertex: } -\frac{b}{2a} = -\frac{40}{2(-16)} = -\frac{40}{-32} = \frac{5}{4} = 1.25$$

To find the h-coordinate of the vertex, we replace t with 1.25 in the original equation $h = -16t^2 + 40t$.

$$h = -16(1.25)^2 + 40(1.25) = 25$$

The vertex is at the point with coordinates $(1.25, 25)$. Now we find the intercepts.

t-intercept:	$0 = -16t^2 + 40t$	Replace h with 0.
	$0 = -8t(2t - 5)$	Factor out a common factor of $-8t$.
$-8t = 0$ or	$2t - 5 = 0$	Use the zero-factor theorem.
$t = 0$	$2t = 5$	
	$t = 2.5$	

The t-intercepts are $(0, 0)$ and $(2.5, 0)$. Note that $(0, 0)$ is also the h-intercept.

b. Because the graph to the right opens down, the maximum height occurs at the vertex of the parabola $(1.25, 25)$, which means that 1.25 seconds after launch, the rocket is at its maximum height of 25 feet.

c. The rocket is on the ground when the height is 0, which are the intercepts $(0, 0)$ and $(2.5, 0)$. This means that the rocket will hit the ground 2.5 seconds after launch.

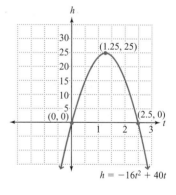

$$h = -16t^2 + 40t$$

Answers Your Turn 4

a.

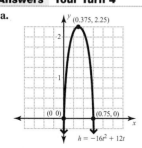

$$h = -16t^2 + 12t$$

b. 2.25 ft.   **c.** 0.75 sec.

Your Turn 4 A cricket leaps straight up with an initial velocity of 12 feet per second. The equation $h = -16t^2 + 12t$ describes the height of the cricket, where h represents the height in feet and t represents the time in seconds from when the cricket jumps until it lands.

a. Graph the equation.

b. What is the maximum height the cricket reaches?

c. Approximate the amount of time the cricket is in the air.

Calculator Tips

Besides being able to graph parabolas on the graphing calculator, you also can find the x-intercepts and the vertex.

Let's work through Example 4 using a graphing calculator. First, press the [Y=] key and enter the equation $y = -16x^2 + 40x$ as Y1. Press [WINDOW] and enter Xmin = −3, Xmax = 3, Xscl = 1, Ymin = −10, Ymax = 30, Yscl = 5. Then press [GRAPH] to see the graph.

Now to find the x-intercepts, press [2nd] [TRACE], which will bring up the CALC menu. Use the arrow keys to highlight ZERO in the CALC menu; then press [ENTER] to select it. This returns you to the graph screen, where you will be prompted for the left bound. Use the left–right arrow keys to move the cursor along the parabola until it is to the left of one of the x-intercepts and press [ENTER]. Next, you will be prompted for the right bound. Again, use the arrow keys to move the cursor to the right of the x-intercept; then press [ENTER].

You will be prompted for a guess. Press [ENTER] and you will see the coordinates of this x-intercept. Follow the same procedure for the other x-intercept.

To find the vertex, we use the MAXIMUM or MINIMUM functions under the CALC menu. Because our graph opens down, the vertex will be a maximum; so we will use the MAXIMUM function. After selecting the MAXIMUM function, you will be prompted for the left bound, then the right bound, then a guess. After entering the left bound and right bound and pressing [ENTER] for the guess, you will see the coordinates of the vertex.

10.5 Exercises   For Extra Help   *MyMathLab*

Note: Exercises marked with a ★ represent challenging exercises.

1. In an equation in the form $y = ax^2 + bx + c$, what determines whether the parabola opens upward or downward?

2. What conditions cause a parabola to have only one x-intercept?

3. Can a parabola have the same point for both its x- and y-intercept? Explain.

4. Suppose the y-intercept for a given quadratic equation is $(0, -5)$. If the vertex is at $(3, -2)$, what can you conclude about the x-intercepts?

5. If the solutions for a quadratic equation are imaginary numbers, what does that indicate about its graph?

6. Given that the x-coordinate of the vertex is $-\dfrac{b}{2a}$, what must be true about a and b if the x-coordinate of the vertex is −1?

For Exercises 7–16, state whether the parabola opens upward or downward. See Objective 1.

7. $y = 2x^2 - x + 1$

8. $y = 3x^2 - 5x + 8$

9. $y = -4x^2 - x + 1$

10. $y = x^2 - 2x + 1$

11. $y = -x^2 - 8x - 9$

12. $y = -2x^2 + 4x + 5$

13. $y = 8x^2 + 2x - 3$

14. $y = x^2 - 2x + 8$

15. $y = -\dfrac{2}{3}x^2 + 4x - 1$

16. $y = \dfrac{2}{5}x^2 - 8x + 1$

For Exercises 17–34, find the x- and y-intercepts. See Example 1.

17. $y = x^2 - x - 2$

18. $y = x^2 - 3x + 2$

19. $y = -x^2 - 2x + 8$

20. $y = -x^2 + 3x + 18$

21. $y = x^2 - 6x + 8$

22. $y = x^2 - x - 30$

23. $y = x^2 + x + 1$

24. $y = x^2 + 3x + 4$

25. $y = x^2 + 10x + 25$

26. $y = x^2 - 12x + 36$

27. $y = 2x^2 + 15x - 8$

28. $y = 3x^2 + 20x + 12$

29. $y = x^2 + 4x + 2$

30. $y = x^2 + 10x - 5$

31. $y = -x^2 - 5x + 1$

32. $y = 3x^2 - 8x + 2$

33. $y = -2x^2 + 3x - 6$

34. $y = -3x^2 - 2x - 5$

For Exercises 35–46, find the coordinates of the vertex and write the equation of the axis of symmetry. See Examples 2 and 3.

35. $y = x^2 - 6x + 1$

36. $y = x^2 + 4x + 5$

37. $y = -x^2 + 6x - 2$

38. $y = -x^2 + 2x + 15$

39. $y = 2x^2 - 8x + 5$

40. $y = 2x^2 + 4x - 5$

41. $y = -3x^2 + 2x + 1$

42. $y = x^2 - x - 2$

43. $y = x^2 + 5x + 16$

44. $y = -x^2 - x + 6$

45. $y = -5x^2 + 3x + 1$

46. $y = 5x^2 - 3x + 2$

For Exercises 47–60: a. State whether the parabola opens upward or downward.
b. Find the x- and y-intercepts.
c. Find the coordinates of the vertex.
d. Write the equation of the axis of symmetry.
e. Graph.

See Example 3.

47. $y = x^2 - 2x - 3$

48. $y = x^2 - 6x + 5$

49. $y = -x^2 + 6x - 8$

50. $y = -x^2 - 2x + 3$

51. $y = x^2 + 6x + 9$

52. $y = x^2 - 4x + 4$

53. $y = x^2 + 4x + 5$

54. $y = x^2 - 4x + 5$

55. $y = -3x^2 + 6x - 5$

56. $y = -2x^2 + 4x - 5$

57. $y = -3x^2 - 6x + 4$

58. $y = 3x^2 - 6x + 1$

59. $y = 2x^2 + 6x + 3$

60. $y = -x^2 + 3x - 1$

★ *For Exercises 61–66, given the vertex and the x- and y-intercepts of a parabola, reconstruct the equation.*

61. Vertex: $(-3, -1)$
Intercepts: $(-2, 0), (-4, 0), (0, 8)$

62. Vertex: $(-1, -9)$
Intercepts: $(-4, 0), (2, 0), (0, -8)$

63. Vertex: $(-1, 0)$
Intercepts: $(-1, 0), (0, 1)$

64. Vertex: $(1, 0)$
Intercepts: $(1, 0), (0, 1)$

65. Vertex: $(-2, 1)$
Intercept: $(0, 5)$

66. Vertex: $(-1, 2)$
Intercept: $(0, 3)$

67. Compare the graphs of each of the following.

$$y = x^2 + x + 1$$
$$y = 2x^2 + x + 1$$
$$y = 4x^2 + x + 1$$
$$y = 6x^2 + x + 1$$

What effect does increasing the value of a, where $a \geq 1$, appear to have on the graph of an equation in the form $y = ax^2 + bx + c$?

68. Compare the graphs of each of the following.

$$y = x^2 + x + 1$$
$$y = 0.5x^2 + x + 1$$
$$y = 0.2x^2 + x + 1$$
$$y = 0.1x^2 + x + 1$$

What effect does decreasing the value of a, where $0 < a \leq 1$, appear to have on the graph of an equation in the form $y = ax^2 + bx + c$?

69. Compare the graphs of each of the following.

$$y = x^2 + x + 1$$
$$y = x^2 + 2x + 1$$
$$y = x^2 + 4x + 1$$
$$y = x^2 + 6x + 1$$

What effect does increasing the value of b, where $b \geq 1$, appear to have on the graph of an equation in the form $y = ax^2 + bx + c$?

70. Compare the graphs of each of the following.

$$y = x^2 + x + 1$$
$$y = x^2 + 0.5x + 1$$
$$y = x^2 + 0.05x + 1$$
$$y = x^2 + 0.005x + 1$$

a. What effect does decreasing the value of b, where $0 < b \leq 1$, appear to have on the graph of an equation in the form $y = ax^2 + bx + c$? (You may need to use the ZOOM feature.)

★ **b.** Explain why decreasing the value of b, where $0 < b \leq 1$, has less and less effect on the graph.

For Exercises 71–74, solve. See Example 4.

71. A toy rocket is launched with an initial velocity of 45 meters per second. The equation $h = -4.9t^2 + 45t$ describes the height of the rocket, where h represents the height in meters and t represents the flight time in seconds.
  a. Graph the equation.
  b. What is the maximum height the rocket reaches?

  c. How long is the rocket in flight to the nearest thousandth of a second?

72. A ball is dropkicked straight up with an initial velocity of 72 feet per second. The equation $h = -16t^2 + 72t$ describes the height of the ball, where h represents the height in feet and t represents the time in seconds from when the ball is kicked until it hits the ground.
  a. Graph the equation.
  b. What is the maximum height that the ball reaches?

  c. How long is the ball in the air?

★ **73.** The equation $y = -0.8x^2 + 3.2x + 6$ models the trajectory of a ball thrown upward and outward from a height of 6 feet. (Assume that $x \geq 0$ and $y \geq 0$.)
  a. Graph the trajectory.
  b. What is the maximum altitude the ball reaches?

  c. How far does the ball travel horizontally to the nearest hundredth of a foot?

★ **74.** In track and field, the javelin toss requires participants to throw a javelin the farthest. Suppose the equation $y = -0.02x^2 + 1.3x + 8$ models the trajectory of one particular throw. (Assume that x and y represent distances in meters and that $x \geq 0$ and $y \geq 0$.)
  a. Graph the trajectory.
  b. What is the maximum altitude the javelin reaches?

  c. How far does the javelin travel to the nearest hundredth of a meter?

Collaborative Exercises   Arch Span

The Gateway Arch, located in St. Louis, Missouri, was built from 1963 to 1965 and is the nation's tallest memorial. The equation $h(x) = -0.0063492063x^2 + 630$ can be used to approximate the height of the arch, where x represents the distance from its axis of symmetry and $h(x)$ represents its height above the ground.
  1. Using the equation, find the maximum height of the structure.

  2. The span of the arch at a given height is the horizontal distance between the two opposing points on the parabola at that height. Find the span of the arch at ground level. (*Hint:* Think of ground level as the x-axis. The height is 0 along the x-axis.)

  3. How does the span at ground level relate to the maximum height of the arch?

  4. The arch has foundations 60 feet below ground level. What is the span of the arch between its foundations?

Of Interest

The shape of the Gateway Arch is not a true parabola, although the equation $h(x) = -0.0063492063x^2 + 630$ approximates its shape well. The actual shape is called an *inverted catenary curve*.

Review Exercises

Exercises 1 and 2 **EXPRESSIONS**

[1.5]   **1.** Evaluate: $-|-2 + 3 \cdot 4| - 4^0$ **[10.2]**   **2.** Complete the square: $x^2 + 5x + $ ▢

Exercises 3–6 **EQUATIONS AND INEQUALITIES**

[7.6]   **3.** Solve by factoring: $x^2 + 2x = 15$ **[10.4]**   **4.** Solve: $x^2 = -20$

[10.3]   **5.** Give the quadratic formula used to solve an equation in the form $ax^2 + bx + c = 0$.   **[10.5]**   **6.** Graph: $y = x^2 + 1$

Chapter 10 Summary

Defined Terms

Section 10.3
Discriminant (p. 718)

Section 10.4
Imaginary unit (p. 724)
Imaginary number
(p. 724)

Complex number (p. 725)
Complex conjugates
(p. 727)

Formulas

The distance formula: To find the distance, d, between two points $(x_1, y_1,)$ and (x_2, y_2),

$$d = \sqrt{(x_2 - x_1)^2 + (y_2 - y_1)^2} \text{ (p. 700)}$$

The quadratic formula: Given an equation in the form $ax^2 + bx + c = 0$,

$$x = \frac{-b \pm \sqrt{b^2 - 4ac}}{2a} \text{ (p. 714)}$$

The discriminant: $b^2 - 4ac$ (p. 718)

Given an equation in the form $y = ax^2 + bx + c$, the x-coordinate of the vertex

of a parabola and the equation of the axis of symmetry: $x = -\dfrac{b}{2a}$ (p. 735)

Procedures, Rules, and Key Examples

Procedures/Rules	Key Examples

SECTION 10.1 The Square Root Principle

Procedures/Rules	Key Examples
Square root principle: If $x^2 = a$, where a is a real number, then $x = \sqrt{a}$ or $x = -\sqrt{a}$. It is common to indicate the positive and negative solutions by writing $\pm\sqrt{a}$.	**Example 1:** Solve. **a.** $x^2 = 36$ $\qquad x = \pm\sqrt{36}$ $\qquad x = \pm 6$ **b.** $(x - 3)^2 = 12$ $\qquad x - 3 = \pm\sqrt{12}$ $\qquad x = 3 \pm 2\sqrt{3}$
To calculate the distance, d, between two points with coordinates (x_1, y_1) and (x_2, y_2), use the following formula: $$d = \sqrt{(x_2 - x_1)^2 + (y_2 - y_1)^2}$$	**Example 2:** Find the distance between $(-8, 3)$ and $(-2, -1)$. $d = \sqrt{(x_2 - x_1)^2 + (y_2 - y_1)^2}$ $d = \sqrt{(-2 - (-8))^2 + (-1 - 3)^2}$ $d = \sqrt{(6)^2 + (-4)^2}$ $d = \sqrt{36 + 16}$ $d = \sqrt{52}$ $d = 2\sqrt{13}$

(continued)

Procedures/Rules	Key Examples

SECTION 10.2 Solving Quadratic Equations by Completing the Square

Given an expression in the form $x^2 + bx$, to complete the square, add the constant term $\left(\dfrac{b}{2}\right)^2$.

Note: The factored form for the completed square is

$$x^2 + bx + \left(\dfrac{b}{2}\right)^2 = \left(x + \dfrac{b}{2}\right)^2.$$

To solve a quadratic equation by completing the square,
1. Write the equation in the form $x^2 + bx = c$.
2. Complete the square.
3. Write the completed square in factored form.
4. Use the square root principle to eliminate the square.
5. Isolate the variable.
6. Simplify as needed.

Example 1: Complete the square; then write in factored form.

a. $x^2 - 10x$

   Completed square: $x^2 - 10x + 25$
   Factored form: $(x - 5)^2$

b. $x^2 + 9x$

   Completed square: $x^2 + 9x + \dfrac{81}{4}$

   Factored form: $\left(x + \dfrac{9}{2}\right)^2$

Example 2: Solve $3x^2 - 9x = 12$.

$$x^2 - 3x = 4 \qquad \text{Divide both sides by 3.}$$

$$x^2 - 3x + \dfrac{9}{4} = 4 + \dfrac{9}{4} \qquad \text{Complete the square.}$$

$$\left(x - \dfrac{3}{2}\right)^2 = \dfrac{25}{4} \qquad \text{Factor.}$$

$$x - \dfrac{3}{2} = \pm\sqrt{\dfrac{25}{4}} \qquad \text{Use the square root principle.}$$

$$x = \dfrac{3}{2} \pm \dfrac{5}{2} \qquad \text{Add } \dfrac{3}{2} \text{ to both sides and simplify.}$$

$$x = \dfrac{3}{2} + \dfrac{5}{2} = 4 \quad \text{or} \quad x = \dfrac{3}{2} - \dfrac{5}{2} = -1$$

SECTION 10.3 Solving Quadratic Equations Using the Quadratic Formula

To solve a quadratic equation in the form $ax^2 + bx + c = 0$, where $a \neq 0$, use the quadratic formula:

$$x = \dfrac{-b \pm \sqrt{b^2 - 4ac}}{2a}$$

Example 1: Solve $3x^2 - 6x + 2 = 0$.

$$x = \dfrac{-(-6) \pm \sqrt{(-6)^2 - 4(3)(2)}}{2(3)}$$

$$x = \dfrac{6 \pm \sqrt{36 - 24}}{6} \qquad \begin{array}{l}\text{In the quadratic formula,} \\ \text{replace } a \text{ with 3, } b \text{ with} \\ -6, \text{ and } c \text{ with 2.}\end{array}$$

$$x = \dfrac{6 \pm \sqrt{12}}{6}$$

$$x = \dfrac{6 \pm 2\sqrt{3}}{6} \qquad \begin{array}{l}\text{Simplify the square} \\ \text{root.}\end{array}$$

$$x = \dfrac{3 \pm \sqrt{3}}{3} \qquad \begin{array}{l}\text{Divide out the common} \\ \text{factor, 2.}\end{array}$$

To determine the number and type of solutions of a quadratic equation in the form $ax^2 + bx + c$, where a, b, and c are rational numbers and $a \neq 0$, evaluate the discriminant $b^2 - 4ac$.

If the **discriminant is positive,** the equation has two real-number solutions. They will be rational if the discriminant is a perfect square and irrational otherwise.

If the **discriminant is 0,** the equation has one rational solution.

If the **discriminant is negative,** the equation has no real-number solutions.

Example 2: Use the discriminant to determine the number and type of solutions for $5x^2 - 7x + 8 = 0$.

$$(-7)^2 - 4(5)(8) \qquad \begin{array}{l}\text{In the discriminant,} \\ \text{replace } a \text{ with 5, } b \text{ with} \\ -7, \text{ and } c \text{ with 8.}\end{array}$$

$$= 49 - 160$$
$$= -111$$

Because the discriminant is negative, the equation has no real-number solutions.

(continued)

Procedures/Rules	Key Examples

SECTION 10.4 Complex Numbers

To write an imaginary number $\sqrt{-n}$ in terms of the imaginary unit i, 1. Separate the radical into two factors, $\sqrt{-1} \cdot \sqrt{n}$. 2. Replace $\sqrt{-1}$ with i. 3. Simplify $\sqrt{n}$.	**Example 1:** Write using the imaginary unit. **a.** $\sqrt{-36} = \sqrt{-1} \cdot \sqrt{36} = 6i$ **b.** $\sqrt{-32} = \sqrt{-1} \cdot \sqrt{32} = i \cdot 4\sqrt{2} = 4i\sqrt{2}$
To add complex numbers, combine like terms.	**Example 2:** Add $(5 - 6i) + (9 + 2i)$. $(5 - 6i) + (9 + 2i) = 14 - 4i$
To subtract complex numbers, change the signs of the second complex number; then combine like terms.	**Example 3:** Subtract $(7 - i) - (3 + 5i)$. $(7 - i) - (3 + 5i) = (7 - i) + (-3 - 5i)$ $\qquad\qquad\qquad\quad = 4 - 6i$
To multiply complex numbers, follow the same procedures for multiplying monomials or binomials. Remember that $i^2 = -1$.	**Example 4:** Multiply $(6 + 5i)(2 - 3i)$. $(6 + 5i)(2 - 3i) = 12 - 18i + 10i - 15i^2$ $\qquad\qquad\qquad\quad = 12 - 8i - 15(-1)$ $\qquad\qquad\qquad\quad = 12 - 8i + 15$ $\qquad\qquad\qquad\quad = 27 - 8i$
To divide complex numbers, rationalize the denominator using the complex conjugate.	**Example 5:** Write $\dfrac{2 - 3i}{4 + 5i}$ in standard form. $\dfrac{2 - 3i}{4 + 5i} = \dfrac{2 - 3i}{4 + 5i} \cdot \dfrac{4 - 5i}{4 - 5i}$ $= \dfrac{8 - 10i - 12i + 15i^2}{16 - 25i^2}$ $= \dfrac{8 - 22i + 15(-1)}{16 - 25(-1)}$ $= \dfrac{8 - 22i - 15}{16 + 25}$ $= \dfrac{-7 - 22i}{41}$ $= -\dfrac{7}{41} - \dfrac{22}{41}i$

SECTION 10.5 Graphing Quadratic Equations

Given an equation in the form $y = ax^2 + bx + c$, if $a > 0$, then the parabola opens upward; if $a < 0$, then the parabola opens downward.	**Example 1:** Determine whether the graph opens up or down. **a.** $y = x^2 - 5x + 3$ $\quad$ The graph opens up because $a > 0$ $(a = 1)$. **b.** $y = -4x^2 + x + 1$ $\quad$ The graph opens down because $a < 0$ $(a = -4)$.
To find the x-intercept(s), replace y with 0 and solve for x. To find the y-intercept, replace x with 0 and solve for y.	**Example 2:** Find the x- and y-intercepts for $y = 2x^2 - 5x + 3$. For the x-intercepts, we replace y with 0 and solve for x. $\qquad\qquad 0 = 2x^2 - 5x + 3$

(*continued*)

SECTION 10.5   **Graphing Quadratic Equations (*continued*)**

We solve by factoring.

$$0 = (2x - 3)(x - 1)$$
$$2x - 3 = 0 \qquad x - 1 = 0$$
$$2x = 3 \qquad\qquad x = 1$$
$$x = \frac{3}{2}$$

x-intercepts: $\left(\dfrac{3}{2}, 0\right)$ and $(1, 0)$

For the y-intercept, we replace x with 0 and solve for y.

$$y = 2(0)^2 - 5(0) + 3 = 3$$

y-intercept: $(0, 3)$

Given an equation in the form $y = ax^2 + bx + c$, to determine the vertex of the corresponding parabola,

1. Find the x-coordinate using the formula $x = -\dfrac{b}{2a}$.

2. Find the y-coordinate by substituting the x-coordinate into the given equation.

Example 3: Find the vertex of $y = x^2 - 8x + 2$.

For the x-coordinate, use $-\dfrac{b}{2a}$, replacing a with 1 and b with -8.

$$x = -\frac{(-8)}{2(1)} = 4$$

For the y-coordinate, replace x with 4 in the original equation.

$$y = (4)^2 - 8(4) + 2$$
$$y = 16 - 32 + 2$$
$$y = -14$$

Vertex: $(4, -14)$

Given an equation in the form $y = ax^2 - bx + c$, the equation of the axis of symmetry is $x = -\dfrac{b}{2a}$.

Example 4: Find the axis of symmetry for $y = 4x^2 - 6x - 1$.

Use $x = -\dfrac{b}{2a}$, replacing a with 4 and b with -6.

$$x = -\frac{(-6)}{2(4)} = \frac{6}{8} = \frac{3}{4}$$

The axis of symmetry is $x = \dfrac{3}{4}$.

Chapter 10 Review Exercises

For Exercises 1–5, answer true or false.

[10.1] **1.** The notation $\pm\sqrt{a}$ indicates both the positive square root of a and the negative square root of a.

[10.3] **2.** The equation $x^2 = -36$ has two real-number solutions.

[10.4] **3.** The set of complex numbers contains both real and imaginary numbers.

[10.4] **4.** The complex conjugate of a complex number $a + bi$ is $a - bi$.

[10.5] **5.** Given an equation in the form $y = ax^2 + bx + c$, if $a < 0$, then the equation's graph is a parabola that opens upward.

For Exercises 6–10, complete the rule.

[10.1] **6.** To calculate the distance between two points with coordinates (x_1, y_1) and (x_2, y_2), use the formula _____.

[10.2] **7.** Given an expression in the form $x^2 + bx$, to complete the square, add the constant term _____.

[10.3] **8.** To solve a quadratic equation in the form $ax^2 + bx + c$, where $a \neq 0$, use the quadratic formula _____.

[10.3] **9.** Given a quadratic equation in the form $ax^2 + bx + c = 0$, where $a \neq 0$, the discriminant is _____.

[10.5] **10.** Given an equation in the form $y = ax^2 + bx + c$, the equation of the axis of symmetry is _____.

Exercises 11–66 EQUATIONS AND INEQUALITIES

[10.1, 10.4] ***For Exercises 11–26, solve and check.***

11. $x^2 = 121$

12. $y^2 = \dfrac{1}{25}$

13. $m^2 = -36$

14. $u^2 = 20$

15. $k^2 + 5 = 30$

16. $n^2 + 7 = 43$

17. $3x^2 = 42$

18. $5x^2 = 125$

19. $4y^2 - 1 = 19$

20. $5h^2 + 24 = 9$

21. $(x + 7)^2 = 25$

22. $(x - 9)^2 = 16$

23. $\left(m + \dfrac{3}{5}\right)^2 = \dfrac{16}{25}$

24. $\left(n - \dfrac{2}{3}\right)^2 = \dfrac{121}{9}$

25. $(2.3x + 1.1)^2 = 1.96$

26. $(0.3x - 2.1)^2 = 0.09$

[10.1] *For Exercises 27–30, determine the distance between the two points. If the distance is an irrational number, write the exact answer; then use a calculator to approximate the distance rounded to three places.*

27. $(4, 3)$ and $(-2, 1)$ **28.** $(6, 2)$ and $(-3, -2)$ **29.** $(-1, -2)$ and $(4, -4)$ **30.** $(0, 0)$ and $(4, 5)$

[10.2] *For Exercises 31–34, complete the square.*

31. $x^2 + 16x$ **32.** $m^2 - 9m$ **33.** $w^2 + \dfrac{2}{5}w$ **34.** $y^2 - \dfrac{1}{3}y$

[10.2] *Exercises 35–46, solve by completing the square.*

35. $m^2 + 8m + 7 = 0$ **36.** $x^2 - 2x - 2 = 0$ **37.** $a^2 + 16a = -9$ **38.** $b^2 + 10b = -21$

39. $y^2 - 4y = 11$ **40.** $b^2 + 12b - 64 = -36$ **41.** $u^2 - 6u - 12 = 100$ **42.** $t^2 - 6t - 5 = -4$

43. $4m^2 - 4m = 7$ **44.** $2b^2 - 5b = 7$ **45.** $25p^2 - 30p = 27$ **46.** $9y^2 + 18y = -3$

[10.3] *For Exercises 47–58, solve using the quadratic formula.*

47. $x^2 - 8x + 16 = 0$ **48.** $h^2 - h - 6 = 0$ **49.** $2t^2 + t - 5 = 0$ **50.** $3m^2 - 2m - 1 = 0$

51. $3r^2 + 8r = 1$ **52.** $9w^2 - 3w = 1$ **53.** $x^2 - 2x - 1 = 0$ **54.** $p^2 + 2p - 5 = 0$

55. $2z^2 - 3z - 1 = 0$ **56.** $3k^2 - 6k - 2 = 0$ **57.** $2v^2 - 3v - 9 = 0$ **58.** $3c^2 + 2c = 6$

[10.3] *For Exercises 59–66, use the discriminant to determine the number of real solutions for the equation. If the solution(s) are real numbers, indicate whether they are rational or irrational.*

59. $b^2 - 4b - 12 = 0$ **60.** $9x^2 + 12x + 4 = 0$ **61.** $6z^2 - 7z + 5 = 0$ **62.** $x^2 - 8x + 5 = 0$

63. $3k^2 + 5k = 0$ **64.** $5x^2 - 3 = 0$ **65.** $4n^2 - 24n + 40 = 0$ **66.** $\dfrac{1}{3}f^2 + 2f + 3 = 0$

Exercises 67–96   EXPRESSIONS

[10.4] *For Exercises 67–72, write the imaginary number using i.*

67. $\sqrt{-100}$ **68.** $\sqrt{-25}$ **69.** $\sqrt{-20}$ **70.** $\sqrt{-60}$ **71.** $\sqrt{-27}$ **72.** $\sqrt{-5}$

[10.4] *For Exercises 73–80, add or subtract.*

73. $(6 + 3i) + (9 - i)$

74. $(5 - 2i) - (3 - 2i)$

75. $(-3 + 4i) + (7 + 2i)$

76. $(-2 - i) + (3 + i)$

77. $(6 - 2i) - (5 - i)$

78. $(10 + 5i) - (6i)$

79. $(-7 - 6i) - (8 + 2i)$

80. $(5i) - (2 - 2i)$

[10.4] *For Exercises 81–90, multiply.*

81. $(10i)(3i)$

82. $(-2i)(i)$

83. $4i(5 - i)$

84. $-3i(2 + 5i)$

85. $(2 + i)(3 - 2i)$

86. $(1 - 4i)(5 - 3i)$

87. $(2 + i)(2 - i)$

88. $(5 - 7i)(5 + 7i)$

89. $(2 - 3i)^2$

90. $(4 + i)^2$

[10.4] *For Exercises 91–96, write in standard form.*

91. $\dfrac{6 + i}{2 - i}$

92. $\dfrac{1 + 2i}{3 + i}$

93. $\dfrac{5}{2 + 3i}$

94. $\dfrac{-2}{6 + 2i}$

95. $\dfrac{3i}{2 - 7i}$

96. $\dfrac{-2i}{4 - i}$

Exercises 97–118 EQUATIONS AND INEQUALITIES

[10.4] *For Exercises 97–106, solve.*

97. $x^2 + 25 = 0$

98. $y^2 + 36 = 0$

99. $m^2 + 12 = 0$

100. $u^2 + 18 = 0$

101. $(x + 4)^2 = -2$

102. $(x - 1)^2 = -20$

103. $h^2 + 11 = 4h$

104. $4p^2 + 8p = -5$

105. $3x^2 + 1 = 2x$

106. $2y^2 - 3y = -5$

[10.5] *For Exercises 107–112:* **a.** *State whether the parabola opens upward or downward.*
b. *Find the x- and y-intercepts.*
c. *Find the coordinates of the vertex.*
d. *Write the equation of the axis of symmetry.*
e. *Graph.*

107. $y = x^2 - 9$

108. $y = x^2 - 4x + 3$

109. $y = -x^2 - 6x - 8$

110. $y = -3x^2 - 4$

111. $y = -2x^2 + 5$

112. $y = x^2 + 2x - 1$

For Exercises 113–118, solve.

[10.1] **113.** A severe thunderstorm watch is issued by the National Weather Service for a square area covering 16,900 square miles. What is the length of each side of the square area?

[10.1] **114.** Using the formula $E = \dfrac{1}{2}mv^2$, where E represents the kinetic energy of an object with a mass m and velocity v in meters per second, find the velocity of an object with a mass of 50 kilograms and 400 joules of kinetic energy.

[10.2] **115.** The length of a toolshed is 4 feet more than its width. If the area is 285 square feet, what are the dimensions of the shed?

[10.1] **116.** A right circular cylinder is to be constructed so that its volume is equal to that of a sphere with a radius of 12 inches. If the cylinder is to have a height of 8 inches, find the radius of the cylinder to the nearest hundredth of an inch.

[10.2] **117.** A ramp is constructed so that it is a right triangle with a base that is 7 feet longer than its height. If the hypotenuse is 17 feet, find the dimensions of the base and height.

[10.5] **118.** An acrobat is launched upward from one end of a lever with an initial velocity of 24 feet per second. The equation $h = -16t^2 + 24t$ describes the height of the acrobat, where h represents the height in feet and t represents the flight time in seconds.
   a. Graph the equation.

   b. What is the maximum height the acrobat reaches?

   c. How long is the acrobat in the air?

Chapter 10 Practice Test

For Extra Help

CHAPTER
Test Prep
VIDEOS

Step-by-step test solutions are found on the Chapter Test Prep Videos available via the Video Resources on DVD, in *MyMathLab*, and on You Tube (search "CarsonElemAlg" and click on "Channels").

For Exercises 1 and 2, use the square root principle to solve and check.

1. $x^2 = 81$

2. $(x - 3)^2 = 20$

For Exercises 3 and 4, find the distance between the two points.

3. $(-1, -3)$ and $(4, -1)$

4. $(0, -2)$ and $(3, 1)$

For Exercises 5–8, solve by completing the square.

5. $x^2 - 8x = -4$

6. $y^2 - 10y = 11$

7. $3m^2 - 6m = 5$

8. $2p^2 - 16p = -36$

For Exercises 9–12, solve using the quadratic formula.

9. $2x^2 + x - 6 = 0$

10. $x^2 - 8x + 15 = 0$

11. $2u^2 + 2u - 7 = 0$

12. $9t^2 - 12t + 1 = 0$

For Exercises 13–17, solve using any method.

13. $u^2 - 16 = -6u$

14. $x^2 = 81$

15. $4w^2 + 6w + 3 = 0$

16. $2x^2 + 4x = 0$

17. $(v + 12)^2 = 16$

18. Write $\sqrt{-18}$ using the imaginary number i.

For Exercises 19–22, perform the indicated operation and write the answer in standard form.

19. $(2 - 3i) + (5i)$

20. $(-4 - 5i) - (2 - 3i)$

21. $(4 + 3i)^2$

22. $\dfrac{2i}{3 - 4i}$

23. Solve: $x^2 + 16 = 0$

24. For $y = x^2 - 8x + 15,$
   a. State whether the parabola opens upward or downward.

   b. Find the x- and y-intercepts.

   c. Find the coordinates of the vertex.

   d. Write the equation of the axis of symmetry.

   e. Graph.

25. A rectangular court for basketball is to be placed in a local park. The length of the court is to be 12 feet more than the width, and the area is to be 3100 square feet. Find the dimensions of the court.

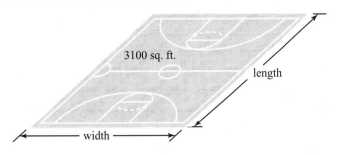

3100 sq. ft.

length

width

Chapters 1–10 Cumulative Review Exercises

For Exercises 1–4, answer true or false.

[2.1] **1.** A solution to an equation in one variable is a number that can replace the variable and make the equation true.

[6.2] **2.** The degree of a monomial is the exponent of the coefficient.

[5.1] **3.** Given a system of two linear equations in two variables, if the graphs are parallel lines, then the system has an infinite number of solutions along either line.

[10.1] **4.** If $x^2 = a$ and a is a real number, then $x = \sqrt{a}$ or $x = -\sqrt{a}$.

For Exercises 5 and 6, fill in the blank.

[7.4] **5.** The expression $a^2 - b^2$ rewritten in factored form is _____.

[8.5] **6.** A complex rational expression contains _____ _____ in the numerator and denominator.

Exercises 7–15   EXPRESSIONS

For Exercises 7 and 8, simplify.

[1.5] **7.** $|8 - 6^2| + 2(4 - 5)^2$

[6.5] **8.** $(3x - 5)(x + 4)$

For Exercises 9 and 10, factor completely.

[7.3] **9.** $6x^2 + 5x - 4$

[7.5] **10.** $18y^4 - 32y^2$

For Exercises 11–14, simplify.

[8.2] **11.** $-\dfrac{9x}{2x^4y^3} \div \dfrac{3y^2}{14x^5y}$

[8.4] **12.** $\dfrac{2}{5x^2} + \dfrac{x - 3}{x^2 + x}$

[9.4] **13.** $(\sqrt{2} - 3)^2$

[10.4] **14.** $\dfrac{2i}{3 - i}$

[9.4] **15.** Rationalize the denominator of $\dfrac{6}{\sqrt{x} - \sqrt{y}}$. Assume that variables represent positive numbers.

Exercises 16–30 **EQUATIONS AND INEQUALITIES**

For Exercises 16–21, solve.

[2.3] **16.** $9x - (2x + 3) = 5 + 3(4x - 1)$

[7.6] **17.** $2x^2 - 5x = 12$

[8.6] **18.** $\dfrac{3}{n} + \dfrac{7}{6n} = \dfrac{5}{n + 1}$

[5.3] **19.** $\begin{cases} 2x + y = 3 \\ 2x - y = 5 \end{cases}$

[9.5] **20.** $\sqrt{x + 45} = \sqrt{x} + 5$

$\begin{bmatrix} \textbf{10.2} \\ \textbf{10.3} \end{bmatrix}$ **21.** $4x^2 + 1 = 6x$

[4.2] **22.** Graph $y = -\dfrac{2}{3}x + 1$.

[4.5] **23.** Write the equation of a line passing through $(1, 3)$ and $(-2, 5)$ in the form $Ax + By = C$, where A, B, and C are integers and $A > 0$.

[5.5] **24.** Graph the solution set of $\begin{cases} 3x + 4y \le 4 \\ x - 4y < 8 \end{cases}$.

For Exercises 25–30, solve.

$\begin{bmatrix} \textbf{2.1} \\ \textbf{2.4} \end{bmatrix}$ **25.** The volume of a cone can be found using the formula $V = \dfrac{1}{3}\pi r^2 h$.

    a. Solve the formula for h.

    b. Suppose the volume of a cone is 15π cubic inches and the radius is 3 inches. Find its height.

[3.2] **26.** Lawyers.com conducted a poll rating prenuptial agreements. If 2731 people were polled and 28% thought that prenuptial agreements made financial sense, how many thought that the agreements made financial sense? If 15% thought that a prenuptial agreement doomed a marriage to fail, how many people predicted marriage failure? (*Source:* Harris Interactive for lawyers.com in *USA Today*)

[3.3] **27.** In the 2008 Summer Olympics, China won the
[5.3] most gold medals, which was fifteen more than
second place United States. If the combined total of gold medals won by the two countries was
87, how many did each country win?

[3.5] **28.** Rudy has a total of $10,000 invested in two plans.
[5.4] Plan A has an APR of 4%, and plan B has an APR
of 8%. If the money invested in plan B earns an
annual income of $320 more than the money in-
vested in plan A, how much did he invest in each
plan?

[7.6] **29.** A rectangular screen is designed so that the
width is 6 feet less than the length. If the area is
112 square feet, find the dimensions of the
screen.

[8.7] **30.** Two hoses that are different sizes are used to re-
fill a huge storage tank with a liquid chemical.
Alone, the larger hose can fill the tank in 2 hours,
whereas the smaller hose alone takes 3 hours to
fill the tank. How long will it take to fill the tank if
both hoses are used together?

Appendix 1 Mean, Median, and Mode

Objectives
1. Find the mean.
2. Find the median.
3. Find the mode.

Objective 1 Find the mean.

Numbers can be used to describe characteristics of data. A number used in this way is called a **statistic**.

Definition *Statistic:* A number used to describe some characteristic of a set of data.

Oftentimes we use statistics, such as averages, which describe the *middle* or *central tendency* in a set of data. One such statistic is the **arithmetic mean**. Many people use the words *average* and *mean* interchangeably. Actually, the mean is only one type of average, called the **arithmetic average**. Other averages are the median and mode, which we will discuss later in this appendix.

Definition *Mean* or *arithmetic average:* The sum of all given numbers divided by the number of numbers.

Procedure **To find the mean, or arithmetic average, of a given set of numbers:**
1. Calculate the sum of all the given numbers.
2. Divide the sum by the number of numbers.

The variable used to indicate the mean is $\bar{x}$.

Example 1 Clerical specialist salaries within a company are listed below. What is the mean salary of a clerical specialist within the company?

$24,000	$21,900
$19,500	$19,500
$20,400	$25,200

Understand: We must calculate the arithmetic average, or mean, salary. We are given a set of salaries.

Plan: Divide the sum of the salaries by the number of salaries in the list.

Execute: $\bar{x} = \dfrac{24{,}000 + 19{,}500 + 20{,}400 + 21{,}900 + 19{,}500 + 25{,}200}{6}$

$\bar{x} = \dfrac{130{,}500}{6}$

$\bar{x} = 21{,}750$

Answer: The mean salary of the clerical specialists at this company is $21,750.

Check: We can verify the calculations by inverse operations. We will leave this to the reader.

Objective ② Find the median.

Another average or statistic that describes the *middle* or *central tendency* of a set of numbers is the **median**.

Definition   *Median:* The middle number in an ordered set of numbers.

Consider the set of numbers: 4, 5, 9, 3, 2, 10, 9

The definition of *median* indicates that the set *must be ordered*. This means that we must put the numbers in order from smallest to largest or largest to smallest. We must also write *each* repetition of a number.

2, 3, 4, **5**, 9, 9, 10
↑
median

Note: Since 5 is the middle number in this ordered set, it is the median.

Notice that the median is easy to find in a set that contains an odd number of numbers. Suppose the set contains an even number of numbers as in 2, 3, 4, 5, 9, 10.

2, 3, 4, 5, 9, 10
↑
median

$$\text{median} = \text{mean of 4 and 5} = \frac{4 + 5}{2} = 4.5$$

Note: The middle of this set of ordered numbers is *between* 4 and 5. In fact, the number exactly halfway between 4 and 5 is the mean of 4 and 5.

Procedure   **To find the median of a set of numbers:**

1. Arrange the numbers in order from least to greatest or greatest to least.
2. Locate the middle number of the ordered set of numbers.

Note: If there are an even number of numbers in the set, the median will be the mean of the two middle numbers.

Example 2   Find the median salary for clerical specialists at the company described in Example 1. The salaries were

$24,000	$21,900
$19,500	$19,500
$20,400	$25,200

Understand: We must find the median of the given salaries.

Plan: Arrange the salaries in order from least to greatest, then locate the middle salary in the ordered list. Because there is an even number of salaries, the median salary will be the mean of the middle two salaries.

Execute: 19,500   19,500   20,400   21,900   24,000   25,200

Note: The median is halfway between 20,400 and 21,900, so we must find the mean of 20,400 and 21,900.

$$\text{median} = \frac{20,400 + 21,900}{2} = \frac{42,300}{2} = 21,150$$

Answer: The median income for the clerical specialists at the company is $21,150.

Check: Verify the calculations using inverse operations.

In Example 2, notice that half of the people have salaries greater than the median salary and half have salaries less than the median salary. Median does not indicate how much greater or less the other salaries might be. Also, it is possible that no one earns exactly the mean or median salary.

Objective ❸ Find the mode.

Another statistic we may consider with data sets is called the **mode**.

Definition   *Mode:* The number that occurs most often in a set of numbers.

Consider the set of numbers: 2, 5, 9, 12, 9, 15, 9, 12

Since 9 is the number that occurs most often, it is the mode of this set. If no number is repeated, then there is no mode. If there is a tie between numbers that occur most often, then we list each as a mode.

Procedure   **To find the mode**

To find the mode of a set of numbers, count the number of repetitions of each number. The number with the most repetitions is the mode.

> **Note:** Ordering the data helps spot repeated numbers and the mode(s). If no numbers are repeated, then there is no mode. If there is a tie between numbers that occur most often, then list each number as a mode.

Example 3   Find the mode.

a. Twelve players in a golf tournament post the following numbers:

$$69, 66, 74, 72, 69, 70, 72, 71, 75, 65, 72, 71$$

Understand: We must find the mode in the set of numbers. The mode is the number that occurs most frequently.

Plan: Order the data and count the number of repetitions of each number. The number with the most repetitions is the mode.

Execute: 65, 66, <u>69, 69</u>, 70, <u>71, 71</u>, <u>72, 72, 72</u>, 74, 75

               twice   twice   three times

                             72 is the number with the most repetitions, so it is the mode.

Answer: The mode is 72.

b. Final grades for a class of 15 students:

$$85, 86, 72, 65, 80, 91, 62, 76, 80, 85, 76, 78, 80, 96, 85$$

Understand: We must find the mode in the set of numbers. The mode is the number that occurs most frequently.

Plan: Order the data and count the number of repetitions of each number. The number with the most repetitions is the mode.

Execute: 62, 65, 72, 76, 76, 78, 80, 80, 80, 85, 85, 85, 86, 91, 96

twice   three times   three times

85 and 80 both occur three times. This
means they are *both* modes.

Answer: 85 and 80 are modes.

c. Birth weights for a group of 6 siblings:

    7.4 pounds   7.5 pounds   8.2 pounds   7.6 pounds   7.8 pounds   8.1 pounds

Understand: We must find the mode in the set of weights. The mode is the weight
that occurs most frequently.

Plan: Order the data and count the number of repetitions of each weight. The weight
with the most repetitions is the mode.

Execute: 7.4 pounds, 7.5 pounds, 7.6 pounds, 7.8 pounds, 8.1 pounds, 8.2 pounds.
Because no weight is repeated, there is no mode.

Answer: There is no mode for this set of weights.

Appendix 1
Exercises   For Extra Help   *MyMathLab*   Math XL PRACTICE   WATCH   DOWNLOAD

1. What is the mean of a set of data?

2. What is the median of a set of data?

3. Explain how to find the median of a set containing an even number of numbers.

4. What is the mode of a set of data?

For Exercises 5–14, find the mean, median, and mode(s).

5. Following is a list of nurse salaries on the surgical floor of a hospital. Find the mean, median, and mode of the salaries.

$25,500	$28,700
$28,700	$27,500
$26,450	$24,200

6. Following is a list of the hourly wages for employees at a packaging plant. Find the mean, median, and mode of the wages.

$11.00	$12.00	$11.00
$15.00	$18.25	$12.50
$13.50	$16.25	$18.00

7. Find the mean, median, and mode for the test scores of students in a history class.

80	92	64	78	88
80	82	74	72	60
55	96	100	71	82
75	82	90	86	58

8. A basketball team has the following final scores. Find the mean, median, and mode of the scores.

76	82	80	78	85	75
78	80	72	70	84	88

9. A marine biologist studying leatherback sea turtles measures and records their lengths. Following is a list of lengths for the last 20 turtles. Find the mean, median, and mode of the lengths.

2.2 m	1.8 m	1.9 m	2.3 m	2.1 m
1.6 m	1.5 m	1.8 m	1.2 m	2.0 m
2.1 m	1.4 m	1.2 m	2.1 m	1.7 m
2.2 m	2.0 m	1.6 m	1.8 m	1.9 m

10. Following is a list of heights of players on a basketball team. Find the mean, median, and mode of the heights.

6.5 ft.	6.75 ft.	6.25 ft.	6.5 ft.	6.8 ft.	6.75 ft.
6.25 ft.	7 ft.	6.75 ft.	6.25 ft.	6 ft.	6.2 ft.

11. Following is a list of rainfall amounts for one month. Each amount is the total of rainfall in one 24-hour period. Calculate the mean, median, and mode.

0.4 in.	0.8 in.	1.2 in.	3.4 in.
0.6 in.	1.5 in.	0.6 in.	1.0 in.
1.4 in.	2.3 in.	1.3 in.	0.5 in.

12. The table lists daily high and low temperatures (°F) for the last two weeks. Find the mean, median, and mode of the high temperatures. Then find the mean, median, and mode of the low temperatures.

	Sun.	Mon.	Tue.	Wed.	Thu.	Fri.	Sat.
High	91	92	95	93	90	89	88
Low	72	74	78	75	72	70	71
High	90	94	96	95	92	90	91
Low	72	74	78	76	73	70	72

13. The table lists each month's electric and gas charges for a family over a two-year period. Find the mean, median, and mode of the electric and gas charges each year.

Month	Year 1	Year 2
January	$158.92	$165.98
February	$147.88	$162.85
March	$125.90	$130.45
April	$108.40	$112.55
May	$87.65	$90.45
June	$114.58	$125.91
July	$145.84	$137.70
August	$142.78	$140.19
September	$90.25	$96.15
October	$104.12	$115.75
November	$136.62	$145.21
December	$158.18	$160.25

14. The table lists water consumption, in cubic feet, by a family over a two-year period. Find the mean, median, and mode for each year.

Month	Year 1	Year 2
January	500	550
February	525	540
March	600	600
April	650	650
May	840	900
June	1100	1430
July	1000	1200
August	1400	1250
September	840	750
October	620	700
November	600	550
December	550	500

Answers

Chapter 1

1.1 Exercises **1.** A collection of objects. **3.** A rational number can be expressed as a ratio of integers, but an irrational number cannot. **5.** true **7.** true **9.** False. All real numbers are either rational or irrational. **11.** {Sunday, Monday, Tuesday, Wednesday, Thursday, Friday, Saturday}
13. {a, e, i, o, u} **15.** {5, 10, 15, 20, ...}
17. {9, 11, 13, 15, ...} **19.** {−6, −5, −4, −3} **21.** rational
23. rational **25.** irrational **27.** rational **29.** rational

31. 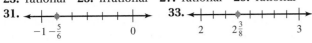 **33.**

35. **37.**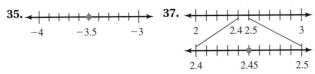

39. 23 **41.** 2 **43.** 5.7 **45.** $3\frac{1}{8}$ **47.** 0 **49.** > **51.** >
53. > **55.** < **57.** > **59.** > **61.** < **63.** = **65.** <
67. = **69.** > **71.** > **73.** < **75.** <

77. $-4.7, -2.56, 5.4, \left|-7\frac{1}{2}\right|, |8.3|$

79. $-0.6, -0.44, 0, |-0.02|, 0.4, \left|1\frac{2}{3}\right|, 3\frac{1}{4}$

1.2 Exercises **1.** Multiply or divide the numerator and denominator by the same nonzero number. **3.** No, 4 is not a prime number. The correct prime factorization is $2 \cdot 2 \cdot 2 \cdot 7$.
5. No, for example, 3 and 7 are whole numbers that are not composite. (They are prime numbers.) **7.** True. Prime numbers are natural numbers, and every natural number is a whole number. **9.** True. Every composite number can be written as a product of prime numbers. **11.** $\frac{3}{10}$ **13.** $\frac{1}{3}$

15. $\frac{3}{4}$ **17.** $\frac{3}{8}$ **19.** $\frac{15}{16}$ **21.** 12 **23.** −35 **25.** 16 **27.** 15

29. $\frac{12}{15}$ and $\frac{10}{15}$ **31.** $\frac{8}{18}$ and $\frac{21}{18}$ **33.** $-\frac{44}{72}$ and $-\frac{51}{72}$

35. $-\frac{9}{144}$ and $-\frac{60}{144}$ **37.** $2 \cdot 2 \cdot 11$ **39.** $2 \cdot 2 \cdot 3 \cdot 3$

41. $2 \cdot 2 \cdot 2 \cdot 2 \cdot 2 \cdot 2$ **43.** $2 \cdot 5 \cdot 5 \cdot 5$ **45.** $\frac{4}{5}$ **47.** $\frac{7}{11}$ **49.** $-\frac{3}{4}$

51. $-\frac{8}{15}$ **53.** Incorrect. You may divide out only factors, not addends. **55.** Incorrect. The prime factorization of 240 should be $2 \cdot 2 \cdot 2 \cdot 2 \cdot 3 \cdot 5$. **57.** $\frac{7}{8}$ **59.** $\frac{12}{23}$ **61.** $\frac{5}{21}$ **63.** $\frac{2}{3}$ **65.** $\frac{19}{50}$

67. $\frac{31}{50}$ **69. a.** 1985 **b.** $\frac{77}{200}$ **71.** $\frac{37}{189}$ **73.** $\frac{6}{73}$ **75.** $\frac{2}{3}$

77. $\frac{202}{435}$ **79.** $\frac{4}{15}$

Review Exercises **1.** {Mercury, Venus, Earth, Mars}
2. It is a rational number because it can be written as a ratio of the integers 8 and 10. **3.**

4. It is an expression because it has no = sign. **5.** 27 **6.** =

1.3 Exercises **1.** The commutative property of addition changes the order, while the associative property of addition changes the grouping. **3.** Adding 0 to a number or an expression does not change the identity of the number or expression. **5.** When adding two numbers that have the same sign, add their absolute values and keep the same sign.
7. To write a subtraction statement as an equivalent addition statement, change the operation symbol from a minus sign to a plus sign and change the subtrahend to its additive inverse.
9. Commutative property of addition. **11.** Additive identity.
13. Additive inverse. **15.** Associative property of addition.
17. Commutative property of addition. **19.** Additive inverse.
21. 21 **23.** −18 **25.** 9 **27.** −9 **29.** 14 **31.** −16

33. $\frac{3}{4}$ **35.** $-\frac{2}{3}$ **37.** $-\frac{2}{3}$ **39.** $\frac{11}{12}$ **41.** $-\frac{3}{4}$ **43.** $-\frac{9}{14}$

45. 0.26 **47.** 6.52 **49.** −4.38 **51.** −28 **53.** 3.18

55. $\frac{29}{24}$ **57.** −5 **59.** 12 **61.** 0 **63.** $-\frac{5}{6}$ **65.** 0.29 **67.** x

69. $-\frac{m}{n}$ **71.** 2 **73.** −4 **75.** −4 **77.** −12 **79.** −9

81. −13 **83.** 7 **85.** −6 **87.** 0 **89.** $\frac{13}{10}$ **91.** $-\frac{18}{35}$ **93.** 0.36

95. −5.75 **97.** −1.6 **99.** −10 **101.** −0.9
103. $877,000,000 **105.** −268.2 N; the negative indicates that the beam is moving downward. **107.** 133.40
109. $25.10 **111.** $-196 - (-208)$; 12 **113. a.** $18.6 - 18.8$
b. −0.2 **c.** The negative difference indicates that the mean composite score in 1989 was less than the score in 1986.
115. 2000; ≈ $16,000

Review Exercises 1. {Washington, Adams, Jefferson, Madison} 2. 7 3. $2 \cdot 2 \cdot 5 \cdot 5$ 4. $\dfrac{6}{7}$ 5. 5 6. $<$

1.4 Exercises
1. positive 3. negative 5. There is no quotient that we can multiply by 0 to get 5. 7. Their product is 1. 9. An odd number of negatives in division will result in a negative answer. 11. Distributive property. 13. Multiplicative identity. 15. Multiplicative property of 0. 17. Commutative property of multiplication. 19. Associative property of multiplication. 21. Commutative property of multiplication. 23. -18 25. -50 27. -54 29. 20 31. 42 33. $-\dfrac{1}{4}$ 35. $\dfrac{1}{2}$ 37. $-\dfrac{4}{9}$ 39. -38 41. 6.72 43. -6.536 45. 54 47. -48 49. -168 51. -144 53. 720 55. -48 57. $\dfrac{3}{2}$ 59. $-\dfrac{2}{5}$ 61. $-\dfrac{1}{5}$ 63. There is no multiplicative inverse of 0. 65. -3 67. 14 69. -2 71. 9 73. 0 75. undefined 77. indeterminate 79. 9 81. $-\dfrac{2}{3}$ 83. $\dfrac{3}{4}$ 85. $-\dfrac{14}{15}$ 87. 2.1 89. 91.8 91. $-46.\overline{6}$ 93. $\dfrac{1}{2}$ 95. $-\$8680$ 97. 612.8 99. -402.5 lb.; the force is downward 101. 169,202.2245 kg 103. -51.2 V 105. 3600 W

Review Exercises 1. irrational
2.

-8 ——— -7.2 — -7

3. $-\dfrac{2}{3}$
4. 6.8 5. -22 6. $-\dfrac{11}{24}$

1.5 Exercises
1. Two cubed. 3. Squaring a number means to multiply the number by itself; finding its square root means to find a number whose square is the given number. 5. We can multiply and then find the square root of the product, or we can find the square root of each factor and then multiply those square roots. 7. Base: 7; exponent: 2; "seven squared." 9. Base: -5; exponent: 3; "negative five cubed." 11. Base: 2; exponent: 7; "additive inverse of two to the seventh power." 13. 81 15. 64 17. -64 19. -125 21. -125 23. 8 25. -1 27. $\dfrac{1}{25}$ 29. $-\dfrac{27}{64}$ 31. 0.008 33. 16.81 35. ± 11 37. No real-number square roots exist. 39. ± 14 41. ± 16 43. 4 45. 12 47. 0.7 49. Not a real number. 51. $\dfrac{8}{9}$ 53. 5 55. 19 57. 0 59. 10 61. -11 63. -4 65. -40 67. -20 69. 17 71. -7 73. -28 75. 26.8 77. -97.4 79. -53 81. $-10\dfrac{4}{5}$ 83. 78 85. 19 87. -29 89. $-\dfrac{253}{300}$ 91. 22 93. 12 95. 0 97. 4 99. 7 101. undefined 103. Associative property of multiplication. The multiplication was not performed from left to right. 105. Distributive property. The parentheses were not simplified first. 107. Mistake: Multiplied before dividing. Correct: 1 109. Mistake: Found the square roots of the subtrahend and minuend in square root of a difference. Correct: 24 111. 77.2 113. 96 115. $\$588.41$ 117. $\$49.41$

Review Exercises 1. $\{0, 1, 2, 3, 4, 5, 6, 7, 8, 9, 10\}$
2.

-1 -$\frac{4}{5}$ 0

3. $-\dfrac{1}{9}$ 4. It is an expression because it has no $=$ sign. 5. 24 6. 3

1.6 Exercises
1. Sum, plus, added. 3. Product, times, twice. 5. Less than 7. Some number plus seven, seven added to some number, the sum of some number and seven, seven more than some number, some number increased by seven. 9. Seven divided by some number, the quotient of seven and some number, the ratio of seven and some number. 11. $4x$ 13. $4x + 16$ 15. $7x - 8$ 17. $-4 \div y^3$ or $\dfrac{-4}{y^3}$ 19. $8p - 4$ 21. $\dfrac{m}{14}$ 23. $x^4 + 5$ 25. $7w - \dfrac{1}{5}$ 27. $-3(n - 2)$ 29. $(4 + n)^5$ 31. $mn - 5$ 33. $4 \div n - 2$ or $\dfrac{4}{n} - 2$ 35. $-27 - (a + b)$ 37. $0.6 - 4(y - 2)$ 39. $(p - q) - (m + n)$ 41. $\sqrt{y} - mn$ 43. $n - 3(n - 6)$ 45. Mistake: Order is incorrect. Correct: $3t - 17$ 47. Mistake: Multiplied x by 9 instead of the sum of x and y. Correct: $9(x + y)$ 49. $w + 5$ 51. $3w$ 53. $\dfrac{1}{2}d$ 55. $42 - n$ 57. $t + \dfrac{1}{4}$ 59. $2w + 2l$ 61. $\dfrac{1}{3}\pi r^2 h$ 63. mc^2 65. $\sqrt{(x_2 - x_1)^2 + (y_2 - y_1)^2}$ 67. Mistake: Could be translated as $3(x + 4)$. Correct: Four more than three times a number. 69. Mistake: Could be translated as $5x - 1$. Correct: Five times the difference of x and one. 71. Mistake: Could be translated as $(n + 5)(n - 6)$. Correct: n plus the product of five and the difference of n and six. 73. One-half the product of the base and height. 75. The product of the length, width, and height. 77. The product of the length and width added to the product of the length and height added to the product of the width and height, all doubled. 79. The ratio of the difference of y_2 and y_1 to the difference of x_2 and x_1.

Review Exercises 1. $2 \cdot 5 + 2 \cdot 2 = 14$ 2. 2 3. $\dfrac{5}{12}$ 4. -28 5. 12 6. -9

1.7 Exercises
1. To evaluate an expression, (1) replace the variables with their corresponding given values and (2) calculate the numerical expression using the order of operations. 3. An expression is undefined when the denominator is equal to 0. 5. A coefficient is the numerical factor in a term. 7. 9 9. 3.2 11. 23 13. $\dfrac{11}{2}$ 15. 4.6 17. -11 19. 3 21. -12 23. -21 25. -48 27. $\dfrac{3}{13}$ 29. a. 25 b. -12 31. a. 2 b. $\dfrac{2}{3}$ 33. -5 35. $-1, 3$ 37. 0 39. $\dfrac{2}{3}$ 41. $6a + 12$ 43. $-32 + 24y$ 45. $\dfrac{7}{16}c - 14$ 47. $0.6n - 1.6$ 49. coefficient: -6 51. coefficient: 1 53. coefficient: -1 55. coefficient: $-\dfrac{2}{3}$ 57. coefficient: $\dfrac{1}{5}$ 59. $8y$ 61. $-5a$ 63. $11x$ 65. $-11r$ 67. $-1.9n$ 69. $-\dfrac{1}{3}b^2$

71. $-26c$ **73.** $5x - 1$ **75.** $-10x + 4y - 1$

77. $-9m + 5n + y - 11$ **79.** $-1.3x - 0.4$ **81.** $\frac{5}{6}c + \frac{5}{2}d + \frac{1}{7}$

83. $\frac{25}{4}a + \frac{18}{35}b^2 + 4$ **85. a.** $14 + (6n - 8n)$ **b.** $14 - 2n$

c. 20

Review Exercises **1.** $-3, -2.5, 4.2, 4\frac{5}{8}, |-6|$ **2.** $\frac{6}{35}$

3. -22 **4.** 50 **5.** 1 **6.** 3

Chapter 1 Review Exercises **1.** false **2.** false
3. false **4.** true **5.** true **6.** true **7.** positive
8. prime **9.** keep **10.** even **11.** {March, May}
12. {a, l, g, e, b, r} **13.** {2, 4, 6, …}
14. {…, $-6, -4, 0, 2, 4, …$} **15.**

16.

17.

18.

19. 6.3 **20.** 8.46 **21.** $2\frac{1}{6}$

22. $4\frac{3}{8}$ **23.** = **24.** < **25.** < **26.** = **27.** $\frac{11}{15}$ **28.** $\frac{3}{8}$

29. 12 **30.** 25 **31.** 5 **32.** -10 **33.** $3 \cdot 17$ **34.** $2 \cdot 2 \cdot 5 \cdot 5$

35. $2 \cdot 2 \cdot 3 \cdot 3 \cdot 3$ **36.** $2 \cdot 2 \cdot 3 \cdot 7$ **37.** $\frac{19}{25}$ **38.** $\frac{1}{2}$ **39.** $\frac{25}{36}$

40. $\frac{2}{3}$ **41.** Associative property of addition.

42. Distributive property. **43.** Commutative property of
multiplication. **44.** Commutative property of addition.
45. Commutative property of multiplication.
46. Associative property of multiplication. **47.** -3.8

48. 15.1 **49.** $\frac{11}{15}$ **50.** $\frac{5}{12}$ **51.** 91 **52.** -48 **53.** -5

54. 10 **55.** -36 **56.** 9 **57.** 52 **58.** 9 **59.** 160 **60.** 1

61. $-\frac{28}{3}$ **62.** -8.6 **63.** -1 **64.** 27 **65.** 7 **66.** -103

67. 3 **68.** 24 **69.** ± 7 **70.** No real-number square roots

exist. **71.** $14 - 2n$ **72.** $\frac{y}{7}$ **73.** $y - 2(y + 4)$

74. $7(m - n)$ **75.** 61 **76.** 52 **77.** -8 **78.** 7 **79.** 5

80. -4 **81.** -6 **82.** $-3, 4$ **83.** $5x + 30$

84. $-15n + 24$ **85.** $2y + \frac{3}{4}$ **86.** $2.7m - 1.26$

87. $-3x - 3y - 15$ **88.** $4y^2 + 7y$ **89.** $2xy$
90. $-2x^3 - 6x^2$ **91.** $-8m$ **92.** $-14x - 6y + 6$
93. $-\$503.59$ **94. a.** -2.94 **b.** loss of $\$0.25$ **95. a.** $\$9494$
b. $\$588$ loss, or $-\$588$ **96.** $-152v$ **97.** ≈ 6.2
98. $\$54,158\frac{1}{3}$, or $\$54,158.33$

Chapter 1 Practice Test
1.

[1.1]

2. 3.67 [1.1] **3.** $2 \cdot 2 \cdot 5 \cdot 5$ [1.2] **4.** $\frac{1}{3}$ [1.2] **5.** 12 [1.2]

6. Distributive property. [1.4] **7.** Commutative property of
addition. [1.3] **8.** 4 [1.3] **9.** $1\frac{17}{24}$ or $\frac{41}{24}$ [1.3] **10.** 0.6 [1.4]

11. $-\frac{5}{4}$ [1.4] **12.** -64 [1.5] **13.** -81 [1.5] **14.** -8 [1.5]

15. 2 [1.5] **16.** 5 [1.5] **17.** 10 [1.5] **18.** $2(m + n)$ [1.6]
19. $3w - 5$ [1.6] **20.** $\$250$ [1.6] **21.** 5.583 [1.6]
22. -11 [1.7] **23.** 10 [1.7] **24.** $-20y - 45$ [1.7]
25. $5.6x + 1.3$ [1.7]

Chapter 2

2.1 Exercises **1.** An equation has an equal sign.
3. (1) Replace the variable in the equation with the value, and
(2) if the resulting equation is true, the value is a solution.
5. no **7.** yes **9.** no **11.** yes **13.** no **15.** yes **17.** yes
19. yes **21. a.** 66 ft. **b.** 6 **c.** $\$53.94$ **23.** ≈ 12.57 km
25. 7000 ft.^{2} **27.** $\$975$ **29. a.** 78 ft.^{2} **b.** 8 bales **c.** $\$36$

31. a. 260 in.^{2} **b.** $\$3900$ **33. a.** 3208 ft.^{2} **b.** $2138\frac{2}{3}$ ft.^{2}

c. 5 **d.** $\$530$ **35.** ≈ 6.9 ft.^{2} **37.** 82.25 ft.^{2} **39.** 47.5 m^3

41. ≈ 1005.3 cm^3 **43.** 83,068,741.$\overline{6}$ ft.^{3} or $83,068,741\frac{2}{3}$ ft.^{3}

45. 64.5 mph **47.** ≈ 54.3 mph **49.** 1747.5 mi. **51.** -52 V
53. 93.2°F **55.** -297.4°F **57.** -60°C

Review Exercises **1.** -51 **2.** -34 **3.** $\frac{1}{3}x + 2$

4. $-4w + 6$ **5.** $8x - 5$ **6.** $5.4y + 5.5$

2.2 Exercises **1.** The addition principle of equality says
that we can add (or subtract) the same amount on both sides
of an equation without affecting its solution(s). **3.** Add 9 to
both sides of the equation. **5.** Use the addition principle of
equality to get the variable terms together on the same side of
the equal sign. **7.** yes **9.** no **11.** no **13.** yes **15.** yes
17. no **19.** yes **21.** yes **23.** 9 **25.** 5 **27.** -7 **29.** 5

31. -10 **33.** -25 **35.** $-\frac{3}{40}$ **37.** $-\frac{11}{24}$ **39.** -8.2 **41.** 4.6

43. -9 **45.** 7 **47.** 13 **49.** 6 **51.** -1 **53.** -2 **55.** 13.2
57. 36 **59.** -10 **61.** 16 **63.** 4 **65.** 9.7 **67.** All real
numbers. **69.** No solution. **71.** No solution. **73.** All real
numbers. **75.** $1947 + x = 2373$; $\$426$ **77.** $16 + x = 42$;
26 mi. **79.** $1741.62 - 1286.65 = 16.82 + 150.88 + 192.71 + x$;
$\$94.56$ **81.** $x + 12.4 + 16.3 + 27.2 = 67.2$; 11.3 cm
83. $x + 19 + 10 = 54$; 25 ft. **85.** $x + 900 + 3 \cdot 1500 +$
$2 \cdot 1245 = 8500$: $\$610$; yes, because $\$610$ is less than all of his

prior weeks. **87.** $x + \frac{1}{3} + \frac{2}{5} = 1; \frac{4}{15}$

Review Exercises **1.** $-\frac{5}{4}$ **2.** -33.2 **3.** -9

4. $-6x + 1$ **5.** $8x^2 - 5x - 9$ **6.** $4x - 9$

2.3 Exercises **1.** We can multiply (or divide) both sides
of an equation by the same amount without affecting its solu-
tion(s). **3.** Multiply both sides of the equation by the multiplicative inverse of the coefficient. **5.** Multiply both sides of
the equation by a common multiple of the denominators.

Using the LCD results in an equation with the smallest integers possible. **7.** 4 **9.** 3 **11.** 16 **13.** -20 **15.** $\dfrac{7}{6}$ **17.** $-\dfrac{4}{3}$

19. 7 **21.** 5 **23.** 5 **25.** -3 **27.** $-\dfrac{3}{2}$ **29.** 16 **31.** 0

33. -9 **35.** -5 **37.** 2 **39.** 16 **41.** 2 **43.** $\dfrac{20}{3}$ **45.** -2

47. 1 **49.** 2 **51.** No solution. **53.** 2 **55.** $-\dfrac{15}{2}$ **57.** All

real numbers. **59.** -9 **61.** No solution. **63.** $\dfrac{15}{8}$ **65.** $\dfrac{48}{13}$

67. 1 **69.** $-\dfrac{2}{3}$ **71.** -8 **73.** 9 **75.** All real numbers.

77. -9 **79.** 12 **81.** 6.5 **83.** Mistake: The minus sign was dropped in front of the 11. Correct: 2 **85.** Mistake: In the second line, the minus sign was not distributed to the 2.

Correct: $\dfrac{13}{3}$ **87.** 14 ft. **89.** 19 cm **91.** $l = 54$ ft., $w = 52$ ft.

93. 110 ft. **95.** 2 in. **97.** ≈ 6374.8 km **99.** ≈ 4.8 m
101. $w = 11$ ft., $x = 2$ ft., $w + 2 = 13$ ft., $y = 23$ ft.
103. -9.5 A **105.** 4 m/sec.^{2} **107.** 945.5 kg
109. 413.6 m/sec. **111.** \$51.10 **113.** 15 ft./sec.

Review Exercises
1. 70 **2.** $3.8x^2 - 7.63x + 1.5$
3. $-\dfrac{9}{2}x - 6$ **4.** 20 **5.** -25 **6.** -10

2.4 Exercises
1. Add 2 to both sides. Because 2 is subtracted from w, adding 2 eliminates the 2, leaving w isolated. **3.** Subtract $2x$ from both sides. Subtracting $2x$ isolates the $3y$ term, which contains the variable we want to isolate.

5. $t = 4u + v$ **7.** $y = -\dfrac{x}{5}$ **9.** $x = \dfrac{b+3}{2}$ **11.** $m = \dfrac{y-b}{x}$

13. $y = \dfrac{8-3x}{4}$ **15.** $Y = \dfrac{mn}{4} - f$ **17.** $c = \dfrac{m - np - 12d}{6}$

19. $y = \dfrac{15 - 5x}{3}$ **21.** $n = c\left(\dfrac{3}{4} + 5a\right)$ **23.** $p = \dfrac{A - P}{tr}$

25. $a = P - b - c$ **27.** $l = \dfrac{A}{w}$ **29.** $d = \dfrac{C}{\pi}$ **31.** $r^2 = \dfrac{3V}{\pi h}$

33. $b = \dfrac{2A}{h}$ **35.** $w = \dfrac{P - 2l}{2}$ **37.** $l = \dfrac{2S - na}{n}$

39. $w = \dfrac{\pi r^2 h - V}{lh}$ **41.** $C = R - P$ **43.** $r = \dfrac{I}{Pt}$

45. $C = nP$ **47.** $r = \dfrac{A - P}{Pt}$ **49.** $t = \dfrac{d}{r}$ **51.** $d = \dfrac{W}{F}$

53. $t = \dfrac{W}{P}$ **55.** $t = \dfrac{v - v_0}{-32}$ or $t = \dfrac{v_0 - v}{32}$

57. $C = \dfrac{5}{9}(F - 32)$ **59.** $m = \dfrac{FR^2}{GM}$ **61.** Mistake: Subtracted the coefficient 7 instead of dividing. Correct: $t = \dfrac{54 - 3n}{7}$

63. Mistake: Multiplied 5 by -2. Correct: $m = \dfrac{5nk + 2}{3}$

Review Exercises
1. $7n + 4$ **2.** $3(x + 2) - 9$ **3.** 0.5
4. $-\dfrac{5}{6}$ **5.** $-\dfrac{3}{4}$ **6.** -80

2.5 Exercises
1. Sum, plus, added. **3.** Of, times, twice.
5. The order of the subtraction is different. **7.** $y + 3 = -8; -11$

9. $6 - x = -3; 9$ **11.** $11m = -99; -9$ **13.** $m \div 4 = 1.6; 6.4$

15. $\dfrac{4}{5}x = \dfrac{5}{8}; \dfrac{25}{32}$ **17.** $7 + 3w = 34; 9$ **19.** $5a - 9 = 76; 17$

21. $8(8 + t) = 160; 12$ **23.** $-3(x - 2) = 12; -2$

25. $\dfrac{1}{3}(g + 2) = 1; 1$ **27.** $3m - 11 = 5 + m; 8$

29. $10 = \dfrac{r}{4} - 1; 44$ **31.** $2a - 3(a + 4) = -3; -9$

33. $(2d - 8) + (d - 12) = 13; 11$

35. $5\left(x + \dfrac{2}{3}\right) = 3x - \dfrac{2}{3}; -2$ **37.** $2x - 4 = 16 + 3x; -20$

39. $\dfrac{2}{5}x = \dfrac{1}{2}x - 2; 20$ **41.** $\dfrac{n - 3}{3} = \dfrac{n - 1}{4}; 9$

43. $-4(2 - x) = 2(4 - 3x) - 6; 1$ **45.** Three added to four times a number is seven. **47.** Six times the sum of a number and four is equal to the product of negative ten and the number. **49.** One-half of the difference of a number and three will result in two-thirds of the difference of the number and eight. **51.** Five-hundredths of a number added to six-hundredths of the difference of the number and eleven is twenty-two. **53.** The sum of two-thirds, three-fourths, and one-half of the same number will equal ten. **55.** Mistake: Subtraction translated in reverse order. Correct: $n - 10 = 40$ **57.** Mistake: Multiplied 5 times the unknown number instead of the difference, which requires parentheses. Correct: $5(x - 6) = -2$ **59.** Mistake: Subtracted the unknown number instead of the sum, which requires parentheses. Correct: $2t - (t + 3) = -6$ **61.** Translation: $P = 2(l + w)$

a. 84 ft. **b.** 61 cm **c.** $37\dfrac{1}{4}$ in. **63.** Translation: $P = b + 2s$

a. 27 in. **b.** 2.7 m **c.** $34\dfrac{1}{4}$ in. **65.** Translation: $I = Prt$

a. \$400 **b.** \$7.50 **c.** \$30 **67.** Translation: $t = \dfrac{e - d}{153.8}$

a. ≈ 49 sec. **b.** ≈ 52 sec. **c.** ≈ 62 sec.

Review Exercises
1. True; 19 is equal to 19. **2.** =
3. > **4.** -5 **5.** $-\dfrac{10}{9}$ **6.** -25

2.6 Exercises
1. Any number that can replace the variable(s) in the inequality and make it true. **3.** Any number greater than -5 and -5 itself. **5. a.** $\{x|x \geq -3\}$ **b.** $[-3, \infty)$
c. **7. a.** $\{h|h < 6\}$ **b.** $(-\infty, 6)$

c. **9. a.** $\left\{n|n < -\dfrac{2}{3}\right\}$

b. $\left(-\infty, -\dfrac{2}{3}\right)$ **c.**

11. a. $\{t|t \geq 2.4\}$ **b.** $[2.4, \infty)$
c. **13. a.** $\{x|-3 < x < 6\}$

b. $(-3, 6)$ **c.**

15. a. $\{n|0 \leq n \leq 5\}$ **b.** $[0, 5]$ **c.**

17. $\{x|x \geq -4\}, [-4, \infty)$ **19.** $\{x|x < 5\}, (-\infty, 5)$

21. $\{x|-4 \le x < 3\}, [-4, 3)$ 23. a. $n > 5$ b. $\{n|n > 5\}$
c. $(5, \infty)$ d. [number line 2–8] 25. a. $z \le -6$

b. $\{z|z \le -6\}$ c. $(-\infty, -6]$ d. [number line −8 to −3]

27. a. $y \ge 2$ b. $\{y|y \ge 2\}$ c. $[2, \infty)$
d. [number line 0–4] 29. a. $x \le 4$ b. $\{x|x \le 4\}$

c. $(-\infty, 4]$ d. [number line −1 to 5] 31. a. $x \ge 6$

b. $\{x|x \ge 6\}$ c. $[6, \infty)$ d. [number line 3–9]

33. a. $m > 8$ b. $\{m|m > 8\}$ c. $(8, \infty)$
d. [number line 4–11] 35. a. $y > 3$ b. $\{y|y > 3\}$

c. $(3, \infty)$ d. [number line 0–6] 37. a. $x > -4$

b. $\{x|x > -4\}$ c. $(-4, \infty)$ d. [number line −5 to 1]

39. a. $a < 1$ b. $\{a|a < 1\}$ c. $(-\infty, 1)$
d. [number line 0–4] 41. a. $f \le -2$ b. $\{f|f \le -2\}$

c. $(-\infty, -2]$ d. [number line −4 to 0] 43. a. $u \le 2$

b. $\{u|u \le 2\}$ c. $(-\infty, 2]$ d. [number line −2 to 3]

45. a. $c > -12$ b. $\{c|c > -12\}$ c. $(-12, \infty)$
d. [number line −14 to −9] 47. a. $w \ge 3$ b. $\{w|w \ge 3\}$

c. $[3, \infty)$ d. [number line 0–5] 49. a. $x \le -13$

b. $\{x|x \le -13\}$ c. $(-\infty, -13]$
d. [number line −15 to −10] 51. a. $n < 7$ b. $\{n|n < 7\}$

c. $(-\infty, 7)$ d. [number line 4–9] 53. a. $l < 10$

b. $\{l|l < 10\}$ c. $(-\infty, 10)$ d. [number line 5–11]

55. a. $t \le 3.\overline{3}$ b. $\{t|t \le 3.\overline{3}\}$ c. $(-\infty, 3.\overline{3}]$
d. [number line 2–5]

57. $n - 4 > 24; n > 28$ 59. $\frac{4}{9}x \le -8; x \le -18$

61. $8y - 36 \ge 60; y \ge 12$ 63. $\frac{1}{2}a + 5 \le 2; a \le -6$

65. $4x - 8 < 2x; x < 4$ 67. $25 \ge 6x + 7; x \le 3$
69. $i \le 10$ ft. 71. $h \ge 15$ in. 73. $r \le 23.89$ in.

75. $t \ge 6\frac{4}{13}$ hr. 77. $R \ge \$1,475,000$ 79. $x \ge 98$

81. $i \le 1.5$ A

Review Exercises 1. $\{1, 2, 3, 4, 5, 6, 7, 8, 9, 10\}$ 2. $\frac{35}{2}$

3. $2(x - 5) + 6$ 4. 16 5. $-\frac{135}{4}$ or $-33\frac{3}{4}$ 6. 1426.208

Chapter 2 Review Exercises 1. true 2. true
3. false 4. false 5. true 6. true 7. identify. 8. linear
9. 1. Simplify a. Distribute b. multiplying; LCD
c. Combine 2. addition 3. multiplication 10. Replace
the variable(s) in the equation with the value. If the resulting
equation is true, the value is a solution. 11. yes 12. no
13. no 14. yes 15. no 16. yes 17. -5 18. 18

19. 6 20. 5 21. -2 22. $\frac{9}{5}$ 23. No solution. 24. All
real numbers. 25. 2 26. 18 27. $\frac{9}{4}$ 28. $-\frac{1}{2}$ 29. $\frac{3}{4}$

30. 5.25 31. $x = 1 - y$ 32. $a^2 = c^2 - b^2$ 33. $d = \frac{C}{\pi}$

34. $m = \frac{E}{c^2}$ 35. $h = \frac{2A}{b}$ 36. $m = \frac{Fr}{v^2}$ 37. $h = \frac{3V}{\pi r^2}$

38. $m = \frac{y - b}{x}$ 39. $w = \frac{P - 2l}{2}$ 40. $h = \frac{2A}{a + b}$

41. $6x = -18; x = -3$ 42. $\frac{1}{2}m - 3 = \frac{1}{4}m - 9; m = -24$

43. $2(v + 4) - 1 = 1; v = -3$
44. $2(y - 1) + 3y = 6y - 20; y = 18$
45. a. $x < -3$ b. $\{x|x < -3\}$ c. $(-\infty, -3)$
d. [number line −4 to 2] 46. a. $x \ge -1$ b. $\{x|x \ge -1\}$

c. $[-1, \infty)$ d. [number line −3 to 3] 47. a. $z \le -1$

b. $\{z|z \le -1\}$ c. $(-\infty, -1]$ d. [number line −3 to 3]

48. a. $v > -4$ b. $\{v|v > -4\}$ c. $(-4, \infty)$
d. [number line −5 to 1] 49. a. $m > 4$ b. $\{m|m > 4\}$

c. $(4, \infty)$ d. [number line 0–6] 50. a. $c \ge -9$

b. $\{c|c \ge -9\}$ c. $[-9, \infty)$ d. [number line −10 to −4]

51. $13 > 3 - 10p; p > -1$ 52. $3 + 2z < 3(z - 5); z > 18$
53. $-\frac{1}{2}x \ge 4; x \le -8$ 54. $-2 \le 1 - \frac{1}{4}k; k \le 12$
55. 19.75 ft.^{2} 56. 102.6 in.^{2} 57. \$23.75 58. 80 in.
59. 7.1 in. 60. 1.2 mi. 61. 15 in. 62. $d \le 3.18$ ft.

63. $t \ge 5\frac{5}{6}$ hr. 64. $R \ge \$825,000$

Chapter 2 Practice Test 1. Equation because there
is an equal sign. [2.1] 2. Nonlinear because there is a variable raised to an exponent other than 1. [2.2] 3. yes [2.1]
4. no [2.1] 5. -3 [2.3] 6. 13.2 [2.2] 7. 3 [2.3]

8. No solution. [2.3] 9. $-\frac{2}{3}$ [2.3] 10. 4 [2.3]

11. $x = \frac{7 - y}{2}$ [2.4] 12. $r = \frac{I}{Pt}$ [2.4] 13. $h = \frac{2A}{b}$ [2.4]

14. $F = \frac{9C + 160}{5}$ or $F = \frac{9}{5}C + 32$ [2.4] 15. Set-builder

notation: $\{x|x \ge 3\}$; Interval notation: $[3, \infty)$;
Graph: [number line 0–7] [2.6]

16. Set-builder notation: $\{x | -1 \le x < 4\}$; Interval notation: $[-1, 4)$; Graph:

$[2.6]$

17. a. $m < 3$ **b.** $\{m | m < 3\}$ **c.** $(-\infty, 3)$

d.

$[2.6]$ **18. a.** $x \ge -8$

b. $\{x | x \ge -8\}$ **c.** $[-8, \infty)$

d.

$[2.6]$ **19. a.** $p > -\dfrac{2}{5}$

b. $\left\{ p | p > -\dfrac{2}{5} \right\}$ **c.** $\left(-\dfrac{2}{5}, \infty \right)$

d.

$[2.6]$ **20. a.** $l \le -\dfrac{5}{3}$

b. $\left\{ l | l \le -\dfrac{5}{3} \right\}$ **c.** $\left(-\infty, -\dfrac{5}{3} \right]$

d.

$[2.6]$

21. $\dfrac{2}{3}n - \dfrac{1}{6} = 2n$; $n = -\dfrac{1}{8}$ $[2.5]$ **22.** $12 + (-7x) = 5$;

$x = 1$ $[2.5]$ **23.** $5(n - 2) - 3 = 10 - 4(n - 1)$;

$n = 3$ $[2.5]$ **24.** $1 - n > 2n$; $n < \dfrac{1}{3}$ $[2.6]$ **25.** 20 ft. $[2.1]$

26. 68.25 in.^{2} $[2.1]$ **27.** 7 cm $[2.1]$ **28.** $530 $[2.1]$
29. $l \le 3$ ft. $[2.6]$ **30.** $x \ge 101$ $[2.6]$

Chapters 1 and 2 Cumulative Review

Exercises 1. true **2.** false **3.** false **4.** true
5. Understand, Plan, Execute, Answer, Check.
6. Grouping symbols, exponents/roots, multiplication/
division, addition/subtraction. **7.** Replace the variables
with their corresponding given values. Calculate the
numerical expression using the order of operations.

8.

9. 102 **10.** -63 **11.** 12

12. -4 **13.** -5 **14.** 148 **15.** 7 **16.** $13x^2 - 13x - 16$

17. $\dfrac{31}{3}$ **18.** 3 **19.** -10 **20.** $r = \dfrac{B - P}{Pt}$

21. $w = \dfrac{3V}{lh}$ **22. a.** $h \ge -9$ **b.** $\{h | h \ge -9\}$

c. $[-9, \infty)$ **d.**

23. a. $m < -5$ **b.** $\{m | m < -5\}$ **c.** $(-\infty, -5)$

d.

24. $3y - 40 = 7y$; -10

25. $12n - (n + 3) = -11 + 2(n - 5)$; -2 **26.** 99 cm^2
27. 22 ft.^{2} **28.** 1376 ft. **29.** $x \le -5$ **30.** $x \le 8$ ft.

Chapter 3

3.1 Exercises 1.
The quantity preceding *to* (longest
side) is written in the numerator, and the quantity following
the word *to* (shortest side) is written in the denominator.
3. $ad = bc$ **5.** To solve proportion problems, (a) set up a
proportion in which the numerators and denominators of the
ratios correspond in a logical manner and (b) solve using

cross products. **7. a.** $\dfrac{4}{15}$ **b.** $\dfrac{6}{11}$ **c.** $\dfrac{2}{1}$ **9.** $\dfrac{1}{2}$ **11.** $\dfrac{2}{3}$

13. $\dfrac{63.5}{1}$; he drives an average of 63.5 miles per hour.

15. $\dfrac{0.125}{1}$; each banana costs 12.5 cents. **17.** $\approx \dfrac{16.47}{1}$; the

price of the stock is $16.47 for every $1 earned in 2009.
19. The 24-oz. container is better because it costs less per
ounce. (The unit ratio of price to quantity is less.)

21. a. $\approx \dfrac{0.80}{1}$ **b.** $\approx \dfrac{0.64}{1}$ **c.** The ratio of convictions to total

cases was greater in 1999 than in 2004. **23. a.** $\approx \dfrac{0.89}{1}$;

women in the 25–34 age group with a bachelor's degree earn
$0.89 for every $1.00 men in the corresponding group earn.

b. $\approx \dfrac{0.60}{1}$ **c.** Younger women with a bachelor's degree are

making more in relation to men with a bachelor's degree than
are older women. **25.** yes **27.** no **29.** no **31.** yes
33. no **35.** 10.5 **37.** -14 **39.** 10.5 **41.** -8 **43.** 115

45. $3\dfrac{3}{5}$ **47.** 4 **49.** 7 **51.** $183\dfrac{1}{3}$ mi. **53.** $1350.65

55. $9450 **57.** ≈ 422.25 pounds **59.** 4.875 in. **61.** $2\dfrac{1}{2}$ tsp.

63. 360 highway miles **65.** ≈ 15.8 min. **67.** 225 lb.
69. $85.91 **71.** ≈ 66 deer **73.** 617,647 **75.** 11.25 cm

77. $a = 4\dfrac{8}{19}$ ft.; $b = 5\dfrac{1}{4}$ ft.; $c = 5\dfrac{17}{19}$ ft. **79.** 300 m

81. 26.4 m

Review Exercises 1.
Yes, because it can be written as a

ratio of integers, $\dfrac{58}{100}$. **2.** $\dfrac{3}{25}$ **3.** 3.06 **4.** 0.452 **5.** 320

6. 0.4 or $\dfrac{2}{5}$

3.2 Exercises 1.
To write a percent as a decimal or frac-
tion, (a) write the percent as a ratio with 100 in the denominator and (b) simplify to the desired form. **3.** multiplication
5. In a word-for-word translation, the division yields a decimal
number that must be written as a percent. When using the
proportion method, the decimal number is multiplied

by 100, which gives the percent. **7.** $0.2, \dfrac{1}{5}$ **9.** $0.15, \dfrac{3}{20}$

11. $0.148, \dfrac{37}{250}$ **13.** $0.0375, \dfrac{3}{80}$ **15.** $0.455, \dfrac{91}{200}$ **17.** $0.\overline{3}, \dfrac{1}{3}$

19. 60% **21.** 37.5% **23.** 83.3% **25.** 66.7% **27.** 96%
29. 80% **31.** 9% **33.** 120% **35.** 2.8% **37.** 405.1%
39. 28 **41.** 3.1 **43.** 5.92 **45.** 103.2 **47.** 50 **49.** 80
51. 62.4 **53.** 20% **55.** 105% **57.** 66.$\overline{6}$% **59.** 80%
61. 43 **63.** $4800; $400 **65.** $898.80 **67.** 5783.5 tg
69. $42.50 **71.** $300 **73.** $\approx 45.2\%$ **75.** $\approx 70\%$
77. $\approx 8.0\%$ **79.** United States $\approx 24.4\%$, China $\approx 2.7\%$,
Russia $\approx 53.6\%$; Russia **81.** 23.7% **83.** 20%
85. $6.22; $130.67 **87.** $94.90; $854.10 **89.** $2500
91. $1950; 75% **93.** $\approx 45.7\%$ **95.** 34.1$\overline{6}$% **97.** $\approx 70.8\%$
99. 2800% **101.** 400,000 **103.** ≈ 2246.77

Review Exercises 1.
-1 **2.** $-3x - 21$
3. $-xy + 4x + 3y$ **4.** $3(n + 9) = n - 11$; -19 **5.** 2.1
6. 10.5 cm

3.3 Exercises **1.** The second number; $5n$. **3.** There is always a difference of 2 between consecutive even or odd numbers. **5. a.** $x + 3$ **b.** $6(x + 3) + 8x = 14x + 18$
7. a. $3w + 4$ **b.** $2w + 2(3w + 4) = 8w + 8$
9. a. $4x - 10$ **b.** $x + (4x - 10) = 90$ **11. a.** $-10, -8; -30$
b. $y + 2, y + 4$ **c.** $y + (y + 2) + (y + 4) = 3y + 6$
13. a. $3n$ **b.** maple: $7.99(3n)$; pecan: $11.99n$
c. $7.99(3n) + 11.99n = 35.96n$ **15. a.** $372 - x$
b. exterior: $16.99x$; interior: $12.99(372 - x)$
c. $16.99x + 12.99(372 - x) = 4x + 4832.28$ **17.** 13, 39
19. 10, 17 **21.** 11, 17 **23.** 6, 8 **25.** 12, 24 **27.** 10, 2
29. Grapefruit: 22 g; orange: 27 g **31.** Big Mac: 580 cal.; large fries: 520 cal. **33.** 25-watt: 215 lumens; 60-watt: 880 lumens
35. length: 70 ft.; width: 50 ft. **37.** length: 12 ft.; width: 4 ft.
39. length: 23 ft.; width: 13 ft. **41.** length: 11 cm; width: 7 cm
43. length: 7.5 ft.; width: 5 ft. **45.** 77°, 103° **47.** 50°, 130°
49. 26°, 64° **51.** 15°, 75° **53.** 73, 74 **55.** 12, 14
57. 92, 93, 94 **59.** 76, 78, 80 **61.** 123, 125, 127
63. 23, 24, 25 **65.** 69, 70, 71 **67.** 15, 17, 19 **69.** 24 $5 bills; 8 $50 bills **71.** 22 shares at $18; 45 shares at $32
73. 220 student tickets; 530 general public tickets
75. 10 $5 bills; 6 $10 bills

Review Exercises **1.** yes **2.** yes **3.** -16 **4.** -8
5. 3 **6.** 6.75 mi.

3.4 Exercises **1.** They will travel the same amount of time. **3.** $45t + 24t = 10$ **5.**

| 1st car | |
| 2nd car | |

Bridge   2nd car catches up

7. a. 165 mi. **b.** $48t$ **c.** $63(t + 2)$ **9.** Sal: $t + 2, 2t$; Jorge: t, $t - 3$ **11. a.** Their times are the same.
b.

c. Time: t, t; distance:

at 58 mph   at 62 mph
Atlanta   Charleston
$58t, 62t$ **d.** 500 mi. **e.** $58t + 62t = 500$ **13. a.** Polly's time is two hours more than Erick's. **b.**

| Polly's distance |
| Erick's distance |

c. They are the same. **d.** Time: $t + 2$; distance: $55(t + 2)$, $68t$
e. $55(t + 2) = 68t$ **15.** 4 hr. **17.** 3.5 hr. **19.** 2 hr.
21. 2 hr. **23.** 9:36 A.M. **25.** Car 1: 22 mph; car 2: 33 mph
27. 4.5 hr.; 247.5 mi. **29.** 5:30 P.M. **31.** 5 P.M.
33. a. 45 mph **b.** 180 mi. **35.** 6 hr. **37.** 4 hr. **39.** 79 mph
41. 3 mph, 67 mph

Review Exercises **1.** Commutative property of multiplication. **2.** 38 **3.** -43 **4.** $8x + 7y + 6$ **5.** $90.\overline{90}$
6. 700

3.5 Exercises **1. a.** Acct. 1: $3600; Acct. 2: $1500, $6800 **b.** $10,000 -$ Known amount **c.** $10,000 - P$
3. $220; $391.88; $0.05 P$; $0.065 (15,000 - P)$ **5.** Acct. 1: $P + 1200, 4000 - P$; Acct. 2: $2P, 6000 - P$ **7.** The sum of the expressions for interest is set equal to the total interest.
9. Soln. 2: 200 ml, $2n$ ml; Comb. Soln.: 120 ml, 300 ml, $3n$ ml

11. 6.5 oz.; 10.725 oz.; $0.125x$ oz.; $0.082(50 - x)$ oz.
13. $4600 at 2%; $13,800 at 4% **15.** $1600 at 2%; $2400 at 3%
17. $1250 at 8%; $1900 at 12% **19.** $1800 in plan 1; $2700 in plan 2 **21.** $940 at 5%; $660 at 4% **23.** $1000 at 5.5%; $4000 at 7% **25.** $4200 at 6%; $8400 at -2% **27.** $4000 in plan A; $6000 in plan B **29.** 6000 L **31.** 52.5 ml

33. $166\frac{2}{3}$ lb. of 26%; $133\frac{1}{3}$ lb. of 35% **35.** 50 g of 6%; 25 g

of 18% **37.** 30 ml **39.** 1.2 qt. **41.** 56 lb. at $1.20; 24 lb at $0.90 **43.** 3 lb. at $1.65; 2 lb. at $1.25

45. $76\frac{2}{3}$ lb. at $0.60; $38\frac{1}{3}$ lb. at $1.20

Review Exercises **1.** $\dfrac{41}{108}$ **2.** -64 **3.** $12x$ **4.** $<$

5. -2 **6.** 72

Chapter 3 Review Exercises **1.** true **2.** true
3. false **4.** false **5.** false **6.** false **7.** 1. 100 **8.** congruent
9. complementary **10.** proportion **11.** $\dfrac{3}{8}$ **12.** $\approx \dfrac{\$0.085}{1 \text{ oz.}}$;

each ounce costs 8.5 cents. **13.** no **14.** yes **15.** yes

16. no **17.** 25 **18.** 22 **19.** $78\frac{3}{4}$ **20.** 9.5625 **21.** -120

22. $21.1\overline{6}$ **23.** 300 mi. **24.** 5 ft. **25.** $9.10 **26.** $2587.20
27. ≈ 220.5 mi. **28.** $270.03 **29.** 12 in. **30.** 43.2 m

31. $x = 12$ ft.; $y = 6$ ft. **32.** $a = 4\frac{5}{7}$ in., $b = 4\frac{2}{3}$ in.,

$c = 13\frac{1}{2}$ in. **33.** $0.15, \dfrac{3}{20}$ **34.** $0.825, \dfrac{33}{40}$ **35.** $0.125, \dfrac{1}{8}$

36. $0.0245, \dfrac{49}{2000}$ **37.** $0.1, \dfrac{1}{10}$ **38.** $0.\overline{3}, \dfrac{1}{3}$ **39.** 40%

40. $33\frac{1}{3}$% or $33.\overline{3}$% **41.** $36\frac{4}{11}$% or $36.\overline{36}$% **42.** 35%
43. 120% **44.** 201.6% **45.** 14.56 **46.** 289.7 **47.** 32.5%
48. 56.25% **49.** 478 **50.** $805.59 **51.** $\approx 59.3\%$
52. $\approx 36.1\%$ **53.** 11, 44 **54.** 2.5, 12.5 **55.** 41, 42
56. Jack: 24; Jill: 26 **57.** 19, 21, 23 **58.** $-2, -1, 0$
59. length: 15 in.; width: 9 in. **60.** length: 45 ft.; width: 15 ft.
61. 160°, 74° **62.** 31°, 59° **63.** 3 regular; 2 control top
64. 10 gal. of interior; 14 gal. of exterior **65.** 3.5 hr.
66. 2.6 hr. **67.** 4 P.M. **68.** 75 mph **69.** $20,000 at 8%; $40,000 at 10% **70.** $10,200 at 7%; $30,600 at -2%
71. 40 L of 90%; 60 L of 40% **72.** 60 oz.

Chapter 3 Practice Test **1.** $\dfrac{2}{5}$ [3.1] **2.** $\dfrac{0.26}{1}$;

1 ounce of cereal costs 26 cents. [3.1] **3.** $3\frac{1}{3}$ [3.1]

4. 2.25 [3.1] **5.** ≈ 49 gal. [3.1] **6.** 12.96 cm [3.1]

7. $0.22, \dfrac{11}{50}$ [3.2] **8.** $0.0\overline{3}, \dfrac{1}{30}$ [3.2] **9.** 320% [3.2]

10. 40% [3.2] **11.** 20% [3.2] **12.** 14.35 [3.2]
13. 175 [3.2] **14.** 185.5, or about 186 [3.2]
15. $\approx 67.6\%$ [3.2] **16.** 12.5% [3.2] **17.** ostrich: 56.4 oz., albatross: 21 oz. [3.3] **18.** 54, 56 [3.3] **19.** 5 km, 11 km [3.3]

20. 102°, 78° [3.3] **21.** 15 shares of Walmart; 22 shares of Target [3.3] **22.** 0.5 hr. or 30 min. [3.4] **23.** 0.625 hr. or 37.5 min. [3.4] **24.** $3000 in plan A; $1500 in plan B [3.5]
25. 30 oz. [3.5]

Chapters 1–3 Cumulative Review Exercises

1. true **2.** false **3.** false **4.** true **5.** itself **6.** LCD
7. $2^3 \cdot 3 \cdot 5$ **8.** 3 **9.** Distributive property. **10.** Set the cross products equal to each other and solve for the variable.
11. $\dfrac{4}{9}$ **12.** 7 **13.** $\dfrac{1}{10}$ **14.** -14 **15.** $\dfrac{11}{10}$ **16.** -26.48
17. $-\dfrac{1}{2}x + 27$ **18.** 9 **19.** $\dfrac{2}{5}$ **20.** All real numbers. **21.** 5
22. $b = \dfrac{2A}{h}$ **23.** $c = P - a - b$ **24.** 6.345 **25.** $-3, -1, 1$
26. $t \geq 88$ **27.** 20 L **28.** 4 P.M. **29.** 3.75 lb. of cashews; 6.25 lb. of peanuts **30.** 300 mi.

Chapter 4

4.1 Exercises **1.** Horizontal-axis coordinate.
3.

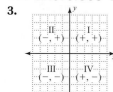

5. $A(4, 3), B(-3, 1), C(0, -2), D(2, -5)$
7. $A(-4, 0), B(2, 4), C(-2, -3), D(4, -4)$
9.

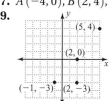

11.

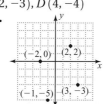

13. II **15.** I

17. IV **19.** III **21.** y-axis **23.** x-axis **25.** linear
27. nonlinear **29.** linear **31.** nonlinear **33.** linear
35. Answers may vary. Some possible answers are $(-1, -5)$, $(0, -3)$, and $(2, 1)$. **37.** $A: (-6, -4); B: (-3, 2); C: (5, 2);$
$D: (8, -4)$ **39. a.** $(-3, 1), (2, 1), (1, -3), (-4, -3)$
b. $(0, 2), (5, 2), (4, -2), (-1, -2)$ **c.** $(x + 3, y + 1)$
41. a.

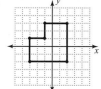

b. 20 units **c.** 21 square units

Review Exercises **1.**

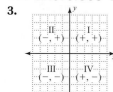

2. 2

3. 12 **4.** 0 **5.** $\dfrac{15}{2}$ **6.** $-\dfrac{9}{10}$

4.2 Exercises **1. a.** Replace the variables in the equation with the corresponding coordinates. **b.** Verify that the equation is true. **3. a.** Choose a value for one of the variables. **b.** Replace the corresponding variable with your chosen value. **c.** Solve the equation for the value of the

other variable. **5.** A minimum of two ordered pairs are needed because two points determine a line. **7.** yes **9.** yes
11. no **13.** yes **15.** no **17.** yes **19.** no **21.** no
23. $(-1, -1), (0, 0), (1, 1)$ **25.** $(-1, -2), (0, 0), (1, 2)$

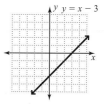

27. $(-1, 5), (0, 0), (1, -5)$ **29.** $(0, -3), (2, -1), (4, 1)$

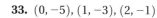

31. $(4, -6), (-4, 2)$ **33.** $(0, -5), (1, -3), (2, -1)$

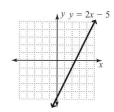

35. $(-1, 6), (0, 4), (1, 2)$ **37.** $(-2, -1), (0, 0), (2, 1)$

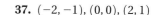

39. $(0, 0), (-3, 2), (3, -2)$ **41.** $(-3, 6), (0, 4), (3, 2)$

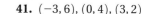

43. $(3, -5), (5, -3), (6, -2)$ **45.** $(1, 4), (3, 0), (5, -4)$

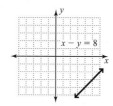
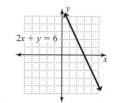

47. $(0, 4), (6, 0), (3, 2)$

49. $(0, 4), (-3, 0), (3, 8)$

51. $(1, -4), (2, 0), (3, 4)$

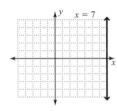

53. $(-1, -5), (0, -5), (1, -5)$

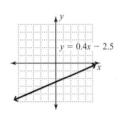

55. $(7, -1), (7, 0), (7, 1)$

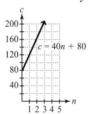

57. $(-1, -2.9), (0, -2.5), (1, -2.1)$

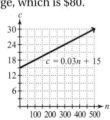

59. As a gets larger, the graph gets steeper.
61. The graph will shift up b units on the y-axis.
63. a. $160   **b.** 4 hr.   **c.**

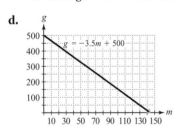

d. It represents the initial charge, which is $80.
65. a. $18   **b.** 500 copies   **c.**

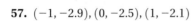

67. a. 412.5 gal.   **b.** ≈ 71.4 min.   **c.** ≈ 143 min.

d.

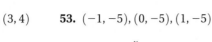

e. It represents the original amount of water in the hot tub, which is 500 gallons.   **f.** It represents the amount of time it takes for the hot tub to empty (0 gallons of water).

69. a. $1200   **b.** 4 yr.   **c.** 8 yr.   **d.**

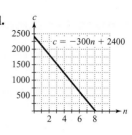

Review Exercises 1. 4   **2.** 31.5 in.   **3.** 5.67 sq. in.
4. $\dfrac{5}{3}$ **5.** $x = \dfrac{v - k}{m}$ **6.**

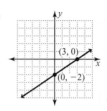

4.3 Exercises 1. The point where a graph intersects the x-axis.   **3. (a)** Replace y with 0 in the given equation.
(b) Solve for x.   **5.** The graph of any equation in the form $y = mx$ is always a line that passes through the origin, which means the x- and y-intercepts are both at $(0, 0)$.
7. $(1, 0), (0, 3)$ **9.** $(-3, 0), (0, 4)$ **11.** $(3, 0)$; no y-intercept   **13.** no x-intercept; $(0, -2)$ **15.** $(4, 0), (0, -4)$
17. $(3, 0), (0, 2)$ **19.** $\left(-\dfrac{10}{3}, 0\right), \left(0, \dfrac{5}{2}\right)$ **21.** $(2, 0), (0, -6)$
23. $\left(-\dfrac{5}{2}, 0\right), (0, 5)$ **25.** $(0, 0)$ for both   **27.** $(0, 0)$ for both
29. $(5, 0)$; no y-intercept   **31.** no x-intercept; $(0, -2)$

33.

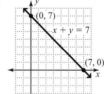

35.

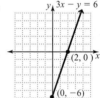

37.

39.

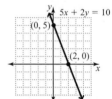

41.

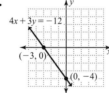

43.

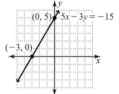

45.

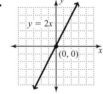

47.

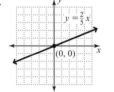

49.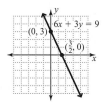

51.

53.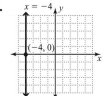

55.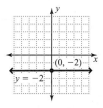

57.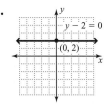

59.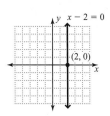

61.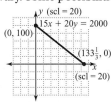

63.

65. d because the x-coordinate of the x-intercept is positive and the y-coordinate of the y-intercept is negative.

67. c because the x-coordinate of the x-intercept is positive and the y-coordinate of the y-intercept is negative.

69. $(-2.7, 0), (2.7, 0), (0, -2.7), (0, 2.7)$ **71. a.** $15x + 20y$

b. $15x + 20y = 2000$ **c.** $\left(133\frac{1}{3}, 0\right), (0, 100)$ **d.** The number of units required to meet the goal if only one size is sold.

e. Answers may vary. Some possibilities are $(20, 85), (40, 70)$, and $(60, 55)$.   **f.**

Review Exercises 1. -2 **2.** undefined

3. $w = \dfrac{P - 2l}{2}$ **4.** $4 + 3x = 5(x + 6); -13$ **5.** On the y-axis.

6. no

4.4 Exercises 1. How steep a line is.   **3.** Downhill because the slope is negative.   **5.** Horizontal because the rise is 0 for any amount of run.

7.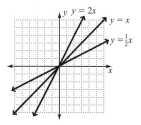

9.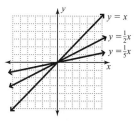

11.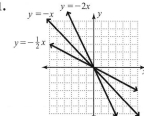

13.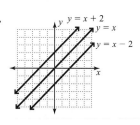

15. $m = 2; (0, 3)$ **17.** $m = -2; (0, -1)$

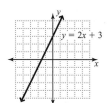 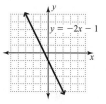

19. $m = \dfrac{1}{3}; (0, 2)$ **21.** $m = \dfrac{3}{4}; (0, -2)$

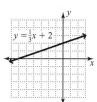 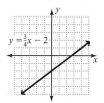

23. $m = -\dfrac{2}{3}; (0, 8)$ **25.** $m = -2; (0, 4)$

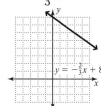 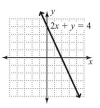

27. $m = 3; (0, 1)$ **29.** $m = -\dfrac{2}{3}; (0, 2)$

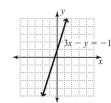 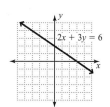

31. $m = -\dfrac{1}{2}$; $(0, -2)$ **33.** $m = \dfrac{3}{2}$; $\left(0, \dfrac{7}{2}\right)$

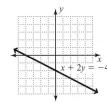

x + 2y = −4

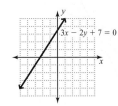

3x − 2y + 7 = 0

35. $m = -\dfrac{3}{2}$; $\left(0, \dfrac{5}{2}\right)$ **37.** $m = 3$; $(0, -5)$

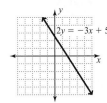

2y = −3x + 5

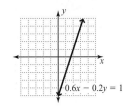

0.6x − 0.2y = 1

39. $y = 2x + 3$ **41.** $y = -3x - 9$ **43.** $y = \dfrac{3}{4}x + 5$

45. $y = -\dfrac{2}{5}x + \dfrac{7}{8}$ **47.** $y = 0.8x - 5.1$ **49.** $y = -3x$

51. $\dfrac{5}{2}$ **53.** $-\dfrac{2}{3}$ **55.** $\dfrac{2}{7}$ **57.** $-\dfrac{4}{3}$ **59.** $\dfrac{2}{3}$ **61.** $-\dfrac{5}{4}$ **63.** 0

65. undefined   **67.** d   **69.** a   **71.** f   **73.** g   **75.** h

77. $m = 2$; $(0, 3)$; $y = 2x + 3$ **79.** $m = -1$; $(0, 2)$;

$y = -x + 2$ **81.** $m = \dfrac{2}{3}$; $(0, -1)$; $y = \dfrac{2}{3}x - 1$

83. $m = -\dfrac{1}{4}$; $(0, 3)$; $y = -\dfrac{1}{4}x + 3$

85. m is undefined; no y-intercept; $x = 2$ **87.** $m = 0$; $(0, 3)$;
$y = 3$ **89.** $y = -2, x = 3$ **91.** $y = 0$

93. a.   **b.** 3 and 0   **c.** They are the same.

95. $0.58\overline{3}, \dfrac{7}{12}$ **97.** $\dfrac{293}{230} \approx 1.27$

Review Exercises 1. $-\dfrac{3}{2}$ **2.** $11x + 7$ **3.** $-2x - 14$

4. $x = \dfrac{C - By}{A}$ **5.**   **6.** $\left(\dfrac{7}{3}, 0\right), \left(0, -\dfrac{7}{5}\right)$

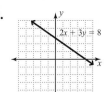

2x + 3y = 8

4.5 Exercises 1. The slope–intercept form would be easiest to use because you are given the values m and b.   **3.** Find the slope.   **5.** Solve for y.   **7.** $y = 3x - 13$ **9.** $y = -x - 2$

11. $y = \dfrac{2}{3}x - 1$ **13.** $y = -\dfrac{3}{4}x - \dfrac{23}{4}$ **15.** $y = -\dfrac{4}{3}x + \dfrac{4}{3}$

17. $y = 2x$ **19.** $2x - 3y = -2$ **21.** $x + 2y = 3$

23. $y = -2x + 5$ **25.** $y = \dfrac{5}{2}x - \dfrac{13}{2}$ **27.** $y = 7x$ **29.** $x = 8$

31. $y = -1$ **33.** $y = 3x + 0.8$ **35.** $y = 2x - 2$

37. $y = -\dfrac{6}{5}x + 3$ **39.** $5x - 2y = 4$ **41.** $3x - 2y = 6$

43. $4x + 11y = -27$ **45.** $2x + 3y = 0$ **47.** $x = -1$
49. $y = -1$ **51. a.** $y = 4x - 14$ **b.** $4x - y = 14$
53. a. $y = -3x - 5$ **b.** $3x + y = -5$

55. a. $y = \dfrac{1}{3}x + \dfrac{1}{3}$ **b.** $x - 3y = -1$ **57. a.** $y = -\dfrac{3}{4}x - \dfrac{15}{4}$

b. $3x + 4y = -15$ **59. a.** $y = -2x + 1$ **b.** $2x + y = 1$

61. a. $y = \dfrac{3}{4}x + 5$ **b.** $3x - 4y = -20$ **63. a.** $y = -\dfrac{1}{2}x - 3$

b. $x + 2y = -6$ **65. a.** $y = -4x + 8$ **b.** $4x + y = 8$

67. a. $y = \dfrac{1}{2}x - 5$ **b.** $x - 2y = 10$ **69. a.** $y = -\dfrac{2}{3}x + \dfrac{5}{3}$

b. $2x + 3y = 5$ **71. a.** $y = -\dfrac{5}{2}x - \dfrac{17}{2}$ **b.** $5x + 2y = -17$

73. a. $y = \dfrac{3}{2}x - \dfrac{17}{2}$ **b.** $3x - 2y = 17$ **75.** parallel

77. perpendicular   **79.** neither   **81.** parallel   **83.** perpendicular   **85.** neither   **87.** perpendicular   **89.** parallel
91. a. $c = 0.3n + 7.75$ **b.** \$19.75

c.   **93. a.**

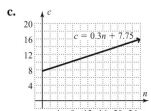

c = 0.3n + 7.75

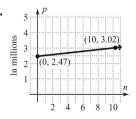

In millions

(10, 3.02)

(0, 2.47)

b. 0.055   **c.** $p = 0.055n + 2.47$ **d.** 2.745 million
e. 3.35 million   **95. a.**   **b.** -1.5

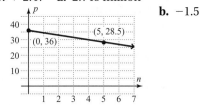

(5, 28.5)

(0, 36)

c. $p = -1.5n + 36$ **d.** $(0, 36)$; the p-intercept indicates the initial price.   **e.** $(24, 0)$; the n-intercept indicates that on the 24th day, the price will be \$0.   **f.** \$24

Review Exercises 1. $<$ **2.** $\dfrac{36}{5}$

3. $x \le -7$
  −10−9−8−7−6 −5−4

4. $x \le \dfrac{4}{3}$ **5.** $m = -\dfrac{3}{4}$; $(0, 2)$
  0 1$\frac{4}{3}$ 2   3

6. $(7, 0)(0, -2)$

4.6 Exercises 1. To determine whether an ordered pair is a solution for an inequality, replace the variables with the corresponding coordinates and see if the resulting inequality is true. If it is, the ordered pair is a solution.   **3.** A solid line indicates equality is included, $\ge$ or $\le$. A dashed line indicates equality is excluded, $>$ or $<$.   **5.** yes   **7.** no

9. no **11.** yes

13. $y \le -x + 1$

15. $y > -2x + 5$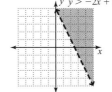

17. $y > 2x$

19. $y > \frac{1}{3}x$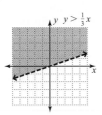

21. $y \ge \frac{3}{4}x + 2$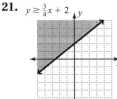

23. $3x + y \le 9$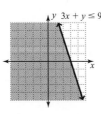

25. $5x - 2y < 10$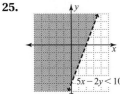

27. $3x + y \ge 8$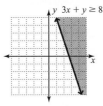

29. $2x + 3y < 9$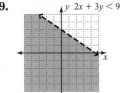

31. $3x + y \ge 0$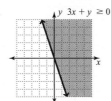

33. $x > -4$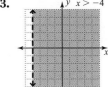

35. $y \le 2$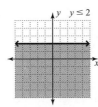

37. $x + 2 \ge 0$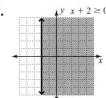

39. $y - 3 < 0$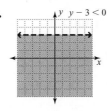

41. a. x rpresents the number of large bottles sold; y represents the number of small bottles sold.

b. $2x + 1.5\,y \ge 280{,}000$
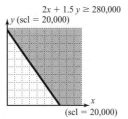

c. The boundary line represents combinations of bottle sizes sold to produce a revenue of exactly $280,000 (breakeven). The shaded region represents combinations that produce a revenue greater than $280,000 (profit). **d.** Answers will vary; a couple combinations are $(80{,}000, 80{,}000)$ and $(20{,}000, 160{,}000)$. **e.** Answers will vary; a couple combinations are $(150{,}000, 0)$ and $(90{,}000, 100{,}000)$. **f.** No, only combinations of whole numbers are possible because we cannot sell fractions of a bottle of hand lotion. **43. a.** $5.5x + 12.5y \le 100$

b. $5.5x + 12.5y \le 100$
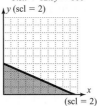

c. The combinations of plant purchases that would cost exactly $100. **d.** The combinations of plants purchased that would cost less than $100. **e.** Answers will vary; some combinations are $(2, 4)$, $(6, 2)$, and $(8, 4)$. **f.** Yes, $(0, 8)$. Because she cannot purchase fractional numbers of plants, the combination must be whole numbers and $(0, 8)$ is the only combination of whole numbers that costs exactly $100.
45. a. $2l + 2w \le 180$ **b.** $2l + 2w \le 180$
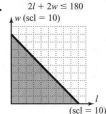

c. The combinations of lengths and widths that yield a perimeter of exactly 180 ft. **d.** The combinations of length and width that yield a perimeter that is less than 180 ft.
e. Answers will vary; some combinations are $(10, 80)$, $(20, 70)$, and $(40, 50)$. **f.** Answers will vary; some combinations are $(10, 10)$, $(20, 20)$, and $(30, 30)$.

Review Exercises **1.** $\{a, e, i, o, u\}$ **2.** 1 **3.** -13
4. 20 ft. by 24 ft. **5.** $\left(\frac{8}{3}, 0\right)$, $(0, -4)$ **6.** $-x + y = 7$

4.7 Exercises **1.** The domain is a set containing all x-values. **3.** Trail along the x-axis and determine what values of x have a corresponding y-value. **5.** Draw or imagine a vertical line through every point in the domain. If each vertical line intersects the graph in at most one point, the relation is a function. **7.** Domain: $\{2, 3, 1, -1\}$; range: $\{1, -2, 4, -1\}$. **9.** Domain: $\{4, 2, 3, -4\}$; range: $\{1, -5, 0\}$. **11.** Domain: $\{0, -3, 2, 8\}$; range: $\{0, -3, 9, -16\}$. **13.** yes **15.** no **17.** yes **19.** no **21.** Yes, every element in the domain is assigned to one element in the range. **23.** No, an element in the domain is assigned to more than one element in the range. (8 is assigned to both 2 and -1.) **25.** Yes, every element in the domain is assigned to one element in the range. **27.** No, an element in the domain is assigned to more than one element in the range. **29.** No, it fails the vertical line test. **31.** Yes, it passes the vertical line test. **33.** No, it fails the vertical line test. **35.** Yes, it passes the vertical line test. **37.** Domain: $\{x \mid 1900 \le x \le 2005\}$ or $[1900, 2005]$; range: $\{y \mid 13 \text{ million} \le y \le 66 \text{ million}\}$ or $[13 \text{ million}, 66 \text{ million}]$; function. **39.** Domain $\{1900, 1910, 1920, 1930, 1940, 1950, 1960, 1970, 1980, 1990, 2000\}$; range: $\{4.7, 5.4, 6.2, 6.9, 8.0, 8.8, 10.4, 11.6, 13.2, 13.6, 14.7\}$; function **41.** Domain: $\{x \mid -4 \le x \le 4\}$ or $[-4, 4]$; range: $\{y \mid -3 \le y \le 2\}$ or $[-3, 2]$; function. **43.** Domain: $\{x \mid x \le 0\}$ or $(-\infty, 0]$; range: all real numbers or $(-\infty, \infty)$; not a function. **45.** Domain: all real numbers or $(-\infty, \infty)$; range: $\{y \mid y \ge -1\}$ or $[-1, \infty)$; function. **47. a.** -5 **b.** -3 **c.** -9 **d.** -4 **49. a.** 7 **b.** 5 **c.** 11 **d.** 17 **51. a.** $\sqrt{3}$ **b.** Not a real number. **c.** $\sqrt{2}$ **d.** $\sqrt{a+3}$ **53. a.** 0 **b.** Not a real number. **c.** 0 **d.** $\sqrt{5}$ **55. a.** -1 **b.** undefined **c.** $-\frac{1}{2}$ **d.** $\frac{1}{a-1}$ **57. a.** -3 **b.** -3.19 **c.** $a^2 - 2a - 3$ **d.** $a^2 - 8a + 12$ **59. a.** -2 **b.** $-\frac{5}{3}$ **c.** $-\frac{7}{3}$ **d.** $\frac{1}{3}a - 2$ **61. a.** $-\frac{3}{4}$ **b.** undefined **c.** $-\frac{1}{6}$ **d.** -6 **63. a.** undefined **b.** $-\frac{1}{\sqrt{3}}$ **c.** Not a real number. **d.** $\frac{1}{\sqrt{a+2}}$ **65. a.** 2 **b.** 4 **c.** 6 **d.** 10 **67. a.** 0 **b.** $-\frac{1}{2}$ **c.** 20 **d.** 7 **69. a.** $8a^2 + 6a + 1$ **b.** $2a^4 + 3a^2 + 1$ **c.** $2a^2 + 11a + 15$ **d.** $8a^2 - 2a$ **71. a.** 3 **b.** 4 **c.** 2 **73. a.** 0 **b.** -2 **c.** undefined

75.
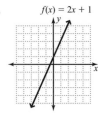
$f(x) = 2x + 1$

77.
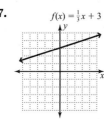
$f(x) = \frac{1}{3}x + 3$

79. $f(x) = -4x + 1$
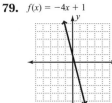

81.
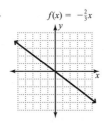
$f(x) = -\frac{2}{3}x$

83.
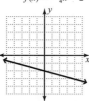
$f(x) = -\frac{1}{4}x - 2$

85. a. \$350 **b.** \$450

c. The cost of producing 30 items is \$750. **87. a.** \$13,500 **b.** \$9000 **c.** The value of the car after ten years is \$3000.

Review Exercises **1.** $-\dfrac{13}{30}$ **2.** $-4x - 2$
3. Commutative property of addition.
4.

5. $y = \dfrac{3}{2}x - 3$ **6.** parallel

Chapter 4 Review Exercises **1.** true **2.** true **3.** false **4.** false **5.** true **6.** true **7. a.** y **b.** x **8. a.** x **b.** y **9.** y-intercept **10.** uphill; downhill **11.** $A(4, 3), B(-2, 4), C(-4, -2), D(2, -1)$ **12.** $A(3, 1), B(0, 4), C(-3, 0), D(0, -2)$ **13.**
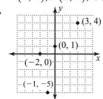
14.

15. II **16.** III **17.** I **18.** IV **19.** yes **20.** yes **21.** no **22.** no **23.** no **24.** no **25.** $(0, 0), (1, -2), (-1, 2)$ **26.** $(0, 0), (1, 1), (-1, -1)$
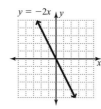
$y = -2x$
$x = y$

27. $(0, 7), (1, 6), (3, 4)$ **28.** $(5, 0), (1, 1), (-3, 2)$
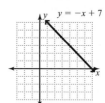
$y = -x + 7$
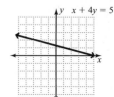
$x + 4y = 5$

29. $(0,2), (3,0), (-3,4)$ **30.** $(0,3), (3,2), (-3,4)$

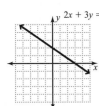
$2x + 3y = 6$

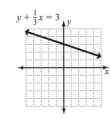
$y + \frac{1}{3}x = 3$

46. $m = -\frac{1}{4}; (0,2)$ **47.** $m = 3; (0,0)$

$y = -\frac{1}{4}x + 2$

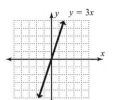
$y = 3x$

31. $(12,0), (0,6)$ **32.** $(0,0)$ for both **33.** $(3.1,0), (0,-2)$

34. $\left(-\frac{15}{2}, 0\right), (0,5)$ **35.** $(-2,0)$, no y-intercept

36. no x-intercept, $(0,7)$ **37.**

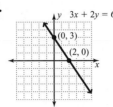
$3x + 2y = 6$
$(0,3)$
$(2,0)$

48. $m = -1; (0,5)$ **49.** $m = \frac{1}{3}; \left(0, -\frac{7}{3}\right)$

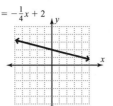
$x + y = 5$

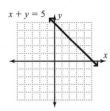
$x - 3y = 7$

38.

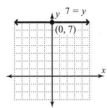
$y = -2x + 1$
$(0,1)$
$\left(\frac{1}{2}, 0\right)$

39. $y = -\frac{1}{5}x + 3$

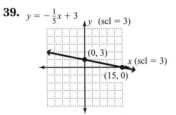
(scl = 3)
$(0,3)$
x (scl = 3)
$(15,0)$

50. $m = \frac{2}{3}; \left(0, -\frac{8}{3}\right)$

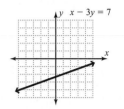
$2x - 3y = 8$

40. $2x = 5y + 10$

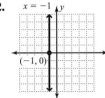
$(5,0)$
$(0,-2)$

41.

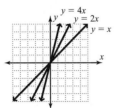
$7 = y$
$(0,7)$

51. 3 **52.** 1 **53.** $-\frac{1}{3}$ **54.** $-\frac{3}{5}$ **55.** 0 **56.** undefined

57. $y = -x + 7$ **58.** $y = -\frac{1}{5}x - 8$ **59.** $y = 0.2x + 6$

60. $y = x$ **61.** $y = -2x + 9$ **62.** $y = x - 9$ **63.** $y = \frac{1}{3}x + \frac{5}{3}$

64. $y = -\frac{2}{5}x + \frac{21}{5}$ **65.** $y = 6.2x + 9.4$ **66.** $y = -0.4x + 1.6$

67. $y = -x + 4, x + y = 4$ **68.** $y = \frac{5}{4}x - \frac{37}{4}, 5x - 4y = 37$

69. $y = -\frac{1}{2}x + \frac{7}{2}, x + 2y = 7$ **70.** $y = -\frac{1}{5}x - \frac{13}{5},$

$x + 5y = -13$ **71.** $y = 2x - 2$ **72.** $y = -\frac{2}{3}x + \frac{17}{3}$

42.

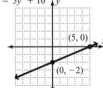
$x = -1$
$(-1,0)$

43.

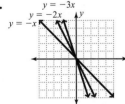
$y = 4x$
$y = 2x$
$y = x$

44.

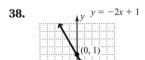
$y = -3x$
$y = -2x$
$y = -x$

45. $m = -2; (0,3)$

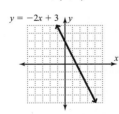
$y = -2x + 3$

73. $y = -\frac{5}{3}x - \frac{11}{3}$ **74.** $y = \frac{5}{2}x + 9$ **75. a.** $\$2200$

b. $(0, 1000)$; This indicates the person's gross pay if he has $\$0$ in sales during a month. **c.**

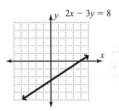
$p = 0.05s + 1000$

76. a.

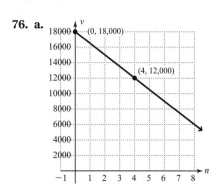

b. -1500 **c.** $v = -1500n + 18{,}000$ **d.** 2010 **e.** 2016
77. yes **78.** yes **79.** yes **80.** no

81.

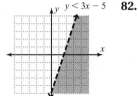

82.

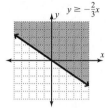

83.

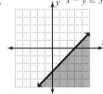

84.

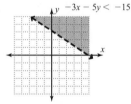

85.

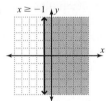

86. a. $6a + 8b \le 12{,}000$ **b.**

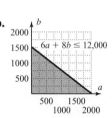

c. Answers may vary; some combinations are $(400, 1200)$, $(800, 900)$, and $(1200, 600)$. **d.** Answers may vary; some combinations are $(500, 100)$, $(1000, 200)$, and $(200, 1100)$.
87. Domain: $\{2, -1, 3, -3\}$; range: $\{3, 5, 6, -4\}$. **88.** Domain: {Volkswagen, Honda, Toyota, Subaru, Nissan}; range: $\{52.2, 49.7, 49.0, 47.8, 45.8\}$. **89.** no **90.** yes **91.** Domain: all real numbers or $(-\infty, \infty)$; range: all real numbers or $(-\infty, \infty)$; it is a function. **92.** Domain: $\{x \mid x \le 0\}$ or $(-\infty, 0]$; range: all real numbers or $(-\infty, \infty)$; it is not a function. **93.** Domain: $\{x \mid -4 \le x \le 5\}$ or $[-4, 5]$; range: $\{1, 2, 3\}$; it is a function. **94. a.** 1 **b.** 3 **c.** 3 **d.** 2
95. a. 0 **b.** 2 **c.** -10 **96. a.** 2 **b.** undefined **c.** $\dfrac{5}{8}$

Chapter 4 Practice Test

1. $A(2, 4)\, B(-3, 2)\, C(0, -5)$ $D(4, -2)$ [4.1] **2.**

[4.1] **3.** I [4.1]

4. no [4.2] **5.** $(4, 0), (0, -2)$ [4.3] **6.** $(5, 0), (0, 2)$ [4.3]

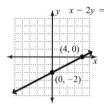
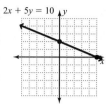

7. $(0, 0)$ [4.3] **8.** $\left(\dfrac{4}{3}, 0\right), (0, -1)$ [4.3]

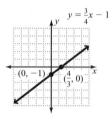
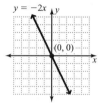

9. $m = \dfrac{3}{4};\, (0, 11)$ [4.4] **10.** $m = \dfrac{5}{3};\, \left(0, -\dfrac{8}{3}\right)$ [4.4]

11. $-\dfrac{10}{3}$ [4.4] **12.** 0 [4.4] **13.** $y = \dfrac{3}{5}x + 4$ [4.4]

14. $y = -\dfrac{2}{5}x + \dfrac{11}{5}$ [4.5] **15.** $2x - 3y = 1$ [4.5]

16. $x - 3y = 10$ [4.5] **17. a.**

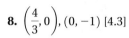

b. 20.229 [4.5] **c.** $y = 20.229x + 848.6$ [4.5]
d. 1071.1 [4.5] **18.** yes [4.6]

19.

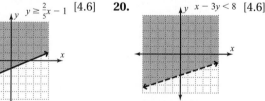

[4.6] **20.**

[4.6]

21. a. $10x + 8y \ge 360$ [4.6] **b.**

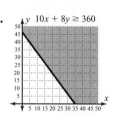

c. Answers may vary. One possible answer is $(0, 45)$. [4.6]
d. Answers may vary. One possible answer is $(10, 35)$. [4.6]
22. yes [4.7] **23.** no [4.7] **24. a.** -2 [4.7]
b. undefined [4.7] **c.** $-\dfrac{9}{7}$ [4.7] **25. a.** Domain:
$\{x \mid -3 \le x \le 3\}$ or $[-3, 3]$; range: $\{-2, 1, 2\}$ [4.7].
b. 2 [4.7]

Chapters 1–4 Cumulative Review **1.** false

2. true **3.** false **4.** false **5.** identity **6.** 100 **7.** $\dfrac{33}{200}$

8. Multiply all the terms by the power of 10 that will clear the decimal from the number with the most decimal places. **9.** -23 **10.** 15 **11.** -18 **12.** -81 **13.** -41
14. 12 **15.** $(3, 0), (0, -2)$ **16.** $(1, 1), (0, 4), (-1, 7)$

17. 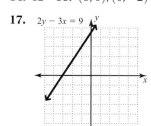 **18.** $m = -\dfrac{1}{3}$ **19.** $y = \dfrac{1}{2}x + \dfrac{1}{2}$

20. 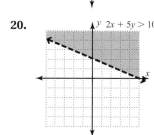 **21.** $\dfrac{85}{6}$ **22.** -12

23. -12.5 **24.** $r = \dfrac{C}{2\pi}$ **25.** 230 **26.** $\approx 1.9\%$ **27.** $36°, 54°$
28. $\approx 136\%$ **29.** 78 quarters, 112 dimes **30.** \$20,000 at 8\%, \$40,000 at 10\%

Chapter 5

5.1 Exercises 1. a. Replace each variable in each equation with its corresponding value. **b.** Verify that each equation is true. **3.** The graphs are parallel. **5.** Yes, when the graphs are identical. **7.** yes **9.** no **11.** yes **13.** no
15. yes **17.** $(4, 1)$ **19.** $(-3, 4)$ **21.** $(3, -2)$ **23.** $(-1, 3)$
25. $(3, -4)$ **27.** $(4, -3)$ **29.** $(2, -3)$ **31.** No solution.
33. All ordered pairs along $2x + 3y = 2$. **35.** $(2, 6)$
37. $(-2, 2)$ **39.** $(-2, 3)$ **41.** $(5, -2)$ **43. a.** inconsistent
b. no solution **45. a.** consistent with independent equations **b.** one solution **47. a.** consistent with dependent equations **b.** infinite number of solutions
49. a. consistent with independent equations **b.** one solution **51. a.** inconsistent **b.** no solution **53. a.** consistent with dependent equations **b.** infinite number of solutions
55. a. consistent with independent equations **b.** one solution **57. a.** 5000 units **b.** \$4000 **c.** $u > 5000$
59. a. The One-Rate plan. **b.** 375 min. **c.** More than 375 min.

Review Exercises 1. $-3y - 35$ **2.** $y = \dfrac{10 - x}{3}$

3. $x = 30 + 6y$ **4.** $\dfrac{40\pi}{3}$ in.^{3} or ≈ 41.89 in.^{3} **5.** $11, 73$
6. $2x + 17$

5.2 Exercises 1. When using the graphing method, if either coordinate in the solution is a fraction, we may have to guess the value. Substitution requires no guessing. **3.** y in the first equation because it can be isolated without using division. **5.** The resulting equation will no longer contain variables and becomes a false equation. **7.** $(4, 2)$
9. $(12, -4)$ **11.** $(-10, -24)$ **13.** All ordered pairs along $y = -2x$ (dependent). **15.** $(-3, 2)$ **17.** $(2, 9)$
19. $(-1, -3)$ **21.** $(3, 4)$ **23.** $(1, 1)$ **25.** $\left(3, -\dfrac{1}{2}\right)$
27. No solution (inconsistent). **29.** $(1, -2)$ **31.** $(4, -2)$
33. $\left(-\dfrac{2}{5}, \dfrac{2}{3}\right)$ **35.** $(2, 5)$ **37.** $(1, 2)$ **39.** Mistake: 3 was
not distributed to $3y$. Correct: $\left(\dfrac{27}{11}, -\dfrac{13}{11}\right)$ **41.** $23, 17$
43. $2, -1$ **45.** $-6, -2$ **47.** Jon is 20; Tony is 7.
49. \$58,600 **51.** Length: 11 m; width: 5 m.
53. Length: 14 ft.; width: 10 ft. **55.** Length: 18 m; width: 9 m.

Review Exercises 1. $5x - 4y$ **2.** $-3x + 21y$ **3.** $9x$
4. $13y$ **5.** 80 **6.** $-\dfrac{3}{2}$

5.3 Exercises 1. The elimination method is advantageous over graphing when the solution involves fractions. The elimination method is advantageous over substitution when no coefficients are 1. **3. a.** x because multiplying only the second equation by -2 allows the x to be eliminated; whereas to eliminate y, we need to multiply each equation by a number. **b.** Multiply $3x + 5y = 7$ by -2 and then add the resulting equation to $6x - 2y = 1$. **5.** Both variables have been eliminated, and the resulting equation is false. **7.** $(6, -8)$
9. $(2, 3)$ **11.** $(-1, -2)$ **13.** $(2, 3)$ **15.** $(3, -1)$
17. $(-4, 1)$ **19.** $(-4, 3)$ **21.** All ordered pairs along $2x + y = -2$ (dependent). **23.** No solution (inconsistent).
25. $(1, 1)$ **27.** $(4, -4)$ **29.** $\left(-\dfrac{2}{5}, \dfrac{2}{3}\right)$ **31.** $(0.2, 0.3)$
33. $(6, 4)$ **35.** $(7, -9)$ **37.** $(-4, -13)$ **39.** $\left(2, \dfrac{5}{2}\right)$
41. $(4, 4)$ **43.** Mistake: Did not multiply the right side of $x - y = 1$ by -9. Correct: $(5, 4)$ **45.** $155, 93$ **47.** $16, 9$
49. 5.5 yr., 9.5 yr. **51.** $126\dfrac{2}{3}°, 53\dfrac{1}{3}°$ **53.** $60°, 30°$ **55.** 22 in.,
8 in. **57.** Germany: 29; United States: 25 **59.** Colts: 29;
Bears: 17 **61.** Nicklaus: 6; Palmer: 4

Review Exercises 1. 13 dimes, 11 quarters **2.** 3:12 P.M.
3. \$500 in checking, \$600 in CD **4.** 160 ml **5.** $(44, 24)$
6. $(4, 8)$

5.4 Exercises 1. $x = 3y$ **3.** elimination
5. a. $x = y + 3$ **b.** CDs: $14x$; DVDs; $20y$ **c.** $14x + 20y = 212$
d. 8 CDs, 5 DVDs **7. a.** $x = y + \dfrac{1}{2}$ **b.** John: $60x$; Susan: $75y$
c. $60x = 75y$ **d.** $2\dfrac{1}{2}$ hr. **9. a.** principal: x, y; interest:

0.07x, 0.09y   **b.** $y = 2x + 250,000$ **c.** $0.07x + 0.09y = 63,500$ **d.** $164,000 in bonds, $578,000 in mutual funds
11. 13 $5 bills, 6 $10 bills   **13.** Rolling Stones: $88 million; Madonna: $72 million   **15.** 27 3-hr. courses, 15 5-hr. courses
17. *Spider-Man 3*: $336.5 million; *Shrek the Third*: $320.7 million
19. checking: $840; savings: $3360   **21.** 78 ft. by 36 ft.
23. 75 ft. by 75 ft.   **25.** 135°, 45°   **27.** 37.5°, 52.5°
29. 4:45 P.M.   **31.** 1:33 P.M. **33.** 10 mph in still water; current is 2 mph   **35.** 470 mph in still air, 30 mph winds   **37.** $12,000 at 4%, $3500 at 3%   **39.** $11,100 at 5%, $2150 at 8%   **41.** $10,253 at 2.9%, $5997 at 6%   **43.** $12,500 at 7%, $7500 at 9%
45. 120 ml   **47.** 200 ml of 15%, 100 ml of 45%   **49.** 11 lb. of Italian Roast, 9 lb. of GCB   **51.** 5 lb. of peanuts, 2 lb. of cashews

Review Exercises
1. $x \le -5$ **2.** $x < 3$ **3.** $x \le 1$

4. 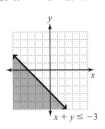   **5.** 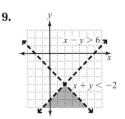   **6.** $(-2, 5)$

5.5 Exercises
1. Select a point in the solution region and verify that it makes both inequalities true.   **3.** Parallel lines with shaded regions that do not overlap.   **5.** $\begin{cases} x < 0 \\ y < 0 \end{cases}$

7. 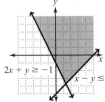   **9.**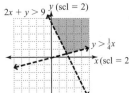

11. 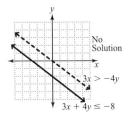   **13.**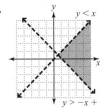

15. 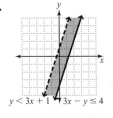   **17.**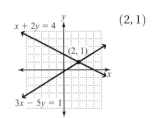

19. 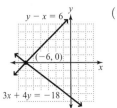   **21.**

23. 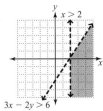   **25.**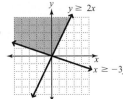

27. 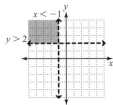   **29.**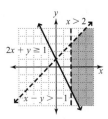

31. $x + y = 3$ should be a dashed line.   **33.** The wrong area is shaded. It should be the region containing the point $(1, 5)$.
35. a. $V + M \ge 100$ $M \ge 500$ **b.** $V \le 800$ $M \le 800$
c. 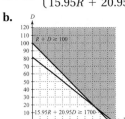   **d.** Answers will vary. Some possible (V, M) pairs are $(500, 700)$, $(600, 500)$, and $(700, 600)$.

37. a. $\begin{cases} R + D \ge 100 \\ 15.95R + 20.95D \ge 1700 \end{cases}$
b. 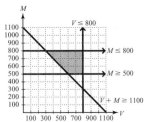   **c.** Answers will vary. Some possible (R, D) pairs are $(30, 90)$, $(40, 80)$, and $(50, 60)$.

Review Exercises
1. 9   **2.** -100 **3.** -1460
4. 14.766   **5.** $9.5x^2$ **6.** $-x - 14$

Chapter 5 Review Exercises
1. false   **2.** false
3. true   **4.** true   **5.** false   **6. 1.** Replace   **7.** consistent
8. parallel   **9.** different, identical   **10. 2.** solution   **11.** yes
12. no   **13. a.** consistent with independent equations
b. one solution   **14. a.** inconsistent   **b.** no solution
15. a. consistent with independent equations   **b.** one solution   **16. a.** inconsistent   **b.** no solution
17. $(-6, 0)$ **18.** $(2, 1)$

19. 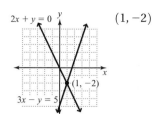 $(1, -2)$

20. 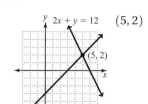 $(5, 2)$

21.  $(-1, -4)$

22. 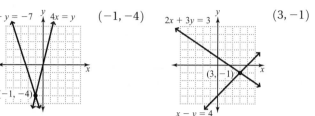 $(3, -1)$

53. 30 ml of 10%, 20 ml of 60% **54. a.** $\begin{cases} x + y \le 16 \\ 250x + 400y \ge 2750 \end{cases}$

b.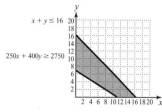

Chapter 5 Practice Test

1. yes [5.1] **2.** no [5.1]

3. $(1, -1)$ [5.2 and 5.3] **4.** $\left(-\dfrac{1}{2}, \dfrac{1}{3}\right)$ [5.2 and 5.3]

5. $(-5, 10)$ [5.2 and 5.3] **6.** No solution (inconsistent). [5.2 and 5.3] **7.** All ordered pairs along $5x + 3y = 4$ (dependent). [5.2 and 5.3] **8.** $(-2, 1)$ [5.2 and 5.3] **9.** $(-1, -2)$ [5.2 and 5.3] **10.** $(-8, 1)$ [5.2 and 5.3] **11.** $(8, 4)$ [5.2 and 5.3] **12.** $(14, 8)$ [5.2 and 5.3]

13. [5.5]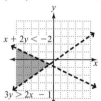

14. [5.5]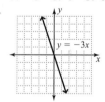

15. 4, 1 [5.4] **16.** 77 waiters, 86 accountants [5.4] **17.** 180 [5.4] **18.** 7 mph [5.4] **19.** 57 at $6, 114 at $10 [5.4] **20.** $4000 at 6%, $8000 at 8% [5.4]

23. $(3, 2)$ **24.** $(4, 9)$ **25.** $(3, -2)$ **26.** $(4, -1)$

27. No solution (inconsistent). **28.** $\left(5, -\dfrac{2}{5}\right)$ **29.** $(4, 6)$

30. $(6, 1)$ **31.** $(1, 3)$ **32.** $(3, 1)$ **33.** $(-2, -7)$

34. $(-1, -1)$ **35.** $(2, -1)$ **36.** $(12, -5)$ **37.** All ordered pairs along $x + 3y = 16$ (dependent). **38.** $(5, -3)$

39. 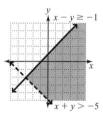 **40.**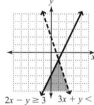

41. 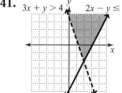 **42.**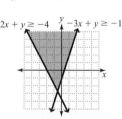

43. 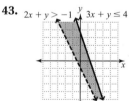 **44.**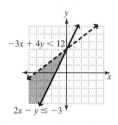

45. 10, 6 **46.** 15°, 75° **47.** Width: 80 ft.; length: 185 ft. **48.** 47 adult, 92 child **49.** 45 black-and-white, 5 color **50.** $3500 at 7%, $1500 at 9% **51.** 5:45 P.M. **52.** 120 mph

Chapters 1–5 Cumulative Review Exercises

1. false **2.** false **3.** true **4.** true **5.** the final amount **6.** the other variable **7.** -3 **8.** 0 **9.** undefined **10.** $(4, 0), (0, 2)$ **11.**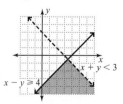

12. Answers may vary. **13.** yes **14.** $\dfrac{1}{2}$ **15.** 625

16. Function, Domain: {Orlando, Tampa–St. Petersburg, West Palm Beach, Austin, Las Vegas, Phoenix, Memphis, Jacksonville, Fort Lauderdale, Kansas City} Range: {18.8, 17.6, 16.6, 15.8, 15.4, 15.2, 14.7, 14.6, 14.5, 14.2} **17.** $\dfrac{7}{2}$ **18.** 2 **19.** $x > -12$

20. $s > 2$ **21.** $r^2 = \dfrac{A}{\pi}$ **22.** $(3, -2)$ **23.** $(-2, 3)$

24. $(2, -1)$ **25.**   **26.** 280

27. 63.5% **28.** $1.14 **29.** 5 nickels, 15 dimes **30.** 247,000 by parent-teacher organizations; 361,000 by public schools

Chapter 6

6.1 Exercises **1.** If a is a real number, where $a \neq 0$ and n is a natural number, then $a^{-n} = \dfrac{1}{a^n}$. **3.** The result is positive. We simplify an expression with a negative exponent by inverting the expression and changing the sign of the exponent: $\left(2^{-3} = \dfrac{1}{2^3} = \dfrac{1}{8}\right)$. **5.** To write a number greater than 1 in scientific notation, **a.** Move the decimal point so that the number is greater than or equal to 1 but less than 10. **b.** Write the decimal number multiplied by 10^n, where n is the number of places between the new decimal position and the original decimal position. **c.** Delete zeroes to the right of the last nonzero digit. **7.** 1 **9.** -8 **11.** -125 **13.** -16 **15.** $\dfrac{81}{256}$

17. -0.008 **19.** 5.29 **21.** $\dfrac{1}{a^5}$ **23.** $\dfrac{1}{8}$ **25.** $-\dfrac{1}{64}$ **27.** $-\dfrac{1}{16}$

29. -125 **31.** $\dfrac{5}{m^2}$ **33.** $-\dfrac{6}{x^7}$ **35.** $\dfrac{n^5}{m^5}$ **37.** $\dfrac{49}{25}$ **39.** b^6

41. 64 **43.** $-\dfrac{1}{7,962,624}$ **45.** $-\dfrac{512}{125}$ **47.** $-160,000$

49. ≈ 0.00039 **51.** Mistake: The expression was interpreted to mean $(-3)^4$, but it means the additive inverse of 3^4. Correct: -81 **53.** Mistake: The negative sign in the exponent was passed to the product instead of the original expression being inverted. Correct: $\dfrac{1}{25}$ **55.** $16,500,000,000$

57. $299,800,000$ **59.** $6,378,000$ **61.** $92,920,000$
63. 0.0000025 **65.** $0.00000000000000000000000000167$
67. 0.00000000295 **69.** $0.0000000000000000000000006645$
71. 2.5×10^{13} **73.** 5.3×10^9 **75.** 2.342×10^7
77. 1.3×10^9 **79.** 5.86×10^{-7} **81.** 9.1094×10^{-31}
83. 6.25×10^{-5} **85.** 5×10^{-8} **87.** $8.95 \times 10^5, 7.2 \times 10^6$, $7.5 \times 10^6, 9 \times 10^7, 1.3 \times 10^8$ **89.** The wavelength of ultraviolet light is about 38.8 times the wavelength of X-rays.

Review Exercises **1.** $-4y + 12$ **2.** -11 **3.** $266\dfrac{2}{3}$
4. $\dfrac{1}{5}$ **5.** slope $\dfrac{2}{3}$; y-intercept: $(0, -2)$ **6.** -65

6.2 Exercises **1.** Find the sum of the exponents on all variables in the monomial. **3.** Find the greatest degree of any of the terms in the polynomial. **5.** Add the coefficients and leave the variables and exponents the same. **7.** monomial
9. monomial **11.** not **13.** not **15.** monomial
17. monomial **19.** c: -5; d: 3 **21.** c: -1; d: 5
23. c: -9; d: 0 **25.** c: 4.2; d: 4 **27.** c: 16; d: 3 **29.** c: 1; d: 1
31. trinomial, 1 **33.** monomial, 3 **35.** No special polynomial name, 3. **37.** binomial, 3 **39.** monomial, 0
41. Not a polynomial. **43.** 4 **45.** 5 **47.** 7 **49.** 4 **51.** 8
53. 60 **55.** -8 **57.** -4.8 **59.** 25 **61. a.** 46 ft. **b.** 26.96 ft.
63. a. 30 units **b.** 43 units **65. a.** ≈ 1273.4 ft.^{3}
b. ≈ 537.2 ft.^{3} **67. a.** 26 million **b.** 45.95 million
c. 40.45 million **d.** 1.3 million; answers will vary, but the prediction is most likely not reasonable because of the increase in world population and unforeseen social factors that might affect travel; it is not likely that the number will decrease this much. **69.** $5x^7 - 3x^5 + 7x^4 - 8x + 14$

71. $-8r^6 + 18r^5 + 7r^3 - 3r^2 + 4r$

73. $-12w^4 - w^3 + 11w^2 + 5w + 20$ **75.** $5x - 3$
77. $-4a + 5$ **79.** $-2x^2 + 6x$ **81.** $2x^2 + \dfrac{5}{3}x$
83. $\dfrac{7}{6}a^2 - \dfrac{19}{15}a$ **85.** $-5x^2 + 14$ **87.** $5y^2 - 9y + 22$
89. $-2k^4 + 6k^3 - 3k^2 + 6k$ **91.** $-10a^4 + 7a^2 + a + 16$
93. $-21v^5 + 8v^3 + 35v^2 - 20$ **95.** $4a - 2b$
97. $-4y^2 + 3y$ **99.** $-9x^2 + 9y$ **101.** $\dfrac{19}{12}a^2 - \dfrac{17}{30}a$
103. $y^6 + 5yz^4 - 6yz - 3z^5 - 7z^3 - 10$
105. $-4w^2z - wz^2 + 12w^2 - 1$

Review Exercises **1.** $-\dfrac{7}{12}$ **2.** -4.3 **3.** -1
4. 2×10^{11} **5.** 12 **6.** 600 L

6.3 Exercises **1.** To add polynomials, combine like terms. **3.** Change the signs of each term in the polynomial.
5. $8x + 1$ **7.** $10y + 12$ **9.** $7x$ **11.** $3x^2 - 2x + 12$
13. $6z^2 - 11z - 2$ **15.** $5r^2 - 7r - 1$ **17.** $9y^2 + 3$
19. $3x^2 + \dfrac{1}{12}x + 1$ **21.** $2r^2 + 1.4r - 7.8$
23. $5a^3 + 3a^2 + 3a - 3$ **25.** $3p^3 - p^2 + 7p + 9$
27. $-8w^4 + 3w^3 - 7w^2 + 13w - 9$
29. $-3a^3b^2 - 2ab^2 - 8a^2b + 4ab + 3b^2 - 8$
31. $-7u^4 + u^2v^3 + 7u^3v + 7u - 13v^2 - 3$
33. $-12mnp - 16m^2n^2 - 12mn^2p + 14m^2n - 31n + 9$
35. $2a^4 - \dfrac{5}{12}a^3b + \dfrac{14}{15}ab^3 - b^4$
37. $1.8a^2bc - 1.8ab^2c^2 - 2.4a^2b^2c^2$
39. $5x - 2$ **41.** $14a - 12$ **43.** $4x + 2$ **45.** $13a^2 - 1$
47. $2x^2 - 9x + 2$ **49.** $2z^2 - 7z + 9$ **51.** $12a^2 + 10a - 15$
53. $-u^3 - 2u^2 - 6u + 8$ **55.** $-11w^5 + 3w^4 + 5w^2 + 2$
57. $-11p^4 + 3p^3 - 13p^2 + 5p - 6$
59. $11v^4 + 3v^3 + 23v^2 - 5v - 9$ **61.** $2xy$
63. $-6xy + 6xz + 2yz$ **65.** $19p^3q^2 + 2pq^2 - pq + 16q^2 + 2$
67. $1.9a^2b + 1.1ab^2 - 6.5a^2b^2$
69. $32.1x + 35.45y + 6.01z$
71. $4.22a + 1.63b + 12.13c$
73. Mistake: Only the sign of the first term in the subtrahend was changed instead of all three terms. Correct: $-4x^2 - 2x + 5$

Review Exercises **1.** -17.25 **2.** 15 **3.** 0.0001
4. 10^7 **5.** $58,900,000$ **6.** 300 m^2

6.4 Exercises **1.** Product rule for exponents.
3. To multiply monomials, **a.** Multiply coefficients. **b.** Add the exponents of the like bases. **c.** Write any unlike variable bases unchanged in the product. **5.** Use $(a^m)^n = a^{mn}$. So $(x^3)^8 = x^{3 \cdot 8} = x^{24}$. **7.** a^7 **9.** 2^{13} **11.** a^5b
13. $6x^3$ **15.** $-8m^3n^2$ **17.** $-\dfrac{5}{28}s^3t^8$ **19.** $2.76a^3b^5c^6$
21. $-6r^5$ **23.** $-12x^4z^9$ **25.** $90x^6y^5z^3$ **27.** $5q^5r^{10}s^{21}$
29. $-\dfrac{1}{4}a^6b^3c^2$ **31.** $w^3x^3y^5$ **33.** Mistake: The exponents of x were multiplied instead of added. Correct: $35x^7y$
35. Mistake: The exponents of x were not added. Correct:
$54x^4y^6z$ **37.** $5w^2$ **39.** $4.5w^3$ **41. a.** height $= \dfrac{1}{2}b$,
top side $= \dfrac{1}{4}b$ **b.** $A = \dfrac{5}{16}b^2$ **43.** 1.2×10^9 **45.** 2.4×10^{14}

47. 1.827×10^5 **49.** 1.1421×10^{-4} **51.** 5.612×10^{-10}
53. $4.2 \times 10^4\,\text{mi}^2$ **55.** $\approx 5.8404 \times 10^8$ mi.
57. $\approx 1.412 \times 10^{27}\,\text{m}^3$ **59.** 2.9817×10^{-19} joule
61. 2.16×10^7 joules **63.** x^6 **65.** 3^8 **67.** h^8 **69.** $-y^6$
71. $x^3 y^3$ **73.** $27x^6$ **75.** $-8x^9 y^6$ **77.** $\frac{1}{64} m^6 n^3$
79. $64p^{24} q^{24} r^{12}$ **81.** $-0.008 x^6 y^{15} z^3$ **83.** $18x^7$ **85.** $r^5 s^4$
87. $100 a^6 b^4$ **89.** $\frac{1}{1024} a^{10} b^{15} c^5$ **91.** $360 a^2 b^3$ **93.** $216 a^{18} b^8$
95. $324 u^6 v^3$ **97.** $3.24 u^6 v^3$ **99.** Mistake: The exponents
were added instead of multiplied. Correct: $25x^6$
101. Mistake: The coefficient was multiplied by 4 instead of
being raised to the 4th power. Also, the exponents were not
multiplied properly. Correct: $16 x^{16} y^4$

Review Exercises **1.** $6x + 8$ **2.** $-1.8 t^5 u^2$ **3.** $-22x$
4. $-4x - 4z$ **5.** -61 **6.** 6

6.5 Exercises **1.** Distributive property. **3.** They differ
only in the sign separating the terms. **5.** The middle term
is missing. **7.** $5x + 15$ **9.** $-12x + 24$ **11.** $28n^2 + 8n$
13. $9a^2 - 9ab$ **15.** $\frac{1}{4} m^2 - \frac{5}{8} mn$ **17.** $-0.3 p^5 + 0.6 p^2 q^2$
19. $20x^3 + 10x^2 - 25x$ **21.** $-21x^5 + 15x^3 - 3x^2$
23. $-r^4 s^4 - 3r^3 s^3 + r^2 s^2$
25. $-4a^5 b^2 + 12a^2 b^3 - 6a^3 b^3 + 2a^4 b^4$
27. $12 a^3 b^3 c^3 - 6a^2 b^2 c^2 + 15abc$
29. $4b^6 - b^4 + \frac{1}{5} b^3 - \frac{1}{2} b^2 + 16b$
31. $-r^4 t^3 + 3.2 r^3 t^4 - 1.64 r^2 t^5 + 0.8 rt^2$
33. $25 x^6 y - 15 x^5 y^2 + 50 x^3 y^9$ **35. a.** $x + 5$ **b.** $2x$
c. $2x(x + 5)$ **d.** $2x^2 + 10x$ **e.** They describe the same area.
37. a. $x + 3$ **b.** $x + 2$ **c.** $(x + 2)(x + 3)$ **d.** $x^2 + 5x + 6$
e. They describe the same area. **39.** $x^2 + 7x + 12$
41. $x^2 - 5x - 14$ **43.** $y^2 - 9y + 18$ **45.** $6y^2 + 19y + 10$
47. $15 m^2 + 11m - 12$ **49.** $12 t^2 - 26t + 10$
51. $15 q^2 + 11qt - 12 t^2$ **53.** $14 y^2 + 34xy + 12 x^2$
55. $a^4 - a^2 b - 2b^2$ **57.** $x^3 - 3x + 2$
59. $3a^3 - 4a^2 - a + 2$ **61.** $6c^3 + c^2 - 11c - 6$
63. $4f^3 - 27fg^2 - 27g^3$ **65.** $27m^3 - 8n^3$ **67.** $4a^3 + 125b^3$
69. $x^4 + 3x^3 y + 2x^2 y^2 + xy^3 - y^4$
71. $x^4 + 14x^3 + 69x^2 + 140x + 100$ **73.** $x - 3$
75. $4x + 2y$ **77.** $4d - 3c$ **79.** $-3j + k$ **81.** $x^2 - 25$
83. $4m^2 - 25$ **85.** $x^2 - y^2$ **87.** $64r^2 - 100s^2$ **89.** $4x^2 - 9$
91. $x^2 + 6x + 9$ **93.** $16t^2 - 8t + 1$ **95.** $m^2 + 2mn + n^2$
97. $4u^2 + 12uv + 9v^2$ **99.** $81w^2 - 72wz + 16z^2$
101. $81 - 90y + 25y^2$ **103.** $10xyz + 2y^2 z$ **105.** $h^2 + 4h$
107. $w^2 + 3w$ **109.** $4x^3 + 24x^2 + 20x$ **111.** $4w^3 + 6w^2$
113. $V = 3.14 r^3 + 6.28 r^2$, or $\pi r^3 + 2\pi r^2$

Review Exercises **1.** 0.45 **2.** -5 **3.** $6{,}304{,}000$
4. x^7 **5.** $r = \dfrac{5 - d}{m}$ **6.** $9, 36$

6.6 Exercises **1.** When dividing exponential forms that
have the same base, subtract the divisor's exponent from the
dividend's exponent and keep the same base. **3.** The decimal
factors and powers of 10 can be separated into a product of two
fractions. This allows us to calculate the decimal division and
divide the powers of 10 separately. **5.** To divide a polynomial

by a monomial, divide each term in the polynomial by
the monomial. **7.** $\dfrac{1}{a^3}$ **9.** $\dfrac{1}{2^5}$ **11.** y^5 **13.** 3 **15.** $\dfrac{1}{4^6}$ **17.** $\dfrac{1}{x^2}$
19. a^2 **21.** $\dfrac{1}{w^3}$ **23.** $\dfrac{1}{a^7}$ **25.** r^{10} **27.** y^8 **29.** $\dfrac{1}{p^3}$ **31.** 1
33. 2.6×10^{-1} **35.** 6×10^{-8} **37.** 7.1×10^{10} **39.** 2.7×10^4
41. 500 sec. or $8\frac{1}{3}$ min. **43.** $\$36{,}862.75$ **45.** $\approx 9.3 \times 10^{-4}$ kg
47. $5x^3$ **49.** $\dfrac{-4}{x^3}$ **51.** $-3m^2 n^4$ **53.** $-\dfrac{8}{5} q^3$ **55.** $\dfrac{4y^4}{3x^5}$ **57.** $\dfrac{4c}{3a^2}$
59. $a + 2b$ **61.** $4x^2 - 2x$ **63.** $3x + \dfrac{2}{x}$ **65.** $xy - 3y^2$
67. $-2c + 8ab$ **69.** $3x^2 + 2x - 1$ **71.** $2x - 4 + \dfrac{3}{x}$
73. $12 uv^3 + \dfrac{4v^4}{u} - 5v$ **75.** $y^2 + y^4 - 3y^5 + \dfrac{1}{y^2}$ **77.** $\dfrac{5mn}{2}$
79. a. $3x - 5 + \dfrac{4}{x}$ **b.** Area $= 18$, length $= 6$, width $= 3$.
81. $x + 4$ **83.** $m - 5$ **85.** $x - 12 + \dfrac{4}{x - 5}$
87. $3x^2 + 15x + 2$ **89.** $2x^2 + 3x + 6 + \dfrac{17}{x - 2}$
91. $p^2 - 5p + 25$ **93.** $3x + 2$ **95.** $2y - 2 + \dfrac{23}{7y + 3}$
97. $4k^2 - 1 + \dfrac{8}{k + 2}$ **99.** $7u^2 - 11u + 12 - \dfrac{30}{3u + 2}$
101. $b^2 - 3b + 5 + \dfrac{3}{b - 3}$ **103.** $y^2 - 2y + 3$
105. a. $x + 4$ **b.** Length $= 7$ ft., height $= 8$ ft.,
volume $= 336$ ft.^{3} **107.** $\dfrac{36 m^4}{n^4}$ **109.** $\dfrac{x^9}{64 y^6}$ **111.** $256 r^9 s^7$
113. m^{16} **115.** $\dfrac{1}{x^{14}}$ **117.** $\dfrac{y^{27}}{x^9}$ **119.** x^{14} **121.** $\dfrac{27 y^6}{x^3 z^6}$
123. $\dfrac{1}{x^{10}}$ **125.** $\dfrac{4 x^4 z^3}{y^5}$ **127.** $\dfrac{2}{3ab^2}$ **129.** $\dfrac{1}{2x^{11}}$

Review Exercises **1.** $2^4 \cdot 3^3 \cdot 7$ **2.** $-18 x^4 y^3$
3. $3xy - 4y$ **4.** $x + \dfrac{1}{2} y$ **5.** 340 ft.^{2} **6.** -24

Chapter 6 Review Exercises **1.** false **2.** false
3. true **4.** false **5.** false **6.** false **7.** combine
8. a. coefficients **b.** exponents **9.** $(a^m)^n = a^{mn}$ **10.** $2ab$
11. $\dfrac{8}{125}$ **12.** -16 **13.** $\dfrac{1}{25}$ **14.** $\dfrac{1}{13}$ **15.** $4{,}500{,}000{,}000$
16. $13{,}800{,}000$ **17.** $0.00000000000000000000001663$
18. $0.000000000000000000000000002006$
19. 1.661×10^{-24} **20.** 9.63×10^{-23} **21.** 3×10^8
22. 6.37×10^6 **23.** yes **24.** yes **25.** no **26.** no
27. c: 6; d: 4 **28.** c: 27; d: 0 **29.** c: -2.6; d: 4 **30.** c: -1; d: 1
31. binomial **32.** monomial **33.** Not a polynomial.
34. No special polynomial name.
35. 9; $-2x^9 + 21x^5 - 19x^3 - 3x + 15$
36. 4; $-j^4 + 22j^2 + 5j - 19$
37. 5; $4u^5 + 13u^4 - 18u^3 - u - 21$
38. 8; $6v^8 + 21v^4 + 16v^3 - v^2 - 2v - 19$
39. $4y^5 + 4y^4 + 8$ **40.** $18 l^5 + 5l^4 - 2l^3 + 6l^2 + 7l + 20$
41. $-2m^3 + 7m^2 + m + 3$ **42.** $-3y^2 + 7y - 15$
43. $a^2 bc - ab^2 c - 6$ **44.** $-11 x^2 yz - 2xyz^2$

45. $-4c^3 - cd^2 + 11cd + 2d + 8$
46. $-8a^5 + 3abc^3 - 8abc + 8$
47. $-3j^4 + 3jk^3 + k^2 + 10jk + 12$
48. $3mn^2 - 8n^2 - 9mn + 7$ **49.** $-4ab - 18a - 19c + 8$
50. $11y - 10$ **51.** $10x^2 + 3x - 4$ **52.** $3m - 8$
53. $-2n^3 + 3n^2 + 4n + 3$ **54.** 9 **55.** $-x^3 + 4x^2 + 17$
56. $-p^3 + 10p^2 - 10p + 9$ **57.** $-16x^3 - 4x - 2$
58. $8x^2 - xy + 6y^2$ **59.** $-3m^2 + 8mn - 2$ **60.** m^5
61. $10a^9$ **62.** $-6x^6y^6$ **63.** $25x^4y^2$ **64.** $72u^{10}$ **65.** $32x^{10}$
66. x^8 **67.** u^7 **68.** $\dfrac{1}{s^6}$ **69.** x^6 **70.** $\dfrac{b^6}{8a^9}$ **71.** $\dfrac{4j^8}{9hk^2}$
72. $\dfrac{-5}{x^2}$ **73.** 7 **74.** 1 **75.** $2x - 42$ **76.** $-16x + 8$
77. $4a^2 - 12a$ **78.** $-12b^3 + 24b^2 + 6b$
79. $-8m^7 - 12m^5 + 4m^4 + 20m^3$
80. $-12a^2bc^2 + 16a^2b^4c^2 - 8a^2b^2c^4 - 32abc^2$
81. $x^2 + 4x - 5$ **82.** $y^2 - 11y + 24$ **83.** $12m^2 + 28m - 5$
84. $15a^2 - 4ab - 4b^2$ **85.** $4x^2 - 1$ **86.** $x^3 - x^2 - x - 15$
87. $3y^3 - 7y^2 + 14y - 4$
88. $a^4 - a^3b - 2a^2b^2 - 3ab^3 - b^4$
89. $16 - x^2$ **90.** $x^2 - 12x + 36$ **91.** $9r^2 - 30r + 25$
92. $36s^2 - 4r^2$ **93. a.** $20x^2 - 28x$ **b.** 96 ft.^2
94. $2.25 \times 10^8 \text{ m}^2$ **95.** $2.94 \times 10^{-4} \text{ V}$ **96.** $x - 5$
97. $-y^2 + 2y - 3$ **98.** $1 + 10st^3 - 2t^4$ **99.** $a^2 - \dfrac{1}{c} + 2ac$
100. $2 + b$ **101.** $\dfrac{2z}{5y^2} - \dfrac{xz}{y} + \dfrac{2}{y}$ **102. a.** $9x - 10$
b. 24 in., 26 in., and 624 in.^{2} **103.** ≈ 4.2 yr. **104.** $z + 5$
105. $3m - 2 + \dfrac{1}{m - 5}$ **106.** $y - 4 - \dfrac{4}{2y + 1}$
107. $2m - 2 + \dfrac{1}{3m - 2}$ **108.** $x^2 - x + 1 - \dfrac{2}{x + 1}$
109. $4x^2 + 6x + 9$ **110.** $2s^2 + 2s - \dfrac{3}{s - 1}$
111. $5y^3 - 3y + 2$ **112.** $4x - 5$

Chapter 6 Practice Test
1. $\dfrac{1}{8}$ [6.1] **2.** $\dfrac{9}{4}$ [6.1]
3. 0.006201 [6.1] **4.** 2.75×10^8 [6.1] **5.** 3 [6.2]
6. 4 [6.2] **7.** 56 [6.2] **8.** $-10x^4 - 6x^3 + 14x^2 + 12$ [6.2]
9. $8x^2 + x - 4$ [6.3] **10.** $5x^4 - 3x^2 - x + 8$ [6.3]
11. $12x^7$ [6.4] **12.** $-2a^6b^4c^7$ [6.4] **13.** $16x^2y^6$ [6.4]
14. $3x^3 - 12x^2 + 15x$ [6.5] **15.** $-24t^5u + 48t^3u^3$ [6.5]
16. $n^2 + 3n - 4$ [6.5] **17.** $4x^2 - 12x + 9$ [6.5]
18. $x^3 - 2x^2 - 5x + 6$ [6.5] **19.** $6n^2 - 7n - 20$ [6.5]
20. x^5 [6.6] **21.** $\dfrac{1}{x^5}$ [6.6] **22.** $\dfrac{x^9y^4}{9}$ [6.6]
23. $4x^3 + 3x$ [6.6] **24.** $x - 4$ [6.6]
25. $5x - 4 + \dfrac{6}{3x - 2}$ [6.6]

Chapters 1–6 Cumulative Review Exercises
1. true **2.** false **3.** true **4.** false **5.** coefficient, coefficient
6. $(0, 0)$ or the origin **7.** 23 **8.** -1 **9.** 60 **10.** $5y^2 - 4y - 1$
11. $6x - 6$ **12.** $2x^2 - 11x - 21$ **13.** $4x^2 - 12x + 9$
14. $m - 2 + \dfrac{1}{3m}$ **15.** $x + 3 + \dfrac{1}{x + 2}$ **16.** $\dfrac{x^6}{27y^9}$ **17.** $\dfrac{4x^6}{y^2}$

18. $-108x^{16}$ **19.**
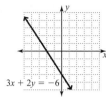
$3x + 2y = -6$

20.
$y > \frac{2}{3}x - 2$

21. -12 **22.** $x < -10$ **23.** $l = \dfrac{P - 2w}{2}$ **24.** $\left(\dfrac{9}{7}, -\dfrac{4}{7}\right)$
25. $(0, -3)$ **26.** 20 lb. **27.** \$33,750 at 13%, \$11,250 at 9%
28. 480 **29.** 53.7% **30.** 27 three-hour classes, 15 five-hour classes.

Chapter 7

7.1 Exercises 1. To list all of the possible factors of a number, we can divide by 1, 2, 3, and so on, writing each divisor and quotient pair as a product until we have all possible combinations. **3.** The factorization for the GCF includes only those prime factors common to all of the factorizations, each raised to its smallest exponent. **5.** Four-term polynomials.
7. 1, 3, 9 **9.** 1, 3, 11, 33 **11.** 1, 2, 3, 6, 9, 18 **13.** 1, 2, 4, 8, 16
15. 1, 2, 4, 11, 22, 44 **17.** 1, 2, 3, 4, 5, 6, 10, 12, 15, 20, 30, 60
19. 1, 2, 4, 7, 8, 14, 28, 56 **21.** 1, 2, 3, 5, 6, 9, 10, 15, 18, 30, 45, 90
23. 3 **25.** 8 **27.** 6 **29.** $2xy$ **31.** $5h^2$ **33.** $3a^4b$
35. $5(c - 4)$ **37.** $4(2x - 3)$ **39.** $x(x - 1)$
41. $6z^4(3z^2 - 2)$ **43.** $3p^3(6 - 5p^2)$ **45.** $3ab(2a - b)$
47. $7uv(2v - 1)$ **49.** $25xy(1 - 2z + 4)$
51. $xy(x + y + x^2y^2)$ **53.** $4ab^2c(7b - 9a)$
55. $4pq(5p + 6 - 4q)$ **57.** $3mnp(n^4p + 6n^2 - 2)$
59. $21a^2b^2(5a - 3b + 4a^4b^2)$
61. $3x(6x^3 - 3x^2 + 10x - 4)$ **63.** $-3(x - 2y)$
65. $-5a(4a + 3)$ **67.** $-4a^3b(3a - 5b^2)$
69. $-4(x^2 + 2x - 4)$ **71.** $-6x^2y^2z(4x + 5yz^3 - 2x^2y^3z)$
73. $(a - 3)(y + 2)$ **75.** $(2m + 3n)(4a + b)$
77. $(4m - 5)(5x - 2)$ **79.** $2(6p + 5q)(r - 2s)$
81. $(x + 2)(b + c)$ **83.** $(m - n)(a - b)$
85. $(x + 2)(x^2 - 3)$ **87.** $(1 - m)(1 + m^2)$
89. $(3x + 5)(y + 2)$ **91.** $(4b - 1)(b + 1)$
93. $(3 + b)(2 - a)$ **95.** $(x + 2y)(3a + 4b)$
97. $(y - s)(x^2 - r)$ **99.** $(w + 3z)(w + 5)$
101. $(s + y)(3t - 2)$ **103.** $(x^2 + y^2)(a - 5)$
105. $2(a + b)(a + 3)$ **107.** $2(2b + 3)(3a + 4)$
109. $3(4x - y)(2z - w)$ **111.** $3a(y - 4)(a + 3)$
113. $2x(x + 3y)(x + 5)$ **115.** $6x(x + 6)$ **117.** $3x(7x + 9)$
119. $12\pi r^2(3r + 2)$

Review Exercises 1. 3,740,000,000 **2.** 4.56×10^7
3. $x^2 + 7x + 10$ **4.** $x^2 - x - 12$ **5.** $x^2 - 8x + 15$
6. $x^2 - 7x$

7.2 Exercises 1. Both are positive. **3.** The number with the greater absolute value is positive, and the other number is negative. **5.** 5 **7.** 2 **9.** 6 **11.** 2 **13.** $(r + 3)(r + 1)$
15. $(x - 7)(x - 1)$ **17.** $(z - 3)(z + 1)$ **19.** $(y + 2)(y + 3)$
21. $(u - 2)(u - 4)$ **23.** $(u + 3)(u - 2)$
25. $(a + 3)(a + 4)$ **27.** $(y - 3)^2$ **29.** $(w - 4)(w + 3)$
31. $(x - 6)(x + 5)$ **33.** $(n - 6)(n - 5)$ **35.** $(r - 3)(r - 6)$
37. $(x - 8)(x + 3)$ **39.** $(p - 9)(p + 4)$ **41.** prime

43. prime **45.** $(p - 3q)(p - 7q)$ **47.** $(a - 9b)(a + 3b)$
49. $(x - 12y)(x - 2y)$ **51.** $(r - 6s)(r + 5s)$
53. $4(x - 3)(x - 7)$ **55.** $2m(m - 1)(m - 6)$
57. $3b(a - 8)(a + 3)$ **59.** $4(x - 3)^2$ **61.** $n^2(n + 2)(n + 3)$
63. $7u^2(u + 5)(u + 1)$ **65.** $6b^2c(a - 2)(a - 4)$
67. $3(x - 2y)(x - 5y)$ **69.** $2b(a - 6b)(a + 3b)$
71. $4x^2y(x^2 - 3xy - 15y^2)$ **73.** $(x - 6)(x + 1)$
75. $(x - 4)(x + 1)$ **77.** 4, 20 **79.** 7, 8, 13 **81.** 6, 10, 12
83. 9, 16, 21, 24, 25 **85.** Length: $h + 4$; width: $h + 2$.
87. Length: $w - 6$; width: $w - 10$.

Review Exercises **1.** 1 **2.** $10x^2 + 17x + 3$
3. $15y^2 - 2y - 24$ **4.** $72n^3 - 60n^2 + 8n$
5. $5x^2y(2y^2 - x - 4y)$ **6.** $4ab^2(2ab^2 - 4 - 3ab)$
7. $(x + 2)(x - 3)$ **8.** $(z + 5)(z - 6)$ **9.** $(3c - d)(5a - b)$
10. $(3c + 2d)(2a - 3b)$

7.3 Exercises **1.** monomial GCF in all of the terms
3. Write a pair of last terms whose product is c. **5.** 2, 4
7. $t, 4t$ **9.** $7, 3x$ **11.** $(2j + 1)(j + 2)$ **13.** $(2y - 5)(y + 1)$
15. $(3m - 4)(m - 2)$ **17.** $(6a + 7)(a + 1)$ **19.** prime
21. $(4a - 3)(a - 4)$ **23.** $(2x + 3)(3x + 5)$
25. $(2d - 3)(8d + 5)$ **27.** $(3p + q)(p + 4q)$
29. $(k + h)(5k - 12h)$ **31.** $(2a - 5b)(6a - 5b)$
33. $(8m - 3n)(m - 3n)$ **35.** $y(8x^2 - 4x - 1)$
37. $3(2y + 1)(2y + 3)$ **39.** $2a(3b - 7)(b - 1)$
41. $2w(3w + 2v)(w + 2v)$ **43.** $(a + 1)(3a + 1)$
45. $(2t - 1)(t - 1)$ **47.** $(3x - 7)(x + 1)$
49. $(y - 2)(2y + 3)$ **51.** $(2a - 1)(5a - 7)$
53. $(4r + 3)(2r - 3)$ **55.** $(5x - 2)(4x - 3)$
57. $(2k + j)(3k + 2j)$ **59.** $(x - 5y)(5x - y)$
61. $(5s - 2t)(2s + t)$ **63.** $(3u - 2v)(2u + 3v)$ **65.** prime
67. $2(k - 2)(2k - 3)$ **69.** $2m(2m + 3)(3m - 2)$
71. $8x(x - 3y)(3x + y)$ **73.** $4x(2x - 3y)(3x + 2y)$
75. Mistake: Using FOIL to check, we see that the first and last terms check but the inner and outer terms combine to give $31x$ instead of $13x$. Correct: $(2x + 1)(3x + 5)$
77. Mistake: Did not factor completely.
Correct: $4(n + 2)(n + 1)$ **79.** $(5x - 2)(3x - 1)$
81. $2w + 1$ and $3w - 2$ **83.** 7, 8, or 13 **85.** 4, 12, or 44
87. 1, 11, 19, or 41

Review Exercises **1.** $9x^2 - 25$ **2.** $x^3 - 8$
3. $4x^2 - 20x + 25$ **4.** $4cd^2(4c^2 - 6cd^2 + 9)$
5. $(b + 5)(a + 3)$ **6.** $(m - 4)(m + 6)$
7. $3x(x - 2)(x + 5)$

7.4 Exercises **1.** The first and last terms are perfect squares, and twice the product of their roots equals the middle term. **3.** conjugates **5.** In the trinomial factor (second term). **7.** $(x + 7)^2$ **9.** $(b - 4)^2$ **11.** Not a perfect square.
13. $(5u - 3)^2$ **15.** $(10w + 1)^2$ **17.** $(y + z)^2$
19. $(2p - 7q)^2$ **21.** $(4g + 3h)^2$ **23.** $4(2t + 5)^2$
25. $(x + 2)(x - 2)$ **27.** $(4 + y)(4 - y)$
29. $(p + q)(p - q)$ **31.** $(5u + 4)(5u - 4)$ **33.** prime
35. $(8m + 5n)(8m - 5n)$ **37.** $2(5x + 4y)(5x - 4y)$
39. $4(x^2 + 25y^2)$ **41.** $(x^2 + y^2)(x + y)(x - y)$
43. $(x^2 + 4)(x + 2)(x - 2)$ **45.** $(n - 3)(n^2 + 3n + 9)$
47. $(x + 3)(x^2 - 3x + 9)$ **49.** $(x - 1)(x^2 + x + 1)$
51. $(m + n)(m^2 - mn + n^2)$ **53.** $(3k - 2)(9k^2 + 6k + 4)$
55. $(3k + 2)(9k^2 - 6k + 4)$ **57.** $(c - 4d)(c^2 + 4cd + 16d^2)$
59. $(5x + 4y)(25x^2 - 20xy + 16y^2)$

61. $(3x - 4y)(9x^2 + 12xy + 16y^2)$
63. $(2p + qz)(4p^2 - 2pqz + q^2z^2)$
65. $2(x + 5)(x - 5)$ **67.** $\left(4x + \dfrac{5}{7}\right)\left(4x - \dfrac{5}{7}\right)$
69. $2u(u + 1)(u - 1)$ **71.** $y^3(y + 4b)(y - 4b)$
73. $2x(25x^2 + 1)$ **75.** $3(y - 2z)(y^2 + 2yz + 4z^2)$
77. $\left(c - \dfrac{2}{3}\right)\left(c^2 + \dfrac{2}{3}c + \dfrac{4}{9}\right)$
79. $2c(2c + d)(4c^2 - 2cd + d^2)$
81. $(2a - b + c)(2a - b - c)$
83. $(4 - 3x + 3y)(4 + 3x - 3y)$
85. $(x + 3y + 3z)(x^2 - 3xy - 3xz + 9y^2 + 18yz + 9z^2)$
87. $(x + y - 4d)(x^2 + 2xy + y^2 + 4dx + 4dy + 16d^2)$
89. 24 **91.** 36 **93.** 25 **95.** 16 **97.** $10(x + 2)(x - 2)$
99. $(2x - 3)(4x^2 + 6x + 9)$

Review Exercises **1.** 1, 2, 3, 4, 6, 9, 12, 18, 36
2. $2 \cdot 2 \cdot 5 \cdot 5$ **3.** $7xy^2(2x^2y - 3xy^2 - 1)$ **4.** $(x + 4y)(x + 3)$
5. $(a + 3b)(a - 6b)$ **6.** $3b(a + 5)(a - 3)$
7. $(3n - 2)(2n + 3)$ **8.** $2xy(2x - 5y)(3x - 2y)$

7.5 Exercises **1.** A monomial GCF. **3.** Sum of squares.
5. Try to factor by grouping when there are four terms. Grouping may also be used as a method to factor trinomials.
7. $3xy(y + 2x)$ **9.** $(x + y)(7a - b)$ **11.** $2(x + 4)(x - 4)$
13. $(a + b)(x + y)$ **15.** $abc(12a^2b + 3abc + 5c^2)$
17. $(x + 3)(x + 5)$ **19.** $(x^2 + 4)(x + 2)(x - 2)$ **21.** prime
23. $(5x - 1)(3x + 2)$ **25.** $a(x + 2)^2$ **27.** $6ab(1 - 6b)$
29. prime **31.** $(x + 7)(x - 7)$ **33.** $(p + 6)(p - 5)$
35. $u(u + 1)(u - 1)$ **37.** $2(b + 3)(b + 4)$ **39.** $3r^2(2 - 5r)$
41. $7(2u - 5)(u + 3)$ **43.** $4(h^2 + 3h + 1)$
45. $5(p + 4)(p - 4)$ **47.** $(3w + 2)(w + 1)$
49. $(2q - 3)(4q + 1)$ **51.** $(3v + 2)(4v + 5)$
53. $2(1 + 5x)(1 - 5x)$ **55.** $20k^2(2 + l)(2 - l)$ **57.** prime
59. $2(5 - t)^2$ **61.** $(x - 2y)(3a + 4b)$ **63.** $(x + 1)(x - 1)^2$
65. $(2x - 7)^2$ **67.** $2(x^2 + 9)(x + 3)(x - 3)$
69. $(a + 2b)(a + b)$ **71.** $(3 + 2m)(3 - 2m)$
73. $x(x^2 + 2y)(x^2 - 2y)$ **75.** prime
77. $(b + 5)(b^2 - 5b + 25)$ **79.** $3(y - 2)(y^2 + 2y + 4)$
81. $2x(3 - y)(9 + 3y + y^2)$ **83.** $l = 2x + 1, w = 3x - 7$
85. $l = 2x, w = 3x + 5, h = 3x + 5$ **87.** $4(5 + 2t)(5 - 2t)$
89. Current: $3i + 4$; resistance: $2r + 5$.

Review Exercises **1.** $x - 4 - \dfrac{4}{2x + 1}$
2. $(6y - 1)(y + 4)$ **3.** 224 ft.^{2} **4.** $4(x + 3) = 20; 2$
5. 17, 18, 19

7.6 Exercises **1.** If a and b are real numbers and $ab = 0$, then $a = 0$ or $b = 0$. **3.** Write the equation in standard form $(ax^2 + bx + c = 0)$. **5.** $-5, -2$ **7.** $-3, 4$ **9.** $-\dfrac{3}{2}, \dfrac{4}{3}$

11. $0, 7$ **13.** $-2, 1, 0$ **15.** $0, 2$ **17.** $0, 4$ **19.** $0, -\dfrac{5}{3}$

21. $-2, 2$ **23.** $-3, 2$ **25.** 3 **27.** $4, -\dfrac{1}{3}$ **29.** $-\dfrac{5}{2}, \dfrac{2}{3}$

31. $-3, 7$ **33.** $1, 2$ **35.** $-3, 3$ **37.** $-7, -2$ **39.** $0, 4, -3$

41. $0, -\dfrac{5}{2}, \dfrac{2}{3}$ **43.** $\dfrac{2}{3}, 4$ **45.** $-4, \dfrac{15}{4}$ **47.** $3, 4$ **49.** $-2, 7$

51. $-\dfrac{7}{2}$ **53.** $-2, -4$ **55.** $3, -6$ **57.** $5, 11$ **59.** $11, 13$

61. 13, 14   **63.** 12 m by 21 m   **65.** 172 ft. by 54 ft.   **67.** 14 ft.
69. Base: 18 cm; height: 24 cm.   **71.** 9 in.   **73.** 10, 12, 14 or
−16, −14, −12   **75.** 1.25 sec.   **77.** 10%   **79.** 15   **81.** 25
83. 37 cm   **85.** 30 blocks   **87.** 45 in.

Review Exercises **1.** No, it is not linear because the
variable x has an exponent other than 1.   **2.** yes
3. $(-10, 0), (0, 4)$ **4.** $3x - y = -7$ **5. a.** 2   **b.** 6
6.

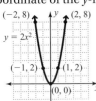

7.7 Exercises **1.** If a vertical line is drawn through the
vertex, the sides of the parabola are mirror images.   **3.** The
y-coordinate of the y-intercept is c.   **5.** $(2, -1)$ **7.** $(3, 0)$

9.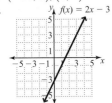

11.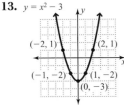

13. $y = x^2 - 3$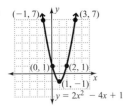

15. $y = -x^2 + 2x$

17.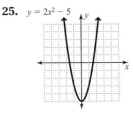

19.

21.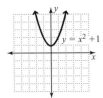

23.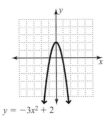

25. $y = 2x^2 - 5$

27.

29.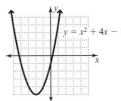

31.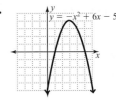

33. $y = 2x^2 - 4x - 3$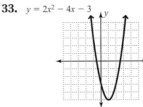

35.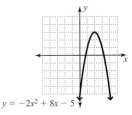

37. $f(x) = 3x^2$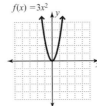

39.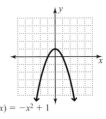

$f(x) = -x^2 + 1$

41.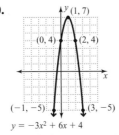

$f(x) = x^2 + 4x - 1$

43.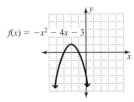

$f(x) = -x^2 - 4x - 3$

45. $f(x) = 3x^2 - 12x + 5$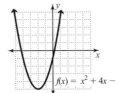

47.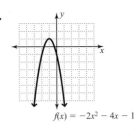

$f(x) = -2x^2 - 4x - 1$

49. A function; domain: all real numbers or $(-\infty, \infty)$; range:
$\{y \mid y \geq 0\}$ or $[0, \infty)$.   **51.** Not a function; domain: $\{x \mid x \geq 0\}$
or $[0, \infty)$; range: all real numbers or $(-\infty, \infty)$.   **53.** A
function; domain: all real numbers or $(-\infty, \infty)$; range:
$\{y \mid y \leq 2\}$ or $(-\infty, 2]$.   **55.** Not a function; domain:
$\{x \mid x \leq -1\}$ or $(-\infty, -1]$; range: all real numbers or
$(-\infty, \infty)$.   **57. a.** $(0, 0)$ **b.** $(0, 2)$ **c.** The original graph is
moved upward c units.   **d.** The original graph moved down-
ward 3 units.   **59. a.** $y = x^2$ opens up, and $y = -x^2$ opens
down. The sign of a indicates whether the parabola opens up
or down.   **b.** yes   **c.** These graphs are "wider."   **d.** The
smaller the absolute value of a, the wider the parabola.

Review Exercises **1.** $3m^3n - 12m^2n^2 + 15mn$
2. $2x^2 - 5x - 3$ **3.** $(3n - 5)(n - 4)$ **4.** $(x + 5)(x - 5)$
5. 0, 3   **6.** −6, 3

Chapter 7 Review Exercises **1.** false   **2.** false
3. false   **4.** true   **5.** true   **6. a.** GCF   **b.** polynomial; GCF
7. a. c, b **8.** $(a + b)(a - b)$ **9.** $(a - b)(a^2 + ab + b^2)$
10. $(a - b)^2$ **11.** 1, 2, 3, 4, 5, 6, 10, 12, 15, 20, 30, 60   **12.** 1,

3, 9   **13.** 1, 3, 9, 27, 81   **14.** 1, 2, 4, 11, 22, 44   **15.** 1, 3, 5, 9, 15, 45   **16.** 1, 7, 49   **17.** 1, 2, 3, 5, 6, 10, 15, 30   **18.** 1, 3, 11, 33   **19.** 3   **20.** 17   **21.** 6   **22.** 5   **23.** $5y^2$ **24.** $4x$
25. $3mn$ **26.** $2k$ **27.** $2(2x - 1)$ **28.** $5m(1 - 7m^2)$
29. $y(y^2 - 1)$ **30.** $abc(1 + abc - a^2b^2c^2)$
31. $xy(x + y + x^2y^2)$ **32.** $21a^2b^2(5a - 3b + 4a^4b^2)$
33. $10k(10k^3 + 12k^4 - 1 + 4k^2)$ **34.** $18ab^2c(b - 2a)$
35. $(x + y)(a + b)$ **36.** $(x + 2)(a + b)$
37. $(y + 2)(y^2 + 3)$ **38.** $(y + 4)(y^3 - b)$
39. $(x + 1)(y + 1)$ **40.** $(x^2 + y^2)(a - 5)$
41. $(2b^2 + 1)(b - 1)$ **42.** $(u - 3)(u + 4v)$
43. $(x - 4)(x + 3)$ **44.** $(x + 5)(x + 9)$
45. $(n - 2)(n - 4)$ **46.** prime   **47.** $(h + 3)(h + 48)$
48. $(y - 12)(y + 2)$ **49.** $4(x - 3)^2$ **50.** $3(m - 9)(m - 2)$
51. prime   **52.** $(2u + 1)(u + 2)$ **53.** $(3m - 1)(m - 3)$
54. $(5k - 12h)(k + h)$ **55.** $2(3a - 4)(a - 2)$
56. $y(8x^2 - 4x - 1)$ **57.** $(3p - q)(p - 4q)$
58. $8(3x + y)(x - 3y)$ **59.** $(v - 4)^2$ **60.** $(u + 3)^2$
61. $(2x + 5)^2$ **62.** $(3y - 2)^2$ **63.** $(x + 2)(x - 2)$
64. $(5 + y)(5 - y)$ **65.** $(x - 1)(x^2 + x + 1)$
66. $(x + 3)(x^2 - 3x + 9)$ **67.** $(3b + 2)(2b - 1)$
68. $4ab(1 - 6b)$ **69.** prime   **70.** $3(x + y)(x - y)$
71. $(x^2 + 9)(x + 3)(x - 3)$ **72.** $7(u - 5)(u + 3)$
73. $(2x - 3y)(4x^2 + 6xy + 9y^2)$ **74.** $2(1 + 5y)(1 - 5y)$
75. $(m - 2n)(3a + 4b)$ **76.** $3(m^2 + 3m + 9)$ **77.** $-3, 4$
78. $-2, 2$ **79.** $2, 3$ **80.** $-3, 7$ **81.** $-3, 3$ **82.** $-6, 3$
83. -3 **84.** $-\dfrac{1}{3}, 4$ **85.** $5x^2 = 2x; 0, \dfrac{2}{5}$
86. $x^2 + (x + 1)^2 = 61; 5, 6$ **87.** $x \cdot 4x = 100; -5, 5$
88. $(w + 6)w = 91; 13$ in. by 7 in.   **89.** $x(x + 1) = 110; 10, 11$
90. $3^2 + 4^2 = c^2; 5$

91.

92.
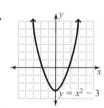

93.
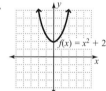

94.
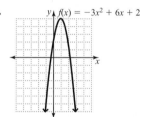

Chapter 7 Practice Test   **1.** 1, 2, 4, 5, 8, 10, 16, 20, 40, 80 [7.1]   **2.** $5m$ [7.1]   **3.** $5(y - 6)$ [7.1]
4. $2y(3x^2 - y)$ [7.1]   **5.** $(x - y)(a - b)$ [7.1]
6. $(m + 3)(m + 4)$ [7.2]   **7.** $(r - 4)^2$ [7.4]
8. $(2y + 5)^2$ [7.3]   **9.** $(q - 2)(3q - 4)$ [7.3]
10. $(3x - 4)(2x - 5)$ [7.3]   **11.** $a(x + 3)(x - 8)$ [7.2]
12. $2n(n + 4)(5n - 1)$ [7.2]   **13.** $(c + 5)(c - 5)$ [7.4]
14. prime [7.4]   **15.** $2(1 + 5u)(1 - 5u)$ [7.4]
16. $-4(x + 2)(x - 2)$ [7.4]
17. $(m + 5)(m^2 - 5m + 25)$ [7.4]
18. $(x - 2)(x^2 + 2x + 4)$ [7.4]   **19.** $-3, 0$ [7.6]
20. $6, -2$ [7.6]   **21.** $\dfrac{3}{2}, -5$ [7.6]   **22.** $8, 9$ [7.6]

23. 9ft., 4 ft. [7.6]
24. [7.7]
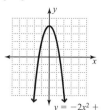

25. [7.7]
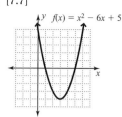

Chapter 1–7 Cumulative Review Exercises
1. false   **2.** true   **3.** false   **4.** false   **5.** $(a + b)(a - b)$
6. a. standard   **b.** factored   **7.** 42   **8.** -89
9. $5x^2 + 4x + 2$ **10.** $18x^4y$ **11.** $48x^7$ **12.** $2y^2 - 15y - 8$
13. $4h + 1 - \dfrac{2}{h}$ **14.** $x + 4$ **15.** $(x + 11)(x - 11)$
16. $10(x^2 + 4)$ **17.** $(x - 4)^2$ **18.** $(2x - 3)(x + 4)$
19. $(x - 5)(x^2 + 5)$ **20.** $\dfrac{17}{12}$ **21.** 13.2   **22.** $-4, 7$
23. $x = \dfrac{7 + 3y}{2}$ **24.** $(3, 0), (0, -6)$ **25.** 0

26.
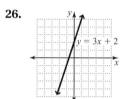

27. a. $\{x \mid x \leq 4\}$ **b.** $(-\infty, 4]$
c.

$-5 \ -4 \ -3 \ -2 \ -1 \ \ 0 \ \ 1 \ \ 2 \ \ 3 \ \ 4 \ \ 5$ **28.** 6, 8, 10

29. \$2500 in 5%, \$5500 in 8%   **30.** $33\dfrac{1}{3}$ ml

Chapter 8

8.1 Exercises   **1.** Replace the variables with given values and then simplify the numerical expression.   **3.** Write an equation that has the denominator set equal to zero and solve the equation.   **5.** y is not a factor in the numerator, $x + y$.
7. a. $\dfrac{16}{45}$ **b.** -1 **c.** 0   **9. a.** 1   **b.** undefined   **c.** -3
11. a. $\dfrac{11}{8}$ **b.** $-\dfrac{1}{7}$ **c.** $-\dfrac{24}{49}$ or ≈ -0.49 **13. a.** $-\dfrac{4}{3}$ **b.** $-\dfrac{3}{2}$
c. $-\dfrac{341}{30}$ or $-11.3\overline{6}$ **15.** 3   **17.** $-3, 3$ **19.** 5, 9   **21.** 0, 3
23. $-1, \dfrac{5}{2}$ **25.** Defined for all real numbers.   **27.** $\dfrac{3x}{4y}$
29. $-\dfrac{4m^2}{n}$ **31.** $\dfrac{3}{2x^2z^2}$ **33.** $-\dfrac{3m^2n}{m + 2n}$ **35.** $\dfrac{1}{3(x + 5)}$ **37.** $\dfrac{3}{5}$
39. $\dfrac{2}{3}$ **41.** $\dfrac{5}{3}$ **43.** $\dfrac{x}{4}$ **45.** m **47.** $\dfrac{5}{x}$ **49.** $\dfrac{x - y}{x + y}$ **51.** $\dfrac{x}{x - 2}$
53. $\dfrac{t - 3}{t + 2}$ **55.** $\dfrac{2a - 3}{3a - 2}$ **57.** $\dfrac{3b - 4}{5b - 1}$ **59.** $\dfrac{x - 4}{x + 2}$ **61.** $\dfrac{p + q}{m - n}$
63. $\dfrac{2u^2 - 1}{u + 1}$ **65.** -1 **67.** -4 **69.** $-y - 2$ **71.** $-\dfrac{1}{w + 1}$

73. $\dfrac{x-2}{x^2-2x+4}$ 75. $\dfrac{x+y}{x}$ 77. $-\dfrac{1}{m^2+3m+9}$

79. Mistake: Placed remaining x in numerator. Correct: $-\dfrac{1}{2x}$

81. Mistake: Divided out a part of a multiterm polynomial. Correct: Can't reduce 83. Mistake: $y-x$ and $x-y$ are not identical factors until -1 is factored out of one of them. Correct: $-x$ 85. a. 23.7, 22.9, 28.3 b. The first two are normal weight, and the third is overweight. 87. $533.\overline{3}\pi, 666.\overline{6}\pi, 666.\overline{6}\pi,$ $833.\overline{3}\pi$ 89. a. $h=\dfrac{3V}{\pi r^2}$ b. ≈ 9 cm 91. $\dfrac{s-3}{s+4}$

Review Exercises 1. 1 2. $-\dfrac{3}{4}$ 3. $\dfrac{9}{20}$ 4. $\dfrac{28}{15}$
5. $(2y+5)(2y-5)$ 6. $(3x-2)(x+5)$

8.2 Exercises 1. To eliminate any common factors.
3. Their product is 1. 5. A fraction with a ratio equivalent to 1.
Examples will vary. 7. $\dfrac{2m^2}{5n^2}$ 9. $\dfrac{s^2}{r^3}$ 11. y^2 13. $-\dfrac{1}{mn^2s}$
15. $\dfrac{9r}{s}$ 17. $\dfrac{mn^2}{2}$ 19. $\dfrac{9}{10}$ 21. $\dfrac{4}{15}$ 23. $(2x-5)(3x+4)$
25. $\dfrac{r^2}{(r+6)(r-3)}$ 27. $\dfrac{q+1}{2q(q-1)}$ 29. $\dfrac{w+4}{w-1}$ 31. $\dfrac{n+1}{n}$
33. 1 35. $\dfrac{(d-5)}{(d+1)}$ 37. $\dfrac{x-2}{x}$ 39. $\dfrac{1}{y}$ 41. $-\dfrac{4y^2}{3x}$
43. $1-3x$ 45. $\dfrac{1}{x^2y^2}$ 47. $\dfrac{5}{4m^2n}$ 49. $\dfrac{y^2}{5z^2}$ 51. $\dfrac{-32x^6}{45}$
53. $\dfrac{4a}{9}$ 55. $\dfrac{8}{5}$ 57. $\dfrac{a-b}{x+y}$ 59. $\dfrac{x+8}{x-8}$ 61. $x-4$
63. $\dfrac{a+2b}{a-2b}$ 65. $\dfrac{1}{k(2k+1)}$ 67. $\dfrac{4x^2+6x+9}{3x-1}$
69. $\dfrac{1}{(x+4)(x+1)}$ 71. $\dfrac{b-4}{c-3}$ 73. $-b-5$
75. $\dfrac{7+y}{y(y+4)}$ 77. $-\dfrac{x^3}{x-3}$ 79. $-\dfrac{x}{y}$

81. Mistake: Did not invert the divisor. Correct: $\dfrac{50}{27}$
83. Mistake: Divided out y, which is illegal because it is not a
factor in $y-2$. Correct: $\dfrac{(x-3)(y-2)}{15y}$ 85. 222 in.
87. 10,560 yd. 89. 288 in. 91. 88 oz. 93. 12.7 T
95. 96,000 oz. 97. 260.48 ft./sec. 99. ≈ 19.03 mi./hr.
101. 939,785,328 km 103. $\approx 5.87 \times 10^{12}$ mi./yr.
105. 255.36£ 107. $665.53 109. 860.82€

Review Exercises 1. 1 2. $-\dfrac{1}{2}$ 3. $2x$ 4. $\dfrac{1}{4}y-4$
5. $8x^2-\dfrac{2}{3}x-4$ 6. $-\dfrac{1}{5}n^2+4$

8.3 Exercises 1. The numerator may not contain like
terms. 3. Write an equivalent addition with the additive
inverse of the subtrahend. 5. $\dfrac{x}{2}$ 7. $\dfrac{a^2}{3}$ 9. $\dfrac{9}{a}$ 11. $-\dfrac{x}{7y^2}$
13. $\dfrac{n+3}{n+5}$ 15. $\dfrac{3r}{r+2}$ 17. $\dfrac{q+1}{q}$ 19. 3 21. 3 23. $\dfrac{11}{w+1}$
25. $\dfrac{3m-1}{m-1}$ 27. $\dfrac{-11t+3}{2t-3}$ 29. $\dfrac{6w-5}{w+5}$ 31. $\dfrac{1}{g+3}$

33. $z-3$ 35. $\dfrac{1}{s+2}$ 37. $\dfrac{y-3}{y}$ 39. $\dfrac{x+2}{x-4}$ 41. $\dfrac{t-1}{t+7}$
43. $3(x+2)$ 45. $\dfrac{1}{x-5}$ 47. $\dfrac{3x-2}{x-5}$ 49. $\dfrac{x-3}{x+5}$

51. Mistake: Combined terms that are not like. Correct: $\dfrac{4y+1}{5}$
53. Mistake: Did not change sign of -5. Correct: 2
55. $\dfrac{5x+9}{2}$ 57. $\dfrac{3x+2}{2}$

Review Exercises 1. $\dfrac{11}{10}$ 2. $\dfrac{11}{24}$ 3. $1\dfrac{6}{35}$ 4. $\dfrac{11}{15}x+\dfrac{7}{12}$
5. $2x^2-12x-\dfrac{11}{24}$ 6. $2y^2-\dfrac{1}{20}y-10$

8.4 Exercises 1. a. Find the prime factorization of each
denominator. b. Write a product that contains each unique
prime factor the greatest number of times that factor occurs in
any factorization—or if you prefer to use exponents, the product
that contains each unique prime factor raised to the greatest
exponent that occurs on that factor in any factorization.
c. Simplify the product found in Step 2. 3. Factor the
denominators. 5. Rewrite the numerator as an equivalent
addition sentence. 7. $mn; \dfrac{2m}{mn}, \dfrac{3n}{mn}$ 9. $a^3b^5; \dfrac{2at}{a^3b^5}, \dfrac{3b^2z}{a^3b^5}$
11. $15a^2; \dfrac{10a}{15a^2}, \dfrac{9}{15a^2}$ 13. $14x^2y^2; \dfrac{6y}{14x^2y^2}, \dfrac{5x}{14x^2y^2}$
15. $90m^3n^3; \dfrac{48}{90m^3n^3}, \dfrac{35mn^2}{90m^3n^3}$
17. $(y+2)(y-2); \dfrac{3y-6}{(y+2)(y-2)}, \dfrac{4y^2+8y}{(y+2)(y-2)}$
19. $12(y-5); \dfrac{9y}{12(y-5)}, \dfrac{14y}{12(y-5)}$
21. $6x(x+3); \dfrac{5x}{6x(x+3)}, \dfrac{12}{6x(x+3)}$
23. $p^2-4; \dfrac{2}{p^2-4}, \dfrac{p^2+p-6}{p^2-4}$
25. $3(u+2)^2; \dfrac{12u}{3(u-2)^2}, \dfrac{2u+4}{3(u+2)^2}$
27. $(t+6)(t-6)(t-1); \dfrac{2t^2-6t+4}{(t+6)(t-6)(t-1)},$
$\dfrac{3t^2+20t+12}{(t+6)(t-6)(t-1)}$
29. $(x+4)(x-1)(x+2); \dfrac{x^2+4x+4}{(x+4)(x-1)(x+2)},$
$\dfrac{x^2-4x+3}{(x+4)(x-1)(x+2)}$
31. $(x-1)(x-2); \dfrac{2x-4}{(x-1)(x-2)}, \dfrac{3}{(x-1)(x-2)},$
$\dfrac{5x^2-5x}{(x-1)(x-2)}$ 33. $\dfrac{-31}{18}$ 35. $\dfrac{8}{3c}$ 37. $\dfrac{5x^2-2}{x^3}$
39. $\dfrac{8y-2}{(y+2)(y-4)}$ 41. $\dfrac{x+10}{(x-2)(x+2)}$
43. $\dfrac{2x+14}{(x+5)(x-5)}$ 45. $\dfrac{10p-8}{p^2-2p+1}$ 47. $\dfrac{-z^2-5z-3}{(z+4)(z-3)}$
49. $\dfrac{2n^2-2n+18}{(n-4)(n+3)}$ 51. $\dfrac{-t^2+t+16}{(t+4)(t+6)}$ 53. $\dfrac{5}{2(c+5)}$

55. $\dfrac{y+2}{y}$ **57.** $\dfrac{r^2-6r-1}{(r+1)(r-1)^2}$ **59.** $\dfrac{q+1}{q(q-3)}$

61. $\dfrac{1}{(w-1)(w-2)}$ **63.** $\dfrac{v^2+16v+51}{(v+4)(v-4)(v+3)}$ **65.** $\dfrac{-3}{t-3}$

67. $\dfrac{4-x}{2x-1}$ **69.** $\dfrac{y+6}{3(y-2)}$ **71.** Mistake: Did not find a common denominator. Correct: $\dfrac{6-x^2}{2x}$ **73.** Mistake: Did not change the sign of 7. Correct: $\dfrac{-x-7}{x+2}$ **75.** Mistake: Added denominators. Correct: $\dfrac{7w}{x}$ **77.** $\dfrac{5t}{6}$ **79.** $\dfrac{4x+6}{x^2+3x}$

81. $\dfrac{4ah+5bh}{8}$ **83.** $\dfrac{6x^2+8x+2}{(2x-1)(2x+1)}$ **85.** xy

87. $\dfrac{3a^2+10a+20}{5a(a+2)}$ **89.** $\dfrac{2a^2+2b^2}{(a+b)^2}$ **91.** $\dfrac{x-5}{3(x+2)}$

Review Exercises 1. $\dfrac{3}{8}$ **2.** $\dfrac{m^2}{108}$ **3.** $4(x+2)(x-2)$

4. $3y(2y+1)(y-5)$ **5.** $\dfrac{x-1}{x+2}$ **6.** $\dfrac{1}{x-3}$

8.5 Exercises 1. $\dfrac{4}{x+3}$ is the numerator. $\dfrac{3}{x}$ is the

denominator. **3.** $\dfrac{\frac{2x}{x-3}}{\frac{4-x}{2x-6}}$ **5.** In this case, multiplying the

numerator and denominator by their LCD, 12, (Method 2) is preferable because doing so requires fewer steps than

simplifying $\dfrac{x}{6}-\dfrac{3}{4}$ and $\dfrac{2}{3}+x$ and then dividing the simplified

expressions (Method 1). (Answers may vary.) **7.** $\dfrac{1}{2}$ **9.** $\dfrac{mp}{nq}$

11. $\dfrac{11}{13}$ **13.** $\dfrac{5a+2}{3a+4}$ **15.** $\dfrac{1-y}{1+y}$ **17.** $\dfrac{4r-3}{2-4r}$ **19.** $\dfrac{x}{3+x}$

21. xy **23.** $\dfrac{x}{x-2}$ **25.** $\dfrac{6(2x+1)}{x(3x-2)}$ **27.** $\dfrac{a}{b(a-b)}$

29. $\dfrac{2y(y-5)}{5}$ **31.** u **33.** $\dfrac{1}{a-1}$ **35.** $\dfrac{1}{3}$ **37.** 2 **39.** $y-1$

41. $\dfrac{1}{n-1}$ **43.** $3-y$ **45.** $\dfrac{x-1}{x}$ **47.** $\dfrac{-2}{x(x+h)}$

49. Mistake: Divided out $\dfrac{1}{2}$, which is not a common factor.

Correct: $\dfrac{2x-1}{2y-1}$ **51.** Mistake: Did not write as an equivalent

multiplication. Correct: $\dfrac{m^2}{18}$ **53.** Mistake: Did not distribute c

in the numerator. Correct: c^2+1 **55.** Mistake: Added the x's

instead of multiplying them. Correct: $\dfrac{18}{x^2}$ **57.** $34\dfrac{2}{7}$ mph

59. $\dfrac{3(5n-2)}{4(n-1)}$ **61. a.** $\dfrac{R_1R_2}{R_2+R_1}$ **b.** $\approx 9.1\,\Omega$ **c.** $8\,\Omega$

Review Exercises 1. -7 **2.** 3 **3.** 2.5 **4.** 4
5. 4.8 mph **6.** 0.09 hr. or 5.4 min.

8.6 Exercises 1. Both sides of the equation are multiplied by the LCD of the rational expressions. **3.** An apparent solution that does not solve its equation. **5.** 4 and -2; substituting either of those numbers for x causes an expression in the given equation to be undefined. **7.** no **9.** yes **11.** yes **13.** 2 **15.** 10

17. 15 **19.** 3 **21.** 1 **23.** 10 **25.** 8 **27.** -3

29. -1 **31.** 1 **33.** 9 **35.** $-2, \dfrac{1}{3}$ **37.** -5

39. $-\dfrac{11}{3}$, $(-1$ is extraneous$)$ **41.** $4, -2$ **43.** $-\dfrac{1}{4}, 5$

45. $-2, 9$ **47.** 1, $(2$ is extraneous$)$ **49.** $0, \dfrac{9}{2}$ **51.** $-3, -\dfrac{2}{3}$

53. -2 **55.** 1 **57.** $6, -1$ **59.** No solution.
61. 5, $(1$ is extraneous$)$ **63.** -3 is the only answer; 3 is extraneous. **65.** In an expression, the LCD is used to combine the numerator over a single denominator. In an equation, the LCD is used to eliminate the denominator.
67. $400\,\Omega$ **69.** 20 and $30\,\Omega$ **71.** ≈ 2.1 ft.
73. $f=15$ mm, $i=20$ mm

Review Exercises 1. ≈ 68.6 mph **2.** $\approx \dfrac{0.084}{1}$; each

ounce costs $\$08.4$ **3.** 6 hr. **4.** $72, 74$ **5.** The length is 11 mm, and the width is 7 mm. **6.** 0.15 hr. $(9$ min.$)$

8.7 Exercises 1. $\dfrac{1}{x}$ **3.** $\dfrac{100}{r}$ **5.** It increases.

7. $4\dfrac{4}{5}$ days **9.** 12 min. **11.** $2\dfrac{2}{5}$ hr. **13.** $2\dfrac{8}{11}$ hr.

15. 6 mph, 4 mph **17.** 340 mph **19.** 6 mph
21. 16 mph, 20 mph **23.** 30 **25.** 4.5 **27.** $\$25.76$
29. 15 gal. **31.** 144.9 lb. **33.** 20 in. **35.** 420 in.^{3}
37. 5 **39.** 14 **41.** 30 psi **43.** 4 A **45.** 660 vps
47. $2.\overline{3}$ cm/sec. **49.** ≈ 4.39 million lb.
51. a.

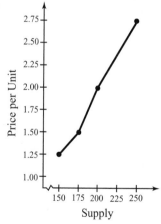

b. Supply and price are directly proportional. As supply increased, so did the price per unit. **c.** Yes; each value in the domain (supply) is paired with one value in the range (price per unit).

Review Exercises **1.** 5, 6 **2.** no

3.
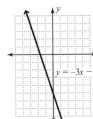

4. $\left(-\dfrac{3}{2}, 0\right), (0, 6)$ **5.** $\dfrac{17}{4}$ **6.** $3a - 4$

Chapter 8 Review Exercises **1.** false **2.** true
3. true **4.** true **5.** true **6. a.** multiplication
b. Divide **7. a.** numerator, denominator **8.** complex
9. a. numerator, denominator **10. 1.** LCD **11. a.** $\dfrac{17}{6}$
b. ≈ 0.077 **c.** $\dfrac{3}{4}$ **12. a.** 3 **b.** 15 **c.** 1.875 **13.** 5

14. $-2, 2$ **15.** $-6, 1$ **16.** $0, -6$ **17.** $\dfrac{x + 5}{3}$ **18.** $\dfrac{2}{3}$

19. $\dfrac{x}{4}$ **20.** $\dfrac{y}{3}$ **21.** $\dfrac{a + b}{a - b}$ **22.** $\dfrac{3}{(x^2 + 9)(x - 3)}$

23. $-(x + y)$ **24.** $-\dfrac{1}{w + 3}$ **25.** $\dfrac{7x^2}{4y^2}$ **26.** $-\dfrac{3}{5mnp^4}$

27. $\dfrac{3}{5}$ **28.** $\dfrac{x^2}{(x + 4)(x - 3)}$ **29.** $\dfrac{n + 1}{n}$ **30.** $\dfrac{2}{r - 1}$

31. $\dfrac{m + 1}{4m(m - 1)}$ **32.** $\dfrac{w + 4}{w - 4}$ **33.** $\dfrac{y^3}{abx}$ **34.** $\dfrac{5}{6}$ **35.** 2

36. $\dfrac{5a}{2}$ **37.** $\dfrac{x + y}{c - d}$ **38.** $\dfrac{5(b + 4)(b - 4)}{6}$ **39.** $\dfrac{2j + 1}{j(2j - 1)}$

40. $\dfrac{1}{(u + 1)^2}$ **41.** $\dfrac{x}{3}$ **42.** $\dfrac{3r}{r + 3}$ **43.** $-\dfrac{1}{3m}$ **44.** $\dfrac{3x + 1}{x + 1}$

45. 1 **46.** $\dfrac{s}{r^2 - s^2}$ **47.** $\dfrac{11x^2 - 21x + 7}{2x + 1}$ **48.** $q - 3$

49. $\dfrac{6c^2}{12c^3}, \dfrac{12}{12c^3}$ **50.** $\dfrac{5x}{x^2y^3}, \dfrac{3y}{x^2y^3}$

51. $\dfrac{4m + 4}{(m - 1)(m + 1)}, \dfrac{4my - 4y}{(m - 1)(m + 1)}$

52. $\dfrac{3h}{4h(h - 2)}, \dfrac{20}{4h(h - 2)}$

53. $\dfrac{x}{(x + 1)(x - 1)}, \dfrac{x^2 + x - 2}{(x + 1)(x - 1)}$

54. $\dfrac{2p + 4}{(p - 2)(p + 2)^2}, \dfrac{p^2 - 2p}{(p - 2)(p + 2)^2}$ **55.** $-\dfrac{23}{36}$

56. $\dfrac{-x - 27}{(x + 3)(x - 3)}$ **57.** $\dfrac{8y - 22}{(y - 2)(y - 4)}$

58. $\dfrac{4x - 3}{(x + 2)(x - 2)}$ **59.** $\dfrac{-3t^2 + 21t}{2(t - 3)^2}$ **60.** $\dfrac{18 - a^2}{(a + 3)(a + 4)}$

61. $\dfrac{15}{2y - 1}$ **62.** $\dfrac{x + 19}{(x + 4)(x - 4)}$ **63.** $\dfrac{ay}{bx}$ **64.** -1

65. $\dfrac{5(x - 2)}{x(3x + 5)}$ **66.** $\dfrac{x}{x - 3}$ **67.** $\dfrac{x}{5(2x - 1)}$ **68.** 1

69. $\dfrac{y^2 + x^2}{7xy}$ **70.** $-h$ **71.** $\dfrac{38}{13}$ **72.** 6 **73.** 1 **74.** 4

75. 3, 4 **76.** -1 **77.** -8, (4 is extraneous) **78.** 23
79. 31.875 min. **80.** 8 yd./sec. and 10 yd./sec. **81.** 30
82. 2 **83.** 10 gal. **84.** 4.8 A

Chapter 8 Practice Test **1.** 7 [8.1] **2.** $-4, 4$ [8.1]
3. $-\dfrac{3}{x - 4}$ [8.1] **4.** $\dfrac{1}{x + y}$ [8.1] **5.** $\dfrac{6f}{f^2g^3}, \dfrac{2g^3}{f^2g^3}$ [8.4]
6. $\dfrac{5x(x + 3)}{(x + 3)^2(x - 3)}, \dfrac{2(x - 3)}{(x + 3)^2(x - 3)}$ [8.4] **7.** $\dfrac{5ab^2x^2}{12}$ [8.2]
8. $-\dfrac{7a^6b^7y^2}{40x^2}$ [8.2] **9.** $-4x$ [8.2] **10.** $\dfrac{12x^2}{(x + 1)(2x + 1)}$ [8.2]
11. $\dfrac{7x}{2x + 3}$ [8.3] **12.** $\dfrac{1}{x + 3}$ [8.3]
13. $\dfrac{x^2 + 5x - 6}{x(x + 2)(x - 2)}$ [8.4] **14.** $\dfrac{9x + 3}{(x - 1)(x + 2)}$ [8.4]
15. $\dfrac{3x + 13}{(x - 4)(x + 1)}$ [8.4] **16.** $\dfrac{x - 2}{2x}$ [8.4] **17.** $\dfrac{2y + 1}{3y - 1}$ [8.5]
18. $\dfrac{2}{x + y}$ [8.5] **19.** $\dfrac{25}{9}$ [8.6] **20.** -1 [8.6]
21. -1, (2 is extraneous) [8.6] **22.** 2, 4 [8.6]
23. 1.2 hr. [8.7] **24.** 6 mph, 4 mph [8.7] **25.** 529.2 N [8.7]

Chapters 1–8 Cumulative Review Exercises
1. false **2.** false **3.** false **4.** denominator **5.** There are
no common factors in the numerator and denominator
besides 1. **6.** multiply, add **7.** complex rational
expression **8.** 21 **9.** 6 **10.** $-3a^{13}$
11. $m^3 - 3m^2 - 17m + 3$ **12.** $(a - 2)(x + 1)$
13. $3(y - 5)(y + 4)$ **14.** $-3(h + 5)(h - 5)$
15. 9 **16.** $2(m + 4)$ **17.** $\dfrac{1}{x + 3}$ **18.** $\dfrac{6x + 14}{x(x + 7)(x - 7)}$
19. $\dfrac{2(y + 3)}{y}$ **20.** 9 **21. a.** $\dfrac{2}{3}$

b.
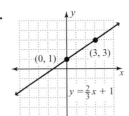

22. $-\dfrac{1}{2}, 7$ **23.** $(1, -2)$ **24.** 2, 6

25. 6 **26.** 2.5 hr. **27.** \$5500 at 9% and \$1500 at 7%

28. 6 oz. **29.** $3\dfrac{3}{7}$ hr. **30.** 6

Chapter 9

9.1 Exercises **1.** Answers will vary. One example of a
number with a rational square root is 9, and 10 has an
irrational square root. Rational roots can be expressed as
whole numbers or fractions with integers in the numerator
and denominator, whereas irrational roots cannot.
3. Squaring a number or its additive inverse results in the
same positive number. **5.** No real number can be squared to
equal a negative number. **7.** ± 6 **9.** ± 11 **11.** ± 14 **13.** ± 15

15. 7 **17.** 19 **19.** -6 **21.** Not a real number. **23.** $\dfrac{5}{6}$

25. $\dfrac{4}{3}$ **27.** Not a real number. **29.** $-\dfrac{13}{5}$ **31.** 0.6 **33.** 0.2

35. Not a real number. **37.** 0.11 **39.** -0.07 **41.** irrational
43. rational **45.** rational **47.** irrational **49.** 1.732
51. 3.464 **53.** -2.449 **55.** 3.317 **57.** 7.141 **59.** 5.385
61. a **63.** x^3 **65.** $8x^2$ **67.** $8ab^3$ **69.** $0.9ab^3$ **71.** $\dfrac{4}{5}rs$
73. $\dfrac{6ab^2}{7m}$ **75.** ≈ -17.709 m/sec. **77.** ≈ 54.8 mi.
79. π or ≈ 3.14 sec. **81.** ≈ 16.4 sec. **83.** 30 mph
85. 13 ft. **87.** 15 N **89. a.** $\approx 416{,}000$ **b.** $\approx 564{,}000$

Review Exercises 1. a. 12 **b.** 12 **2.** x^8
3. $-45m^4n^3$ **4.** $12x^4 - 20x^3 + 4x^2$ **5.** $21y^2 + 23y - 20$
6. $2x^2 - 17x - 9$

9.2 Exercises 1. Because we cannot easily determine
the square roots of 3 and 12, using the product rule for square
roots leads to a radicand that is a perfect square, 36.
3. The radicand, 28, has a perfect square factor, 4. **5.** 8
7. 60 **9.** $8m$ **11.** $9x^3$ **13.** $6y^2$ **15.** $10p^4$ **17.** $12k^5$
19. $5x^3$ **21.** $120yz^2$ **23.** $-28g^2t^2$ **25.** $78w^2xz$ **27.** $24c^5y^8$
29. $-32x^5y^5z^7$ **31.** $2\sqrt{3}$ **33.** $2\sqrt{6}$ **35.** $3\sqrt{7}$ **37.** $12\sqrt{3}$
39. $2a^2\sqrt{2}$ **41.** $2m\sqrt{5m}$ **43.** $4p\sqrt{3q}$ **45.** $2xy^2\sqrt{30y}$
47. $15k^2m^3\sqrt{17m}$ **49.** $8x^2y^3\sqrt{10}$ **51.** $-6a^5b^4\sqrt{6a}$
53. $-24x^6y^3\sqrt{3xy}$ **55.** Mistake: The x was written out of the
radical. Correct: $4\sqrt{2x}$ **57.** Mistake: Did not find the square
root of 4. Correct: $2x^2\sqrt{3}$ **59.** 28 **61.** $20a^2b^5$ **63.** $90\sqrt{3}$
65. $6\sqrt{5}$

Review Exercises 1. $-13x^4 + 7x$ **2.** $2x^3$ **3.** $\dfrac{r^4}{2}$
4. $(x-3)(x-4)$ **5.** $(3y-7)(2y+3)$
6. $(n-6)(n+6)$

9.3 Exercises 1. a. 24 contains a perfect square factor, 4.
b. 16 is a perfect square, and the radicand contains a fraction.
c. The denominator needs to be rationalized. **3.** Multiply
both the numerator and denominator of the fraction by
the same square root that appears in the denominator or
by the square root of a number that makes a perfect square
in the denominator. Simplify. **5.** 2 **7.** 2 **9.** 2 **11.** 4
13. $2\sqrt{6}$ **15.** $3\sqrt{2}$ **17.** x **19.** $4a^3$ **21.** $2m^4\sqrt{2n}$ **23.** $\dfrac{1}{3t}$
25. $\dfrac{2a\sqrt{3}}{b^3}$ **27.** $\dfrac{3}{8}$ **29.** $-\dfrac{4}{9}$ **31.** $\dfrac{a^3}{10}$ **33.** $\dfrac{y}{5x}$ **35.** -8
37. $\dfrac{2}{3}$ **39.** $-\dfrac{3\sqrt{3}}{10}$ **41.** $\dfrac{4}{x^3}$ **43.** $\dfrac{4}{mn}$ **45.** $\dfrac{2a\sqrt{3}}{3}$
47. Mistake: Divided the radicands without evaluating the
square root. Correct: 2 **49.** Mistake: Divided the radicands
without evaluating the root. Correct: 5 **51.** $\dfrac{\sqrt{3}}{3}$ **53.** $\dfrac{3\sqrt{2}}{4}$
55. $\dfrac{3\sqrt{a}}{a}$ **57.** $\dfrac{6\sqrt{7}}{7}$ **59.** $\dfrac{\sqrt{mn}}{n}$ **61.** $\dfrac{x\sqrt{14}}{10}$ **63.** $\dfrac{\sqrt{7}}{7}$
65. $x\sqrt{3x}$ **67.** $\dfrac{\sqrt{14}}{7}$ **69.** $\dfrac{\sqrt{ab}}{b^2}$ **71.** Mistake: Multiplied by
2 instead of $\sqrt{2}$. Correct: $\dfrac{\sqrt{6}}{2}$ **73.** $2x^3$ **75.** $5x^3y$
77. $T = \dfrac{\pi\sqrt{9.8L}}{4.9}$ **79. a.** $s = \dfrac{\sqrt{3Vh}}{h}$ **b.** ≈ 745 ft.
81. a. $V_{rms} = \dfrac{V_m\sqrt{2}}{2}$ **b.** $\dfrac{163\sqrt{2}}{2}V$ **c.** $\approx 115.3\ V$
83. a. $2\sqrt{3}$ **b.** $3\sqrt{2}$ **c.** $\dfrac{6\sqrt{5}}{5}$

85. a. The graphs are the same. The functions are the same.
b. $f(x) = g(x)$ **87. a.** $\dfrac{1}{\sqrt{x}}$ is in quadrant I, and $-\dfrac{1}{\sqrt{x}}$ is its
mirror image in quadrant IV. **b.** The negative sign causes
the graph to reflect across the x-axis.

Review Exercises 1. 0.1 **2.** 71 **3.** $2x + 10y$
4. $-2t^3 + 10t^2 - 5$ **5.** $9y^2 - 2y - 3$
6. $-5n^3 + 10n^2 - 14$

9.4 Exercises 1. The radicands must be identical.
The coefficients can be different. **3.** The processes are similar in that we multiply every term in the first parentheses by
every term in the second parentheses (FOIL). The results differ in that the product of the binomials is a trinomial, whereas
the product of the radical expressions is a two-term radical
expression. **5.** $6\sqrt{3}$ **7.** $3\sqrt{7}$ **9.** $2\sqrt{10}$ **11.** $-10\sqrt{2}$
13. $-12\sqrt{x}$ **15.** $-\dfrac{7\sqrt{5}}{12}$ **17.** $\dfrac{23\sqrt{2}}{18}$ **19.** $-0.2\sqrt{n}$
21. $3\sqrt{5} + 3\sqrt{3}$ **23.** $5\sqrt{x} - 12\sqrt{y}$ **25.** $-3\sqrt{3}$ **27.** $10\sqrt{2}$
29. $-15\sqrt{3}$ **31.** $\sqrt{5}$ **33.** 0 **35.** $46\sqrt{3}$ **37.** $-20\sqrt{2}$
39. 0 **41.** $9\sqrt{3}$ **43.** $-4\sqrt{3}$ **45.** $-24\sqrt{3r}$ **47.** $-\dfrac{\sqrt{7}}{30}$
49. $\dfrac{11\sqrt{5}}{12}$ **51.** $17x\sqrt{6x}$ **53.** $-17t^2\sqrt{2t}$ **55.** Mistake:
Added radicands. Correct: $8\sqrt{2}$ **57.** Mistake: Subtracted
radicands. Correct: $9\sqrt{6x} - \sqrt{2x}$ is simplest form.
59. $25\sqrt{3}$ **61. a.** $(42 + 2\sqrt{5})$ ft. **b.** ≈ 46.5 ft.
c. $87.89 **63.** $2 + \sqrt{6}$ **65.** $\sqrt{10} - 5\sqrt{2}$
67. $2\sqrt{70} + 2\sqrt{30}$ **69.** $12\sqrt{6} - 6\sqrt{42}$ **71.** $8\sqrt{14} - 40\sqrt{3}$
73. $24\sqrt{2} + 8\sqrt{6}$ **75.** $6x + 9\sqrt{x}$ **77.** $30\sqrt{n} - 60n\sqrt{2}$
79. $2 - \sqrt{3}$ **81.** $\sqrt{10} - \sqrt{5} + 4\sqrt{2} - 4$
83. $\sqrt{10} - \sqrt{14} + \sqrt{15} - \sqrt{21}$ **85.** $-4 - 3\sqrt{15}$
87. $3x\sqrt{2} + 5\sqrt{6x} - 4\sqrt{3x} - 20$ **89.** $7a - 36\sqrt{2ab} + 10b$
91. 4 **93.** -13 **95.** $25x - 49$ **97.** $9 - 2\sqrt{14}$
99. $9 + 6\sqrt{2}$ **101.** $9n + 24\sqrt{n} + 16$
103. $72x - 60\sqrt{2xy} + 25y$ **105. a.** $(30\sqrt{2} + 18\sqrt{3})$ ft.^{2}
b. ≈ 73.6 ft^2. **107.** $3\sqrt{2} - 3$ **109.** $8 + 4\sqrt{3}$
111. $5\sqrt{3} - 5\sqrt{2}$ **113.** $-1 - \sqrt{5}$ **115.** $\dfrac{3 + \sqrt{3}}{2}$
117. $\dfrac{-6 - 8\sqrt{3}}{13}$ **119.** $\dfrac{4\sqrt{21} - 4\sqrt{6}}{5}$ **121.** $\dfrac{64 + 8\sqrt{6}}{29}$
123. $\dfrac{6y - 6\sqrt{y}}{y - 1}$ **125.** $\dfrac{3t - 6\sqrt{tu}}{t - 4u}$ **127.** $\dfrac{\sqrt{2xy} + 2y\sqrt{3}}{x - 6y}$
129. $\dfrac{15\sqrt{2} - 10\sqrt{3}}{3}$

Review Exercises 1. 5 **2.** $-6, 6$ **3.** 2, 3
4. 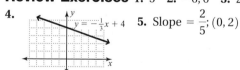 **5.** Slope $= \dfrac{2}{5}; (0, 2)$

6. Yes, it passes the vertical line test.

9.5 Exercises 1. Some of the answers may be
extraneous. **3.** The principal square root of a number cannot
equal a negative. **5.** Subtract $3x$ from both sides to isolate the
radical. This allows us to use the squaring principle of equality
to eliminate the radical. **7.** 4 **9.** No real-number solution.
11. 17 **13.** 11 **15.** 6 **17.** $-\dfrac{1}{2}$ **19.** No real-number solution.

21. 124 **23.** 55 **25.** 30 **27.** -1 **29.** 5 **31.** 6 **33.** 4, 5

35. 12, (5 is extraneous) **37.** 3, 1 **39.** 5, (1 is extraneous)

41. $\dfrac{1}{2}, \left(-\dfrac{3}{8} \text{ is extraneous}\right)$ **43.** No real-number solution.

$\left(-1 \text{ and } \dfrac{2}{9} \text{ are extraneous}\right)$. **45.** 1, 9 **47.** 27, (3 is extraneous)

49. 4 **51.** 5, $\left(\dfrac{13}{9} \text{ is extraneous}\right)$ **53.** Mistake: Did not check

to see that 81 is extraneous. It is extraneous because the principal square root of a number cannot be negative. Correct: no real-number solution **55.** Mistake: x still is not isolated.
Correct: 4 **57.** 9.8 m **59.** 2.45 m **61.** 1.44 ft. **63.** 144 ft.
65. 112.5 ft. **67.** 253.125 ft. **69.** 4 N **71.** 6 N
73. a. $\approx \$539{,}000{,}000$ **b.** 2005 **75. a.** 25, 3, 4
b.

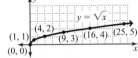

c. No, the x-values must be 0

or positive because real square roots exist only when $x \geq 0$.
The y-values must be 0 or positive because, by definition, the
principal square root is either 0 or positive. **d.** Yes, because
it passes the vertical line test. **77.** The graph becomes
steeper. **79.** The graph rises or lowers according the value of
the constant.

Review Exercises 1. 81 **2.** -0.008 **3.** $\dfrac{625}{16}$ **4.** x^8
5. n^{24} **6.** y^4

9.6 Exercises 1. $5^3 = 125$ **3.** The index, n, must be
odd because even roots are positive if $a > 0$ and undefined if
$a < 0$. **5.** You cannot take the fourth root of a negative
number. **7.** 4 **9.** -5 **11.** 3 **13.** 5 **15.** 6 **17.** 4 **19.** 2
21. x^2 **23.** r^6 **25.** -4 **27.** Not a real number. **29.** 5
31. 2 **33.** 3 **35.** x **37.** m^2 **39.** $4y^2$ **41.** $2\sqrt[3]{4}$
43. $2\sqrt[5]{2}$ **45.** $x^2\sqrt[3]{x^2}$ **47.** $2x^2\sqrt[3]{3x^2}$ **49.** $2\sqrt[5]{2x^4}$ **51.** $\dfrac{3}{4}$
53. $-\dfrac{3}{5}$ **55.** $\dfrac{2\sqrt[3]{4}}{5}$ **57.** $\dfrac{\sqrt[5]{3}}{2}$ **59.** $\sqrt{16}, 4$ **61.** $\sqrt[4]{-81}$, not a
real number **63.** $\sqrt[5]{-32}, -2$ **65.** $(\sqrt{4})^3, 8$
67. $-(\sqrt[4]{16})^3, -8$ **69.** $\dfrac{1}{(\sqrt{4})^3}, \dfrac{1}{8}$ **71.** $-\dfrac{1}{(\sqrt[4]{16})^5}, -\dfrac{1}{32}$
73. $\left(\sqrt[4]{\dfrac{81}{16}}\right)^3, \dfrac{27}{8}$ **75.** $\left(\sqrt[5]{\dfrac{243}{32}}\right)^2, \dfrac{9}{4}$ **77.** x **79.** $h^{2/21}$
81. $m^{2/5}$ **83.** $k^{2/7}$ **85.** $\dfrac{1}{n^{5/4}}$ **87.** $\dfrac{1}{s^{1/2}}$ **89.** $t^{1/5}$ **91.** $x^{3/8}y^{1/2}$
93. $\dfrac{x^{1/4}}{y^{1/2}}$ **95.** Mistake: Not in simplest form.

Correct: $2x\sqrt[3]{2x}$ **97.** Mistake: Misinterpreted the rational
exponent. Correct: 512 **99.** 2.714 **101.** 1.861 **103.** 4.327
105. 6.266 **107.** 216 **109.** 36 **111.** 4096 **113.** 0.020
115. \$1263.18 **117.** \$871.92 **119.** $-2\sqrt[3]{7}$
121. $15\sqrt[4]{2} - 5\sqrt[3]{2}$ **123.** $3\sqrt[3]{2} - 5\sqrt[3]{9}$
125. $2\sqrt[3]{2} + 4\sqrt[4]{4} - 12$ **127.** $2\sqrt[4]{4} + 6\sqrt[4]{8} + 9$

Review Exercises 1. $4(x + y)(x - y)$ **2.** $\dfrac{6}{x - 5}$

3. $2(n + 8) = 18; 1$ **4.** $-\dfrac{1}{2}, 5$ **5.** 24 **6.** $(-6, 2)$

Chapter 9 Review Exercises 1. true **2.** true
3. true **4.** false **5.** true **6.** Multiply **7.** radicands
8. coefficients **9.** conjugate **10.** extraneous **11.** ± 11
12. ± 7 **13.** 13 **14.** -7 **15.** Not a real number. **16.** $\dfrac{1}{5}$
17. irrational **18.** rational **19.** $9x$ **20.** $5mn^2$ **21.** $\dfrac{5}{4}a^2b^5$
22. $\dfrac{9x^2y}{8a^4}$ **23.** 6 **24.** $4xy^2$ **25.** $10x^2$ **26.** $6m^2$ **27.** $4w^2z^2$
28. $10xyz^2$ **29.** $2\sqrt{5}$ **30.** $4\sqrt{3}$ **31.** $3\sqrt{2}$ **32.** $2\sqrt{30}$
33. $x^2b\sqrt{6b}$ **34.** $x^3\sqrt{10}$ **35.** $6u^2v^2\sqrt{u}$ **36.** $2x^2\sqrt{3}$ **37.** 2
38. $\dfrac{\sqrt{11}}{6}$ **39.** $\dfrac{3}{11}$ **40.** $\dfrac{m^2}{9}$ **41.** $\dfrac{1}{4m^2}$ **42.** $\dfrac{h^2}{2}$ **43.** $\dfrac{5\sqrt{2}}{4}$
44. $\dfrac{3}{2y}$ **45.** -3 **46.** $\dfrac{a\sqrt{17}}{c}$ **47.** $\dfrac{\sqrt{2}}{2}$ **48.** $\sqrt{3}$ **49.** $\dfrac{2\sqrt{7}}{7}$
50. $\dfrac{x\sqrt{10}}{2}$ **51.** $\dfrac{\sqrt{51}}{3y}$ **52.** $\dfrac{\sqrt{3}}{3}$ **53.** $2\sqrt{3}$ **54.** $6\sqrt{2}$
55. $-2\sqrt{6}$ **56.** $-\dfrac{\sqrt{2}}{6}$ **57.** $4\sqrt{3}$ **58.** $2\sqrt{5} + 5\sqrt{3}$ **59.** $4\sqrt{5}$
60. $8\sqrt{3}$ **61.** $-2\sqrt{2}$ **62.** $21\sqrt{2}$ **63.** $18\sqrt{5}$
64. $3\sqrt{11} - 2\sqrt{6}$ **65.** $\dfrac{-2\sqrt{6}}{35}$ **66.** $\dfrac{16\sqrt{2}}{55}$ **67.** $\sqrt{15} + \sqrt{10}$
68. $\sqrt{6} - 3\sqrt{2}$ **69.** $\sqrt{21} + 14$ **70.** $2\sqrt{70} + 2\sqrt{30}$
71. $6\sqrt{6} - 54$ **72.** 60 **73.** $\sqrt{10} + \sqrt{14} - \sqrt{15} - \sqrt{21}$
74. $\sqrt{10} + 2\sqrt{2} - 4\sqrt{5} - 8$ **75.** $2x\sqrt{3} - \sqrt{6x} + 2\sqrt{2x} - 2$
76. $5a - 3b$ **77.** $16r - s$ **78.** 7 **79.** $3 + 2\sqrt{2}$
80. $7 - 2\sqrt{10}$ **81.** $9t - 12\sqrt{t} + 4$ **82.** $16x + 8\sqrt{xy} + y$
83. $10\sqrt{15}$ **84.** $18\sqrt{7}$ **85.** $-4\sqrt{2} - 4\sqrt{3}$ **86.** $\dfrac{4 - \sqrt{3}}{13}$
87. $2 + \sqrt{3}$ **88.** $\sqrt{6} + 2$ **89.** 49 **90.** No real-number
solution. **91.** 10 **92.** -1 **93.** $\dfrac{17}{3}$ **94.** 35 **95.** -3 **96.** 5
97. -2, (-5 is extraneous) **98.** 1, 2 **99.** 0, 4 **100.** $\dfrac{3}{16}$
101. a. $8\sqrt{5}$ mph **b.** ≈ 17.9 mph **102.** 312.5 ft. **103.** π sec.
104. ≈ 1.089 m **105. a.** $\dfrac{\sqrt{10}}{2}$ sec. **b.** ≈ 1.58 sec. **106.** 4 ft.
107. 5 ft. **108.** 2 **109.** 4 **110.** Not a real number. **111.** -2
112. 6 **113.** 2 **114.** $2x$ **115.** $2y$ **116.** $\dfrac{4}{5}$ **117.** $-\dfrac{1}{2}$
118. $2 - 5\sqrt[4]{8}$ **119.** $2\sqrt[3]{18} + 14\sqrt[3]{12} + 49$ **120.** $\sqrt{25}, 5$
121. $\sqrt[3]{8}, 2$ **122.** $-(\sqrt[5]{32})^2, -4$ **123.** $-(\sqrt[4]{16})^3, -8$
124. $\sqrt[3]{-27}, -3$ **125.** $\left(\sqrt{\dfrac{25}{9}}\right)^3, \dfrac{125}{27}$ **126.** $\dfrac{1}{(\sqrt[3]{64})^4}, \dfrac{1}{256}$
127. $\dfrac{1}{(\sqrt[3]{-125})^2}, \dfrac{1}{25}$ **128.** $n^{3/5}$ **129.** $\dfrac{1}{h^{3/20}}$ **130.** $\dfrac{1}{x^{2/5}}$
131. $y^{1/5}$ **132.** \$4077.71 **133.** \$1805.88 **134.** \$3274.01

Chapter 9 Practice Test 1. Not a real number. [9.1]
2. 10 [9.1] **3.** $6x^2y$ [9.1] **4.** $2\sqrt{5}$ [9.2] **5.** 4 [9.2]
6. $3xy^2\sqrt{3y}$ [9.2] **7.** $\dfrac{\sqrt{5}}{7}$ [9.3] **8.** 2 [9.3] **9.** $\dfrac{1}{2x}$ [9.3]
10. $12\sqrt{2}$ [9.4] **11.** $-4\sqrt{5}$ [9.4] **12.** $10\sqrt{6} + 45$ [9.4]
13. $\sqrt{15} + \sqrt{6} - \sqrt{10} - 2$ [9.4] **14.** $4 - 2\sqrt{3}$ [9.4]
15. $60\sqrt{2}$ [9.2] **16.** $\dfrac{5\sqrt{3}}{3}$ [9.3] **17.** $\sqrt{5} - \sqrt{2}$ [9.4]
18. 26 [9.5] **19.** 1, (-7 is extraneous) [9.5]
20. a. 30 mph [9.5] **b.** 400 ft. [9.5] **21.** 2 [9.6]
22. -2 [9.6] **23.** $2x$ [9.6] **24.** 6 [9.6] **25.** -8 [9.6]

Chapters 1–9 Cumulative Review Exercises

1. false   2. false   3. false   4. $ab + ac$ 5. $0, 0$
6. $a^2 + b^2 = c^2$ 7. 11 8. $4y^2 - 11y - 2$
9. $25x^2 - 60x + 36$ 10. $(2x + 1)(3x + 5)$

11. $4x(3x - 1)(x - 2)$ 12. $m(m + 1)(m - 1)$ 13. $\dfrac{1}{x}$
14. $\dfrac{x + 1}{x - 7}$ 15. $4m^2$ 16. $2x^2\sqrt{3}$

17. $20 + 5\sqrt{2} - 4\sqrt{6} - 2\sqrt{3}$ 18. $6 + 5\sqrt{3}$ 19. $\dfrac{5\sqrt{3}}{3}$

20. 10 21. $2, 3$ 22. $\dfrac{1}{2}$ 23. $(6, -8)$ 24. $-4, -2$

25. 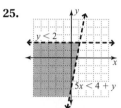   26.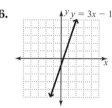

27. 825 28. 10 ft.   29. 60 mph   30. 17 small, 34 large

Chapter 10

10.1 Exercises
1. Because you can square both a negative and positive number to get a.   3. $3 + 2\sqrt{5}, 3 - 2\sqrt{5}$
5. $x = \pm\sqrt{\dfrac{b + c}{a}}$ 7. ± 4 9. ± 14 11. $\pm\dfrac{2}{3}$ 13. ± 0.6
15. $\pm 2\sqrt{3}$ 17. $\pm 5\sqrt{2}$ 19. No real-number solution.
21. ± 10 23. ± 7 25. ± 1 27. ± 6 29. ± 3 31. ± 5
33. $\pm\sqrt{11}$ 35. $\pm\dfrac{3}{5}$ 37. $\pm\dfrac{2\sqrt{6}}{3}$ 39. $\pm\dfrac{3\sqrt{2}}{4}$ 41. ± 1
43. $\pm\sqrt{2}$ 45. $\pm\sqrt{3}$ 47. $\pm\dfrac{2\sqrt{3}}{3}$ 49. ± 1.2 51. $-7, 1$
53. $-\dfrac{9}{2}, \dfrac{3}{2}$ 55. $7 \pm 2\sqrt{3}$ 57. $\dfrac{1 \pm 2\sqrt{10}}{4}$ 59. $\dfrac{-5 \pm 3\sqrt{2}}{2}$
61. No real-number solution.   63. $0, \dfrac{4}{3}$ 65. $-\dfrac{1}{8}, \dfrac{7}{8}$
67. $-5, -9.\overline{6}$ 69. $1.5\overline{3}, 3.4$ 71. Mistake: Gave only the positive square root. Correct: ± 3 73. Mistake: Changed -2 to 2. Correct: no real-number solution   75. $\pm\sqrt{90} \approx \pm 9.487$
77. $\pm\sqrt{15} \approx \pm 3.873$ 79. $2 \pm \sqrt{3} \approx 0.268, 3.732$ 81. 13
83. 5 85. $\sqrt{29} \approx 5.385$ 87. $\sqrt{145} \approx 12.042$
89. $2\sqrt{13} \approx 7.211$ 91. $4\sqrt{2} \approx 5.657$ 93. $\sqrt{5} \approx 2.236$
95. 14 in.   97. Width: 20 ft.; length: 60 ft.   99. 305 m
101. 6 ft.   103. 18 in.   105. 4 m/sec.   107. $\dfrac{3}{4}$ sec.
109. 4.9 sec.   111. $90\sqrt{2} \approx 127.3$ ft.   113. $2\sqrt{341}; 36.9$ in.

Review Exercises
1. 36 2. $x^2 + 8x + 16$
3. $y^2 - 14y + 49$ 4. $(x + 3)^2$ 5. $3x$ 6. -4

10.2 Exercises
1. Add the square of half of b. (Wording may vary.)   3. Because $x^2 - 6x + 8$ is easy to factor, factoring is better as it requires fewer steps than completing the square.
5. a. $x^2 + 10x + 25$ b. $(x + 5)^2$ 7. a. $y^2 - 6y + 9$
b. $(y - 3)^2$ 9. a. $d^2 - 12d + 36$ b. $(d - 6)^2$
11. a. $x^2 - 5x + \dfrac{25}{4}$ b. $\left(x - \dfrac{5}{2}\right)^2$ 13. a. $y^2 + 9y + \dfrac{81}{4}$

b. $\left(y + \dfrac{9}{2}\right)^2$ 15. a. $v^2 + \dfrac{1}{6}v + \dfrac{1}{144}$ b. $\left(v + \dfrac{1}{12}\right)^2$
17. a. $s^2 - \dfrac{3}{4}s + \dfrac{9}{64}$ b. $\left(s - \dfrac{3}{8}\right)^2$ 19. $-2, -8$ 21. $-6, 8$
23. $4, -5$ 25. $3, 6$ 27. $-3, 5$ 29. $-2 \pm \sqrt{13}$
31. $3 \pm 2\sqrt{5}$ 33. $-\dfrac{5}{2}, 1$ 35. $-2, \dfrac{3}{2}$ 37. $-\dfrac{3}{2}, \dfrac{2}{3}$ 39. $-\dfrac{4}{3}, \dfrac{7}{2}$
41. $-\dfrac{3}{2}, 1$ 43. $\dfrac{3 \pm \sqrt{19}}{2}$ 45. $\dfrac{-3 \pm \sqrt{3}}{2}$ 47. $\dfrac{-1 \pm \sqrt{41}}{10}$
49. $\dfrac{2 \pm \sqrt{10}}{2}$ 51. Did not divide the right side of the equation by 2. Correct: $\dfrac{1 \pm \sqrt{3}}{2}$ 53. 12 ft. by 14 ft.
55. Base: 12 ft.; height: 5 ft.   57. Base: 8 ft.; height: 6 ft.
59. 15 in.   61. a. 5 in.   b. Length: 30 in.; width: 20 in.

Review Exercises
1. $x^2 + 8x + 16$
2. $(3x - 5)(2x + 1)$ 3. $(x - 5)^2$ 4. 1 5. $\dfrac{1}{2}$ 6. $\pm 2\sqrt{3}$

10.3 Exercises
1. Follow the procedure for solving a quadratic equation by completing the square.   3. When a quadratic equation is easily factored.   5. The radicand.
7. $a = 1, b = -4, c = 9$ 9. $a = 2, b = -7, c = -5$
11. $a = 0.5, b = -1, c = 2.4$ 13. $a = \dfrac{1}{2}, b = \dfrac{2}{3}, c = -8$
15. $-6, -3$ 17. $-8, 2$ 19. $-4, 1$ 21. $0, 3$ 23. 5
25. $\dfrac{1 \pm \sqrt{5}}{2}$ 27. $-\dfrac{1}{3}, 1$ 29. $-\dfrac{1}{2}, -\dfrac{2}{3}$ 31. $\dfrac{-5 \pm \sqrt{10}}{3}$
33. $\dfrac{9 \pm \sqrt{17}}{8}$ 35. $\dfrac{3 \pm \sqrt{105}}{12}$ 37. No real-number solutions.   39. $-0.2, 0.1$ 41. $-3, \dfrac{1}{2}$ 43. $\pm\dfrac{3}{2}$
45. $\dfrac{3 \pm \sqrt{13}}{4}$ 47. $\dfrac{20 \pm \sqrt{382}}{6}$ 49. Mistake: Did not evaluate $-b$. Correct: $-1, \dfrac{5}{2}$ 51. Mistake: Ignored the negative radicand. Correct: no real-number solutions.
53. Square root principle; $\pm 4\sqrt{5}$.   55. Factoring; 4.
57. Quadratic formula; $2 \pm \sqrt{3}$.   59. Square root principle; $-1, -9$.   61. Factoring; $0, -4$.   63. Two rational.
65. Two irrational.   67. One rational.   69. No real-number solutions.   71. One rational.   73. a. 1 b. $c < 1$ c. $c > 1$
75. a. 9 b. $a < 9$ c. $a > 9$ 77. a. ± 8 b. $b > 8$
c. $0 < b < 8$ 79. $x^2 + 6(x + 1) = 33; 3, 4$
81. $x^2 + (x + 1)^2 = (x + 2)^2; 3, 4, 5$ 83. $w(2w - 3) = 350$; width $= 14$ in., length $= 25$ in.   85. a. $20x = 1.25x^2 + 5x; 12$ ft.
b.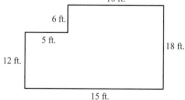

87. $0.5n^2 + 2.5n = 4.5n; 4$ units   89. a. 2003 b. 2006
91. a. 45 b. 25

Review Exercises
1. $-x + 7$ 2. $-x - 6$
3. $-6x^2 + 3x$ 4. $y^2 - 49$ 5. $2\sqrt{5}$ 6. $3 - \sqrt{3}$

10.4 Exercises 1. $\sqrt{-1}$ 3. No, the set of complex numbers contains both the real and imaginary numbers. For example, 2 is complex but not imaginary. 5. We subtract complex numbers just like we subtract polynomials—by writing an equivalent addition and changing the signs in the second complex number. 7. $4i$ 9. $i\sqrt{2}$ 11. $2i\sqrt{2}$ 13. $2i\sqrt{5}$ 15. $3i\sqrt{3}$ 17. $5i\sqrt{5}$ 19. $3i\sqrt{5}$ 21. $7i\sqrt{3}$ 23. $1 + 5i$ 25. $8 - 4i$ 27. $-6 + 5i$ 29. $10 + 2i$ 31. $-13 - 2i$ 33. 0 35. $8 - 4i$ 37. $2 - 4i$ 39. -6 41. 16 43. $12 + 10i$ 45. $12 - 6i$ 47. $7 + i$ 49. $45 + 7i$ 51. 13 53. $8 + 6i$ 55. $\dfrac{6}{5} - \dfrac{3}{5}i$ 57. $-\dfrac{8}{25} + \dfrac{6}{25}i$ 59. $5 - 2i$

61. $\dfrac{2}{5} + \dfrac{1}{5}i$ 63. $\dfrac{10}{17} + \dfrac{11}{17}i$ 65. $\pm 4i$ 67. $\pm 2i\sqrt{5}$

69. $-4 \pm 4i$ 71. $3 \pm 2i\sqrt{2}$ 73. $1 \pm 2i$ 75. $-1 \pm \dfrac{1}{2}i$

77. $-1 \pm i\sqrt{7}$ 79. $2 \pm i\sqrt{3}$ 81. $\pm \dfrac{7}{5}i$ 83. $\dfrac{3}{4} \pm \dfrac{\sqrt{31}}{4}i$

85. The equation indicates that the profit will reach 20 million in $\dfrac{3}{2} \pm \dfrac{\sqrt{23}}{2}i$ years, which are imaginary numbers of years and therefore impossible. 87. a. ≈ 0.27 sec. and 1.6 sec. b. The ball will not reach 25 feet because $\dfrac{15}{16} \pm \dfrac{\sqrt{127}}{16}i$ are not real numbers.

Review Exercises 1. -2 2. $(3, 0), (0, 2)$ 3. 6

4.   5.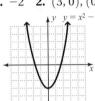

6.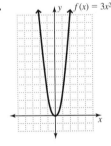

10.5 Exercises 1. The sign of a. 3. Yes, if it has a vertex at the origin. 5. There are no x-intercepts.
7. upward 9. downward 11. downward 13. upward 15. downward 17. $(2, 0), (-1, 0), (0, -2)$
19. $(-4, 0), (2, 0), (0, 8)$ 21. $(4, 0), (2, 0), (0, 8)$ 23. No x-intercepts, $(0, 1)$. 25. $(-5, 0), (0, 25)$ 27. $\left(\dfrac{1}{2}, 0\right), (-8, 0),$
$(0, -8)$ 29. $(-2 + \sqrt{2}, 0), (-2 - \sqrt{2}, 0), (0, 2)$
31. $\left(\dfrac{5 + \sqrt{29}}{-2}, 0\right)\left(\dfrac{5 - \sqrt{29}}{-2}, 0\right), (0, 1)$ 33. No x-intercepts, $(0, -6)$. 35. $(3, -8), x = 3$ 37. $(3, 7), x = 3$
39. $(2, -3), x = 2$ 41. $\left(\dfrac{1}{3}, \dfrac{4}{3}\right), x = \dfrac{1}{3}$ 43. $(-2.5, 9.75),$
$x = -2.5$ 45. $(0.3, 1.45), x = 0.3$ 47. a. upward
b. $(-1, 0), (3, 0), (0, -3)$ c. $(1, -4)$ d. $x = 1$

e.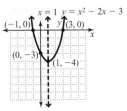

49. a. downward
b. $(2, 0), (4, 0), (0, -8)$
c. $(3, 1)$
d. $x = 3$

e.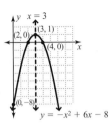

51. a. upward
b. $(-3, 0), (0, 9)$
c. $(-3, 0)$
d. $x = -3$

e.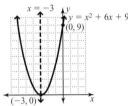

53. a. upward
b. no x-intercepts, $(0, 5)$
c. $(-2, 1)$
d. $x = -2$

e.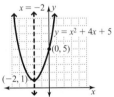

55. a. downward
b. no x-intercepts, $(0, -5)$
c. $(1, -2)$
d. $x = 1$

e.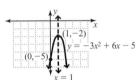

57. a. downward b. $\left(\dfrac{3 + \sqrt{21}}{-3}, 0\right), \left(\dfrac{3 - \sqrt{21}}{-3}, 0\right)$ or $(\approx -2.53, 0), (\approx 0.53, 0); (0, 4)$ c. $(-1, 7)$ d. $x = -1$
e.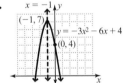

59. a. upward b. $\left(\dfrac{-3 + \sqrt{3}}{2}, 0\right), \left(\dfrac{-3 - \sqrt{3}}{2}, 0\right)$ or $(\approx -0.63, 0), (\approx -2.37, 0); (0, 3)$ c. $(-1.5, -1.5)$
d. $x = -1.5$ e.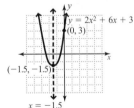

61. $y = x^2 + 6x + 8$ 63. $y = x^2 + 2x + 1$
65. $y = x^2 + 4x + 5$ 67. The graph appears to narrow as a becomes larger. 69. The graph shifts to the left, and the vertex moves downward.

71. a.

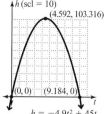

$h = -4.9t^2 + 45t$

b. 103.316 m   **c.** ≈ 9.184 sec.

73. a.

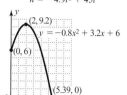

$y = -0.8x^2 + 3.2x + 6$

b. 9.2 ft.   **c.** ≈ 5.39 ft.

94. $-\dfrac{3}{10} + \dfrac{1}{10}i$ **95.** $-\dfrac{21}{53} + \dfrac{6}{53}i$ **96.** $\dfrac{2}{17} - \dfrac{8}{17}i$ **97.** $\pm 5i$

98. $\pm 6i$ **99.** $\pm 2i\sqrt{3}$ **100.** $\pm 3i\sqrt{2}$ **101.** $-4 \pm i\sqrt{2}$

102. $1 \pm 2i\sqrt{5}$ **103.** $2 + i\sqrt{7}$ **104.** $-1 \pm \dfrac{1}{2}i$

105. $\dfrac{1}{3} \pm \dfrac{\sqrt{2}}{3}i$ **106.** $\dfrac{3}{4} \pm \dfrac{\sqrt{31}}{4}i$

107. a. upward
b. $(-3, 0), (3, 0), (0, -9)$
c. $(0, -9)$
d. $x = 0$
e.

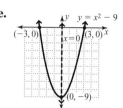

Review Exercises

1. -11 **2.** $\dfrac{25}{4}$ **3.** $-5, 3$

4. $\pm 2i\sqrt{5}$ **5.** $x = \dfrac{-b \pm \sqrt{b^2 - 4ac}}{2a}$ **6.**

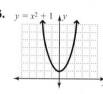

$y = x^2 + 1$

108. a. upward
b. $(1, 0), (3, 0), (0, 3)$
c. $(2, -1)$
d. $x = 2$
e.

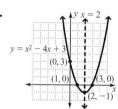

$y = x^2 - 4x + 3$

Chapter 10 Review Exercises

1. true   **2.** false
3. true   **4.** true   **5.** false   **6.** $d = \sqrt{(x_2 - x_1)^2 + (y_2 - y_1)^2}$
7. $\left(\dfrac{b}{2}\right)^2$ **8.** $x = \dfrac{-b \pm \sqrt{b^2 - 4ac}}{2a}$ **9.** $b^2 - 4ac$

10. $x = -\dfrac{b}{2a}$ **11.** ± 11 **12.** $\pm \dfrac{1}{5}$ **13.** $\pm 6i$ **14.** $\pm 2\sqrt{5}$

15. ± 5 **16.** ± 6 **17.** $\pm\sqrt{14}$ **18.** ± 5 **19.** $\pm\sqrt{5}$

20. $\pm i\sqrt{3}$ **21.** $-12, -2$ **22.** $5, 13$ **23.** $-\dfrac{7}{5}, \dfrac{1}{5}$ **24.** $-3, \dfrac{13}{3}$

25. $-1.09, 0.13$ **26.** $6, 8$ **27.** $2\sqrt{10} \approx 6.325$
28. $\sqrt{97} \approx 9.849$ **29.** $\sqrt{29} \approx 5.385$ **30.** $\sqrt{41} \approx 6.403$

31. $x^2 + 16x + 64$ **32.** $m^2 - 9m + \dfrac{81}{4}$ **33.** $w^2 + \dfrac{2}{5}w + \dfrac{1}{25}$

34. $y^2 - \dfrac{1}{3}y + \dfrac{1}{36}$ **35.** $-7, -1$ **36.** $1 \pm \sqrt{3}$

37. $-8 \pm \sqrt{55}$ **38.** $-7, -3$ **39.** $2 \pm \sqrt{15}$ **40.** $-14, 2$

41. $-8, 14$ **42.** $3 \pm \sqrt{10}$ **43.** $\dfrac{1 \pm 2\sqrt{2}}{2}$ **44.** $-1, \dfrac{7}{2}$

45. $-\dfrac{3}{5}, \dfrac{9}{5}$ **46.** $\dfrac{-3 \pm \sqrt{6}}{3}$ **47.** 4 **48.** $-2, 3$

49. $\dfrac{-1 \pm \sqrt{41}}{4}$ **50.** $-\dfrac{1}{3}, 1$ **51.** $\dfrac{-4 \pm \sqrt{19}}{3}$ **52.** $\dfrac{1 \pm \sqrt{5}}{6}$

53. $1 \pm \sqrt{2}$ **54.** $-1 \pm \sqrt{6}$ **55.** $\dfrac{3 \pm \sqrt{17}}{4}$ **56.** $\dfrac{3 \pm \sqrt{15}}{3}$

57. $-\dfrac{3}{2}, 3$ **58.** $\dfrac{-1 \pm \sqrt{19}}{3}$ **59.** Two rational.   **60.** One
rational.   **61.** No real-number solutions.   **62.** Two irrational.
63. Two rational.   **64.** Two irrational.   **65.** No real-number
solutions.   **66.** One rational.   **67.** $10i$ **68.** $5i$ **69.** $2i\sqrt{5}$
70. $2i\sqrt{15}$ **71.** $3i\sqrt{3}$ **72.** $i\sqrt{5}$ **73.** $15 + 2i$ **74.** 2
75. $4 + 6i$ **76.** 1 **77.** $1 - i$ **78.** $10 - i$ **79.** $-15 - 8i$
80. $-2 + 7i$ **81.** -30 **82.** 2 **83.** $4 + 20i$ **84.** $15 - 6i$
85. $8 - i$ **86.** $-7 - 23i$ **87.** 5 **88.** 74 **89.** $-5 - 12i$

90. $15 + 8i$ **91.** $\dfrac{11}{5} + \dfrac{8}{5}i$ **92.** $\dfrac{1}{2} + \dfrac{1}{2}i$ **93.** $\dfrac{10}{13} - \dfrac{15}{13}i$

109. a. downward
b. $(-4, 0), (-2, 0), (0, -8)$
c. $(-3, 1)$
d. $x = -3$
e.

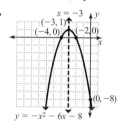

$y = -x^2 - 6x - 8$

110. a. downward
b. no x-intercepts, $(0, -4)$
c. $(0, -4)$
d. $x = 0$
e.

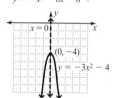

$y = -3x^2 - 4$

111. a. downward
b. $\left(-\dfrac{\sqrt{10}}{2}, 0\right), \left(\dfrac{\sqrt{10}}{2}, 0\right), (0, 5)$
c. $(0, 5)$
d. $x = 0$
e.

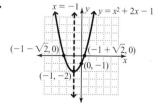

$y = -2x^2 + 5$

112. a. upward   **b.** $(-1 + \sqrt{2}, 0), (-1 - \sqrt{2}, 0), (0, -1)$
c. $(-1, -2)$ **d.** $x = -1$ **e.**

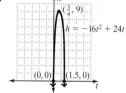

$y = x^2 + 2x - 1$

113. 130 mi.   **114.** 4 m/sec.   **115.** 15 ft. by 19 ft.
116. ≈ 16.97 in.   **117.** 8 ft., 15 ft.
118. a.

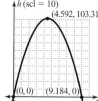

$h = -16t^2 + 24t$

b. 9 ft.   **c.** 1.5 sec.

Chapter 10 Practice Test

1. ± 9 [10.1]
2. $3 \pm 2\sqrt{5}$ [10.1]   **3.** $\sqrt{29}$ [10.1]   **4.** $3\sqrt{2}$ [10.1]
5. $4 \pm 2\sqrt{3}$ [10.2]   **6.** $-1, 11$ [10.2]   **7.** $\dfrac{3 \pm 2\sqrt{6}}{3}$ [10.2]

8. $4 \pm i\sqrt{2}$ [10.2]   **9.** $-2, \dfrac{3}{2}$ [10.3]   **10.** $3, 5$ [10.3]

11. $\dfrac{-1 \pm \sqrt{15}}{2}$ [10.3]   **12.** $\dfrac{2 \pm \sqrt{3}}{3}$ [10.3]

13. $-8, 2$ [10.2, 10.3]   **14.** $-9, 9$ [10.1]

15. $-\dfrac{3}{4} \pm \dfrac{\sqrt{3}}{4}i$ [10.2, 10.3]   **16.** $0, -2$ [10.2, 10.3]

17. $-16, -8$ [10.1]   **18.** $3i\sqrt{2}$ [10.4]   **19.** $2 + 2i$ [10.4]
20. $-6 - 2i$ [10.4]   **21.** $7 + 24i$ [10.4]

22. $-\dfrac{8}{25} + \dfrac{6}{25}i$ [10.4]   **23.** $\pm 4i$ [10.4]   **24. a.** upward

b. $(5, 0), (3, 0), (0, 15)$ **c.** $(4, -1)$ **d.** $x = 4$

e. [10.5]

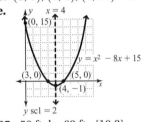

25. 50 ft. by 62 ft. [10.2]

Chapters 1–10 Cumulative Review Exercises

1. true   **2.** false   **3.** false   **4.** true   **5.** $(a + b)(a - b)$
6. rational expressions   **7.** 30   **8.** $3x^2 + 7x - 20$
9. $(3x + 4)(2x - 1)$ **10.** $2y^2(3y + 4)(3y - 4)$

11. $-\dfrac{21x^2}{y^4}$ **12.** $\dfrac{5x^2 - 13x + 2}{5x^2(x + 1)}$ **13.** $11 - 6\sqrt{2}$

14. $-\dfrac{1}{5} + \dfrac{3}{5}i$ **15.** $\dfrac{6\sqrt{x} + 6\sqrt{y}}{x - y}$ **16.** -1 **17.** $4, -\dfrac{3}{2}$

18. 5   **19.** $(2, -1)$ **20.** 4   **21.** $\dfrac{3 \pm \sqrt{5}}{4}$
22.

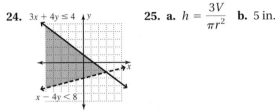

23. $2x + 3y = 11$

24.

25. a. $h = \dfrac{3V}{\pi r^2}$ **b.** 5 in.

26. 765 people thought it made financial sense; 410 people thought it doomed a marriage to fail.   **27.** China: 51; United States: 36.   **28.** \$4000 at 4%, \$6000 at 8%   **29.** 14 ft. by 8 ft.
30. $\dfrac{6}{5}$, or $1\dfrac{1}{5}$ hr.

Appendix

1. The sum of all given numbers divided by the number of numbers.   **3.** 1. Arrange the numbers in order from least to greatest or greatest to least.   2. Find the mean of the middle two numbers.   **5.** mean = \$26,841.$\overline{6}$; median = \$26,975; mode = \$28,700   **7.** mean = 78.25; median = 80; mode = 82   **9.** mean = 1.82 m; median = 1.85 m; modes = 1.8 m and 2.1 m   **11.** mean = 1.25 in.; median = 1.1 in.; mode = 0.6 in.   **13.** year 1: mean = \$126.76; median = \$131.26; no mode year 2: mean = \$131.95: median = \$134.08; no mode

Glossary

Absolute value: A given number's distance from 0 on a number line.

Additive inverses: Two numbers whose sum is 0.

Area: The total number of square units that fill a figure.

Axis of symmetry: A line that divides a graph into two symmetrical halves.

Base: The number that is repeatedly multiplied.

Binomial: A polynomial containing two terms.

Circumference: The distance around a circle.

Coefficient: The numerical factor in a term.

Coefficient of a monomial: The numerical factor in a monomial.

Complementary angles: Two angles are complementary if the sum of their measures is 90°.

Complex conjugates: The complex conjugate of a complex number $a + bi$ is $a - bi$.

Complex number: A number that can be expressed in the form $a + bi$, where a and b are real numbers and i is the imaginary unit.

Complex rational expression: A rational expression that contains rational expressions in the numerator or denominator.

Congruent angles: Angles that have the same measure. The symbol for congruent is $\cong$.

Conjugates: Binomials that differ only in the sign separating the terms.

Consistent system of equations: A system of equations that has at least one solution.

Constant: A symbol that does not vary in value.

Contradiction: An equation that has no real-number solution.

Degree of a monomial: The sum of the exponents of all variables in a monomial.

Degree of a polynomial: The greatest degree of any of the terms in the polynomial.

Diameter: The distance across a circle through its center.

Direct variation: Two variables, y and x, are in direct variation if $y = kx$, where k is a constant.

Discriminant: The radicand $b^2 - 4ac$ in the quadratic formula.

Domain: The set of all input values (x-values) for a relation.

Equation: A mathematical relationship that contains an equal sign.

Exponent: A symbol written to the upper right of a base number that indicates how many times to use the base as a factor.

Expression: A constant, a variable, or any combination of constants, variables, and arithmetic operations that describes a calculation.

Extraneous solution: An apparent solution that does not solve its equation.

Factored form: A number or an expression written as a product of factors.

Factors: If $a \cdot b = c$, then a and b are factors of c.

Formula: An equation that describes a mathematical relationship.

Fraction: A quotient of two numbers or expressions a and b having the form $\frac{a}{b}$, where $b \neq 0$.

Function: A relation in which every value in the domain is paired with (or assigned to) exactly one value in the range.

Greatest common factor (GCF): The largest natural number that divides evenly into all numbers of a given set of numbers.

Identity: An equation that has every real number as a solution (excluding any numbers that cause an expression in the equation to be undefined).

Imaginary number: A number that can be expressed in the form bi, where b is a real number and i is the imaginary unit.

Imaginary unit: The number represented by i, where $i = \sqrt{-1}$ and $i^2 = -1$.

Inconsistent system of equations: A system of equations that has no solution.

Inequality: A mathematical relationship that contains an inequality symbol ($\neq$, $<$, $>$, $\leq$, or $\geq$).

Inverse variation: Two variables, y and x, are in inverse variation if $y = \dfrac{k}{x}$, where k is a constant.

Irrational number: Any real number that is not rational.

Least common denominator (LCD): The least common multiple of the denominators of a given set of fractions.

Least common multiple (LCM): The smallest number that is a multiple of each number in a given set of numbers.

Like radicals: Two radical expressions with identical radicands and indexes.

Like terms: Constant terms or variable terms that have the same variable(s) raised to the same exponents.

Linear equation: An equation in which each variable term contains a single variable raised to an exponent of 1.

Linear equation in one variable: An equation that can be written in the form $ax + b = c$, where a, b, and c are real numbers and $a \neq 0$.

Linear inequality: An inequality containing expressions in which each variable term contains a single variable with an exponent of 1.

Lowest terms: Given a fraction $\dfrac{a}{b}$ and $b \neq 0$, if the only factor common to both a and b is 1, then the fraction is in lowest terms.

Monomial: An expression that is a constant, a variable, or a product of a constant and variable(s) that are raised to whole-number powers.

Multiple: A multiple of a given integer n is the product of n and an integer.

Multiplicative inverses: Two numbers whose product is 1.

nth root: The number b is the nth root of a number a if $b^n = a$.

Percent: A ratio representing some part out of 100.

Perimeter: The distance around a figure.

Polynomial: A monomial or an expression that can be written as a sum of monomials.

Polynomial in one variable: A polynomial in which every variable term has the same variable.

Prime factorization: A factorization that contains only prime factors.

Prime number: A natural number that has exactly two different factors, 1 and the number itself.

Proportion: An equation in the form $\dfrac{a}{b} = \dfrac{c}{d}$, where $b \neq 0$ and $d \neq 0$.

Quadratic equation in one variable: An equation that can be written in the form $ax^2 + bx + c = 0$, where a, b, and c are real numbers and $a \neq 0$.

Quadratic equation in two variables: An equation that can be written in the form $y = ax^2 + bx + c$, where a, b, and c are real numbers and $a \neq 0$.

Radical equation: An equation containing at least one radical expression whose radicand has a variable.

Radius: The distance from the center of a circle to any point on the circle.

Range: The set of all output values (y values) for a relation.

Ratio: A comparison of two quantities using a quotient.

Rational expression: An expression that can be written in the form $\dfrac{P}{Q}$, where P and Q are polynomials and $Q \neq 0$.

Rational number: Any real number that can be expressed in the form $\dfrac{a}{b}$, where a and b are integers and $b \neq 0$.

Real numbers: The union of the rational and irrational numbers.

Relation: A set of ordered pairs.

Scientific notation: A number expressed in the form $a \times 10^n$, where a is a decimal number with $1 \leq |a| < 10$ and n is an integer.

Set: A collection of objects.

Similar figures: Figures with congruent angles and proportional side lengths.

Slope: The ratio of the vertical change between any two points on a line to the horizontal change between these points.

Solution: A number that makes an equation true when it replaces the variable in the equation.

Solution for a system of equations: An ordered set of numbers that makes all equations in the system true.

Supplementary angles: Two angles are supplementary if the sum of their measures is 180°.

System of equations: A group of two or more equations.

Terms: Expressions that are the addends in an expression that is a sum.

Trinomial: A polynomial containing three terms.

Unit ratio: A ratio with a denominator of 1.

Variable: A symbol that can vary in value.

Vertex of a parabola: The lowest point on a parabola that opens up or the highest point on a parabola that opens down.

Volume: The total number of cubic units that fill a space.

x-intercept: A point where a graph intersects the x-axis.

y-intercept: A point where a graph intersects the y-axis.

Index of Applications

Statistics/Demographics

Transportation

Index

POWERS AND ROOTS

n	n^2	n^3	$\sqrt{n}$	$\sqrt[3]{n}$	$\sqrt{10n}$	n	n^2	n^3	$\sqrt{n}$	$\sqrt[3]{n}$	$\sqrt{10n}$
1	1	1	1.000	1.000	3.162	51	2,601	132,651	7.141	3.708	22.583
2	4	8	1.414	1.260	4.472	52	2,704	140,608	7.211	3.733	22.804
3	9	27	1.732	1.442	5.477	53	2,809	148,877	7.280	3.756	23.022
4	16	64	2.000	1.587	6.325	54	2,916	157,464	7.348	3.780	23.238
5	25	125	2.236	1.710	7.071	55	3,025	166,375	7.416	3.803	23.452
6	36	216	2.449	1.817	7.746	56	3,136	175,616	7.483	3.826	23.664
7	49	343	2.646	1.913	8.367	57	3,249	185,193	7.550	3.849	23.875
8	64	512	2.828	2.000	8.944	58	3,364	195,112	7.616	3.871	24.083
9	81	729	3.000	2.080	9.487	59	3,481	205,379	7.681	3.893	24.290
10	100	1,000	3.162	2.154	10.000	60	3,600	216,000	7.746	3.915	24.495
11	121	1,331	3.317	2.224	10.488	61	3,721	226,981	7.810	3.936	24.698
12	144	1,728	3.464	2.289	10.954	62	3,844	238,328	7.874	3.958	24.900
13	169	2,197	3.606	2.351	11.402	63	3,969	250,047	7.937	3.979	25.100
14	196	2,744	3.742	2.410	11.832	64	4,096	262,144	8.000	4.000	25.298
15	225	3,375	3.873	2.466	12.247	65	4,225	274,625	8.062	4.021	25.495
16	256	4,096	4.000	2.520	12.649	66	4,356	287,496	8.124	4.041	25.690
17	289	4,913	4.123	2.571	13.038	67	4,489	300,763	8.185	4.062	25.884
18	324	5,832	4.243	2.621	13.416	68	4,624	314,432	8.246	4.082	26.077
19	361	6,859	4.359	2.688	13.784	69	4,761	328,509	8.307	4.102	26.268
20	400	8,000	4.472	2.714	14.142	70	4,900	343,000	8.367	4.121	26.458
21	441	9,261	4.583	2.759	14.491	71	5,041	357,911	8.426	4.141	26.646
22	484	10,648	4.690	2.802	14.832	72	5,184	373,248	8.485	4.160	26.833
23	529	12,167	4.796	2.844	15.166	73	5,329	389,017	8.544	4.179	27.019
24	576	13,824	4.899	2.884	15.492	74	5,476	405,224	8.602	4.198	27.203
25	625	15,625	5.000	2.924	15.811	75	5,625	421,875	8.660	4.217	27.386
26	676	17,576	5.099	2.962	16.125	76	5,776	438,976	8.718	4.236	27.568
27	729	19,683	5.196	3.000	16.432	77	5,929	456,533	8.775	4.254	27.749
28	784	21,952	5.292	3.037	16.733	78	6,084	474,552	8.832	4.273	27.928
29	841	24,389	5.385	3.072	17.029	79	6,241	493,039	8.888	4.291	28.107
30	900	27,000	5.477	3.107	17.321	80	6,400	512,000	8.944	4.309	28.284
31	961	29,791	5.568	3.141	17.607	81	6,561	531,441	9.000	4.327	28.460
32	1,024	32,768	5.657	3.175	17.889	82	6,724	551,368	9.055	4.344	28.636
33	1,089	35,937	5.745	3.208	18.166	83	6,889	571,787	9.110	4.362	28.810
34	1,156	39,304	5.831	3.240	18.439	84	7,056	592,704	9.165	4.380	28.983
35	1,225	42,875	5.916	3.271	18.708	85	7,225	614,125	9.220	4.397	29.155
36	1,296	46,656	6.000	3.302	18.974	86	7,396	636,056	9.274	4.414	29.326
37	1,369	50,653	6.083	3.332	19.235	87	7,569	658,503	9.327	4.431	29.496
38	1,444	54,872	6.164	3.362	19.494	88	7,744	981,472	9.381	4.448	29.665
39	1,521	59,319	6.245	3.391	19.748	89	7,921	704,969	9.434	4.465	29.833
40	1,600	64,000	6.325	3.420	20.000	90	8,100	729,000	9.487	4.481	30.000
41	1,681	68,921	6.403	3.448	20.248	91	8,281	753,571	9.539	4.498	30.166
42	1,764	74,088	6.481	3.476	20.494	92	8,464	778,688	9.592	4.514	30.332
43	2,849	79,507	6.557	3.503	20.736	93	8,649	804,357	9.644	4.531	30.496
44	2,936	85,184	6.633	3.530	20.976	94	8,836	830,584	9.695	4.547	30.659
45	2,025	91,125	6.708	3.557	21.213	95	9,025	857,375	9.747	4.563	30.882
46	2,116	97,336	6.782	3.583	21.148	96	9,216	884,736	9.798	4.579	30.984
47	2,209	103,823	6.856	3.609	21.679	97	9,409	912,673	9.849	4.595	31.145
48	2,304	110,592	6.928	3.534	21.909	98	9,604	941,192	9.899	4.610	31.305
49	2,401	117,649	7.000	3.659	22.136	99	9,801	970,299	9.950	4.626	31.464
50	2,500	125,000	7.071	3.684	22.361	100	10,000	1,000,000	10.000	4.642	31.623

PROBLEM-SOLVING OUTLINE

1. **Understand** the problem.
   a. Read the question(s) (not the whole problem, just the question at the end) and write a note to yourself about what you are to find.
   b. Read the whole problem, underlining the key words.
   c. If possible or useful, make a list or table, simulate the situation, or search for a related example problem.

2. **Plan** your solution by searching for a formula or translating the key words to an equation.

3. **Execute** the plan by solving the equation/formula.

4. **Answer** the question. Look at the note about what you were to find and make sure you answer that question. Include appropriate units.

5. **Check** results.
   a. Try finding the solution in a different way, reversing the process, or estimating the answer and make sure the estimate and actual answer are reasonably close.
   b. Make sure the answer is reasonable.

USEFUL FORMULAS

Perimeter of a rectangle: $P = 2l + 2w$

Circumference of a circle: $C = \pi d$ or $C = 2\pi r$

Area of a parallelogram: $A = bh$

Area of a triangle: $A = \dfrac{1}{2}bh$

Area of a trapezoid: $A = \dfrac{1}{2}h(a + b)$

Area of a circle: $A = \pi r^2$

Surface area of a box: $SA = 2lw + 2lh + 2wh$

Volume of a box: $V = lwh$

Volume of a pyramid: $V = \dfrac{1}{3}lwh$

Volume of a cylinder: $V = \pi r^2 h$

Volume of a cone: $V = \dfrac{1}{3}\pi r^2 h$

Volume of a sphere: $V = \dfrac{4}{3}\pi r^3$

Distance, d, an object travels given its rate, r, and the time of travel, t: $d = rt$

The temperature in degrees Celsius given degrees Fahrenheit: $C = \dfrac{5}{9}(F - 32)$

The temperature in degrees Fahrenheit given degrees Celsius: $F = \dfrac{9}{5}C + 32$

The profit, P, after cost, C, is deducted from revenue, R: $P = R - C$

The Algebra Pyramid:
A Guide to the Development of Algebra Topics

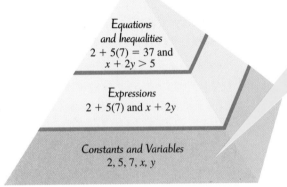

Constants and variables are the foundation of algebra. The following diagram summarizes the entire complex number system (all constants). These constants, along with variables such as x or y, are the basic building blocks that give meaning to the expressions, equations, and inequalities.

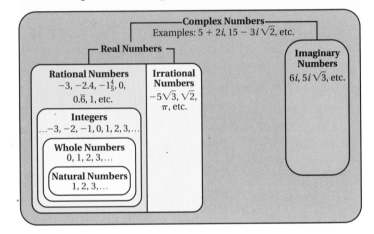

Expressions relate constants and variables using symbols such as operations of arithmetic to describe a calculation. We **evaluate** or **rewrite** expressions. To evaluate, we replace variables with numbers and perform the calculation. To rewrite, we use rules to write an alternative (and usually simpler) form.

The following list outlines the development of expressions in this text:

Chapter 1	Order of Operations
	Translating Word Phrases to Expressions
	Evaluating and Rewriting Expressions
Chapter 6	Exponent Rules and Scientific Notation
	Adding, Subtracting, Multiplying, and Dividing Polynomials
Chapter 7	Factoring Polynomials
Chapter 8	Simplifying, Adding, Subtracting, Multiplying, and Dividing Rational Expressions
Chapter 9	Simplifying, Adding, Subtracting, Multiplying, and Dividing Square Roots and Radical Expressions

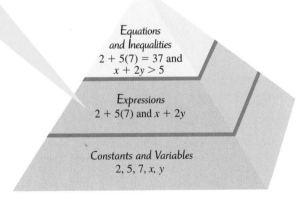